DEDICATIONS

To all the future health care professionals learning Anatomy, Physiology, and Disease. May your chosen professions be as personally rewarding as ours have been.

—Your Travel Guides Bruce, Jeff, and Karen

To those closest to me, who share this wonderful journey through life: my wife Patty, my sons Joshua and Jeremy, and my brothers and sister. And to Jen and Kristi Jan, who, through my sons, have become part of this family. Finally, to the memory of my Mom and Dad, who taught me the importance of education.

—Bruce

To the most wonderful family a person could ever wish for, and a better one than I probably deserve: Patty, Zack, Emily, Sara, Mom, and Dad, and, of course, Rusty. And to my past teachers and professors; here's proof that underachievers sometimes do hit their stride!

—Jeff

I dedicate this book to my family who have always supported me, no matter where life has taken me; not only my "real" family—my late father, Ed, my mother Pat, brother Eddie, sister-in-law Sheila, and assorted aunts, uncles, and cousins, who really had no choice but to be part of my life—but also those members of my extended family who have inexplicably chosen to be part of my life, giving me the gift of their friendship. I couldn't have done it without them.

—Karen

Contents

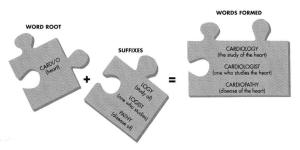

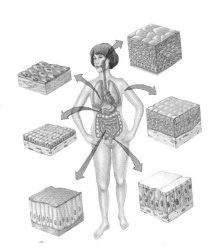

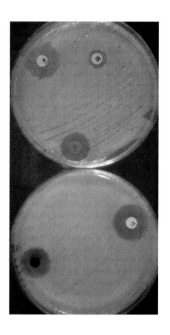

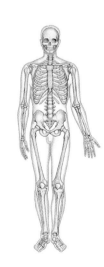

Chapter 7 **The Muscular System:**
Movement for the Journey 196

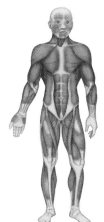

Chapter 8 **The Integumentary System:**
The Protective Covering 228

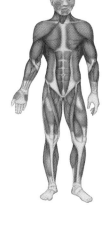

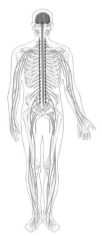

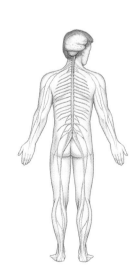

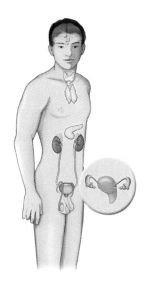

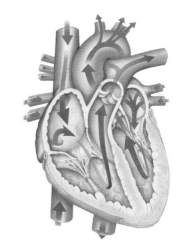

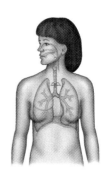

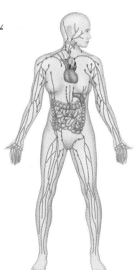

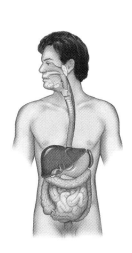

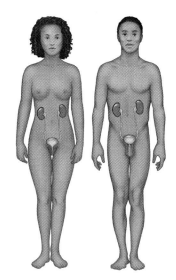

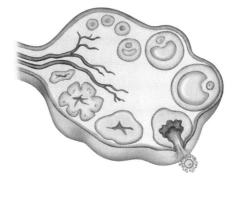

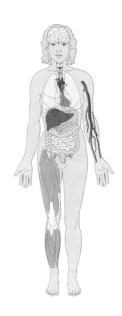

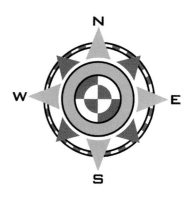

ABOUT THE AUTHORS

Bruce Colbert is the Director of the Allied Health Department at the University of Pittsburgh at Johnstown. He holds a Master's in Health Education and Administration, has authored five books, written several articles, and given over 175 invited lectures and workshops at both the regional and national level. Many of his workshops provide teacher training involving techniques to make the health sciences engaging and relevant to today's students. In addition, he does workshops on developing effective critical and creative thinking, stress and time management, and study skills. He is an avid basketball player, even after three knee surgeries, which may indicate he has some learning difficulties.

Jeff Ankney is the Director of Clinical Education for the University of Pittsburgh at Johnstown Respiratory Care program where he is responsible for the development and evaluation of hospital clinical sites. In the past, Jeff has served as a public school teacher, Assistant Director of Cardiopulmonary Services, Program Coordinator of Pulmonary Rehabilitation, and a member of hospital utilization review and hospital policy committees. He is a consultant on hospital management and pulmonary rehabilitation concerns. Jeff was also the recipient of the American Cancer Society Public Education Award. He is a pretty fair fly fisherman and wing shot, although his springer spaniel, Rusty, would probably disagree.

Karen Lee is an Associate Professor in the Biology Department at the University of Pittsburgh at Johnstown, where she teaches all the anatomy courses, including Anatomy and Physiology for Nursing and Allied Health students. She presents regularly at scientific conferences and has published several articles on crustacean physiology and behavior. An active member of the Council on Undergraduate Research, she is also the chair of the University of Pittsburgh at Johnstown's annual undergraduate symposium. For the last fifteen years she has sung bass in women's barbershop choruses and quartets and more recently has been harmonizing with any singer or guitar player who will let her near the microphone.

PREFACE

We have made every effort to put together an Anatomy, Physiology, & Disease book that students would actually enjoy reading. This book is supplemented with a DVD and a dedicated website that reinforces the concepts and allows for a visual and interactive learning experience.

Anatomy, physiology, and *pathology* are critical concepts you must master to succeed in the health professions. ***Anatomy, Physiology, & Disease: An Interactive Journey for Health Professionals***, along with the accompanying *Study Success Companion*, integrated DVD, and website, are written in a manner that will enhance learning of the material versus mass memorization of facts. Too often students adopt the strategy of memorizing massive amounts of information and simply storing it in their short-term memories. Then they literally repeat this information during their exams. However, memorization alone does not help learners master the material and make the lasting connections that will help them *thrive* as a health care practitioner.

In the course of a day as a health professional, you will be exposed to a variety of diseases involving all parts of the body, and therefore you must truly *learn* the workings and interrelatedness of all the body systems and functions.

So what have we done to facilitate learning the material? First, we have placed study skills and stress management tips in a *Study Success Companion* in the back of your book to help you along your journey through this class and beyond. Second, we have strived to make anatomy, physiology, and pathology "come alive" by using an engaging writing style to make it seem as if we are sitting next to you talking about the concepts. We hope you consider us assistants to the most important guide through this journey: your teacher.

Humor, where appropriate, and analogies that compare the human body to everyday things to which you can relate, have been interwoven throughout the text. Finally, we have added special features in a unique fashion tailored to the visual learning styles and relevant learning that today's students require.

We have worked hard and sincerely hope these features will help make studying ***Anatomy, Physiology, & Disease: An Interactive Journey for Health Professionals*** a positive experience. Have a safe and happy journey!

ACKNOWLEDGMENTS

To Jill Rembetski, who acted as our guide through the development and completion of this textbook. To all those at Pearson/Prentice Hall who were so helpful and friendly in this monumental process. A special thanks goes to Mark Cohen and Melissa Kerian for their enthusiastic belief and support in the vision and execution of this unique project. Thank you to a great group of reviewers and contributors who helped shape this project with their thoughtful insights. Finally, special thanks to Jan Snyder for putting up with two quirky authors and helping in so many ways.

REVIEWERS

Carmen Carpenter, CMA, RN, MS
Chair, Allied Health Sciences/Medical Assisting
South University
West Palm Beach, Florida

Kris A. Hardy, CMA, RHE, CDF
Program Director, Medical Assisting
Brevard Community College
Cocoa, Florida

William J. Havins, BUS, CPhT
Instructor, Pharmacy Technology
Central New Mexico Community College
Albuquerque, New Mexico

Barbara Klomp BA (Psy), R.T.(R)
Adjunct Faculty
Macomb Community College
Clinton Township, Michigan

Mark Lafferty, Ph.D., M.Ed.
Program Coordinator, Exercise Science
Delaware Technical and Community College
Wilmington, Delaware

Rosann Turner, RN, MSN
Instructor, Nursing
Shelton State Community College
Tuscaloosa, Alabama

Wendy Zamora
Professor
Kaplan University

CONTRIBUTORS

Instructor's Resource Manual

Jennifer Esch, BS, PA-C
Instructor, Medical Assisting
Bryant and Stratton College
Milwaukee, Wisconsin

Debra S. McKinney, RN, BSN
Director, Practical Nursing Program
TESST College of Technology
Alexandria, Virginia

Test Bank

Nina Beaman, MS, RNC, CMA
Program Area Coordinator, Allied Health
Bryant and Stratton College
Richmond, Virginia

Deborah J. Bedford, CMA
Former Program Coordinator/Instructor, Medical
   Assisting
North Seattle Community College
Seattle, Washington

Joy Harden, RN, BSN, CNOR, MBA
Former Director, Health Services Education
Virginia College
Mobile, Alabama

Ruth Ann O'Brien, MHA, RRT, CPFT
Director, Respiratory Therapy Program
Miami-Jacobs Career College
Dayton, Ohio

Dominick Squicciarini, MPH
Senior Instructional Designer, Health Sciences
Kaplan University
Fort Lauderdale, Florida

PowerPoint Lecture Outlines

Ursula E. Cole, CMA (AAMA), CCS-P
Medical Program Coordinator/Medical Curriculum
  Chair
Indiana Business College
Terre Haute, Indiana

Debra S. McKinney, RN, BSN
Director, Practical Nursing Program
TESST College of Technology
Alexandria, Virginia

Companion Website

Margaret N. Anfinsen, RN, MSN
Associate Program Manager, Allied Health Studies
Southwest Florida College
Fort Myers, Florida

Deborah J. Bedford, CMA
Former Program Coordinator/Instructor, Medical
  Assisting
North Seattle Community College
Seattle, Washington

Minda Brown, RMA
Instructor, Medical Assisting
Pima Medical Institute
Colorado Springs, Colorado

Student DVD

Margaret N. Anfinsen, RN, MSN
Associate Program Manager, Allied Health Studies
Southwest Florida College
Fort Myers, Florida

Jennifer Esch, BS, PA-C
Instructor, Medical Assisting
Bryant and Stratton College
Milwaukee, Wisconsin

Joy Harden, RN, BSN, CNOR, MBA
Former Director, Health Services Education
Virginia College
Mobile, Alabama

Mary Warren-Oliver, BA
Clinical Coordinator, Medical Assisting
Gibbs College
Vienna, Virginia

Richard Silberman, M.D.
Adjunct Professor, School of Continuing Education
University of Wisconsin-Milwaukee
Milwaukee, Wisconsin
Adjunct Professor
Bryant & Stratton College
Milwaukee, Wisconsin

SPECIAL FEATURES DIRECTORY LISTING

Chapter 1

In-Text Features

Learning Hint: Using the Margins of This Book 6
Learning Hint: Combining and Forming Medical Terms 7
Learning Hint: General Hints on Forming Medical Terms 8
Clinical Application: Metabolic Syndrome, or Syndrome X 13
Clinical Application: "Breaking" a Fever 14
Amazing Body Facts: Bizarre Signs and Symptoms 17
Clinical Application: The Vital Sign of Pulse 18
Clinical Application: The Geriatric Patient 19

DVD Interactive Exercises

Video on Medical Specialties, 1-1
Video on Vital Signs and Pulse Sites, 1-2
Videos on Proper Hand Washing and Gloving, Gowning, Goggles, and Masks for Infection Control, 1-3
Animation on Cervical Diving Injuries, 1-4
Interactive Puzzles and Games with Medical Terminology and Disease Concepts, 1-5

Companion Website

Professional Profiles:
• Medical Assisting
• Medical Records and Health Information Technician
• Medical Transcriptionist
Related Internet Links
Additional Review Questions

Chapter 2

In-Text Features

Clinical Application: Do You Know Your Left from Your Right? 37
Clinical Application: Central Versus Peripheral Cyanosis 38
Clinical Application: Hernias 42
Clinical Application: The Central Landmark: The Spinal Column 43
Amazing Body Facts: Psoas Test 45

DVD Interactive Exercises

Additional Video Information on Body Positions, 2-1
Videos on Ultrasound, MRIs and Other Diagnostic Imaging, 2-2
Interactive Drag-and-Drop Exercise to Reinforce Learning of Body Cavities, 2-3
Interactive Drag-and-Drop Exercise to Reinforce Learning of Anterior and Posterior Body Regions, 2-4
Interactive Games and Puzzles, 2-5

Companion Website

Professional Profiles:
• Radiologic Technology
• Surgical Technology

DVD Interactive Exercises

Videos on Taking a Proper Patient History and Performing Physical Assessment/Exams, 5-1
Videos on Blood Sampling and Performing Glucose Monitoring, 5-2
Video on Blood Test Basics, 5-3
Video on Performing a Dipstick Urinalysis, 5-4
Video on Complete Urinalysis, 5-5
Video on Occult Testing Procedure, 5-6
Videos on Electrocardiogram Testing, Electrode Placement, and Fetal Heart Monitoring, 5-7
Video on Arthroscopic Procedures, 5-8
Interactive Games and Puzzles, 5-9

Companion Website

Professional Profile:
• Physician Assistant
Related Internet Links
Additional Review Questions

Chapter 6

In-Text Features

Amazing Body Facts: Bone Composition 158
Amazing Body Facts: Red Bone Marrow and Red Blood Cells 162
Amazing Body Facts: Aging and Bone Building 164
Amazing Body Facts: The Flexibility of Cartilage 166
Amazing Body Facts: Wear and Tear and Disease 169
Learning Hint: Number of Vertebrae 178
Clinical Application: When to See a Doctor 184

DVD Interactive Exercises

Interactive Drag-and-Drop Exercise: Labeling the Parts of a Bone, 6-1
Animations on Joint Classification and Movement, 6-2
Animations of Various Types of Skeletal Body Movements, 6-3
Interactive Drag-and-Drop on Labeling the Major Bones of the Skeletal System, 6-4
Interactive Drag-and-Drop on Labeling the Bones of the Skull, 6-5
Animation on How Bone Fractures Heal, 6-6
Video on Arthritis and Osteoporosis, 6-7
Interactive Games and Puzzles, 6-8

Companion Website

Professional Profile:
• Radiologic Technologists
Related Internet Links
Additional Review Questions

Chapter 7

In-Text Features

Clinical Application: Muscle Tone 200
Amazing Body Facts: Muscles 201

Amazing Body Facts: What Makes a Muscle Tear Good? 201
Learning Hint: Muscle Names 205
Applied Sciences: Kinesiology 210
Applied Sciences: Interrelatedness of Neuromuscular System 215
Amazing Body Facts: Rigor Mortis 216
Learning Hint: Smooth Muscle Regulation of Blood Pressure 218
Applied Science: Maintaining a Core Body Temperature 220
Applied Science: A Useful Application of a Deadly Toxin 222

DVD Interactive Exercises

Interactive Drag-and-Drop Exercise on Locating the Anterior and Posterior Muscles of the Body, 7-1
Video of Various Massage Therapy Techniques, 7-2
Interactive 3-D Visual Tour of Labeled Muscle Groups of Specified Regions: Head and Neck, Upper Limb, Forearm and Hand, Shoulder and Arm, Trunk and Abdomen, Pelvis, Hip and Thigh, Lower Limb and Foot, 7-3
Video on Intramuscular (IM) Injection Techniques, 7-4
Interactive Drag-and-Drop Exercise on Myofibril Structures, 7-5
Videos on Muscle Atrophy and Muscular Dystrophy, 7-6
Animation of Cellular Muscle Contraction, 7-7
Interactive Games and Puzzles, 7-8

Companion Website

Professional Profiles:
- Kinesiology
- Physical Therapy
- Occupational Therapy
- Massage Therapy
Related Internet Links

Chapter 8

In-Text Features

Amazing Body Facts: A Bothersome Fact of Life 233
Amazing Body Facts: Keloids 241
Clinical Application: Medicine Delivery via the Integumentary System 242
Clinical Application: Assessing Peripheral Perfusion 245
Applied Science: Forensics and Hair 247

DVD Interactive Exercises

Interactive Drag-and-Drop Exercises: Three Layers of Skin and the Integumentary System, 8-1
Videos on the Changes that Occur with Aging to the Integumentary System, 8-2
Animation of Pressure Sore Formation, Prevention, and Treatment, 8-3
Animation on Wound Repair and the Formation of Scar Tissue, 8-4
Video on Intradermal Drugs and Subcutaneous Injections, 8-5
Interactive Drag-and-Drop Exercise on the Anatomical Structures of a Hair Follicle, 8-6
Videos and Animations on Diseases of the Skin: Decubitus Ulcers, Eczema, and Skin Cancer, 8-7
Videos on Topical Wound/Burn Care, 8-8
Interactive Games and Puzzles, 8-9

Companion Website
Professional Profiles:
- EMT-Paramedic
- Nursing

Related Internet Links
Additional Review Questions

Chapter 9

In-Text Features
Amazing Body Facts: Faster than a Speeding Bullet 266
Applied Science: Fugu 271
Clinical Application: Epilepsy 272
Clinical Application: Bioterrorism 277
Learning Hint: Directional Terms 278
Clinical Application: Spinal Blocks, Epidural Anesthesia, and Lumbar Punctures (Spinal Taps) 283
Clinical Application: Knee Jerk Reflex 285
Clincal Application: Carpal Tunnel Syndrome 286
Amazing Body Facts: Why Not a Smooth Brain? 288
Learning Hint: Gray Versus White Matter 298
Learning Hint: Mnemonic Devices 301
Clinical Application: Spastic Versus Flaccid Paralysis 309

DVD Interactive Exercises
Animation of Multiple Sclerosis, 9-1
Animation of Neurochemical Synaptic Transmission, 9-2
Videos of Epidural Placement, 9-3
3-D Animation of Brachial and Lumbosacral Plexus, Spinal Cord, and Cervical Spine Injuries, 9-4
Animation of Neuroreflex Arc, 9-5
Video of Carpal Tunnel Syndrome (CTS), 9-6
Positron Emission Tomography (PET) Scan of the Brain, 9-7
Drag-and-Drop Exercise: Brainstem and Subarachnoid Space, 9-8
Video on Glasgow Coma Scale Assessment, 9-9
Animation on Coup/Contracoup Injuries, 9-10
Video on Alzheimer's Disease, 9-11
Videos on Assessing Pain and Pain Management, 9-12
Video on Parkinson's Disease, 9-13
Videos of Related Nervous System Disorders, 9-14
Interactive Games and Puzzles, 9-15

Companion Website
Professional Profiles:
- Pharmacy
- Electroneurodiagnostician

Related Internet Links
Additional Review Questions

Chapter 10

In-Text Features

Amazing Body Fact: Lesser Known Endocrine Glands 324
Clinical Application: Childbirth and Positive Feedback 329
Learning Hint: Hormone Names 332
Clinical Application: Stature Disorders 336
Clinical Application: Cortisol Busters and Weight Loss 347

DVD Interactive Exercises

3-D Animation and Drag-and-Drop Exercise on the Endocrine System, 10-1
Videos on the Effects of Aging on the Endocrine System, 10-2
Animation and Video Spotlighting the Pathology of Diabetes and How to Monitor
Blood Glucose Levels, 10-3
Videos on the Importance of Insulin and Monitoring Glucose, 10-4
Interactive Games and Puzzles, 10-5

Companion Website

Professional Profiles:
• Phlebotomist
• Dietician
Related Internet Links
Additional Review Questions

Chapter 11

In-Text Features

Amazing Body Fact: Why We Can See in the Dark 363
Clinical Application: Snellen Charts 364
Amazing Body Facts: Eye Dominance 365
Clinical Application: Life Without Glasses 366
Applied Science: Sound Conduction 367
Clinical Application: Heat and Cold Therapy 375
Clinical Application: Cataract Surgery 378

DVD Interactive Exercises

Interactive Drag-and-Drop Exercise of the Eye Structures and
3-D Animation of the Eye, 11-1
Pathophysiologic Spotlights on Eye Disorders such as Cataracts, Macular Degeneration,
and Conjunctivitis, 11-2
Animation of the Workings of the Middle Ear, 11-3
Video on Tympanic Membrane and Thermometer Measurements, 11-4
Interactive Drag-and-Drop Exercise of the Ear Structures, 3-D Animation of the Ear,
and Animations of the Child and Adolescent Ear, 11-5
Videos on Ear Disorders such as Otitis Media and Hearing Tests, 11-6
Videos on Heat and Cold Therapy Procedures, 11-7
Videos on Pain Perception and Pain Scales, 11-8
Videos on Opthalamic and Otic Medications and Their Delivery, Along with Ear and
Eye Irrigation, 11-9
Interactive Games and Puzzles, 11-10

Companion Website

Professional Profiles:
- Ophthalmology and Opticians
- Audiologist

Related Internet Links
Additional Review Questions

Chapter 12

In-Text Features

Applied Science: Color Coded Blood 390
Learning Hint: Arteries or Veins? 390
Learning Hint: The Tricuspid Valve 392
Amazing Body Facts: Pulmonary Arteries and Veins 392
Amazing Body Facts: How Does it Keep Going and Going and Going? 394
Clinical Application: Angiography 395
Clinical Application: Cardiac Enzymes 399
Clinical Application: ECG/EKG 404
Clinical Application: CPR 406
Clinical Application: Heart Attacks 407
Amazing Body Facts: What's in a Drop of Blood? 408
Amazing Body Facts: Coconut Juice Blood Transfusions and Artificial Blood 413
Clinical Application: Umbilical Cord Blood and Leukemia 415
Clinical Application: Regulation of Blood Pressure 417
Clinical Application: Taking a Blood Pressure 418

DVD Interactive Exercises

Animation on the Chambers of the Heart, Interactive Drag-and-Drop Exercises Labeling Various Parts of the Heart, 12-1
Animation of the Heart Contraction and Related Blood Flow, 12-2
Animations and Videos on Heart Attacks and Angina, 12-3
Videos and Animations of CAD, Coronary Heart Disease, and Heart Catheterization, 12-4
Animations and Videos on Performing ECGs/EKGs, Electrode Placement, and Dysrhythmia, 12-5
Videos on Automated External Defibrillator (AED), Defibrillation, and Cardioversion, 12-6
Drag-and-Drop Exercise on Blood Types and a Video on Administering Blood, 12-7
Videos and Animations on Sickle Cell Anemia and Leukemia, 12-8
3-D Views of the Heart and Blood Vessels of the Body and Interactive Exercise on Labeling the Circulatory System, 12-9
Videos on Peripheral Artery Assessment and Carotid Artery Assessment, 12-10
Animations and Videos on Congenital Heart Disease and Aging of the Cardiovascular System, 12-11
Animation and Videos on Shock, Bleeding Control, Administration of Digoxin, Dopamine, and Nitroglycerin, and Starting an IV Line, 12-12
Interactive Games and Puzzles, 12-13

Companion Website

Professional Profile:
- Cardiovascular Technology

Related Internet Links

Additional Review Questions

Chapter 13

In-Text Features

DVD Interactive Exercises

Companion Website

Professional Profiles:
- Respiratory Therapist
- Perfusionist

Related Internet Links

Additional Review Questions

Chapter 14

In-Text Features

Clinical Application: How You Acquire Immunity to Pathogens 507

DVD Interactive Exercises

Video on Lymphatic Drainage Massage Therapy, 14-1
3-D Illustration of the Lymphatic System and Internactive Drag-and-Drop Exercise of
the Lymphatic System, 14-2
Video on Skin Cancer, 14-3
Video on Proper Hand Washing Technique, 144
Video on Leukemia, 14-5
Animation showing T Cell Destruction by HIV, 14-6
Animation and Video on HIV/AIDS, 14-7
Animation on the Treatment of Severe Allergic Reactions with an EpiPen and
Video on Allergic Rhinitis, 14-8
Interactive Games and Puzzles, 14-9

Companion Website

Professional Profiles:
• Nuclear Medicine
• Pharmacy
Related Internet Links
Additional Review Questions

Chapter 15

In-Text Features

Clinical Application: Sublingual Medication 526
Clinical Application: PEG Tube 536
Learning Hint: Emulsifiers 540
Learning Hint: The Ending ASE 540
Clinical Application: Lactose Intolerance 540
Clinical Application: Colonostomy 544

DVD Interactive Exercises

3-D Animation of the Digestive System, 15-1
Animation and Videos Concerning Oral Care and Assessment Along with
Cavity Prevention, 15-2
Video and an Interactive Drag-and-Drop Exercise of the Digestive System and
Intestinal Wall, 15-3
Animations of Carbohydrates, Lipids, and Proteins, 15-4
Videos on Nutrition and Enteral Feedings, 15-5
Animation on the topic of GERD, 15-6
Videos on the topics of Anorexia, Bulimia, and Obesity, 15-7
Interactive Games and Puzzles, 15-8

Companion Website

Professional Profiles:
- Dental Assistants and Hygienists
- Medical Profession: Dietician

Related Internet Links
Additional Review Questions

Chapter 16

In-Text Features

DVD Interactive Exercises

Companion Website

Professional Profile:
- Ultrasound Technician

Related Internet Links
Additional Review Questions

Chapter 17

In-Text Features

DVD Interactive Exercises

Animation on Cellular Division, 17-1
Videos on the Human Genome Project, 17-2
Animation showing Fertilization, 17-3
3-D Animation of the Female Reproductive System with Drag-and-Drop Labeling Exercise, 17-4
Animation on Oogenesis with Drag-and-Drop Labeling Exercise, 17-5
3-D Animation of the Male Reproductive System with Drag-and-Drop Labeling Exercise, 17-6
Animation on Spermatogenesis with Drag-and-Drop Labeling Exercise, 17-7
Video on Erectile Dysfunction, 17-8
Videos on the Vasectomy Procedure, Fetal Life, Labor, Infant Delivery, the Placenta, and Post-Partum Assessment, 17-9
Video on Breast Cancer, 17-10
Interactive Games and Puzzles, 17-11

Companion Website

Professional Profile:
• Doula or Midwife
Related Internet Links
Additional Review Questions

Chapter 18

In-Text Features

Applied Science: Tomatoes Were Once Thought to be Poisonous 645
Applied Science: Forensics and History 645
Applied Science: If Bones Could Speak 646
Amazing Body Facts: You're Not Getting Older, You're Getting Smarter 649
Clinical Application: Age- and Activity-Related Diets and Nutritional Needs 657
Applied Science: Antibiotics 660

DVD Interactive Exercises

Animation of the Cause and Effect of Lead Poisoning, 18-1
Animation of General Information on Poisons, 18-2
Animation on Geriatric Evaluation, Multi-System Problems, and a Variety of Geriatric Issues, 18-3
Video on Nutrition and Aging, 18-4
Video on Cancer Overview and Skin Cancer, 18-5
Videos on Emotional, Social, and Spiritual Well-Being, Caring, and Empathy, 18-6
Videos on Osteoporosis and Carpal Tunnel Syndrome, 18-7
Video on Muscle Atrophy, 18-8
Video on Children and Obesity, 18-9
Videos on Smoking, Passive Smoke, and Smoking Cessation, 18-10
Video on Nutrition and Eating Disorders Such as Anorexia and Bulimia, 18-11
Videos on Macular Degeneration, Cataracts, and Audiology, 18-12
Videos on SIDS and Major Life Transitions of Infancy, 18-13
Video on AIDS and Other Sexually Transmitted Diseases, 18-14
Interactive Games and Puzzles, 18-15

Companion Website

Professional Profiles:
- Criminalist
- Mental Health Professionals
- Physician Assistant

Related Internet Links

Additional Review Questions

Study Success Companion

In-Text Features

A User's Guide to the Features of This Book

We have designed this textbook to be fun, interesting, and rich in features to aid your understanding of this challenging topic. Here is a quick guide to what makes this text different from others. We hope that the special highlights of this book enhance your learning experience as the journey of your health care career unfolds.

SPECIAL FEATURES MAKE LEARNING FUN

Medical Terminology Guides

While this is not a medical terminology book, understanding and pronouncing medical words is critical to your success. You will find proper pronunciations and word part analysis in the margins where key terms are introduced. Audio pronunciation of terms is contained on both your DVD as well as the student website at **www.prenhall.com/colbert**.

> **endocytosis** *(en DOE sigh TOE sis)*
>   **endo** = *within*
>   **cyt/o** = *cell*
>   **osis** = *condition*
> **phagocytosis** *(FAG oh sigh TOE sis)*
>
> **pinocytosis** *(PIE no sigh TOE sis)*
>   **phag/o** = *eating*
>   **pin/o** = *to drink*
> **exocytosis** *(Ex oh sigh TOE sis)*
>   **ex/o** = *outside*
> **vesicle** *(VESS ih kle)*

Clinical Application

DO YOU KNOW YOUR LEFT FROM YOUR RIGHT?

By now, it should be clear that a precise, standardized language with directional terms is needed to study anatomy and physiology and apply it in a health care setting. Something as simple as left and right can become critical. For example, suppose you are a surgical technologist and are ordered to put a tag around a patient's right leg to designate it as the leg to be amputated in an upcoming surgery. If you approach the patient from the bottom of the bed and place the tag on the leg on YOUR right side, you have erroneously placed it on the patient's left leg, and this could have disastrous results. The take-home message is that left and right *always* refer to the patient's left and right, *not yours*.

Clinical Applications

These highlight boxes show the relevance of what you are learning and how that knowledge is needed in clinical practice. This feature includes topics such as aging, major diagnostic studies, and therapeutics.

Pathology Connection

After discussions of Anatomy & Physiology concepts, a special section on related disease conditions will be presented. This will make the connection as to <u>why</u> the understanding of the A&P concept discussed is so important to the pathologic process. This will aid the student in understanding abnormal diagnostic test results and why certain treatments are effective for a specific disease.

Applied Science

A USEFUL APPLICATION OF A DEADLY TOXIN

Botulism is a potentially deadly disease caused by food poisoning with the *Clostridium botulinum* bacteria. Science has found a way to utilize the poison generated by this bacteria for medical and cosmetic treatment. Small amounts of botulinus toxin are injected into facial muscles to stop previously untreatable facial twitching. The toxin basically paralyzes the muscles. The same toxin is used to treat wrinkles without the use of surgery and is known as Botox injections.

Applied Sciences

Instead of having a separate non-integrated chemistry or physics chapter, the sciences are presented in context.

Amazing Body Facts

These are "that's awesome" kinds of facts to give an appreciation of just how wonderful the human design is. For example, nerve impulses can travel at speeds of up to 426 feet per SECOND!

Amazing Body Facts

MAGNETOTAXIS: SOME BACTERIA RESPOND TO MAGNETIC FIELDS

Believe it or not, some bacteria can sense and respond to a magnetic field. These types of bacteria are sensitive to Earth's magnetic field and orient themselves to this force! This ability to move in response to magnetic forces is called *magnetotaxis*. This discovery was made by Richard Blakemore as he observed bacteria living in sulfide-rich mud from a lake. As he changed the position of the mud, the bacteria would reorient themselves to Earths's magnetic field. Upon further examination, Blakemore determined that these bacteria possessed particles of iron oxide, a magnetic metal compound that is stored in a cell structure called *magnetosome*.

Learning Hint

MNEMONIC DEVICES

A mnemonic device is a tool used to help you memorize long lists. It can be very useful in anatomy. To make a mnemonic device, take the first letter of each part of the list you are trying to memorize and make it into a sentence. For example, the five great lakes in order from west to east are Superior, Michigan, Huron, Erie, and Ontario. The mnemonic device used to remember the right order is **S**am **M**ade **H**arry **E**at **O**nions, much easier to remember than the lakes themselves. An example for the cranial nerves is this one: **O**n **O**ld **O**lympus **T**owering **T**ops **A** **F**inn **V**ith **G**erman **V**alked **A**nd **H**opped.

Learning Hints

We present helpful hints to facilitate learning difficult concepts. These may sometimes be amusing stories or other learning aids.

Test Your Knowledge

After a concept is fully developed within a chapter, a "Test Your Knowledge" section will insure you understand what was just covered before moving on and running the risk of getting really lost on your journey.

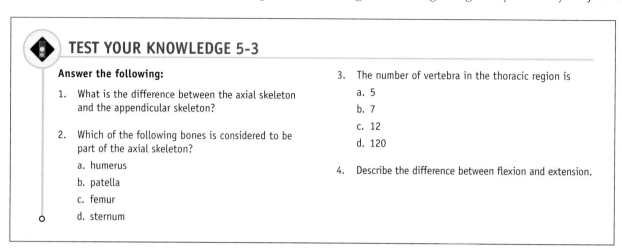

TEST YOUR KNOWLEDGE 5-3

Answer the following:

1. What is the difference between the axial skeleton and the appendicular skeleton?

2. Which of the following bones is considered to be part of the axial skeleton?
   a. humerus
   b. patella
   c. femur
   d. sternum

3. The number of vertebra in the thoracic region is
   a. 5
   b. 7
   c. 12
   d. 120

4. Describe the difference between flexion and extension.

Quick Trip Through Some Diseases of...

This chart provides a valuable quick reference for diseases related to a given body system chapter. The chart includes the cause, signs and symptoms, diagnostic tests, and treatment for each disease.

A Quick Trip Through Some Diseases of the Gastrointestinal System

DISEASE	ETIOLOGY	SIGNS AND SYMPTOMS	DIAGNOSTIC TEST(S)	TREATMENT
Cleft palate	Congenital anomaly in which the roof of the mouth has a split or fissure.	Obvious structural deformity making nursing difficult, allowing milk to be aspirated into the lungs from the nasal cavity.	Physical examination, imaging studies.	Surgical correction.
Crohn's disease	Form of chronic inflammatory bowel disease affecting the ileum and/or colon. Also called regional ileitis/enteritis.	Pain, cramps, diarrhea, bloating, weight loss.	Physical exam and history, radiologic studies.	Anti-inflammatory medications, such as prednisone, surgical intervention if severe.
Esophageal stricture	Narrowing of the	Difficulty swallowing,	Patient history, physi-	Surgical repair or

 Pharmacology Corner

This section includes a variety of medications used in the treatment of diseases that are related to a given body system of the chapter.

Snapshots from the Journey

→ The body can assume many different positions, and to standardize the study of anatomy, scientists often reference the anatomical position. In the anatomical position, the person stands with face and toes forward, hands at sides, and palms facing forward. Other positions, such as the prone, supine, and Fowler positions, are used in health care for assessment and treatments.

→ The body can be divided by the use of planes into different sections. For example, the transverse, or horizontal plane divides the body into superior and inferior sections. The median, or midsagittal, plane divides the body into equal right and left halves, and the frontal, or coronal, plane divides the body into anterior and posterior sections.

→ Directional terms, such as internal and external, proximal and distal, superficial and deep, central and peripheral, help us to navigate the body.

→ It is important to always remember that directions such as right and left are referenced from the *patient's* perspective and NOT yours.

→ The body has several cavities that house anatomical structures (mainly organs). For example, the cranial cavity houses the brain, the thoracic cavity houses the heart and lungs, the abdominopelvic cavity houses the digestive and reproductive organs, and the spinal cavity houses (guess what) the spinal cord.

→ The body has many specific regions. For example, the umbilical region is found around your naval, or belly button, and the femoral region is located in the upper inner thigh area.

→ The directional terms, anatomical landmarks, body regions, and body cavities are all important to know so that health care professionals can communicate in specific terms that leave no room for confusion.

Snapshots from the Journey

This is a concise review summary of key points covered within each chapter.

RAY AND MARIA'S STORIES

Through the use of our two patients within each system chapter, Ray a quadriplegic and Maria a diabetic, the text case studies pull together many of the points discussed within the chapter to show the inter-relatedness of what you are learning. Getting to know these patients well will help you understand how a disease impacts on a variety of body systems.

MARIA'S STORY

Maria has always wanted to have children but she has read horror stories about diabetes and pregnancy. Her doctor has told her that if she is healthy, she could try to get pregnant when the time is right. There are significant risks, however, including a high birth weight baby.

a. What characteristic of diabetes mellitus would contribute to increased birth weight?

b. How can the condition be prevented?

c. What other conditions might Maria need to take care to prevent?

ANCILLARIES GUIDE YOUR JOURNEY

Study Success Companion

Your Study Success Companion will help you establish a good foundation for your trip. This appendix includes study skills and stress management techniques to help you as you journey through Anatomy, Physiology, & Disease. It also contains topics such as the Metric System in case you do not have this background and need a self taught mini-refresher.

Student Workbook

This supplemental workbook contains even more practice and reinforcement opportunities and helps you prepare for quizzes and exams.

INTEGRATED MULTIMEDIA BRINGS CONCEPTS ALIVE

DVD Interactive Exercises

At the completion of selected sections, we prompt readers to visit the DVD to view animated or video presentations of what is described in the book as well as interactive exercises. The DVD contains:

- Animations and short videos
- Interactive games and puzzles
- Audio pronunciation glossary

Companion Website

Located at **www.prenhall.com/colbert** the companion website complements the textbook and DVD. The features include:

- A variety of health care career video profiles and corresponding Internet links for each chapter
- Quizzes in multiple question formats that offer immediate scoring and feedback
- Audio glossary in which key terms are pronounced
- Links to updates on news items related to healthcare from *The New York Times*
- Related Web links (i.e. American Heart Association)

www.prenhall.com/colbert
Not only do the skeletal muscles facilitate movement but, integrated with the nervous system, they provide support for posture while standing or sitting. Promoting balance and posture, along with proper muscle function, is one of the responsibilities of physical therapists. Physical therapists perform many therapies, such as range of motion (ROM) exercises, to ensure full muscle movement. Occupational therapists assist patients in utilizing and adapting their muscle function to perform activities of daily living and improving their quality of life. Massage therapists work directly on the muscles to aid in their relaxation and optimal functioning. To learn more about these professions, please go to the Web site for this chapter.

TEACHING ANCILLARIES BRING OUT THE BEST IN INSTRUCTORS

This text offers a rich array of ancillary materials to benefit instructors and help infuse a spark in the classroom. The full complement of supplemental teaching materials is available to all qualified instructors from your Pearson Health Science sales representative.

Online Course Management Systems

Instructors wishing to facilitate on-line courses will be able to access OneKey, an integrated on-line resource that brings a wide array of resources together in one convenient place for both students and faculty. OneKey features everything you and your students need for out-of-class work, conveniently organized to match your syllabus. OneKey's online course management solution features interactive modules, text and image PowerPoint presentationss, animations, videos, case studies, and more. OneKey also provides course management tools so faculty can customize course content, build online tests, create assignments, enter grades, post announcements, and communicate with students. OneKey content is available in Blackboard, WebCT, and the nationally hosted version CourseCompass platform. Please contact your Pearson Health Science sales representative for a demonstration or go online to **www.prenhall.com/onekey**.

Instructor's Resource Manual

This manual contains a wealth of material to help faculty plan and manage the anatomy and physiology course. It includes many teaching tips, individual and team activities and games, ethical dilemmas, outlines, learning objectives, concept maps, answers to the chapter review activities, worksheets, handouts and a complete 3,000 question test bank. The IRM also guides faculty how to assign and use the text-specific Companion Website, **www.prenhall.com/colbert**, and the DVD that accompanies the textbook.

Instructor's Resource DVD

Packaged along with the Instructor's Resource Manual, this cross-platform DVD provides many resources in an electronic format. First, the DVD includes the complete 3,000 question test bank that allows instructors to generate customized exams and quizzes. Second, it includes a comprehensive, turn-key lecture package in PowerPoint format. The lectures contain discussion points along with embedded color images from the textbook as well as bonus illustrations, animations, and videos to help infuse an extra spark into classroom experience. Instructors may use this presentation system as it is provided, or they may opt to customize it for their specific needs.

PowerPoint content to support instructors who wish to use a classroom response system ("clickers") is provided. Visit **www.prenhall.com/crs** for more information. Finally, the IRDVD contains a complete image library that includes every photograph and illustration from the textbook. These images are provided in three formats: with labels, with leader lines only, and unlabeled.

A COMMITMENT TO ACCURACY

It is vital that the content of this book and all ancillary resources be completely accurate, matching the need for precision in today's health care environment. In order to attain the highest level of accuracy possible throughout this educational program, we put an extensive development process into place. No fewer than a dozen content experts have read each page of chapter for accuracy, including all test questions, the website, the student DVD, and all other ancillary resources.

While our intent is for all books to be error-free, it is possible for some mistakes to get through our system. Pearson/Prentice Hall takes this issue seriously and therefore welcomes any and all feedback that you can provide along the lines of helping us enhance the accuracy of this book. If you identify any errors that need to be corrected in a subsequent printing, please send them to: Pearson Health Science Editorial; Health Professions Corrections; 1 Lake St.; Upper Saddle River, NJ, 07458.

ANATOMY, PHYSIOLOGY, & DISEASE

Learning the Language

Imagine the many potential problems you would encounter traveling to a foreign country where you do not speak the language. To maximize the success and safety during your journey, one of the most important preparatory steps is to develop a basic understanding of the native language before you actually arrive. Every profession seems to have a language all its own. The health care profession is no exception. It has a highly specialized language complete with numerous medical abbreviations.

The language upon which health professions and the study of anatomy, physiology, and disease is based is medical terminology. Therefore, this chapter lays the foundation of learning the native language (medical terminology) of medicine. In addition, this chapter introduces the key concepts of anatomy, physiology, and disease so future chapters can build on this foundation. Finally, as in all journeys there are road signs to make your trip easier. Our road signs are the special features presented by identifiable icons.

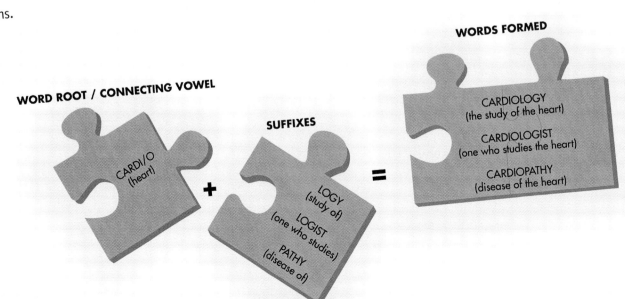

WORD ROOT / CONNECTING VOWEL

CARDI/O
(heart)

+

SUFFIXES

LOGY
(study of)

LOGIST
(one who studies)

PATHY
(disease of)

=

WORDS FORMED

CARDIOLOGY
(the study of the heart)

CARDIOLOGIST
(one who studies the heart)

CARDIOPATHY
(disease of the heart)

Chapter

1

LEARNING OBJECTIVES

At the end of your journey through this chapter, you will be able to:

→ Understand the terms anatomy, physiology and disease and their various related areas

→ Relate the importance and purpose of medical terminology to anatomy, physiology and disease

→ Construct and define medical terms using word roots, prefixes, and suffixes

→ Explain the concept and importance of homeostasis

→ Contrast the metabolic processes of anabolism and catabolism

→ Relate signs and symptoms to the disease process

→ Discuss disease concepts related to the body's defense mechanism

→ Contrast routes of transmission of disease and appropriate preventative measures

MULTIMEDIA APPLICATIONS

DVD Interactive Exercises

→ Video on medical specialties, 1-1

→ Videos on vital signs and pulse sites, 1-2

→ Videos on proper hand washing and gloving, gowning, goggles and masks for infection control, 1-3

→ Animation on cervical diving injury, 1-4

→ Interactive puzzles and games with medical terminology and disease concepts, 1-5

www.prenhall.com/colbert

→ Professional Profiles:
  • Medical Assisting
  • Medical Records and Health Information Technician
  • Medical Transcriptionist

→ Related Internet Links

→ Additional Review Questions

Pronunciation Guide

 Correct pronunciation is important in any journey so that you and others are completely understood. Here is a "see and say" Pronunciation Guide for the more difficult terms to pronounce in this chapter.

anabolism (ah NA bow lizm)

anatomy (ah NA tom ee)

catabolism (ka TA bow lizm)

diagnosis (dye ahg NOH siss)

epidemiology
   (EP uh dee me ALL oh jee)

etiology (ee tee ALL oh jee)

homeostasis (HOH mee oh STAY siss)

idiopathic (ID ee oh PATH ick)

macroscopic anatomy
   (MAK roh scop ic ah NA tom ee)

microscopic anatomy
   (MY kroh scop ic ah NA tom ee)

metabolism (me TA bow lizm)

nosocomial (NOHS oh KOH me al)

organism (OR gan iz em)

pathology (path ALL oh jee)

physiology (fiz ee ALL oh jee)

prognosis (prog NOH siss)

syndrome (SIN drohm)

WHAT IS ANATOMY AND PHYSIOLOGY?

You're probably so accustomed to hearing the words anatomy and physiology (A&P) used together that you may not have given much thought to what each one means and how they differ. They each have unique meanings. Let's take a closer look.

Anatomy

anatomy *(ah NA tom ee)*

Anatomy is the study of the internal and external structures of plants, animals, or for our focus, the human body. The human body is an amazing and complex structure that can perform an almost limitless number of tasks. To truly understand how something works, it is important to know how it is put together. Leonardo da Vinci, in the 1400s, correctly drew the human skeleton and could be considered one of the earliest anatomists (one who studies anatomy). The word anatomy is from the Greek language and literally means "to cut apart," which is exactly what you must do to see how something is put together. For example, the study of the arrangement of the bones that comprise the human skeleton, which is the anatomical framework for our bodies, is considered skeletal anatomy.

microscopic anatomy
   (MY kroh scop ic ah NA tom ee)
   micro = *small*
   scope = *instrument to examine*
macroscopic anatomy
   (MAK roh scop ic ah NA tom ee)
   macro or gross = *large*
   cyto = *cells*
   histo = *tissues*
   ology = *the study of*

Just as we can subdivide biology into more specific concentrations, such as cell biology, plant biology, and animal biology, we can also broadly divide anatomy into **microscopic anatomy (fine anatomy)** and **macroscopic anatomy (gross anatomy)**. Microscopic anatomy is the study of structures that can be seen and examined only with magnification aids such as a microscope. The study of cellular structure (cytology) and tissue samples (histology) are examples of microscopic anatomy.

Gross or macroscopic anatomy represents the study of the structures visible to the unaided or naked eye. For example, the study of the various bones that make up the human body is gross anatomy. Viewing an X-ray of the arm to determine the type and location of a broken bone is considered an examination of gross anatomy.

Physiology

Physiology focuses on the function and vital processes of the various structures making up the human body. These physiological processes include muscle contraction, our sense of smell and sight, how we breathe, and so on. We will focus on each of these processes in their respective chapters. Physiology is closely related to anatomy because it is the study of how an anatomical structure such as a cell or bone actually functions. Physiology deals with all the vital processes of life. It is more complex, and has many subspecialties. Human physiology, animal physiology, cellular physiology, and neurophysiology are just some of the specific branches of physiology.

physiology *(fiz ee ALL oh jee)*
  physi/o = *relationship to nature*
  logy = *study of*

Putting It All Together

In summary, anatomy focuses on structure and how something is put together, whereas physiology is the study of how those different structures work together to make the body function as a whole. For example, anatomy would be the study of the structure of the red blood cells (RBCs), and physiology would be the study of how the RBCs carry vital oxygen throughout our body. Closely related to anatomy is morphology, which is a science that deals with the form and structure of organisms (that is, individual living systems, such as people, animals, and plants). For example, the morphology of cells would be a description of the various shapes of cells. Figure 1–1 ■ shows the morphology of deformed RBCs (sickle shaped) that are present in the disease sickle cell anemia. Because of the anatomical deformity, the physiological process of effectively carrying oxygen is adversely affected.

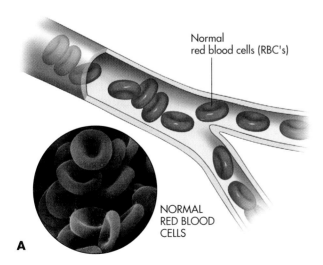
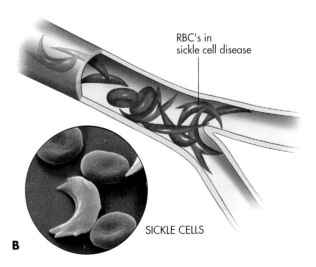

FIGURE ■ 1–1

A. Normal red blood cells (RBCs) are flexible and donut shaped and move with ease through blood vessels. **B.** The anatomical distortion of the *structure* of RBCs in sickle cell anemia affects its normal *function* to carry oxygen. In addition, the sickle cells lose their ability to bend and pass through the small blood vessels, thereby causing blockages to blood flow.

Learning Hint

USING THE MARGINS OF THIS BOOK

Notice that the margin notes present a breakdown of the medical terms discussed in the text. Sometimes you may already know the term and may not need to refer to the margin note, but it is always there to help reinforce the word. On occasion, you may even see a short story on the word origin when it is of interest or helps to further explain the term to make it easier to learn.

You will notice on your journey that the design of a structure is often related to its function. For example, as you will learn in Chapter 6, the type of joint located between bones is dictated by the functions of those bones: hinge joints are located at the knees where back and forth bending movement is required, while a ball and socket joint of the hip provides for a greater range of motion.

Therefore, it makes sense to combine these two sciences into anatomy and physiology (A&P). Understanding human anatomy and physiology forms the foundation for all medical practice because medical treatment attempts to bring the body's structure and function back towards normal A&P.

WHAT IS DISEASE?

disease *(dih ZEEZ)*
   dis = *undone from, not; literally, disease means "not at ease"*

pathology *(path ALL oh jee)*
   path/o = *disease*

etiology *(ee tee ALL oh jee)*

idiopathic *(ID ee oh PATH ick)*
nosocomial *(NOHS oh KOH me al)*
epidemiology
   (EP uh dee me ALL oh jee)

Disease, simply put, is a condition in which the body fails to function normally. The word **disease** literally means not (dis) at ease. If we consider the body at ease when normal anatomy and physiology exists, anything that upsets the normal structure or functioning could be called disease. The official medical term for the study of disease is **pathology**. Pathology is the branch of medicine that studies the characteristics, causes, and effects of disease. Another closely related term is pathophysiology, which is the science that investigates abnormal body function.

In order to effectively treat a disease, the cause (or **etiology**) of the illness must be determined. If the cause of a disease cannot be determined, it is called an **idiopathic** disease. A **nosocomial** infection is an infectious disease you acquire while in a medical facility. The study of the transmission, frequency of occurrence, distribution and control of a disease is called **epidemiology**. For example, an epidemiologist determined that there are nearly 62 million cases of the common cold each year in the United States.

While *communicable* and *contagious* may sound like the same thing, there is a fine difference. A **communicable disease** can be spread in a variety of ways, such as person-to-person or insect-to-person. In contrast, a **contagious disease** is one that is readily transmitted from one person to another such as the common cold.

The Centers for Disease Control and Prevention (CDC) tracks disease worldwide. If a disease occurs in a specific population or region, it is called **endemic**. If the disease occurs in large numbers over a specific region, it is called an **epidemic**. If the disease spreads country or worldwide, it is called a **pandemic**. You will learn more terms associated with disease as you further explore this chapter.

1-1 Medical Specialties

There are numerous medical specialists (ologists), and just by going to the Yellow Pages of the phone book, you can see firsthand the vast array that exists—from anesthesiologists to urologists. Many of these specialties are discussed in the upcoming chapters that deal with specific body systems, but if you're so excited that you can't wait to know more, go to the DVD to view a video on medical specialties and learn a few more word roots along the way.

TEST YOUR KNOWLEDGE 1-1

Indicate whether the following examples are gross anatomy or microscopic anatomy by putting a G or M in the space provided.

1. _____ Viewing an X-ray to determine the type of bone break

2. _____ Classifying a tumor to be cancerous by cell type

3. _____ Viewing bacteria to determine what disease is present

4. _____ Examining the chest for any obvious deformities

5. _____ The type of anatomy studied by a histotechnologist or cytotechnologist

Complete the following or choose the best answer:

6. John was diagnosed with idiopathic respiratory disease. What does this mean?

   _____

7. A study was conducted on the frequency of occurrence of a particular infection in South America. The scientists performing the study were called _____. The cause or _____ of the infection was determined to be a type of bacteria.

8. The spread of avian flu throughout the world would lead to a/an:
   a. endemic   c. birdemic
   b. epidemic   d. pandemic

THE LANGUAGE

Even if you're traveling to another city within your home country, you will most likely need to learn a few things about the language its citizens speak. For example, think of the many different names people use to identify a sandwich made on a long skinny roll. Your "sub" might be someone else's "grinder" or "hoagie." Keep in mind that there may also be small regional differences in language and abbreviations.

Medical Terminology

As stated earlier, the language of A & P and disease is based primarily on medical terminology. Understanding medical terminology may seem like an overwhelming task because, on the surface, there appears to be so many terms. In reality, there are only a relatively few root terms, prefixes, and suffixes, but they can be put together in a host of ways to form numerous terms.

Each medical term has a basic structure upon which to build, and this is called the word root. For example, *cardi* is the word root for terms pertaining to the heart. Rarely is the word root used alone. Instead, it is combined with prefixes and suffixes that can change its meaning. Prefixes come before the word root, while suffixes come after the word root. The suffix **ology** means "study

> *Learning Hint*
>
> **GENERAL HINTS ON FORMING MEDICAL TERMS**
>
> While you can learn the various word roots, prefixes, and suffixes, it gets confusing trying to put them correctly together. In most instances, the medical definition indicates the last part of the term first, especially when suffixes are used. For example, an inflammation of the stomach is gastritis, not itisgastro, and one who studies the stomach is a gastrologist, not an ologistgastro. When using prefixes, you usually put the parts together in the order you say the definition. For example, slow heart rate is bradycardia, not cardiabrady. As with general rules, there are exceptions, but with practice it will become familiar to you.

card/i = *heart*
logy or ology = *study of*
tachy = *fast*
ologist = *one who studies*

of," and therefore, we can combine cardi and ology to form **cardiology**, which is the study of the heart. The prefix *tachy* means "fast," and can be placed in front of the word root to form **tachycardia**, which means a fast heart rate.

Often you will be given a **combining form**, which is the word root and a connecting vowel (usually o), to make it easier to pronounce and link with possible suffixes. For example, the combining form for heart is cardi/o. Figure 1–2 ▦ shows the components of a medical term.

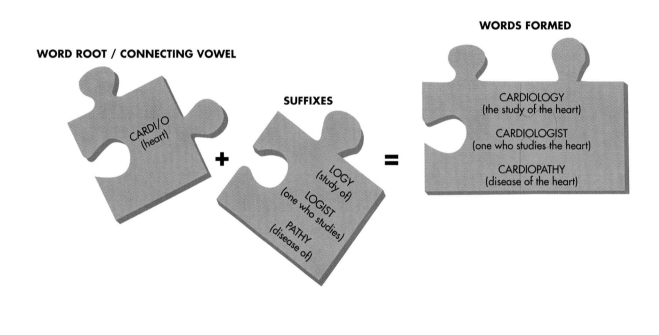

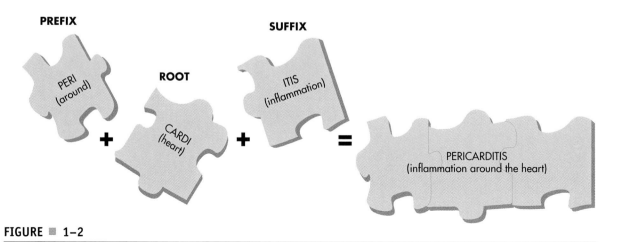

FIGURE ▦ **1–2**

How prefixes and suffixes can be combined with a word root to form many medical terms.

Table 1–1 presents some common combining forms to get you started. Then let's add some common prefixes that can be placed before the word roots to alter their meaning (see Table 1–2).

TABLE 1–1 Common Combining Terms

WORD ROOT/ COMBINING FORM	MEANING
abdomin/o	abdomen
aden/o	gland
angi/o	vessel
arthr/o	joint
cardi/o	heart
col/o	colon
cyan/o	blue
cyt/o	cell
derm/o	skin
erythr/o	red
gastr/o	stomach
glyc/o	sugar
hemat/o, hem/o	blood
hepat/o	liver
hist/o	tissue
hydr/o	water
leuk/o	white
mamm/o	breast
nephr/o	kidney
neur/o	nerve
oste/o	bone
path/o	disease
phag/o	to swallow
phleb/o, ven/o	vein
rhin/o	nose

TABLE 1–2 Common Prefixes

PREFIX	MEANING
a or an	without
acro	extremities
brady	slow
dia	through
dys	difficult
electro	electric
endo	within
epi	upon or over
hyper	above normal
hypo	below normal
macro	large
micro	small
peri	around
sub	under, below
tachy	fast

Finally, let's add some common suffixes (Table 1–3) and then see what kinds of words we can form with just these few parts.

Using Tables 1–1, 1–2, and 1–3, look at all the terms you can make from just the one word root, cardi/o. *Cardiology* is the study of the heart, and a cardiologist is a specialist who studies the heart. *Bradycardia* is a slow heart, *tachycardia* is a fast heart, and an *electrocardiogram* is an electrical recording of the heart. If your heart were enlarged due to inflammation (*carditis*), you may develop *cardiomegaly*, which would mean you have a disease of the heart (*cardiopathy*). The Tin Man from *The Wizard of Oz* thought he had no heart (*acardia*), but later realized that he had a heart all along.

TABLE 1–3 Common Suffixes

SUFFIX	MEANING
-al, ic	Pertaining to or related to
-algia	Pain
-cyte	Cell
-ectomy	Surgical removal of, or excision
-gram	Actual record
-graphy	Process of recording
-ist	One who specializes
-itis	Inflammation of
-megaly	Enlargement of
-ologist	One who studies
-ology	Study of
-oma	Tumor
-otomy	Cutting into
-ostomy	Surgically forming an opening
-pathy	Disease
-penia	Decrease or lack of
-phobia	Fear of
-plasty	Surgical repair
-scope	Instrument to view or examine

Abbreviations

Abbreviations are used extensively in the medical profession. They are useful in simplifying long, complicated terms for disease, diagnostic procedures, and therapies that require extensive documentation. For now, review Table 1–4 for some common abbreviations you may have heard in a health care setting or on television.

TABLE 1–4 Common Medical Abbreviations

ABBREVIATIONS	MEANING
A&P	anatomy and physiology
ACLS	advanced cardiac fife support
b.i.d.	twice a day
BM	bowel movement
BP	blood pressure
CA	cancer
CAD	coronary artery disease
CBC	complete blood count
CPR	cardiopulmonary resuscitation
CVA	cerebral vascular accident (stroke)
CXR	chest X-ray
Dx	diagnosis
GI	gastrointestinal
ICU	intensive care unit
IM	intramuscular
IV	intravenous
MI	myocardial infarction (heart attack)
NPO	Latin *nil per os*, which means "nothing by mouth"
P.O.	orally
p.r.n.	when needed
Q	every
SOB	shortness of breath
STAT	Latin *statim*, which means "immediately"
t.i.d.	three times a day
ER/ED	emergency room/emergency department

Note: ER was popularized by the television show of the same name. However, in actuality it is really a whole department and not just a room, so most prefer the abbreviation ED, which stands for Emergency Department.

Of course you will learn many more terms and abbreviations as we explore the upcoming chapters and become fluent in conversational medical language. This will help you to avoid using lay terms (common, everyday terms) to describe medical and anatomical concepts. Now you know that the correct term for "getting a nose job" is **rhinoplasty** (surgical repair of the nose).

rhinoplasty *(RYE noh plass tee)*
rhin/o = *nose*
plasty = *surgical repair*

TEST YOUR KNOWLEDGE 1-2

Define the medical terms:

1. acrocyanosis

    _____

2   gastritis

    _____

3. rhinoplasty

    _____

4. bradycardia

    _____

5. mammogram

    _____

6. cytomegaly

    _____

Give the correct medical term:

7. inflammation of the kidneys

    _____

8. removal of the stomach

    _____

9. enlarged heart

    _____

10. disease of the bones

    _____

11. specialist who studies the nerves

    _____

The Metric System

Whereas medical terminology represents the written and spoken language for understanding anatomy and physiology, the metric system is the "mathematical language" of medicine. For example, blood pressure is in millimeters of mercury (mm Hg), and organ size is usually measured in centimeters (cm). Medications and fluids are given in milliliters (ml), and weight is often measured in kilograms (kg). What exactly does it mean when you are taught that normal cardiac output is 6 liters per minute? You can now see why you must be familiar with the metric system in order to truly understand anatomy and physiology and medicine. While the metric system may seem complicated if you are not familiar with it, it really isn't if you have a basic understanding of math.

There are two major systems of measurement in our world today. The United States Customary System (USCS) is used by the general population of the United States. The International System of Units (SI) is used almost everywhere else, as well as by US health care professionals. The SI system is also known as the **metric system** and is based on the power of 10. The metric system is also the system used by drug manufacturers.

The USCS system is based on the British Imperial System and uses several different designations for the basic units of length, weight, and volume. We commonly call this the **English system**. For example, in the English system, volumes can be expressed as ounces, pints, quarts, gallons, pecks, bushels, or cubic feet. Distance can be expressed in inches, feet, yards, and miles. Weights are measured in ounces, pounds, and tons. This may be the system most Americans are familiar with, but it is not the system of choice used throughout the world and within the medical profession because the English system has no common base and is therefore very cumbersome to use. It is difficult to know the relationship between each unit of measure because they are not based in an orderly fashion according to the powers of 10 as in the metric system. For example, how many pecks are in a gallon? Just what is a peck? How many inches are in a mile? These all require extensive calculations and memorization of certain equivalent values, whereas in the metric system, you simply move the decimal point the appropriate power of 10.

If you want to learn more about the metric system, please refer to the student *Study Success Companion* at the end of this text for a detailed explanation on how to use the metric system in the health care setting.

ANATOMY, PHYSIOLOGY, AND DISEASE CONCEPTS YOU WILL ENCOUNTER ON YOUR JOURNEY

In this section, we take a closer look at some additional concepts related to the study of A&P and pathology. You will learn more about these as you journey through the chapters in this text.

Metabolism

If you travel to other countries, you will see many different cultures and customs. Even though each culture is unique, they all share certain similarities. The same can be said in anatomy, physiology and disease. We all share certain functions that are vital to survival. All humans, for example, need food in order to produce complex chemical reactions necessary for growth, reproduction, movement, and so on. **Metabolism** refers to all of the chemical operations going on within our bodies. Metabolism requires various nutrients or fuel to function and produces waste products much like a car consumes gas for power and produces waste, or exhaust. Metabolism, for now, can be thought of as "all the life-sustaining reactions within the body." Fever is a common disease process that will speed up metabolism.

> ## *Clinical Application*
>
> ### METABOLIC SYNDROME, OR SYNDROME X
> A disturbing new syndrome is affecting nearly one quarter of the United States adult population, known as Metabolic Syndrome, or Syndrome X. A patient with this syndrome exhibits at least three of the following five common conditions: high blood sugar levels (hyperglycemia), high blood pressure (hypertension), abdominal obesity, high triglycerides (a lipid substance in the blood), and low level of HDL (which is sometimes referred to as "good cholesterol"). Individuals who exhibit this syndrome are at an increased risk for diabetes, heart attacks, and/or strokes. The main causes of the syndrome are poor diet and lack of exercise.

metabolism *(me TA bow lizm)*

anabolism *(ah NA bow lizm)*
ana = *up, as in build up*

catabolism *(ka TA bow lizm)*
cata = *down, as in tear down*

Metabolism is further subdivided into two opposite processes. **Anabolism** is the process by which simpler compounds are built up and used to manufacture materials for growth, repair, and reproduction, such as the assembly of simple amino acids to form complex proteins. This is the building phase of metabolism. **Catabolism** is the process by which complex substances are broken down into simpler substances. For example, the breakdown of food into simpler chemical building blocks for energy use is a catabolic process. An abnormal and extreme example of catabolism is a starvation victim whose body "feeds upon itself," actually consuming the body's own tissues.

Homeostasis

For the body to remain alive, it must constantly monitor both its internal and external environment and make the appropriate adjustments. In order for cells to thrive, they must be maintained in an environment that provides a proper temperature range, balanced oxygen levels, and adequate nutrients. Heart rate and blood pressure must also be monitored and maintained within a certain range or set point for optimal functioning depending upon the body activity. **Homeostasis** is the physiological process that monitors and maintains a stable internal environment or equilibrium. Survival depends upon the body's ability to maintain homeostasis. Homeostatic regulation refers to the adjustments made in the human organism to maintain this stable internal environment.

homeostasis
(HOH mee oh STAY siss)
home/o = *unchanging*
stasis = *standing still; as you will soon see, stasis is not an accurate term because homeostasis is actually a dynamic state of equilibrium*

The thermostat in your house functions like a homeostatic mechanism. A temperature is set and then maintained by a sensor that monitors the internal environmental temperature and either heats the house if the sensor registers too cold or cools the house if the sensor registers too hot. There is a continuous feedback loop from the sensor to the thermostat to determine what action is needed. Because the feedback loop opposes the stimulus (cools down if too hot, heats up if too cold), it is referred to as a **negative feedback loop.**

The body also relies on negative feedback loops that continually sense the internal and external environment and signal the body to make adjustments to maintain homeostasis (see Figure 1–3 ■). For example, the hypothalamus in the brain represents the body's thermostatic control. If the hypothalamus senses a very cold environment, it opposes this cold stimulus (negative feedback loop) and requests physiological processes to create or conserve heat within the body to maintain an internal temperature near 98.6°F. For example, the body begins to shiver, and this increased muscular activity generates heat. In addition, since most heat loss is through peripheral areas (head, arms, and legs), the body decreases the size of the peripheral blood vessels (vasoconstriction), causing the blood to be farther away from the skin surface where the heat would be lost to the cold environment. This keeps the blood closer to the core of the body, where it is warmer. Of course, we can assist the body by wearing a heavy coat and hat, which would remove much of the

Clinical Application

"BREAKING" A FEVER

It is believed that most fevers are the body's way of making an inhospitable environment for a pathogen to survive. Why is it that when someone begins sweating after a prolonged fever, the fever is said to be "breaking?" A fever sets the hypothalamus to a higher set point temperature. The body increases metabolism to generate more heat to reach this now-higher set temperature. Once whatever is causing the fever is gone, the hypothalamus set temperature is turned back down to the true normal. The body must now rapidly get rid of the excess heat by the cooling process of evaporation through sweating.

stress of the cold environment, or simply by getting out of the cold and into a warmer environment.

Conversely, if you are in the desert and the temperature is 120°F, the body senses this as too hot and stimulates physiological processes to cool you down. These processes include sweating (because evaporation is a cooling process) and enlarging the peripheral vessels (peripheral vasodilation) in order to dissipate the body heat into the external environment.

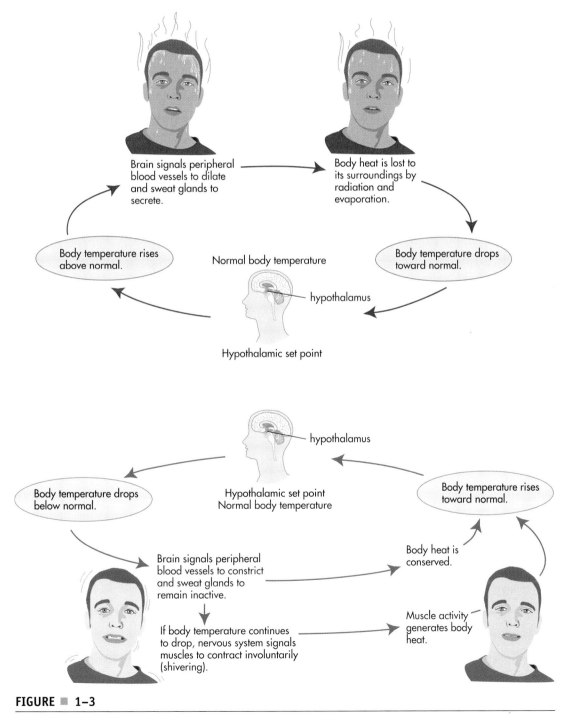

FIGURE ■ 1–3

The homeostatic control of normal body temperature (37°C or 98.6°F).

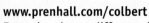
www.prenhall.com/colbert
Even time has a different designation in the medical professions than in everyday life. In health care, military time is used,
which is a 24-hour clock. For times after 12:00 noon, you add 12 to the
time. For example, for 1:00 p.m., the time is 13:00. Please visit the website if you are not familiar with military time, for a discussion and demonstration of its use.

In health care practice, if a patient presents with a very high temperature, he or she may need to be rapidly cooled with ice packs or baths to reduce the body temperature. Much of health care practice is just that— assisting the body through therapy and treatments so it can return to homeostasis.

Your body is also capable of **positive feedback**, which increases the magnitude of a change. This process is also known as a "vicious cycle." Positive feedback is not a way to regulate your body, because it increases a change away from the ideal set point. In most cases, positive feedback is harmful if the vicious cycle cannot be broken, but sometimes positive feedback is necessary for a process to run to completion.

A good example of necessary positive feedback is the continued contraction of the uterus during childbirth. When a baby is ready to be born, a signal, not well understood at this time, tells the hypothalamus to release the hormone oxytocin from the posterior pituitary. Oxytocin increases the intensity of uterine contractions. As the uterus contracts, the pressure inside the uterus caused by the baby moving down the birth canal increases the signal to the hypothalamus. More oxytocin is released, and the uterus contracts harder. Pressure gets higher inside the uterus, the hypothalamus is signaled to release more oxytocin and the uterus contracts yet harder. This cycle of ever-increasing uterine contractions due to ever-increasing release of oxytocin from the hypothalamus continues until the pressure inside the uterus decreases—that is, until the baby is born.

DISEASE CONCEPTS

Ideally, the body works to make things function smoothly and in balance. However, sometimes things happen to alter those functions. Eating habits, smoking, inherited traits, trauma, environmental factors, and even aging can alter the body's balance and lead to disease. Again, in general terms, disease is a condition in which the body fails to function normally.

Think back to a time when you were sick. You may have had a fever, cough, nausea, dizziness, joint aches, or a generalized weakness. These are examples of

1-2 Go to the DVD to learn more about pulse sites and to view videos
on the proper way to take the various vital signs.

signs and **symptoms** of disease. While the terms signs and symptoms are often used interchangeably, each has its own specific definition. Signs are more definitive, objective, measurable indicators of an illness. Fever and change in the size or color of a mole are examples of signs. **Vital signs** are the signs vital to life, and include pulse (heart rate), blood pressure, body temperature, and respiratory rate. The vital sign standard values can change according to the patient's age and sex. In some health care settings, weight is now being added to the list of vital signs. Because of variations in people, we often use a clinical range of acceptable values when measuring vital signs.

Symptoms, on the other hand, are perceived by the patient and are therefore more subjective and difficult to measure. A perfect example of a symptom is pain. Tolerance to pain varies among individuals, so an equal amount of pain such as a controlled pin prick applied to a number of people could be perceived as a light, moderate, or intense level of pain depending on each individual's perception. Other symptoms can include dizziness and itchiness. In spite of the fact that symptoms are hard to measure, they are still very important in the diagnosis of disease. Sometimes a disease exhibits a set group of signs and symptoms that may occur at about the same time. This specific grouping of signs and symptoms related to a specific disease is known as a **syndrome**. For example, the signs and symptoms of Down Syndrome would include mild to moderate mental retardation, sloping forehead, low set ears, short broad hands and often, cardiac valvular disease. Signs, symptoms, and syndromes are further explained throughout the rest of the textbook as they relate to the pathology of the various body systems.

Amazing Body Facts

BIZARRE SIGNS AND SYMPTOMS!

Here are some strange signs and symptoms that may be indications of diseases. To keep lawyers from making a ton of money on this book, please note that there are other signs, symptoms, and tests to determine specific diseases. Please do not use this list of oddities as a sole diagnostic tool! Void where prohibited, you must be 18 years of age to play, cannot be used with other coupons, and all other restrictions apply.

1. Generalized itching skin can be an indication of Hodgkin's disease.

2. Sweating at night may indicate tuberculosis.

3. A desire to eat clay or starchy paste may indicate an iron deficiency in the body.

4. Breath that smells like fruit-flavored chewing gum may be an indication of diabetes.

5. A magenta colored tongue is indicative of a riboflavin deficiency.

6. A patient with profound kidney disease often doesn't have moons on his or her fingernails.

7. A "hairy" tongue may mean that a patient's normal mouth flora has died from improper use of antibiotics.

8. Spoon-shaped fingernails may point to an iron deficiency in the body.

9. Brown linear streaks on the fingernails of fair-skinned people may indicate melanoma (skin cancer).

The word *gnosis* is Greek for knowledge. *Dia* means through or complete, and therefore, *diagnosis* literally means "know through or completely." *Pro* is a prefix meaning before or in front of, and *prognosis* literally means the foreknowledge or predicting of the outcome of a disease.

diagnosis *(dye ahg NOH sis)*
dia = *thorough or complete*
gnosis = *knowledge*

Discovering as many signs and symptoms as possible can help to **diagnose** a disease. A diagnosis is an identification of a disease determined by studying the patient's signs, symptoms, history, and results of diagnostic tests. The diagnostic procedure is begun by first determining why the individual is seeking medical help. This is known as determining their **chief complaint (CC)**.

Although the individual may have several medical problems, which may be of equal or more importance, the chief complaint is what brought that person to now seek help. It may be something as simple as a sore throat or as involved as shortness of breath with concurrent radiating chest pain. Regardless, it is important to accurately record their chief complaint. Expanding upon this, the next step is to discuss and document the history of the present illness. Ask open ended questions that allow the patient to fully explain what they feel. When did the symptoms first appear? What did you do to cause them? How long do they last? What makes them go away?

Getting the medical history can help in determining the etiology, or cause, of the disease. The **prognosis** is the prediction of the outcome of a disease. Hopefully, your prognosis for doing well in this course is excellent.

prognosis *(prog NOH siss)*
pro = *before or in front of*
gnosis = *knowledge*

Clinical Application

THE VITAL SIGN OF PULSE

The pulse is commonly taken by applying slight finger pressure over the radial artery located in each wrist (on the thumb side) and counting the number of beats in a 60-second period (please see Figure 1–4 ■). The normal heart rate for an adult is 60 to 100 beats per minute, child rate is approximately 70 to 120, and a newborn's rate is 90 to 170 beats per minute. If an adult has a heart rate of 165 beats per minute, what medical term would you use to describe that condition?

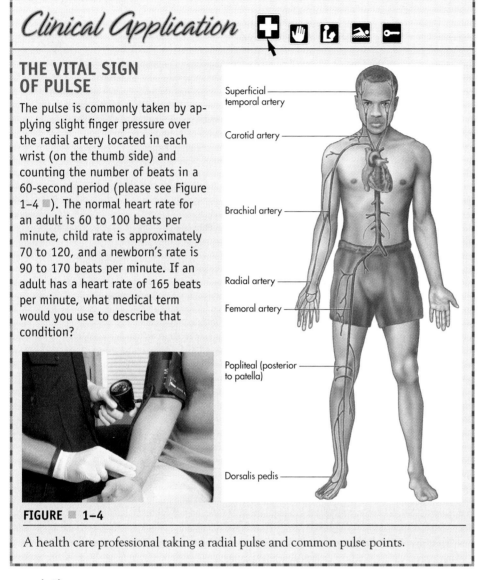

Superficial temporal artery

Carotid artery

Brachial artery

Radial artery

Femoral artery

Popliteal (posterior to patella)

Dorsalis pedis

FIGURE ■ 1–4

A health care professional taking a radial pulse and common pulse points.

exacerbation
(ecks ASS er bay shun)

It is also helpful to determine if the chief complaint was gradual or of a sudden onset. Quite often, symptoms gradually develop from a disease process that may have been there for some time. These often are *chronic* conditions as opposed to *acute* conditions that exhibit a rapid onset of signs and symptoms. One of the problems with chronic conditions is that due to their usually gradual onset, older patients often attribute them to "just getting older" and often ignore them as an indicator of disease. As these conditions worsen these people can no longer ignore these conditions and they seek help. By that time, treating the disease may be more complicated or difficult due to its severity.

The signs and symptoms of a chronic disease may disappear at times and this period is known as **remission** of the disease. **Relapses** are recurrences of the disease. If the signs and symptoms acutely "flare up," this is known as an **exacerbation** of the disease. Diseases that have a prognosis of death are referred to as **terminal** diseases.

TEST YOUR KNOWLEDGE 1-3

Answer the following questions:

1. Check which of the following are vital signs:
   a._____ pulse
   b._____ pain
   c._____ blood pressure
   d._____ age
   e._____ indigestion
   f._____ respiratory rate
   g._____ body temperature

2. Which of the following is the medical term for the cause of a disease?

   a. prognosis

   b. diagnosis

   c. etiology

   d. syndrome

3. Which of the following is the medical term for the outcome of a disease?

   a. prognosis

   b. diagnosis

   c. etiology

   d. syndrome

THE BODY'S DEFENSE SYSTEM

Most microorganisms that enter the body are harmless, but some can be **pathogenic** (disease producing). Some microorganisms are pathogenic only when they enter through openings known as portals of entry or grow in an uncontrolled manner. For example *Escherichia coli (E. coli)* exists normally in your intestinal tract. It breaks down waste products and helps with the production of vitamin K. However, if *E. coli* enters your blood stream, it can cause a severe and potentially fatal infection.

pathogenic
(pa thoe JEN ick)

Body Barriers

Your body has a series of barriers and defenses to protect itself from infection. The first defense is your skin. As long as your skin is unbroken, it provides a shield from penetration by harmful organisms. Your skin is somewhat acidic in nature, so it provides an environment that many harmful organisms don't like.

Sometimes an opening occurs on the skin's surface, allowing germs through that first barrier. Then the second barrier comes into action and the **immune response** is triggered in the form of microscopic cells that either attack and "eat" those harmful invaders or release powerful chemicals that disintegrate parts of the invading germ.

An **inflammatory response** occurs whenever the tissues in your body are injured. This response can occur as a result of physical injury, intense heat, irritation from certain types of chemicals, or a bee sting. In addition, the inflammatory response can also occur as a reaction to the invasion of germs as in an infection. See Figure 1-5 ■. The four signs/symptoms of an inflammatory response are: redness, increased temperature at the affected site, swelling (edema), and pain. The inflammatory response is normally protective and attempts to isolate the injured area and increase blood flow to restore normal function.

Clinical Application

THE GERIATRIC PATIENT

The "senior citizen" population continues to grow rapidly, but just because people are living longer it doesn't mean that they are necessarily living better. Throughout this text you will learn that there are many differences between the various age groups of patients that you will deal with. For example, as people age, their bodies undergo changes in body composition and function, nutritional and drug dosage requirements, and mental functions. A role reversal may also occur, in which a parent gradually becomes dependent upon an adult child. This change may create relationship problems among family members, with the health care professional often being pulled into the situation. There is the potential for you to make a difference in these situations by educating patients and family members or directing them to the proper professionals or organizations.

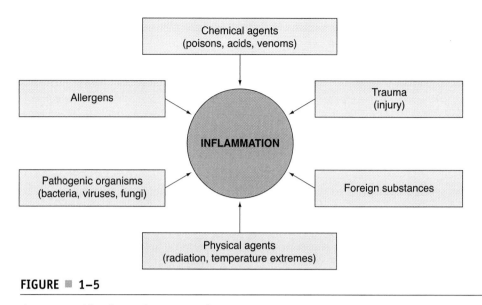

FIGURE ■ 1–5

Agents capable of stimulating an inflammatory response.

If you have been attacked by a certain germ before, a **specific immune response** is activated and your body creates substances to specifically target and destroy a germ that had previously been in your body and caused problems. Closely related to the immune response is an **allergic** or **hypersensitivity reaction**. During an allergic reaction, the immune system goes too far, causing tissue damage and impairing normal function.

Understanding Routes of Transmission

Germs can gain entrance to your body in a variety of ways. The four main routes of transmission are: vectors, contact transmission, common vehicle, and airborne spread. Let's look at each one individually.

Although rarely occurring in hospitals, the vector route of transmission is quite common in the community. A **vector borne** transmission is one in which the organism is carried by an insect or other animal. This can occur in one of two ways. The insect may have the pathogen living inside it, and when it bites you, it infects your blood with that organism. This is known as a **biological vector**. An example of a biological vector is the spread of malaria through the bite of an infected mosquito. Alternatively, the insect may have that organism *on* its body, and spread it to you by landing on an open wound. This is known as a **mechanical vector**. If a fly walks over cow feces, and then lands on an open bowl of potato salad sitting in the warn sun at your picnic, that would be considered disease spreading via mechanical vector.

Contact transmission can also happen in two ways: direct or indirect contact. An example of **direct contact** is when a patient has a bowel movement in bed and the feces come into contact with an open wound on the patient's buttocks. Direct contact can also occur when a health care provider doesn't wash his or her hands or "glove up" before tending to an open wound. An example of **indirect contact** is when medical equipment is not properly sterilized before it is used on the next patient, and germs from the previous patient are transferred to the next. An example at home would be touching a door knob that had been previously handled by someone with the flu.

Common vehicle transmission is one you may hear about in the news. This occurs when consumable goods (such as blood or blood products, IV fluids, food such as ground meat, vegetables or seafood) become contaminated. This sets up the potential for a major epidemic and is readily recognizable by the sudden occurrence of many infections caused by the same pathogen.

Airborne transmission is a result of the spread of droplets that contain the pathogen. Examples include the spread of tuberculosis by sneezing or even laughing, or pathogens growing in an air conditioner and being spread when the air conditioner is turned on, as happened in the case of the outbreak of Legionnaires' Disease in 1977 in Philadelphia.

Standard Precautions

This cycle of the infection, beginning with the creation of a source of infection, continuing with the transportation of the pathogen, and then ending with the entry into the body, is called the **chain of infection**. We can stop the spread of infectious disease simply by breaking this chain at some point. This can be done by something as simple as properly washing your hands.

Tracing the history of healthcare practices over time illustrates our attempts to improve our ability to prevent the spread of infection. However, in the not so distant past our focus was very disease specific and not generalized to the health care provider and his or her interaction with each patient. The realization has come that infectious agents are not always recognized and labeled as such and that they may be present and can be spread without our awareness.

Because of this fact, Standard Precautions have been developed. These provide the most basic protection for both the health care provider and the patient. Standard Precautions assumes that everyone, including the provider and receiver of healthcare, family members, and visitors could be carrying a communicable disease.

The single most important practice to reduce the transmission of infectious agents in healthcare settings is hand hygiene. Hand hygiene includes effective handwashing with plain or antiseptic-containing soap and water, or the use of alcohol-based products that do not require the use of water.

Personal protective equipment (PPE) is a variety of barriers and respirators used alone or in combination to protect mucous membranes, airways, skin, and clothing from contact with infectious agents. These items include gloves, isolation gowns, face protection (masks, goggles, and face shields), and respiratory protection with specialty masks. These items are specific to different types of communicable disease and their methods of spreading.

Another important measure that prevents transmission of hepatitis and human immunodeficiency virus is the prevention of needlesticks and other sharps-related injuries. These include measures to handle needles and other sharp devices in a manner that will prevent injury to the user and to others who may encounter the device during and after a procedure.

See Figure 1 – 6 ▪ for a Standard Precautions guide. As you can see by the chart, there are a variety of ways to stop the spread of infection. Once again, the single most important thing that you can do to stop the chain of infection is to wash your hands before and after working with each patient! See Appendix G for more information on Standard Precautions and for a list of procedures.

Standard Precautions Guides

Procedure	Wash Hands	Gloves	Gown	Mask	Eyewear
Talking to patient					
Adjusting IV fluid rate or noninvasive equipment	●				
Assess patient without touching blood, body, fluids, mucous membranes	●				
Assess patient including contact with blood, body fluids, mucous membranes	●	●			
Drawing blood	●	●			
Inserting venous access	●	●			
Suctioning	●	●	If splattering is likely	If splattering is likely	If splattering is likely
Handling soiled waste, linen, other materials	●	●	If they are extensively soiled or splattering is likely	If they are extensively soiled or splattering is likely	If they are extensively soiled or splattering is likely
Intubation	●	●	●	●	●
Inserting arteriole access	●	●	●	●	●
Endoscopy	●	●	●	●	●
Operating and other procedures producing extensive splattering of blood or body fluids	●	●	●	●	●

FIGURE ■ 1–6

Standard Precautions Guidelines.

 1-3 To view a video on proper hand washing and gloving techniques, please go to your DVD for this chapter. In addition, you can see videos on proper gowning, goggles, and mask techniques to prevent the spread of infection.

TEST YOUR KNOWLEDGE 1-4

Choose the best answer:

1. A region of a state had an outbreak of hepatitis. When reviewing the histories of the victims, it was noted that they had received either whole blood or blood products within the past 9 months. Upon further investigation, it was determined that all of the blood and blood products had come from one organization's blood lab that had failed to test their donors for hepatitis. This transmission of infection to others is known as:

   a. Direct contact

   b. Indirect Contact

   c. Common Vehicle

   d. Vector

Complete the following:

2. A 27-year-old female presents to the Emergency Department of a local hospital. She has a cough and complains of shortness of breath. An elevated temperature is noted along with an increased pulse rate. She states that her daughter had stayed home from school for the past few days and was diagnosed with an upper respiratory viral infection.

   a. What is her chief complaint?

   _____

   b. What is the etiology of her chief complaint?

   _____

   c. What follow up questions would you ask about her chief complaint?

   _____

   d. Are there any possible routes of transmission involved? Which one(s)?

   _____

   e. What would you do to keep from spreading this infection to other patients?

   _____

SUMMARY

Snapshots from the Journey

→ Anatomy is the study of the actual internal and external structures of the body, and physiology is the study of how these structures normally function. Pathology is the study of the disease processes by which abnormal structures and abnormal body functions can occur.

→ Medical terminology is the language of medicine and combines word roots, prefixes, and suffixes to construct numerous medical terms to describe conditions, locations, diagnostic tools, and so on.

→ The metric system is the mathematical language of medicine based on the powers of 10. If you require more practice with this system, please go to your student *Study Companion Guide* at the end of this text for a simplified review.

→ Metabolism refers to all of the chemical operations going on within the body and can be broken down into two opposite processes. The building phase of metabolism is anabolism, in which simpler compounds are built up and used to manufacture materials for growth, reproduction, and repairs. The tearing down phase is catabolism, in which complex substances are broken down into simpler substances, such as food broken down for energy use.

→ The body tries to maintain a balanced or stable environment called homeostasis. The body must constantly monitor the environment and make changes to maintain this balance, often through the use of negative feedback loops.

→ A change in objective measurable values such as temperature (signs) and subjective patient perceptions (symptoms) can indicate a disease is present. Vital signs are pulse, respirations, temperature and blood pressure.

→ The body's defenses are set up to guard against pathogenic organisms invading through portals of entry. There are various routes of transmission for pathogens to gain entry and these include vectors, contact transmission, common vehicle and airborne spread.

Case Study

A 66-year-old Asian male involved in a vehicular accident is taken to the ICU with dyspnea and abdominal pain. He has acrocyanosis, tachycardia, and a past medical history of cardiopathy. He weighs 150 pounds and is 5 feet 6 inches tall. His chest X-ray shows an enlarged heart. His facial injuries will require future rhinoplastic surgery. An electrocardiogram and lower GI series is ordered.

a. Where exactly in the hospital was the patient taken?

_____

b. Describe the patient's color, heart rate, and breathing.

_____

c. What is the medical term for what the X-ray showed?

_____

d. What future facial surgery will he need?

_____

RAY AND MARIA'S STORIES

As you journey through the chapters, you will meet two common cases that will demonstrate the inter-relatedness of the various systems. We first meet Ray who suffered a high cervical injury and is a ventilator-dependent quadriplegic. Ray's treatment will require a solid knowledge of the body's various systems in order to give him the best possible care. In an upcoming chapter you will meet Maria who is a diabetic and will also have multi-system involvement. You will be responsible for Ray's and Maria's care throughout the book by utilizing the knowledge you gain from each new chapter.

RAY'S STORY

Ray glanced at the strategically placed clock and calendar on the wall. 3:00 pm, September 15. His buddies were on their way to the Caribbean for a week of cave diving. He was supposed to be on that plane with them. Instead, he was lying in a hospital bed. Since high school the four of them had been inseparable. Every spare dime they made went into their annual extreme vacation. They had been hang gliding over the Grand Canyon, rafting on the Amazon, sky diving and parasailing in Florida and snowboarding down a glacier in Alaska. How ironic that a dip in a backyard swimming pool would put an end to all that. Not only wouldn't he be skiing or hang gliding any time soon, he couldn't even mark the red X's on the calendar to keep track of the passing days. A nurse did that for him. In fact, nurses or his mom or his brother did everything for him. He couldn't even scratch his own nose, or even tell anybody when it itched.

He was just fooling around with his 6-year-old cousin at the end of the summer, one last swim before his aunt and uncle closed their pool. He had done a back flip off the diving board dozens of times, but this time he miscalculated. He knew as soon as he hit the water that he was sinking too fast and tried desperately to correct, but he couldn't get his head around fast enough. His last thought as the bottom of the pool rose to meet him was that he was going to mess up his face.

When he regained consciousness several days later, his mother and the doctor gently explained what had happened to him. He had struck the bottom of the pool on the point of his chin, jamming and twisting his neck. His neck was broken, the first two vertebrae shattered and his spinal cord irreversibly damaged. He was paralyzed from the neck down and would be dependent on a ventilator for the rest of his life.

a. Explain how the relationship of anatomy to physiology has worked against Ray.

_____

b. What is the etiology of Ray's current condition and his diagnosis?

_____

c. How would you state his prognosis? What future complications do you expect to see?

_____

1-4 Please go to your DVD to view an animation of a diving accident with related cervical injury.

REVIEW QUESTIONS

Multiple Choice

1. Which of the following is an example of microscopic anatomy?
   a. viewing an X-ray
   b. examining the shape of an organ during an autopsy
   c. classifying a type of bacterial cell
   d. watching how the pupils in the eyes react to light

2. Acromegaly means which of the following?
   a. a large stomach
   b. enlarged extremities
   c. an inflamed stomach lining
   d. a large acrobat

3. The breakdown of sugar in the body for energy is called:
   a. anabolism
   b. catabolism
   c. dogabolism
   d. hyperbolism

4. Which of the following is a measurement system based on the power of 10?
   a. English system
   b. British Imperial system
   c. metric system
   d. weights and measures system

5. The cause of a disease is referred to as the:
   a. prognosis
   b. diagnosis
   c. pathology
   d. etiology

Fill in the Blank

1. Ted's knee injury occurred at last night's football game. Today his doctor wants to make a small incision and use a device to "look around the joint" to assess the damage. What is the term for this device? _____

2. _____ is the study of the structures of the body, and _____ is the study of the functions of these structures.

3. For years, Ali never learned to swim because of her unnatural fear of the water, which is called _____

4. Pulse and temperature represent two _____ signs of the body.

5. Raheem had blood work done that showed a normal number of white blood cells (WBCs) and red blood cells (RBCs). What are the respective medical terms for these cell types?

   _____

6. If a cause of a disease cannot be determined, it is referred to as a(n) _____ disease.

Short Answer

1. Explain the difference between diagnosis and prognosis.

   _____

   _____

   _____

   _____

2. Knowing that difficulty swallowing is called dysphagia, what do you think the function of a phagocyte is?

   _____

   _____

   _____

   _____

3. Contrast negative and positive feedback loops.

   _____

   _____

   _____

   _____

4. Describe one example of homeostasis in your body.

_____

_____

_____

_____

5. Contrast the terms endemic, epidemic, and pandemic.

_____

_____

_____

_____

Suggested Activities

1. Using a medical dictionary, find five new medical terms and give their definition and pronunciation.

_____   _____

_____   _____

_____   _____

_____   _____

_____   _____

2. Make up 3 x 5 inch note cards with five word roots discussed in this chapter and see how many medical words you can make using either prefixes or suffixes in the tables. For example, the word root arthr/o can be used to make the following: arthritis, arthralgia, arthroscope, and arthroplasty. Confirm that you made a real word by looking it up in a medical dictionary.

1-5 Now that you have completed your journey through this chapter, please go to the DVD for interactive games and puzzles concerning the medical terms and concepts contained in this chapter. By playing the games you will reinforce your learning of medical terminology in a fun way.

T H E
HUMAN BODY
Reading the Map

Now that we have a basic understanding of the native language and some basic anatomy, physiology, and disease concepts, how will we successfully navigate through an unfamiliar city or country? We must, of course, study maps to plan our visit and know where we are going. The same effort is required to learn the "terrain" of the human body. This chapter provides the major external map of the human body that serves as a guide for future chapters, which map in detail the internal regions. We'll use medical directional terms and body locations as our foundation as we journey together through the human organism. Isn't it ironic that if there is one thing we should know better than anything else, it should be our own bodies? Like the old saying says, by the end of our journey through this textbook, you will know your entire body like the back of your hand.

Chapter 2

LEARNING OBJECTIVES

At the end of your journey through this chapter, you will be able to:

→ List and describe the various body positions

→ Define the body planes and associated directional terms

→ Locate and describe the body cavities and their respective organs

→ List and describe the anatomical divisions of the abdominal region

→ Identify and locate the various body regions

→ Explain the various imaging techniques to view the body

→ Describe situations in which body position can help or hinder the disease process

MULTIMEDIA APPLICATIONS

DVD Interactive Exercises

→ Additional video information on body positions, 2-1

→ Videos on ultrasound, MRIs, and other diagnostic imaging, 2-2

→ Interactive drag-and-drop exercise to reinforce learning of body cavities, 2-3

→ Interactive drag-and-drop exercise to reinforce learning of anterior and posterior body regions, 2-4

→ Interactive games and puzzles, 2-5

www.prenhall.com/colbert

→ Professional Profiles:
- Radiologic Technology
- Surgical Technology

→ Related Internet Links

→ Additional Review Questions

Pronunciation Guide

 Correct pronunciation is important in any journey so that you and others are completely understood. Here is a "see and say" Pronunciation Guide for the more difficult terms to pronounce in this chapter.

abdominopelvic cavity
   (ab dom ih noh PELL vik)

antecubital (an tee CUE bi tal)

buccal (BUCK al)

caudal (KAWD al)

cephalic (seh FAL ik)

coronal plane (KOH roh nal)

cranial (KRAY nee al)

crural (CRUR al)

distal (DISS tal)

dorsal (DOR sal)

gluteal (GLOO tee al)

mediastinum (me dee ah STY num)

midsagittal plane (mid SAJ ih tal)

pleural cavities (PLOO ral)

superficial (SOO per FISH al)

thoracic cavity (tho RASS ik KAV ih tee)

transverse (tranz VERS)

---

supine position *(sue PINE)*

Trendelenburg
   (trin DELL in berg)

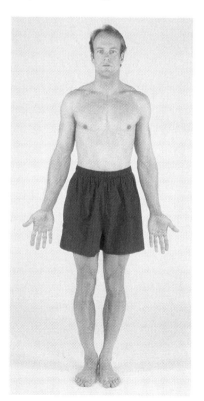

FIGURE ■ 2-1

The anatomical position.

THE MAP OF THE HUMAN BODY

When reading a map, you need certain universal directional terms, such as north, east, south and west. Maps of specific regions can include more details about that region, making it easier to explore. Likewise, scientists have created standardized body directional terms and split the body into distinct regions, sections, and cavities so that we can more clearly and rapidly locate and discuss anatomical features. Having certain anatomical landmarks on the body also provides needed points of reference for assessment and surgical procedures. For example, the spinal cord is a major anatomical landmark for many structures in the center of our bodies.

If a patient states, "I have pain in my stomach," what does that really tell you? Location of pain can help in determining what is wrong with a patient. It is helpful to know the type of pain (dull, sharp, or stabbing) and *exactly* where in that region the pain is located to help determine its cause. For example, pain in the general stomach area can indicate a variety of problems, such as an ulcer, heart attack, appendicitis, indigestion, or liver problems. Knowing the exact region can help a clinician better determine the exact problem.

Body Positions

The body can assume many positions and therefore have different orientations. To standardize the orientation for the study of anatomy, scientists developed the **anatomical position.** The anatomical position, as shown in Figure 2-1 ■, is a human standing erect, face forward, with feet parallel and arms hanging at the side, and with palms facing forward.

Other body positions that are important to discuss because of clinical assessments and treatments in health care are the **supine, prone, Trendelenburg** and

Fowler's positions. The supine position is laying face *upward*, or on your back. The prone position is laying face *downward*, or on your stomach. When a patient is in the Trendelenburg position, the head of the bed is lower than the feet. Fowler's position is sitting in bed with the head of the bed elevated 45 to 60 degrees. This position is often used in the hospital to facilitate breathing and for comfort of the bedridden patient while eating or talking. See Figure 2-2 ■ for these body positions.

anatomical position
(an ah TOM ih kal)

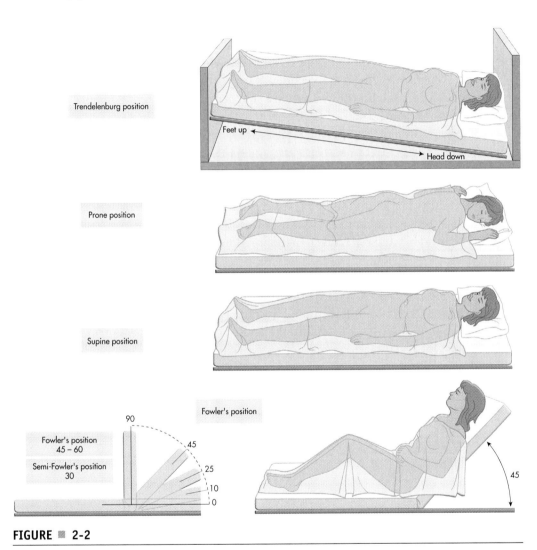

Trendelenburg position
Feet up
Head down

Prone position

Supine position

90
Fowler's position
45 – 60
Semi-Fowler's position
30
45
25
10
0
Fowler's position
45

FIGURE ■ 2-2

Common patient positions.

Pathology Connection: Body Positions and A&P Changes

A patient may be placed in certain positions for treatment. For example, placing patients with secretions in the bases of their lungs in the Trendelenburg position helps drain those segments of the lungs. While this is therapeutic, certain precautions must be taken. Because the head is lower than the heart, gravity

 2-1 There are even more body positions, including the lateral, Sims, dorsal recumbent, and lithotomy positions. For more information on this subject and to view videos, please go to your DVD for this chapter.

increases the blood flow and therefore the intracranial pressure. This position may be contraindicated in patients with cerebral injury or bleeding. Also, patients are more prone to aspirate (take in) vomitus into their lungs and therefore patients should not eat within 2-4 hours of being placed in the Trendelenburg postion.

Another example: when a patient experiences severe heart failure, the neck veins become filled with extra blood due to the backup of fluid into the venous system. This is called **Jugular Venous Distension** (JVD). If the patient experiences engorged neck veins while in the upright sitting position, this indicates severe heart failure. The physiologic reason is that the pressure of the "backed-up" venous blood has become greater than the effects of gravity.

Sometimes when a person quickly arises from a seated position he or she becomes weak and dizzy. This may be a sign of **orthostatic hypotension**. When the person is seated, the blood pressure may be low, but the brain is still receiving adequate blood flow (perfusion) for normal functioning. When the person stands, however, the heart has to pump harder against gravity to send the blood to the brain. If the heart cannot compensate, the pressure becomes even lower and the person experiences weakness and/or dizziness due to lack of cerebral perfusion. These kinds of postural changes in blood pressure will be a problem for Ray, our spinal cord patient introduced to you in the Common Case Study at the end of Chapter 1, due to a dysfunction of his nervous system when propped up in his wheelchair.

Some people sleep better when they prop themselves up with several pillows in bed. Such patients are experiencing **orthopnea**, in which it is easier to breathe in a more upright position than lying flat. This is because when a person sits upright, gravity assists the diaphragm in the downward movement needed for inspiration. You'll learn more on how the diaphragm functions in Chapter 13 on the respiratory system.

TEST YOUR KNOWLEDGE 2-1

Complete the following:

1. Try standing in the anatomical position.

2. Give the best body position (prone, supine, or Fowler's position) for the following circumstances:

   a. Getting a back massage _____

   b. Eating in a hospital bed _____

   c. Watching television in bed _____

   d. Watching the stars at night _____

3. A person experiencing orthopnea would breathe better in which position?

   _____

4. What position would be contraindicated for someone who just had eye surgery?

   _____

Body Planes and Directional Terms

Sometimes it is necessary to divide the body or even an organ or tissue sample into specific sections to further examine it. A **plane** is an imaginary line drawn through the body or organ to separate it into specific sections. For example, in Figure 2-3 ■, we see the **transverse plane** or **horizontal plane,** dividing the body into top (**superior**) and bottom (**inferior**) sections. This can also be called **cross-sectioning** the body. Cross-sectioning is often done with tissue and organ samples to further examine internal structures.

transverse *(tranz VERS)*

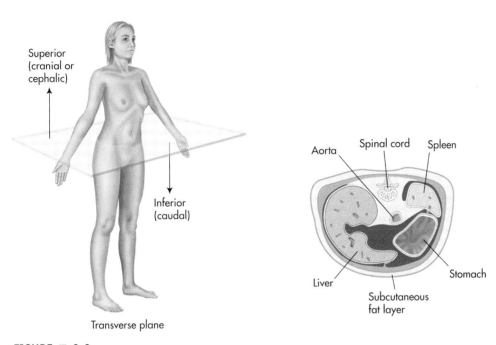

FIGURE ■ **2-3**

Transverse plane and a cross-sectional view of the upper abdominal region.

Notice in Figure 2-3 that certain directional terms can be used to describe areas divided by the transverse plane. One more analogy that relates to a map is the concept of a reference point. If you were traveling from Colorado to Florida, you would have to travel in a southeasterly direction. Colorado is your starting point and serves as your reference point. However, if you were traveling from Florida to Colorado, you would travel in a northwesterly direction because Florida is now your point of reference. In Figure 2-3, you can see that superior (**cranial** or **cephalic**) means toward the head or upper body and inferior (**caudal**) means away from the head or toward the lower part of the body. Any body part can be either superior or inferior depending upon your reference point. For example, the knee is superior to the ankle if the ankle is the reference point. Turning this around, the ankle is inferior to the knee if the knee is the reference point. Two other terms from this illustration are **cranial**, which refers to the skull, and **caudal,** which refers to body parts near the tail (tailbone).

cranial *(KRAY nee al)*
  crani/o = *skull*
cephalic *(seh FAL ik)*
  cephal/o = *toward the head*
caudal *(KAWD al)*
  caud/o = *tail*
-ic and -al are adjective endings that mean "pertaining to"

midsagittal (mid SAJ ih tal)
medial (MEE dee al)
lateral (LAT er al)

The **median plane,** or **midsagittal plane,** divides the body into right and left halves. Figure 2-4 ■ shows this plane and the directional terms associated with it. **Medial** refers to body parts located near the middle or midline of the body. **Lateral** refers to body parts located away from midline (or on the side). If a technician were to section and examine an organ for disease, he or she might make a midsagittal cut (cut the organ into equal right and left halves) in order to examine the internal parts of the organ or might simply make several sagittal (vertical or lengthwise) cuts to slice the organ into smaller sections for closer examination.

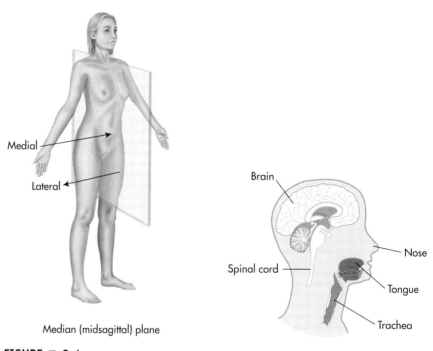

FIGURE ■ **2-4**

Midsagittal or median plane along with a sagittal view of the head.

coronal (KOR oh nal)

The **frontal plane,** or **coronal plane,** divides the body into front and back sections. **Anterior** and **ventral** refer to body parts toward or on the front of the body, and **posterior** and **dorsal** refer to body parts toward or on the back of the body. Figure 2-5 ■ demonstrates the coronal plane and associated directional terms. Remember, if during your trip you stop at the beach, you will know it is not safe to swim if you see a shark's *dorsal* fin sticking out of the water.

Additional Directional Terms

proximal (PROK sim al)
distal (DISS tal)

There are some additional directional terms that are important in health care. **Proximal** refers to body parts close to a point of reference of the body. This is contrasted by **distal,** which refers to body parts away from a point of reference. For example, using your shoulder as a reference point, your elbow is proximal to your shoulder, while your fingers are distal to your shoulder. **External** means on the outside, and **internal** refers to structures on the inside. Did you know that your external skin is actually the body's largest organ? Most other organs are located internally within body cavities.

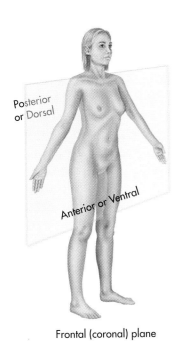

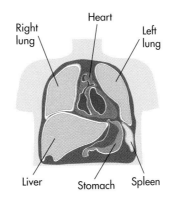

FIGURE ■ 2-5

Frontal or coronal plane along with a coronal view of the chest and stomach.

Clinical Application ✚ ✋ 👣 🏊 🔑

DO YOU KNOW YOUR LEFT FROM YOUR RIGHT?

By now, it should be clear that you need a precise, standardized language with directional terms to study anatomy and physiology and apply it in a health care setting. Something as simple as left and right can become critical. For example, suppose you are a surgical technologist and are ordered to put a tag around a patient's right leg to designate it as the leg to be amputated in an upcoming surgery. If you approach the patient from the bottom of the bed and place the tag on the leg on your right side, you have erroneously placed it on the patient's left leg, and this could have disastrous results. Left and right always refer to the patient's left and right, not yours. See Figure 2-6 ■.

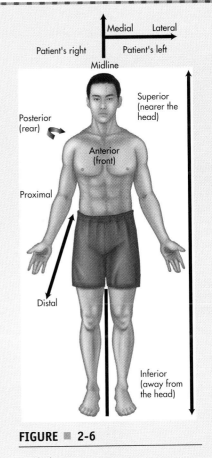

FIGURE ■ 2-6

Body location terms.

embolism *(EM bow lizm)*

An **embolism** is a sudden obstruction of a blood vessel by debris that can include blood clots (thrombi), plaques, bacteria, cancer cells, fat from bone marrow, and air bubbles.

Superficial means toward or at the body surface. When a clinician draws blood from you, he or she looks for superficial veins that are easy to see and easy to access with the needle. **Deep** means away from the body surface. The large veins in your legs are deep veins and are more protected than superficial veins because injury to them can be more critical to survival than can injury to a smaller, superficial blood vessel. These deep leg veins are a common site for blood clots (thrombi) to form that can then break away and travel to areas such as the lungs (pulmonary emboli) or brain (cerebral emboli) and block vital blood flow. **Central** refers to locations around the center of the body (torso and head), and **peripheral** refers to the extremities (arms and legs) or surrounding or outer regions. Table 2-1 provides a summary of directional terms.

Clinical Application

CENTRAL VERSUS PERIPHERAL CYANOSIS

cyanosis *(sigh ah NOH siss)*

  cyano = *blue*

  osis = *condition of*

Cyanosis is a condition of bluish colored skin that is usually the result of low levels of oxygen in the blood. Peripheral cyanosis presents as bluish fingers and toes and may indicate the need for oxygen therapy depending on the condition of the patient. Peripheral cyanosis is sometimes difficult to detect in people with dark skin. Central cyanosis is much more serious and presents as bluish discoloration of the torso and inside the mouth. See Figure 2-7 ■, which illustrates central and peripheral cyanosis.

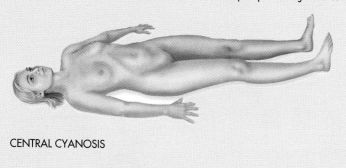

CENTRAL CYANOSIS

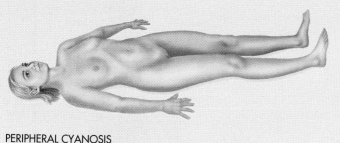

PERIPHERAL CYANOSIS

FIGURE ■ **2-7**

Contrast of central versus peripheral cyanosis.

www.prenhall.com/colbert

Surgical technologists must have a command of medical directional terms used during surgical procedures. If you are interested in learning more about this profession, please visit the website for this chapter.

TABLE 2-1 Directional Terms

DIRECTIONAL TERM	MEANING	USE IN A SENTENCE
Proximal	near point of reference	The wrist is *proximal* to the fingers.
Distal	away from point of reference	The shoulder is *distal* to the fingers.
External	on the outside	The *external* defibrillator is used on the outside of the chest.
Internal	on the inside	He received *internal* injuries from the accident.
Superficial	at the body surface	The cut was only *superficial*.
Deep	under the body surface	The patient had *deep* wounds from the chainsaw.
Central	locations around center of body	The patient had *central* chest pain.
Peripheral	surrounding or outer regions	The patient had *peripheral* swelling of the feet.
Medial	toward the midline	The nose is *medial* to the eyes.
Lateral	toward the sides	The ears are *lateral* to the eyes.

TEST YOUR KNOWLEDGE 2-2

Answer the following questions:

1. Give the opposite directional term:

   a. superior _____

   b. posterior _____

   c. caudal _____

   d. ventral _____

   e. distal _____

   f. external _____

   g. superficial _____

   h. peripheral _____

   i. medial _____

2. A spinal tap is performed on the _____ portion of the body.

3. The plane that divides the body into upper and lower regions is called the _____ plane.

4. Cutting an organ into two equal halves (right and left) requires a _____ incision.

5. A scratch on the surface of the skin is called a _____ wound.

6. The wrist is _____ to the hand and _____ to the elbow.

7. The nose is _____ to the mouth.

8. A pain in your side can also be referred to as _____ pain.

9. If your hands and feet are swollen with fluid (edema), you are said to have _____ edema.

10. A cerebral embolus can often originate in the _____ veins of the legs.

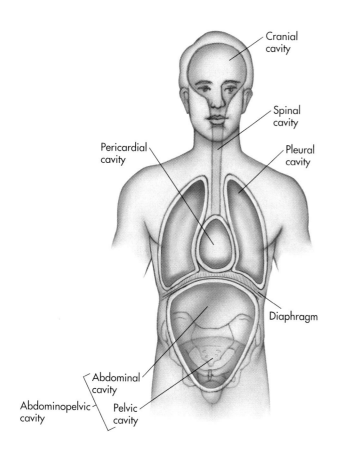

Body Cavities

The body has two large spaces or cavities that house and protect organs. Located in the back of the body are the dorsal cavities and in the front, the ventral cavities. Figure 2-8 ■ illustrates these cavities. The larger anterior cavity is subdivided into two main cavities called the **thoracic cavity** and **abdominopelvic cavity.** These cavities are physically separated by the large, dome-shaped muscle called the diaphragm, which is used for breathing. The thoracic cavity contains the heart, lungs, and large blood vessels. The heart has its own small cavity called the pericardial cavity. The abdominopelvic cavity contains the digestive organs, such as the stomach, intestines, liver, gallbladder, pancreas, and spleen in the upper or abdominal portion. The lower portion, called the pelvic cavity, contains the urinary and reproductive organs and the last part of the large intestine. A posterior or dorsal cavity is located in the back of the body and consists of the **cranial cavity,** which houses the brain, and the **spinal cavity,** which contains the spinal cord.

There are also smaller body cavities that designate specific areas, and these are further explored in upcoming chapters. For example, the nasal cavity is the space behind the nose, the oral or buccal cavity is the space within the mouth, and the orbital cavity houses the eyes.

thoracic cavity
  (thoh RASS ik)
abdominopelvic cavity *(ab dom ih noh PELL vik)*
cranial cavity
  (KRAY nee al)

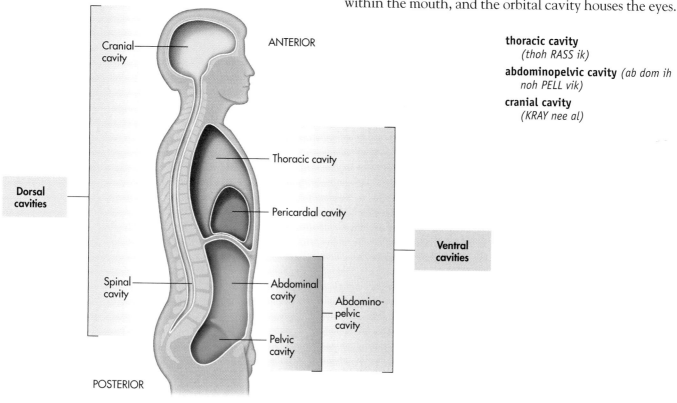

FIGURE ■ 2-8

Main body cavities.

TEST YOUR KNOWLEDGE 2-3

Identify the major body cavity in which the following organs are located:

1. heart _____

2. spinal cord _____

3. stomach _____

4. lungs _____

5. reproductive organs _____

6. brain _____

Body Regions

The abdominal region houses a number of organs. Anatomists have divided this region as shown in Figure 2-9 ■. Understanding directional terms assists in locating the regions. For example, the **epigastric** region (*epi*, above; *gastric*, stomach) is located superior to the umbilical region. The right and left **hypochondriac** regions are located on either side of the epigastric region, and contain the lower ribs. The centrally located **umbilical** region houses the navel

epigastric *(ep ih GAS trik)*
  epi = *above*
  gastric = *stomach*
hypochondriac
  (high poh KON dree ak)
  hypo = *below*
  chondriac = *refers to ribs*
umbilical = *navel*

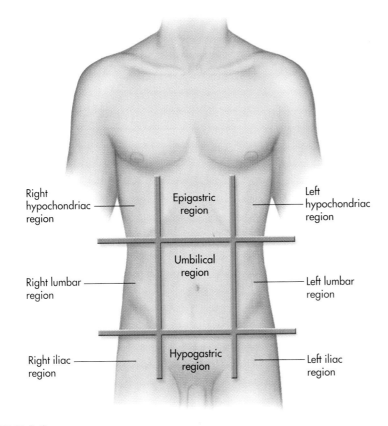

Right hypochondriac region

Epigastric region

Left hypochondriac region

Right lumbar region

Umbilical region

Left lumbar region

Right iliac region

Hypogastric region

Left iliac region

FIGURE ■ 2-9

The nine divisions of the abdominal region.

lumbar = *lower back*
hypogastric *(high poh GAS trik)*
inguinal = *referring to groin*

or belly button. You may not remember your umbilical cord being cut as a newborn, but your belly button is a reminder that it occurred. Lateral to this region are the right and left lumbar regions at the level of the **lumbar** vertebrae. The **hypogastric** region lies inferior to the umbilical region and is flanked by the right and left iliac or inguinal regions. The **inguinal** region is where the thigh meets the trunk and is also called the groin region.

2-3 To reinforce the concept and location of body cavities and their locations, please go to your DVD for an interactive drag-and-drop labeling exercise.

Clinical Application

HERNIAS

You may have heard of an umbilical (belly button) bulge (hernia), or inguinal hernia, and now you know exactly where such hernias are located. Just what is a hernia? A hernia is a tear in the muscle wall that allows a structure (usually an organ) to protrude through it. Sometimes this can be a minor nuisance, but a hernia can also be very dangerous if the blood flow is restricted to the portion of the organ that is protruding. Restricted blood flow can lead to death of the tissue and to serious consequences. Death of a tissue is called necrosis. Figure 2-10 ▒ shows an inguinal and umbilical hernia with a protrusion of the intestines.

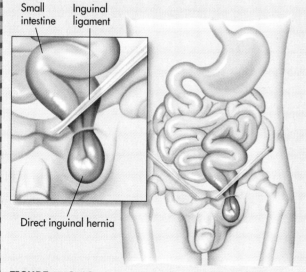

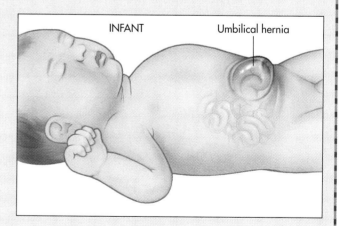

FIGURE ▒ **2-10**

Illustrations of inguinal and umbilical hernias.

Clinical Application

THE CENTRAL LANDMARK: THE SPINAL COLUMN

The spinal or vertebral column is a major, centrally located anatomical landmark and has five sets of vertebrae (spinal bones) labeled for the body region (see Figure 2-11 ■). The seven cervical (C) vertebrae are located in the neck; the 12 thoracic (T) vertebrae are located in the chest; the five lumbar (L) vertebrae are located in the lower back; and the five fused sacral (S) vertebrae (sacrum) are located near the final coccyx vertebra (tailbone). For example, the T5 vertebra is used to help locate on a chest X-ray the area where the right and left lung begin to branch. You'll learn even more about the spinal column and cord in Chapter 6, "The Skeletal System," and Chapter 9, "The Nervous System."

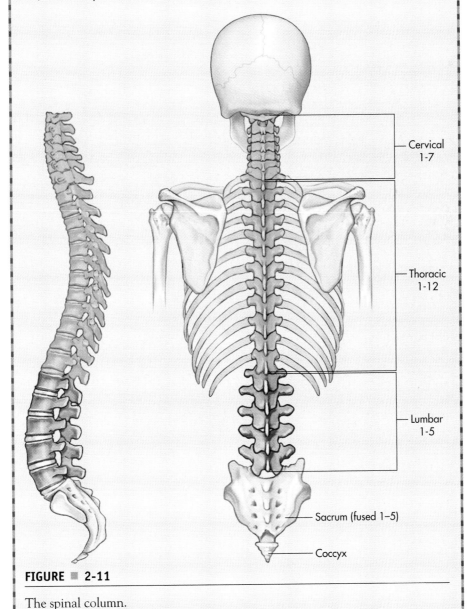

Cervical 1-7

Thoracic 1-12

Lumbar 1-5

Sacrum (fused 1–5)

Coccyx

FIGURE ■ 2-11

The spinal column.

A more practical way for health professionals to compartmentalize the abdominal region is to separate it into anatomical quadrants. Figure 2-12 ■ illustrates these quadrants, which are helpful in describing the location of abdominal pain. Knowing the organs located in the quadrant where the pain occurs can provide a clue to what type of problem the patient has. For example, tenderness in the right lower quadrant (**RLQ**) can be a symptom of appendicitis, because that is where the appendix is located. RUQ (right upper quadrant) pain may mean a liver or gallbladder problem.

2-4 Go to the DVD for more interactive practice with drag-and-drop exercises focused on understanding and locating the various anterior and posterior body regions.

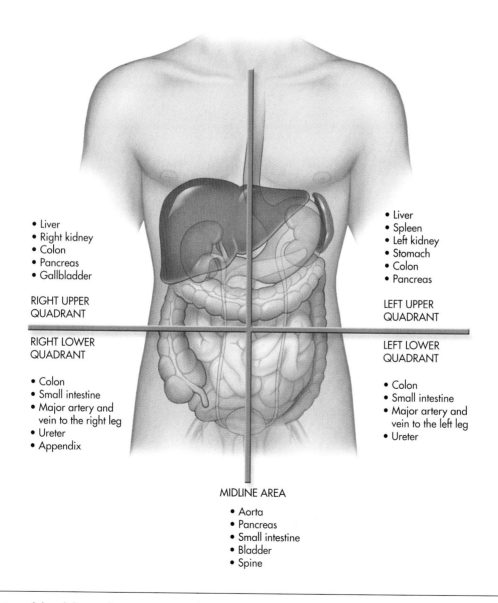

• Liver
• Right kidney
• Colon
• Pancreas
• Gallbladder

RIGHT UPPER QUADRANT

RIGHT LOWER QUADRANT

• Colon
• Small intestine
• Major artery and vein to the right leg
• Ureter
• Appendix

• Liver
• Spleen
• Left kidney
• Stomach
• Colon
• Pancreas

LEFT UPPER QUADRANT

LEFT LOWER QUADRANT

• Colon
• Small intestine
• Major artery and vein to the left leg
• Ureter

MIDLINE AREA

• Aorta
• Pancreas
• Small intestine
• Bladder
• Spine

FIGURE ■ 2-12

The clinical division of the abdominal region into quadrants with related organs and structures.

There are additional body regions that further aid in locating areas and structures. For example, what if you were asked to obtain an axillary temperature on an infant? Just where is the brachial or femoral pulse? What part of the body does carpal tunnel syndrome affect? See Figure 2-13 ■ for other common body regions and parts that are discussed in later chapters. In addition, review Table 2-2 for further practical examples of the medical importance of the various body regions.

Amazing Body Facts

PSOAS TEST

The psoas (*SOH as*) test—with its strange name—is one way to help determine if a patient has appendicitis. The patient is placed in a supine position and instructed to raise his or her right leg while the practitioner places a hand on the patient's right thigh and gives a slight opposing downward force. If the patient has appendicitis, he or she will usually experience pain in the right lower quadrant.

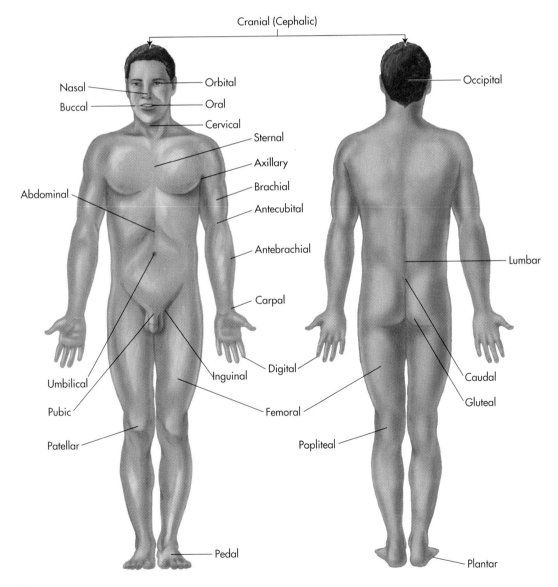

FIGURE ■ 2-13

Anterior and posterior body regions.

TABLE 2-2 Examples of Body Regions and Their Locations

BODY REGION	LOCATION	MEDICAL EXAMPLE
Antebrachial	forearm	between the wrist and elbow
Antecubital	depressed area in front of elbow	area used to draw blood or start an IV
Axillary	armpit	can be used to take temperature
Brachial	upper arm	used to take blood pressure
Buccal	cheek	check buccal region for central cyanosis
Carpal	wrist	carpal tunnel syndrome
Cervical	neck	cervical collar needed for neck injuries
Digital	fingers	place digital oxygen sensors
Femoral	upper inner thigh	femoral pulse indicates adequate circulation to legs
Gluteal	buttocks	the buttock is an injection site
Lumbar	lower back	lumbar pain often occurs on long car trips
Nasal	nose	medications can be given by nasal spray
Oral	mouth	oral route is most common route for medications
Orbital	eye area	orbital injury can cause damage to sight
Patellar	knee	patellar injuries are very common in sports
Pedal	foot	people with heart problems may have pedal edema (swelling)
Plantar	sole of foot	plantar warts can be painful
Pubic	genital region	the pubic region is often checked for body lice
Sternal	breastbone area	the sternal area is used for CPR
Thoracic	chest	the thoracic area is used to listen to heart and lung sounds

TEST YOUR KNOWLEDGE 2-4

Fill in the blank with the appropriate medical term for the body region. These can have more than one answer.

1. People who chew smokeless tobacco or snuff are more susceptible to _____ cancer.

2. Antiperspirant sprays are usually used in the _____ region.

3. Belly button rings are usually found in the _____ region.

4. If you sit too long at your desk, you can develop _____ pain.

5. During physicals, your reflexes are checked with a little rubber hammer that taps your _____ region.

RADIOLOGY: THE SCIENCE OF VIEWING THE BODY

While Chapter 5 is devoted to diagnostic testing, diagnostic imaging fits well within this chapter on body positions and locations. Positions are important in the radiologic sciences for obtaining a proper internal view of body structures. **X-rays** are a form of high-energy radiation that penetrate the body and give a two-dimensional view of the bones, air, and tissues in the body. The standard X-ray (much like a photograph) can be computer-enhanced to give much greater detail and contrast and to allow for a more realistic three-dimensional view. For example, if a standard chest X-ray showed a golf ball–sized tumor in the lung, you would have no idea of its actual depth because it would look flat, like a quarter. **Computed tomography** (CT) scanning uses a narrowly focused X-ray beam that circles rapidly around the body. The computer constructs thin-slice images and combines them to give much greater detail and allow for a more three-dimensional view, much like a loaf of sliced bread gives a better idea of the total shape of the loaf than does a single slice. The CT scan reveals the true depth of the quarter-shaped tumor shown on the regular X-ray. A **magnetic resonance imager** (MRI) produces even greater detail of tissue structures, even down to individual nerve bundles. Another advantage of the MRI is the absence of radiation.

X-rays (Radiograph or Roentgenogram)

The most common type of radiologic diagnostic modality is the X-ray. An X-ray is produced by passing X-ray beams through a specific area of the body and then exposing a photographic film to those rays that successfully pass through that area. As a result, the developed film or digital image will possess an image of varying degrees of dark and light, depending on the densities found in the body that the X-rays passed through. Less dense areas, such as the lungs which contain a lot of air, allow the X-rays to pass through easily and show up on the film as a dark (**radiolucent**) region. Bones are much denser and allow fewer X-rays to pass through them. As a result, the denser bones appear as light (**radiopaque**) structures on the film. The densities of the other body parts are somewhere between these examples, and vary in their lightness or darkness on the film. In general there are four densities that can be found in the body:

1. *Air* is the least dense, and therefore the most radiolucent, showing up as black on an X-ray film.
2. *Tissue/Fat* densities can depend on the thickness of the tissue and/or the amount of fat present. The thicker the layer, the more radiation is absorbed, and the lighter the image.
3. *Water* density can be represented by blood, or by edema as a result of tissue injury or inflammatory processes. Water density is a mid-range density: more dense than air and less dense than bone.
4. *Bone/metal* is the highest density, absorbing the greatest amount of radiation. Since less radiation passes through and onto the film, the film image is white or lighter than the other densities.

One of the problems with an X-ray film is that it is a one-dimensional view of a specific area. As a result, the patient has to be positioned differently in relationship to the X-ray machine and film depending on what the physician is looking for. For example, it would be difficult to see a small lung tumor located right behind the heart if you were looking at a frontal view of the patient. Every attempt should be made to position the patient's area to be X-rayed as close to the film as possible. This will help insure that the structures appear as clear and as close to actual size as possible on the X-ray. The following are the three main views normally used in the clinical setting:

- **Posteroanterior (PA)** This standard position places the patient in an upright position with the chest placed in front of the X-ray film. The X-ray beam travels from the machine, through the patient's posterior region, out the patient's anterior, and onto the film. The shoulders are commonly rotated forward to move the shoulder blades away from the lungs if there is a question about any lung pathology. Generally, the distance between the X-ray machine and the film is six feet.

- **Anteroposterior (AP)** In this position, the patient's posterior is against the X-ray film. The X-ray beam travels through the patient's anterior region, out the posterior, and onto the film. The distance between the X-ray machine and the film is 48 inches. This is the position commonly seen in the hospital where a portable X-ray machine is brought to the patient's room if he or she is too ill to be taken to the Radiology department. The patient is usually in a sitting position on the bed.

- **Lateral:** This position is often done as a complement to the PA image. The lateral view is done to eliminate interfering organs or structures in front of other structures of concern, such as a tumor. The object in question is placed as close to the film as possible. For example, if there is a suspected lesion in the left lung behind the heart, the left side of the patient is placed next to the film. This would be a left lateral view. This view prevents the heart from obstructing the view of the suspected tumor and, in conjunction with a PA X-ray, gives a more 3D sense of the tumor.

Computerized Tomography (CT or CAT Scan)

As previously stated, one of the problems with an X-ray is that it provides a one-dimensional picture. The CT scan provides a high-resolution series of cross sectional "slices" (much like a loaf of sliced bread) of the body. As a result, it creates a three-dimensional view of structures in the body as well as tissue structures within solid organs . This allows the clinician to determine the exact location, size, and shape of a suspected tumor in the body, which would not be possible with only an X-ray. The downside to a CT scan is the high level of radiation exposure required. A single CT scan is the equivalent of hundreds of chest X-rays in terms of radiation exposure.

Magnetic Resonance Imaging (MRI)

Instead of utilizing radiation, this diagnostic tool uses magnetic energy to produce cross-sectional images of body structures. The clarity of these images is generally much better than those from X-rays and is often better than a CT scan. However, there are some drawbacks. Since an intense magnetic field is used, certain steel aneurysm clips or prosthetic heart valves may be adversely affected, even to the point of shifting their position in the body. Patients must also be completely still or there will be a loss in resolution. Open MRIs have been developed to decrease the feelings of claustrophobia experienced by some individuals in a closed MRI which takes place inside a very small tunnel.

Ultrasound (Sonography)

Much more sensitive than a regular X-ray, ultrasound uses sound waves to distinguish structures in the body, much like SONAR or fish finders in fishing boats. Ultrasound studies also allow for body activities to be viewed in real time. Due to this feature, the actions of heart valves can be observed, or fetal development and movement can be monitored, for example. No radiation exposure is involved in these studies. In the past, there had been some concern about potential damage to a developing fetus due to the sound waves used during these studies, but no evidence has been found that this is a hazard. Figure 2-14 ■ contrasts X-ray, CT, MRI, and ultrasound.

 2-2 Go to the DVD to view videos concerning ultrasound, CT, MRI, and other diagnostic imaging techniques used to visualize the interior of the body.

 www.prenhall.com/colbert
If you are interested in learning more about the profession of radiologic technology, please visit this chapter's website.

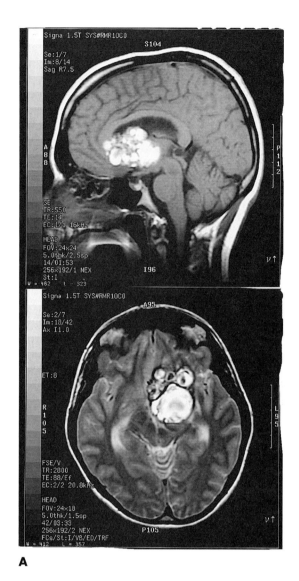

A

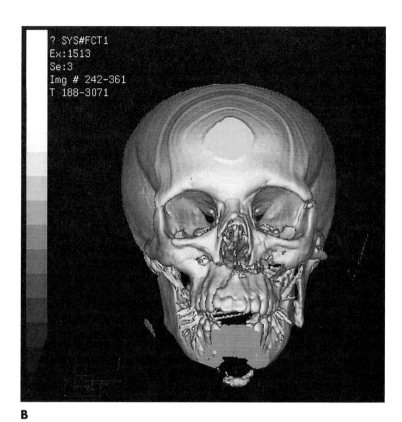

B

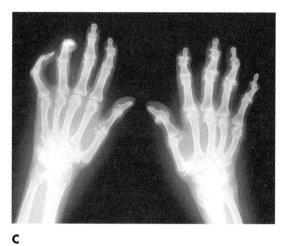

C

D

FIGURE ■ 2-14

Contrasts of X-ray, CT, MRI and ultrasound. **A.** MRI head showing large hemorrhagic lesion. (Courtesy of Teresa Resch); **B.** 3D CT scan, multiple facial fractures. (Courtesy of Teresa Resch); **C.** X-ray showing typical joint changes associated with osteoarthritis. (Source: Getty Images/Stone Allstock); **D.** Ultrasound, left kidney, and spleen. (Courtesy of Teresa Resch).

SUMMARY

Snapshots from the Journey

→ The body can assume many different positions, and to standardize the study of anatomy, scientists often reference the anatomical position. In the anatomical position, the person stands with face and toes forward, hands at sides, and palms facing forward. Other positions, such as the prone, supine, and Fowler's positions, are used in health care for assessment and treatment.

→ The body can be divided into different sections along different planes. For example, the transverse or horizontal plane divides the body into superior and inferior sections. The median or midsagittal plane divides the body into equal right and left halves. The median or sagittal plane divides the body into right and left parts. The frontal or coronal plane divides the body into anterior and posterior sections.

→ Directional terms such as internal and external, proximal and distal, superficial and deep, central and peripheral help us to navigate the body.

→ It is important to remember that directions such as right and left are referenced from the patient's perspective and *not* yours.

→ The body has several cavities that house anatomical structures (mainly organs). For example, the cranial cavity houses the brain, the thoracic cavity houses the heart and lungs, the abdominopelvic cavity houses the digestive and reproductive organs, and the spinal cavity houses (guess what) the spinal cord.

→ The body has many specific regions. For example, the umbilical region is found around your navel, or belly button, and the femoral region is located in the upper inner thigh area.

→ The directional terms, anatomical landmarks, body regions, and body cavities are all important to know so that health care professionals can communicate in specific terms that leave no room for confusion.

→ Detailed images of the internal structures of the body can be obtained from radiologic studies such as X-rays, CT scans, MRIs, and ultrasound.

Case Study

A 50-year-old female patient presents with sternal pain radiating to the left brachial area. Peripheral cyanosis is noted in the digital areas, and she exhibits pedal edema. No epigastric pain is noted. She reports that she became dizzy and fell, bruising the right orbital region, and she received superficial cuts to the right patellar region. The physician orders an IV to be started in the left antecubital region. Please answer the following questions in common lay terms.

a. Where would you suggest placing a bandage?

_____

b. Where did her pain begin?

_____

c. Where does the pain move to?

_____

d. Does she have stomach pain?

_____

e. Where will the IV be started?

_____

f. What part of her body is swollen?

_____

RAY'S STORY

Ray's quadriplegia will require extensive radiologic studies throughout the rest of his life. Given the following scenarios, give your best diagnostic answer.

a. Since this was initially a neck injury, on what specific portion of the spinal column would you focus your studies?

_____

b. Due to being bedridden and inactive, Ray will be breathing in a monotonous shallow manner even with the assistance of a mechanical ventilator and therefore not fully exercising his respiratory system. This can lead to lung collapse and pneumonia. Which body cavity would you focus your study?

_____

c. Ray will be more prone to accidental falls while being assisted in daily activities of living. What could happen that would require radiologic studies?

_____

REVIEW QUESTIONS

Multiple Choice

1. A massage therapist would ask you to assume which position for a back massage?
   a. prone
   b. supine
   c. Fowler's
   d. lotus

2. Which of the following is *not* in the abdominopelvic cavity?
   a. stomach
   b. liver
   c. reproductive organs
   d. heart

3. Carpal tunnel syndrome occurs in what region of the body?
   a. head
   b. cheek
   c. armpit
   d. wrist

4. The midsagittal plane divides the body into:
   a. top and bottom
   b. front and back
   c. upper and lower
   d. left and right

5. An organ contained in the RLQ would be:
   a. appendix
   b. heart
   c. lungs
   d. brain

6. The type of radiologic diagnostic tool that can give you real time action images:
   a. CT scans
   b. chest X-rays
   c. PET scans
   d. ultrasound

Fill in the Blank

1. A standard position in which a person stands erect, face forward, with feet parallel, arms at sides, and palms forward is called the _____ position.

2. The _____ position is laying face upward and on your back.

3. The mouth is located _____ to the nose, whereas the nose is located _____ to the mouth.

4. The organ found in the cranial cavity is the _____.

5. _____ indicates blueness of the extremities and therefore affects the peripheral areas of the body.

6. A person who gets weak and dizzy when he or she quickly stands may be suffering from _____.

Short Answer

1. List the organs found in the abdominal cavity.

_____

_____

_____

_____

2. Contrast the differences between the prone, supine, and Fowler's positions.

_____

_____

_____

_____

3. List and describe two specific body regions that are found on the lower extremities.

_____

_____

_____

4. List and describe the location of the nine abdominal regions using directional terms.

_____

_____

_____

Suggested Activities

1. Using a white t-shirt, draw and label the abdominal quadrants and the related organs.

2. Play Pin the Tail on the Donkey by guiding the blindfolded person using only medical directional terms.

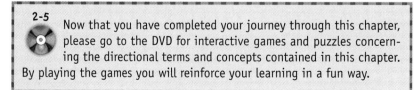

2-5 Now that you have completed your journey through this chapter, please go to the DVD for interactive games and puzzles concerning the directional terms and concepts contained in this chapter. By playing the games you will reinforce your learning in a fun way.

T H E
CELLS

The Raw Materials and Building Blocks

As you continue your journey, chances are that you will visit a city. A city is a complex combination of structures and systems. A brick or cement block is a basic structure upon which many of the buildings are constructed. The human body is also a complex combination of structures and systems, and a cell is the basic building block upon which the body is built. Just as there are different types, shapes, and sizes of building blocks, there are different types, shapes, and sizes of cells. Blood cells, skin cells, nerve cells, and so on, are all different. You will learn more about each kind of cell in later chapters. Although a cement block is a basic building block, it could not exist without sand, lime, and water, the components necessary to make cement. As this chapter explains, cells also consist of component parts, tiny cell structures called **organelles** that are needed to perform specific functions to keep the cell alive.

On a more involved level, cells of a similar type form tissue that works together in an organ, while organs perform specific functions to create a system. For example, cardiac cells form heart tissues, which form the heart organ, which is part of the circulatory system. Finally, systems work together to form a functioning organism, such as the amazing human body! When you think about it, we are very much like a city. A city, like the human body, needs transportation systems, control systems, systems to import food and water and export waste, and heating and cooling systems. A breakdown at any level (cellular, tissue, organ, or system) can have dire pathological consequences.

Chapter 3

LEARNING OBJECTIVES

At the end of your journey through this chapter, you will be able to:

→ List and describe the various parts of a cell and explain their function and pathology

→ Explain the process of cellular mitosis

→ Describe cancer growth and staging

→ Describe the types of active and passive transport within cells

→ Explain cellular respiration and enzyme function and dysfunction

→ Differentiate between bacteria, viruses, fungi, and protozoa and understand how these pathogens cause disease

MULTIMEDIA APPLICATIONS

DVD Interactive Exercises

→ Animation of cellular parts, cell structure, and cell division, 3-1

→ Microbiology and proper hand washing and gloving techniques, 3-2

→ Interactive games and puzzles, 3-3

www.prenhall.com/colbert

→ Professional Profiles:
  • Cytology
  • Lab Technician

→ Related Internet Links

→ Additional Review Questions

Pronunciation Guide

 Correct pronunciation is important in any journey so that you and others are completely understood. Here is a "see and say" Pronunciation Guide for the more difficult terms to pronounce in this chapter.

benign (bee NINE)

capsid (CAP sid)

centrioles (SEN tree oles)

centrosomes (SEN tre soams)

chromatin (CROW ma tin)

cilia (SILL ee ah)

cytoplasm (SIGH toe plazm)

deoxyribonucleic acid (dee AUK see RYE bow NEW clee ick ass id)

diabetes mellitus (DIE ah BEE teez Mel LITE us)

endocytosis (en DOE sigh TOE sis)

endoplasmic reticulum (EN doh PLAZ mic ri TIH cue lum)

exocytosis (EX oh sigh TOE sis)

flagella (flah GELL ah)

fungi (FUN jie)

Golgi apparatus (GOAL jee app ah RA tuss)

hypercholesterolemia (HI per koh LESS tur ul EE me ah)

lysosomes (LIE seh soams)

malignant (mah LIG nant)

metastasis (meh TASS tuh sis)

mitochondria (my teh CAHN dree ah)

mycelia (my SEE lee ah)

organelles (ore ga NELLS)

organism (OR gan iz em)

osmosis (ahz MOE sis)

phagocytosis (FAG oh sigh TOH sis)

phenylketonuria (FIN ill KEE toe new ree ah)

pinocytosis (pin oh se TOH sis)

protozoa (pro toe ZOE ah)

ribonucleic acid (rie bow new KLEE ic)

ribosomes (RIE beh Soams)

vesicle (VESS ih kle)

OVERVIEW OF CELLS

Composed of chemicals and structures, **cells** make up all living things. Some organisms are composed of only a single cell. Practically all of the cells in our body are microscopic in size, ranging from about one third to one thirteenth the size of the dot on this exclamation point! Equally amazing is that certain nerve cells can be two feet in length or longer! When we refer to cells as the "building blocks" of our bodies, we immediately think of brick-shaped objects. However, cells can be flat, round, threadlike, or irregularly shaped. While the approximately 7.5 *trillion* cells found in the human body vary in size, shape, and purpose, they normally work together to allow for proper functioning of processes necessary for life, such as digestion, respiration, reproduction, movement, and production of heat and energy. Figure 3–1 ■ shows examples of various cell types found within the human body.

Applied Science

ATOMS AND MOLECULES

Although cells are composed of small structures called organelles, these organelles are composed of even smaller substances. **Atoms**, which are the tiny building blocks of all matter, combine to form molecules such as water, sugar, and proteins, which are then used to build cellular structures and facilitate cellular functions. The four basic types of molecules found in cells are proteins, carbohydrates, lipids, and nucleic acids.

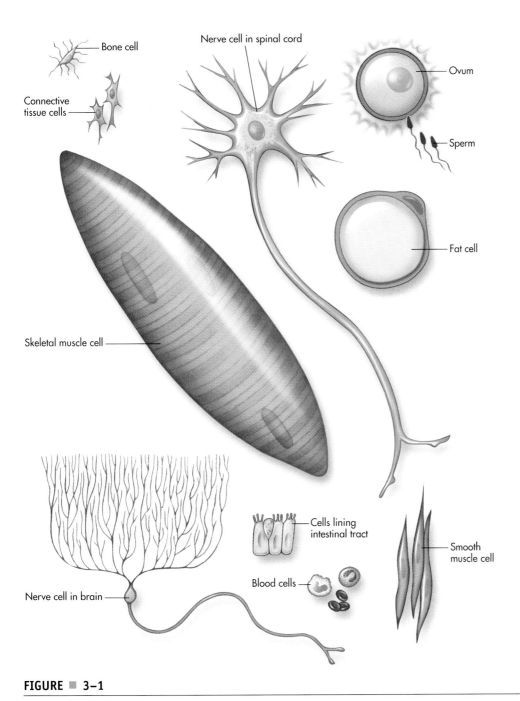

Bone cell

Connective
tissue cells

Nerve cell in spinal cord

Ovum

Sperm

Fat cell

Skeletal muscle cell

Cells lining
intestinal tract

Blood cells

Smooth
muscle cell

Nerve cell in brain

FIGURE ▪ 3–1

Various types of cells within the human body.

CELL STRUCTURE

Even though cells in our bodies can vary greatly in size, shape, and function, they
share certain common traits. To better understand them, let's look at cells as miniature cities with a variety of systems, structures, and organizations that are necessary for proper function. Almost all human cells possess a nucleus (except mature
red blood cells), organelles, cytoplasm, and a cell membrane. Each component of
a cell has a special purpose. See Figure 3–2 ▪, which represents a typical cell with
its major components. We will now discuss the individual components.

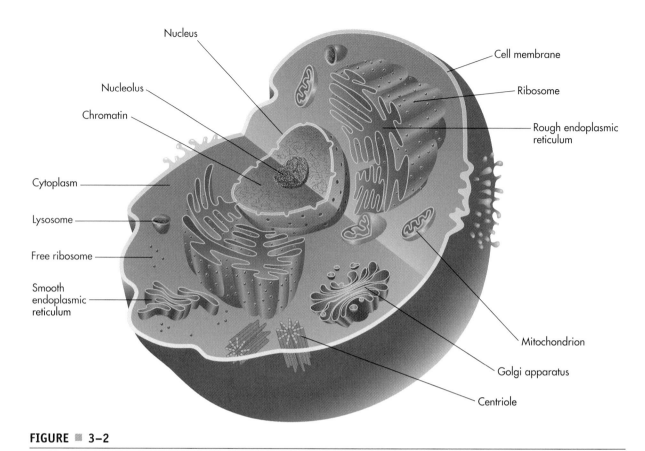

Nucleus

Nucleolus

Chromatin

Cytoplasm

Lysosome

Free ribosome

Smooth endoplasmic reticulum

Cell membrane

Ribosome

Rough endoplasmic reticulum

Mitochondrion

Golgi apparatus

Centriole

FIGURE ▪ 3–2

Cellular components.

Cell Membrane

The **cell membrane** is also called a *plasma membrane* since it surrounds the cytoplasm of the cell and acts as a protective covering. We can think of the **cell membrane** as the city limits. This is a defined boundary that possesses a definite shape and actually holds the cell contents together. Each cell, regardless of its shape or function, must have a cell membrane in order to maintain its integrity and survive.

For a city to thrive, people and materials must be able to travel into and out of the city. A cell membrane is responsible for allowing materials into and out of the cell. The membrane allows only certain things in or out. Because it chooses what may pass through, we call it a *selectively permeable* (or *semi-permeable*) membrane. In fact, cell membranes have specialized structures to help regulate the movement of some substance into and out of the cell. These structures (either channels, tunnels through the membrane, or carriers, which carry substances across the membrane) are essential for the health of cells. We will describe them in more detail later in this chapter.

In addition, the cell membrane has identification markers on it to show that it comes from a certain person, much like *PA* is an "identification marker" that tells us Pittsburgh, PA is in Pennsylvania. If a foreign cell shows up in an individual (such as in a transplanted organ), the body signals an attack on that cell or group of cells. For all that it is responsible for, the cell membrane is only 3/10,000,000 of an inch thick. Please see Figure 3–3 ▪, which depicts components of a cell membrane.

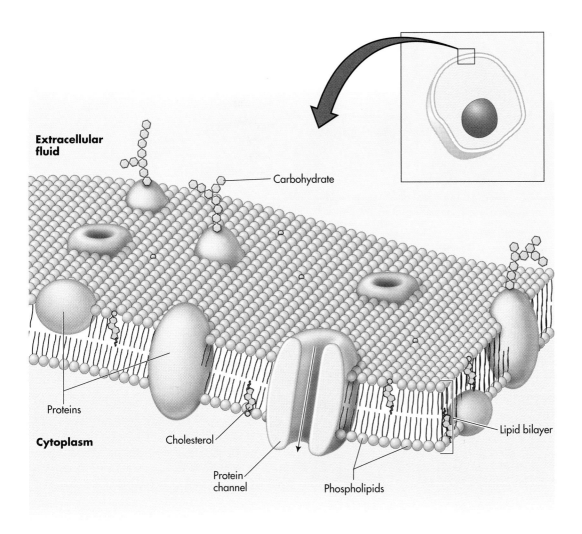

Extracellular fluid

Carbohydrate

Proteins

Cytoplasm

Cholesterol

Protein channel

Phospholipids

Lipid bilayer

FIGURE ▦ 3–3

The cell membrane.

Transport Methods

If we think of the cell membrane as the city limits or a boundary, how then do things get across this boundary? It can be done in two broad ways. **Passive transport** and **active transport** describe the movement of materials across the cell membrane. Passive transport, as the words suggest, requires no extra form of energy to complete. It is similar to having a yo-yo in your hand and letting go. The yo-yo simply falls as it unwinds from the string. No input of energy is needed. Active transport requires some addition of energy to make it happen. Throwing your yo-yo requires an addition of energy from your arm to make it happen.

Learning Hint

CONCENTRATION GRADIENT

If a substance moves from higher concentration to lower concentration, it is said to be moving *with* the concentration gradient (difference). For example, if you are in a long grocery store line "concentrated" with people and three new lanes open up, the people will quickly move from the highly concentrated line to the lower concentrated areas. Soon, you will notice, all lines will equalize, demonstrating that even people move with a concentration gradient and reach equilibrium.

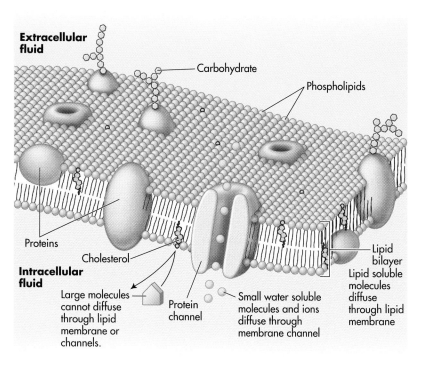

Extracellular fluid

Carbohydrate

Phospholipids

Proteins

Cholesterol

Intracellular fluid

Large molecules cannot diffuse through lipid membrane or channels.

Protein channel

Small water soluble molecules and ions diffuse through membrane channel

Lipid bilayer

Lipid soluble molecules diffuse through lipid membrane

FIGURE ■ 3–4

Two examples of diffusion.

PASSIVE TRANSPORT

Passive transport can be further divided into four types:

- diffusion
- osmosis
- filtration
- facilitated diffusion

Diffusion is our most common means of passive transport by which a substance of higher concentration travels to an area of lesser concentration. The difference between these two concentrations is called the **concentration gradient**. This is like dumping a packet of powdered drink mix into a pitcher of water. The water gradually assumes the color and flavor of the powder until the entire contents of the container are the same color and taste (nature likes a nice, equal balance!). Another example may be one of your classmates overusing perfume or cologne. Once in the classroom, the smell diffuses from high concentration on the individual to low concentration throughout the classroom. He or she may need to be reminded of the old saying that perfume should never announce your presence before your arrival.

Diffusion is necessary in the transportation of oxygen from the lungs and into the blood. It is also needed to transport the waste (carbon dioxide) from the blood to the lungs and eventually out into the air. This vital process is further discussed in Chapter 13, "The Respiratory System." See Figure 3–4 ■ for examples of diffusion.

Osmosis is another form of passive transport, in which water travels through a selectively permeable membrane until concentrations of a substance are equalized. The **solute** is the substance that is dissolved in the water. Remember that nature likes things balanced and equal. Water tends to travel across a membrane from areas that have a low concentration of a solute to areas that have a higher concentration of the solute until the concentration is the same on both sides of the membrane. Keep in mind that the water is moving with *its* concentration gradient. See Figure 3–5 ■ for a visual description. This ability of a substance to "pull" water toward an area of higher concentration of the solute is called **osmotic pressure.** The greater the concentration of the solute, the greater the osmotic pressure it exerts to bring in water.

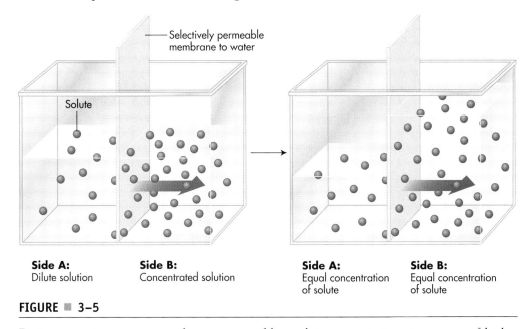

Side A: Dilute solution **Side B:** Concentrated solution **Side A:** Equal concentration of solute **Side B:** Equal concentration of solute

FIGURE ■ 3–5

During osmosis, water moves from an area of low solute concentration to an area of higher solute concentration.

Filtration differs from osmosis in that pressure is applied to force water and its dissolved materials across a membrane. Filtration caused by pressure is similar to a crowd of people being pushed through the turnstiles during rush hour, or the effect you get when you squeeze the trigger on a squirt gun. The major supplier of force in the body is the pumping of the heart, which forces blood flow into the kidneys where filtration takes place. This concept is expanded in Chapter 16, "The Urinary System." For now, see Figure 3–6 ■, which illustrates the process of filtration.

Facilitated diffusion (also called **carrier mediated passive transport**) is a variation of diffusion in which a substance is helped in moving across the membrane, like an usher helping you to your seat at a ball game. The "usher" in this case is a carrier protein. Glucose is one substance that is often helped across the cell in this way. These carriers are very specific. Only glucose, for example, can be carried by its special carrier. You may wonder, if it is helped, how can it be considered *passive* transport? Think of it as a situation in which the glucose was already moving in an attempt to cross the membrane, and it conveniently encounters an already revolving door. Once it steps into the door, it is quickly "pushed" along until it comes out on the other side of the membrane. See Figure 3–7 ■, which illustrates facilitated diffusion.

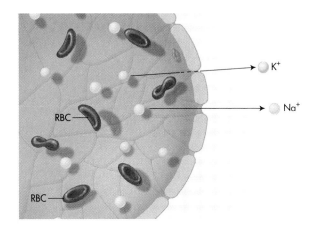

FIGURE ■ 3–6

The process of filtration in the kidneys, where smaller solutes such as the electrolytes sodium and potassium pass through the membrane, while the larger blood protein and cells normally do not.

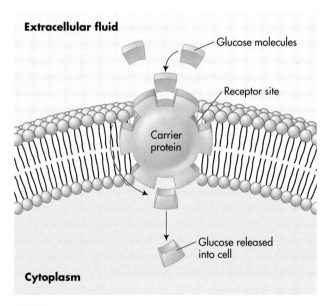

FIGURE ■ 3-7

Facilitated diffusion.

 Pathology Connection: Cystic Fibrosis (CF): A Membrane Channel Abnormality

Cystic fibrosis is a relatively common (1 in 3,000 Caucasian births) incurable, fatal genetic disorder caused by the malformation of membrane channels for chloride and sodium ions. In people with CF, sodium and chloride do not diffuse across the cell membrane as they normally would, because the channels are not normal. This makes the fluid around the cells extremely salty (excess sodium and chloride) and causes excessively thick mucus in the respiratory, digestive, and reproductive systems, clogging the organs. People with CF have difficulty breathing, nutritional deficits due to decreased absorption of nutrients, and increased risk of respiratory infections, diabetes, and infertility (especially males). While CF cannot be cured, treatments are available, such as nutritional supplements, antibiotics to prevent pneumonia, and mucus thinning drugs. Until recently, many CF patients did not survive into their 20's. With treatment available today, the average life span of a CF patient is longer than 35 years. CF can be diagnosed by prenatal or postnatal genetic testing, or by testing the amount of sodium in sweat.

Pathology Connection: Diabetes Mellitus: a Disorder of Facilitated Diffusion

Diabetes mellitus (DM), usually known simply as **diabetes**, is a common health problem. The chief symptom of diabetes, high blood sugar, is caused by a problem with facilitated diffusion. Remember when we discussed facilitated diffusion, we used glucose as an example. Glucose can't get into cells without facilitated diffusion. For glucose to get into cells by facilitated diffusion, a hormone, **insulin**, must be present. In DM, either insulin is absent or the cells are insensitive to insulin, so glucose doesn't get into cells like it should. This causes two problems. First, lots of glucose hangs around in the bloodstream, causing big osmotic problems for cells. Second, cells need glucose to make energy, so cells can't make as much energy as they need. This causes tremendous health problems. There will be a more detailed description of diabetes mellitus in the section on cellular respiration later in the chapter, and more information on diabetes throughout the book.

diabetes mellitus
(DIE ah BEE teez mel LITE us)

TEST YOUR KNOWLEDGE 3-1

Fill in the blanks:

1. This form of passive transportation is like combining drink mix and a pitcher of water: _____.

2. During osmosis, water travels across a semipermeable membrane from an area of _____ concentration of solute to an area of _____ concentration of solute.

3. _____ allows only certain sizes of particles to pass through.

4. The act of removing carbon dioxide from the blood to the lungs is achieved through _____.

5. Glucose is often transported by _____.

6. Diabetes mellitus may be caused by a deficiency in this hormone _____.

7. Cystic fibrosis occurs when a _____ is malformed.

ACTIVE TRANSPORT

Active transport can be broken down into three types:

→ active transport pumps
→ endocytosis
→ exocytosis

Active transport pumps (also called carrier mediated active transport) require the addition of an energy molecule called **adenosine triphosphate (ATP)** to move a substance. (We'll meet ATP again later in the chapter.) Like facilitated diffusion, active transport pumps use a protein carrier to move a specific substance across the cell membrane. Unlike facilitated diffusion, active transport pumps require energy to run because the cell is trying to move a substance into an area that already has a high concentration of that substance (up the concentration gradient). It's kind of like trying to put six pounds of sugar into a five pound bag! It can be done, but you have to apply a lot of energy. A common example in our cells is the need to transport potassium (K^+). Cells contain a good amount of potassium. The only way to get more into a cell is to apply energy to "push" it in.

endocytosis *(en DOE sigh TOE sis)*
   endo = *within*
   cyt/o = *cell*
   osis = *condition*
phagocytosis *(FAG oh sigh TOE sis)*

pinocytosis *(PIE no sigh TOE sis)*
   phag/o = *eating*
   pin/o = *to drink*
exocytosis *(Ex oh sigh TOE sis)*
   ex/o = *outside*
vesicle *(VESS ih kle)*

Endocytosis is utilized by the cells for the *intake* of liquid and food when the substance is too large to diffuse across the cell membrane. The cell membrane actually surrounds the substance with a small portion of its membrane, forming a chamber, or **vesicle,** which then separates from the membrane and moves into the cell. If a solid particle is being transported, we call it **phagocytosis.** White blood cells do it to bacteria to prevent infections in our bodies. If the intake involves liquid, it is called **pinocytosis.**

Exocytosis occurs when the cells needs to transport substances out of itself. Some cells may internally produce a substance needed outside the cell. Once this substance is made, it is surrounded by a membrane forming a **vesicle** (bladder or sac) and moves to the cell membrane. This vesicle becomes a part of the cell membrane and expels its load out of the cell. For further explanation of active transport, see Figure 3–8 ▦.

Table 3–1 summarizes all of the methods of transport.

TABLE 3–1 Methods of Cellular Transportation

CELLULAR TRANSPORTATION METHODS	DESCRIPTION
Passive Transportation (no energy required)	
Diffusion	Moving a substance from an area of high concentration to an area of low concentration
Filtration	Pressure is applied to force water and dissolved materials across a membrane
Osmosis	Water travels across a membrane from areas that have a low concentration of a solute to areas that have a higher concentration of solute until the concentration is the same on both sides of the membrane
Facilitated diffusion	Substance is assisted via carrier in a direction it was already traveling from an area of high concentration to an area of low concentration
Active Transport (energy required)	
Active transport pumps	Require additional energy in form of ATP to move substances against the concentration gradient (from low concentration to high concentration) using a carrier molecule
Endocytosis	Ingesting substances that are too large to diffuse across the cell membrane via vesicles
Phagocytosis	Form of endocytosis in which solid particles are being brought into the cell
Pinocytosis	Form of endocytosis in which liquid is being brought into the cell
Exocytosis	Transportation of material outside of the cell via vesicles

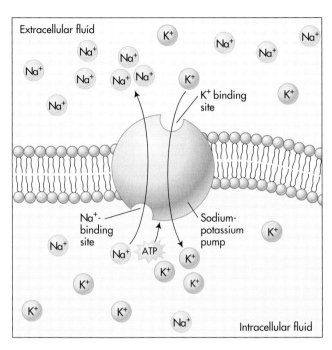

ACTIVE TRANSPORT PUMP

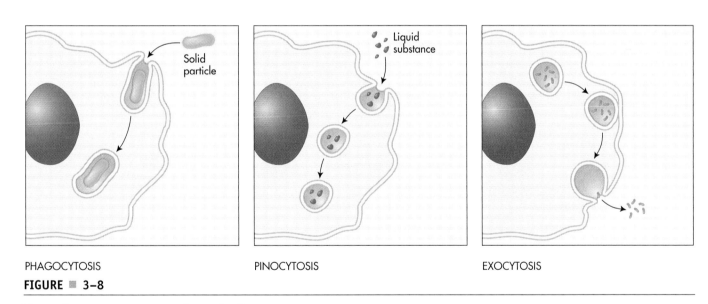

PHAGOCYTOSIS   PINOCYTOSIS   EXOCYTOSIS

FIGURE ▨ 3–8

Types of active transport in and out of cells.

Pathology Connection: Familial Hypercholesterolemia: A Disorder of Endocytosis

Hypercholesterolemia is a condition in which blood cholesterol is too high. For many people the condition can be cured by more nutritious food choices and increased exercise. However, some people have high cholesterol due to a genetic condition called **familial hypercholesterolemia**. Familial hypercholesterolemia causes high levels of the "bad" cholesterol known as Low Density Lipoprotein (LDL) in blood. In people without familial hypercholesterolemia, LDL in blood is bound to cholesterol. The LDL and cholesterol move into cells via endocytosis. Once inside, the cell uses the cholesterol to make other lipids. In people

hypercholesterolemia
(HI per koh LESS tur ul EE me ah)

with familial hypercholesterolemia, LDL doesn't move into cells and stays in the blood. This causes two problems. There is too little cholesterol getting into the cells so cells must make more cholesterol. In addition, the LDL that should be going into cells instead hangs around in the blood causing plaques in blood vessels, like a buildup of rust in pipes. These plaques lead to clots, strokes, and heart attacks. There are two forms of familial hypercholesterolemia: severe and moderate. People with severe familial hypercholesterolemia often have heart attacks or strokes as children, and this form is often fatal in children or teenagers. There is no effective treatment for the severe form. The moderate form also leads to heart attacks and strokes, but usually not until mid-life. Moderate familial hypercholesterolemia can be treated with diet modifications and cholesterol lowering drugs such as the statin group of drugs. It is estimated that 1 in 500 Americans have the moderate form.

TEST YOUR KNOWLEDGE 3-2

Complete the following:

1. Differentiate between phagocytosis and pinocytosis.

   _____

   _____

   _____

2. Explain how familial hypercholesterolemia causes heart problems.

   _____

   _____

   _____

3. Tell whether the following processes are active or passive:
   a. endocytosis _____
   b. facilitated diffusion _____
   c. osmosis _____
   d. phagocytosis _____
   e. filtration _____

Cytoplasm

Living organisms require balanced environments in which to thrive. Humans require the right mixture of oxygen and nitrogen; sea creatures require the right balance of salt and water; a chick embryo requires albumen, or "egg white," in which to develop. Likewise, the internal parts of a cell require a special environment, called **cytoplasm,** in order to survive.

Nucleus and Nucleolus

The **nucleus** has been described as the "brain of the cell." In our case, we will consider it to be City Hall: a control center. City Hall dictates the activity of the city departments much as the nucleus dictates the activities of the organelles in the cell. Since City Hall is crucial to the city's function, security is important to protect it from attack or damage. That is why metal detectors are installed at the doors. The nucleus of a cell is similar. It is surrounded by a double-walled nuclear membrane. Even though this membrane is composed of two layers, it has large pores that allow certain materials to pass in and out of the nucleus.

Somewhere in City Hall are blueprints of the city showing the buildings, streets, water and gas lines, and so on. The nucleus also contains "blueprints"— and "building codes" too. **Chromatin** is the material found in the nucleus that contains **DNA (deoxyribonucleic acid)**. DNA contains the specifications (blueprints), for the creation of new cells. Chromatin eventually forms chromosomes, which contain **genes.** Genes determine our inherited characteristics. (Remember Aunt Pearl saying how much you look like your Uncle Elmer?)

A spherical body called the **nucleolus,** made up of dense fibers, is found within the cell nucleus. Its major function is to synthesize the **ribonucleic acid (RNA)** that forms **ribosomes**. Now we have the blueprints, but what about the materials that we need? Who's in charge of getting these materials together? That's where **ribosomes** and **centrosomes** come into play. Figure 3–9 ■ shows the cell membrane, cytoplasm, nucleus, and nucleolus. We will continue to build the cell's infrastructures as we discuss their specific functions.

deoxyribonucleic acid *(dee AUK see RYE bow NEW clee ick ass id)*

ribosomes *(RIE beh soams)*

centrosomes *(SEN tre soams)*

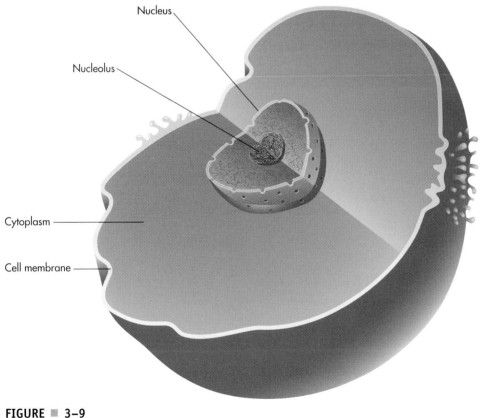

FIGURE ■ 3–9

The cell membrane, cytoplasm, nucleus, and nucleolus.

Ribosomes

Ribosomes are organelles that are found on the endoplasmic reticulum or floating around in the cytoplasm. Ribosomes are made of RNA and are the site for production of enzymes and other proteins that are needed for cell repair and reproduction. Following our analogy, consider ribosomes the manufacturing plant for building materials used for remodeling and repair.

Centrosomes

centrioles *(SEN tree oles)*

In a city, completely new structures often are needed to replace old ones. Therefore, the cell needs a building contractor to build new structures. The **centrosomes** are specialized regions that fill this need. Centrosomes contains **centrioles** that are involved in the division of the cell. Cellular division, or reproduction, will be discussed once we cover all the cell parts. Centrioles are tubular shaped and usually found in pairs. See Figure 3–10 ■, which now adds the nucleus, nucleolus, ribosomes, and centrosomes to the cell.

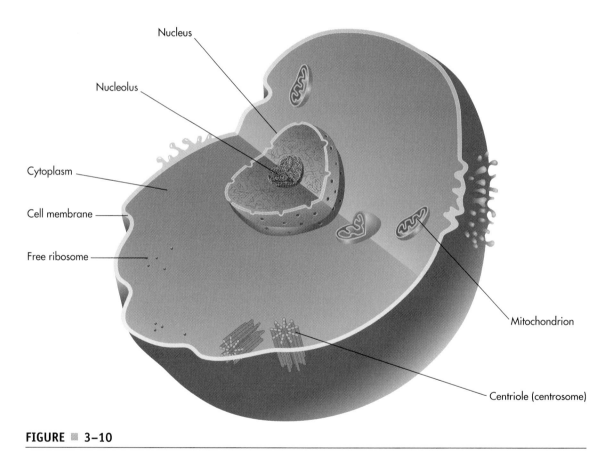

FIGURE ■ 3–10

The cell membrane, cytoplasm, nucleus, nucleolus, ribosomes, centriole, and mitochondria.

Mitochondria

Imagine what it would be like if we had no electricity in our city. Things would come to a standstill. The **mitochondria,** tiny bean-shaped organelles, act as power plants to provide up to 95 percent of the body's energy needs for cellular repair, movement, and reproduction. As a city's need for power increases, more power plants are built. Similarly, if a given cell type is very active and needs more power, there are a larger number of mitochondria in that cell. For example, liver cells, which are quite active, can have up to 2,000 mitochondria in each cell. In contrast, sperm cells "swim" with a tail (flagellum), so they need only a single mitochondria coiled around the tail for energy. Special enzymes in the mitochondria help to take in oxygen and use it to produce energy. While our city uses electricity for most of its power, the cell uses adenosine triphosphate (ATP), which is created by the mitochondria (shown in Figure 3–10).

mitochondria
(my teh CAHN dree ah)

Endoplasmic Reticulum

Although there are paths, walkways, and sidewalks in our city, the main structure for travel is a road system. The **endoplasmic reticulum** is a series of channels set up in the cytoplasm that are formed from folded membranes. The endoplasmic reticulum has two distinct forms. One has a sandpaper-like surface, the result of ribosomes on its surface, which we call the *rough* endoplasmic reticulum, and it is responsible for the synthesis of protein. Once the protein is synthesized, it is sent to the Golgi apparatus for processing. The second form has no ribosomes on its surface, making it appear smooth, so we call it the *smooth* endoplasmic reticulum. (Complex stuff, huh?) The smooth endoplasmic reticulum synthesizes lipids (fats) and steroids. Think of this as a series of dirt and paved roads with butcher shops and food processing plants along the way (see Figure 3–11 ▦).

endoplasmic reticulum
(EN doh PLAZ mic ri TIH cue lum)

Golgi Apparatus

Cities have factories with their own fleet of trucks. The **Golgi apparatus** is very similar to these factories. This organelle looks like a bunch of flattened, membranous sacs. Once the Golgi apparatus receives protein from the endoplasmic reticulum, it further processes and stores it as a shippable product, much like a packaging plant does. Not only does it prepare the protein for shipping, a part of the Golgi apparatus surrounds the protein with a vesicle that then separates itself from the main body of this organelle. That portion, with its load, then travels to the cell membrane where it releases (secretes) the protein! This is an example of exocytosis. Cells that constitute organs with a high level of secretion or storage (like the digestive system) contain higher numbers of the Golgi apparatus. Salivary glands and pancreatic glands, for example, are made of cells containing many Golgi apparati. Some proteins made in the Golgi apparatus stay in the cell and become lysosomes.

Golgi apparatus
(GOAL jee app ah RA tuss)

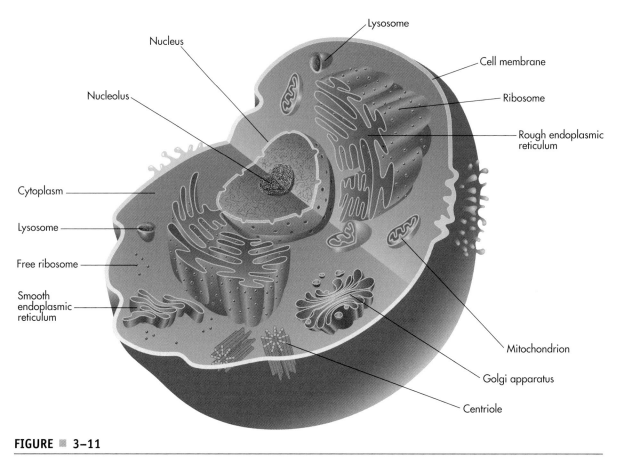

FIGURE ▪ 3–11

The cell membrane, cytoplasm, nucleus, nucleolus, ribosomes, centriole, mitochondria, endoplasmic reticulum, Golgi apparatus, and lysosomes.

Lysosomes

All cities create waste that must be removed. Cells are no different. **Lysosomes** are vesicles containing powerful enzymes that take care of cleaning up intracellular debris and other waste. Lysosomes are multitalented. They also aid in maintaining health by destroying unwanted bacteria through the process of phagocytosis. Again, please see Figure 3–11, which contains all the organelles.

 3-1 For an animation of the parts, structure of a cell, and cell division, please go to your DVD for this chapter.

Other Interesting Parts

Cells need a way to move things around or move themselves from place to place. Vesicles, or chambers, often created by the Golgi apparatus, can be thought of as little trucks. They can be loaded up with substances, travel to another site in the cell or cell membrane, and then drop off their loads. Similar in some ways to the skeleton, the **cytoskeleton** is a network of microtubules and interconnected filaments that provide shape to the cell and allow the cell and its contents to be mobile. Suppose we could build cities that float on large bodies of water. How would we propel them to new locations? Certain cells can

cytoskeleton
(SIGH toe SKELL eh ton)

solve this problem through the use of **flagella,** or whip-shaped tails that move some cells in a fashion similar to that of a tadpole.

flagella *(flah GELL ah)*

The future may offer new ways of transporting people and materials. For example, perhaps cities will install moving sidewalks, similar to those in some airports. Certain cells have solved such transport problems already through the use of **cilia.** Cilia are short, microscopic, hair-like projections located on the outer surface of some cells. They begin a wavelike motion that carries particles in a given direction. This is one way our lungs stay clean from the dust particles and germs that we inhale every day. The action is comparable to when a band member jumps into a crowd and is moved by the wave action of the audience's arms. This behavior is NOT recommended by the authors.

cilia *(SILL ee ah)*

Pathology Connection: Organelle Disorders

All organelles can malfunction. Therefore, there are many disorders caused by abnormalities in organelles. Many are widespread, affecting some or all of the cells in the body. Here are just a few examples of organelle disorders.

- **Tay Sachs, a disorder of lysosomes:** Lysosomal storage disorders occur when lysosomes are missing one of their enzymes. With the enzyme missing or broken, some molecules will not break down and will accumulate in cells. There are 40 different kinds of lysosomal storage disorders, all are genetic, and most lack effective therapies. Tay Sachs disease is the best known of the lysosmal storage diseases. It was, until recently, most common in Jews of central and eastern European descent. Tay Sachs is caused by a missing enzyme in the lysosomes of cells in the nervous system. Because of the missing enzyme, glycoproteins accumulate in cells in the nervous system, causing inflammation and eventually destroying the cells. The symptoms of Tay Sachs—mental regression, dementia, and paralysis—appear during the first year of a patient's life. There is no treatment, so the disease is generally fatal within two or three years of onset. Tay Sachs is diagnosed by the appearance of a cherry red spot on the back of patient's eyes and by abnormality in the startle reflex. The availability of a genetic test has decreased the incidence of Tay Sachs in recent years.

- **Cigarettes and paralyzed cilia:** As you learned earlier, your lungs stay clean because of the action of cilia. One of the side effects of smoking is to paralyze cilia. If cilia cannot move, they cannot keep the lungs clean. This condition may lead to many of the lung problems seen in smokers and is the primary etiology for Chronic Obstructive Pulmonary Disease (COPD), which affects more than 12 million people in the United States. Passive smoking (the exposure of nonsmokers to cigarette smoke) also increases the risk of COPD-related diseases.

Pathology Connection: Diabetes Mellitus (Revisited): A Disorder Tied to Cellular Respiration

Diabetes mellitus (DM) is a disorder characterized by blood glucose levels that are much higher than normal because the glucose cannot get into cells from the bloodstream as a result of either an insulin deficit in the body or a cellular insensitivity to insulin. Insulin, a hormone secreted by the pancreas, is necessary for facilitated diffusion to carry glucose into cells. There are two types of DM: type I, which is also known as Insulin Dependent Diabetes Mellitus (IDDM),

Applied Science

CELL ENERGY AND ATP: THE ENERGY MOLECULE

We all know that we need to eat to obtain energy, but how does energy get from food to cells? In simple terms, the body takes in food and breaks it down (digestion). During this process, energy is released from the food. Now, cells can't use this energy directly; only food converted to glucose (simple sugar) can be used to make energy. Your cells use glucose during a series of chemical reactions called **cellular respiration**. During cellular respiration, glucose is combined with oxygen and is transformed in your mitochondria into the high-energy molecule called **adenosine triphosphate (ATP)**. During cellular respiration, glucose is "burned" in the presence of oxygen, making water, carbon dioxide, and lots of energy. The equation

$$C_6H_{12}O_6 + 6O_2 \rightarrow 6CO_2 + 6H_2O + ATP \text{ (lots)}$$

Glucose + oxygen (Yields)$\xrightarrow{}$ carbon dioxide + water + energy

can be used to represent cellular respiration. Once the glucose is used up and energy is made, carbon dioxide and water are produced as waste products. In order to make energy for your cells, you must have glucose (from food) and abundant oxygen. Now you know why you breathe! You need to bring in oxygen to make energy and you to need to exhale to get rid of the waste product carbon dioxide.

The point of cellular respiration is to make energy in the form of ATP. ATP is made up of a base, a sugar, and three (hence, *tri*phosphate) phosphate groups. The phosphate groups are held together by high-energy bonds. When a bond is broken, a high level of energy is released. Energy in this form can be used by the cells. When a bond is used, ATP becomes **ADP (adenosine diphosphate)**, which has only two phosphate groups. ADP now is able to go and pick up another phosphate and form a high-energy bond so energy is stored and the process can begin again! Life is good.

juvenile, or early onset diabetes, and type II, which is also known as Non Insulin Dependent Diabetes Mellitus (NIDDM), or late onset diabetes.

The two types of diabetes are very different diseases. Type I is an autoimmune disorder. For some reason, the immune system attacks the cells in the pancreas, the beta cells, which make insulin. They are destroyed and stop making insulin. Type II is due to insulin resistance. That is, the pancreas makes plenty of insulin, but for some reason, your body's cells stop responding to it. The cells act as if there isn't enough insulin, even when there is enough. Type II is caused by obesity and is usually associated with high cholesterol, high lipids, and high blood pressure. Whether you have type I or type II diabetes, the ultimate problem is the same: too much glucose in the blood and too little glucose in cells.

The symptoms of diabetes—weight loss, excessive thirst and excessive urination, and chronic long term problems like heart disease and kidney failure—are caused by the glucose abnormalities. In order for cells to make ATP in the mitochondria, glucose must be able to get into cells. If glucose doesn't get into cells, the cells will look for other sources of glucose in order to make enough ATP to survive. Cells will begin to sacrifice body tissues like muscles in order to get energy from alternate sources, such as fats and proteins. Body chemistry will be abnormal due to the breakdown of body tissues caused by the cells' desperate attempts to get fuel.

The extra glucose in the blood causes an abnormal concentration gradient between the inside and outside of cells. Your body desperately tries to get rid of the extra glucose by filtering extra blood through your kidneys. Your kidneys produce excessive amounts of urine. This causes water loss, which can lead to dehydration. People with untreated diabetes are often thirsty, driven to drink water to replace the water lost during urination.

Both types of diabetes are treatable. Type I is treated with daily insulin injections, but there is no cure. Patients should stick to a strict diet to help keep blood glucose at healthy levels. Type II is often treatable and even reversible. The most important treatments are adoption of healthy diet, good exercise habits, and weight loss. Some patients with type II can treat their disease with diet and exercise alone. Drugs which stimulate the pancreas may also treat the disease, as can cholesterol lowering medications. Some patients with type II may eventually become insulin dependent, but some can keep their disease from progressing to that point. We will discuss many more details of diabetes as we continue our journey through the body.

Pathology Connection: PKU, an Enzyme Disorder

Phenylketonuria (PKU) is a genetic condition most common in Caucasians descended from Ireland, Scotland, and some Scandinavian countries. People with PKU are missing an enzyme called phenylalanine hydroxylase. Without this enzyme, the amino acid phenylalanine builds up in cells, especially in the nervous system, leading to progressive mental retardation in untreated individuals. In addition, individuals with PKU have light pigmentation of skin, hair, and eyes, abnormalities of posture and gait, and epilepsy.

The primary treatment for PKU is a low phenylalanine diet, typically avoiding high protein foods. Recently, however, adhering to this diet has become a bit more complicated. Next time you pick up a can or bottle of diet soda, read the warning on the side of the can. Aspartame, the sweetener in NutraSweet, is made of two amino acids. One of them is phenylalanine! People with PKU must avoid consuming products containing Aspartame.

In children, coming "off diet" can lead to cognitive deficits. In adults, a significant percentage of patients "off diet" report depression or anxiety and other neurological changes. PKU is diagnosed by blood tests and genetic tests. Untreated PKU is rare in the U.S. since all states currently screen every newborn for PKU.

Clinical Application

ENZYMES: MAKING REACTIONS HAPPEN

In order for your cells to be able to do anything, some chemicals must be broken down and others must be made. You need building materials to build organelles and to make energy. Any of these processes require chemical reactions to take place in the cell. The problem is that these reactions are usually very slow—so slow that by the time the reaction happens, it would be too late for your cells to use the building materials. To solve this problem, cells have special proteins called **enzymes**. Enzymes speed up the rate of chemical reactions making them fast enough for your cells to use the materials. Enzymes are like carriers molecules, in that they are not used up and they are very specific. Only certain reactions can be sped up (catalyzed) by certain enzymes.

TEST YOUR KNOWLEDGE 3-3

Choose the best answer:

1. What organelle most closely follows the given analogy?
   a. city hall _____
   b. road system _____
   c. power plant _____
   d. packing and shipping plant _____
   e. the "cell propeller" _____

Answer the following:

2. What is a lysosomal storage disorder?
   _____

3. Explain why enzymes are important.
   _____
   _____

4. Why is it so important for glucose to get into cells?
   _____
   _____

Mitosis

Cellular reproduction is the process of making a new cell. Cellular reproduction is also known as **cell division,** because one cell divides into two cells when it reproduces. Cells can only come from other cells. When cells make identical copies of themselves without the involvement of another cell, that is called **asexual reproduction.** Most cells are able to reproduce themselves asexually, whether they are animal cells, plant cells, or bacteria.

The cells that make up the human body are a type of cell known as a **eukaryotic cell.** Eukaryotic cells have a nucleus, cellular organelles, and usually, several chromosomes in the nucleus. (Reminder: the genetic material of the cell, DNA, is bundled into "packages" of chromatin known as chromosomes.) Since chromosomes carry all the instructions for the cells, all cells must have a complete set after reproduction. These instructions include how the cell is to function within the body and blueprints for reproduction. No matter whether a cell has one chromosome, like bacteria, or 46 chromosomes, like humans, all the chromosomes must be copied before the cell can divide.

Let's start off with the simple cellular division of a bacterium. Bacterial cells, which do not have a nucleus or organelles, are called **prokaryotic cells,** and reproduce very easily, through a process known as **binary fission**. Bacterial cells simply copy their DNA, divide up the cytoplasm, and split in half!

Now let's go to the more complex cellular reproduction. Eukaryotic cells, like yours, must go through a more complicated set of maneuvers in order to reproduce. Not only do your cells have to duplicate all 46 of their chromosomes, but they also have to make sure that each cell gets all of the chromosomes and all of the right organelles. The process of sorting the chromosomes, so that each new cell gets the right number of copies of all of the genetic material, is called **mitosis.** Mitosis is the only way that eukaryotic cells can reproduce asexually.

eukaryotic cell
(you care ee AH tic SELL)

mitosis *(my TOE sis)*

mitotic *(my TOT ick)*
cytokinesis
(site oh keh KNEE sess)

The Cell Cycle

The total life of a eukaryotic cell can be divided into two major phases known as the **cell cycle.** Most of the cell cycle is devoted to a phase known as **interphase.** During interphase, a cell is not dividing, but is performing its normal functions along with stockpiling needed materials and preparing for division by also copying DNA and making new organelles. Only a brief portion of the cell cycle, the **mitotic phase,** is devoted to actual cell division. The mitotic phase is divided into two major portions. Mitosis is the division and sorting of the *genetic material,* while **cytokinesis** is the division of the *cytoplasm.*

As you might imagine, the cell cycle is very tightly controlled to avoid overproduction of cells. (The overproduction of cells can sometimes lead to cancer, which will be discussed later in the chapter.) At several places in the cell cycle there are checkpoints, spots where cells must wait for a signal to tell them to keep dividing. If something is wrong with the cell or there is some reason the cell should not reproduce, the cell will not get a signal and will stop dividing. If the timing is good, the cell will get the go-ahead signal and will divide without interruption.

Amazing Body Facts

BACTERIAL REPRODUCTION
Bacteria can reproduce so rapidly that they can double their population every hour! No wonder a bacterial infection can get out of hand so fast.

Mitosis, the division of the genetic material, is the most complicated part of cell division for a eukaryotic cell. What further complicates mitosis is that the process is further divided into four specific phases. As you'll soon see, the four phases are based on the position of the chromosomes relative to the new cells. They are: **prophase, metaphase, anaphase,** and **telophase.** To put all these processes and phases in perspective, please refer to the flowchart in Figure 3-12 ▪.

THE PHASES OF MITOSIS

Let's proceed step by step through the four phases of mitosis using Figure 3-13 ▪ as a reference.

1. **Prophase (pro = *before*)** The nucleus disappears, the chromosomes become visible, and the centrioles move toward the sides of the cell; a set of chromosomal anchor lines or guide wires (the **spindle**) forms.
2. **Metaphase (meta = *between*)** The chromosomes line up in the center of the cell.
3. **Anaphase (an = *without*)** The chromosomes split and the spindles pull them apart.
4. **Telophase (telo = *end*)** The chromosomes go to the far end of the cell, the spindle disappears, and the nuclei reappear.

During or directly after telophase, cytokinesis happens and the cell divides in half. The original cell was the mother cell that has now formed into two new identical daughter cells. Thus mitosis, asexual reproduction in eukaryotic cells, results in two new daughter cells identical to the original mother cell.

MITOSIS IN YOUR BODY

Mitosis (asexual cellular reproduction) serves many purposes in your body. First, your body uses mitosis any time your cells need to be replaced. Many tissues, like bone, epithelium, skin, and blood cells, all replace themselves on a regular basis. Mitosis also accomplishes repair and regeneration of damaged tissue. If you cut your hand, the skin is replaced, first by collagen, but eventually by the original type of tissue. Mitosis increases in cells near the injury so that the damaged or destroyed cells can be replaced. A broken bone is replaced in much the same way.

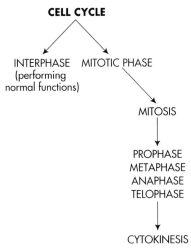

FIGURE ▪ **3–12**

Flow Chart of the cell cycle.

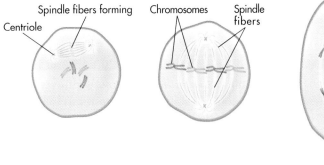

Prophase   Metaphase   Anaphase   Telophase and Cytokinesis

FIGURE ▪ **3–13**

Phases of mitosis.

MITOSIS VERSUS MEIOSIS

These words sound and look alike and are therefore often confused. Remember, meiosis produces gametes or sexual cells which contain half of the chromosomes because the sexual union of male and female will contribute the other half. Mitosis (*I reproduce myself*) is asexual and produces exact copies of the cell and the full complement of chromosomes since no union is needed.

meiosis *(my OH sis)*

Growth is also accomplished by mitosis. Lengthening of bones as you grow, increases in muscle mass due to exercise, and most other ways that tissue gets bigger are all due to mitosis of cells in the tissues or organs. Without mitosis, your body would not be able to grow or replace old or damaged cells.

Sexual reproduction is different from mitosis. **Meiosis** is the term for the division of cells in order for reproduction to occur. During meiosis, four cells are produced from one cell. These cells are not identical to the mother, because they have half as many chromosomes. These cells are the eggs and sperm which will be used for reproduction. This process will be discussed in detail in Chapter 17, "The Reproductive System."

TEST YOUR KNOWLEDGE 3-4

Choose the best answer:

1. Cells are reproducing themselves during this portion of the cell cycle:
   a. metaphase
   b. mitotic phase
   c. meiotic phase
   d. manic phase

2. During this phase of mitosis the nucleus disappears and the spindle appears:
   a. prophase
   b. metaphase
   c. anaphase
   d. telophase

3. During this phase of mitosis the chromosomes begin to pull apart:
   a. prophase
   b. metaphase
   c. anaphase
   d. telophase

4. Which of the following is not a function of asexual reproduction?
   a. tissue repair
   b. replacement of cells
   c. tissue growth
   d. cloning humans

 Pathology Connection: Cancer: Mitosis Run Amok

When the body is healthy, cells grow in an orderly fashion: they grow at the appropriate rate in the appropriate number, shape, and alignment. If everything is working as it should, the control system prevents cells from reproducing too fast. Sometimes conditions are altered in the body (either internally or externally) that trigger changes in the way the cells grow. Cell growth can become wild and uncontrolled, leading to too many cells being produced and resulting in a lump, or tumor. Generally, tumors are classified as either benign or malignant. **Benign** tumors typically grow slowly and push healthy cells out of the

benign *(bee NINE)*

way. Usually, benign tumors are not life-threatening. **Malignant,** or cancerous, tumors grow rapidly and invade rather than push aside healthy cell tissue.

Cancer actually means "crab" (remember your Zodiac signs?) and is a good description of cancer cells in that they spread out into healthy tissue like the legs and pincers of a crab. Cancerous tumors also differ from benign tumors in that parts of a cancerous tumor can break off and travel through the blood system or the lymphatic network and start new tumors in other parts of the body. This breaking off and spreading of malignant cells is called **metastasis.** One reason that lung cancer is so deadly (more women die from it than from breast cancer) is that it can **metastasize** for a fairly long time before it is even diagnosed in the lungs. By the time it is discovered, tumors may be growing in the liver and brain and in other parts of the body, making survival very difficult.

Cancer prognosis is often determined by the stage of the cancer at diagnosis. The most general cancer stages are determined by the amount of metastasis. Stage I cancer has not spread; stage II cancer has spread to nearby tissues; stage III cancer has spread to the lymphatic system; and stage IV cancer has spread to distant organs. Staging can also be done using the T, N, M classification system, which takes into account the extent of the primary Tumor, the amount of lymph Nodes invaded by the cancer, and the extent of tumor Metastasis. More recently, scientists have begun to develop specific staging systems for specific cancers. These systems are developed after examination of the outcomes of thousands of cases looking for specific characteristics of tumors that better predict outcome.

Cancer is diagnosed by a variety of techniques, including imaging (MRI, CT scan, and X-ray), blood tests, and **biopsy**, which is the surgical examination of the abnormal tissue.

Cancer can be treated by any method that attacks cancer cells. The most common treatments are **chemotherapy**, which uses chemicals to kill rapidly dividing cells; radiation therapy, which uses energy to target cancer cells; surgery, which removes cancer cells from the body; and biological or immunotherapy, which trains the body's natural defenses to fight the cancer cells. Typically more than one treatment will be used to attack a cancer. Success of treatment depends on stage at diagnosis and specific characteristics of the cancer itself. Some cancers are more responsive to treatment than others.

malignant *(mah LIG nant)*

metastasis *(meh TASS tuh sis)*

chemotherapy
(KEE moh THAIR ah pee)

MICROORGANISMS

As in any city, you have a large and diverse population. While the vast majority of the population are good citizens who provide positive contributions to the city, there are some who can cause problems. The same can be said for the world of microorganisms. The following are the main "microcitizens" of our city:

→   bacteria

→   viruses

→   fungi

→   protozoa

Bacteria

bacteria *(back TEER ee ah)*
pathogen *(PATH oh jenn)*
   path/o = *disease*
   gen = *producing*

When you hear the word **bacteria,** you probably first think of an organism that produces disease, called a **pathogen.** You are somewhat correct in assuming so because bacteria make up the largest group of pathogens. Some bacterial pathogens even release toxic substances in your body. Bacteria, which are prokaryotes, grow rapidly and reproduce by splitting in half, sometimes doubling as rapidly as every 30 minutes!

normal flora *(FLOOR ah)*

You will learn throughout this book that bacteria are often harmless and, in fact, essential for life. These bacteria live within or on us, and are part of what is called our **normal flora.** For example, certain bacteria in your intestine help to digest food, and some help to synthesize vitamin K, which helps clot blood so you don't bleed to death when you get a cut or scrape. See Figure 3–14 ■, which shows examples of various types of bacteria.

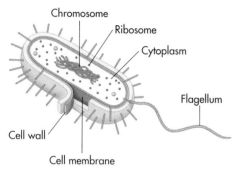

Chromosome
Ribosome
Cytoplasm
Flagellum
Cell wall
Cell membrane

BACILLUS (ROD-SHAPED)

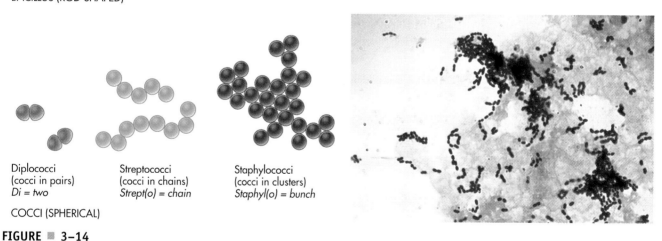

Diplococci
(cocci in pairs)
Di = two

Streptococci
(cocci in chains)
Strept(o) = chain

Staphylococci
(cocci in clusters)
Staphyl(o) = bunch

COCCI (SPHERICAL)

FIGURE ■ **3–14**

Types of bacteria.

Viruses

capsid *(CAP sid)*

An even more basic pathogen is a **virus** (see Figure 3–15 ■). Viruses (from a Latin term meaning "poison") are infectious particles that have a core containing genetic material (codes to replicate) surrounded by a protective protein coat called a **capsid.** Some viruses have an additional layer, or membrane, surrounding the capsid. Viruses are interesting because they cannot grow, "eat," or reproduce by themselves. They must enter another cell (host cell) and hijack that cell's parts, energy supply, and materials to do all cellular activities! In addition, each virus must target specific cells in the body to claim as hosts. For example, the viruses that cause a cold target the cells found in the respiratory system, and the viruses that cause herpes target cells that are found in the tissues of our nervous system. Most of the upper respiratory infections that people get are caused by viruses and, like all viral infections, they do not respond to antibiotics.

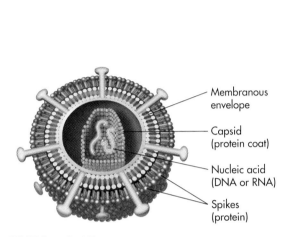
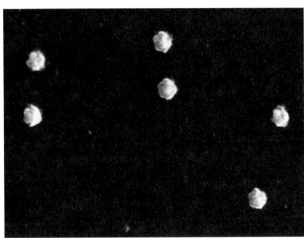

Membranous envelope

Capsid (protein coat)

Nucleic acid (DNA or RNA)

Spikes (protein)

FIGURE ▪ 3–15

A virus.

It is interesting to note (or disturbing for those of you who worry about everything) that viruses can stay dormant in the body and become active once again later in life. This is true for people who have had chicken pox. Those viruses stay in the body and may later become active and cause a potentially painful skin condition called *shingles*, caused by the herpes zoster virus. It is also interesting to note that the actual virus is relatively easy to kill by itself, but once it becomes part of a cell, it is hard to kill without harming the individual cell.

Fungi

Fungi, the plural form of fungus, can be either a one-celled or multi-celled organism. These plantlike organisms have tiny filaments, called **mycelia,** which travel out from the cell to find and absorb nutrients. Like bacteria, fungi, such as edible mushrooms, can be good, but in certain situations they can also cause problems (see Figure 3–16 ▪).

Fungi also can spread through the release of **spores.** Normally, people are not affected by fungi, but if the body has a problem with its immune system, then fungi have a chance to cause an opportunistic infection. If tissue becomes damaged, fungi can more easily create an infection. Wind can pick up and carry spores that can be inhaled, potentially causing lung infections. Interestingly, you can inhale fungi spores and still not develop an infection. It is estimated that up to 80 percent of the population in certain regions of the United States have tested positive for the inhalation of certain fungi! Examples of fungal infections are athlete's foot and a mouth fungus called thrush or candidiasis.

fungi *(FUN jie)*
mycelia *(my SEE lee ah)*

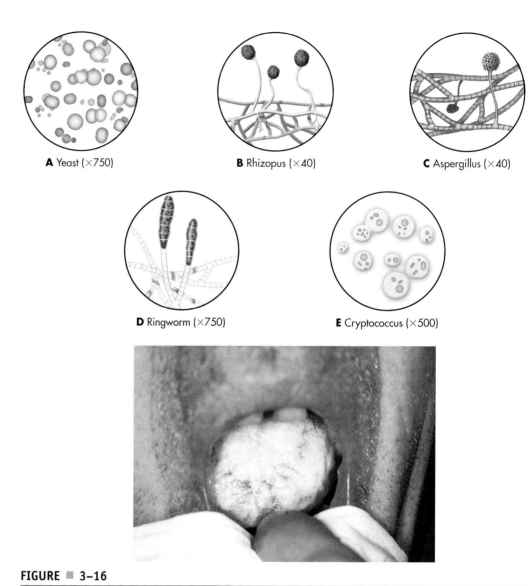

FIGURE ■ 3–16

Types of fungi and a fungal infection of the tongue. (Photo Source: Courtesy of Jason L. Smith, MD.)

Protozoa

protozoa *(pro toe ZOE ah)*

Protozoa are one-celled animal-like organisms that can be found in water, such as ponds, and in soil. Disease caused by these microorganisms can result from swallowing them (such as by drinking contaminated water) or from being bitten by insects that carry them in their bodies (such as malaria-carrying mosquitoes). See Figure 3–17 ■.

 Pathology Connection: How Microorganisms Cause Disease

All microorganisms, bacteria, viruses, fungi and protozoa can cause disease, but the disease mechanism and treatments are different for each type of organism. Any microorganism that causes disease is called a pathogen. The treatment and course of any infection is determined by the type of pathogen involved.

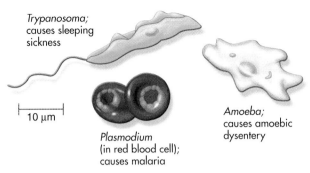

Trypanosoma; causes sleeping sickness

10 μm

Plasmodium (in red blood cell); causes malaria

Amoeba; causes amoebic dysentery

FIGURE ■ 3–17

Protozoa.

BACTERIA

While bacteria can cause disease directly by destroying infected tissue, many bacteria cause disease due to the release of toxins into the host body. Toxins can cause many problems including, destruction of body tissues, destruction of blood cells, inhibition of ribosomes, fluid loss, high fever, decreased blood pressure, increased blood clotting, fluid in lungs, and paralysis. Signs and symptoms of bacterial infection may include high fever; rapid pulse and breathing; abnormal, often foul-smelling discharge from infected area; and pain and swelling at site of infection. Other symptoms depend on the location of the infection. Bacterial infections are treated with **antibiotics**, chemicals which can kill the prokaryotic bacteria without harming eukaryotic cells. Most antibiotics are produced naturally by other microorganisms. See Table 3-2 for common bacterial pathogens and their associated disease.

anti = *against*
bios = *life*

TABLE 3–2 Common Bacterial Pathogens and Associated Diseases

BACTERIAL PATHOGENS	DISEASES
Gram-positive cocci	
Staphylococcus aureus	Skin infections, food poisoning
Streptococcus pyogenes	Pharynigitis (strep throat)
Streptococcus pneumoniae	Lobar pneumonia
Gram-positive bacilli	
Clostridium tetani	Tetanus
Clostridium botulinum	Botulism
Bacillus anthracis	Anthrax
Gram-negative cocci	
Neisseria gonorrhoeae	Gonorrhea
Gram-negative bacilli	
Salmonella typhimurium	Salmonellosis
Legionella pneumophila	Legionellosis (Legionnaires' disease)
Pseudomonas aeurginosa	Urinary tract infections, burn infections

VIRUSES

Viruses are not cells at all, but intracellular parasites composed of a nucleic acid (either DNA or RNA) with a lipid or protein coat. Viruses must use the materials of a cell to make more viruses. Viruses cause disease in two ways: directly, by shutting down a cell or destroying the cell outright (cell ruptures to release viruses), or indirectly, by making a good environment for the development of secondary bacterial or fungal infections. For example, influenza rarely kills people. The leading cause of death due to the flu is bacterial pneumonia, which can easily infect lung tissue damaged by the flu virus.

Signs and symptoms of viral infection include low grade fever (although it may also be high sometimes), muscle aches, and general fatigue, though some viral infections may cause no symptoms. Viral infections are complicated in that they may be latent, hiding in the body without causing symptoms and then activating years after the infection. Some viral infections may also become chronic, causing low level symptoms for weeks, months, or even years. There are very few treatments for viral infections. Viruses are not prokaryotes, so antibiotics do not kill viruses. In fact, since viruses are using the eukaryotic cell's machinery, antiviral drugs usually have side effects because they affect the host cell. Most viruses are destroyed by the immune system within a few days of infection. The treatments for most viral infections are rest, fluids, and management of symptoms to keep the patient comfortable. Please see Table 3-3 for common pathogenic viruses and their associated diseases.

3-2 If you would like more information on microbiology, pathogenic organisms, and ways to prevent the spread of disease, please go to your DVD. In addition, a video demonstrates the proper handwashing and gloving techniques to minimize the spread of infection. Don't forget the interactive games created just for this chapter!

FUNGI

Most fungal infections are caused by the inhalation or ingestion of fungal spores, or the entrance of spores through open wounds. Spores are tiny bodies resistant to environmental changes, meaning they can stay dormant until conditions are just right. Most fungal spores do not cause disease in otherwise healthy individuals, though fungal infections of the skin, such as athlete's foot and jock itch, are common. Many fungal infections are opportunistic, causing disease in individuals with compromised immune systems or other underlying disease. Symptoms of fungal infection vary widely depending on the location of the infection. Fungal infections are often difficult to treat. Some antifungal drugs are highly toxic and many fungal infections are resistant to treatment.

PROTOZOANS

Most protozoan infections are caused by ingestion of contaminated water or by insect bites. Many of these protozoans are parasites, actually taking up residence in the human body and living off its cells. Most protozoans that infect humans cause disease. Symptoms of protozoan infections vary widely depending on the type of protozoan infection. Many are very serious diseases which cause long term debilitating illness, such as malaria, which is transmitted by mosquitoes. Others are relatively mild illnesses, like "beaver fever" caused by Giardia, a protozoan that lives in streams and water supplies contaminated by fecal matter.

TABLE 3–3 Viral Pathogens and their Diseaeses

VIRUSES	DISEASES
DNA viruses	
Herpes viruses	
Herpes simplex 1	Cold sores and fever blisters
Herpes simplex 2	Genital herpes
Varicella-Zoster	Chicken pox and shingles
Hepatitis B	Hepatitis ("serum hepatitis")
Epstein-Barr	Infectious mononucleosis
RNA viruses	
Influenza A, B, C	Influenza ("flu")
Hepatitis A, C, D	Infectious hepatitis
Rhinovirus	Common cold
Human immunodeficiency virus	HIV infection/AIDS

Clinical Application

ANTIBIOTICS' RESISTANCE IN BACTERIA

Prior to the discovery of antibiotics in the 1930's, most people who got bacterial infections died of their infections. Those who did survive were often left with long-term effects. Diseases that are easy to deal with today, like strep throat or simple wound infections, killed thousands of people every year. Many deaths during wars were caused by the bacterial infections of the wounds rather than the wounds themselves. When antibiotics were discovered they were considered a miracle cure and were used carelessly. As the drugs became readily available, doctors began to prescribe them whenever a patient wanted them, whether they had a bacterial infection or a viral infection. Farmers began to add antibiotics to livestock feed. At the time these practices seemed harmless. Now we know that overuse of antibiotics was a big mistake. Today, many bacteria cannot be killed by antibiotics that used to work without fail. Many bacteria have evolved resistance to the drugs. Some bacteria are even resistant to many different antibiotics. Bacterial infections, like strep pneumonia and tuberculosis, are not always killed by antibiotics that used to work on them in the past. Many bacterial infections must be treated by several antibiotics before the infection is killed. Think about the last time you or someone you know had to take an antibiotic. Did the first one work? Did the second? Thirty years ago, penicillin was the drug of choice for strep infections. Now doctors rarely prescribe penicillin because bacteria are no longer sensitive. Sensible use of antibiotics and the development of new antibiotics is the only way to combat this important public health problem.

SUMMARY

Snapshots from the Journey

→ All living organisms are made of one or more cells. Cells are the fundamental units of living organisms.

→ Even though cells are the fundamental units, they are composed of a variety of parts that are necessary for proper cellular function. These small parts are called organelles.

→ Substances can cross the cell membrane via passive or active transport.

→ Passive transport can occur through diffusion, facilitated diffusion, osmosis, or filtration.

→ Active transport can occur through active transport pumps, endocytosis, or exocytosis.

→ Problems with transport across cell membranes can cause serious disorders such as cystic fibrosis and diabetes.

→ One form of cellular movement is through the use of flagella.

→ Cilia found in the lungs aid in the removal of foreign particles in the airways through rhythmic movement.

→ When organelles do not work properly, serious problems result, often in every cell of the body.

→ Cells must have the energy molecule ATP in order to power the cell. ATP is made in the mitochondria by a process called cellular respiration. In order to make ATP, cells must have glucose and oxygen.

→ Diabetes mellitus (DM) results when glucose cannot get into cells. Cells cannot make energy and excess glucose builds up in the blood causing many problems.

→ Most chemical reactions in the body could not happen without the help of proteins called enzymes. Missing enzymes usually cause major problems.

→ Cells and tissues grow, are replaced and are repaired by asexual reproduction. Cells make identical copies of themselves. This takes place all over your body whenever tissues grow or are repaired.

→ Some tissues, like epidermis, blood, and bone, replace themselves continually, always by asexual reproduction.

→ The life of the cell can be depicted as a cell cycle. Only about 10% of the cell cycle is devoted to mitosis. The rest of the time the cell is in interphase, preparing to divide and carrying on day-to-day cellular activities.

→ Asexual reproduction in eukaryotic cells is accomplished by a relatively complex process called mitosis and cytokinesis. Mitosis, the division of the genetic material, takes place in four phases: prophase, metaphase, anaphase, and telophase. Cytokinesis is the division of the cytoplasm and organelles. Mitosis produces two daughter cells, identical to each other.

→ Meiosis is the sexual reproduction of cells. If an organism is going to reproduce sexually it must use specialized cells called gametes with only half the typical number of chromosomes for that organism. The chief difference between mitosis and meiosis is that mitosis is asexual and produces the exact number of chromosomes while meiosis is sexual and combines two cells with half of the needed chromosomes.

→ Cancer results when mitosis is not regulated and cells divide out of control, spreading to other tissues.

→ Not all bacteria are bad. In fact, the body needs bacteria to survive.

→ A virus is not technically a one-celled organism; it needs another cell to replicate.

→ Fungi can be single-celled or a multi-celled organism and can cause infections in the body. Spores can be immune to a harsh environment, thus allowing fungi to spread.

→ Protozoa are one-celled and can cause disease through ingestion or through insect bites.

→ Many microorganisms cause disease. Each has unique characteristics and treatments. Antibiotics can be used *only* to treat illnesses caused by bacteria.

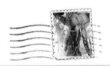

Case Study

Given the following mini-scenarios, identify what type of microorganism may be the causative agent.

a. Two young boys complain of a stomach ache and severe diarrhea after drinking pond water.

_____

b. Julia is a 13-year-old with a compromised immune system due to an inherited disease. Two days after returning home from a school field trip to a mushroom factory, Julia complains of shortness of breath and is diagnosed with a respiratory infection.

_____

c. Bob has had a stubborn cold for three days and is given an antibacterial agent. However, he doesn't respond to the treatment and the cold persists.

_____

MARIA'S STORY

Maria is a 35-year-old diabetic who now takes good care of herself. Diagnosed with autoimmune type I at 12 years old, Maria rebelled against the dietary restrictions and daily insulin injections and blood monitoring. During adolescence she had wild blood sugar swings, hypoglycemic episodes that would cause her to black out one day and be off the chart with hyperglycemia the next. She ignored all the "scare tactics" her doctors and parents tried to lay on her. She figured it was no big deal. She'd deal with her diabetes when she was good and ready.

In early 20's, Maria's retinas began to hemorrhage. Luckily the problem was caught early enough to save her eyesight. Then a minor cut on her leg turned into an ulcer that had to be treated surgically. Alarmed, she did some Web surfing and realized that she was flirting with disaster. She talked to her doctor and a nutritionist and began to stick to her diet and exercise program and to monitor her blood glucose level daily.

For several years her blood glucose was relatively stable, marked by occasional periods of hyperglycemia or hypoglycemia. When she turned 30, things began to unravel. Even when she followed her treatment plan rigorously her blood sugar was sometimes difficult to control. She tried an insulin pump to stabilize her blood sugar. Nothing worked. She passed out several times in public and then began to have other problems. She had several skin infections and began to have sharp pain in her hands and feet. Her kidneys began to fail. In an attempt to control her diabetes, her doctor suggested a pancreas transplant.

a. List and explain two problems the lack of insulin causes for cells.

_____

_____

_____

b. In what future system chapters will you expect to see Maria discussed?

_____

_____

_____

_____

REVIEW QUESTIONS

Multiple Choice

1. The cell membrane can best be described as:
   a. permeable to all materials
   b. nonpermeable
   c. rigid
   d. semipermeable

2. All of the following are passive forms of transport *except*:
   a. facilitated diffusion
   b. exocytosis
   c. osmosis
   d. filtration

3. With a greater concentration of a solute, what will happen to osmotic pressure?
   a. it will become less
   b. it will become greater
   c. it will remain the same
   d. there is no relation between osmotic pressure and the concentration of solute.

4. Cystic fibrosis is caused by:
   a. a missing carrier protein
   b. a malformed membrane channel
   c. a hole in the cell membrane
   d. too little insulin

5. Where are the specifications for the creation of new cells found?
   a. DNA
   b. RNA
   c. ATP
   d. ADP

6. Which microorganism can cause disease?
   a. bacteria
   b. fungi
   c. virus
   d. all of the above

7. Antibiotics are used to treat illness caused *only* by this type of pathogen:
   a. virus
   b. bacteria
   c. fungus
   d. protozoa

Matching

1. Match the following organelles with their function:

   _____nucleus   a. the "powerhouse" of the cell

   _____cell membrane   b. processing, packaging, and shipping of materials

   _____Golgi apparatus   c. the gatekeeper of the cell

   _____mitochondria   d. the "brain" of the cell

   _____cytoplasm   e. the internal environment

   _____lysosome   f. sanitation engineers

2. Match the disorder with its cause:

   _____diabetes mellitus   a. uncontrolled mitosis

   _____cystic fibrosis   b. missing lysosomal enzyme

   _____cancer   c. missing cellular enzyme

   _____Tay Sachs disease   d. malformed membrane channel

   _____PKU   e. failed facilitated diffusion

Short Answer

1. List and describe the four methods of passive transport.

   _____

   _____

   _____

   _____

2. Why do viruses need cells?

   _____

   _____

   _____

   _____

3. How does passive transport differ from active transport?

   _____

   _____

   _____

   _____

4. Discuss how cells can provide motility for themselves.

   _____

   _____

   _____

   _____

5. Explain the importance of enzymes to cells.

_____

_____

_____

_____

6. Explain the stages of cancer.

_____

_____

_____

_____

7. Compare and contrast bacterial, viral, and fungal diseases.

_____

_____

_____

_____

Suggested Activities

1. Research the various cell types within the body and see how many you can list.

2. Research five sexually transmitted diseases (STDs) and list the causative microorganism for each.

3. List the ways that we abuse antibiotics and ways that we could slow down antibiotics resistance in bacteria.

 3-3 Now that you have completed your journey through this chapter, please go to the DVD for interactive games and puzzles concerning the medical terms and concepts contained in this chapter. By playing the games you will reinforce your learning in a fun way.

TISSUES AND SYSTEMS

The Inside Story

Previously, we discussed cells as the basic building blocks of the body. In this chapter, we explore **tissues,** which are collections of similar cells. We can consider cells to be individual employees. At the next level, we can think of tissue as a group of employees who have the same or similar educational background, such as radiologic technicians, at a hospital. A combination of tissues designed to perform a specific function or several functions is called an **organ**. This could be compared to an X-ray department in the hospital, where specific functions such as X-rays, CT scans, MRIs, and barium swallows (a procedure, not a bird!) are performed. Organs that work together to perform specific activities, often with the help of accessory structures, form what we call **systems**. We can compare a system to a hospital that provides a service (health care) to the citizens of our city through the combination of all of the departments: laboratory, nursing, respiratory care, physical therapy, dietetics, housekeeping, and so on. Of course, all those systems combine to make a living **organism**, in this case the wondrous human body. This chapter provides an overview of tissues, organs, and systems, which later chapters will expand upon.

Chapter 4

LEARNING OBJECTIVES

At the end of your journey through this chapter, you will be able to:

→ Explain the relationship between cells, tissues, organs, and systems

→ List and describe the four main types of tissues

→ Identify and describe the various body membranes

→ Differentiate the three main types of muscle tissues

→ Describe the main components of nerve tissue

→ List and describe the main functions of the body systems

→ Provide general examples of how pathologic conditions can impact on cells, tissues, organs and body systems

MULTIMEDIA APPLICATIONS

DVD Interactive Exercises

→ Video on skin cancer, 4-1

→ Interactive exercises: Body systems identification and labeling of main components, 4-2

→ Interactive games and puzzles, 4-3

www.prenhall.com/colbert

→ Professional Profile:
  • Histotechnology

→ Related Internet Links

→ Additional Review Questions

Pronunciation Guide

Correct pronunciation is important in any journey so that you and others are completely understood. Here is a "see and say" Pronunciation Guide for the more difficult terms to pronounce in this chapter.

anorexia (an oh REX ee ah)

atherosclerosis
   (ATH ur oh sklur OH sis)

cuboidal (cue BOYD al)

cutaneous membranes (cue TAY nee us)

epithelial tissue (ep ih THEE lee al)

genitourinary
   (gen i toe YOUR in air EE)

glia (GLEE ah)

meninges (men IN jeez)

neuroglia (noo ROH glee ah)

neurons (NOO ron)

parietal (pah RYE eh tal)

serous membrane (SEER us)

skeletal muscle (SKELL eh tal)

squamous (SKWAY muss)

stratified (STRAT ih fied)

striated muscle (STRY ate ed)

synovial membrane (sin OH vee al)

transitional (tran ZISH ion al)

visceral (VISS er al)

OVERVIEW OF TISSUES

Just as there are many different types of cells with various functions and re-sponsibilities, tissues come in different shapes and sizes, again with the struc-ture dependent upon the function. Let's begin to explore the different tissue types.

Tissue Types

Tissue is a collection of similar cells that act together to perform a function. Imagine individual cells as bricks. Placing these bricks (cells) in a specific pat-tern creates the functional wall (tissue) of a building. There are many different types of tissues depending upon the required function. The four main types of tissues are

- epithelial
- connective
- muscle
- nervous

Let's look at these major types of tissue and their subdivisions.

EPITHELIAL TISSUE

Similar in purpose to the plastic wrap we use to keep food fresh, epithelial tis-sue not only covers and lines much of the body but also covers many of the parts found in the body. The cells in this form of tissue are packed tightly to-gether, forming a sheet that usually has no blood vessels in it.

We can further classify epithelial cells by their shape and arrangement. These cells can be flat or scale-like (**squamous**), cube-shaped (**cuboidal**), column-like (**columnar**), or stretchy and variably shaped (**transitional**). If these cells are arranged in a single layer and are all the same type of cell, we classify them as

squamous *(SKWAY muss)*
cuboidal *(cue BOYD al)*
transitional *(tran ZISH ion al)*

simple. If they are arranged in several layers, we say they are **stratified,** and they are named by the type of cell that is on the outer layer (such as stratified columnar). An exception to that rule is **pseudostratified** columnar epithelium, which looks stratified but is not. The function required of the cell dictates which type of cell formation is utilized. For example, simple squamous cells are utilized in the lungs because of their flat, thin design, which makes for easy transfer of oxygen from the lungs to the blood. Figure 4–1 ■ shows the types and locations of epithelial tissues. These are further discussed in later chapters.

stratified *(STRAT ih fied)*

pseudo = *false*

Membranes   Generally, **membranes** are sheet-like structures found throughout the body that perform special functions. Although membranes can be classified as organs, we will discuss them along with tissues for ease of explanation.

Amazing Body Facts

SKIN AND VITAMIN PRODUCTION

You know that you get vitamins and minerals from the foods you eat and the supplements you take, but did you know that your skin produces vitamin D when you are exposed to sunlight?

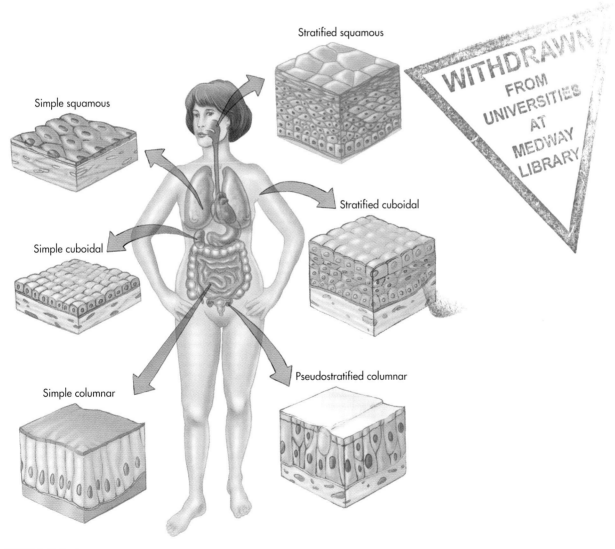

Stratified squamous

Simple squamous

Stratified cuboidal

Simple cuboidal

Simple columnar

Pseudostratified columnar

FIGURE ■ **4–1**

Types and locations of epithelial tissues.

epithelial *(ep ih THEE lee al)*

Epithelial membranes possess a layer of epithelial tissue and a bottom layer of a specialized connective tissue. Epithelial membranes are classified into three general categories, as you can see in Table 4–1.

Figure 4–2 ▦ shows the location of the serous and mucous membranes of the body.

TABLE 4–1 Types of Epithelial Membranes

1. cutaneous	Functions like a tarp placed over a boat
	The main organ of the integumentary system, commonly known as your skin
	Makes up approximately 16 percent of the total body weight
	Skin is your largest, visible organ
2. serous	A two layered membrane with a potential space in between
	Comprised of the parietal and visceral layers
parietal	Lines the wall of the cavities in which organs reside
	Produces serous fluid, which reduces friction between different tissues and organs.
visceral	Wraps around the individual organs
	Also produces serous fluid, which reduces friction between different tissues and organs
3. mucous	Lines openings to the outside world, such as your digestive tract, respiratory system, and urinary and reproductive tracts
	Called mucous membranes because they contain specialized cells that produce mucus. Mucus can act as a lubricant like the oil in a car. Mucus also serves several other important purposes besides grossing you out, as you will see in future chapters.

serous membranes *(SEER us)*

parietal *(pah RYE eh tal)* = *"wall"; therefore, the parietal membrane lines the wall of the cavity.*

visceral *(VISS er al)*
  viscer/o = *organs; therefore, visceral membranes enclose organs.*

  muc/o; myx/o = *combining forms indicating relative to mucus.*

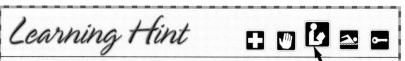

Learning Hint

To get a visualization of the importance of serous fluid consider the following: Without this friction-reducing fluid, each beat of your heart and every breath you take would be uncomfortable. The effect is similar to running water over a sheet of plastic placed on the grass. You can run, jump onto the plastic, and slide. Imagine how that would feel trying to do a body slide without the water running over the plastic!

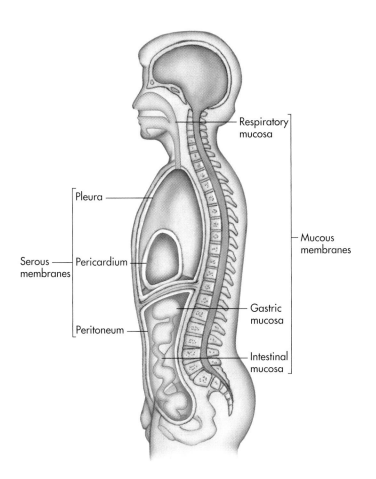

FIGURE ▦ 4–2

Location of serous and mucous membranes.

CONNECTIVE TISSUE

Connective tissue is the most common tissue and is found throughout the body more than any other form. That is because it is found in organs, bones, nerves, muscles, membranes, and skin. Connective tissue's job is to hold things together and provide structure and support. Fine, delicate webs of loosely connected tissue (**areolar tissue**) hold organs together and help to hold other connective tissues together. Fat (**adipose tissue**) is also a connective tissue. Although we always seem to want to lose fat, we truly need some fat in our bodies for proper functioning. Connective tissue can also be more densely packed and form strong cordlike structures similar to wire cables on suspension bridges. Tendons and ligaments are composed of dense connective tissue. The skin also needs to be densely packed to form a protective barrier, and it therefore uses dense connective tissue. Because connective tissue is so versatile and found throughout the body, we will discuss it in more depth in the relevant chapters. Please see Figure 4–3 ▦, which illustrates the various types of connective tissues and shows some of the places where they are found.

Amazing Body Facts 🞣 ✋ 🧍 🏊 🔑

BLOOD AND LYMPH AS CONNECTIVE TISSUES

Even though blood and lymph are fluid, they are considered to be connective tissue because they are a liquid mixture comprised of a group of cells that have specialized functions. These fluid connective tissues contain specialized cells and dissolved proteins suspended in a watery substance. We expand upon these two very important tissues in later chapters.

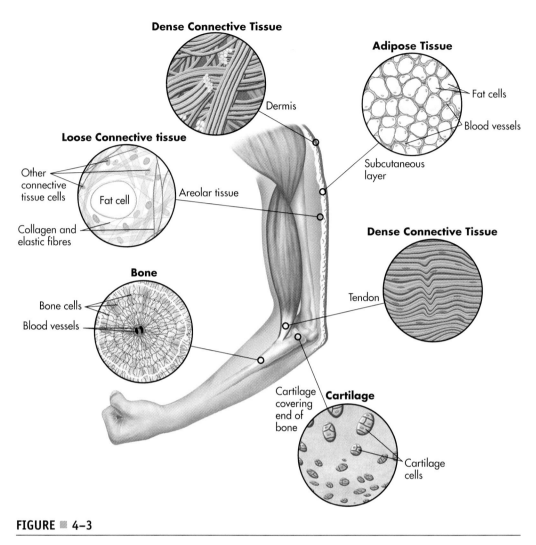

FIGURE 4–3

Types and locations of connective tissues.

synovial *(sin OH vee al)*

A membrane associated with connective tissue is the **synovial membrane.** This important membrane type is found in the spaces between joints and produces a thick, colorless, and slippery substance called synovial fluid, which greatly reduces friction when joints move. Imagine runners without synovial membranes: their knees just might burst into flames during track meets! Figure 4–4 shows a synovial joint and membrane.

MUSCLE TISSUE

Muscle tissue provides the means for movement by and in the body. This form of tissue has the ability to shorten itself (contractility). There are three types of muscle tissue: **skeletal, cardiac,** and **smooth.**

Learning Hint

MUCOUS ISN'T ALWAYS MUCUS

A rose by any other name is still a rose might work well in Shakespeare's plays, but mucous isn't always mucus. Although they sound the same, mucous and mucus are two different things. *Mucous* is an adjective that describes the type of membrane that produces *mucus,* the actual substance.

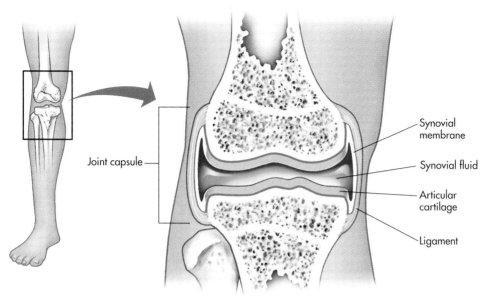

FIGURE ▪ 4–4

The synovial joint.

Skeletal Muscle   Skeletal muscle (often described as **striated** because of its striped appearance) is attached to bones and causes movement by contracting and relaxing. It also surrounds certain openings of the body, such as the mouth, and controls the size of the opening. Unless you are a ventriloquist, it is very hard to speak clearly without moving your lips. The cells that make up this tissue type are long and fiber-like with many nuclei in each cell. Our brain thinks about moving or speaking and causes the correct muscle to contract or relax as necessary. Since this is a conscious effort, we call these muscles *voluntary* muscles.

striated *(STRY ate ed)*

Cardiac Muscle   Cardiac muscle is found in the walls of the heart. Our hearts beat without our conscious control. Since the heart contracts and relaxes without conscious thought, this muscle type is considered *involuntary* muscle tissue. The cells in this tissue type interlock with each other, promoting more efficient contraction, as you will learn in Chapter 12, "The Cardiovascular System."

Smooth Muscle   Smooth muscle tissue forms the walls of hollow organs, such as in our digestive system (which is why it is often called *visceral* tissue) and blood vessels. Since we don't have to consciously think about digesting food, we consider this muscle type to also be involuntary muscle tissue. Cells forming this tissue are not as long and fibrous as skeletal muscle, and each cell has only one nucleus. Figure 4–5 ▪ shows the various types of muscle tissue.

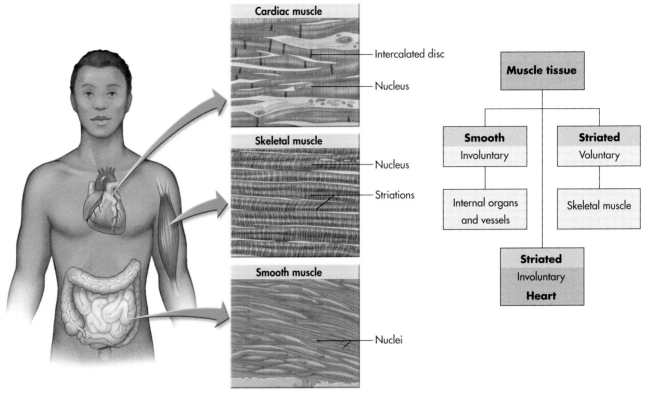

FIGURE 4–5

Labeled diagram and flowchart of the three muscle tissue types.

NERVOUS TISSUE

Nerve tissue acts as a rapid messenger service for the body, and its messages can cause actions to occur. There are two types of nerve cells. **Neurons** are the conductors of information, and **glia** (sometimes called **neuroglia**) cells function as support by helping to hold the neurons in place. The branchlike formations that make up part of the neuron, called **dendrites**, receive sensory information. The trunk-shaped structure, called the axon, transports information *away* from the cell body. Ray, our spinal cord injury patient and common case study, has permanently damaged the axons running up and down his spinal cord, which carry information between the body and the brain. Many nerves have an insulating layer called the myelin sheath, which is further discussed in Chapter 9, "The Nervous System." Figure 4–6 shows the two types of nerve cells.

Medical Controversy: To Vaccinate or Not

Even membranes can be involved in pathologic processes. The membranes associated with covering the brain and spinal cord are called **meninges**. **Meningitis** (inflammation of the meninges) is a potentially life threatening process that can be caused by bacteria or viruses. The bacterial form of meningitis (for which *meningococcus* is one of the main causative agents) can be spread via droplets as a result of sneezing or coughing or by saliva contact of an infected person. This is why college students or military personnel in crowded situations are at higher risk. It is also believed that stressful situations and alcohol / tobacco consumption may increase the chance for infection.

neurons *(NOO ron)*

> **neuro** = *nerves*
>
> **glia** *(GLEE ah)* = *glue, hence the name for cells that holds the nerve cells together*
>
> **dendr/o** = *tree, hence the name of the branching dendrite structure*

meninges *(men IN jeez)*

meninges (men IN jeez)

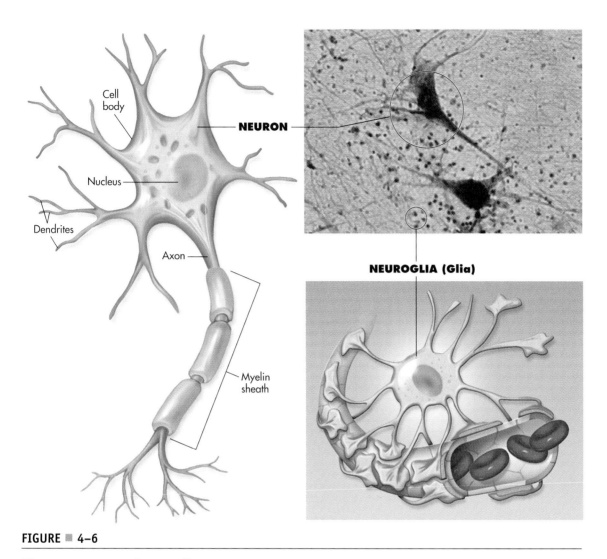

FIGURE 4–6

The two main types of nerve cells.

Once you are infected, you become a **carrier** of the disease. That doesn't necessarily mean that you will develop the disease. Your immune system may actually destroy and remove that pathogen. Some individuals may develop the disease once the infecting agent gets into the blood stream or the nervous system (meninges). Symptoms can range from non-specific headaches or fever, nausea, neck stiffness, and a skin rash to hearing loss, neurologic / brain damage, kidney failure, and loss of limbs through necessary surgical removal. There is approximately a 10% fatality rate in this form of meningitis. Although most of the media attention seems to focus on events on college campuses, those occurrences represent less than 3% of the cases in this country.

Enter the vaccine and the controversy. There has been some major debate about mandatory vaccination of individuals in the higher risk groups such as college students. Only about 25% of the cases in the United States are caused by a pathogen covered by the vaccine; approximately 75% of the US cases are caused by pathogens it does not cover. While the chances for getting bacterial meningitis is low (1-3 people per 100,000) and the chance of developing severe disability from the disease or dying is even less, that is of little consolation to patients with amputated legs and fingers or families of deceased patients.

So just get a vaccination and be done with it, right? Well, it's not that simple. Receiving a vaccination may expose you to potential adverse reactions. These reactions can include: headaches, dizziness, vomiting, convulsions, and even death. In one United Kingdom study, it was found that one adverse reaction occurred for every 907 vaccinations given.

And now the question becomes, "What should I do?" The decision to use certain vaccines not required by law should be weighed by their benefits versus their potential danger. The choice is yours; choose wisely. You alone must make that decision, but base it on sound scientific evidence, not media hype or political musings. Are you a member of a higher risk group? Please see Table 4-2 which lists universally recommended vaccinations.

TABLE 4-2   Universally Recommended Vaccinations

POPULATION	VACCINES
All young children	Measles, mumps, and rubella
	Diphtheria-tetanus toxoid and pertussis
	Poliomyelitis
	Haemophilus influenzae type B
	Hepatitis B
	Varicella
Previously unvaccinated or partially vaccinated adolescents	Hepatitis B
	Varicella
	Measles, mumps, and rubella
	Tetanus-diphtheria toxoid
All adults	Tetanus-diphtheria toxoid
All adults aged >65 years	Influenza
	Pneumococcal

Pathology Connection: Blood Sugar and Tissue Damage

You've probably listened to people talking about diabetes and the high levels of "sugar" in their blood or urine. But did you know that diabetes also affects tissues in the body? As you learned in Chapter 3 in the Common Case Study, Maria, a diabetic, doesn't produce enough insulin to allow glucose (*the* principal nutrient) into the body's cells. Since the cells don't have their main source of food, they resort to "eating" fats first, and later, the proteins found in the body. As the body uses up its protein, it is harder for the body to create more, and as a result tissues start to break down. It is more difficult and sometimes almost impossible for wounds to heal,

www.prenhall.com/colbert
The allied health profession that specializes in the study of tissues is histotechnology. To learn more about this professional career, please visit the book's companion website for this chapter.

and infections are harder to fight. That is why Maria developed a leg ulcer after a minor injury. In addition, blood circulation to tissue is decreased as lipids are released from fat tissues and are deposited around the inside wall of the blood vessels, impairing blood flow. This is a condition called **atherosclerosis**.

Tissue destruction and blood flow occlusion can cause diabetics to become susceptible to tissue death and gangrene, thus potentially leading to the loss of toes, feet, and even legs as their condition worsens. More life-altering effects of diabetes can be found in Chapter 10, and throughout the chapters in which we follow Maria's common case study.

atherosclerosis
(ATH ur oh sklur OH sis)
sclerosis = *hardness*

TEST YOUR KNOWLEDGE 4-1

Complete the following:

1. List and describe the four types of epithelial cells.

    _____

    _____

    _____

2. Lubrication for joints is produced by which type of membrane?

    _____

3. Why is connective tissue so prominent in the body?

    _____

    _____

    _____

4. Explain the difference between the terms mucous and mucus.

    _____

    _____

    _____

5. When considering diabetes, which substance is prohibited from entering cells of the body? How can this potentially lead to tissue damage?

    _____

    _____

    _____

Clinical Application

MELANOMA

Melanoma, a skin cancer that effects the epithelial membrane, has had a rapid increase in the rate of incidence in the past several decades. Melanoma typically appears as a brown or black irregular patch on the skin, or a change in an existing wart or mole. Treatment consists of surgical removal of the melanoma and the surrounding area and a possible follow-up with chemotherapy.

melan/o = *black*
oma = *tumor*

Melanoma is significant for two main reasons. First, it has a very high mortality rate, and in fact is the cancer with the most rapidly increasing mortality rate. While causing only 3% of all skin cancers, it accounts for approximately 75% of all skin cancer deaths. It is the most common form of cancer in people ages 25 to 29. About 60,000 new cases are reported annually. Second, it is one of the more preventable cancers. The culprit? Excessive sun exposure and tanning! Although the classic patient has fair skin, blue eyes, and blonde or red hair, more darkly pigmented individuals are also at risk, but at a somewhat lower rate. The effects of blistering sunburns have an additive effect. This means that each childhood sunburn you had increases your chance of developing melanoma in adulthood. Protection from excessive sun and early detection are keys to survival.

4-1 To view a video on skin cancer, see your DVD.

ORGANS

As mentioned earlier, a hospital department is made up of employees who work together to perform specific functions. Similarly, an **organ** is the result of two or more types of tissues organizing in such a way as to accomplish a task that the tissues cannot do on their own. Your heart, lungs, stomach, liver, and kidneys are all examples of organs found in your body. Some organs occur singly, such as the heart, and some occur in pairs, such as the lungs. It is interesting to note that we can survive quite well with only one healthy organ from a paired group (such as a lung or kidney).

It is important to understand that there are *vital* organs and organs that are not vital. Vital organs are the ones that you can't live without. Your heart, brain, and lungs are vital organs. Organs that you can live without include your appendix, spleen, and gallbladder. Some vital organs come in pairs, so if one is damaged or removed, you can still survive. Lungs and kidneys are a good example. Table 4–3 provides a quick reference to the various organs of the body, their medical terminology combing form, and related medical specialty.

TABLE 4–3 Systems and Organs of the Human Body

BODY SYSTEM	ORGANS IN THE SYSTEM	COMBINING FORM	MEDICAL SPECIALTY
Integumentary	Skin	dermat/o, cutane/o	Dermatology (*DUR mah TALL oh jee*)
	Hair	trich/o	
	Nails	ung/o	
	Sweat glands	sud/o, hidr/o	
	Sebaceous glands	seb/o	
Musculoskeletal	Muscles	my/o, muscul/o	Orthopedics (*OR thoh PEE dik*s)
	Bones	oste/o	
	Joints	arthr/o	Rheumatology (*ROO ma TALL oh jee*)
Endocrine	Thyroid gland	thyr/o	Endocrinology (*EN doh krin ALL oh jee*)
	Pituitary gland	pituit/o	Internal medicine
	Testes	test/o, orchi/o	
	Ovaries	ovari/o, oophor/o	Gynecology (*GUY neh KOL oh jee*)
	Adrenal glands	Adren/o	
	Pancreas	pancreat/o	
	Parathyroid glands	parathyroid/o	
	Pineal gland	pineal/o	
	Thymus gland	thym/o	
Cardiovascular	Heart	cardi/o	Cardiology (*car dee ALL oh jee*)
	Blood	hemat/o, hem/o	Hematology (*HEE mah TALL oh jee*)
	Arteries	arteri/o	Internal medicine
	Veins	phleb/o, ven/o, veni/o	
Lymphatic and Immune	Spleen	splen/o	Immunology (*IM yoo NALL oh jee*)
	Lymph	lymph/o	
	Thymus gland	thym/o	

TABLE 4–3 Systems and Organs of the Human Body (*continued*)

BODY SYSTEM	ORGANS IN THE SYSTEM	COMBINING FORM	MEDICAL SPECIALTY
Respiratory	Nose	nas/o, rhin/o	Otorhinolaryngology (*OH toh RYE noh lair in GALL oh jee*)
	Pharynx	pharyng/o	Thoracic (*tho RASS ik*) surgery
	Larynx	laryng/o	
	Trachea	trache/o	
	Lungs	pneum/o	Pulmonology (*pull mon ALL oh jee*)
	Bronchial tubes	bronch/o	Internal medicine
Gastrointestinal	Mouth	or/o	Gastroenterology (*GAS troh EN ter ALL oh jee*)
	Pharynx	pharyng/o	Internal medicine
	Esophagus	esophag/o	
	Stomach	gastr/o	
	Small intestine	enter/o	
	Colon	col/o colon/o	Proctology (*prok TOL oh jee*)
	Liver	hepat/o	procto = *anus*
	Gallbladder	cholecyst/o	
	Pancreas	pancreat/o	
Urinary	Kidneys	nephr/o, ren/o	Nephrology (*neh FROL oh jee*)
	Ureters	ureter/o	Urology (*yoo RALL oh jee*)
	Bladder	cyst/o, vesic/o	uro = *urine*
	Urethra	urethr/o	
Reproductive	Ovaries	oophor/o	Gynecology
	Uterus	uter/o, hyster/o	gynec/o = *woman*
	Fallopian tubes	salping/o	Obstetrics (*ob STET riks*)
	Vagina	vagin/o	
	Mammary glands	mamm/o aden/o	
	Testes	orchid/o	
	Prostate	prostat/o	
	Urethra	urethr/o	
Nervous	Brain	encephal/o	Neurology (*noo RAL oh jee*)
	Spinal cord	myel/o, spin/o	Neurosurgery (*noo roh SIR jer ee*)
	Nerves	neur/o	
Senses	Eye	ocul/o, ophthalm/o	Ophthalmology (*off thal MALL oh jee*)
	Ear	ot/o	Otolaryngology (*OH toh LAIR in GALL oh jee*)

SYSTEMS

A body **system** is formed by organs that work together to accomplish something more complex than what a single organ can do on its own. Take the heart, for example. Even if your heart is functioning perfectly, you would die without all of the other parts that make up the cardiovascular system. Much like a road system in the city, you need the arteries, veins, and blood to get vital oxygen and nutrients to the cells and to remove the waste products produced by those cells.

Although we discuss the body systems separately, it is extremely important that you understand that all of the body systems are interrelated, often depending on each other for proper functioning.

Skeletal System

Throughout your journey, you will see many different kinds of buildings. The one thing all of the buildings have in common is some sort of skeletal structure for support. Most people think that the skeleton's only job is to provide support and structure to the body, much like the framework of a house, but it does much more. The skeleton protects organs such as the brain; in combination with muscles, it provides movement; it acts as a storage vault for a variety of minerals, such as calcium and phosphorus; *and* it also produces blood cells! That's pretty impressive stuff! The main components of this system are bones, joints, ligaments, and cartilage. Please see Figure 4–7 ■, which shows the main components of the skeletal system.

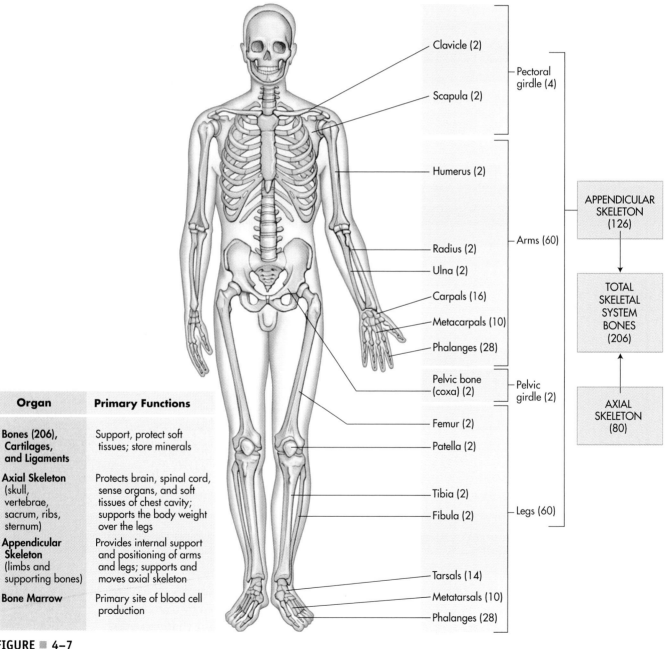

Organ	Primary Functions
Bones (206), Cartilages, and Ligaments	Support, protect soft tissues; store minerals
Axial Skeleton (skull, vertebrae, sacrum, ribs, sternum)	Protects brain, spinal cord, sense organs, and soft tissues of chest cavity; supports the body weight over the legs
Appendicular Skeleton (limbs and supporting bones)	Provides internal support and positioning of arms and legs; supports and moves axial skeleton
Bone Marrow	Primary site of blood cell production

FIGURE ■ 4–7

The skeletal system.

Muscular System

"All of this dry reading makes me thirsty. I think I'll go get something to drink." The muscular system is responsible for getting you up and over to that refrigerator (see Figure 4–8 ■). This voluntary action is made possible by skeletal muscles that are attached to your bones. Two general classifications of muscle are *voluntary* (of which we just had an example) and *involuntary*. Involuntary muscles perform without consciously being told to do so. The smooth muscle found in the walls of organs, often called *visceral muscle*, and the muscle found in the heart, called cardiac muscle, are examples of involuntary muscles. Smooth muscle is also found in blood vessels and airways, where it helps control the diameter of these passageways. So, the main parts of the muscle system are skeletal, smooth, and cardiac muscles.

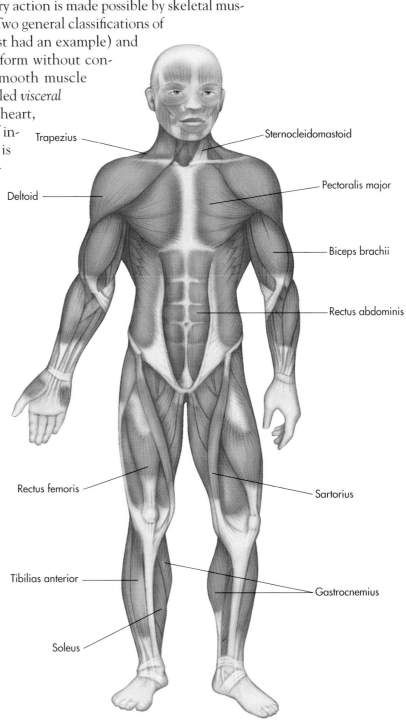

Trapezius — Sternocleidomastoid

Deltoid — Pectoralis major

Biceps brachii

Rectus abdominis

Rectus femoris — Sartorius

Tibilias anterior — Gastrocnemius

Soleus

Organ	Primary Functions
Skeletal muscles (700)	Provide skeletal movement, control openings of digestive tract, produce heat, support skeletal position, protect soft tissues

FIGURE ■ 4–8

The muscular system.

Integumentary System

All cities must have first responders, such as paramedics, police, firefighters, and safety inspectors, who cover and protect the city and its inhabitants. The body's first line of protection is your skin. Skin is the main part of the integumentary system. Besides protecting your body from invasion, the integumentary system also helps to regulate body temperature through sweating, shivering, and changes in the diameter of blood vessels in the skin. Much of the sensory information received from the outside world (heat, cold, pain, pressure, etc.) comes from sensors in the skin. Glands in the skin help to lubricate and waterproof the skin and also inhibit the growth of unwanted bacteria. The main components of this system include skin, hair, sweat glands, sebaceous glands, and nails (see Figure 4–9 ■).

Nervous System

Much like the activities of City Hall, the nervous system is the rapid messenger system of the body that both receives and sends messages for activities to occur. The messages conducted by the nervous system are stimulated by the body's internal and external environments. This is important not only so we may experience the world around us but to also protect us from harm. The nervous system also monitors what is going on inside the body. How do we know when we are hungry or when we have had enough to eat? This information is

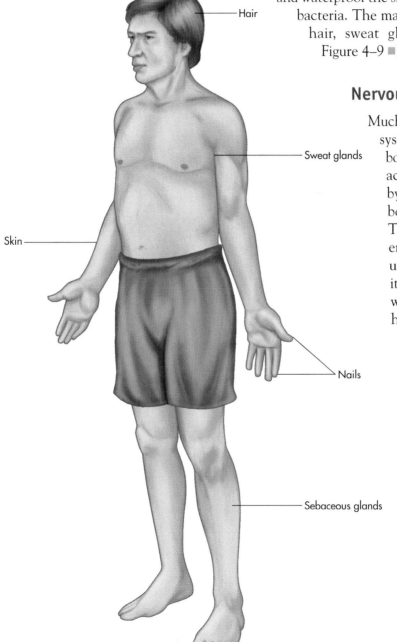

Organ	Primary Functions
Skin	Largest organ of the body
Epidermis	Covers surface, protects underlying tissues
Dermis	Nourishes epidermis, provides strength, contains glands
Hair Follicles	Produce hair
Hair	Provides sensation, provide some protection for head
Sebaceous glands	Secrete oil that lubricates hair
Sweat Glands	Produce perspiration for evaporative cooling
Nails	Protect and stiffen tips of fingers and toes
Sensory Receptors	Provide sensations of touch, pressure, temperature, pain

FIGURE ■ 4–9

The integumentary system.

obtained through **sensations**, which are conscious (and unconscious) feelings or an awareness of conditions that occur inside and outside of the body. These sensations are caused by stimulation of our sensory receptors. So, then, the three main functions of the nervous system are sensory (receiving messages), processing and interpreting messages, and motor (acting on those messages). The main parts of the nervous system are the nerve cells (glial cells and neurons); the spinal cord (damaged in Ray) along with its spinal fluid; peripheral nerves; and, of course, the brain. The sensory organs include the eyes (sight), nose (smell), tongue (taste), skin (touch) and ears (hearing *and* balance). Chapter 11 covers the special senses in more detail. See 4–10 ■, which depicts the nervous system.

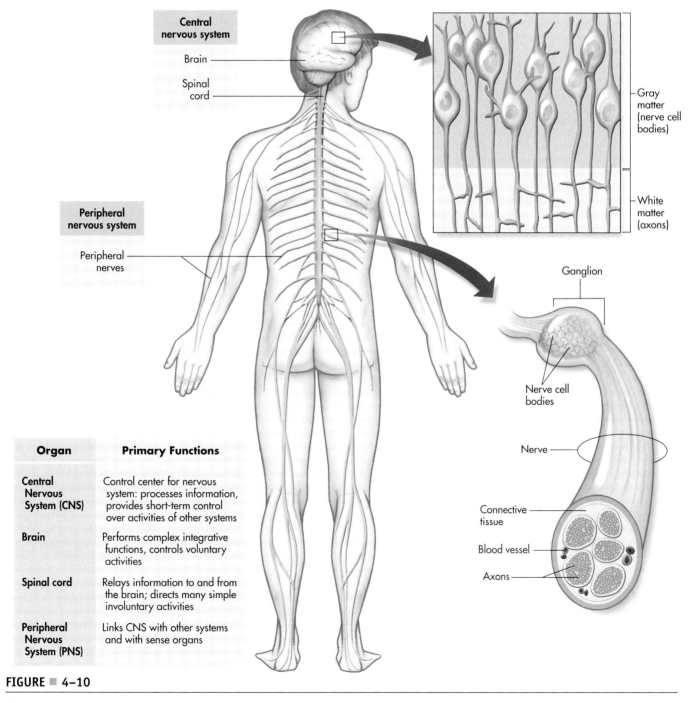

Organ	Primary Functions
Central Nervous System (CNS)	Control center for nervous system: processes information, provides short-term control over activities of other systems
Brain	Performs complex integrative functions, controls voluntary activities
Spinal cord	Relays information to and from the brain; directs many simple involuntary activities
Peripheral Nervous System (PNS)	Links CNS with other systems and with sense organs

FIGURE ■ 4–10

The nervous system.

Endocrine System

While not as quick-acting as the nervous system, the endocrine system also acts as a control center for virtually all of the body's organs (see Figure 4–11 ▪). This control is accomplished through **endocrine glands** that release chemical substances called **hormones** that are circulated through the cardiovascular system. The endocrine system helps to regulate the body's metabolic processes that utilize carbohydrates, fats, and proteins, and it plays an important role in the rate of growth and reproduction.

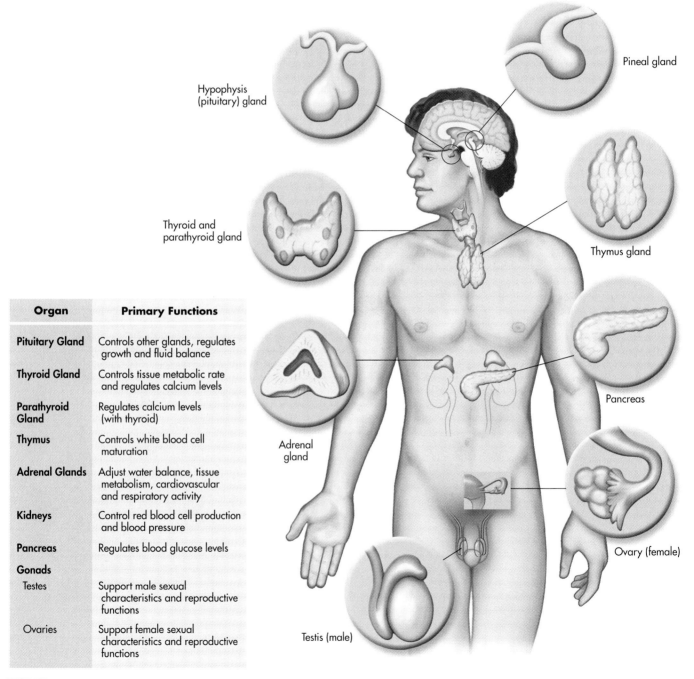

Organ	Primary Functions
Pituitary Gland	Controls other glands, regulates growth and fluid balance
Thyroid Gland	Controls tissue metabolic rate and regulates calcium levels
Parathyroid Gland	Regulates calcium levels (with thyroid)
Thymus	Controls white blood cell maturation
Adrenal Glands	Adjust water balance, tissue metabolism, cardiovascular and respiratory activity
Kidneys	Control red blood cell production and blood pressure
Pancreas	Regulates blood glucose levels
Gonads	
Testes	Support male sexual characteristics and reproductive functions
Ovaries	Support female sexual characteristics and reproductive functions

FIGURE ▪ 4–11

The endocrine system.

In addition, the endocrine system helps to regulate the fluid and electrolyte balances of the body. If that weren't enough, the hormones produced by the endocrine system help you to deal with general stress and the stresses produced by infection and trauma! The main parts of the endocrine system are the hypothalamus, the pineal, pituitary, thyroid, parathyroid, thymus, adrenal glands, the pancreas, and the gonads (testes in males and ovaries in females), plus a variety of hormones. In our common case study, Maria's diabetes is caused by a dysfunctional pancreas, which does not produce enough of the hormone insulin.

Cardiovascular System

Often referred to as the circulatory system, the cardiovascular system, shown in Figure 4–12 ■ , is the main transportation system to each cell of our body, much like the streets, sidewalks, and subways of our city. Through this system, water, oxygen, and a variety of nutrients and other substances necessary for life are transported to the cells, and waste products are transported away from the cells. Also like our city, these routes can become clogged or blocked, causing major problems. Imagine what happens to a busy four-lane highway if two of the lanes are shut down because of construction or an accident. The traffic slows and pressure builds up due to the congestion, much like the blood flow does when the arteries become partially obstructed. The buildup of pressure (hypertension) can be very dangerous. The main components of this system are the heart, arteries, veins, capillaries, and—of course— blood.

Heart

Major veins
(in blue)

Major arteries
(in red)

Organ	Primary Functions
Heart	Pumps blood, maintains blood pressure
Blood Vessels	Distribute blood around the body
Arteries	Carry blood from heart to capillaries
Capillaries	Site of exchange between blood and interstitial fluids
Veins	Return blood from capillaries to heart
Blood	Transports oxygen and carbon dioxide, delivers nutrients, removes waste products, assists in defense against disease

FIGURE ■ 4–12

The cardiovascular system.

Pathology Connection: Septicemia

Many pathologic conditions involve only one body system. However, as you journey through this text you will begin to visualize the interrelatedness of the systems and how a disease in one system will affect one or more of the other systems.

Septicemia (also called **sepsis** or **blood poisoning**) is a condition in which a pathogen is present in the blood. Since blood is needed throughout all of the body's systems, the potential for a multi-system infection is a very real danger. For example, a perforation in the stomach or intestine (as caused by surgical procedures, trauma, knife wound, etc.) can allow gastric contents to spill into the abdominal cavity, which can lead to infection of that region. Blood that is perfusing (flowing through) that area may pick up some of the bacteria and spread it to other organs. Once in other organs, the bacteria can continue to colonize, thus potentially affecting function(s) of that organ. The term "sepsis syndrome" is often used when there is a decrease in blood perfusion to organs, with other systemic signs. "Septic shock" occurs when there is a drop in blood pressure due to the decrease in organ perfusion. If not quickly and effectively treated, the patient can die from multi-system failure, often referred to as **MODS** (multiple organ dysfunction syndrome). The mortality rates in these cases increase as the number of involved organ systems increases, and can approach 100% if MODS continues for more than four days, depending on the patient.

Some signs/symptoms that can be related to sepsis/sepsis syndrome include: fever, chills, tachypnea, tachycardia, skin lesions, diffused redness over areas of the skin (erythema), wide spread erythema (erythroderma), hypoxemia, and changes in mental status. Both Ray and Maria are at increased risk for sepsis due to their chronic conditions.

Respiratory System

Think about a time you went swimming and stayed a little too long underwater, and you will remember how important our respiratory system is! We all know the old concept of being able to live weeks without food and days without water, but think about how long you would last without oxygen. Without conscious effort, your lungs move approximately 12,000 quarts of air a day! Our respiratory system, much like the ventilation system in an office building, not only supplies us with fresh oxygen but performs several other important functions as well. Our lungs eliminate the carbon dioxide created as a result of cellular metabolism. The respiratory system filters, warms, and moistens air as it is inhaled. The mucous lining of the airway helps trap foreign particles and germs. This system also helps to maintain the proper acid-base balance of the blood and aids in the elimination of ingested alcohol. The main parts of the respiratory system are the pharynx, larynx, trachea, bronchial tubes, and lungs (see Figure 4–13 ■).

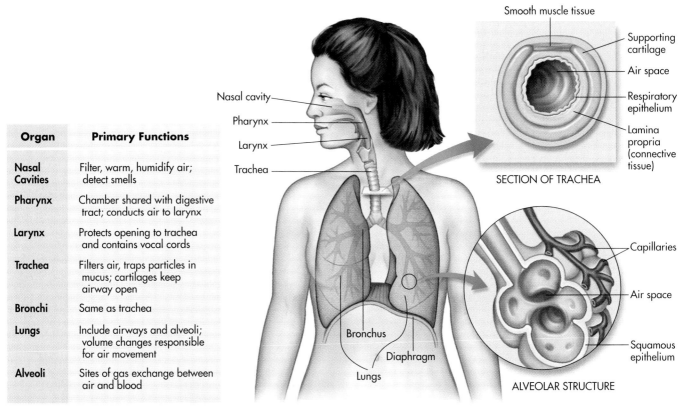

Organ	Primary Functions
Nasal Cavities	Filter, warm, humidify air; detect smells
Pharynx	Chamber shared with digestive tract; conducts air to larynx
Larynx	Protects opening to trachea and contains vocal cords
Trachea	Filters air, traps particles in mucus; cartilages keep airway open
Bronchi	Same as trachea
Lungs	Include airways and alveoli; volume changes responsible for air movement
Alveoli	Sites of gas exchange between air and blood

FIGURE ▪ 4–13

The respiratory system.

Lymphatic and Immune System

Much like the storm drain system of our city, this very important—but often forgotten—system is responsible for helping to maintain proper fluid balance in our body and to protect it from infection. Excess fluid that may collect in places it shouldn't in the body is brought back into the lymphatic system, cleaned and processed, and then recirculated. Special structures called **lymph nodes** act as filters to capture unwanted infective agents. Lymph vessels and ducts, lymph nodes, the thymus gland, tonsils, and the spleen are the major parts of the immune system. In addition, the immune portion of the lymphatic system produces specialized infection-fighting white blood cells called **lymphocytes**. The immune system is the police force of the human body, patrolling for harmful invaders (see Figure 4–14 ▪).

lymphocytes *(LIMF oh sights)*

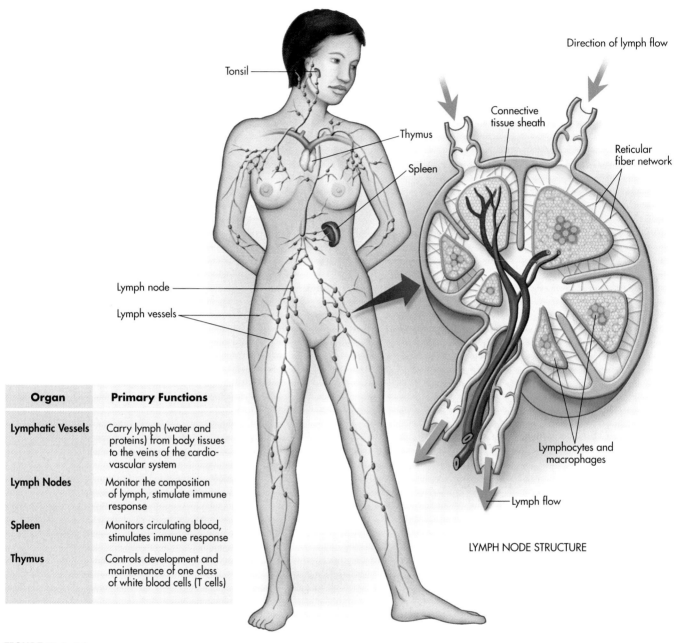

Organ	Primary Functions
Lymphatic Vessels	Carry lymph (water and proteins) from body tissues to the veins of the cardio-vascular system
Lymph Nodes	Monitor the composition of lymph, stimulate immune response
Spleen	Monitors circulating blood, stimulates immune response
Thymus	Controls development and maintenance of one class of white blood cells (T cells)

FIGURE ■ 4–14

The lymphatic system.

Gastrointestinal (Digestive) System

The **digestive system** (often called the GI system, short for *gastrointestinal*) breaks down raw materials (food), both mechanically and chemically, into usable substances. Please see Figure 4–15 ■. Once these usable substances are created, this system absorbs them for transportation to the cells of the body. Materials that aren't used, as well as cellular waste, are transported out of the body by this system, much like the waste disposal and sewage system of our city. The main parts of the digestive system are the mouth, pharynx, esophagus, stomach, intestines, accessory organs, and anal canal.

Organ	Primary Functions
Salivary Glands	Provide lubrication, produce buffers and the enzymes that begin digestion
Pharynx	Passageway connected to esophagus
Esophagus	Delivers food to stomach
Stomach	Secretes acids and enzymes
Small Intestine	Secretes digestive enzymes, absorbs nutrients
Liver	Secretes bile, regulates blood chemistry
Gallbladder	Stores bile for release into small intestine
Pancreas	Secretes digestive enzymes and buffers; contains endocrine cells
Large Intestine	Removes water from fecal material, stores waste

Labels on figure: Tongue, Salivary glands (Sublingual, Submandibular), Parotid gland, Parotid duct, Pharynx, Esophagus, Liver, Gallbladder, Spleen, Stomach, Ascending colon, Cecum, Appendix, Rectum, Pancreas, Descending colon, Transverse colon, Small intestine (duodenum, jejunum and ileum)

FIGURE ▪ 4–15

The digestive system.

Pathology Connection: Body Image: Obesity and Anorexia

It seems, at times, that we are a country of extremes. Anyone who watches or reads the news knows that the U.S population is becoming increasingly overweight every year. The situation is so pervasive and severe that the current generation of adolescents potentially will have a shorter lifespan than their parents. On the other hand, we also are facing an epidemic of individuals trying to be super thin, trying to emulate their image of what a super model or movie star should look like. While there is absolutely nothing wrong with proper dieting and exercising, some individuals develop a body image that is unrealistic. These individuals starve themselves or vomit after eating to attempt to keep from forming body fat, which they usually only *think* they have. **Bulimia** is a condition in which an individual goes on eating binges, overeats and then attempts to get rid of the food by either vomiting of using laxatives to keep from gaining weight. **Anorexia nervosa** is the term used for the condition in which there is a progressive and severe weight loss even as the patient denies that there is a problem. Anorexics, rather than purging what they eat, often avoid eating or eat too little food to sustain a healthy weight. Figure 4–16 ▪ shows an individual with this condition. These conditions are both a mental and a physical health issue.

FIGURE ▪ 4–16

This is an example of an individual with a body image problem in which her self perception is of being overweight.

Urinary System

While the digestive system plays a large role in the elimination of certain digested waste, the **urinary system** plays an important role in the elimination of waste products such as forms of nitrogen. In addition, electrolytes, drugs and other toxins, and excessive water are removed. This system is crucial for maintaining the proper balance of water you have in your body and regulating your blood pressure. The urinary system helps regulate the number of red blood cells and the acid-base and electrolyte balance of blood. The main parts of the urinary system are the kidneys, ureters, urinary bladder, and urethra (see Figure 4–17 ■).

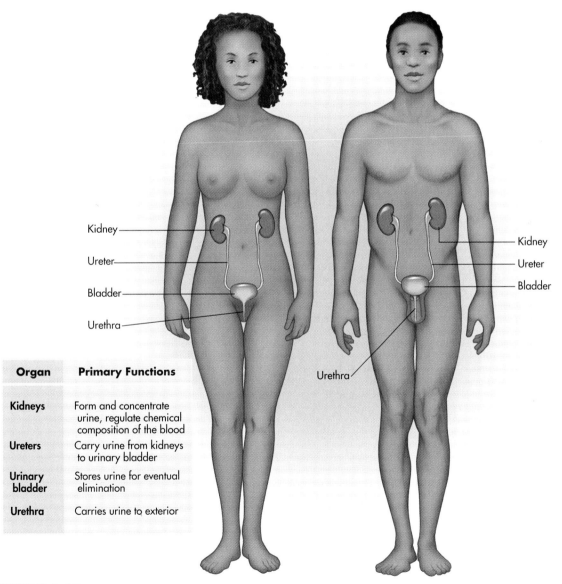

Organ	Primary Functions
Kidneys	Form and concentrate urine, regulate chemical composition of the blood
Ureters	Carry urine from kidneys to urinary bladder
Urinary bladder	Stores urine for eventual elimination
Urethra	Carries urine to exterior

FIGURE ■ 4–17

The male and female urinary systems.

Reproductive System

We build new cities or new buildings to accommodate growing needs or to replace worn out structures. The reproductive system does the same thing. Quite simply, without this system, we would not exist. The reproductive system is often combined with the urinary system to create the **genitourinary,** or **GU** system. Humans require a male and female to produce offspring. The male is needed to provide sperm that contains certain genetic traits of that individual, while the female provides an egg with her traits and a place for the fertilized egg to grow to maturity. The main female parts of this system are the ovaries, eggs, fallopian tubes, uterus, and vagina. For men, the main parts are the testes, sperm, and penis (see Figure 4–18 ■).

4-2 Now that we have briefly discussed the body systems, go to your DVD for an interactive activity identifying the various systems along with their major components.

TEST YOUR KNOWLEDGE 4-2

List the correct system for the following activities:

1. Exchanges carbon dioxide for oxygen

   _____

2. Eliminates nitrogen, drugs, and excessive water from the body

   _____

3. Main storage for calcium

   _____

4. Maintains body temperature and provides much of the sensory information from the external world

   _____

5. Protects the body from invading pathogens

   _____

6. Moves blood through the body

   _____

7. Converts food to energy

   _____

8. With the help of the sun, produces vitamin D

   _____

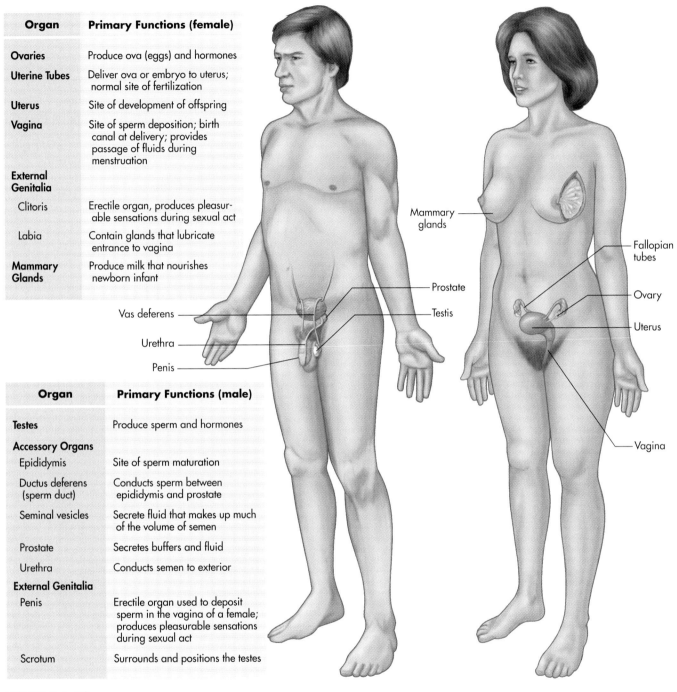

Organ	Primary Functions (female)
Ovaries	Produce ova (eggs) and hormones
Uterine Tubes	Deliver ova or embryo to uterus; normal site of fertilization
Uterus	Site of development of offspring
Vagina	Site of sperm deposition; birth canal at delivery; provides passage of fluids during menstruation
External Genitalia	
Clitoris	Erectile organ, produces pleasurable sensations during sexual act
Labia	Contain glands that lubricate entrance to vagina
Mammary Glands	Produce milk that nourishes newborn infant

Organ	Primary Functions (male)
Testes	Produce sperm and hormones
Accessory Organs	
Epididymis	Site of sperm maturation
Ductus deferens (sperm duct)	Conducts sperm between epididymis and prostate
Seminal vesicles	Secrete fluid that makes up much of the volume of semen
Prostate	Secretes buffers and fluid
Urethra	Conducts semen to exterior
External Genitalia	
Penis	Erectile organ used to deposit sperm in the vagina of a female; produces pleasurable sensations during sexual act
Scrotum	Surrounds and positions the testes

FIGURE ▪ 4–18

The male and female reproductive systems.

SUMMARY

Snapshots from the Journey

→ Cells are the basic building blocks of the body.

→ Tissues are collections of similar cells that act together to perform a function. The four main types of tissues are epithelial, connective, muscle, and nervous.

→ Membranes are sheet-like structures found throughout the body; they perform specific functions.

→ The four major membrane types are cutaneous, serous, mucous, and synovial.

→ Tissues that combine to perform a specific function or functions are called an organ.

→ Organs that work together, often with the help of accessory structures, to perform specific activities, are called a system.

→ There are 11 major body systems: skeletal, muscular, integumentary, nervous, endocrine, cardiovascular, respiratory, lymphatic/immune, gastrointestinal, urinary, and reproductive. Even though these are distinct systems, they are interrelated, and their relationships are highlighted in upcoming chapters.

→ A disease process can affect more than one body system at a time often making it more difficult to treat. Septicemia is one such example. If left unchecked, it can lead to multiple organ dysfunction syndrome (MODS).

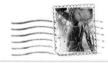

A 73-year-old male presents to an emergency department of a local hospital. Initial assessment reveals the following:

- afebrile
- mild tachypnea with mild shortness of breath
- acrocyanosis
- mild tachycardia
- history of smoking
- history of diabetes
- moderately overweight

Based on the information given, have members of the group identify which system or systems of the body they would want to further investigate to determine why this individual has come to the emergency department. You have had some medical terminology already but may need additional help from the text, medical dictionary, or website. Compile a list of specialists or health care professionals to whom you might refer this patient, and explain why.

3. What is the purpose of synovial fluid?

_____

_____

_____

_____

4. Contrast the three types of muscle tissues and identify where they are found.

_____

_____

_____

Suggested Activities

1. Choose one system to research. Identify all the tissue types and membranes found within that system and describe the individual tissue and membrane functions.

2. Create five to ten multiple-choice quiz questions related to this chapter and have a quiz show contest.

4-3 Now that you have completed your journey through this chapter, please go to the DVD for interactive games and puzzles concerning the medical terms and concepts contained in this chapter. By playing the games you will reinforce your learning in a fun way.

BASIC DIAGNOSTIC TESTS

What Do The Tests Tell Us?

Sometimes a general map just isn't enough to understand and enjoy various parts of your trip. City maps, dining guides, tourist highlights, and brochures on geologic/historic points of interest are all important tools that help you zero in on specific areas of interest. The same thing can be said about diagnostic tests, which help to provide us with a more accurate view of the patient's overall condition and disease state including diagnosis, progression, or improvement. To make sure we have a trouble-free journey, we need a reliable means of transportation. A quick check of our car (oil, gas, or windshield washer fluid) can be compared to a general physical examination, but there are times when we may need to check the more involved car systems that require special equipment such as the transmission, brakes, or engine compression. Medical diagnostics are like those more advanced diagnostics for your car. These tests can be used to verify the obvious or to discover where the signs and symptoms are pointing. In this chapter, we will provide you with an overview of some of the common diagnostic tests used in health care, including quick reference tables and lists of possible causes for abnormal test results of some common clinical tests. While we haven't covered any of the specific body systems yet, these tests will lay a foundation to better understand diagnostics as they relate to each upcoming body system.

Chapter

5

LEARNING OBJECTIVES

At the end of your journey through this chapter, you will be able to:

→ List and describe common diagnostic tests often used in clinical settings

→ Determine which diagnostic test(s) should be chosen to determine a patient's suspected disease state

→ Recognize abnormal findings of several basic diagnostic tests

→ Determine conditions that may produce abnormal test results

MULTIMEDIA APPLICATIONS

DVD Interactive Exercises

→ Videos on taking a proper patient history and performing physical assessment/exams, 5-1

→ Videos on blood sampling and performing glucose monitoring, 5-2

→ Video on blood test basics, 5-3

→ Video on performing a dipstick urinalysis, 5-4

→ Video on complete urinalysis, 5-5

→ Video on occult testing procedure, 5-6

→ Videos on electrocardiogram testing, electrode placement and fetal heart monitoring, 5-7

→ Video on arthroscopic procedures, 5-8

→ Interactive games and puzzles, 5-9

www.prenhall/colbert

→ Professional Profile:
  • Physician Assistant

→ Related Internet Links

→ Additional Review Questions

Pronunciation Guide

Correct pronunciation is important in any journey so that you and the others are completely understood. Here is a "see and say" Pronunciation Guide for the more difficult terms to pronounce in this chapter.

anemia (uh NEE me uh)

dyspnea (DISP nee uh)

eosinophil (EE oh SIN oh fill)

erythrocytes (eh RITH roh site)

leukocytes (LOO koh sites)

leukocytosis (LOO koh sigh TOE sis)

leukopenia (LOO koh PEE nee uh)

occult (uh CULLT)

polycythemia
(PAUL ee sigh THEE me uh)

polymorphonuclear
(PAUL ee MORE foh NEW klee air)

polysomnography
(PAUL ee sahm NOG graf ee)

urea (you REE ah)

OVERVIEW OF DIAGNOSTIC TESTING

While a good physical examination with a solid patient history (including family/occupational information) and review of current medications are very important, they aren't always enough to make a correct diagnosis. A headache can be caused by infected sinuses, stress, tumors, eye strain, medication side effects, and so on. **Dyspnea** (difficulty breathing) can be caused by asthma, heart problems, stress, anxiety, smoking or inhalation of other toxic substances, or tumors, to name just a few. To help determine the specific cause of various signs or symptoms, appropriate medical testing comes into play. Another aspect of diagnostic testing is to confirm the effectiveness of a prescribed treatment or medication.

dyspnea *(DISP nee uh)*

To thoroughly cover all of the types of diagnostic testing available, you need a textbook much larger than this one. This chapter provides an overview of a variety of common diagnostic tests that aid in determining a patient's disease state. In addition, each system chapter of this textbook will have more information on "system specific" diagnostic tests. Not only will this information aid in your understanding of these tests, but it will help you to explain to your patient the "whys and hows" of their ordered tests, when appropriate. Note that the normal values listed in this chapter for the various tests may sometimes vary slightly from other reference sources. Different hospital labs utilize different *but similar* test values, and that is okay. These values are all within acceptable ranges for making a correct diagnosis.

5-1 Please go to your DVD if you would like to view videos on taking a proper patient history and performing physical assessment/exams.

One major caution: *A diagnosis is based on the patient, not solely on the test numbers!* As you spend more time in the clinical setting, you will see a variety of exceptions to the rules. Each patient is different. Here's one more very important caution: *One abnormal test result does not a diagnosis make.* Though we struggle daily to maintain homeostasis, there are occasions when our own diagnostic test values may be abnormal even when we are healthy!

BLOOD TESTING

The one type of test that most of us have had at some time in our lives is the common blood test. Blood is a vital fluid that communicates with all parts of our bodies. Therefore, the information gained from analyzing the various components of blood can greatly aid in diagnosis and treatment.

Blood Basics

Blood is composed of a liquid portion called **plasma** (which is about 90% water and originates from the intestines and organs) and a variety of cells (which are mainly produced by bone marrow) called **red blood cells (erythrocytes)**, **white blood cells (leukocytes)**, and **platelets (thrombocytes)**. These cell types are not of a uniform size. Leukocytes are the largest, thrombocytes are the smallest, and erythrocytes are in the middle. Their numbers also vary, normally, with one leukocyte for every 30 thrombocytes and every 500 erythrocytes.

erythrocytes *(eh RITH roh sites)*
leukocytes *(LOO koh sites)*

Each of these three main blood cell types has distinct responsibilities. Red blood cells have the main responsibility of transporting oxygen from the lungs to all of the body's cells. White blood cells are our protectors against infection. Platelets help blood to clot so we don't bleed to death from minor cuts and abrasions. Figure 5-1 ▓ shows the major cell types found in the blood.

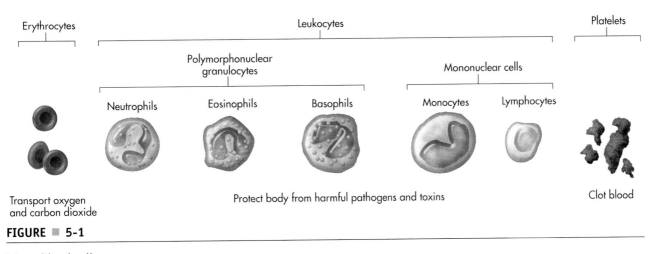

FIGURE ▓ 5-1

Major blood cell types.

Blood samples are usually obtained from the veins, which are the vessels you see in your skin that have a bluish tint to them. Another way to obtain a blood sample is from a pin prick of your skin, usually on the finger tip. Diabetics often do this when they are measuring their blood glucose (sugar) levels. See Figure 5-2 ▓ for an example of a device that measures blood glucose in this manner. For special testing to determine the amount of oxygen you have in your body, or how well you are breathing, blood samples are taken from your arteries, usually

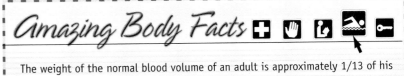

Amazing Body Facts

The weight of the normal blood volume of an adult is approximately 1/13 of his or her total body weight!

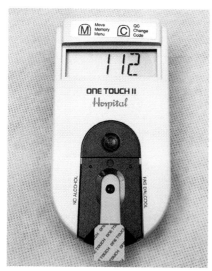

FIGURE ■ 5-2

After the patient pricks his or her skin a droplet of blood is placed on the absorbent test strip, which is then inserted into the monitor. The patient's blood glucose level is then numerically displayed.

anemia *(uh NEE me uh)*
polycythemia
    (PAUL ee sigh THEE me uh)
    poly = *many*
    cyt/o = *cells*
    hemia = *blood*
leukocytosis *(LOO koh sigh TOE sis)*
leukopenia *(LOO koh PEE nee uh)*
    penia = *to decrease*

5-2   Videos on blood sampling and performing glucose monitoring are on your DVD.

the ones in your arms. You can't see arteries since they are deeper than veins, but you can feel them. As a result, it takes special skill to obtain an arterial blood sample.

General Blood Disorders

Numerous potential blood disorders can be categorized by *blood cell type* and by the *amount* of those major blood cell types. Red blood cell disorders can be classified as either **anemias** or **polycythemias**.

In cases of anemia, there are fewer than normal amounts of erythrocytes for a given volume of blood. This could be a result of blood loss caused by a bleeding injury, lower than normal production of red blood cells, or some disorder that is destroying erythrocytes at a higher than normal rate. *Polycythemia,* on the other hand, is a condition in which there is a higher than normal amount of erythrocytes for a given blood volume. In Chapter 13, "The Respiratory System," you will learn that a form of polycythemia can be caused by chronically insufficient amounts of oxygen in your blood.

Problems with white blood cells can also be determined through testing. **Leukocytosis** occurs when there is a higher than normal number of white blood cells. This can occur when there is an infection or your patient has leukemia, for example. **Leukopenia** is a situation when there are lower than normal amounts of white blood cells. In an interesting twist, this can occur if an individual has a long term (*chronic*) infection and the body just can't keep up with sufficient white blood cell production. Of course, there are other reasons why a patient may have leukopenia, as you will see in future chapters.

Thrombocytopenia is a condition in which the number of platelets is below normal. This can lead to a decreased blood clotting ability. Patients with thrombocytopenia are in danger of experiencing massive bleeding.

Specific Blood Tests

When venous blood is withdrawn, it is placed in a tube that is centrifuged (spun around) at a very high speed. This separates the blood into two major components, as shown in Figure 5-3 ■. The heavier cells are forced to the bottom of the tube and are referred to as the **formed elements**. The upper level, which is lighter in weight and color, is the plasma or liquid portion of the blood, containing numerous dissolved substances. Please refer back to Figure 5-3 as you read through the upcoming sections.

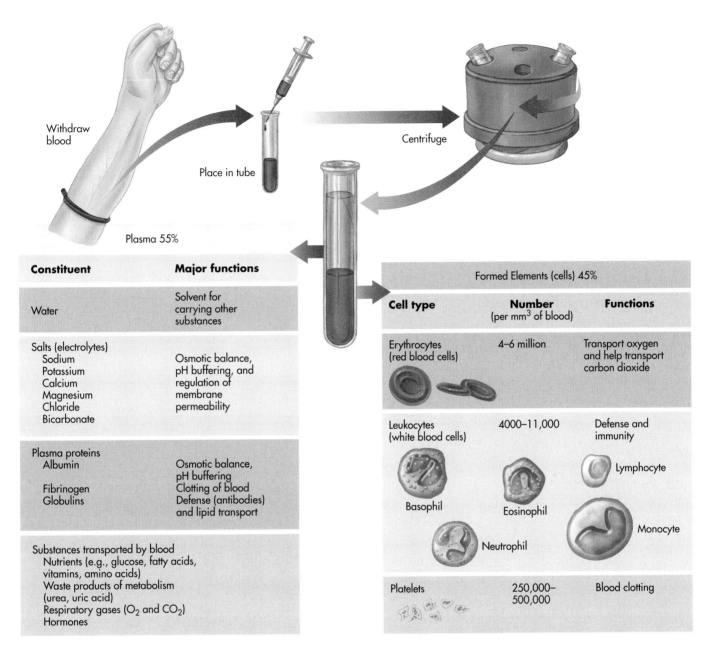

FIGURE ■ 5-3

As blood is spun around in a centrifuge, the heavier cells (RBCs, WBCs and platelets) form the bottom layer with plasma remaining as the top layer.

COMPLETE BLOOD COUNT (CBC)

This is one of the most commonly ordered tests. Due to the number of things that this test looks at, it is useful in determining a patient's diagnosis, prognosis, their response to medication or treatment, or if they have recovered from the diagnosed disease. Among other things, a typical CBC looks at the following:

- RBC (red blood cell count)
- Hct (hematocrit)
- Hgb (hemoglobin)
- WBC (white blood cell count)
- Diff (differential white blood cell count)
- Platelet Count

hemoglobin *(HEE moh GLOH binn)*

Whereas WBCs are mostly involved with protecting us from infection, erythrocytes are mainly responsible for transporting oxygen from the lungs and depositing it into body tissue. This is accomplished by a substance found in red blood cells (RBCs) called **hemoglobin**. RBCs are manufactured in the red bone marrow and are shaped like a biconcave, or hollow on both sides, disk. Some RBCs are also manufactured in the yellow marrow. RBCs are flexible to better fit through tiny capillaries when they have to.

Normally, millions of RBCs die each day, so millions have to be created daily to replace them. RBCs are measured by the number of them contained in a cubic millimeter (mm^3) of blood. The normal amount found in that mm^3 varies between males and females:

Normal RBC values
Men: $4.6 - 6.2 \times 10^6$ / mm^3
Females: $4.2 - 5.4 \times 10^6$ / mm^3

Anemia is a condition in which there is a decrease in the normal RBC amount. It is *not* a disease, but rather a sign of some disease or cause such as:

- Blood loss
- Dietary insufficiency (iron, some vitamins, folic acid)
- Decreased RBC production
- Increased RBC destruction

As stated before, polycythemia is a general term for an excess of RBCs in the blood. Some of the causative agents of this condition are:

- Dehydration
- Diarrhea (severe)
- High altitude
- Over production by bone marrow

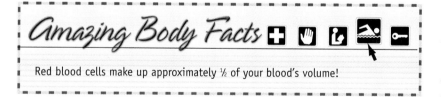

Red blood cells make up approximately ½ of your blood's volume!

Hct (Hematocrit) This test is used to determine the amount of red blood cells packed in a specific volume of blood. A tube two thirds full of blood is spun in a centrifuge so the RBCs are separated from the rest of the blood. The height of the packed RBCs at the bottom of the tube is measured and recorded in terms of the volume of RBCs in 100 ml. of blood. Again, please see Figure 5-3 ■. This result is written as a percentage of the total amount of blood that was centrifuged.

Normal Hct values
Men = 40 – 54%
Women = 38 – 47%

Higher than normal percentages of Hct can be caused by:

- Severe dehydration (loss of fluid concentrates the cells)
- Shock

Lower than predicted values can be a result of:

- **Hemolytic** reactions (incompatible blood transfusions and infections)
- Acute massive hemorrhage (blood loss)

hemolytic
  hem/o = *blood*
  lytic = *to break down*

Hgb (Hemoglobin)   Erythrocytes contain a substance called **hemoglobin** that is needed to transport oxygen from the lungs to all of the tissues in the body. Normally, every gram of hemoglobin can carry 1.34 ml. of oxygen. Hemoglobin also acts as a buffer to help maintain the acid-base balance of the blood. Hemoglobin levels are measured in grams per deciliter to determine if the individual is anemic or not.

Normal Hgb values
Women: 12 – 16 g. / dl
Men: 13.5 – 17.5 g. / dl
Newborn: 14 – 16 g. / dl.

Decreases in the level of hemoglobin can be caused by:

- Anemia
- Excessive fluid intake
- Hemorrhage (severe)
- Pregnancy

An increased level of hemoglobin can be a result of:

- Chronic obstructive pulmonary disease
- High altitude

WBC (White Blood Cells)   Several types of leukocytes (white blood cells) help to protect us from infection. They can be divided into two main groups. The first group is called **granulocytes**, since they possess granules in their cytoplasm. They also have multi-lobed nuclei, and because of this are also known as **polymorphonuclear leukocytes**, or—more commonly—PMNs. This group consists of basophils, eosinophils, and neutrophils.

polymorphonuclear
  (PAUL ee MORE foh NEW klee air)

Finding out the number of *total* WBCs in the blood is important when looking for a diagnosis or prognosis for your patient. As you will see in a little while, increases or decreases of *specific* types of WBCs can aid us in determining a *specific* disease.

Normal WBC values
Men: 4.5 – 11 x 10^3 / mm^3
Women: 4.5 – 11 x 10^3 / mm^3

Leukocytosis is diagnosed when the total WBC count is greater than 10, 000/mm^3. It is usually a result of one type of WBC increasing in number. Leukocytosis can occur when there is an acute infection, but sometimes it occurs with no clinical disease. Other potential causes include but are not limited to:

- Circulatory disease
- Coma
- Hemorrhage
- Leukemia
- Malignancy (esp. in the GI tract, bone /liver)
- Menstruation
- Steroid therapy
- Stress (pain / excitement)

Leukopenia is diagnosed when the total number of WBCs falls below 4,000/mm^3. Some possible causes can be:

- Alcoholism
- Viral infections
- Drugs (such as epinephrine)
- Fever
- Severe electric shock
- Tissue necrosis
- Toxins
- Trauma / tissue injury

As previously stated, leukopenia can occur in a patient with chronic infection(s) and whose body is so "worn out" that it just can't produce enough WBCs anymore. A WBC count less than 500 is a panic value requiring immediate attention

DIFF (Differential Leukocyte Count)   It is important to remember that there are several different types of WBCs that circulate through the body in varying amounts.

Each of these types performs a special function and will increase in percentage if the following situations exist:

Neutrophils: Combat bacterial infection, inflammation, stress
Lymphocytes: Fight viral infections including measles, chicken pox, infectious mononucleosis
Eosinophils: Respond to asthma, allergic conditions, parasitic invasions
Monocytes: Respond to severe and chronic infections
Basophils: Respond to inflammation and blood disorders

Normal Diff values
Neutrophils make up about 60 - 70% of all circulating WBCs
Lymphocytes comprise about 20 - 40%
Eosinophils are around 1 - 4%
Monocytes are 2-6%
Basophils are 0.5 – 1%

By monitoring the varying amounts of the specific types of WBCs and comparing that to the total amount of WBCs we may see which type of disorder is causing the problem; one more piece of the puzzle. For example, an **eosinophil** count above 4% may indicate an allergic reaction.

eosinophil *(EE oh SIN oh fill)*

TEST YOUR KNOWLEDGE 5-1

Choose the best answer:

1. The largest of the blood cells are:
   a. leukocytes
   b. touristcytes
   c. erythrocytes
   d. thrombocytes

2. The general blood cell type that aids in fighting infection is:
   a. WBC
   b. CBC
   c. RBC
   d. NBC

3. A disorder involving fewer than normal red blood cells per a given volume of blood:
   a. analogue
   b. leukopenia
   c. anemia
   d. RBC

4. An increase in hemoglobin levels can be caused by:
   a. excessive fluid intake
   b. anemia
   c. severe hemorrhage
   d. high altitude living

5. This type of blood cell is responsible for the transportation of oxygen to the cells of the body:
   a. WBC
   b. CBC
   c. RBC
   d. NBC

Platelet Count   Also known as thrombocytes, **platelets** are the smallest of the formed elements in your blood. They can be disk-shaped, oval, flattened, or round. They are important in helping blood to clot and maintaining vascular integrity. They can also pick up, store, and release certain vasoactive substances. In normal situations, 2/3 of all of your platelets circulate in the blood, while 1/3 are found in the spleen. A platelet count is important to help see how well anticoagulant therapy is working, if there is bone marrow failure (as in leukemia), or if a patient has thrombocytopenia, and to diagnose a variety of conditions such as a bleeding disorder due to liver disease.

Normal Platelet values
150, 000 – 350,000 / mm^3

Situations in which there are higher than normal numbers of platelets can include:

- Cancer/Heart disease
- High altitude living
- Iron deficiency
- Splenectomy
- Trauma
- Tuberculosis
- Vitamin K deficiency
- Vomiting
- Excessive sweating
- Fever

Conditions that can lead to a decrease in platelet numbers include:

- Blood transfusions
- Bone marrow lesions
- Cancer chemotherapy
- Infections/pneumonia

PT (PROTHROMBIN TIME OR PRO TIME)

Pro Time is a timed test to measure the blood's ability to clot. This is an indicator of levels of prothrombin, a protein produced by the liver that converts to thrombin during the clotting process. A body needs vitamin K to produce prothrombin. Since prothrombin is produced by the liver, low blood levels of this substance may be indicative of liver disease.

Normal PT: 10 – 14 seconds

Increases in PT indicate that the blood will not clot as quickly as normal. For example, individuals on anticoagulant therapy such as Coumadin will have an increased PT. Most cardiac patients are kept at a PT time that's 2 to 2 1/2 times normal to inhibit clots from forming in the cardiovascular system.

Decreases in PT indicate an increase in the blood's ability to clot. This can be a result of consumption of excessive amounts of green, leafy vegetables, which increases the absorption of vitamin K, thus increasing the amount of prothrombin. Drugs, depending on which type, can either increase or decrease PT.

PTT (PARTIAL THROMBOPLASTIN TIME)

PTT is another test for the ability of blood to clot. This one looks at the intrinsic thromboplastin system, and is often used to monitor the effectiveness of heparin (an anti-coagulation drug) therapy. Conditions that cause changes to PTT are similar to the ones that cause changes to PT.

Normal PTT: 30 – 45 seconds

Table 5-1 provides a quick reference to blood test abnormalities and their possible causative conditions or situations.

TABLE 5–1 Blood Testing Results

Abnormal Test Result(s)	Possible Causative Conditions / Situations
WBCs, decreased	Alcoholism, bone marrow depression, viral infections
WBCs, increased	Circulatory disease, coma, drugs (such as anesthetics, quinine), fever, hemorrhage, leukemia, malignancy, menstruation, moderate physical activity, newborns, serum sickness, severe electric shock, steroid therapy, stress (pain / excitement), tissue necrosis, toxins, trauma / tissue injury, uremia, allergies
RBCs, decreased	Blood loss, dietary insufficiency, decreased RBC production, increased RBC destruction, Hodgkin's disease, leukemia, multiple myeloma, pernicious anemia, rheumatic fever, Addison's disease
RBCs, increased	Dehydration, diarrhea (severe), exercise, high altitude, poisoning (acute), secondary polycythemia, pulmonary fibrosis
Hct, decreased	Anemia, cirrhosis, hemolytic reactions (incompatible transfusions, infections, drugs / chemicals, burns), hyperthyroidism, leukemia, massive hemorrhage
Hct, increased	Erythrocytosis, dehydration (severe), polycythemia, shock
Hgb, decreased	Anemia, cirrhosis of the liver, excessive fluid intake, hemolytic reactions, hemorrhage (severe), hyperthyroidism, pregnancy
Hgb, increased	Chronic obstructive pulmonary disease (COPD), congestive heart failure, high altitude living, conditions of hyperconcentration of blood
Platelet count, decreased	Allergic conditions, anemia, blood transfusions, bone marrow lesions, cancer chemotherapy, chemical exposures, toxic drug effects, infections, pneumonia
Platelet count, increased	Asphyxiation, anemia (post hemorrhagic) Cancer, cirrhosis, exercise/excitement, heart disease, high altitude living, iron deficiency, polycythemia vera, Rheumatoid arthritis, splenectomy, trauma, tuberculosis, Winter (!)

Blood Chemistry

In addition to the blood cells we have discussed, there are also numerous chemicals in the blood. Looking back at Figure 5-3, note the plasma portion of the centrifuged blood. That is where you will find these numerous chemicals. Disease processes can alter the amounts of many of these chemicals, and by measuring their values, they aid us in making a diagnosis. So many substances can be measured in the blood that it would be impossible to discuss them all in this chapter. The following are common ones often found in the results of a physician-ordered patient's blood chemistry.

BUN (Blood Urea Nitrogen) No, BUN isn't something you eat but rather an important blood chemistry measurement. **Urea** is formed by the liver and is a very significant non-protein nitrogenous end waste product of protein catabolism in the body. Once urea is produced, it is transported to the kidneys via the blood for eventual excretion in the urine. While this is a diagnostic test often used to determine renal function, Table 5 – 2 shows that there are other causes for abnormal BUN levels.

urea *(you REE ah)*

Normal BUN Values: 7 – 18 mg / dl or 2.5 – 6.3 mmol/L

TABLE 5-2 BUN Testing Results

ABNORMAL TEST RESULT(S)	POSSIBLE CAUSATIVE CONDITIONS / SITUATIONS,	SIGNS / SYMPTOMS (S/S)
Increased BUN	Acute myocardial infarction, chronic gout, dehydration, diabetes, drugs, excessive protein consumption / protein catabolism, GI hemorrhage, infection, pregnancy(late stage), renal function impairment, shock, tissue trauma(severe).	Confusion, convulsions, disorientation, many more.
Decreased BUN	Diet (low protein / high carbohydrate), drugs, fluid overload (as in intravenous therapy), liver failure, malnutrition, nephrotic syndrome, overhydration	Edema, many more disease specific

Electrolytes   Electrolytes are crucial for proper cellular function throughout the body. As a result, the body must maintain normal concentrations of the various electrolytes. In severe imbalances, life-threatening heart arrhythmias can occur. The amount of water in the body (too much or too little) can affect electrolyte concentrations. In the clinical setting we have patients with conditions that will affect their fluid balances, and thus their electrolyte balances. Dehydration may result from not drinking enough water, from diarrhea or from vomiting, while excessive intravenous (IV) therapy may add too much water to the patient. Table 5-3 shows the possible causes of abnormal values of some common electrolytes.

Normal Electrolyte Values
Calcium (Ca^{++}): 4.5.-5.4 mEq/L
Chloride (Cl^-): 95 – 103 mEq/L
Potassium (K^+): 3.8 – 5.0 mEq/L
Sodium (Na^+): 136 – 142 mEq/L

Enzymes   Enzymes are complex proteins that create changes, often speeding up changes or processes, without changing themselves. They can be found in the majority of the body's cells. Because of this, the levels of certain enzymes are often noted in the blood when they are released following an injury to or the death of cells. This fact is particularly useful in the case of heart attacks. Cardiac tissue death is a common result of a heart attack. As a result, particular enzymes are released over a specific period of time. These enzyme levels can be measured through a series of blood tests. You will learn more about this in Chapter 12, "The Cardiovascular System."

5-3   To view a video on the basics of blood testing, please refer to your DVD.

TABLE 5–3 Electrolyte Testing Results

ABNORMAL TEST RESULT(S)	POSSIBLE CAUSATIVE CONDITIONS / SITUATIONS,	SIGNS / SYMPTOMS (S/S)
Calcium increased	Diuretic therapy, excessive consumption of antacids or milk, hyperparathyroidism, malignant tumors, vitamin D intoxication	Anorexia, constipation, hyporeflexia, lethargy, mental deterioration, renal stones, weakness
Calcium decreased	Pregnancy, hypoparathyroidism, vitamin D deficiency	Convulsions, cramping of muscles, mental disturbances, paresthesia
Chloride increased	Renal tubular acidosis	Breathing rapid and deep, disorientation, weakness
Chloride decreased	Excessive vomiting	Breathing depressed, muscle hypertonicity, tetanus
Potassium increased	Muscle tissue damage, renal failure	Diarrhea, irritability, nausea, ventricular fibrillation, weakness
Potassium decreased	Chronic stress, diuretic therapy, diarrhea, endocrine disorder	Cardiac arrhythmias, hypotension, malaise, muscle weakness
Sodium increased	Dehydration	Dry mucous membranes, dry tongue, flushed skin, intense thirst
Sodium decreased	Burns, excessive water intake, loss of gastrointestinal secretions, excessive sweating	Abdominal cramping, confusion, muscle twitching, perfusion decrease, seizures, vasomotor collapse

TEST YOUR KNOWLEDGE 5-2

Choose the best answer:

1. Situations that can lead to higher than normal numbers of platelets include all *except*:
   a. high altitude living
   b. Vitamin K deficiency
   c. exercise
   d. massive blood transfusions

2. Platelets can be found in all of the following shapes *except*:
   a. oval
   b. disk-shaped
   c. rhomboid
   d. round

3. A severe imbalance of this blood chemistry value can lead to potential life threatening cardiac arrhythmias:
   a. BUN
   b. electrolytes
   c. platelets
   d. PTT

4. Increased hemoglobin levels can be a result of:
   a. high altitude living
   b. anemia
   c. low altitude living
   d. pregnancy

5. These values are often measured over time to determine if the patient had a heart attack (often a sign of tissue injury or death):
   a. erythrocytes
   b. enzymes
   c. leukocytes
   d. erythropoietin

URINE TESTING

Although it is 95% water, urine is a complex fluid containing thousands of dissolved substances. Keep in mind that urine is the end product of metabolism, with all of its waste products, that is performed by *trillions* of cells in your body! As a result of this activity and the body's need to maintain proper fluid balance, you produce approximately 1 to 1 1/2 liters of urine daily.

The chief excretory systems of the body are the skin, lungs, and kidneys, with the kidneys being responsible for urine production. This is a very important job. Kidneys regulate the dilution or concentration of urine and regulate the excretion of sodium, among other substances.

Most of the substances found in urine are also found in your blood, but in different amounts. Some blood substances, such as glucose, will "spill" over into your urine once the substance reaches a certain concentration in the blood. This is why diabetics like Maria sometimes have glucose in their urine. If your kidneys couldn't function properly, death could occur in a matter of days. And yes, normal urine is sterile. You *can* (ugh) drink it.

In general, it is best to use urine that is freshly voided in the morning for testing purposes. This urine usually is more concentrated, (sort of like adding extra packets of drink mix to a pitcher of water) so it should show more abnormalities. The following are some of the urine tests you might see.

5-4 To view a video on performing a dipstick analysis, please refer to your DVD.

Dipsticks

More than a name you call someone or a device to measure the amount of motor oil in your engine, a **dipstick test** doesn't require an accredited laboratory to be performed, and often can be "read" at home by the patient. A dipstick looks a lot like a strip of blotter paper and is impregnated with different types of reactive chemicals, depending on the type of test. Either the tip of the dipstick changes color, or the color just changes shades, depending on the concentration of the substance that is being measured. See Figure 5-4 ■ for an example of a dipstick and how it is read.

Different dipsticks can quickly determine the presence of the following properties of urine:

Bilirubin
Glucose
Hemoglobin
Ketones
Leukocyte esterase
Nitrite
pH Protein

FIGURE ■ 5-4

Once a fresh urine sample is passed over and absorbed by the dipstick, color changes will occur. By matching the dipstick colors to the reference chart, various substances found in the sample can be determined.

Specific Urine Tests

A urine sample can yield a variety of valuable diagnostic information.

Specific Gravity (SG)   A specific gravity (SG) test looks at the kidneys' ability to concentrate urine. The weight of a given amount of urine is compared to a weight of distilled water (which has a weight of 1.000). As we said before, urine is composed of water *plus* minerals, salts, and a variety of other compounds that would make it heavier than 1.000.

Normal Specific Gravity Values
1.003 – 1.035 (usually 1.010 – 1.025)
1.025 - 1.030+ (if urine is concentrated)
1.001 -1.010 (if urine is diluted.)

Specific gravity of urine depends on how well or poorly the patient is hydrated, the volume of urine being excreted, or the amount of solids mixed in the urine. The results can be a reflection of how well the kidneys are functioning.

Concentration   If there is suspicion of renal disease, this test is often performed because it measures the kidneys' ability to concentrate urine when liquid is withheld from the patient for several hours. Healthy kidneys can produce urine that has a greater than 1.020 concentration in this situation. Any concentration less than this indicates renal failure. However, a normal finding does not always indicate healthy kidneys.

Normal Concentration 1.020 – 1.025 (Depending on the test performed)

Patients who take diuretics before their urine concentration is measured can create a false reading that wouldn't provide a true picture of the kidneys' function. There are also numerous other factors that can interfere with these test results, including:

Acute renal disease
Bone disease
Chronic liver disease
Hypercalcemia
Hyperparathyroidism
Potassium deficiency
Low salt or protein diets

Urine Color   Normally a color range of straw to amber is acceptable and is usually due to the pigments in food that is consumed or from the metabolism of bile. Abnormal pigments (often a result of disease) can change the color of urine. Although clinically there are many colors and shades of color in urine, some of the major ones can be found in the upcoming Table 5-4.

Urine Odor   Odor from fresh, normal urine is a result of volatile acids and isn't offensive. Certain foods, such as asparagus, create a distinctive odor. An ammonia scent usually comes from stale urine that has bacterial activity and the decomposition of urea. Infected urine has a very disagreeable odor, depending on how intense the infection is. A sweet smell is usually indicative of diabetic ketosis;

however, an amino acid metabolic disorder in some infants produces a smell much like maple syrup.

Urine pH   The acidity of your urine is mainly controlled by your kidneys in their attempt to maintain a homeostatic pH in your body.

Normal pH range 4.6 – 8 (Average: 6 which is slightly acidic)

An individual with lung disease in which CO_2 is retained in the lungs and blood, a condition called **respiratory acidosis**, will have excessively acidic urine. Ray, our quadriplegic, may have problems with respiratory acidosis due to the paralysis of his diaphragm. Carbon dioxide acts as an acid, so acid has to be excreted through the urine to maintain homeostasis in the body. Other potential causes of excessive acidic urine can include:

Dehydration
Diabetes (uncontrolled)
Diarrhea
High-protein diets
Sleep disorders that affect breathing
Starvation

Urine that is alkaline in nature (pH greater than 7) can be caused by pulmonary diseases that cause **hyperventilation**, a condition in which too much CO_2 is lost, leaving too little acid in the body. As a result, the body holds on to as much acid as possible, trying to bring itself back to homeostasis. Other potential causes of alkaline urine include:

Chronic renal failure
Renal tubular acidosis
Salicylate (aspirin) intoxication
Urinary tract infections (UTI)

TURBIDITY

If urine appears cloudy it could be an indicator of the presence of bacteria, pus, or red (sometimes white) blood cells. However, normal urine can be cloudy due to rapid cooling which allows for the formation of precipitates, ingestion of some foods (especially greasy/fatty foods), or vaginal contamination from samples taken from women (a rather common occurrence).

Normal Turbidity: clear to slightly hazy

SUGAR (GLUCOSE)

This test can be used to detect diabetes, confirm the diagnosis of diabetes, or determine the effectiveness of the treatment for diabetes. Note that sugar in urine, known as glucosuria or glycosuria, is not always abnormal. Situations of emotional stress, diet, or a low reabsorption rate by the kidneys with normal blood levels of glucose may cause high readings. You will find that in most cases, sugar in the urine of your patient is a result of diabetes mellitus.

Normal Urine Glucose: < 0.5 mg/dl

PROTEIN (ALBUMIN)

Healthy patients pass, at most, only trace amounts of protein in their urine. If there is *continual* excretion of protein (which is a rather large molecule) in the urine, that is often indicative of renal disease. Maria's kidney disease may have been detected by the presence of protein in her urine. It's kind of like trying to throw a rock through a screen door; the only way it will get through is if the screen becomes damaged. (Assuming you just don't open the door and throw it in!)

Normal Urine Protein Amount: 50 -80 mg / 24 hrs.

Some situations that can lead to a temporary proteinuria can include:

Bathing or swimming in cold water
Eating large amounts of protein
Severe emotional stress
Violent / intense exercise

Renal diseases that can lead to chronic proteinuria include:

Kidney stones
Nephritis / nephrosis
Polycystic kidney
Tuberculosis or cancer of the kidney

KETONE BODIES (ACETONE)

This is an important test, especially with the continued popularity of high fat and protein / low carbohydrate diets. Ketones are released as a result of the metabolism of fat and fatty acids. Normally, ketones are produced by the liver and are completely metabolized so only minute amounts would be in urine.

Normal Urine Ketone Bodies: Negative

If the body runs out of carbohydrates to burn, it switches to fats. **Ketonuria** is usually associated with diabetes, but consider the effects of diet on your patients as well as the following:

ketonuria *(kee toh NEW ree ah)*

Anorexia
Diarrhea
Drugs (i.e. insulin)
Fasting
Fever
Post anesthesia
Prolonged vomiting
Starvation

 5-5 To view a video on the urinalysis procedure, please refer to your DVD.

BACTERIA

Even small numbers of bacteria may indicate a urinary tract infection (UTI), which when untreated may lead to serious kidney disease. Patient's with abnormal bladder function or who need a urinary catheter to empty his bladder like Ray the quadriplegic patient, are at high risk for developing UTI's.

Normal Urine Bacteria: Negative

alkaptonuria
(AL cap toh NEW ee ah)

SEDIMENT EXAMINATION

Usually, healthy urine contains a small number of cells and other formed elements. By examining the sediment of urine a diagnosis or prognosis can be made without the need for a biopsy/surgery, in some cases. See Table 5-4 for a summary of urine testing results.

TABLE 5–4 Urine Testing Results

ABNORMAL TEST RESULT(S)	POSSIBLE CAUSATIVE CONDITIONS / SITUATIONS
Color	**Straw**: Normal with specific gravity less than 1.010 except in diabetics with a higher specific gravity due to high sugar levels
	Yellow: Normal with a specific gravity of 1.001 – 1.019 with an output of 1 – 1.5 liters
	Amber: Normal but indicates a higher specific gravity greater than 1.020 with an output of less than 1 liter.
	Black: Alkaptonuria, hemoglobin, melanin, Lysol poisoning
	Brown: Addison's disease, blood/hemoglobin, bilirubin, drugs, rhubarb ingestion, drugs (phenolic class), possible melanotic tumor
	Clear/nearly clear: Alcohol ingestion, chronic interstitial nephritis, diabetes insipidus, diuretics, large fluid intake, nervousness/excitement, reduced perspiration, untreated diabetes mellitus
	Orange: Concentrated urine, decreased fluid intake, drugs (e.g. pyridium), excessive sweating, fever
	Red: Beets, blood/hemoglobin, drugs (e.g. certain laxatives, Dilantin, Pyridium), methemoglobin, and oxyhemoglobin
Concentration, decreased	Kidney Failure, overhydration (IV overload)
Concentration, increased	Dehydration, excessive sweating Ketone bodies (acetone), positive anorexia, diabetes, diarrhea, drugs (e.g. ether, insulin, isopropyl alcohol)
pH, acidic, more than normal	Dehydration, diabetes (uncontrolled), diarrhea, high-protein diets, pulmonary diseases with retained CO_2 (causes respiratory acidosis forming acidic urine), starvation
pH, alkaline	Chronic liver failure, pulmonary diseases with hyperventilation, pyloric obstruction, renal tubular acidosis, salicylate intoxication, urinary tract infection (UTI)
Protein (albumin), increased	<u>Temporary increase:</u> Bathing or swimming in cold water, high-protein diet, severe emotional stress, Strenuous exercise.
	<u>Chronic increase:</u> Abdominal tumors, acute infections, anemia (severe) ascites, cancer of the kidney, cardiac disease, convulsive disorders, fever, hyperthyroidism, intestinal obstruction, kidney stones, liver disease, poisoning (e.g. lead, mercury, opiates, turpentine), nephritis, polycystic kidney, trauma, tuberculosis of the kidney
Specific gravity, decreased	Overhydration
Specific gravity, increased	Dehydration
Sugar, increased	Diabetes, emotional stress, diet, low reabsorption rate of kidneys
Turbidity (increased)	Foods (e.g. fat/greasy), precipitates (rapidly cooled specimens), vaginal contamination

FECAL TESTING

The examination of feces (also known as stools) aids in the detection of bleeding or obstruction of the gastrointestinal system, parasites, ulcerative colitis, gall stones, and a variety of other gastrointestinal disorders. A stool examination looks at: consistency, color, odor, pH, shape, size, and the presence or absence of the following: blood, bacteria, food residue, mucus, parasites (and their eggs), pus, and tissue particles.

Let's look at some aspects of a fecal examination.

Amount, Consistency, Form, and Shape

Stool consistency is dependent on diet but normally is plastic and soft with a definite shape—usually the same caliber as the colonic lumen. Depending on diet, there may be indigestible seeds and vegetable/fruit skin fragments present. A pasty stool is often associated with high fat content. Patients with cystic fibrosis have a greasy, tar-like/buttery texture stool. A bulky, frothy stool is often related to celiac disease and a condition called sprue. Diarrhea is watery due to its rapid passing through the intestine before water can be reabsorbed.

Normal Amount of Stool Produced : 100 – 200 g / day

Shape can be changed to a narrow, ribbon-like structure possibly due to spastic bowel, decreased elasticity, rectal narrowing or stricture, or a partial obstruction. Small, ball-like stools are often a result of chronic moderate constipation, while excessively hard stools are often a result of excessive fluid absorption due to slow movement through the colon.

Stool pH

Normal pH is neutral or slightly alkaline and is dependent upon diet and bacterial fermentation. Carbohydrates tend to cause an acidic pH, while proteins tend to cause an alkaline pH.

Stool Color

Normal stool color is brown. After eliminating dietary or food dye as a cause for abnormal color, changes in color may be indicative of a disease process or drug intake. Color changes with some of their possible causative factors are as follows:

Yellow/yellow-green: Severe diarrhea, breast fed infant, bowel sterilization due to antibiotics.
Green: Severe diarrhea, antibiotic therapy, ingestion of chlorophyll-rich vegetables
Tan/clay-colored: Common bile duct blockage, pancreatic insufficiency, excessive fat intake
Black: Usually due to upper GI bleeding, high meat diet, ingestion of iron, charcoal, or bismuth
Red: Bleeding from the lower GI tract. If streaking is on the outer surface, consider hemorrhoids or anal pathology; if blood is throughout the stool, consider a problem higher up the GI tract.

Blood in Stool

A healthy individual passes 2.0 – 2.5 ml. of blood daily into the gastrointestinal, or GI, tract. If an individual passes more than 2.8 ml. of blood into their GI tract over a given 24 hour period, that is indicative of some form of GI pathology. The most common causes for this are hemorrhoids and anal fissures. If there is **occult** blood (blood that is hidden) in the feces, that may be a result of bleeding somewhere in the upper GI region. A gastric ulcer would cause this, for example. Testing for blood in the feces is also one of the ways to detect colon cancer. A false reading of this test can be a result of the ingestion of red meat, as can large doses of vitamin C or inorganic iron. Some drugs, such as aspirin or steroids, can cause greater than normal blood loss to the GI tract.

Conditions that can lead to blood in the stool include:

Cancer (e.g. colon, gastric)
Colitis (ulcerative)
Diverticulitis
Gastric ulcers
Gastritis

occult *(uh CALLT)*

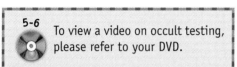

5-6 To view a video on occult testing, please refer to your DVD.

Mucus In the Stool

Normally there is no mucus in the stool. Mucus is secreted as a result of parasympathetic stimulation. Mucus that is translucent and gelatinous can be a result of:

Emotional excitation
Excessive straining during bowel movements
Mucus colitis
Spastic constipation

Mucus with blood may indicate:

Inflammation of the rectal canal
Neoplasm (in some cases, up to 3 – 4 liters of mucus/ 24 hrs.)

Mucus with a blood / pus mixture may indicate:

Cancer of the colon (ulcerative)
Colonitis (ulcerative)
Diverticulitis (acute, in rare cases)
Dysentery (bacterial)
Tuberculosis (intestinal, in rare cases)

TEST YOUR KNOWLEDGE 5-3

Choose the best answer:

1. Urine turbidity can be a result of all of these *except*:
   a. rapid cooling
   b. bacteria
   c. bacon ingestion
   d. overhydration

2. This measuring device is impregnated with reactive chemicals that cause either a color change or a change in color shade when exposed to a substance.
   a. dipstick
   b. needle gauge
   c. test tube
   d blotter paper

3. Which of the following is *not* a normal color of urine:
   a. straw
   b. yellow
   c. amber
   d. orange

4. Starvation will change urine's pH to be more:
   a. acidic
   b. alkaline
   c. neutral
   d. no change

5. Tan or clay-colored feces can be indicative of:
   a. severe diarrhea
   b. upper GI bleed
   c. excessive fat intake
   d. hemorrhoids

CEREBRAL SPINAL FLUID (CSF) TESTING

This fluid is found in the ventricles of the brain and the central canal of the spinal cord. Cerebral spinal fluid serves many purposes that include: acting as a shock absorber, helping to regulate intracranial pressure that can affect the transporting of nutrients and waste produces in the brain, and influencing brain functions, such as glucose levels, leading to hunger sensations and eating behaviors. CSF can be tested for the different type and amounts of cells found in the sample, pathogens, substances such as protein, sugar, and chloride along with their concentration, as well as its general appearance and consistency.

Normal CSF Values:

Normal Daily Production: 500 ml

Normal Circulation Around Brain and Spinal Cord: 150 – 200 ml

CSF Color

Normally CSF fluid is clear and colorless, like water. Appearance changes can be due to:

Hemorrhage / Inflammatory disease (e.g. meningitis)
Micro-organisms (e.g. bacteria, yeast)
Trauma (like Ray's injury)
Tumors

CSF Cell Counts

Normal CSF does not contain white blood cells. If there are white blood cells present, the percentage of each WBC type is counted and compared to the total number of WBCs found.

Normal CSF Cell Counts: 0-5/μl

Usually, increased numbers of white blood cells are a result of an inflammatory disease, hemorrhage, neoplasms, or trauma. If WBCs reach above 500 (mostly neutrophils), then expect a purulent (pus) infection.

High numbers of neutrophils in CSF can be associated with:

Cerebral abscess
Emboli (bacterial endocarditis)
Meningitis (bacterial, early viral, aseptic, mycotic)
Syphilis
Tuberculosis (early)

If there are high numbers of mononuclear cells, it could be indicative of:

Guillain – Barré syndrome
Multiple sclerosis
Syphilis of the central nervous system
Tuberculosis meningitis
Tumor / abscess
Viral infection (e.g. poliomyelitis)

CULTURE AND SENSITIVITY (C&S) TESTING

We fight bacterial infections through the use of antibiotics. Fungi are dealt with through the use of antifungal agents. Remember, not all pathogens are the same and not all antibiotics or antifungal agents are the same. Some antibiotics are labeled as "broad spectrum" drugs, which means that they do a pretty good job of killing off a wide variety of bacteria. Due to the ever increasing problem of drug resistant bacterial, the ideal objective is to first determine the specific organism by growing it (culturing), and then find the best drug (or several acceptable ones) for a specific pathogen—one that will kill it most effectively with as little toxic effect to the patient as possible.

To perform a **culture and sensitivity (C&S) test**, a patient sample is first taken from the suspected infected area. Areas of interest can include the lungs, wounds, throat, eyes, ears, blood, urine, stool, and so forth. The sample is then placed in a nutrient growth medium and allowed to grow over a period of time so colonies of that organism can increase. Certain types of organisms can grow certain shaped or colored colonies. Some of these organisms are subjected to a staining agent applied by a laboratory technologist. Some types of organisms will accept the stain, whereas others are resistant to staining. Then a microscopic examination is made to further determine the type of organism. These are some of the ways to help determine the identity of the organism.

To determine a pathogen's sensitivity to a variety of antibiotics, that organism is placed on a growth medium and allowed to colonize. What makes this different from just culturing is that small discs of several types of antibiotic (or antifungal) drugs are placed throughout the culture medium (see Figure 5-5). The "weakest" drug for that pathogen can be identified by observing how close the organism colonizes near it. In some cases, the colony grows right over a disc, indicating no effect against that pathogen. The most powerful drug disc(s) in that nutrient medium won't allow any colonization to occur near them at all. Those are the ones you consider when determining which drugs are best for getting rid of an infection.

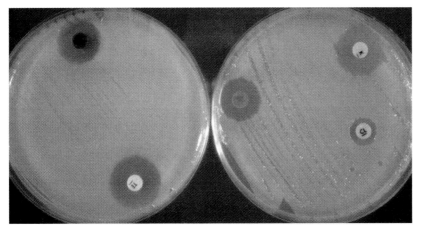

FIGURE 5-5

Note the different sizes of the ring around each disc that was placed in the bacterial growth medium. The larger the ring of non-growth, the more effective that drug is against that specific pathogen.

Here are some things to consider. Normal flora may grow in these colonies. Make sure the specimen that is comprised of normal flora came from where that organism is normally found. For example, the organism *E-coli* is part of the normal flora in your intestinal tract. But if it is found in a blood sample, that is indicative of a potentially fatal situation. Also, verify if the patient may be on some form of antibiotic treatment before a C&S test is performed. That may alter findings.

CARDIAC DIAGNOSTICS

Due to the importance of the heart, several tests can be done to determine its level of function. Malfunctioning of the heart can affect all of the body systems.

Note
The terms ECG and EKG both refer to the same procedure. ECG is from the English word "electro-cardiogram," whereas EKG is from the German word "elektrokardio-gramm."

Electrocardiogram (ECG/EKG)

Later in this book you will learn that all muscle movement is initiated by electrical impulses. The heart is a muscle, so it needs electrical impulses for it to correctly contract allowing blood to be efficiently pumped. We can monitor how these electrical impulses travel through the heart through the use of an **electrocardiogram (ECG / EKG)**. Electrical wires, known as **leads**, are attached at specific locations on a patient's torso to detect any electrical activity occurring in the heart. That electrical activity is converted to either an image seen on the screen of an ECG monitor or as a tracing on a strip of paper.

5-7 To view videos on ECG testing, electrode placement, and fetal heart monitoring, please see your DVD.

Holter Monitor

Sometimes heart irregularities don't occur all day long, or occur only during certain activities. A simple ECG taken in a doctor's office or in the clinical setting may miss that abnormal cardiac event. A **Holter monitor**, which attaches to a belt or shoulder strap and utilizes ECG leads, is used to record all cardiac activity, usually during a 24 hour period (although it may even be a longer time). The patient also keeps a written diary that includes a list of activities plus the time the activities were done. Clinicians then compare the activities to the tracings provided by the Holter monitor to determine any cardiac pathology.

Stress Testing

Patients may develop heart disease that is not recognizable in its early stages. If the individual is placed in a situation where there is physical exertion, symptoms of the disease (such as chest discomfort/pain, shortness of breath, or light-headedness) may exhibit themselves. This situation can be created in a controlled environment during a **stress test**. A normal stress test involves walking on a treadmill at varying speeds and inclines over a given period of time. The patient is monitored by an ECG to determine any arrhythmias, and blood pressures are taken throughout the test. Radioactive dyes may also be injected into the bloodstream and followed via a monitor to determine if there is sufficient blood perfusing the heart muscle. Low amounts of blood flow through cardiac muscle may indicate clogged cardiac arteries that may need to be opened (balloon angioplasty) or replaced (coronary artery grafting or bypass). In cases when the patient is unable to walk, a drug is injected intravenously to cause the heart to beat rapidly, mimicking exercise.

"SCOPES"

"Scope" is a general clinical term incorrectly used to describe a procedure in which a lit, tubular device, usually with a magnifying lens, is inserted into a region of the body to obtain a better view. **Endoscope** is the correct general term for the device used, and to be correct, "scope" is replaced with "scopy" when referring to the actual procedure. So an endoscope is used to perform an **endoscopy**. There are a variety of specialized scopes in the health field. Here is a sampling of some of them.

endoscope (EN doe skope)
  endo = within
  scope (skopein) = to examine
endoscopy (en DOS koh pee)

A common one that most of us have had utilized on us, especially as children, is the **otoscope**. An otoscope is often used to examine the external ear canal (auditory canal) and the eardrum (tympanic membrane) when there is a suspicion of an inner ear infection or a ruptured eardrum. If done properly, it is a painless procedure.

ot/o = ear

A flexible **bronchoscope** is comprised of a magnifying lens (often called a "head") and a tube approximately the thickness of a pencil. There are several channels within this tube. These channels provide for a light source, the instillation of fluids such as saline or drugs, suctioning of airway secretions / fluids, and little forceps that can take a small tissue sample. This is a somewhat uncomfortable procedure, so the patient is usually mildly sedated and numbing agents are administered.

The **gastroscope** is used to examine the interior of the stomach. Since this is somewhat uncomfortable, the patient is mildly sedated

A **laparoscope** is used to examine the abdominal region, and has revolutionized surgery. Laparoscopic surgery allows for much smaller incisions and quicker patient recovery time. Since this is a surgical procedure, a higher level of sedation and anesthesia is used.

If you look at Figure 5-6 ■ you will see a **cystoscope**. This scope is used for viewing the internal anatomy of the bladder.

cyst/o = bladder

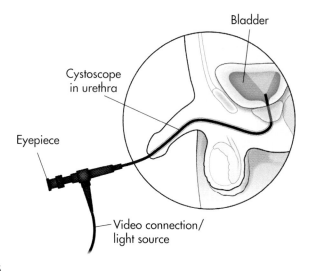

FIGURE ■ **5-6**

A cystoscope is utilized to view the internal anatomy of a bladder. Note that it is a flexible tube with a light providing illumination.

Colonoscopy is the term for visual examination of the colon. Often used to look for sources of bleeding, growths, or cancerous tumors, this procedure requires a level of sedation that is less than laparoscopic surgery.

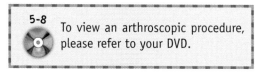

5-8 To view an arthroscopic procedure, please refer to your DVD.

PULMONARY FUNCTION TESTING (PFT)

A basic **pulmonary function test** (PFT) looks at the flow and volume of air going both in and out of your lungs to determine if there is a pathologic occurring. Lower than normal inhaled volumes may be indicative of a restrictive problem. Lower than normal flows may indicate an obstructive disease.

Basic PFTs are regularly used to determine the degree of lung involvement in several diseases such as asthma, emphysema, chronic bronchitis, and cystic fibrosis as well as determining the severity of neuromuscular diseases.

More complex PFTs involve the inhalation of gases to see how effective lungs are at transferring gases to the blood system.

Often the success and accurate results of PFTs are dependent on the effort made by the patient to breathe as instructed. That is why good explanations and coaching are important when conducting the test.

POLYSOMNOGRAPHY

polysomnography
(PAUL ee sahm NOG gra fee)

Polysomnography, or sleep studies, is an area of rapid growth in the clinical setting. Sleep disorders can be a result of certain patient conditions such as stress or depression. They can also be a causative agent in a variety of conditions such as heart disease, memory difficulties, and personality changes and can be a major factor in automobile and industrial accidents.

A basic polysomnographic study looks at several things while a person is asleep:

- The flow of air in and out of the nose / mouth
- Eye movement to determine the stage of sleep the patient is in
- Movements of muscles needed to breathe
- Monitoring of brain waves to determine the stage of sleep the patient is in
- ECG to monitor heart rate and any beat irregularities
- Monitoring of oxygen levels in the blood

By monitoring these areas it can be determined if there is a structural problem in the airway that blocks the flow of air as a person sleeps or if there is a problem with the brain telling the body to breathe during periods of sleep.

TEST YOUR KNOWLEDGE 5-4

Choose the best answer:

1. White blood cell counts are important when determining:
   a. infection
   b. age
   c. coronary artery blockage
   d. race

2. This test is used to determine a specific pathogenic organism and the drug(s) most effective in combating it:
   a. PTT
   b. BUN
   c. C & S
   d. PT

3. A testing device used to record heart activities during a 24 hour period is a(n):
   a. EMG
   b. Holter monitor
   c. Hanasen monitor
   d. GMC

4. This device has revolutionized abdominal surgery in that a large incision isn't needed and, as a result, recovery time is shortened:
   a. bronchoscope
   b. endoscope
   c. otoscope
   d. laparoscope

5. This patient effort dependent test measures flows and volumes of air breathed in and out of the lungs to determine the degrees of obstruction and /or restriction:
   a. LSMFT
   b. PFT
   c. STD
   d. C&S

SUMMARY

Snapshots from the Journey

→ Diagnostic tests help to provide a better picture of your patient's overall condition.

→ These tests help to determine the diagnosis, prognosis and improvement of a disease condition as well as aid in determining the effectiveness of a treatment.

→ Usually, it is best to not base a diagnosis solely on one abnormal diagnostic test.

→ Even healthy individuals can occasionally produce an abnormal test result.

→ Blood is composed of a liquid portion called plasma, a variety of specialized cells such as leukocytes, erythrocytes, and platelets, and a variety of other substances, all of which can be analyzed to determine your patient's condition.

→ Excessive or lower than normal levels of the various blood cell types can be indicative of disease.

→ A complete blood count (CBC) is a common diagnostic test.

→ White blood cell analysis can provide evidence of infection.

→ Urine is 95% water but also contains thousands of dissolved substances that can be analyzed.

→ The best urine samples are obtained in the morning upon awakening due to its high concentration.

→ Fecal examination is important in the determination of gastrointestinal bleeding.

→ Cerebral spinal fluid (CSF) regulates intracranial pressure, transports nutrients and waste in the brain, influences brain function and acts as a shock absorber.

→ Culture and sensitivity (C & S) testing identifies the specific organism and the drug (or drugs) that is most effective against it.

→ Diagnostic tools that are valuable for cardiac assessment include: electrocardiogram (EKG / ECG), Holter monitor, and stress testing.

→ A variety of scoping devices play important roles in diagnostics, and surgical procedures such as the: otoscope, endoscope, bronchoscope, and laparoscope.

→ Pulmonary function testing (PFT) is utilized as an aid in determining lung function and disease.

→ Polysomnography testing monitors several parameters during sleep to determine if a sleep disorder is present.

RAY AND MARIA'S STORIES

 Given our common cases in this textbook deal with a diabetic and a quadriplegic patient, select two diagnostic tests that could assess and monitor their conditions or associated problems along with a justification for each test and what results you might expect.

Maria, diabetic patient

Diagnostic test 1 _____

Justification _____

Expected results _____

Diagnostic test 2 _____

Justification _____

Expected results _____

Ray, quadriplegic patient

Diagnostic test 1 _____

Justification _____

Expected results _____

Diagnostic test 2 _____

Justification _____

Expected results _____

REVIEW QUESTIONS

Multiple Choice

1. Polysomnography is a diagnostic test used to determine:
   a. sleep disorders
   b. blood chemistry
   c. many bacteria
   d. cell counts

2. Mucus found in a fecal sample may be a result of:
   a. a normal finding
   b. emotional excitation
   c. gelatin ingestion
   d. runny nose

3. A value measured to determine the effectiveness of heparin therapy is:
   a. PTT
   b. MRI
   c. BUN
   d. DOG

4. A specific gravity test looks at the kidneys' ability to concentrate:
   a. RBCs
   b. enzymes
   c. urine
   d. bacteria

5. This test is used to determine the extent of lung damage from diseases such as emphysema, chronic bronchitis, or cystic fibrosis:
   a. BUN
   b. EKG
   c. C & S
   d. PFT

6. High-protein diets or dehydration can cause urine to be more:
   a. acidic
   b. alkaline
   c. neutral
   d. thin

Fill in the Blank

1. _____ is the term used for a condition of excess RBCs in the blood.

2. A group of blood chemistry values used to determine if a patient had a heart attack are the _____ levels.

3. The body organs chiefly responsible for the regulation of the excretion of sodium and the concentration or urine are the _____.

4. A _____ is a diagnostic tool used to quickly determine the level of glucose in a patient's urine.

5. A _____ is utilized to place physical exertion on the heart and is often helpful in determining early stages of heart disease.

Short Answer

1. List the things that basic polysomnography looks at to determine if there is a possible sleep disorder.

_____

_____

_____

_____

2. Explain why diagnostic tests are important when diagnosing a patient's condition.

_____

_____

_____

_____

3. Explain why it is important not to treat a patient based solely on diagnostic test values.

_____

_____

_____

_____

Suggested Activities

1. Contact your local hospital to arrange a tour of their clinical laboratory.

2. Do a Web search to discover the effects of the overuse of antibiotics in humans.

3. Do a Web search on the usage and possible effects of antibiotics administered to animals that we consume.

5-9 Now that you have completed your journey through this chapter, please go to the DVD for interactive games and puzzles concerning medical terms and concepts contained in this chapter. By playing the games you will reinforce your learning of diagnostic testing in a fun way.

T H E SKELETAL System

The Framework

As we continue our journey, we may stop to visit friends or relatives and perhaps stay overnight. Their house will protect us from the elements and provide a safe place to sleep. Our hosts will offer us a good breakfast in the morning from the stored goods within. Although we cannot see the framework that holds up the house, without it everything would collapse. We can think of the human body as a house. The wood framework is the skeleton, composed of bones that provide shape and strength. That is why we are beginning the system chapters with the body's own framework, upon which the muscles and skin are layered, much like stone or siding. Mounted on the framework are hinges that allow doors to swing open and shut and windows that glide to open or close. These are analogous to the joints and muscles attached to the skeleton that allow for body movement. What's the first thing you think of when someone asks about the function of the skeleton? From our analogy, you would probably say, "to provide support and allow us to move." But the skeleton does so much more. As explained in this chapter, the bones that make up your skeleton also protect the soft body parts, produce blood cells, and act as a storage unit for minerals and fat—much like our hosts' home, which has stored goods within for that scrumptious breakfast! In this chapter, we discuss the makeup and importance of the 206 bones in the adult skeleton, as well as cartilage, ligaments, and joints. And now it's time to "bone up" on the skeletal system.

Chapter 6

 Correct pronunciation is important in any journey so that you and others are completely understood. Here is a "see and say" Pronunciation Guide for the more difficult terms to pronounce in this chapter.

appendicular skeleton
  (app en DIK yoo lahr)
arthritis (ahr THRYE tiss)
articulation (ahr TICK you LAY shun)
axial skeleton (AK see al)
cancellous bone (CAN cell us)
diaphysis (dye AFF ih siss)

epiphyseal plate (eh PIFF ih SEE al)
epiphysis (eh PIFF ih siss)
hemopoiesis (HEE moh poy EE sus)
medullary cavity (MED you lair ee)
osseous tissue (OSS see us)
ossification (OSS siff ih KAY shun)
osteoarthritis
  (OSS tee oh ahr THRYE tiss)

osteocyte (OSS tee oh site)
osteons (OSS tee ons)
periosteum (pair ee OSS tee um)
synovial fluid (sin OH vee al)
trabeculae (tra BECK you lay)
vertebrae (VER teh bray)

SYSTEM OVERVIEW: MORE THAN THE "BARE BONES" ABOUT BONES

The skeleton has many more uses than just scaring people on Halloween. It is a wondrous structure that serves more functions than simply providing a framework for the human body. It also produces blood cells, provides protection for organs, helps us to breathe, acts as a warehouse for mineral storage, and, along with the muscular system, allows for movement.

Amazing Body Facts

Approximately 50% of a bone's weight comes from stored minerals, about 30% from protein, and the remaining 20% from water.

General Bone Classification

The primary components of the skeleton are bones. Although they may seem lifeless and are composed of some nonliving minerals such as calcium and phosphorus, they are very much alive, and are constantly building and repairing themselves. This is somewhat ironic because the word *skeleton* is derived from the Greek word meaning "dried-up body."

We can classify bone types according to their shape:

- long bones
- short bones
- flat bones
- irregular bones

Long bones are longer than they are wide and are found in your arms and legs. The femur, for example, is the longest and strongest bone in the body. *Short bones* are fairly equal sized in width and length, similar to a cube, and are mostly found in your wrists and ankles. *Flat bones* are thinner bones that can be either flat or curved and are plate-like in nature. Examples of flat bones are the skull, ribs, and breastbone (**sternum**). *Irregular bones* are like the parts of a jigsaw puzzle. These are the odd-shaped bones needed to connect to other bones. Some examples of irregular bones are the hip bones and the **vertebrae** that make up your spine. See Figure 6-1 ▪ for a view of the various bone shapes.

sternum *(STER num)*

vertebrae *(VER teh bray)*

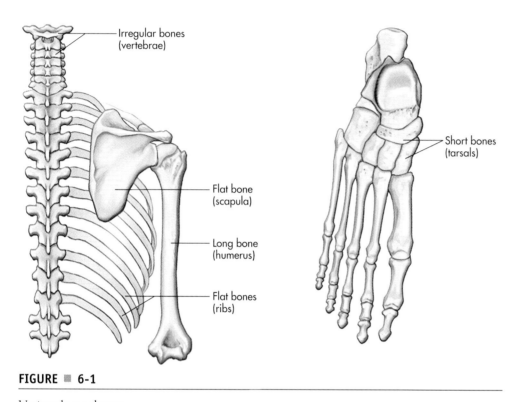

FIGURE ▪ **6-1**

Various bone shapes.

Basic Bone Anatomy

Let's look at the overall construction of a bone by examining a long bone in Figure 6-2 ▪. Bone is covered with **periosteum,** which is a tough and fibrous connective tissue. This cover contains blood vessels, which transport blood and nutrients into the bone to nurture the bone cells. It also contains lymph vessels and nerves. In addition, the periosteum provides anchor points for ligaments and tendons, which we discuss later. Note in Figure 6-2 that both ends of the long bone are larger than the middle portion. Each bone end is called an **epiphysis.** The region between or "running through" the two ends is called the **diaphysis,** or shaft.

The hollow region in the diaphysis is called the **medullary cavity** (sometimes called the **medullary canal**) and acts as a storage area for bone marrow, like a well-stocked kitchen cabinet. There are two kinds of bone marrow: red and yellow. **Red marrow** makes blood cells. **Yellow marrow** has a high fat content. In emergencies—for instance, in the event of massive blood loss—when you need more red blood cells, some of the yellow marrow can convert to red bone marrow to help in red blood cell production.

periosteum *(pair ee OSS tee um)*
  peri = *around*
  osteum = *bone*
epiphysis *(eh PIFF ih siss)*
  epi = *over or upon*
diaphysis *(dye AFF ih siss)*
  dia = *through*

medullary cavity *(MED you lair ee)*

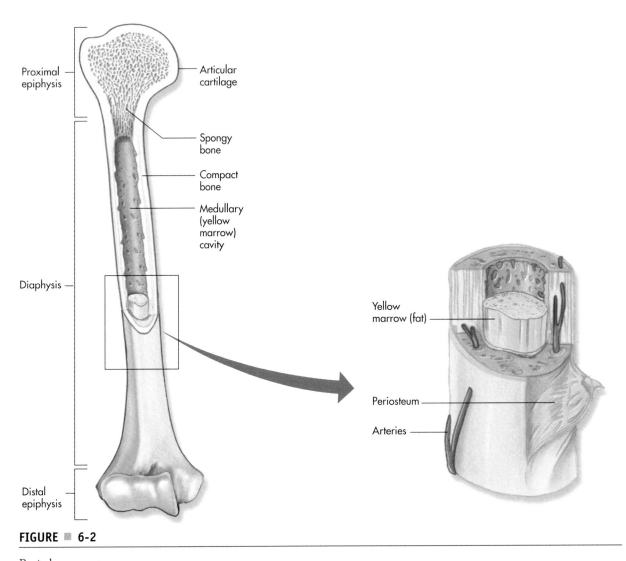

FIGURE ■ 6-2

Basic bone anatomy.

Bone Tissue

There are two types of bone tissue: compact and spongy. **Compact bone** is a dense, hard tissue that normally forms the shafts of long bones and the outer layer of other bones. Microscopic examination reveals that the material of compact bone is tightly packed. This makes for a dense and strong structure. This material forms microscopic cylindrical-shaped units called **osteons,** or **Haversian systems**. Each unit has mature bone cells (**osteocytes**) forming concentric circles around blood vessels. The area around the osteocytes is filled with protein fibers, calcium, and other minerals. The osteons run parallel to each other, with blood vessels laterally connecting with them to ensure sufficient oxygen and nutrients for the bone cells.

Spongy (or **cancellous**) bone is different than compact bone. Instead of Haversian systems, spongy bone tissue has bars and plates called **trabeculae.** Irregular holes between the trabeculae give the bone a spongy appearance (porous).

osteons *(OSS tee ons)*
osteocytes *(OSS tee oh sites)*

cancellous *(CAN cell us)*
trabeculae *(tra BECK you lay)*

Spongy bone is lined with **endosteum**, a tissue similar to periosteum. Spongy bone serves two purposes: it helps make the bones lighter in weight and it provides a space for red bone marrow, which produces red blood cells (see Figure 6-3 ■).

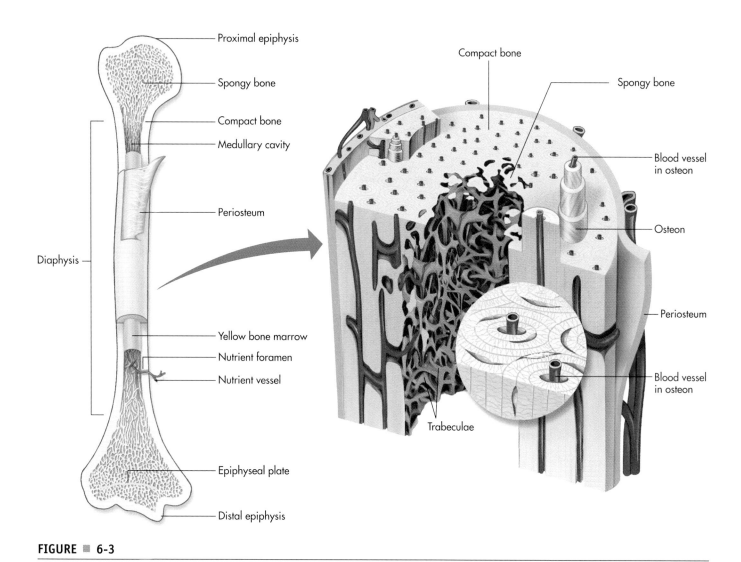

FIGURE ■ 6-3

Comparison of compact and spongy bone.

Surface Structures of Bones

Bone is not perfectly smooth, but includes a variety of projections, bumps, and depressions. Generally, projecting structures act as points of attachment for muscles, ligaments, or tendons, while grooves and depressions act as pathways for nerves and blood vessels. Both projecting structures and depressions can work together as joining or articulation points to form joints such as the ball and socket joint in the hip. Table 6-1 lists many of these bone features.

TABLE 6-1 Bone Features

BONE SURFACE STRUCTURES	DESCRIPTIONS
Projecting Structures and Processes	
Condyle	A large, rounded knob, usually articulating with another bone
Crest	A narrow ridge
Epicondyle	An enlargement near or superior to a condyle
Facet	A small, flattened area
Head	An articulating end of a bone that is rounded and enlarged
Process	A prominent projection
Spine	A sharp projection
Trochanter	Located only on the femur; a larger version of a tubercle
Tubercle	A knoblike projection
Depressions and Openings	
Foramen	A passageway through a bone for blood vessels, nerves, and ligaments; a hole
Fossa	Either a groove or shallow depression
Meatus	A tube or tunnel-like passageway through bone
Sinus	A hollow area

hemopoiesis *(HEE moh poy EE sus)*
  heme = *blood*
  poiesis = *to make*

Amazing Body Facts

RED BONE MARROW AND RED BLOOD CELLS

The areas of bone with red bone marrow, which produces red blood cells, are found in the skull, clavicles (collar bones), vertebrae of the spinal column, sternum (breast bone), ribs, and pelvis, and in the spongy bone that makes up the epiphysis of the long bones. The production of red blood cells, known as **hemopoiesis**, is truly an incredible process! Since red blood cells last only about 120 days, red blood cell production is a constant job in order to maintain the trillions (give or take a few) of red blood cells contained in the human body. As a result, it has been calculated that approximately 3 million new red blood cells are created every second! What's more, your body can step up production to 10 times that rate in cases of severe blood loss. If the red marrow can't maintain the needed production, some of the yellow marrow can be converted to red marrow to assist. White blood cells, which fight infection, are also made in the marrow, but can be made in both red and yellow marrow.

TEST YOUR KNOWLEDGE 6-1

Complete the following:

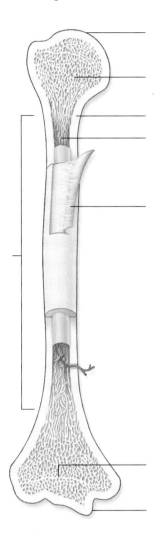

1. Label the diagram:
   a. diaphysis
   b. proximal and distal epiphysis
   c. periosteum
   d. spongy bone
   e. compact bone
   f. medullary cavity
   g. epiphyseal plate

2. Where are the locations and what is the purpose of red bone marrow?

   _____

   _____

   _____

3. Discuss three functions of bone in your body.

   _____

   _____

   _____

Choose the best answer:

4. Mature bone cells in compact bone are called
   a. marrowcytes
   b. osteocytes
   c. riflecytes
   d. monocytes

5. The end of a long bone is called the
   a. epiphysis
   b. periosteum
   c. diaphysis
   d. knob

Bone Growth and Repair

Ossification, or **osteogenesis**, is the formation of bone in the body. Bones grow *longitudinally* in order to lengthen (which makes you taller), and they grow *horizontally* (wider and thicker) so they can more efficiently support body weight

ossification *(OSS siff ih cay shun)*
oste/o = *bone*

and any other weight we support when we work or play. There are four types of cells involved in the formation and growth of bone:

- osteoprogenitor cells
- osteoblasts
- osteocytes
- osteoclasts

Osteoprogenitor cells are non-specialized cells found in the periosteum, endosteum, and central canal of compact bones. Non-specialized cells (bone stem cells) can turn into other types of cells as needed. **Osteoblasts** are the cells that actually form bones. They arise from the non-specialized osteoprogenitor cells and are the bone cells that secrete a matrix of calcium and other minerals that give bone its typical characteristics. **Osteocytes** are mature bone cells that were originally osteoblasts. In other words, osteoblasts surround themselves with a matrix of calcium that help them become the mature osteocytes. Bone is built up or formed by osteoprogenitor cells becoming osteoblasts, which in turn surround themselves with a mineral matrix to become full-blown osteocytes or bone cells.

Not only does the body constantly build new bone, but it also constantly tears down old bone. This is the job of the **osteoclast.** It is believed that osteoclasts originate from a type of white blood cell called a **monocyte** that is found in red bone marrow. As we already learned, the bones of the body store both calcium and phosphorus in the osteoblasts. However, if the body suddenly needs more of these minerals for other functions, the osteoclast will return the stored minerals back to the blood stream. The osteoclasts' two jobs are to tear down bone material and help move calcium and phosphate into the blood. You can think of osteoblasts and osteoclasts as employees of a house remodeling company: the osteoblasts ("b" as in builder) are masons laying down brickwork to make new exterior walls, and the osteoclasts ("c" as in clearing away) are tearing out the inside to remodel! As explained shortly, the job the osteoclasts do is very important for bone growth and repair.

Bone development and growth begins in the womb through intramembranous and endochondral ossification. The process starts at 8 weeks after conception, when bone begins to replace cartilage.

progeny = *offspring*

blast = *immature stage of cell development*

> **6-1** For a drag-and-drop exercise in labeling parts of the bone, go to your DVD for this chapter.

clast = *causing breakage into parts*

Amazing Body Facts

AGING AND BONE BUILDING

Even in adults, bone continues to be broken down and rebuilt. In fact, about 10% of the body's bone is torn down and rebuilt each year! Your bones continue to increase in mass well into your 20s. The osteoclasts break down and remove worn-out bone cells and deposit calcium into the blood. They last for approximately three weeks. Osteoblasts then pull the calcium out of the blood, surround themselves with that mineralized matrix we talked about, and mature into osteocytes. Because of this continual breakdown of old and creation of new bone, adults actually need more calcium in their diets than children do! The process of breaking down and rebuilding bone continues well into a person's 40s, so there normally is no net gain or loss of bone mass.

This continuing process allows your body to sculpt bone into shapes that accommodate the body's activity. For example, exercise, such as running or weight lifting, causes calcium to stay in the bone, making it thicker, denser, and stronger than those of a sedentary person (couch potato) or an individual who experiences extended weightlessness such as an astronaut. Continuous or repeated actions or postures tend to cause bone to be resculpted. For example, due to constant squatting, a certain pattern of bumps forms on the bones of the hips, shins, and knees. As a result of this pattern, it was determined that Neanderthal man squatted rather than sat.

Intramembranous ossification occurs when bone develops between two sheets composed of fibrous connective tissue, such as in the development of your skull. Cells from connective tissue turn into osteoblasts and form a matrix similar to the trabeculae of spongy bone, while other osteoblasts create compact bone over the surface of the spongy bone. As discussed previously, once the matrix surrounds the osteoblasts, they become osteocytes. This is how the bones of the skull develop.

The majority of your skeletal bones are created through **endochondral ossification**, in which shaped cartilage is replaced by bone. Refer to Figure 6-4 ■. In this situation, which begins several months before birth, periosteum surrounds the diaphysis of the "cartilage" bone as the cartilage itself begins to break down. Osteoblasts come into this region and create spongy bone in an area that is then called the **primary ossification center**. Meanwhile, other osteoblasts begin to form compact bone under the periosteum. Here is where the osteoclasts come into play. Their job is to break down the spongy bone of the diaphysis to create the medullary cavity. After birth, the epiphyses on your long bones continue to grow. However, around birth, secondary ossification of this area begins, with spongy bone forming and not breaking down. A thin band of cartilage forms an **epiphyseal plate** (often called the **growth plate**) between the primary and secondary ossification centers. This is the growth zone of your bones. The plate is important because, as long as it exists, the length and width of the bone will increase. As explained in Chapter 10, "The Endocrine System," hormones control the growth of bones. Eventually the plates become ossified, thereby stopping bone growth, usually at the end of puberty.

intra = *within*

endochondral
    endo = *within*
    chondral = *referring to cartilage*
epiphyseal plate *(eh PIFF ih SEE al)*

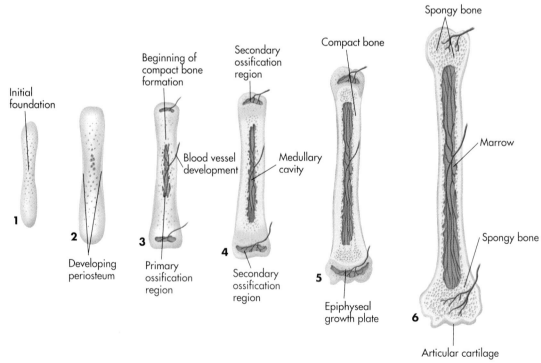

FIGURE ■ 6-4

Endochondral ossification of long bone.

Pathology Connection: Osteoporosis

As time goes on, bone mass also changes with age. In our 50s, the skeleton begins to change: the breakdown of bone becomes greater than the formation of new bone. At the cellular level, the osteoclasts are tearing down more bone than the osteoblasts are forming. As a result, we see total bone mass beginning to gradually decrease. A microscopic examination of bones going through this process reveals increasing holes in the bone. This bone is lighter in weight and weaker than healthy bone, thereby making it more prone to breakage. Men appear to lose less than 25% of their bone mass, whereas women can experience a 35% loss on average. Hormonal changes during menopause increase the rate of bone loss, and the risk of bone loss in many women. Thin Caucasian and Asian women are at the highest risk. For both men and women, the risk of developing thin bones may be hereditary. This condition of decreasing bone density, known as **osteoporosis,** is a serious problem. It is estimated that close to 4 million Americans suffer with this disease, and over 6 million related doctor visits per year are made.

Is there any way to stop or at least slow down the loss of bone mass? The early intervention of preventative practices is crucial for all people. Simple dietary changes such as adding more calcium, fluoride, and vitamin D may help in maintaining bone density as will the elimination of smoking and decrease of caffeine consumption. Smoking and caffeine are believed to aid in the depletion of calcium. Research is showing that exercise, especially weight bearing and impact forms, improve bone density. This helps people of all ages, even those in their 80s and 90s. The use of medications that increase bone mass, such as alendronate, may also be considered. For some post-menopausal women, hormone replacement therapy (HRT) may be appropriate, but the use of HRT is controversial.

osteoporosis
  (OSS tee oh poor OH siss)
  oste/o = *bone*
  porosis = *condition of porous nature*

Cartilage

cartilage *(KAR tih lij)*

Cartilage is a special form of dense connective tissue that can withstand a fair amount of flexing, tension, and pressure. Cartilage plays many roles throughout the body. The flexible parts of your nose and ears are cartilage. (Imagine how many people would have broken off ears and noses if it weren't for cartilage!) When you were born, two areas in the middle of your skull were made of membranes somewhat similar to cartilage. These "soft spots," called **fontanels**, slowly became bone during your first two years of life.

Cartilage also makes a flexible connection between bones. For example, the cartilage between the breast bone and ribs allows your chest to flex and give so you don't break your ribs when you run into things or collide with another player during a football game. Something as simple as taking a deep breath could become a major struggle if it weren't for this flexibility.

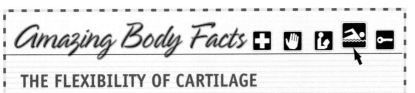

THE FLEXIBILITY OF CARTILAGE

Did you know that before you were born your skeleton was entirely made of cartilage but through development most of it became bone? Your ears and nose will always remain cartilage, and therefore are called permanent cartilage. In some cases, our ears and nose may get larger as we get older because permanent cartilage can continue to grow.

This amazing tissue also acts as a cushion between the bones. As you can see in Figure 6-5 ■, **articular cartilage** is located on the ends of bones and acts as a shock absorber, preventing the bone ends from grinding together as they move. In addition at this location, a small sac, called the **bursa,** contains a lubricant called **synovial fluid.** Even with cartilage and synovial fluid protecting the area between bones, joints can wear out and become inflamed, resulting in a condition called **arthritis,** or **osteoarthritis.**

bursa *(BER sah)*
synovial *(sin OH vee al)*
arthritis *(ahr THRYE tiss)*
   arthr/o = *joint*
   itis = *inflammation of*
osteoarthritis
   (OSS tee oh ahr THRYE tiss)
   oste/o = *bone*

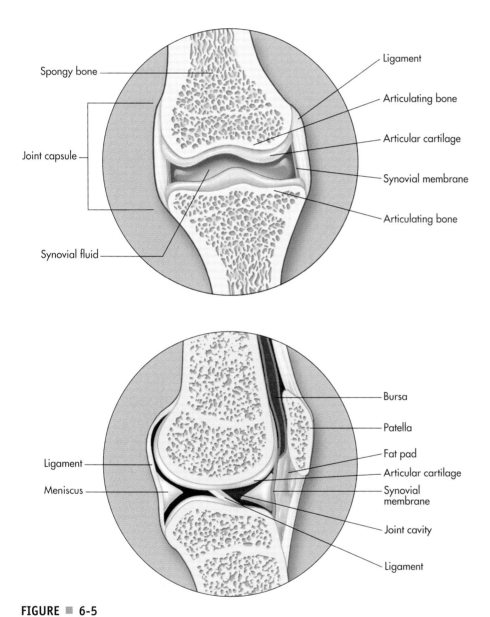

FIGURE ■ 6-5

Articular cartilage and synovial joint.

Joints and Ligaments

Without **joints,** the body could not move. Two or more bones joined together form a joint, or an **articulation.**

    Freely moving joints have to be held together and yet still be moveable. This is accomplished through the use of another specialized connective tissue called a **ligament.** (Again, see Figure 6-5.) Ligaments are very tough, whitish bands that connect from bone to bone and can withstand pretty heavy stress. Do not confuse ligaments with **tendons.** While ligaments hold bone to bone, tendons are cordlike structures that attach muscle to bone. There are several types of joints, and each works in a specific way.

articulation
    (AHR tick you lay shun)

TEST YOUR KNOWLEDGE 6-2

Choose the best answer:

1. A term that can be used to describe the formation of bone is:
   a. ossification
   b. periosteum
   c. bonafide
   d. osteoclasts

2. These cells actually form bones:
   a. osteoclasts
   b. osteons
   c. generator cells
   d. osteoblasts

3. Another name for the "growth plate" is:
   a. tectonic plate
   b. epiphyseal plate
   c. diaphysis plate
   d. periosteum plate

4. This special connective tissue composes your ears and nose:
   a. tendons
   b. ligaments
   c. cartilage
   d. cartridge

5. This lubricant helps to prevent wear between the joints:
   a. pleural fluid
   b. synovial fluid
   c. mucous
   d. petroleum jelly

6. These structures attach bone to bone:
   a. ligaments
   b. tendons
   c. cords
   d. articulations

    Joints are classified either by function or structure. In terms of function, joints can be immobile, move a little, or move freely. For example, skull sutures are immobile, the pubic symphysis (line of fusion between two bones) between your pelvic bones moves a little, and your elbow moves freely. If we characterize joints by structure, we divide them based on the type of connective tissue that links the bones together. **Fibrous joints** are held together by short connective tissue strands. They are either immobile or slightly movable. The sutures in your skull are fibrous joints. **Cartilaginous joints** are held together by cartilage disks. The pubic symphysis and the joints between your ribs and sternum are cartilaginous joints, which are either immobile or slightly movable. Finally, **synovial joints** are joined by a joint cavity lined with a synovial membrane and filled with synovial fluid, which decreases friction in the joint. All synovial joints are freely moving. Synovial joints are constructed in various ways that determine how they can move.

 For some excellent animations on joint classification and how they move, go to your DVD for this chapter.

 To see a video on the types of movement made by the human body, go to your DVD for this chapter.

- *Pivot joints* (which act like a turnstile) have a circular portion of one bone that spins inside a ring-shaped portion of the other. Pivot joints are the type of joint found in your neck and forearm. They can only rotate.

- *Ball and socket joints* usually consist of a spherical bone articulating with a cup-shaped socket on the other bone. Ball and socket joints are found in your hips and shoulders and can perform all types of movement, including rotation. These allow the greatest movement.

- *Hinge joints* are found in your knees and elbows. Typically one bone is in the shape of a cylinder and the other a trough. They can either open or close.

- *Gliding joints* are the flat, or slightly curved, plate-like bones found in your wrists and ankles. Gliding joints slide back and forth.

- *Saddle joints* have a bone shaped just like a saddle and another bone similar to a horse's back. This joint type is found in the base of your thumb. Saddle joints rock up and down and side to side.

- *Ellipsoidal joints* (also called *condyliod joints*) provide two axes of movement through the same bone, like the joint formed at the wrist with both the radius and ulna. The knuckles of your fingers are also ellipsodial joints.

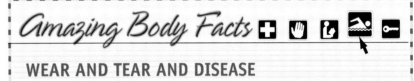

Amazing Body Facts

WEAR AND TEAR AND DISEASE

When considering chronic disorders of the skeletal system, most involve the joints and not the shafts of the bones. This is due to the wear and tear of contact areas over time. One major exception to this is osteoporosis which affects actual bone integrity.

To better visualize these various joints, look at Figure 6-6 ▪.

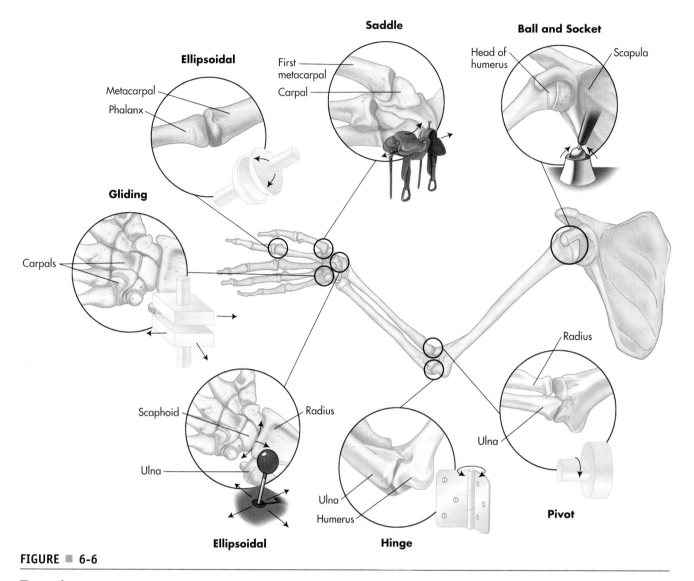

FIGURE ■ 6-6

Types of joints.

Movement Classification

Since joints allow for various types of movement, these individualized movements can also be classified, as you can see in the examples in Figure 6-7 ■. **Flexion** occurs when a joint is bent, decreasing the angle between the involved bones, as when the leg is bent at the knee. **Extension** is a result of straightening a joint so the angle between the involved bones increases, as occurs with a kicking motion. Ballerinas utilize **plantar flexion** when they dance on their toes. **Dorsiflexion** occurs when the foot is bent up toward the leg. If the joint is forced to straighten beyond its normal limits, **hyperextension** occurs.

Abduction and **adduction** can be confusing. Abduction means to move *away* from the body's midline (think, "**B**e gone!" as you move your arm up and away to swat a bee). A**dd**uction means to move *toward* the midline of the body. To remember adduction, think of your a**dd**ress, where packages and mail come *to* you. Then when you move your arm back toward yourself after swatting at a bee, you are a**dd**ucting your arm.

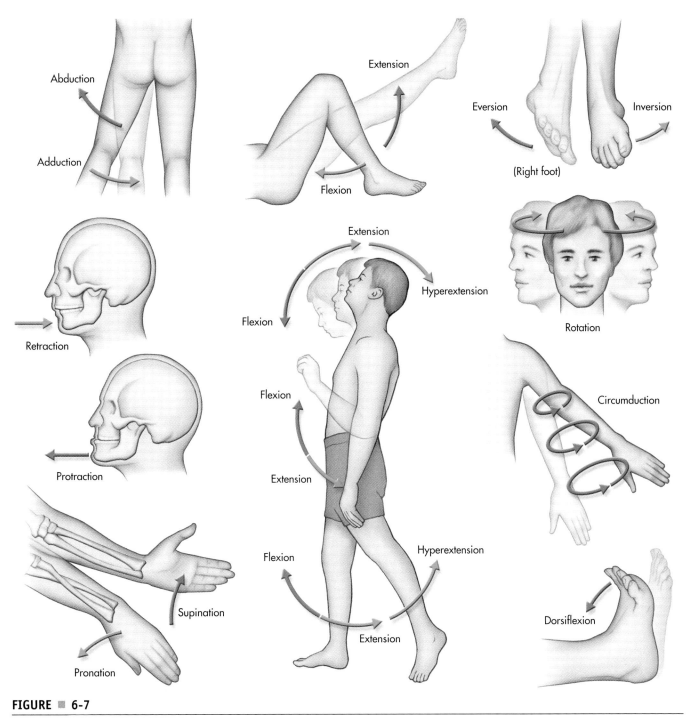

FIGURE ■ 6-7

Classification of joint movements.

Inversion results when the sole of one foot is turned inward so it points to the other foot, while **eversion** is the opposite: the foot is turned outward, pointing away from the opposite foot. **Supination** occurs when the hand is turned to the point where the palm faces upward; **pronation** turns the palm downward. Although you may not have heard of **circumduction**, you have seen this combination of movements in the circular arm movement that a softball pitcher utilizes.

Protraction is the motion of drawing a part forward. **Retraction** is the motion of drawing backward. Figure 6-7 shows the protraction and retraction of the jaw. The movements are analogous to a turtle sticking his head out (protracting) and drawing it back in (retracting) to the shell.

Finally, **rotation** is when a bone "spins" on its axis. An example is when your head rotates (looking left and right) before you cross the street.

Pathology Connection: Common Joint Disorders

All good things must come to an end, and the same can be said about the health of your skeletal system. In general, as the body ages, the cartilage and bones deteriorate. Although this is a natural process that we will all encounter, we can slow the process down in many cases.

As the body ages, the chemical composition of cartilage changes. The bluish tint and flexibility of young skeletal cartilage changes to a more brittle, opaque yellow form. Calcification or hardening of cartilage leads to brittleness. Articular cartilage, once it becomes brittle, doesn't function as well as young, healthy cartilage. **Arthritis** is a general term used to describe an inflammatory process of a joint or joints. There are literally dozens of disorders that cause swollen, painful, stiff joints and, as such, are classified as arthritis. The related tendons and ligaments also become less flexible, causing a decrease in the range of motion in joints. **Osteoarthritis** is a degenerative condition that is a result of simple "wearing out" of a joint from a sports injury, trauma, repetitive motion, obesity, or just the aging process. **Rheumatoid arthritis** is a chronic, systemic disease that is believed to be an autoimmune condition affecting the connective tissue of the body. This condition appears to affect mostly the small joints of the hands and feet, but also can affect the elbows and knees. A third form of arthritis is **septic arthritis**, the result of the infective process of a pathogen that was introduced to a joint from a penetrating wound, or a blood-borne pathogen. For a list of the various treatments for these forms of arthritis, please go to the Quick Trip table at the end of the chapter.

Once considered a disease of royalty due to the belief that it was caused by excessive amounts of alcohol and rich foods, **gout** is a form of arthritis. Gout is a result of the body's decreased ability to control uric acid production or its excretion, thus causing high uric acid levels in the blood. Once the uric acid levels increase sufficiently, it crystallizes and is deposited in the connective tissue found throughout the body. If these crystals wind up in synovial fluid, sharp joint pain can occur. Primary gout may be an inherited disease due to a metabolic defect. Excessive ingestion of protein (as in meats) can lead to high uric acid levels. Secondary gout may be a result of chronic renal disease, drugs such as Chlorothiazide (used to treat hypertension), lead poisoning, or leukemia. Although any joint may be affected, gout commonly affects the big toe! If not treated (sometimes by simply reducing protein intake), gout can lead to a chronic disability.

Although bursitis and tendonitis (tendinitis) are different disorders, they are similar due to their predisposing factors, symptoms, and treatment regimens. **Bursitis** is an inflammation of one or more bursae (the fluid filled sacs that "oil" up your joints for friction-free movement). **Tendonitis** is an often painful inflammation of the tendon itself or the tendon-muscle attachment. Even though bursitis can be caused by an infection of the bursae, both bursitis and tendonitis can be commonly caused by congenital defects, skeletal misalignment, repetitive movements, rheumatic (inflammatory) diseases, sports activities, and even our old friend, gout! Bursitis' resulting pain can present gradually or suddenly and often limits joint movement. Tendonitis commonly presents in the shoulder, producing joint pain during movement and also at night, when it can be severe enough to disturb sleep. The best "treatment" for both of these conditions is prevention through improving your strength and flexibility, warming up properly before exercise, and avoiding repeated stress and strain in sporting activities. Treatment for both includes rest of the affected joint(s), analgesics, and application of cold and moist heat. (Watch out for potential skin damage when using extreme temperatures!) The injection of corticosteroids into the affected joint may be beneficial in reducing inflammation.

A common sport injury involving the shoulder is a **torn rotator cuff**. The group of muscles that hold the head of the humerus into the shoulder socket is known as the rotator cuff. Sports such as tennis, baseball, and basketball tend to stress this region and, as a result, tears can occur in the tendons that hold those muscles in place. When these tears happen, there is a snapping sound and acute pain radiating from that area. In addition, there is difficulty abducting (raising) the affected arm. Surgery is often required to correct the tears, with rest and immobilization needed for three to four weeks. Active rehabilitation is often needed to restore full shoulder mobility and function.

Sudden and unusual movements during exercise, such as twisting an ankle, can cause a sprain. A **sprain** is an injury to the ligaments in a joint. A grade I sprain is mild. The ligaments swell and feel tender, but they are not torn. You can treat a grade I sprain at home. Pain medication such as ibuprofen or acetaminophen can help to reduce swelling and pain. The following steps (called RICE) should also be followed to treat a grade I sprain:

Rest
Ice for 48 hours
Compression of the injury with an elastic wrap
Elevation of the sprained area above the heart

A grade II sprain is more serious. A ligament is partially torn, so the joint swells and becomes very painful. When a grade II sprain happens, bruising occurs around the sprained joint and it becomes difficult to move the joint without severe pain. RICE may help a grade II sprain, but it may also need additional medical treatment.

A grade III sprain is even more serious. A ligament is completely torn, so the joint becomes impossible to move without severe pain. A grade III sprain may require a cast and physical therapy.

A hard fall or sudden blow to a joint can cause **joint dislocation**. When a joint becomes dislocated, its bones no longer line up properly. Dislocation to a joint can also damage the ligaments and nerves within the joint. It may be swollen, bruised, and very painful. A dislocated joint is an emergency that should always be treated by trained medical professionals. During treatment, a doctor may use **closed reduction**, a process of applying gentle pressure to move the joint back into place. The doctor may then use a splint or sling to immobilize the joint and allow it to heal.

arthroscopy *(are THROSS koh pee)*

Arthroscopy is a procedure used to diagnose and treat many joint injuries. Knee injuries are commonly treated with arthroscopy. An **arthroscope** is a snake-like instrument that contains a long tube, fiber optic light source, and an eyepiece. During arthroscopy, the tube of the arthroscope is inserted into the joint through a small incision in the skin. The light shines through the tube into the joint. A small television camera attached to the eyepiece projects images from within the joint onto a television screen. The projected images can help a physician to diagnose any injuries to the joint. These images can also be used to assist the surgeon during joint repair surgery.

THE SKELETON

axial skeleton
  (AK see al)

appendicular skeleton
  (app en DIK yoo lahr)

Anatomically, the skeleton can be divided into two main sections. The **axial skeleton** includes bones of the bony thorax, spinal column, hyoid bone, bones of the middle ear, and skull. This part of the skeletal system protects the organs of the body and is composed of 80 bones. The **appendicular skeleton** is, as its name implies, the region of your appendages (arms and legs), as well as the connecting bone structures of the hip and shoulder girdles, and contains 126 bones (see Figure 6-8 ■). Interestingly, nearly half of the total number of your bones can be found in your hands and feet! Your ten fingers are made up of 28 bones called **phalanges**. Two phalanges are found in each thumb. Three phalanges are found in each finger.

www.prenhall.com/colbert
Professional Profile
Radiologic Technologist is a title for a variety of health care professionals who utilize radiation to make images of bones and other body parts. Radiographers prepare patients for X-rays, take the "picture," and develop the X-ray film so physicians can read (analyze) them. For more information about these areas of health care, visit the companion website for links to the American College of Radiology and the American Society of Radiologic Technologists.

Special Regions of the Skeletal System

The skeleton consists of many different special regions. These distinctions make it easier to locate and discuss the hundreds of bones and associated components in this system. In this section, we make a quick tour of each region, beginning at the top.

THE SKULL

The skull protects and houses the brain and has openings needed for our sensory organs, such as the eyes, nose, and ears. The skull also contains the **oral cavity** or **mouth**, which is a common passageway for both the digestive and respiratory systems. The maxilla and mandible (moveable bone) are the bones of the mouth. The skull contains fibrous connective tissue joints called **suture lines** that hold the bony plates of the skull together. While these joints are not actually moveable, they do provide some degree of flexibility, which is important for

6-4 For more detailed views of the skeletal system (appendicular and axial) and drag-and-drop labeling exercises, go to your DVD for this chapter.

6-5 For an interactive drag-and-drop exercise on naming the bones of the skull, please go to your DVD.

absorbing shock from a blow to the head, thus decreasing the chance of a skull fracture. Figure 6-9 ■ shows the bones of the skull in greater detail.

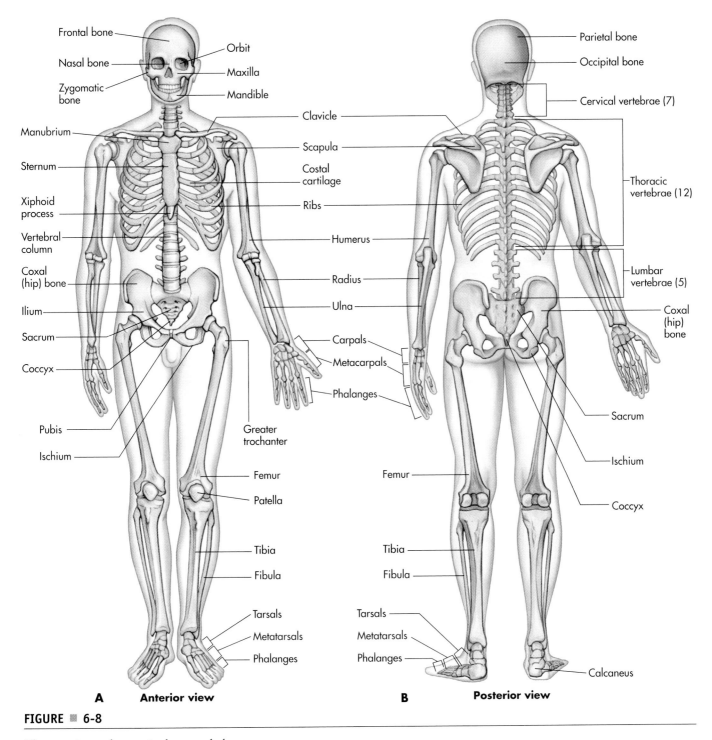

A **Anterior view**

B **Posterior view**

FIGURE ■ **6-8**

The anterior and posterior human skeleton.

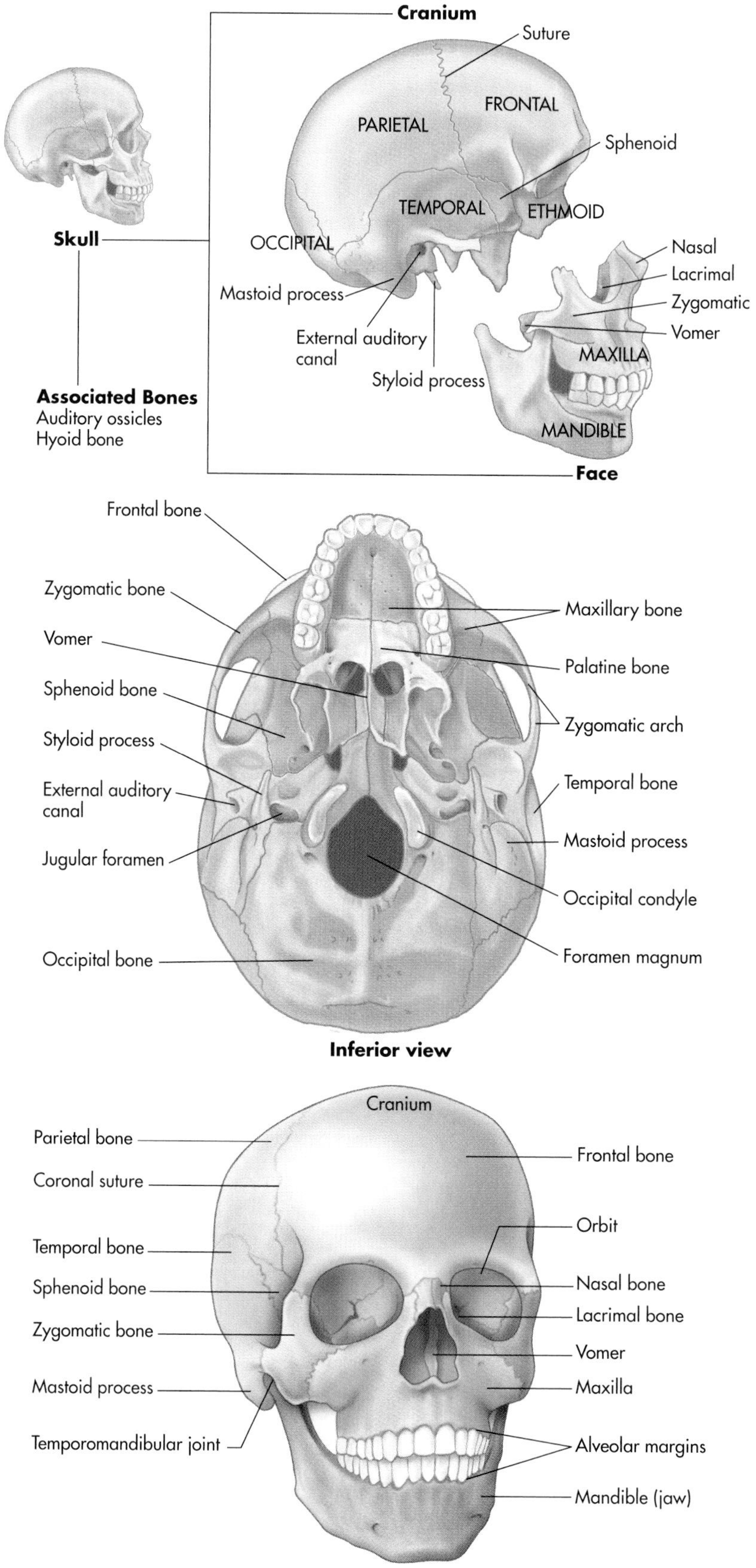

Cranium
Suture
FRONTAL
PARIETAL
Sphenoid
TEMPORAL
ETHMOID
OCCIPITAL
Nasal
Lacrimal
Zygomatic
Vomer
Mastoid process
MAXILLA
External auditory canal
Styloid process
MANDIBLE
Skull
Associated Bones
Auditory ossicles
Hyoid bone
Face

Frontal bone
Zygomatic bone
Vomer
Sphenoid bone
Styloid process
External auditory canal
Jugular foramen
Occipital bone
Maxillary bone
Palatine bone
Zygomatic arch
Temporal bone
Mastoid process
Occipital condyle
Foramen magnum

Inferior view

Cranium
Parietal bone
Coronal suture
Temporal bone
Sphenoid bone
Zygomatic bone
Mastoid process
Temporomandibular joint
Frontal bone
Orbit
Nasal bone
Lacrimal bone
Vomer
Maxilla
Alveolar margins
Mandible (jaw)

FIGURE ■ 6-9

Bones of the skull.

THE BONY THORAX

The bones of the chest form a thoracic "cage" that provides support and protection for the heart, lungs, and great blood vessels (see Figure 6-10 ▦). This cage is flexible because of cartilaginous connections that allow for movement during the process of breathing. The **sternum**, or **breastbone**, is the anatomical location for conducting compressions of the heart during cardiopulmonary resuscitation (CPR). The sternum is composed of three distinct areas. The **manubrium** is the superior portion, and the **body** is the largest, central portion. The **xiphoid** is the final and inferior portion that ossifies (hardens) by age 25 and can be broken off if CPR is improperly performed. During cardiac compressions, the heart is compressed anteriorly by the body of the sternum and posteriorly by the bones of the vertebral column.

The thoracic cage consists of 12 pairs of elastic arches of bone called ribs. The ribs are attached by cartilage to allow for their movement when we breathe. The true ribs are pairs 1 to 7 and are called **vertebrosternal** because they connect anteriorly to the sternum and posteriorly to the thoracic vertebrae of the spinal column. Pairs 8 to 10 are called the false ribs, or **vertebrocostal**, because they connect to the costal cartilage of the superior rib and again posteriorly to the thoracic vertebrae. Rib pairs 11 and 12 are called the **floating ribs** because they have no anterior attachment.

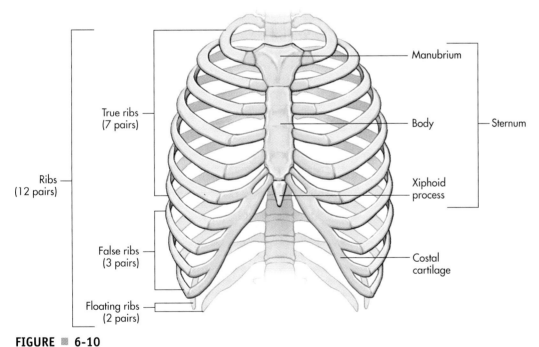

FIGURE ▦ 6-10

The bony thorax.

THE SPINAL COLUMN

The **spinal column** (**vertebral column**) protects the spinal cord which is the superhighway for information coming to and from the central nervous system. The individual bones, or vertebrae, are numbered and classified according to the body region where they are located (see Figure 6-11 ▦). For example, there are seven vertebrae found in the cervical or neck region, and they are numbered C-1 though C-7 respectively.

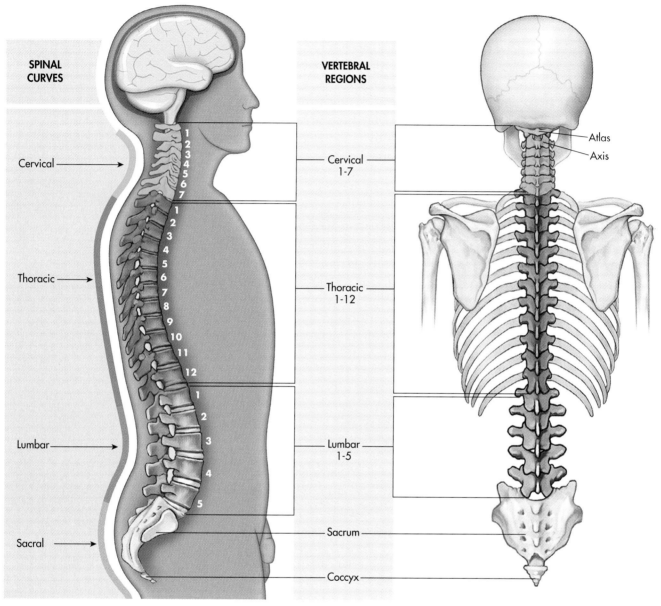

FIGURE ■ 6-11

The spinal column.

NUMBER OF VERTEBRAE

To remember the number of vertebrae in each region, think of 7 days in a week for cervical and 12 months in a year for thoracic. Finally, the lumbar region has the same number of vertebrae as digits on your hand (5).

As seen in Figure 6-11, there are seven vertebrae in the neck region (cervical) where Ray our quadriplegic is injured, twelve in the upper back (**thoracic**), five in the lower back (**lumbar**), five fused vertebrae in the mid-buttock region (**sacral**), and three to five small bones at the very end (**tailbone - coccygeal**). At birth, the vertebral column is concave to the front, like a fetal position (primary curvature) but bends in the opposite direction as the infant starts to raise and hold her head up as well as starts to walk. In other words, there will be secondary curvatures by the time a child is 2. From 2 years onward, the vertebral column develops a secondary curvature in the neck, a primary curvature in the upper back, a secondary curvature in the lower back, and a primary curvature in the mid-buttocks and tailbone regions.

Pathology Connection: Spinal Column Abnormalities

A "slipped disk" or, more correctly, a **herniated disk**, is a common disorder found in individuals less than 45 years of age due to poor posture and/or poor lifting techniques, and in the geriatric population due to degenerative joint disease. The vertebral column is separated by pads of cartilage called **invertebral disks**. Herniation occurs when the soft central part of the intervertebral disk is forced through the outer covering. The herniated portion then may press on a spinal nerve root, causing lower back pain. Sometimes this pain will radiate down the sciatic nerve to the buttock, leg, and foot on one side of the body. Although this condition can occur anywhere along the spine, such as the cervical area which produces a tingling of the arm on the affected side, the lower back region appears to be the most common region. Treatment for this condition can include rest, application of heat, special exercises, analgesics and muscle relaxants, proper lifting techniques training, and in severe cases, surgery.

Another potential spinal problem involves the spine's curvature. If the body is not in balance, whether due to congenital deformity, trauma, poor posture, or disease, these curvatures may be exaggerated, leading to a number of curvature abnormalities. **Kyphosis** is a condition in which the upper portion of the spine exhibits a posterior curve, producing what is commonly called a humpback or a hunchback. **Scoliosis** is a lateral (left or right) curvature of the spine. **Lordosis** is an anterior curve of the lower back and is commonly called swayback. Figure 6-12 ▪ illustrates these conditions. It is important to note that these conditions present at various levels of severity; some individuals may go through life never knowing that they have an abnormal curvature of the spine. The key to successful treatment is early detection. Spinal corrections can be accomplished through exercise, good nutrition with weight control, bracing, corrective shoes (so legs are of equal length), or surgery.

UPPER AND LOWER EXTREMITIES

The appendicular region consists of the arms and legs. Since these areas perform most of the body movement, the greatest number of sport-related injuries occur here. Please see Figure 6-13 ▪, which shows the bones of the upper and lower extremities. These figures show the bony landmarks of the limbs and girdles, which will be beneficial in learning and understanding where the muscles discussed in Chapter 7 will attach.

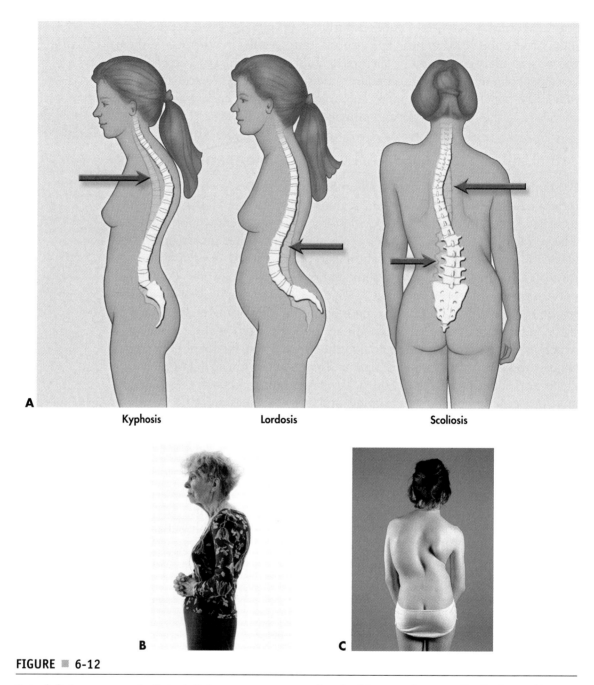

Kyphosis   Lordosis   Scoliosis

FIGURE ■ 6-12

Spinal disfigurements. **A.** Spinal disfigurements compared to healthy spinal curves. **B.** Kyphosis. (Source: Phototake NYC.) **C.** Scoliosis. (Source: Photo Researchers, Inc.)

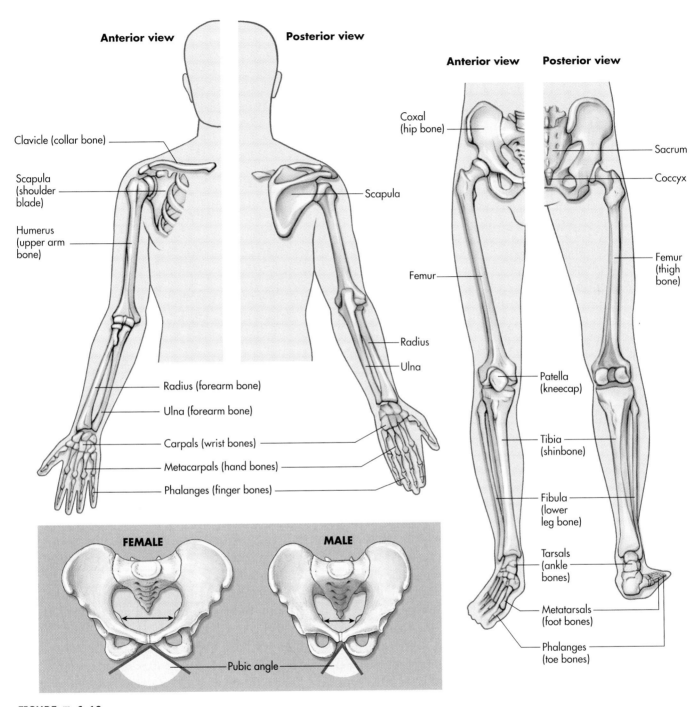

FIGURE ■ 6-13

Bones of the upper and lower extremities.

Notice that the pelvic girdle in women is different than in men. Women have a greater pubic angle, which facilitates childbirth, and also a relatively broad girdle to support the extra weight of the child. This difference can be used to identify the sex of a skeleton, such as in a murder case or an archeological find. The pelvis consists of several parts, including the ilium, ischium, and pubis as shown in Figure 6-8.

TEST YOUR KNOWLEDGE 6-3

Answer the following:

1. What is the difference between the axial skeleton and the appendicular skeleton?

   _____

   _____

2. Which of the following bones is considered to be part of the axial skeleton?
   a. humerus
   b. patella
   c. femur
   d. sternum

3. The number of vertebra in the thoracic region is:
   a. 5
   b. 7
   c. 12
   d. 120

4. Describe the difference between flexion and extension.

   _____

   _____

   _____

5. This form of arthritis is a result of an excessive build-up of uric acid.

   _____

 ### Pathology Connection: Bone Fractures and Healing

Chances are that you have broken a bone or know someone who has broken a bone in the past, but did you know that not all breaks are the same? A **fracture** is any break in a bone. Although there are a variety of fractures, the following are the more common ones. Please refer to the illustrations for further clarification.

A **hairline** fracture, which looks like a piece of hair on the X-ray, is a fine fracture that does not completely break or displace the bone. A **simple** or **closed** fracture is a break without a puncture to the skin. An individual in an accident who has a bone that is severely twisted may receive a **spiral** fracture. **Greenstick** fractures are incomplete breaks, which more often occur in children because they have softer, more pliable bones (like sapling branches) than adults (like seasoned twigs). If a bone is crushed to the point that it becomes fragmented or splintered, that is classified as a **comminuted** fracture. A fracture in which the bone is pushed through the skin is referred to as a **compound** or **open** fracture. These fractures are particularly nasty because deep tissue has the potential to be exposed to bacteria once the bone is set into place, and, hence, the chance for infection in addition to the break is increased. See Figure 6-14 ▪ for examples of common fractures.

Bones take several weeks to heal. They can heal normally only if the ends of the bones are touching. If the bones are not touching (are poorly aligned), the bone must be **set (reduced)**. If the misalignment is slight, closed reduction may be sufficient. During a **closed reduction**, force is exerted on the broken bone to bring the bones into alignment. In more severe fractures, **surgical** or **open reduction** is necessary. During a surgical reduction, pins, screws, plates, or other hardware are used to fix the bones in place. In either case, the broken bone must often be immobilized to keep the ends of the bone in proper alignment. A cast or splint is applied to immobilize the bone.

Traction may be used to treat fractures of long bones. Traction aligns broken bones and holds them in place. It uses heavy weights to exert a pulling force on the injured bones. The force from the weights keeps the broken bones in the correct position and allows them to heal normally

 6-6 For an animation on how bones heal, visit your DVD for this chapter.

In the first few hours after the injury, a **hematoma**, or **blood clot**, forms around the broken bone and inflammation sets in. For the next 3-4 weeks, a soft callus forms, replacing the hematoma and bridging the gap between the broken ends of the bone. A soft callus starts as hyaline cartilage ,which eventually has some bone cells within it. Capillaries invade the site, ready to vascularize the healing bone. From week 4 to week 12 after the injury, a bony callus forms, replacing the soft callus via endochondral ossification. (Remember endochondral ossification? It's the same process that causes bone development in embryos and allows your long bones to get longer after birth.) Usually the bony callus will contain excess bone. Remodeling is necessary after bony callus formation to make the repaired bone match the rest of the bone.

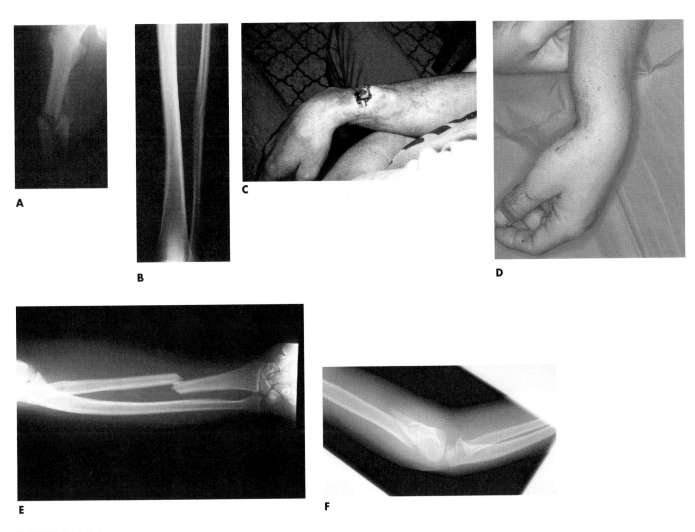

FIGURE ▦ 6-14

A. Femur, AP view, comminuted fracture. **B.** Tibia, simple, transverse fracture. **C.** Open fracture of the wrist. **D.** Displaced fracture of the distal radius. (Source: Charles Stewart & Associates.) **E.** X-ray of complete fracture of the radius. (Source: James Stevenson/Science Photo Library/Photo Researchers, Inc.) **F.** Fractured humerus. (Source: Charles Stewart & Associates.)

MAINTAINING GOOD BONE HEALTH

Even though bone mass loss is a natural process of aging, it can be slowed by a healthy lifestyle. It is important to consume the proper amount of dietary calcium to build strong bones in the first place. Proper calcium intake during the formative years, including the teenage years, and continued calcium consumption as the body ages, is crucial. Vitamin D is important because it allows your body to absorb ingested calcium from the digestive tract. As previously discussed, exercise (especially weight-bearing forms) also plays a vital role in developing and maintaining bones, so stay active. And, surprisingly, the excessive use of caffeine and cigarette smoking can reduce bone density. Individuals who, for a lifetime, consume two cups of black caffeinated coffee per day tend to show an increased bone loss, while cigarette smokers have shown a loss of 5% to 8% bone mineral density.

While osteoporosis and arthritis are big concerns, there are many more potential disorders of bones and joints. As you can see in Tables 6-2 and 6-3, these disorders can generally be classified by the following causative agents: congenital, degenerative, nutritional, secondary disorders, infection, inflammation, trauma, and tumors.

Clinical Application

WHEN TO SEE A DOCTOR

We have discussed a lot of pathological conditions relating to bones and the skeleton in this chapter, but when do you need to see a doctor and when do you take care of things yourself? Anyone who has been involved in sports or physical labor knows all about the bumps, bruises, and strains that go with those activities. So, in general, when should you see a doctor? Here are a few general guidelines:

- Any obvious bone deformity or misalignment of a joint
- Pain that continues beyond 10 days
- An injury that does not improve after 5 to 7 days
- Continued swelling and/or inflammation of a joint or injury
- Any indication of an infection such as a fever, pus, red streaks visible in tissue, or the swelling of lymph glands

Of course, these are guidelines, always use common sense to err on the safe side.

osteomalacia
(OSS tee oh mah LAY she ah)
  malacia = *softening*
chondrosarcoma
(KON droe sar KOE ma)
  chondr/o = *cartilage*
  oma = *tumor*
  sarcoma = *tumor of connective tissue*

TABLE 6-2 Bone Disorders

CLASSIFICATION	EXAMPLE(S)
Congenital disorders	Abnormal curvature of the spine (kyphosis, lordosis, scoliosis), cleft palate, clubfoot, osteogenesis imperfecta, achondroplasia
Degenerative disorders	Osteoporosis
Infection	Osteomyelitis
Nutritional disorders	Osteomalacia (vitamin D deficiency), rickets (vitamin D deficiency), scurvy (vitamin C deficiency)
Secondary disorders	Endocrine system dysfunction: gigantism, pituitary dwarfism
Trauma	Bruises, fractures
Tumors	Chondrosarcomas, myelomas, osteosarcomas

TABLE 6-3 Joint Disorders

CLASSIFICATION	EXAMPLE(S)
Degenerative disorders	Osteoarthritis
Infection	Gonococcal arthritis, rheumatic fever, septic arthritis, viral arthritis
Inflammation	Bursitis, arthritis
Secondary disorders	Immune system dysfunction: rheumatoid arthritis; metabolic dysfunction: gout
Trauma	Ankle and foot injuries, dislocations, hip fractures, knee injuries

6-7 For more information about bone diseases, go to your DVD, which features videos concerning arthritis and osteoporosis.

See the following Quick Trip table for information on the different forms of arthritis as well as other diseases of the skeletal system.

A Quick Trip Through Some Diseases of the Skeletal System

DISEASE	ETIOLOGY	SIGNS AND SYMPTOMS	DIAGNOSTIC TEST(S)	TREATMENT
Arthritis	A general term for the inflammation of a joint / synovial membrane.			
Osteoarthritis	A degenerative condition in which a joint wears out, often a result of simple aging, sports injuries, trauma, obesity or a family history.	Painful swelling of the affected joint, wearing away of the articular cartilage, joint develops a crooked deformation.	Visual and radiologic examination, fluid culture for infective agents, patient history.	Rest, exercises (non-weight-bearing), analgesics, anti-inflammatories, steroid injections into affected site, surgical intervention/joint replacement.

continued

A Quick Trip Through Some Diseases of the Skeletal System (*Continued*)

DISEASE	ETIOLOGY	SIGNS AND SYMPTOMS	DIAGNOSTIC TEST(S)	TREATMENT
Rheumatoid Arthritis	An insidious, chronic, systemic, autoimmune disease affecting connective tissue in the body, and mostly the small joints of the hands and feet, can also affect the elbows and knees. Three times as many women as men develop this condition.	Stiffness, swelling, and fluctuating pain of multiple joints (usually symmetrical), pronounced joint deformities and destruction, synovial membrane inflammation, fibrous adhesions obliterate the joint spaces, other systemic effects involving connective tissue.	Visual and radiologic examination, fluid culture for infective agents, patient history, antibody screening (rheumatoid factor).	Aspirin (effective only for patients with mild disease) non-steroidal anti-inflammatory drugs, corticosteroids, the cancer drug Methotrexate in low doses, drugs that suppress the immune system specifically for autoimmune disorders (such as Humira and Enbrel) rest, range of motion exercises, surgical intervention (in advanced cases).
Septic Arthritis	Originates from an infective organism either from a penetrating joint wound or blood-borne pathogen.	Pain, swelling of affected joint, filling of the joint space with an inflammatory exudate destroying the joint and replacing the space initially with fibrous tissue and then bone.	Visual examination and imaging, fluid culture for infective agents, patient history.	Preventative aseptic techniques, aggressive, early as possible treatment with antibiotics.
Bursitis	Repetitive movement, strain, congenital defect, rheumatic diseases.	Pain on movement, painful inflammation of affected site, swelling, limited range of motion due to pain.	Visual examination and imaging, patient history.	Joint rest, moist heat / cold therapy (although cold may aggravate the situation), analgesics, anti-inflammatory medications, corticosteroid injection at the affected site, draining if severe. Range of motion exercises once pain subsides to restore full joint function.
Cruciate ligament tears	Trauma such as twisting of the leg, anterior or posterior blow when leg is planted or bearing weight. A common contact sport injury.	Pain, limited stability and mobility.	Physical examination and imaging, joint stability testing.	Rest, immobilization, surgical intervention.

A Quick Trip Through Some Diseases of the Skeletal System (*Continued*)

DISEASE	ETIOLOGY	SIGNS AND SYMPTOMS	DIAGNOSTIC TEST(S)	TREATMENT
Gout	A metabolic disease with joint involvement. Overproduction of uric acid, leading to crystallization and deposition of the excessive uric acid in synovial fluid. Excessive protein ingestion precipitates the condition. Possibly an inherited disease. 95% of cases are male; onset is usually after 30 years of age. Secondary gout may be caused by leukemia, lead poisoning, chronic renal diseases, or drug ingestion such as Chlorothiazide. Considered a form of arthritis by some.	Excruciating pain in a joint, commonly the big toe. Involved joint is inflamed, hot, tender, and in more severe cases, swollen. In advanced stages: hypertension, kidney stones, renal dysfunction, severe back pain, disability, ulceration, and unremitting pain.	Visual examination, blood testing for excessive uric acid.	Dietary restrictions of protein consumption, especially organ meats such as liver and seafood (high in purines that are precursors to uric acid). Rest and immobilization of the affected joint, blood uric acid monitoring, analgesics, anti-inflammatory medications, and drugs designed specifically for gout management such as colchicine (utilized for past 1,500 years!), probenecid, and allopurinol. Low-fat dairy products seem to help prevent attacks.
Kyphosis	Causative agents include: congenital defect, trauma, diseases (e.g. polio). In adults: aging process, osteoporosis, tuberculosis, steroid therapy, arthritis, and tumor growth.	An exaggerated curve of the upper back (commonly called "humpback"). May lead to backache, fatigue, dyspnea, tenderness, muscle weakness, pulmonary insufficiency.	Visual examination and imaging.	Age and severity dependant: exercise, bracing, surgery, electrical stimulation, weight control.
Osteomalacia	Decreased or impaired mineralization of bone material due to insufficient vitamin D. May be due to improper diet, lack of sufficient sunlight, or a malabsorption condition.	Bone pain, loss of height, deformity of weight bearing bones.	Bone scan, visual examination.	Correct the nutritional deficiency.

continued

A Quick Trip Through Some Diseases of the Skeletal System (*Continued*)

DISEASE	ETIOLOGY	SIGNS AND SYMPTOMS	DIAGNOSTIC TEST(S)	TREATMENT
Osteomyelitis	An acute or chronic inflammatory condition of the bone, commonly caused by the bacteria *Staphylococcus aureus*. The infection usually originated from a wound, the skin or throat. More common in children than adults, more common in boys than girls. In adults, may be a result of trauma, surgery (especially if any bone had been manipulated).	In children: most commonly occurs in the long bones and may affect the growth plates. In adults: most commonly occurs in the vertebrae and pelvis. Begins with sudden pain, swelling, heat, and tenderness of the affected site, high fever, chills. Nausea and malaise are also possible.	Visual examination, culture for suspected infecting agent.	Preventive therapy: decrease the possibility of infection. Maintain a healthy immune system. Antibiotics (aggressive therapy), possible surgical debridement.
Osteoporosis	A metabolic problem with slow, progressive bone implications that lead to decreased bone mass. Usually a condition related to aging but may be related to diseases that decrease mobility. Also found in postmenopausal and estrogen-deficient women, regardless of age.	Decreasing bone density, compression fractures of the spine, wrist fractures, height loss up to 4 – 5 inches, thoracic and lumbar spine pain radiating around the trunk, kyphosis, decreased thoracic cavity size with decreasing exercise tolerance. In some cases, bone breaks that were attributed to falls actually occurred first and then the fall.	Imaging, bone densitometry.	Early initiation of preventative practices such as diet with sufficient calcium. Smoking and caffeine may aid in depleting calcium. Regular (daily is ideal) exercise, especially weight-bearing and impact sports / exercises, supplements (calcium, vitamin D, fluoride), drugs that increase bone mass such as alendronate. Estrogen replacement is somewhat controversial but may have a place.
Plantar fasciitis (runner's heel)	Repetitive impact trauma to the heel affecting the thick and fibrous connective tissue that extends the length of the foot beginning at the heel. Additional pain if pre-existing heel spur is involved. Predisposing factors can include: sudden increases in activity, increase in body weight, high arches, poor supportive shoes, and flat feet.	Intermittent pain that worsens as one walks following sitting or standing for a period of time.	Radiologic examination.	Rest, analgesics, anti-inflammatories, application of ice, utilization of padding for the heel, surgical intervention in some cases.

A Quick Trip Through Some Diseases of the Skeletal System (*Continued*)

DISEASE	ETIOLOGY	SIGNS AND SYMPTOMS	DIAGNOSTIC TEST(S)	TREATMENT
Rickets	A pediatric condition caused by the lack of vitamin D, leading to impaired calcium absorption and bone calcification. In adults, a deficiency of vitamin D may lead to bone softening (osteomalacia).	Weak, deformed bones.	Visual and radiology examination, patient history.	Dietary correction to increase vitamin D, supplemental vitamin D, sunlight exposure (which increases vitamin D production).
Scoliosis	Congenital defect, trauma, poor posture, obesity.	A lateral curvature of the spine, usually affecting the chest or lower back regions. Symptoms include backache, fatigue, dyspnea, pulmonary insufficiency, and sciatica. Symptoms are dependent upon the age of patient, severity of the condition, and timeliness of the treatment.	Visual and radiologic examination.	Age and severity dependant: exercise, bracing, surgery, electrical stimulation, weight control.
Tendonitis (tendinitis)	Repetitive movement or calcium deposits at the affected site, rheumatic diseases, congenital defects. Often occurs in conjunction with bursitis.	Inflammation of the involved tendon and/or the connective tissue that attaches that tendon to the muscle or bone. Pain on movement and localized pain especially at night.	Patient history, visual and radiologic examination.	Analgesics, joint rest, cold/moist heat therapy, corticosteroid injection at affected site, other anti-inflammatory medications.

℞ Pharmacology Corner

There are several medication options when treating joint pain. Analgesic creams often containing capsaicin can help to ease minor joint pain. For more severe cases, corticosteroids are injected directly into the joints. Pain medication can include non-steroidal anti-inflammatory drugs (NSAIDs) such as naproxen and ibuprofen that reduce the inflammation and thus reduce the pain and swelling. Pain medication such as acetaminophen and aspirin can help to ease the pain in general. The cancer drug Methotrexate in smaller doses has been used to treat rheumatoid arthritis, as has cyclosporine (an anti-rejection drug) and several new classes of immunosuppressants such as Remicade, Humira, and Enbrel (see Chapter 14 for more details).

In addition to proper diet and exercise, bone density in diseases such as osteoporosis can also be improved with calcium supplements. Interestingly some calcium based antacids used for heartburn can help improve bone density. Alendronate is a more powerful prescribed calcium medication that can be taken once a month. Vitamin D supplements help bone health by improving the body's ability to absorb calcium. Because Vitamin D is a fat soluble vitamin, care must be taken to not overdose or it may cause liver damage. Estrogen replacement therapy, while controversial, does seem to improve bone density in women.

SUMMARY

Snapshots from the Journey

→ In addition to providing support and protection for the body, the skeleton also produces blood cells and acts as a storage unit for minerals and fat.

→ The 206 bones of the skeleton can be classified according to their shapes: long bones, short bones, flat bones, and irregular bones.

→ Bone is covered with periosteum, which is a tough, fibrous connective tissue. In long bones, each bone end is called an epiphysis, and the shaft is called the diaphysis. The hollow region within the diaphysis is called the medullary cavity and stores yellow marrow.

→ Compact bone is a dense, hard tissue that normally forms the shafts of long bones or is found as the outer layer of the other bone types. Spongy bone is different in that it contains irregular holes that make it lighter in weight and provides a space for red bone marrow, which produces red blood cells.

→ Ossification is the formation of bone in the body. Osteoprogenitor cells are non-specialized cells that can turn into osteoblasts, which are the cells that actually form bones. Osteocytes are mature bone cells that were originally osteoblasts. Osteoclasts originate from a type of white blood cell called a monocyte, found in red bone marrow. Osteoclasts break down bone material and help move calcium and phosphate into the blood.

→ A thin band of cartilage forms an epiphyseal plate (often referred to as the growth plate), and as long as it exists, the length and width of the bone will increase.

→ Cartilage is a special form of dense connective tissue that can withstand a fair amount of flexing, tension, and pressure and makes a flexible connection between bones, as between the breastbone and ribs. It also acts as a cushion between bones.

→ Various types of joints join two or more bones and provide various types of movement. The point at which they join is called an articulation. Ligaments are very tough, whitish bands of connective tissue that connect from bone to bone to hold the joint together and can withstand heavy stress.

→ With the exception of osteoporosis, most chronic diseases of the skeleton involve the joints.

→ The skeleton can be divided into two main sections. The axial skeleton includes bones of the bony thorax, spinal column, hyoid bone, bones of the middle ear, and skull. The appendicular skeleton is the region of your appendages (arms and legs) as well as the connecting bone structures of the hip and shoulder girdles.

→ Although they are different diseases, bursitis and tendonitis have similar treatment regimens that include rest of the affected joint(s), analgesics, and application of cold and moist heat. (Watch out for potential skin damage when using extreme temperatures!) The injection of corticosteroids into the affected joint may be beneficial in reducing inflammation.

→ As we age, the chemical composition of cartilage changes, causing it to become more brittle. Articular cartilage that ages or becomes injured can lead to arthritis, which is an inflammatory process of the joint or joints. Gout is a form of arthritis and is a result of higher than normal levels of urea in the body. Bone mass also gradually decreases with age, beginning in a person's 50s. Even though this is a natural process of aging, it can be slowed by a healthy lifestyle.

→ Although bumps and bruises are a fact of life and often not a big deal, the general guideline concerning when to see a doctor include: any obvious bone deformity or misalignment of a joint, pain that continues beyond 10 days, an injury that does not improve after 5 to 7 days, continued swelling and / or inflammation of a joint or injury, any indication of an infection such as a fever, pus, red streaks visible in tissue, or the swelling of lymph glands and of course, common sense.

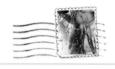

Case Study

A somewhat frail 76-year-old female visits her physician's office for an annual check-up. Her social history shows she smokes a pack of cigarettes a day and she is a heavy coffee drinker. She has had several fractured bones in the last five years that required medical attention. During initial examination, measurements show that the patient has lost approximately an inch of height over the past year. She has also lost several pounds but states she still wears the same size clothes.

a. What possible bone disease do you think she is exhibiting?

_____

_____

_____

b. Describe the bone changes in this condition on a macro and cellular level.

_____

_____

_____

c. What treatments and/or lifestyle changes would you suggest?

_____

_____

_____

RAY'S STORY

You are a member of the discharge planning team responsible for developing a home care plan for our quadriplegic, Ray. Thinking back to Ray's condition, remember that he can only breathe with mechanical assistance, cannot move himself without help from others, needs assistance eating, and spends most of his time lying in a bed. What suggestions would you make for his skeletal, bone, and joint health? Which profession(s) should be involved in maintaining or improving his "skeletal health"?

_____

_____

_____

_____

REVIEW QUESTIONS

Multiple Choice

1. Your elbow is an example of which type of joint?
   a. hinge joint
   b. ball-and-socket joint
   c. gliding joint
   d. fibrous joint

2. The sternum is the correct medical term for which bone?
   a. shin bone
   b. breastbone
   c. shoulder blade
   d. collar bone

3. The end of a long bone is the:
   a. diplodicus
   b. epiphysis
   c. condylcorn
   d. perla

4. As long as this exists, your bones will increase in length and width:
   a. Torger center
   b. ossifier
   c. Mantoux membrane
   d. epiphyseal plate

5. The aging process, excessive caffeine, and cigarette smoking can each contribute to this bone disease:
   a. rheumatoid arthritis
   b. osteoporosis
   c. scoliosis
   d. de-ossification

6. This group of muscles is responsible for holding the head of the humerus in the shoulder joint socket:
   a. bundle branch
   b. rotator cuff
   c. ball flexor
   d. ossiforator

7. A "slipped disk" is more correctly called a:
   a. displaced disk
   b. degenerated disk
   c. herniated disk
   d. infarcted disk

Fill in the Blank

1. Name three large appendicular bones: _____, _____, and _____.

2. List three places where cartilage is found in the body: _____, _____, and _____.

3. _____ is a liquid found in joints that keeps them lubricated.

4. The specialized cells that constantly rebuild bone are called _____.

5. These specialized cells are needed to tear down bone _____.

6. The condition of a lateral curvature of the spine is called _____.

7. _____ is an inflammation of the sac(s) that provide(s) the lubricating fluid for the joints in the body.

Short Answer

1. Describe the difference in function between tendons and ligaments.

_____

_____

_____

_____

2. List three functions of the skeletal system.

_____

_____

_____

_____

3. What happens to our skeletal system as we age?

_____

_____

_____

_____

4. What are the functions of cartilage?

_____

_____

_____

_____

5. Discuss the general guidelines used to help determine when you should see a doctor following a bone or joint injury:

_____

_____

_____

_____

Suggested Activities

1. Using art paper, cut out, label, and assemble bones to create a full-sized skeleton.

2. Act out various range-of-motion exercises and perform and identify different joint movements.

3. Visit the library and research how the human skeleton has evolved, also noting the differences between male and female skeletons, posture changes, and changing shapes of various bones.

 6-8 Now that you have completed your journey through this chapter, please go to the DVD for interactive games and puzzles concerning the medical terms and concepts contained in this chapter. By playing the games you will reinforce your learning of the skeletal system in a fun way.

T H E
MUSCULAR
System

Movement for the Journey

As we continue our journey of exploration, we obviously need a transportation method to reach our destination. We can go by a plane, train, or automobile. However, no matter what transportation system we use, we must utilize the body's muscular system to get to the vehicle. While the skeletal system provides the framework for the human body, the body also needs a system that allows movement, or locomotion, which is the job of the muscular system. We are most familiar with external movements that use our external muscles to walk, run, or lift objects. However, internal movement within the body is necessary to transport food, air, waste products, and body fluids such as blood. For example, if you drink some bad water on our journey, the smooth muscles in your digestive tract will rapidly pass it through your system to be expelled in the form of urgent diarrhea. Different types of specialized muscles within the muscular system allow for both external and internal movement. This chapter defines and contrasts the different muscle types needed for external and internal body movement.

Chapter 7

LEARNING OBJECTIVES

At the completion of your journey through this chapter, you will be able to:

→ Differentiate the three major muscle types

→ Discuss the functions of tendons and ligaments

→ Explain the difference between voluntary and involuntary muscles

→ Describe the various types of skeletal muscle movement

→ Identify and explain the components of a muscle cell

→ Describe the chemical activities required for muscle movement

→ Describe the process of neuromuscular transmission

→ Contrast the activity of cardiac, smooth, and skeletal muscles

→ Discuss disorders of the muscular system

MULTIMEDIA APPLICATIONS

DVD Interactive Exercises

→ Interactive drag-and-drop exercise on locating the anterior and posterior muscles of the body, 7-1

→ Videos of various massage therapy techniques, 7-2

→ Interactive 3-D visual tour of labeled muscle groups of specified regions: head and neck; upper limb, forearm, and hand; shoulder and arm; trunk and abdomen; pelvis, hip and thigh; lower limb and foot, 7-3

→ Video on Intramuscular (IM) injection techniques, 7-4

→ Interactive drag-and-drop exercise of myofibril structures, 7-5

→ Videos of muscle atrophy and muscular dystrophy, 7-6

→ Animation of cellular muscle contraction, 7-7

→ Interactive games and puzzles, 7-8

www.prenhall.com/colbert

→ Professional Profiles:
- Kinesiology
- Physical Therapy
- Occupational Therapy
- Massage Therapy

→ Related Internet Links

Pronunciation Guide

 Correct pronunciation is important in any journey so that you and others are completely understood. Here is a "see and say" Pronunciation Guide for the more difficult terms to pronounce in this chapter. In addition, your DVD contains an audio pronunciation guide.

acetylcholine (ah SEET ul KOE leen)

actin (ak TIN)

adenosine triphosphate
   (ah DEN oh sin)

ataxia (ah TAK see ah)

atrophy (AT roh fee)

diaphragm (DYE ah fram)

electromyography
   (ee LEK troh my OG rah fee)

fibromyalgia (fie bro my AL jeuh)

flaccid (FLAS sid)

flexion (FLEK shun)

glycogen (GLIE co jin)

Guillian-Barré syndrome
   (Gey yan bar RAY)

hypertrophy (high PER troh fee)

intercalated discs (in TER cow LATE ed)

muscular dystrophy
   (MUSS kyoo lahr DISS troh fee)

myalgia (my AL jee ah)

myasthenia gravis (my as THEE nee ah)

myofibril (my oh FIE bril)

myosin (MY oh sin)

rigor mortis (RIG or MORE tiss)

sarcomeres (SAR koh mearz)

sphincters (SFING ters)

tetanus (TET ah nus)

tonus (TONE us)

OVERVIEW OF THE MUSCULAR SYSTEM

Because of the numerous functions they must perform, muscles come in many shapes and sizes. The structure of the muscle matches its function, as you will shortly see.

Types of Muscles

contractile tissue = *tissue that has the ability to contract, draw together, or shorten*

Muscle is a general term for all **contractile tissue**. The term muscle comes from the Latin word *mus*, which means "mouse," because the movement of muscles looks like mice running around under our skin. The contractile property of muscle tissue allows it to become short and thick in response to a nerve impulse and then to relax back to its original length once that impulse is removed. This alternate contraction and relaxation is what causes movement. Muscle cells are elongated and resemble fibers such as those in rope. Muscle tissue is constructed of bundles of these muscle fibers. These fibers are approximately the diameter of human hair. Under the direction of the nervous system, all of the muscles provide for motion of some type for your body.

The body has three major types of muscles: **skeletal, smooth,** and **cardiac.** Let's begin with a general description and comparison of these three muscle types and then get more specific about each type.

Skeletal muscles are **voluntary** muscles, or under conscious control, and derive their name from the attachment of the bones to the skeletal system. The fibers in skeletal muscles appear to be striped and are therefore called **striated** (striped) muscle. These muscles allow us to perform external movements—running, lifting, or scratching, for example. These are the muscles we try to develop through exercise and sports and also so we look good at the beach.

Unlike skeletal muscle, smooth muscle is **involuntary** and not under our conscious control. It is also called **smooth muscle** because it does not have the striped appearance of skeletal muscles. This involuntary muscle is found within certain organs, blood vessels, and airways. Because it is the muscle of organs, it is sometimes called **visceral muscle**. Smooth muscle allows for the internal movement of food (peristalsis) in the case of the stomach and other digestive organs. Smooth muscle also facilitates the movement of blood by changing the diameter of the blood vessels (vasoconstriction and vasodilation), and also the movement of air by changing the diameter of the airways found in our lungs.

The third type of muscle is the specialized **cardiac muscle**, which has a striated appearance. This muscle type is found solely in the heart. It makes up the walls of the heart and causes the heart to contract. These contractions cause the internal movement (circulation) of blood within the body. Fortunately, cardiac muscle, like smooth muscle, is an involuntary muscle. Imagine if we had to think each time in order for our heart to beat. Figure 7-1 ▪ contrasts the three types of muscles found within the body.

All muscle has certain characteristics in common. One of these is the ability to be stretched. This is called extensibility. Think of yoga instructors and how flexible they can be to pose in yoga positions. It is fortunate for muscles to have extensibility to some degree. For example, if you swallow a large piece of food by accident, your visceral or smooth muscle in your esophagus and the sphincter muscle at the top of your stomach will stretch to let it pass.

We will now explore each of these types of muscles in further depth.

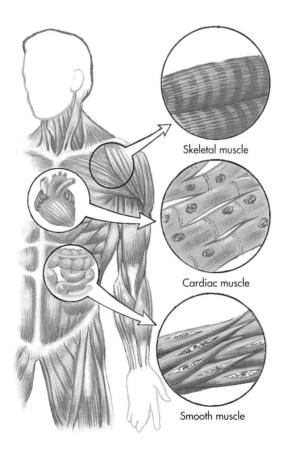

Skeletal muscle

Cardiac muscle

Smooth muscle

FIGURE ▪ **7-1**

The three types of muscle: skeletal, cardiac, and smooth.

hypertrophy
   hyper = *greater than normal*
   trophy = *growth or development*
flaccid *(FLAS sid)*
atrophy *(AT roh fee)*
   a = *without, lack of*

Clinical Application

MUSCLE TONE

Have you ever had a cast on for an extended period of time? When it is removed, the arm or leg is much smaller and weaker than the limb without the cast. Why does this occur? Normally, all muscles exhibit muscle tone (**tonus**). Tonus is the partial contraction of a muscle with a resistance to stretching. Athletes who exercise regularly have increased muscle tone, making their muscles more pronounced, and also increased flexibility due to muscle stretching. The muscle fibers in an athlete increase in diameter (**hypertrophy**) and become stronger. Hypertrophy refers to increased growth or development. When muscles are not used, they begin to lose their tone and become **flaccid** (soft and flabby). For example, if a patient is required to remain in bed (bedfast) for an extended period of time, his or her muscles waste away (**atrophy**) from the lack of use. One of the reasons patients get out of bed as soon as possible is to prevent atrophy from occurring. Patients who are confined to bed, like our quadriplegic patient Ray, often have problems with muscle atrophy. If skeletal muscle is damaged, it can regenerate itself. However, if the damage is extensive, then a scar forms.

Pathology Connection: Myopathy

Myopathy is the general term for muscle disease or disorders. There are many causes of myopathy including injury, genetics, nervous system disorders, medication, and cellular abnormalities. Symptoms of myopathy include weakness, cramping, stiffness, and spasm. Treatment for myopathy depends on the cause. Specific myopathies will be discussed at appropriate places in the chapter.

TEST YOUR KNOWLEDGE 7-1

Choose the best answer:

1. The biceps muscle is an example of a:
   a. smooth muscle
   b. cardiac muscle
   c. skeletal muscle
   d. dinosaur muscle

2. Smooth muscle is found in all the following *except:*
   a. airways
   b. digestive system
   c. blood vessels
   d. heart

3. Which types of muscles are striated?
   a. smooth and cardiac
   b. cardiac and skeletal
   c. skeletal and smooth
   d. smooth only

4. A decrease in the size of a muscle due to disuse or disease is called:
   a. myopathy
   b. hypertrophy
   c. atrophy
   d. amyotrophy

Skeletal Muscles

Skeletal muscles are attached to bones of the skeleton and provide movement for your body. As you learned in Chapter 6, "The Skeletal System," **tendons** are fibrous tissues that attach skeletal muscle to bones, and **ligaments** attach bone to bone. Some muscles can attach to a bone or soft tissue *without* a tendon. Such muscles use a broad sheet of connective tissue called an **aponeurosis**. This type of connection is found, for example, in some facial muscles.

Skeletal muscle is also known as voluntary muscle because we can control its movement by conscious thought. The numerous skeletal muscles found throughout the body are responsible for movement, giving the body its shape and form, maintaining our body posture, and heat generation. See Figure 7-2 , which shows some of the major muscles found in the human body.

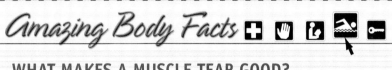

MUSCLES

- Muscles make up almost half the weight of the entire body.

- There are 650 different muscles in the human body.

- The size of your muscles depends on how much you use them and how big you are. Ice skaters, for example, have large leg muscles.

- The little muscles around the eye can contract 100,000 times a day.

- Individual elongated muscle cells can be up to 12 inches, or 30 centimeters, in length.

- At about the age of 40, the number and diameter of muscle fibers begin to decrease, and by age 80, 50% of the muscle mass may be lost. Exercise and good nutrition helps to decrease this loss.

7-1 To reinforce your learning, please visit your DVD for a drag-and-drop exercise on locating the anterior and posterior muscles of the body.

Pathology Connection: Strains and Tears

Muscle **strains** are injuries caused by overstretching the tendons or the muscles themselves. Injuries can range from slight overstretch (a **pulled muscle**) to complete muscle tear or tendon rupture. Injuries may be acute, resulting from trauma, or chronic, typically resulting from overuse or disease. Symptoms of muscle or tendon injuries vary with severity. A mild strain, with no tear of tendon or muscle fibers, results in mild pain and perhaps some stiffness. A moderate strain, with some tearing of muscle or tendon fibers, will cause more severe pain as well as bruising and obvious weakness. A severe strain, usually a complete muscle or tendon tear, will result in severe pain, swelling, extensive bruising, and often complete loss of movement.

WHAT MAKES A MUSCLE TEAR GOOD?

Some muscle tears are actually good for you. When muscle groups are exercised, they develop small tears. The healing process actually leads to increased muscle growth and mass. This is most efficiently accomplished if you give that muscle group a rest after exercise.

Diagnosis of muscle or tendon injuries is accomplished by physical examination, imaging (MRI, X-ray), and medical history. Treatment of acute tendon or muscle strains varies with severity. Moderate and severe strains should be treated by a physician. In the early stages of the injury (during the first 72 hours) the standard treatment is *r*est, *i*ce, *c*ompression, and *e*levation (abbreviated RICE).

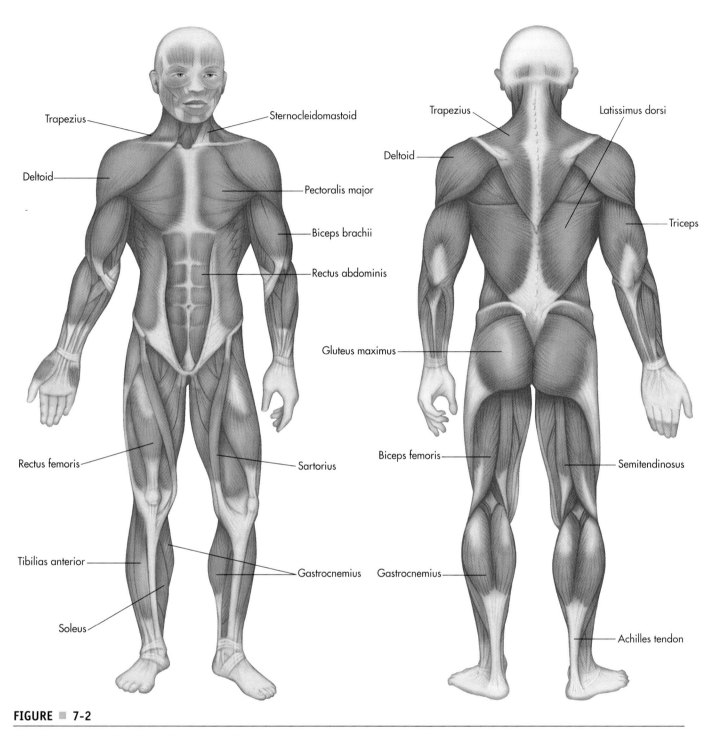

FIGURE ■ 7-2

Anterior and posterior view of major muscles.

Sometimes another letter is added to form PRICE, where P stands for *protection*. Acetaminophen (such as Tylenol) or one of the NSAIDs, (non-steroidal anti-inflammatory drugs, ibuprophen, such as Advil) may be given for pain relief, though their efficacy for promoting healing is questionable. After the first 72 hours, the injury should be rehabilitated, either by gradual increase in activity or by physical therapy. Heat may be applied at this stage. If after several weeks therapy has not shown any improvement, surgery may be necessary. Muscles or tendons that are obviously torn may be treated surgically at the time of diagnosis. Strains are often slow to heal, particularly tendon injuries. Keep in mind that skeletal muscle and dense regular connective tissue have only moderate ability to repair themselves.

Chronic tendon injuries are typically called **tendinitis** (or **tendonitis**). This name is incorrect, however, because there is rarely inflammation present in patients who present with symptoms of chronic tendon damage. More physicians are referring to these chronic injuries, often due to overuse (repetitive motion) or to untreated acute injuries, as tendinosis. However, the term tendinitis is so engrained that it will probably continue to be used. For patients who do not have inflammation of the tendons but are still experiencing discomfort, treatment usually includes (P)RICE, and the use of pain medications. Physical therapy may also be prescribed to improve form and strengthen the muscles around the affected tendon, and the patient will likely be monitored to make sure the condition does not get worse.

Tendinosis is a degenerative disease leading to breakdown and scarring of tendons that appears to be caused by the failure of tendons to repair themselves after injury. Symptoms of tendinosis include pain, tenderness, and stiffness, and often do not appear until the disease is advanced. Several tendons seem more susceptible to tendinosis than others, including the rotator cuff, Achilles tendon, tibialis posterior tendon, patellar tendon, and tendons of the lateral elbow. Several factors affect the development of tendinosis, including age, gender, skeletal anatomy, types of occupational equipment used, and systemic disease (like rheumatoid arthritis or diabetes mellitus). Maria, our diabetes patient, is at a slightly higher risk of tendon damage because of impaired wound healing. Treatment includes (P)RICE, rehabilitation, physical therapy, steroids (short term only), lasers, ultrasound, and extracorporeal shock wave therapy. Surgery may be used as a last resort. Unfortunately, none of these treatments have proven to be very effective. Long term prognosis for chronic tendon injuries is not as good as for acute injuries.

> **7-2** With physical activity and the daily stress of life, muscles often become sore and fatigued. Massaging techniques help to stimulate blood flow and relax tense muscles. To understand more about various massage techniques, please go to your DVD to view videos showing several different methods of massage.

A common running-related inflammatory injury of the extensor muscles and surrounding tissues of the lower legs is called **shin splints**. This condition can be treated with rest or reduction of intensity of the exercise along with ice, anti-inflammatory medication, and modification of footwear.

SKELETAL MUSCLES OF SPECIFIC BODY REGIONS

Many times on a journey, we need a roadmap for reference. These roadmaps are often big maps of an entire state. However, there are also inserts of specific cities that give much greater detail. Think of Figure 7-2 as our "state map" of the anterior and posterior major muscles. The following series of "city maps" will provide you with greater detail.

Facial Skeletal Muscles   Please see Figure 7-3 ■, which shows the facial skeletal muscles.

Anterior and Posterior Trunk Skeletal Muscles   Now take an in-depth look at the muscles of the anterior and posterior trunk of the body in Figure 7-4 ■.

7-3   To view interactive 3-D labeled animations of more specific muscles of body regions in greater detail, please go to your DVD for this chapter. The distinct regions covered are the head and neck; upper limb, forearm, and hand; shoulder and arm; trunk and abdomen; pelvis, hip, and thigh; and lower limb and foot. In addition, you can play an interactive drag-and-drop exercise to label various muscles.

7-4   To view actual examples of intramuscular (IM) injection techniques, please go to your DVD.

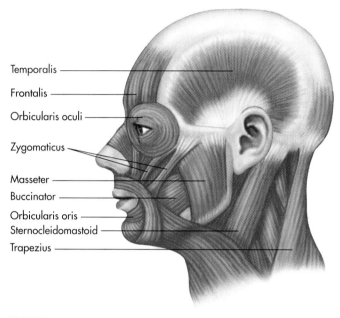

FIGURE ■ **7-3**

Skeletal facial muscles.

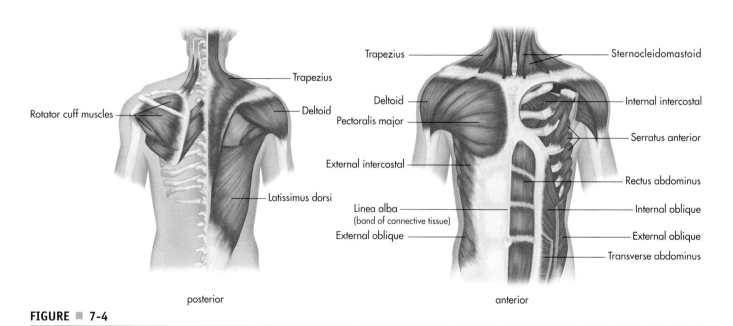

posterior

anterior

FIGURE ■ **7-4**

Skeletal muscles of the anterior and posterior trunk.

Skeletal Muscles of the Arm and Shoulder   Moving out to the peripheral area of the body, we now zoom in on the skeletal muscles of the hand, arm, and shoulder in Figure 7-5 ■.

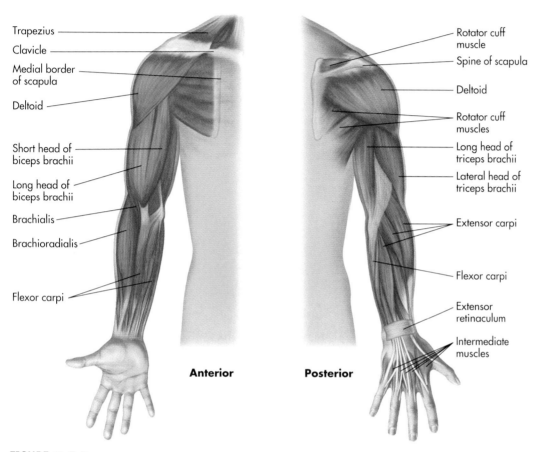

Trapezius
Clavicle
Medial border of scapula
Deltoid
Short head of biceps brachii
Long head of biceps brachii
Brachialis
Brachioradialis
Flexor carpi

Rotator cuff muscle
Spine of scapula
Deltoid
Rotator cuff muscles
Long head of triceps brachii
Lateral head of triceps brachii
Extensor carpi
Flexor carpi
Extensor retinaculum
Intermediate muscles

Anterior   **Posterior**

FIGURE ■ 7-5

Skeletal muscles of the shoulder, arm, and hand.

Learning Hint

MUSCLE NAMES

Muscles can be named based on any of the following criteria:

- Muscle location (Example: *biceps brachii* is in the arm. Brachii = *arm*.)
- Number of origins (Example: *biceps brachii* has two origins. Biceps = *two heads*.)
- Action (Example: *adductor longus* adducts the thigh)
- Size (Example: *gluteus maximus*. Maxiumus = *biggest*.)
- Location of attachments (Example: *brachioradialis; radialis* refers to the radius)
- Shape (Example: deltoid is triangular. Delta = *triangle*.)
- Combination (Example: *pectoralis major*. Pectoral = *shoulder*, major = *big*.)

Skeletal Muscles of the Legs   We finish our tour with the skeletal muscles of the hip and leg in Figure 7-6 ■. Also see Table 7-1, which describes some of the major skeletal muscles along with their location and function.

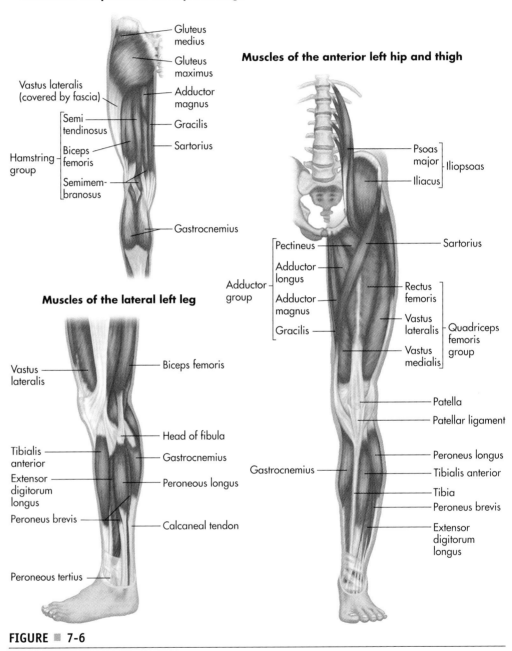

Muscles of the posterior left hip and thigh

- Gluteus medius
- Gluteus maximus
- Vastus lateralis (covered by fascia)
- Adductor magnus
- Semi tendinosus
- Gracilis
- Biceps femoris
- Sartorius
- Hamstring group
- Semimem-branosus
- Gastrocnemius

Muscles of the anterior left hip and thigh

- Psoas major
- Iliacus
- Iliopsoas
- Pectineus
- Sartorius
- Adductor longus
- Adductor group
- Rectus femoris
- Adductor magnus
- Vastus lateralis
- Gracilis
- Quadriceps femoris group
- Vastus medialis
- Patella
- Patellar ligament
- Gastrocnemius
- Peroneus longus
- Tibialis anterior
- Tibia
- Peroneus brevis
- Extensor digitorum longus

Muscles of the lateral left leg

- Vastus lateralis
- Biceps femoris
- Tibialis anterior
- Head of fibula
- Extensor digitorum longus
- Gastrocnemius
- Peroneus brevis
- Peroneous longus
- Calcaneal tendon
- Peroneous tertius

FIGURE ■ 7-6

Skeletal muscles of the hip and leg.

TABLE 7-1: Major Muscle Groups

MUSCLE NAME	MUSCLE LOCATION	MUSCLE FUNCTION
Biceps brachii	Anterior upper arm	Flexes arm
Triceps brachii	Posterior upper arm	Extends arm
Orbicularis oculi	Encircles eye	Closes eyelid
Masseter	Jaw or mandible	Closes jaw
Sternocleidomastoid	From sternum (breast bone) and clavicle (collar bone) to temporal bone	Flexes and rotates head
Pectoralis major	Chest	Flexes the chest area
Deltoid	Shoulder	Moves arms; also an IM injection site
Latissimus dorsi	Mid-lower back and connects to anterior portion on shoulder	Helps to adduct and rotates arm and extends arm
Trapezius	Between shoulder and neck	Provides movement for scapula and extends neck
Intercostals	Between ribs	Lifts and lowers the ribs to assist breathing
Diaphragm	Floor of thoracic cavity	Primary muscle of normal breathing
Gluteus maximus	Buttocks	Abducts and rotates thigh; also an IM injection site
Hamstrings	Posterior portion of thigh	Flexes lower leg
Quadriceps	Anterior portion of thigh	Extends lower leg
Tibialis anterior	Front of lower leg	Dorsiflexes foot
Gastrocnemius	Posterior portion of calf	Flexes foot
Soleus major	Posterior portion of calf	Flexes foot

Pathology Connection: Fibromyalgia Syndrome

Myalgia means pain or tenderness in a muscle. Myalgia is a symptom of many disorders, including acute injury and viral infection. This type of myalgia is usually treated using RICE and over-the-counter or prescription pain medication or by treating the viral infection that is causing the myalgia. **Fibromyalgia Syndrome (FMS)** is a chronic pain syndrome characterized by pain of at least three months duration, bilateral tenderness, fatigue, sleep disorders, depression, anxiety, and exercise intolerance. It is found in about 2% of the population of the U.S. and Canada. The majority of fibromyalgia patients are women. FMS is also common in patients with rheumatological disorders. FMS may be either primary, with FMS only, or secondary, with pain related to another disorder. The cause is unknown, but there is no inflammation involved. Evidence suggests that FMS may be caused by hyperactive stress response or by a sensory or neurological problem that leads to increased sensitivity to pain. FMS can be distinguished from other types of chronic pain by the location of pain in 11 of 18 designated

myalgia *(my AL jee ah)*
fibromyalgia
  (fie bro my AL jee ah)
  algia = *pain*

tender points. Other disorders that cause chronic pain should always be ruled out. There is no definitive treatment for FMS. Symptom management is often inadequate. Some patients are helped by antidepressants (even without evident depression), antiepileptic medications, exercise, or pain relievers.

SKELETAL MUSCLE MOVEMENT

The body requires several different types of movement for various tasks. This movement is accomplished through the coordination of the contraction and relaxation of various muscles. All muscles have four common characteristics. They are contractibility, excitability or irritability, extensibility, and elasticity. Contractibility is when a muscle shortens or contracts. Excitability or irritability is the ability to respond to certain stimuli by producing electric signals called impulses. Extensibility is the muscle's ability to be stretched. Elasticity is the muscle's ability to return to its original length when relaxing.

Contraction and Relaxation

Movement of the body is a result of the contraction (shortening of the muscle fibers) of certain muscles. Consider the act of bending your arm so your fingers touch your shoulder. To really learn the concept, actually bend your arm and touch your fingers to your shoulder while resting your other hand on your biceps muscle. In order to do this, your forearm is drawn to your shoulder as a result of the contraction of your biceps. Did you feel the shortening and bulging of the biceps? Muscles, either by themselves or in muscle groups that cause movement, are known as **agonists** or **primary movers.**

agonist *(AG on ist)*

The chief muscle causing the movement is the primary mover, and in this example it is the biceps muscle. Typically, as your muscle contracts, one of the bones will move (lower forearm) while the other (humerus) will remain stationary. The end of the muscle that is attached to the stationary bone is the **point of origin,** and in this example, it is at the shoulder area. The muscle end that is attached to the moving bone is the **point of insertion;** in this example, it is near the elbow (see Figure 7-7 ■). So the action of a muscle is to move the insertion toward the origin.

Other muscles can assist this movement, such as some of the muscles in the hands and wrist. These are called **synergistic** muscles because they assist the primary mover. To straighten out that same arm requires you to contract the triceps muscles. Your biceps must also relax, but this is a passive process. You do not relax your biceps, but simply stop contracting it and it relaxes. Since these muscles cause movement in the opposite direction when they contract, they are called **antagonists.** This brings us to an important concept. All movement is a result of contraction of primary movers and relaxation of *opposing* muscles. In our example, you cannot forcefully contract the biceps muscles and straighten out your arm. Try it. Again, see Figure 7-7 for the illustration of the antagonistic muscles of the biceps and triceps.

synergistic *(SIN er GIS tic)*

One very important skeletal muscle that controls our breathing is the **diaphragm.** This dome-shaped muscle separates the abdominal and thoracic cavities and is responsible for performing the major work of bringing atmospheric air into our lungs. Exactly how this process occurs is discussed in detail in

Chapter 13, "The Respiratory System." The diaphragm is unique in that it is under both voluntary and involuntary control. You don't have to think each time you breathe, but you can voluntarily change the way you breathe. Figure 7-8 ▦ shows the major muscles of breathing.

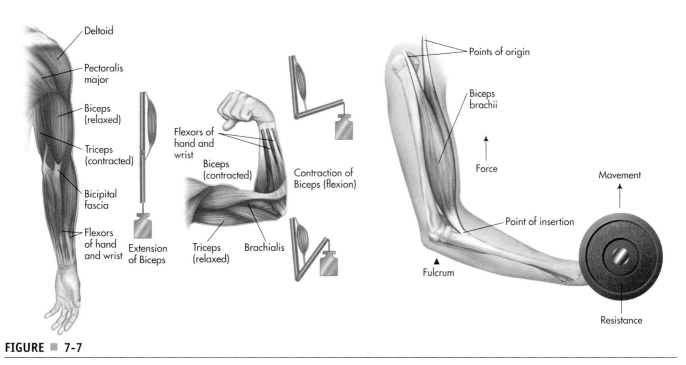

FIGURE ▦ 7-7

Coordination of antagonistic muscles to perform movement.

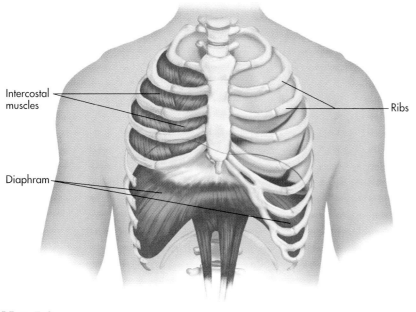

FIGURE ▦ 7-8

The diaphragm: the major muscle of breathing.

Movement Terminology

Certain terms are utilized to describe the direction of body movement. In Chapter 6, we discussed movement as it relates to joints in the skeletal system. In this chapter, we briefly discuss movement as it relates to muscles. **Rotation** describes circular movement that occurs around an axis. Rotation occurs, for example, when you turn your head from left to right or right to left. **Abduction** means to move *away* from the midline of the body. When you raise your arm to point when giving directions, you are performing abduction. **Adduction** occurs when you produce a movement that moves *toward* the midline of the body. When you bring your arm back down to your side from pointing, you are performing adduction.

abduction *(ab DUK shun)*
  ab = *away, as in abduct or abnormal*
adduction *(add DUK shun)*
  ad = *toward*

extension *(eks TEN shun)*
flexion *(FLEK shun)*
ataxia *(ah TAK see ah)*
  a = *without*
  tax/o = *coordination*

    Extension is a term used for *increasing* the angle between two bones connected at a joint. Extension is needed when you kick a football. In this situation, extension occurs when your leg straightens out during the kick. The muscle that straightens the joint is called the **extensor muscle.** Improved extensibility is accomplished by stretching. **Flexion** is the opposite of extension. In this situation, you *decrease* the angle between two bones. Flexion occurs when you bend your legs to sit down. Flexion and rotation occur when you get your arm into position to arm wrestle. The muscle that bends the joint is called the **flexor muscle.** In this case a picture is worth a thousand words or at least the 124 words used to explain these concepts. Figure 7-9 ■ illustrates these movements. Some muscular diseases can cause **ataxia**, which is a condition of irregular muscle movement and lack of muscle coordination.

Applied Science

KINESIOLOGY

Kinesiology is the study of muscles and movement. A kinesiologist is one who studies movement and can employ therapeutic treatment (kinesiotherapy) by specific movements or exercises. Go to the companion website to learn more about the profession of kinesiology.

TEST YOUR KNOWLEDGE 7-2

Give the correct body movement term for the following activities:

1. Looking right and left at a stop sign _____

2. Doing a split _____

3. Patting yourself on the back for labeling all these activities correctly _____

4. The first movement in curling a weight _____

5. Returning the weight from the curled position to your side _____

Fill in the blank:

1. Overstretch of a muscle or tendon is called a(n) _____.

2. Chronic tendon injury is properly called _____.

3. Fibromyalgia is a chronic inflammatory condition. True or false? _____

4. Muscles may be named by the location of their _____.

Flexion

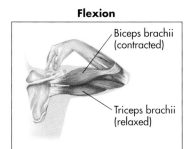

Biceps brachii
(contracted)

Triceps brachii
(relaxed)

A.

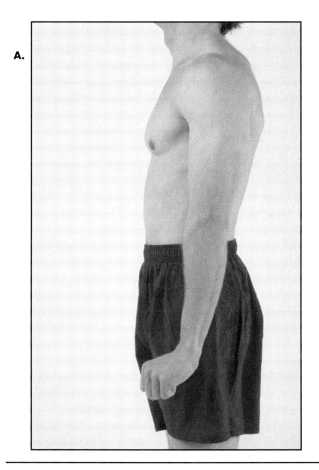

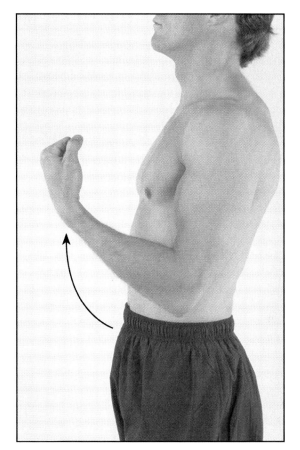

Extension

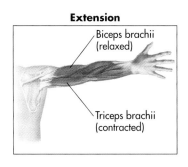

Biceps brachii
(relaxed)

Triceps brachii
(contracted)

B.

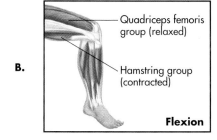

Quadriceps femoris
group (relaxed)

Hamstring group
(contracted)

Flexion

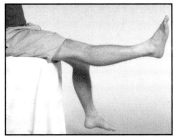

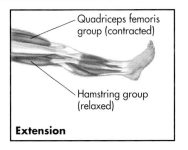

Quadriceps femoris
group (contracted)

Hamstring group
(relaxed)

Extension

FIGURE ▪ 7-9

The types of skeletal movement. **A.** Flexion and extension of left forearm. **B.** Flexion and extension of the leg.

MUSCULAR MOVEMENT AT THE CELLULAR LEVEL

Exactly how is skeletal muscular contraction and relaxation accomplished? How does the muscle tissue cause a coordinated and smooth contraction? Let's look in more detail at how muscles work.

The Functional Unit of the Muscle

We have talked about muscles on a macro (very large) scale. For example, how do those large biceps you've developed from working out contract when you touch your shoulder? Now, let's explore the makeup of the individual muscle fiber to learn exactly how this contraction takes place. As stated previously, muscle consists of elongated cells called muscle fibers, which can be up to 12 inches, or 30 centimeters, in length. Each muscle fiber is encased in a cell membrane called the **sarcolemma,** and contains units called **myofibrils.** The muscle myofibril is sort of like a strand of metal, and these strands of metal can be bundled together to form a cable, (see Figure 7-10A ▪).

In order for contraction to take place, each fiber must possess many functional contractile units called **sarcomeres.** Each fiber has the ability to contract because of the makeup of the sarcomere. Each sarcomere has two types of threadlike structures called thick and thin **myofilaments.** The thick myofilaments are made up of the protein **myosin,** and the thin ones are primarily made up of the protein **actin.** The sarcomere has the actin and myosin filaments arranged in repeating units separated from each other by dark bands called **Z lines,** which give the striated appearance to skeletal muscle (see Figure 7-10A and B ▪). Each myofibril is made up of several sarcomeres.

Note in Figure 7-10C ▪ that the contraction of a muscle causes the two types of myofilaments to slide toward each other, shortening each sarcomere and therefore the entire muscle. Picture a tube sliding within a tube, such as on a trombone. This sliding filament action and corresponding contraction requires that temporary connections, or crossbridges, be formed between the thick filament heads (myosin heads) and the thin filaments (actin) to pull the sarcomere together. Once the crossbridges form, the myosin heads rotate and pull the actin toward the center of the sarcomere. When the sarcomere relaxes, the filaments return to their resting or relaxed position. Visualize a raised drawbridge, where cars cannot pass. In order to be functional, the cross-connection—or lowering of the drawbridge—must occur, similar to the crossbridges needed to be formed for a muscle contraction.

myofibril *(my oh FIE bril)*
  my/o = *muscle*

sarcomeres *(SAR koh mearz)*

myosin *(MY oh sin)*

actin *(AK tin)*

7-5 To perform an interactive drag-and-drop exercise on labeling myofibril structures, please go to your DVD.

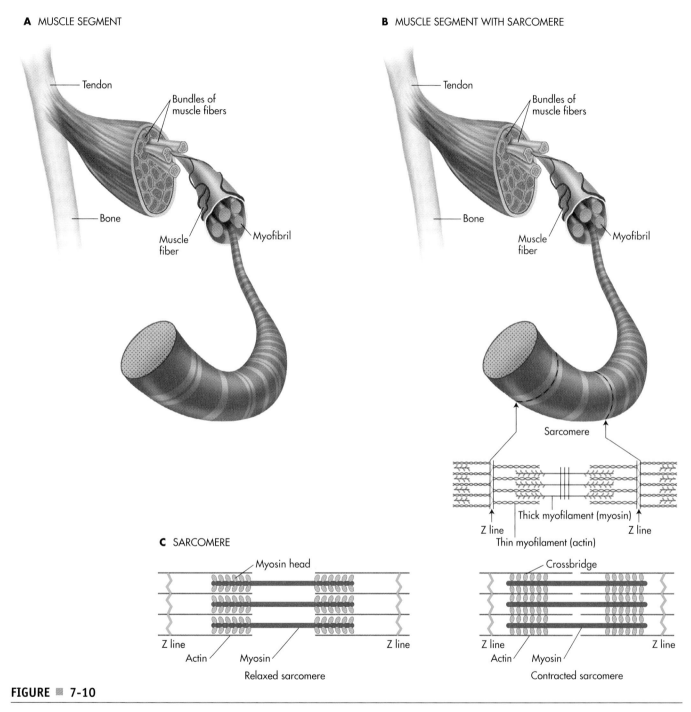

A MUSCLE SEGMENT

- Tendon
- Bundles of muscle fibers
- Bone
- Muscle fiber
- Myofibril

B MUSCLE SEGMENT WITH SARCOMERE

- Tendon
- Bundles of muscle fibers
- Bone
- Muscle fiber
- Myofibril
- Sarcomere
- Thick myofilament (myosin)
- Z line
- Z line
- Thin myofilament (actin)

C SARCOMERE

- Myosin head
- Z line
- Actin
- Myosin
- Z line
- Relaxed sarcomere

- Crossbridge
- Z line
- Actin
- Myosin
- Z line
- Contracted sarcomere

FIGURE ■ 7-10

A. The muscle segment; **B.** The muscle segment with sarcomere, and **C.** relaxed and contracted sarcomeres.

 Pathology Connection: Duchenne's Muscular Dystrophy

Duchenne's muscular dystrophy (DMD) is a genetic, incurable myopathy. It is the most common childhood form of muscular dystrophy. Because the gene for DMD is carried on the X chromosome, the disease is seen overwhelmingly in boys. It is relatively common, with an incidence of 1 in 3,500 live male births. The disease is caused by a mistake in the gene for a protein called dystrophin. Scientists think dystrophin's function is to hold muscle fibers together

muscular dystrophy
(MUSS kyoo lahr DISS troh fee)
dys = *difficult*
trophy = *growth or nourishment*

during contraction. Without functional dystrophin, muscle fibers degenerate. All types of muscle: smooth, cardiac and skeletal are affected.

Symptoms are typical of myopathy, mainly muscle weakness, in the early stages of the disease. As the disease progresses, muscles become progressively weaker, scarred, and filled with fatty deposits as the muscle fibers disappear. DMD is progressive. Most boys are diagnosed around age 4 and are wheelchair-bound by age 10. Average life span is 17 years. Many DMD patients have significant cardiac and smooth muscle abnormalities as well. Death is usually due to respiratory or cardiac failure. A milder, late-onset form of muscular dystrophy, called Becker's muscular dystrophy (BMD), is diagnosed in patients in their 20s and 30s. It leads to debilitating myopathy, but progresses more slowly.

DMD is diagnosed by physical examination. Its diagnostic symptoms include gait abnormality in toddlers, progressive myopathy, and pseudohypertrophy in the calves. Biochemical tests for muscle enzymes, genetic testing, muscle biopsy, and an **electromyogram** (EMG) can confirm diagnosis based on physical exam. An electromylogram is a test in which a muscle group is stimulated with an electrical impulse that causes a contraction. The strength of the contraction can then be recorded. There is no effective treatment for DMD, so treatment is mainly symptom management, including assistive devices and palliative care. Physical therapy should not be used because it increases the breakdown of muscle fibers. There are some treatments in the experimental stage that may offer hope for decreasing progression of the disease, including steroids, immunosuppressants, nutritional supplements, gene therapy, and stem cells, but none of these is widely available to patients. Research is focusing on increasing muscle repair to keep pace with muscle damage.

electromyogram
(ee LEK troh MY oh gram)
electr/o = *electricity*
my/o = *muscle*
gram = *recording*

7-6 To view videos on muscle atrophy and muscular dystrophy, please go to your DVD for this chapter.

IMPORTANT INGREDIENTS: ATP AND CALCIUM

In the previous analogy of the drawbridge, consider that a toll must be paid in order for the bridge to lower and connect. The body's toll is the energy molecule **adenosine triphosphate** (ATP) and **calcium** (Ca), which are needed for contraction and relaxation. ATP provides the energy to help the myosin heads form and break the crossbridges with actin. When the muscle is relaxed, calcium is stored away from the actin and myosin in the **sarcoplasmic reticulum** (SR), which is a specialized series of interconnecting tubules and sacs that surround each myofibril. When the muscle is stimulated, calcium is released from the SR and allows actin, myosin, and ATP to interact, which causes the contraction. When calcium returns to the SR, the crossbridges are broken and the muscle relaxes.

When the nervous system tells the muscle to contract by releasing the neurotransmitter acetylcholine (ACh), the signal causes the muscle fiber to open **sodium ion channels**. Sodium ions flow through these channels into the muscle fibers. This causes the muscle fibers to become excited, and calcium is released from the sarcoplasmic reticulum. The calcium, now free in the cytoplasm of the muscle fiber, helps myosin bind with actin in the presence of ATP. The calcium is then pumped back into the sarcoplasmic reticulum for storage and the muscle relaxes.

Just to recap, for muscle to contract, ACh must be released from the nervous system to excite the muscle. ACh causes sodium channels to open and sodium to flow into the muscle. The sodium excites the muscle and causes calcium to be released from the sarcoplasmic reticulum. The calcium allows the crossbridges to form between the myosin heads and the actin filaments. ATP is needed to allow the myosin head to let go of actin and reattach, causing the actin filaments to slide along the myosin filaments. This sliding shortens the sarcomerem and therefore shortens the muscle. This shortening is muscle contraction.

Applied Science

INTERRELATEDNESS OF NEUROMUSCULAR SYSTEM

As you will discover in Chapter 9, contraction of a skeletal muscle requires the coordination of both the muscular and nervous systems. A **motor unit** is a motor neuron plus all the muscle fibers it stimulates. The initiation of a skeletal muscular contraction requires an impulse from a motor neuron of the nervous system to trigger a release of a neurochemical transmitter called **acetylcholine** (ACh), which opens the sodium ion channels and sets the process of muscle contraction into motion. Acetylcholine is a chemical neurotransmitter that diffuses across the synaptic cleft. This all occurs at the neuromuscular junction. This is the junction between the motor neuron's fiber that transmits the impulse and the sarcolemma (the muscle cell membrane). The surface of the muscle is studded with sodium channels that are **ligand-gated** (requiring a "special key" to open). A ligand-gated channel opens or closes when a molecule binds to a receptor that is part of the channel, like a key fitting into a lock. In the case of skeletal muscles, the ligand (key) is the neurotransmitter acetylcholine (ACh). Acetylcholine is released from the terminal end of a motor neuron, diffuses across the synaptic cleft, and binds to the surface of the skeletal muscle, opening sodium channels and causing sodium to flow into the cell. The sodium flowing in leads to calcium release, and the muscle contracts. Like all chemical synapses, the neuromuscular junction must be cleaned up at the terminal end of transmission. The enzyme responsible for cleaning up the synapse is called **acetylcholinesterase**.

acetylcholine *(ah SEET ul KOE leen)*

Pathology Connection: Myasthenia Gravis

Myasthenia Gravis (MG), which literally means *grave muscle weakness*, is an autoimmune disorder in which the immune system attacks and destroys a large number of acetylcholine receptors at the neuromuscular junction. The remaining receptors are blocked, and there are also modifications of the muscle fiber at the neuromuscular junction. Motor neurons (nerve cells that cause movement) continue to release acetylcholine, but because of the damage, the muscle cannot respond. Eye muscles are typically the first muscles affected, but some patients initially experience difficulty chewing, swallowing, or talking. This disorder, like most autoimmune disorders, is progressive, though the course

Amazing Body Facts

RIGOR MORTIS

Have you ever heard of a dead body rising from a table or showing signs of movement? This may sound like the opening for a movie about zombies or the "undead." Actually, it is a normal physiological process called **rigor mortis** that can be explained by science and not by science fiction.

When a body dies, all the stored calcium cannot be pumped back into the sarcoplasmic reticulum. Therefore, excess calcium remains in the muscles throughout the body and causes the muscle fibers to shorten (contract) and stiffen the whole body. In addition, ATP is not present in a dead body to break the cross bridges. This stiffening process of the entire body is termed rigor mortis.

rigor mortis
(RIG or MORE tiss)

7-7 Cellular muscular contraction may be hard to visualize. Please go to your DVD to see an animation of cellular-level muscle contraction that visually pulls all of these concepts together.

of the disease varies widely among patients. The muscle weakness fluctuates, worsening with activity and improving with rest.

MG is a rare disorder, but is most common in women under 40 years old and in men over 50. Because it is rare, MG is hard to diagnose and is often not diagnosed during the first year of the disease. Many disorders cause muscle weakness, but the fluctuating weakness seen in MG is unique. Blood tests for acetylcholine receptor antibodies (molecules that attack the receptors), and electromyography (muscle recordings) are the best way to diagnose MG. There is no specific definitive treatment for MG. Treatment typically includes acetylcholinesterase (AChE) inhibitors (allows for more Ach to be available), corticosteroids, immunosuppressant and plasma exchange.

Pathology Connection: Tetanus

tetanus *(TETT ah nus)*

spasm = *A muscle spasm or cramp is an involuntary, sudden and violent muscle contraction*

Tetanus (also known as Lock Jaw) is a muscle disorder caused by an untreated bacterial infection of a wound. It was long thought that you got it from stepping on a rusty nail, but actually the bacteria reside in the soil and the puncture allows for their entry. Therefore, a new clean nail is just as effective. The bacteria (*Clostridium tetani*) release a toxin that keeps muscle constantly contracted. Only a tiny amount of toxin, less than a milligram, is needed to cause the disease. (Remember our discussion of bacterial pathology in Chapter 3.) Tetanus symptoms include muscle spasm, rigid paralysis, stiffness, and pain, usually starting in the jaw. Spasm can be initiated with minimal stimuli, such as loud noises or even turning on a light. Patients may also have a fever. Symptoms appear 3 days to 3 weeks post-injury. Symptoms are usually progressive, and can eventually paralyze the diaphragm. Diagnosis of tetanus is by physical examination and laboratory testing to rule out other disorders. Treatment includes cleaning of the wound that is the source of the bacteria, injection or IV tetanus antitoxin, sedation, ventilator support, and pain management. Most patients must be treated in the Intensive Care Unit (ICU). Recovery takes several weeks and many patients have long-term problems.

Because of widespread vaccination against tetanus, most clinicians in the U.S. and other developed countries will never see a case of tetanus. Tetanus would otherwise be rather common, as its infectious bacteria is widespread in soil around the world. Prior to vaccination, there were 500 cases per year in the US. Today there are 800,000 tetanus deaths per year worldwide, mainly in countries without vaccination programs. Current recommendations are that children be vaccinated several times for tetanus and that adults be vaccinated every 10 years.

Several other disorders, classified as neuromuscular disorders, also cause paralysis of skeletal muscle, due to abnormalities of the nervous system. These include Guillaine-Barré syndrome, post-polio syndrome, and Lou Gehrig's disease. These will be described in Chapter 9.

www.prenhall.com/colbert
Not only do the skeletal muscles facilitate movement, but integrated with the nervous system, they provide support for posture while standing or sitting. Promoting balance and posture, along with proper joint and muscle function, is one of the responsibilities of **physical therapists**. Physical therapists perform many therapies, such as range of motion (ROM) exercises, to ensure full muscle movement. **Occupational therapists** assist patients in utilizing and adapting their muscle function to perform activities of daily living and improving their quality of life. **Massage therapists** work directly on the muscles to aid in their relaxation and optimal functioning. To learn more about these professions, please go to the website for this chapter.

TEST YOUR KNOWLEDGE 7-3

Fill in the blank:

1. The region between two Z lines is called a
   _____.

2. The thick myofilament needed for muscle contraction is composed of _____.

3. The thin myofilament needed for muscle contraction is composed of _____.

4. The two ingredients needed for cross bridges to form and break are _____ and
   _____.

5. Muscular dystrophy causes fragile muscle fibers due to this missing or damaged protein
   _____.

6 _____ is caused by a toxin created by bacteria found in the soil.

VISCERAL OR SMOOTH MUSCLE

We've introduced the concept of smooth muscle earlier in this chapter; now let's take a closer look. **Visceral muscle,** or **smooth muscle,** is found in organs (except the heart) such as the stomach and other digestive organs, and in the blood vessels and bronchial airways. The ability of smooth muscle to expand and contract plays a vital role in many of the body's internal workings. For example, blood pressure can be affected by whether the blood vessels get larger in diameter (**vasodilate**) or get smaller in diameter (**vasoconstrict**). Vasodilation can lead to decreases in blood pressure due to smooth muscle relaxation in the vessel that allows it to enlarge. The enlarged vessel has less resistance to flow, and the blood pressure therefore goes down. Conversely, vasoconstriction can cause increased blood pressure due to the smooth muscle contraction that restricts the blood vessel.

vasodilate *(VASE oh DIE late)*
vasoconstrict *(VASE oh con STRICT)*

As another example, during an asthma attack, smooth muscles in the airways of the lungs constrict, making it difficult to get air in and out of the lungs. This is what causes the wheezing sound heard during an attack.

sphincters *(SFING ters)*

A special type of smooth muscle, called a **sphincter**, is found throughout your digestive system. Sphincters usually occur between major digestive organs, such as between the esophagus and stomach. These donut-shaped muscles act as doorways to let materials in and out by alternately contracting and relaxing. For example, the sphincters of the stomach act like doors that open up to allow food in from the esophagus and also to allow partially digested food out into the small intestine. Have you ever swallowed a large amount of bread or stuffing and had it get stuck on the way down to your stomach? This is a painful reminder that there is a sphincter that must relax and open to allow food to enter your stomach. The muscles of the digestive system are discussed in greater depth in Chapter 15, "The Gastrointestinal System."

Learning Hint

SMOOTH MUSCLE REGULATION OF BLOOD PRESSURE

In considering blood pressure, visualize a large highway. If one lane is taken away (vasoconstriction), the same number of cars must now fit through one less lane, leading to traffic congestion (increase in pressure). If you open up another lane (vasodilate), you relieve some of this pressure.

Smooth (visceral) muscles are involuntary muscles and do not contract as rapidly as skeletal muscles. Skeletal muscles, once stimulated, can contract 50 times faster than smooth muscle. Because of their slower activity and lower metabolic rate, smooth muscles receive only moderate amounts of blood. Once injured, smooth muscle rarely repairs itself; instead it forms a scar.

CARDIAC MUSCLE

Cardiac muscle forms the walls of the heart. The contraction of cardiac muscle squeezes blood out of the chambers of the heart, causing the blood to circulate through your body. Cardiac muscle is involuntary muscle. Cardiac muscle fibers are somewhat shorter than the other muscle types. Since the heart must work constantly, the cardiac muscles must receive a generous blood supply to get enough oxygen and nutrition, as well as to get rid of waste. In fact, cardiac muscle has a richer supply of blood than any other muscle in the body. The cardiac muscle fibers are connected to each other by **intercalated disks.** Because of this connection, as one fiber contracts, the adjacent one contracts, and so on. This is similar to the domino effect or the human wave at a football stadium if done correctly. A wave of contraction occurs, allowing blood to be squeezed out of the heart and into the body. This directed wave is important for a full and effective emptying of the blood within the heart. Imagine if everyone squeezed the tube of toothpaste in the middle: think of all the wasted toothpaste that would be left in the tube and how happy the toothpaste manufacturers would be. See Figure 7-11 ■.

intercalated *(in TER cow LATE ed)*

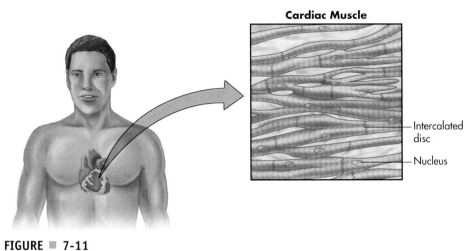

Cardiac Muscle

— Intercalated disc

— Nucleus

FIGURE ▪ 7-11

Heart and intercalated disks.

Cardiac muscle does not regenerate after severe damage; damage leads to tissue death, such as occurs in a severe heart attack. If the blood supply going to the heart from the coronary arteries is blocked, cardiac muscle damage can occur, causing scarring of the heart. Scar tissue does not help the healthy muscles of the heart to contract. If the scarred area is extensive, the remaining cardiac muscle may not be sufficient to pump blood efficiently. An individual with scarred cardiac muscle may have a severely diminished cardiac output, which could lead to severe disability or even death.

MUSCULAR FUEL

Muscle, like all tissue, needs fuel in the form of nutrients and oxygen in order to survive and function. The body stores a carbohydrate called **glycogen** in the muscle. Glycogen is always on reserve waiting to be converted to a usable energy source. When needed, the muscle can convert glycogen to **glucose,** During cellular respiration, which requires oxygen, energy is released from the glucose and converted to ATP, which powers the muscle. For very intense exercise, the body uses the process of **fermentation** to break down glucose and provide ATP. This process is less efficient than using glucose and can provide energy for only a short time, but it does not require oxygen. A byproduct of this process is **lactic acid**, which builds up in the muscles, causing temporary muscle fatigue, or weakness. After exercise is completed, more oxygen is needed to convert the lactic acid back to glucose. This additional oxygen is known as the body's **oxygen debt.** The body overcomes this debt by increasing the rate of breathing and heart rate until it has returned to normal.

Muscles with very high demands (such as leg muscles) also store fat and use it as energy. The release of energy also produces heat, and this is why strenuous or prolonged exercises can overheat our bodies.

The higher-demand muscles not only use fat as an energy source, but have a much richer blood supply than do less demanding muscles. These muscles are

glycogen *(GLIE co jin)*

glucose *(GLOO kohs)*

Applied Science

MAINTAINING A CORE BODY TEMPERATURE

Not only do muscles produce movement, but they help maintain posture, stabilize joints, and produce body heat. Producing heat is important in maintaining the body core temperature. As the energy-rich ATP is used for muscle contraction, three-fourths of its energy escapes as heat. This process helps to maintain body temperature by producing heat when muscles are utilized. This is why your temperature rises when exercising and also why you shiver when you are very cold. Shivering is your body's way of saying it is too cold and it needs to generate a lot of heat via many muscle contractions (shivering). In turn, this increases the body temperature.

needed for endurance, such as required by long-distance running. The richer blood supply carries extra oxygen to hardworking muscles, giving those muscles a darker color.

Some muscles, such as those in the hand, have fewer heavy demands placed on them and need only a small supply of blood. These muscles utilize the local blood supply for glucose and the glycogen stored within. They therefore have a lighter color. These muscles are faster but do not have the endurance capabilities that heavily used muscles have. Next time you take a long walk, keep pumping your hand. While the hand can move faster than the leg muscles, it will tire more quickly.

Another example can be found in a chicken. Because its breast and wing muscles are not heavily used, those parts contain white meat. Its legs, however, endure constant use, and the meat is therefore dark. By contrast, a woodcock, a migratory bird that must fly long distances (endurance), has dark breast meat. Now you know why a chicken's breast meat is white. When is the last time you saw a chicken flying overhead?

A Quick Trip Through Diseases and Conditions of the Muscular System

DISEASE	ETIOLOGY	SIGNS AND SYMPTOMS	DIAGNOSTIC TEST(S)	TREATMENT
Strains (tears)	Acute injury.	Varies with severity; may include pain, stiffness, bruising, weakness, loss of function.	Examination, radiologic studies, patient history.	Depends on severity, (P)RICE, pain relievers, heat, physical therapy, surgery.
Tendinosis (tendinitis)	Chronic overuse or disease.	Varies with severity; may include pain, stiffness, weakness.	Examination, radiologic studies, patient history.	(P)RICE, physical therapy, anti-inflammatory drugs, lasers, ultrasound, shock waves.
Shin splints	Repetitive lower body exercise such as running. Can be due to faulty foot mechanics or footwear.	Pain in the tibia region.	Patient history and physical exam.	Ice packs, anti-inflammatory medications, decrease of exercise intensity, avoidance of hills and hard surfaces, change in footwear.

A Quick Trip Through Diseases and Conditions of the Muscular System (*Continued*)

DISEASE	ETIOLOGY	SIGNS AND SYMPTOMS	DIAGNOSTIC TEST(S)	TREATMENT
Cramps/spasms	Sudden, severe involuntary muscle contraction.	Can be result of prolonged physical activity, excessive fluid/electrolyte loss, menstruation.	Patient history and physical exam. Blood work for electrolyte levels.	Rest from specific task, rehydration, passive stretching, dietary electrolyte replacement.
Fibromyalgia Syndrome	Unknown, but may be neurological.	Chronic pain, bilateral tenderness, fatigue, sleep disorders, depression, exercise intolerance.	Location of pain confined to "tender points."	Symptom management not very effective, some helped by antidepressants, exercise, pain relievers, anti-epileptic medications.
Duchenne's Muscular Dystrophy (DMD)	Genetic defect in muscle protein, causes disintegration of muscle fibers.	Muscle weakness in early stages; later, significant progressive muscle weakness including skeletal, cardiac, and smooth muscles.	Physical exam, biochemical and genetic tests, Electromyogram (EMG), muscle biopsy.	No real treatment, disease is invariably fatal in adolescence or young adulthood. Symptom management and palliative care. Some experimental treatments in development.
Mitochondrial Myopathy	Defect in ATP production in mitochondria.	Progressive muscle weakness, often accompanied by hearing loss, diabetes mellitus, heart problems, nervous system disorders and other biochemical abnormalities.	Genetic tests, biochemical tests, EMG, muscle biopsy (for ragged red fibers).	No consensus on treatment, some drugs can decrease symptoms but nothing can stop progression.
Myasthemia Gravis	Autoimmune attack at neuromuscular junction.	Progessive, flucuating muscle weakness, often starting with facial or eye muscles.	Blood tests, EMG.	Steroids, immunosuppressant drugs, plasma exchange, acetlycholinesterase inhibitors.
Tetanus (Lock Jaw)	Bacterial infection; *Clostridium tetani*.	Progressive muscle spasm, paralysis, stiffness and pain, especially in jaw.	Physical exam, lab tests to rule out other disorders, history of wound.	Wound hygiene, anti-biotics, tetanus antitoxin, sedation, ventilator support, pain management.

℞ Pharmacology Corner: Skeletal Muscle Drugs

Although there are many types of muscles, this corner will focus on drugs related to skeletal muscles. Cardiac and smooth muscles will be discussed in their respective chapters. At some times in our lives, we have all pulled or strained muscles. The most common drugs prescribed to treat the pain and resulting discomfort are anti-inflammatories and pain medication. The anti-inflammatories are of the non-steroidal type due to the many potential side effects of steroid use, and are known generally as Non-Steroidal Anti-Inflammatory Drugs (NSAIDs). Examples include ibuprofen (Advil) and naproxen (Alleve). Common pain medications include aspirin and acetaminophen (Tylenol).

Sometimes the actual muscle itself must be rested in order to heal. This mild relaxation can be achieved by a muscle relaxant. Skeletal muscle relaxants also help to relieve musculoskeletal pain or spasm. They are used to treat spasticity in stroke patients and in patients with multiple sclerosis or spinal cord injuries. Examples include Flexeril and Parafon Forte, both of which tell the brain to decrease muscle tone.

There are certain situations when we want to clinically induce paralysis. For example, during surgery we don't want to have the patient moving all around the table! Here, neuromuscular blocking drugs actually cut off the communication between the brain and muscles. Examples include succinlycholine (Anectine) and pancuronium (Pauvulon).

Clinical paralysis is a very precise science because an overdose of these medications can stop the heart or breathing. Even though the patient is completely paralyzed, he or she can still feel pain and are conscious and aware. This is why it is imperative to give other drugs that sedate the patient.

Applied Science

A USEFUL APPLICATION OF A DEADLY TOXIN

Botulism is a potentially deadly disease caused by food poisoning with the *Clostridium botulinum* bacteria. Science has found a way to utilize the poison generated by this bacterium for medical and cosmetic treatment. Small amounts of *botulinus* toxin are injected into facial muscles to stop previously untreatable facial twitching. The toxin basically paralyzes the muscles. The same toxin is used to treat wrinkles without the use of surgery and is known as Botox injection.

SUMMARY

Snapshots from the Journey

→ The three main types of muscles are skeletal, smooth, and cardiac.

→ Skeletal muscle is striated (striped) voluntary muscle that allows movement, stabilizes joints, and helps maintain body temperature.

→ Smooth muscle is nonstriated involuntary muscle found in the organs of the body and linings of vessels; it facilitates internal movement within the body.

→ Cardiac muscle is involuntary, striated muscle found only in the heart.

→ All movement is a result of contraction of primary movers and relaxation of opposing muscles.

→ Muscles usually attach to bones via tendons.

→ Large muscles consist of many single muscle fibers comprised of myofibrils. The smallest functional contractile unit is called a sarcomere.

→ Each sarcomere unit contains the two threadlike contractile proteins: myosin and actin.

→ Muscles contract as the actin and myosin protein filaments, in the presence of ATP and calcium, form crossbridges that cause the filaments to slide past each other, thereby causing the muscle to contract or shorten.

→ There is an interrelation between the nervous and muscular systems in which the motor neuron of the nervous system initiates the activity of muscle contraction through the release of a neurotransmitter transmitter chemical called acetylcholine (ACh).

→ There are many common diseases and conditions of the muscles, and because the nervous system is so closely related, there are also many common neuromuscular diseases.

MARIA'S STORY

From previous chapters you remember Maria, who is an insulin-dependent diabetic. Recently, on the advice of her doctor, she began an exercise program. She started slowly, walking for 20 minutes per day. After a month of walking she decided to take an aerobics class. Two weeks into the class, one of her knees became painful and stiff. Her doctor explained to her that, as a diabetic, Maria is at a risk for tendinosis.

a. Do Maria's symptoms fit the symptoms of tendinosis?

_____

b. Why might diabetics be at higher risk for tendinosis?

_____

c. Is this a new injury or a more long standing problem?

_____

RAY'S STORY

Ray is devastated. His mother finally brought him a mirror so he could look at his body. He couldn't believe his eyes. He had been injured for a few months and now looked like a skeleton. He had always taken pride in his appearance, working out regularly. He had even entered a Mr. Muscles contest once on a dare from his buddies.

How would you explain to Ray what happened to his body (or appearance)?

_____

_____

_____

_____

REVIEW QUESTIONS

Multiple Choice

1. Another name for voluntary muscle is:
   a. skeletal
   b. smooth
   c. cardiac
   d. nonstriated

2. Which structure does *not* contain smooth muscle?
   a. blood vessels
   b. heart
   c. digestive tract
   d. bronchi

3. Most skeletal muscles attach to bones via:
   a. ligaments
   b. joints
   c. flexors
   d. tendons

4. The state of partial skeletal muscle contraction is known as:
   a. homeostasis
   b. muscle tone
   c. partialus contractus
   d. flexerus

5. Cardiac muscle:
   I. is a voluntary muscle
   II. has intercalated disks to assist contraction
   III. regenerates after injury
   VI. lines the blood vessels

   a. I only
   b. I and II
   c. II only
   d. I, II, III, VI

6. A disorder characterized by multiple points of tenderness and fatigue is:
   a. pompe disease
   b. fibromyalgia
   c. tendinosis
   d. Duchenne muscular dystrophy

Fill in the Blank

1. A sudden or violent muscle contraction is a _____.

2. Partial or total loss of voluntary muscle use is _____.

3. The neurotransmitter that is necessary for skeletal muscle contraction is _____.

4. The body stores a carbohydrate called _____ in the muscle; it can be converted to a usable energy source.

5. _____ means pain or tenderness in the muscle.

6. Chronic injury to a tendon is known as _____.

7. This molecule is missing in Duchenne muscular dystrophy _____.

8. Which neuromuscular disorder is caused by a bacterial toxin? _____.

Short Answer

1. List the three major muscle types and give an example of each.

_____

_____

_____

_____

2. Contrast the terms *hypertrophy* and *atrophy* and give an example of how each situation could occur.

_____

_____

_____

_____

3. Explain how vasoconstriction and vasodilation affect blood pressure.

_____

_____

_____

_____

4. Explain the steps needed in a skeletal muscle contraction.

_____

_____

_____

_____

5. Define myopathy and explain two disorders which cause myopathy.

_____

_____

_____

_____

6. Muscle disorders may be caused by malfunction of organelles or structural breakdown of muscle fibers. List and briefly explain these disorders.

_____

_____

_____

_____

Suggested Activities

1. Pick a major muscle group and discuss how your life would be different if that group could not function properly due to a disease or accident.

2. Bodybuilding requires an extensive knowledge of muscles and muscle groups. Demonstrate five different exercises and the different muscles they would develop.

3. Pair off with a partner and perform various muscle movements showing rotation, abduction, adduction, extension, and flexion. See if your partner can accurately classify each motion.

7-8 Now that you have completed your journey through this chapter, please go to the DVD for interactive games and puzzles concerning the medical terms and concepts contained in this chapter. By playing the games you will reinforce your learning of the muscular system in a fun way.

T H E
INTEGUMENTARY *System*

The Protective Covering

You learned in Chapter 6 that the skeletal system is like the framework of a building or house. But the framework is just one part of a building; the integrity of a house wouldn't last very long without shingles on its roof, siding of some sort, and windows, all of which help prevent the environment from damaging the main structures and inner workings. Like a house, the human body must be sheltered from the environment: that is the job of the integumentary system. Your skin forms a protective barrier to shield your body from the elements, guard against pathogens, and perform several other vital functions. Think how important your skin is to your well-being and your ability to enjoy your journey fully. Without your skin, you would be unable to regulate your body temperature and would be uncomfortable in any environment. We paint our houses to protect them from the elements; likewise, we apply sun screen when we spend a day at the beach. While your skin is an important organ, there are also several other accessory components such as nails, hair, and glands —hence the name integumentary *system*.

Chapter 8

LEARNING OBJECTIVES

At the end of your journey through this chapter, you will be able to:

- → Discuss the functions of the integumentary system
- → List and describe the layers of the skin
- → Explain the healing process of skin
- → Describe the structure and growth of hair and nails
- → Explain how the body regulates temperature through the integumentary system
- → Describe various skin diseases, causative agents, and their related treatments

MULTMEDIA APPLICATIONS

DVD Interactive Exercises

- → Interactive drag-and-drop exercises: three layers of skin and the integumentary system, 8-1
- → Videos on the changes that occur with aging to the integumentary system, 8-2
- → Animation of pressure sore formation, prevention, and treatment, 8-3
- → Animation on wound repair and the formation of scar tissue, 8-4
- → Video on intradermal drugs and subcutaneous injections, 8-5
- → Interactive drag-and-drop exercise of the anatomical structures of a hair follicle, 8-6
- → Videos and animations on diseases of the skin: decubitus ulcers, eczema, and skin cancer, 8-7
- → Videos on topical wound/burn care, 8-8
- → Interactive games and puzzles, 8-9

www.prenhall.com/colbert

- → Professional Profiles:
  - EMT-Paramedic
  - Nursing
- → Related Internet Links
- → Additional Review Questions

Pronunciation Guide

Correct pronunciation is important in any journey so that you and others are completely understood. Here is a "see and say" Pronunciation Guide for the more difficult terms to pronounce in this chapter.

alopecia (al oh PEE she ah)

apocrine (APP oh crin)

carotene (CARE eh teen)

corium (CORE ee um)

ecchymosis (ek ee MOH sis)

eccrine (EKK rin)

epidermis (ep ih DER miss)

epithelial cells (ep ih THEE lee al)

keratin (KAIR ah tin)

keratinization
  (KAIR ah tin eye ZAY shun)

lesion (LEE zhun)

lunula (LOO nyoo lah)

melanin (MELL an in)

melanocytes (mell AN oh sights)

pustule (PUS tyool)

sebaceous gland (see BAY shuss)

seborrheic keratosis
  (SEB oh REE ik KERR ah TOH sis)

sebum (SEE bum)

scabies (SKAY beez)

squamous cells (SKWAY muss)

stratum corneum
  (STRAY tum core NEE um)

subcutaneous fascia
  (sub cue TAY nee us FAY she ah)

tinea (TIN ee ah)

vesicles (VES ih koolz)

SYSTEM OVERVIEW

integument = *a covering*

In this section, we will look at the functions and parts of the **integumentary system**. This system is the protective covering of the body and is the most exposed system. Get ready for this chapter to "get under your skin."

Integumentary System Functions

The integumentary system includes the skin and its accessory components of hair, nails, and associated glands. The skin is your first line of defense protecting you from disease, and intact skin is the best protection from most infections. Surprisingly, some diseases of the integumentary system are actually caused by diseases from other body systems! Measles would be a perfect example. This is a viral disease of the respiratory system that causes a rash on your skin.

Your integumentary system also performs several vital functions besides protecting you from an invasion of disease-producing pathogens. This system helps keep the body from drying out, provides a natural sunscreen, acts as storage for fatty tissue necessary for energy, and, with the aid of some sunshine, produces vitamin D (needed to help your body utilize phosphorus and calcium for proper bone and tooth formation and growth). In addition, the skin provides sensory input (pleasant and unpleasant sensations involving pressure and temperature, for example) to your brain and helps regulate your body temperature.

The Skin

Your skin is quite a large organ, easily weighing twice as much as your brain, approaching 20 pounds in an average adult. In fact, the skin is your largest organ. It covers an area of about 20.83 square feet on an adult-sized body. A closer examination of a cross section of skin reveals three main layers of tissue:

- **epidermis**
- **dermis**
- **subcutaneous fascia** (also called the **hypodermis** layer because it lies *under* the dermis)

As we discuss these three layers, please refer to Figure 8-1 ▦ for further clarification.

epidermis *(ep ih DER miss)*
  epi = *upon*
  dermis = *true skin*
subcutaneous fascia
  (sub cue TAY nee us FAY she ah)
  sub = *under*
  cutane/o = *skin*
  fascia = *band*

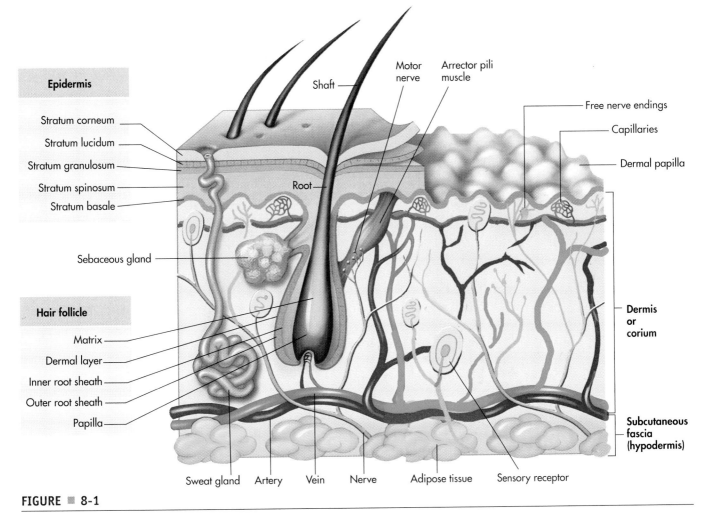

FIGURE ▦ **8-1**

The three layers of the skin.

EPIDERMIS

The epidermis is the layer of skin that we normally see. It is made up of several even smaller layers of stratified squamous epithelium. The epidermis is interesting for several reasons. First, it is **avascular** and contains no nerve cells. Second,

avascular *(ay VASS cue ler)*
  a = *without*
  vascular = *referring to vessels*

stratum corneum
   (STRAY tum core NEE um)
   stratus = *to spread out*
   kerat/o = *hard or horny*

melanocytes *(mell AN oh sights)*
   melan/o = *black, extremely dark hue*
melanin *(MELL an in)*
carotene *(CARE eh teen)*

the cells on the surface of this layer are constantly shedding, being replaced with new cells that arise from the deeper region called the **stratum basale** or basal layer of the epidermis in a process that takes 2 to 4 weeks. In fact, the outermost surface of skin is actually a layer of dead cells called **stratum corneum,** which are characteristically flat, scaly, keratinized (hardened) epithelial cells. Cells are born in the basal layer. This layer is also called the *stratum germinativum.* As cells are pushed toward the surface, they die and fill up with the protein, keratin. Through everyday activities such as bathing, drying off, and moving around, the body sloughs off *500 million* cells a day, equaling about *one and a half pounds* of dead skin a year! This continuous replacement of cells is very important because it allows your skin to quickly repair itself in cases of injuries.

Specialized cells called **melanocytes** are located deep in the epidermis and are responsible for skin color. Melanocytes produce **melanin,** which is the actual substance that affects skin color. An interesting note is that all people possess about the same number of melanocytes. The variations of skin color are a result of the amount of melanin that is produced and how it is distributed. This is obvious when you are exposed to the ultraviolet rays of the sun. In order to protect your skin, melanocytes produce more melanin and, voila!, you've got a tan. You will also note that in time you will develop a tan line demonstrating that the production of melanin occurs only as needed, in the areas where needed. Regardless of an individual's skin tone or color, all skin types respond to sun exposure. For some of us, the melanin locates in patches on the skin, forming *freckles.* **Carotene,** which is another form of skin pigment, gives a yellowish hue to skin. Individuals with a pinkish hue derive that color from the hemoglobin in their blood.

8-1 To perform an interactive drag and drop labeling exercise of the three layers of skin and the integumentary system, see your DVD.

People with a condition called *albinism* have very little pigment in their skin, hair, or eyes. This is because they have inherited genes that do not allow body cells to produce the usual amounts of melanin. People with albinism may have vision problems, due to the lack of pigment in their eyes. They may also be at a greater risk for skin cancer.

Pathology Connection: Skin Color and Disease

There are times when skin color can indicate an underlying disease. Although carotene gives a yellowish hue to skin, that is normal. In a situation where liver disease exists, the body can't excrete a substance call **bilirubin**. As bilirubin builds up in the body, **jaundice** occurs, giving the skin a deeper yellow color. The changes in color are not as apparent on individuals with darker skin, but the yellowish color is easily seen in the whites of the eyes. Although bronze skin is associated with healthy outdoor living, individuals with a malfunctioning adrenal gland may have the same color due to excessive melanin deposits in the skin. Excessive bruising (black and blue marks called **ecchymosis**) could indicate skin, blood, or circulatory problems as well as possible physical abuse.

ecchymosis *(ek ee MOH sis)*

DERMIS

The layer right below, or inferior to, the epidermis layer is the thicker dermis layer, often called the **corium.** This layer of dense, irregular, connective tissue is considered the "true skin" and contains the following:

corium *(CORE ee um)*

- capillaries (tiny blood vessels)
- collagenous and elastic fibers
- involuntary muscles
- nerve endings
- lymph vessels (transport fluids from tissue to the blood system)
- hair follicles
- sudoriferous glands (sweat)
- sebaceous glands (oil)

Small dermal papillae project from the surface of the dermis (stratum germinativum) and anchor this layer to the epidermal layer. These ridges are arranged so they provide the maximum resistance to slipping when grasping and holding objects. Fingerprints, toe prints, and other unique skin patterns also arise from this layer and are used in identification. Nerve fibers are located in the corium so the body can sense what is happening in the environment. Because this layer also possesses blood vessels, this is where your blush comes from when you get embarrassed!

The collagenous and elastic fibers of this layer help your skin flex with the movements that you make. Without that ability, your skin would eventually tear from all of your moving around. In addition to flexing, these fibers allow your skin to return to its normal shape when at rest. In older people and people regularly exposed to high levels of sunlight, the skin's firmness and ability to recoil to normal decreases. To better understand this process, try this experiment. With one of your hands resting palm down, take the thumb and index finger of your other hand and gently pinch and pull up the skin on the back of your resting hand. Let go and observe how quickly the skin recoils to normal. Try this experiment with an older person and observe how much slower his or her skin returns to normal. Another example of the skin's resilience due to the collagenous and elastic fibers is when it returns to normal after an injury that caused swelling.

Amazing Body Facts

A BOTHERSOME FACT OF LIFE

Sometimes the sebaceous glands within the skin become blocked. As a result, sebum stagnates and is exposed to air, drying it out. When this occurs, the sebum turns black, creating the infamous *blackhead!* To make matters worse, if that blackhead becomes infected, a *pimple* is formed (medically known as a *pustule*). The best thing to do is keep those areas clean through gentle washing with soap and water and let nature take its course.

A somewhat related condition is folliculitis which is usually a staphylococcus bacterial infection and inflammation of hair follicles. This condition presents with the formation of small pustules that surround the follicle.

This brings us to the topic of cleaning your skin. The skin is generally too dry for microbial growth, except for certain areas. Most skin bacteria is associated with hair follicles or sweat glands, so cleaning the areas of your skin that are hair covered and/or contain a high concentration of sweat glands is especially important. In addition, washing your hands frequently is the best way to prevent spread of disease. However, be careful not to wash too frequently with too hot water and/or too harsh soap, and avoid aggressive drying. This is important because excessive cleaning has the potential to dry out your skin and remove the antibacterial layer of sebum. In fact, aggressive washing with a soap that isn't pH balanced (the same acidity as your skin) can cause skin to lose its antibacterial abilities for up to 45 minutes after washing. As with most things in life, moderation is the key.

SWEAT AND SEBACEOUS GLANDS

The sweat and sebaceous glands are part of the dermis. Sweat glands are distributed over the entire skin surface, with large numbers under the arms, in the palms, on the soles of the feet, and on the forehead. The perspiration they generate is excreted through pores. There are two main types of sudoriferous or sweat glands: apocrine and eccrine glands (see Figure 8-2 ■). **Apocrine** sweat glands secrete at the hair follicles in the groin and anal region as well as the armpits. These glands become active around puberty and are believed to act as a sexual attractant. Located all over your skin, and in people of all ages, **eccrine** glands are important in the regulation of body temperature. Eccrine glands are found in greater numbers on your palms, feet, forehead, and upper lip. Sweat glands can be activated by heat, pain, fever, or nervousness. Your body has approximately 3,000,000 sweat glands, generating an average fluid loss via sweating of 500ml per day! It is interesting to note that sweat is 99% water and by itself does not have a strong odor. However, if it is left on the skin, bacteria degrade substances in the sweat into chemicals that give off strong smells, commonly called body odors. Ingestion of certain foods, like garlic, can also affect the odor of sweat.

apocrine *(APP oh crin)*

eccrine *(EKK rin)*

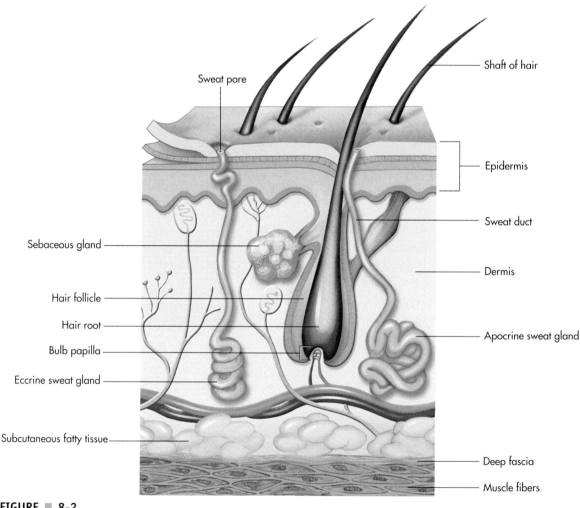

FIGURE ■ 8-2

Sweat and sebaceous glands.

Sebaceous glands play an important role by secreting oil, or **sebum**, that keeps the skin from drying out. Because sebum is somewhat acidic in nature, it also helps destroy some pathogens on the skin's surface.

8-2 Please go to your DVD to view a video on the changes that occur with aging to the integumentary system.

Sebaceous glands are usually found in hair-covered areas, where they are connected to hair follicles. The glands deposit sebum on the hairs, and bring it to the skin surface along the hair shaft. Sebaceous glands are also found in areas that do not have hair follicles, such as eyelids, penis, labia minora, and nipples. In these areas the sebum transverses ducts that terminate in sweat pores on the surface of the skin. A specialized form of sebaceous gland located at the rim of the eyelids called **meibomiam glands** (or tarsal glands) secrete sebum into the tears that coat the eye, slowing evaporation.

SUBCUTANEOUS FASCIA

Finally, the innermost layer is the subcutaneous fascia, or **hypodermis,** which is composed of elastic and fibrous connective tissue and fatty tissue. Within this layer, **lipocytes,** or fat cells, produce the fat needed to provide padding to protect the deeper tissues of the body and act as insulation for temperature regulation. Fat is also necessary as an efficient store for energy. The hypodermis is also the layer of skin that is attached to the muscles of your body.

hypodermis *(high poh DER miss)*
  hypo = *below, under*
lipocyte *(lip OH sight)*
  lip/o = *fat*
  cyte = *cell*

The Pathology Connection: Infections of the Skin

In this section we will discuss a variety of additional skin conditions that are caused by infection. **Herpes** is a lifelong viral infection that produces an inflammation of the skin in the form of clusters of small, fluid filled blister–like sacs called **vesicles**. Although this is a lifelong infection, there are periods of remission (periods where there are no noticeable signs of infection) and exacerbation (periods of "flare ups"). Exacerbation is usually caused by decreased immunity as a result of stress or other diseases. See Figure 8-3 ■ for an example of herpes.

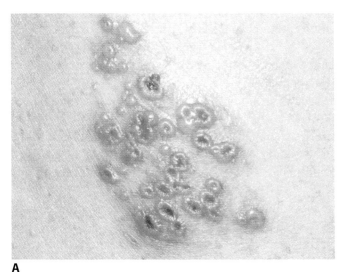

A

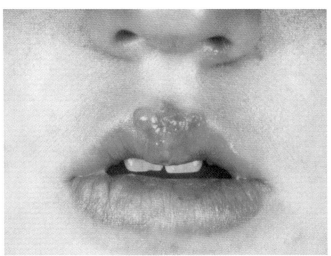

B

FIGURE ■ **8-3**

Herpes types. **A.** Shingles and **B.** typical cold sores or fever blisters.

Herpes varicella is also called **chickenpox**, and is a highly contagious common childhood infection that can be spread by airborne particles or direct contact. Vesicles produced by Herpes varicella can be found on the face, trunk and extremities. It can be widespread or limited in nature and can cause intense itching. There is now a vaccine given to children for chickenpox. This vaccine decreases the risk of contracting the disease and if it is contracted, the disease is less severe without possible deadly complications. For those who reach adulthood without having the disease and have a negative titer, which means bloodwork shows the person has never been exposed to chickenpox, the vaccine may be worth consideration even in older patients.

Herpes zoster is commonly known as **shingles**, and is a painful condition that creates lesions that appear to follow a spinal nerve pathway. It is believed that shingles is the dormant, adult version of chickenpox that the patient initially had exhibited in childhood. Shingles develop in the adult when the immune system cannot keep *herpes zoster* in check as a result of stress, disease, trauma, or the ordinary aging process. Approximately 50% of individuals over 80 has had at least one episode of shingles. Although the blisters and rashes are usually found on the trunk in a midline fashion, following the course of a sensory nerve, they are often located on the face and can cause conjunctivitis. In either case, these lesions are extremely painful, producing sharp, stabbing sensations that may last from 10 days to several weeks.

Herpes simplex type 1 forms the "cold sore" or "fever blister" that is found around the mouth or nose following the common cold or febrile situations. *Herpes simplex type 2* is the causative agent of genital herpes. In its active stage, this virus produces blisters in the genital area. The lesions created by *herpes simplex type 1* and *type 2* are the same and clinically indistinguishable. *Herpes simplex type 2* is highly contagious when in its active stage, although it can also spread in its remission stage. It is spread by direct contact. Although it is mainly spread by sexual contact, it is not limited only to sexual transmission. *Herpes simplex type 2* can also be transmitted from infected mother to newborn child during vaginal delivery.

The drug acyclovir may be used to prevent outbreaks by people who have genital herpes. It can also be used to treat the symptoms of chickenpox and shingles. However, it cannot cure herpes.

Warts, or **verruca**, are caused by the papillomavirus, which causes a hypertrophy of the keratin cells in the skin. Warts present in a variety of shapes, sizes, numbers, and appearances. See Figure 8-4 ■ for examples of warts.

Common warts are normally found on children's hands and fingers. These normally painless and harmless warts are spread as a result of scratching and direct contact. Often they will disappear on their own. Adults should be concerned with wart-like lesions that form on their skin that could possibly be a skin cancer. These should be watched for any change in character such as size, shape or color.

Plantar warts are found on the soles of the foot and tend to grow inward, exhibiting a smooth surface. The fact that this type of wart grows inward tends to make walking painful if the wart is located on a pressure point. The first line of treatment is with a topical application of salicylic acid, but this is usually temporary and they come back. These warts can be removed surgically or cryogenically (frozen).

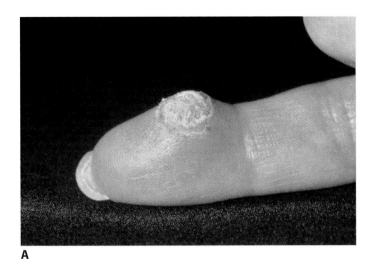
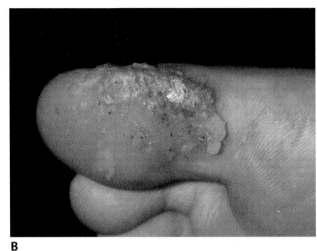

A

B

FIGURE ■ 8-4

Examples of warts. **A.** common wart and **B.** plantar wart.

Genital warts are highly contagious and sexually transmitted. Evidence indicates a strong correlation between genital warts and cervical cancer. The evidence is strong enough that there currently is a controversy over requiring a papillomavirus vaccine for teenage girls.

Fungi can also cause various skin disorders. **Tinea** is a general term relating to fungal skin infections that are usually located in warm, moist regions of the body. Fungal infections exhibit cracking and weeping skin with an itch. Often fungal infections can reoccur to the point of being a chronic condition. There are many fungal infections, usually named for the region in which they are found. See Figure 8-5 ■ for examples of fungal infections.

Tinea pedis, or athlete's foot, is the most common fungal skin infection. It is highly contagious, spreading by direct contact of contaminated surfaces such as shower and locker room floors and towels. Since the areas between the toes are warm and maintain moisture, this is the most common location for tinea infection (see Figure 8-5A). If left untreated, the fungus can spread over the entire foot.

Tinea cruris, also known as jock itch, mainly affects men. It is a fungal infection of the groin and scrotal areas. Warm weather with increased perspiration, increased physical activity, and tight fitting shorts/pants/ undergarments tend to aggravate this condition.

Tinea corporis is somewhat misleadingly known as ringworm. This condition presents on the smooth skin of the legs, arms, and body as red, ring-shaped structures with a pale center. This can resemble the shape of a worm under the skin, but no worm is involved; a parasitic fungus is responsible for the infection.

Tinea unguium is a condition in which a fungal infection is involved with either finger nails or toe nails (see Figure 8-5B). These infections can be difficult to treat due to the infection's location under the affected nail. If this disease is left untreated, the nail will become overgrown and thickened with a white and brittle appearance.

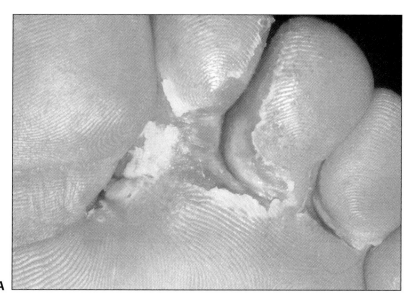

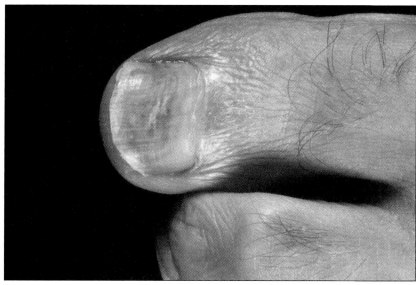

FIGURE ▪ 8-5

Examples of fungal infections. **A.** Athlete's foot (*tinea pedia*), **B.** Nail fungus (*tinea unguium*).

8-3 Often patients who are bedridden for a prolonged period of time develop skin breakdown or pressure sores. To view an animation of pressure sores formation, prevention, and treatment, please go to your DVD.

Cellulitis, which is caused by the bacterium staphylococcus, is an inflammatory condition of the skin and subcutaneous tissue, producing a red, swollen, and painful area. Often the source of this infection is a bedsore, ulcer, or wound. If left untreated, cellulitis can lead to life-threatening situations such as endocarditis or even septicemia.

You probably have heard a lot about this next condition if you listen to the news or spend a fair amount of time outdoors. Named after the town of Lyme,

Connecticut, **Lyme disease** involves multiple systems of the body as a result of a bacterial infection that is spread via the bite of the deer tick. Flu-like symptoms, malaise, joint inflammation, fever, and chills are some of the symptoms of this disease. If left untreated, Lyme disease can lead to neurologic and cardiovascular problems as well as arthritis. So why are we discussing this disease in the integumentary section of this book? A common early sign of this condition is a rash that looks like a red circle with a lighter center resembling a "bull's eye." This rash appears between a few days and several weeks following the tick bite. A blood test will confirm the actual infection.

TEST YOUR KNOWLEDGE 8-1

Complete the following:

1. List the three main layers of skin.

   _____

   _____

   _____

2. List four of the functions of your integumentary system.

   _____

   _____

   _____

   _____

Choose the best answer:

3. Which cells are responsible for your normal skin color?
   a. monocytes
   b. jandicytes
   c. eyecytes
   d. melanocytes

4. The two main types of sudoriferous glands are
   a. apocrine and pelicine
   b. appeltine and eccrine
   c. eccrine and apocrine
   d. sudacrine and melocrine

5. This form of herpes causes the painful condition known as shingles:
   a. varicella
   b. zoster
   c. simplex type I
   d. simplex type II

6. This form of wart can make walking painful due to its inward growth and position over a pressure point:
   a. plantar
   b. common
   c. genital
   d. cruris

HOW SKIN HEALS

Just as storms can damage the exterior of a house by high winds tearing off shingles or siding, lightning burning portions of the exterior, or ice causing damage to the roof, there are many ordinary objects that can damage skin. As a result, the body has developed ways to repair itself when injury threatens its first line of defense—the skin.

If skin is punctured and the wound damages blood vessels in the skin, as shown in Figure 8-6 ■, the wound fills with blood. Blood contains substances that cause it to clot. (For more information about clotting, see Chapter 12.) The top part of the clot, which is exposed to air, hardens to form a scab. This is nature's bandage, forming a barrier between the wound and the outer environment to prevent pathogens from entering. So don't pick your scabs.

Next, the inflammatory process kicks in, with the white blood cells entering to destroy any pathogens that may have entered when the wound occurred. At about the same time, cells called **fibroblasts** (cells that can develop into connective tissue) come in and begin pulling the edges of the wound together. The basal layer of the epidermis begins to hyperproduce new cells for the repair of the wound. If the wound is severe enough, a tough scar composed of collagen fibers may form. Scars usually don't contain any accessory organs of the skin or any sense of feeling. Scar production can be greatly minimized if stitches, adhesive strips, or a specialized glue are used to draw the margins of the wound together before the healing process begins. Ideally, the wound starts to heal from the *inside*, working its way to heal the wound toward the outside. This aids in preventing pathogens from becoming trapped between a healed surface and the deeper layers of skin, where they could develop into a major pocket of infection. Diabetic patients such as Maria often have difficulty with wound healing due to decreased blood flow and WBC activity to that area.

fibroblast

> **fibr/o** = *fibers, or fibrous tissue*
>
> **blast** = *immature cellular development*

8-4 For an animation on wound repair and the formation of scar tissues, visit the DVD for this chapter.

Pathology Connection: Burns

Skin burns present special problems for healing. We naturally think that burns are caused by heat, and that is true. However, burns can also be caused by chemicals, electricity, or radiation. When assessing the damage caused by burns, there

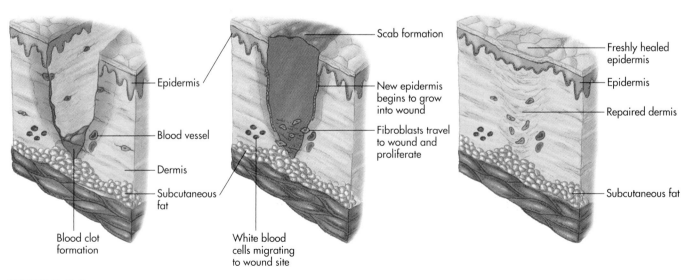

Epidermis

Blood vessel

Dermis

Subcutaneous fat

Blood clot formation

Scab formation

New epidermis begins to grow into wound

Fibroblasts travel to wound and proliferate

White blood cells migrating to wound site

Freshly healed epidermis

Epidermis

Repaired dermis

Subcutaneous fat

FIGURE ■ 8-6

Wound repair.

are two factors to consider: the depth of the burn and the size of the area damaged by the burn.

The depth of a burn relates to the layer or layers of skin affected by the burn. A **first-degree burn** has damaged only the epidermis, and is classified as a **partial thickness burn** (that is, a burn that does not affect the entire depth of the skin). With a first-degree burn there is skin redness and pain but no blistering. The pain usually subsides in 2 to 3 days with no scarring. The damaged layer of skin usually sloughs off in about a week or so. Sunburn is a classic example of a first-degree burn.

Second-degree burns involve the entire depth of the epidermis and a portion of the dermis. These burns are still considered partial-thickness burns because they do not go all the way through the skin. Such burns cause pain, redness, and blistering. The extent of blistering is directly proportional to the depth of the burn. Blisters continue to enlarge even after the initial burn. Excluding any additional complications such as infection, these blisters usually heal within 10 to 14 days, but burns reaching deeper into the dermis require anywhere from 4 to 14 weeks to heal. Scarring in second-degree burn cases is common.

Small second-degree burns may be treated at home. For a burn without open blisters, the patient should run cool water over the burn. They should dry the burn carefully, in order to avoid breaking the blisters, and apply a dry, sterile dressing to the burn. They may take ibuprofen to help with the pain. A burn with open blisters should be covered with a dry, sterile bandage.

Amazing Body Facts

KELOIDS

A **keloid** can be considered a "scar gone wild." It is a mass of scar tissue that has a raised, firm, irregular shape (see Figure 8-7■). A keloid often occurs as a result of trauma or surgical incision, and can even form following an ear piercing. The formation of this mass is a result of an overproduction of connective tissue at the affected site. The formation of keloids appears to be more prominent in the black population. Surgical removal would not make sense (because more keloids would form), so treatment usually consists of steroid injection at the lesion site and cryotherapy (freezing).

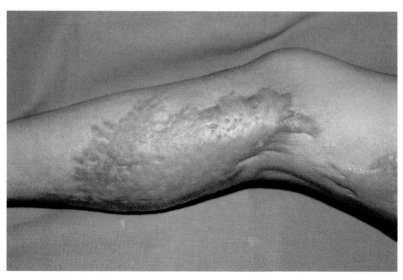
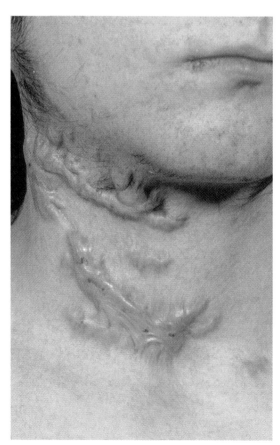

FIGURE ■ 8-7

Examples of keloids. (Custom Medical Stock Photo, Inc.)

Clinical Application

MEDICINE DELIVERY VIA THE INTEGUMENTARY SYSTEM

A variety of medicines can be delivered via the integumentary system. Medicines can be applied to adhesive patches that are placed on the skin where it is slowly absorbed into the bloodstream. These are called **transdermal patches.** Nicotine (for smoking cessation), nitroglycerin (for vasodilation in the heart), birth control compounds, and pain medication, for example, can be delivered in this manner. If a more rapid response is required, the cardiac drug nitroglycerin can be placed under the tongue (**sublingually**), where it is rapidly absorbed into the bloodstream because of the high vascularization of the mucosa in that area.

Of course, the other method of drug delivery via the integumentary system is a little more painful but very effective. Injection is used when a drug can't be taken by mouth or the digestive system may alter the desired effects of the drug. Medication can be injected utilizing a syringe and needle to deliver the medication either **subcutaneously** (under the skin) or **intradermally** (into the skin). Other routes of injection also include intramuscular (IM), intraspinal, and intravenous (IV) routes.

www.prenhall.com/colbert

Emergency medical technicians (EMTs) and paramedics often deal with burn victims as a result of structural or automobile fires. To learn more about emergency professions, please go to the companion website for information and to view a video.

To avoid infection, it should not be rinsed with water. Larger second-degree burns, or burns with open blisters, may need to be treated by a doctor. The doctor may prescribe intravenous fluids to replace those lost by your body, and antibiotics to fight infection. Stronger pain medicine may also be prescribed.

Third-degree burns affect all three of the skin layers and are therefore classified as **full-thickness burns**. Here the surface of the skin has a leathery feel to it and varies in color: black, brown, tan, red, or white. Initially, the victim will feel no pain at the affected site because pain receptors are destroyed by third-degree burns. Also destroyed are the sweat and sebaceous glands, hair follicles, and blood vessels. Third degree burns can be life threatening, but improvements in burn care have reduced the number of mortalities. The doctor may prescribe antibiotics to fight infection and intraveneous fluids. Skin grafting may be necessary.

Fourth-degree burns are burns that penetrate to the bone. These burns are full-thickness burns that also destroy tendons and muscle. The victim feels no pain at the affected site. Fourth-degree burns on arms or legs may require amputation of the affected area.

A clinician can estimate the extent of the area covered by the burn by using the "rule of nines." As you can see in Figure 8-8 ■, the body is divided into the following regions and given a percentage of body surface area value: head and neck, 9%; *each* upper limb, 9%; *each* lower limb, 18%; front of trunk, 18%; back of trunk and buttocks, 18%; perineum (including the anal and urogenital region), 1%. These regions can also be further divided for smaller burn areas, as you can see in Figure 8-8.

The clinical concerns for burn patients relate to the functions of the skin already discussed:

- bacterial infection
- fluid loss
- heat loss

Patients with third-degree burns need prompt medical treatment. The patient should be taken to a burn center or hospital emergency room. Burn patients generally receive intravenous therapy, which provides replacement fluids, antibiotics, and pain medications. Burns are cleaned, treated with antibiotic

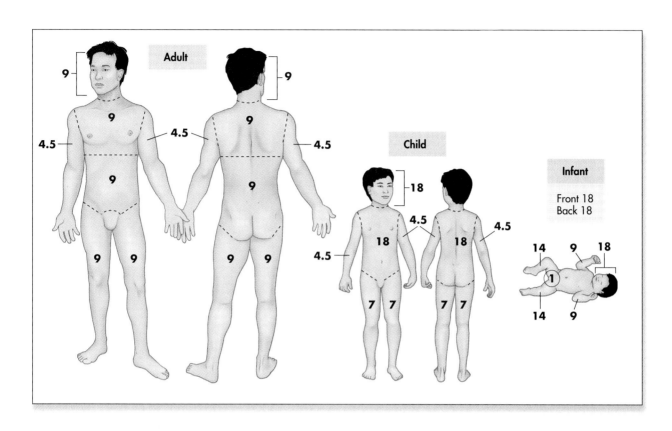

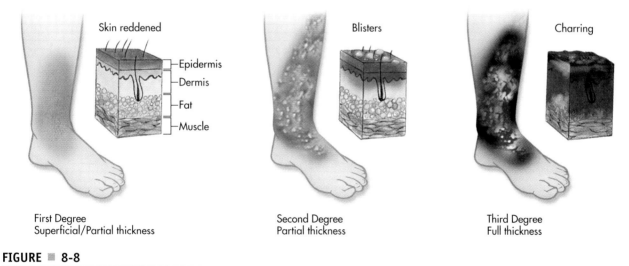

First Degree	Second Degree	Third Degree
Superficial/Partial thickness	Partial thickness	Full thickness

FIGURE ■ 8-8

Assessing the degree of the burn.

cream, and covered with dry, sterile dressings. The burn patient may be given a tetanus shot. The patient may also be placed in a hyperbaric chamber—a sealed structure in which pure oxygen is supplied at a higher than normal atmospheric pressure. Exposure to oxygen helps to speed healing.

Severe burns, such as third-degree burns, require healing at an intensity that the body can't normally achieve on its own. Damaged skin must be removed as soon as possible to allow the process of skin grafting to begin. In this process, healthy skin is placed over damaged areas so it may begin to grow.

auto = *self*

Ideally, it is best to use the patient's own skin, known as **autografting**, because it generally eliminates the chances of tissue rejection. However, the destruction may be severe enough to require tissue from a donor (**heterografting**). Grafting usually requires repeat surgeries because large areas cannot be done all at once, and often the grafts don't "take." Other options may include growing sheets of skin in a laboratory from cells of the patient or utilizing synthetic materials that act as skin.

 8-5 For videos on intradermal drugs and subcutaneous injections, please visit your DVD.

NAILS

keratinized *(KAIR ah tin ized)*
kerat/o = *hard or horny*

Specialized epithelial cells originating from the **nail root** form your nails (see Figure 8-9 ▦). The nail root is also called the **germinal matrix**. As these cells grow out and over the **nail bed** (actually a part of the epidermis), they become **keratinized,** forming a substance similar to the horns on a bull that is the same protein that fills the cells of the stratum corneum. This process occurs as cells dry and shrink, are pushed to the surface, and become filled with a hard protein called keratin. The **cuticle** is a fold of tissue that covers the nail root. The portion that we see is called the **nail body**. The **sterile (nail) matrix** is a proximal layer of cells that helps the nail body to stick to the nail bed. Nails normally grow about 1 millimeter every week. The pink color of your nails comes from the vascularization of the tissue under the nails, while the white half-moon shaped area, or **lunula,** is a result of the thicker layer of cells at the base.

lunula *(LOO nyoo lah)*
luna = *moon*

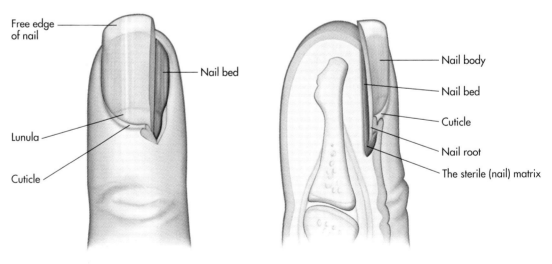

 FIGURE ▦ 8-9

Structures of the fingernail.

HAIR

Body hair is normal and served important purposes in our evolutionary past as well as today. Hair helps to regulate body temperature, as you will see in the next section, and it functions as a sensor to help detect things on your skin, such as bugs or cobwebs. Eyelashes help to protect your eyes from foreign objects, and hair in the nose helps to filter out gross particulate matter.

Hair is composed of a fibrous protein called **keratin,** just like your fingernails and toenails. This is also the same protein that fills the dead cells on the surface of the epidermis. The hair that you see is called the shaft. The shaft is covered by a protective layer of flat cells called the **cuticle.** The **cortex** layer lies beneath the cuticle. The cortex is made of twisted proteins and the melanin pigments that give hair its color. Coarse hair has another layer of cells beneath the cortex, called the **medulla.** Each hair has a root that extends down into the dermis to the **follicle** (see Figure 8-10 ■). The follicle is formed by epithelial cells, which have a rich source of blood provided by the dermal blood vessels. As a result, cells divide and grow in the base of the follicle. As new cells continually form, the older cells are crowded out and pushed upward toward the skin's surface. As these old cells are pushed away from the blood source (which provides their nourishment), they die, becoming keratinized in the process. So basically, the hair you see on the individual next to you is a bunch of dead cells. This isn't so bad considering that if your hairs were alive, your haircuts would be extremely painful, with a good chance of you bleeding to death if you got more than a light trim! The old, popular belief that shaving or frequent cutting of hair makes it grow quicker or thicker is wrong. Neither shaving nor cutting does anything to affect the rate of cell growth at the base of those hair follicles.

If you look at Figure 8-10 ■, you will note that there is a **sebaceous gland** associated with the hair follicle. The sebaceous gland secretes **sebum,** an oily substance that coats the follicle and works its way to the skin's surface to waterproof and lubricate the skin and hair. Sebum is somewhat antibacterial, so it aids in decreasing infections on your skin. Sebum production decreases with age. As a result, older individuals exhibit drier skin and more brittle hair.

Like skin color, your hair color is dependent on the amount and type of melanin you produce. Generally, the more melanin, the darker the hair. White hair occurs in the absence of melanin. Red hair is a result of an altered melanin that contains iron. Gray hair is usually due to age-related decreases in melanin production.

You may be extremely envious of someone with a head full of curly hair or of someone with absolutely straight hair. Don't be envious of the person—just envy his or her hair shafts! Flat hair shafts produce curly hair, and round hair shafts produce straight hair.

keratin *(KAIR ah tin)*

follicle *(FALL ih kle)*

sebaceous *(see BAY shuss)*
 seb/o = *tallow (animal fat)*
sebum *(SEE bum)*

ASSESSING PERIPHERAL PERFUSION

The pink color of the nail bed is clinically significant in that it can aid in the assessment of perfusion (blood flow) to the extremities and can be a determinant of oxygenation. If you pinch one of your fingernails straight down with the thumb and index finger of your other hand for 5 seconds and release that pinch, you will note that the nail bed changes from a blanched white color back to pink in a matter of seconds. This shows good perfusion as a result of the blood rushing back into the nail bed. In cases of poor perfusion, it takes longer for the nail bed to "pink-up." Normally, it takes less than 3 seconds for the nail to return to pink from the blanched white state. If that time is greater than 3 seconds, perfusion to the extremities is considered sluggish. If the refill time is greater than 5 seconds, there is clearly an abnormal situation occurring. This is called capillary refill time.

Diabetes is a disease that may cause a condition of reduced blood flow to the extremities known as **peripheral vascular disease (PVD).** An increase in the time required to re-perfuse the nail bed may be an indication of PVD. Blood clots or vascular spasms can decrease blood flow and thereby extend refill time, as can hypothermia, which naturally constricts blood vessels in the periphery to conserve heat.

In addition, nail beds can change colors under certain conditions. For example, as the level of oxygen decreases in the tissue, the nail beds become bluish in tint. Of course, one can always paint their nails different colors!

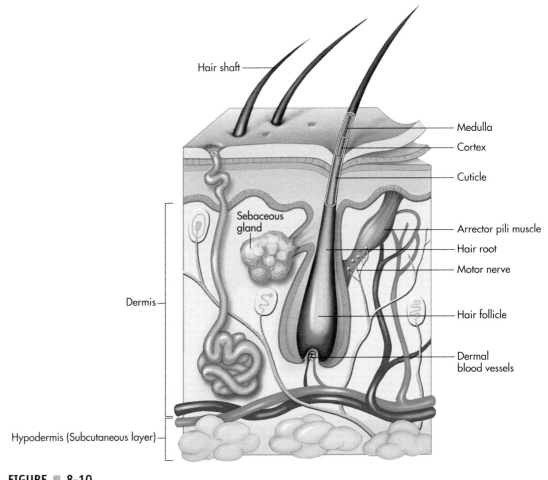

Diagram of a hair follicle.

FIGURE ■ 8-10

Alopecia is the term for any type of hair loss and can be acute or chronic. Some forms of alopecia, such as male pattern baldness, do not represent a disease, but are inherited traits. Hair loss can also be a result of chemotherapy for cancer treatments, hormonal imbalance, infections, severe emotional or physical stress, or side effects of some other medications.

The Pathology Connection: Lice and Scabies

One of the "rites of passage" in school used to be having your scalp checked for head lice. Lice are tiny insect parasites. **Pediculosis** is the term for an infestation with lice that is a result of direct contact with a person infested with lice or sharing objects that may be infested. Please note that lice know nothing about the socioeconomic status of their victims. *Anyone* can become infested.

8-6 To perform an interactive drag-and-drop exercise on the hair follicle, please go to your DVD.

There are several types of lice infestations. **Head lice** is the common type that you are checked for in school. **Body lice** infestations are often a result of poor personal hygiene. These lice can be carriers of disease such as typhus.

Pubic lice, often called "crabs," are spread through sexual contact.

Treatment for all forms of lice is the same: bathing or shampooing with a medicated shampoo specifically designed to eliminate lice and their eggs (known as "nits"). All bedding, towels and clothing, hats and combs / hairbrushes must be thoroughly cleaned or discarded to prevent reinfestation.

Another little fellow that can cause problems is a tiny mite that burrows into the skin to lay eggs, causing a condition known as **scabies**. These mites are transmitted via direct contact from an infected individual, and usually lodge themselves in the folds of the skin (wrist, underarms, groin, under breasts, etc.). Symptoms of scabies include vesicles, pustules, and intense itching of the affected area. The eggs usually hatch within 3 to 5 days. The young mature on the skin surface within 2 to 3 weeks, and if not destroyed, they mate and start the whole process over again. A specially formulated cream is used on the skin to kill these mites. All infected individuals must be treated for the cycle of infestation to be broken. For examples of lice and scabies, see Figure 8-11 ■.

Applied Science

FORENSICS AND HAIR

An interesting sidelight concerning hair is its ability to tell a pathologist if an individual ingested certain drugs or other substances, such as lead or arsenic. Trace amounts of ingested substances can become part of the hair's composition. As a result, analysis of a hair sample can reveal what and how long ago something was ingested. The longer the length of hair, the longer the record of what that individual consumed.

The life span of hair is dependent on location. Eyelashes last around 3 to 4 months, whereas the hair on the scalp lasts for about 3 to 4 years. Teachers with difficult classes usually pull their hair out sooner!

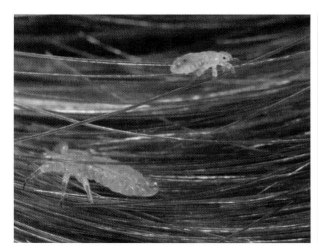 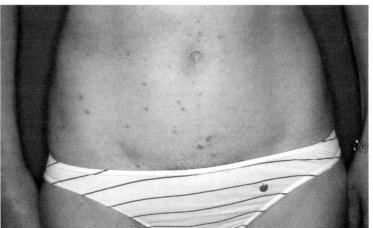

FIGURE ■ 8-11

Examples of lice and scabies.

TEST YOUR KNOWLEDGE 8-2

Choose the best answer:

1. The blood clot that forms in a wound that is exposed to air becomes a:

   a. scar

   b. scab

   c. hematoma

   d. keloid

2. When assessing the skin damage caused by a burn, the two main factors to assess are:

   a. temperature and depth

   b. depth and odor

   c. area of damage and depth

   d. odor and color

3. The white, half-moon shaped area of your fingernail is called:

   a. cuticle

   b. lunula

   c. keratin

   d. lingula

4. Corium is another term for:

   a. epidermis

   b. contrasta

   c. corneum

   d. dermis

5. A skin formation that can be considered a "scar gone wild" is a:

   a. scab

   b. hematoma

   c. keloid

   d. lunula

TEMPERATURE REGULATION

The integumentary system plays a major role in the regulation of body temperature. It is amazing how we can stay in a relatively "tight" range of body temperature while we do a variety of things in a variety of environments. Of course, clothes do help tremendously, but still it is critical to have a properly functioning temperature regulator like our integumentary system. This is accomplished through a complex series of activities.

Part of temperature regulation is accomplished by changes in the size of blood vessels in your skin. As your temperature rises, your body signals the blood vessels in your skin to get larger in diameter. This is called **vasodilation.** It is the body's attempt to get as much "hot" blood as possible exposed to a cooler surrounding environment, so the heat radiates away from the body. In addition, sweat glands excrete water (as well as some waste products such as nitrogenous wastes and sodium chloride) onto the skin's surface. As the water evaporates, cooling occurs. As long as you stay hydrated so you can produce sweat, this system works pretty well. Hydration is especially important before and during activities to avoid feeling thirsty. Thirst indicates the body is already dehydrating. It's pretty easy to dehydrate, especially when you realize that you can have up to 3,000 sweat glands on a square inch of skin on your hands and feet and you can potentially excrete up to 12 liters of sweat in a 24 hour period! The risk of dehydration can be serious.

vasodilation *(vaz oh DYE lay shun)*
vas/o = *vessel*
dilation = *enlargement*

Conversely, if you were in a cold environment and needed to warm up, your blood vessels would become smaller in diameter, or **vasoconstrict.** Vasoconstriction forces blood *away* from the skin and back toward the core of the body where the heat is. The perfect visual explanation of vasoconstriction is a skinny little kid running around a pool on a chilly summer day, shivering, with blue lips. The lips are blue because of vasoconstriction. Rings have a tendency to slide off your fingers in cold weather more readily than in hot weather because of vasoconstriction.

Body temperature regulation is also aided by the hairs on your skin. Muscles in your skin called **arrector pili** are attached to your hairs, and when those muscles contract, they make your hairs stand erect. The constriction of those muscles shows up as goose flesh or goose bumps when you are chilled. When the hair stands up, pockets of still air are formed right above the skin, creating a dead air space that insulates the skin from the cooler surrounding environment. This is also how goose down (feather) clothing works in protecting against winter's cold. Interestingly, older homes, built before the now-common and effective practice of placing insulation in the walls and ceilings, actually insulated the homes through the creation of dead air (nonmoving) spaces between the outer and inner walls. See Figure 8-12 ■, which illustrates how your integumentary system regulates body temperature with the aid of your nervous system. Patients with spinal cord injuries such as Ray have poor control of their body temperatures. This is because their nervous systems can't communicate well with their bodies to control vasoconstriction and dilation in response to temperature change.

vasoconstrict *(vaz oh kon STRIKT)*

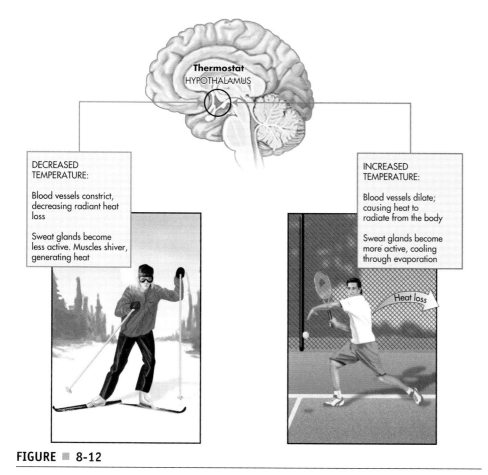

FIGURE ■ **8-12**

Integumentary regulation of body temperature.

Pathology Connection: Skin Lesions

A **lesion** is a pathologically altered piece of tissue that can include a wound or injury or a single infected patch of skin. The color of a lesion is usually different than that of normal skin. Figure 8-13 ■ shows a variety of lesion types.

A **macule** is a discolored spot on the skin; freckle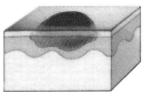

A **wheal** is a localized, elevation of the skin that is often accompanied by itching; uticaria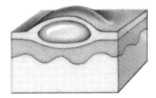

A **papule** is a solid, circumscribed, elevated area on the skin; pimple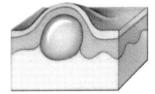

A **nodule** is a larger papule; acne vulgaris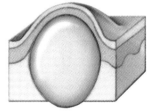

A **vesicle** is a small fluid filled sac; blister. A bulla is a large vesicle varicella (chickenpox)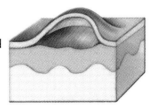

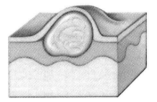 A **pustule** is a small, elevated, circumscribed lesion of the skin that is filled with pus; whitehead

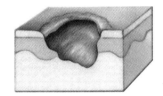 An **erosion** or ulcer is an eating or gnawing away of tissue; decubitus ulcer

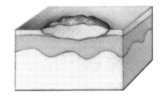 A **crust** is a dry, serous or seropurulent, brown, yellow, red, or green exudation that is seen in secondary lesions; eczema

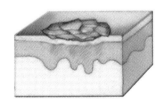 A **scale** is a thin, dry flake of cornified epithelial cells such as psoriasis

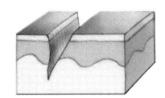 A **fissure** is a crack-like sore or slit that extends through the epidermis into the dermis; athlete's foot

FIGURE ■ 8-13

Various types of skin lesions.

A Quick Trip Through Some Diseases of the Integumentary System

DISEASE	ETIOLOGY	SIGNS AND SYMPTOMS	DIAGNOSTIC TEST(S)	TREATMENT
Abrasion	Mechanical removal of skin.	Loss of skin surface integrity, redness, swelling, inflammation.	Visual examination.	Proper cleansing technique, removal of any foreign matter, antiseptic, bandage if necessary.
Acne	A metabolic condition, allergies, various drugs or endocrine disorders are possible causative agents.	Inflammation of hair follicles / sebaceous glands especially on the face, neck, chest, upper back / shoulders. Can form blackheads, cysts, nodules, pustules, and pimples.	Visual examination.	Mild: proper cleansing techniques and OTC treatments Severe: proper cleansing techniques including prescribed medications, antibiotics, steroids and/ or all-trans retinoic acid (tretinoin).
Basal cell carcinoma	New growth or malignant tumor.	Begins as small, shiny papule, enlarges to form a whitish border around a central depression or ulcer that may bleed.	Biopsy.	Surgical removal, radiation, or cryosurgery.
Bed sores (decubitus or pressure ulcer)	Tissue injury (resultant of unrelieved pressure placed upon a specific area).	A red, inflamed, crater-like lesion usually located over a bony prominence These are classified in stages 1-4, with 4 eroding to the bone.	Visual inspection, culturing of site for infection.	Preventative measures such as turning at least every two hours and padding. Treat infection of the sore.
Boil (furuncle)	Bacteria (Staphylococcus).	Inflammation, a localized encapsulated pus filled lesion, painful affected site carbuncles are a large abscess composed of several furnuncles.	Visual examination, site culture.	Proper antiseptic cleansing techniques, antibiotics, application of warm moist heat. Depending on severity, may require draining.
Burns (thermal)	Heat or radiation or varying intensities and duration.	Depending on the intensity and duration: reddening of the affected surface, penetrating additional skin layers as the severity increases. Color is dependent on the severity, with a deepening red to black.	Visual examination.	Dependent upon the severity.
Cellulitis	Bacteria (Streptococcus and Staphylococcus).	Inflammation of skin and subcutaneous tissue, red and swollen, painful.	Visual examination, site culture.	Antibiotics (oral or intravenously, dependant on the severity).

continued

A Quick Trip Through Some Diseases of the Integumentary System (*Continued*)

DISEASE	ETIOLOGY	SIGNS AND SYMPTOMS	DIAGNOSTIC TEST(S)	TREATMENT
Contusion (bruise)	Blunt force or some form of skin injury without the skin surface breaking.	Pain, swelling, discoloration.	Visual inspection. Imaging may be needed to check for more severe injury.	Cold applications, firm bandage to impede swelling, elevation when possible, heat application, massage.
Dermatitis (contact)	Contact with an allergen (extreme range of potential agents including, soaps, cosmetics, metals, drugs, plastics, etc.).	Small, reddish lesions to larger vesicles, weepy and crusted areas, itching possible.	Visual examination.	Avoidance of the causative agent. Medication to decrease inflammatory process.
Dermatitis (general)	Stress, habitual scratching, dry skin such as winter itch.	Skin blotches, rash with itching and redness.	Patient history, visual examination.	Removal from stressful situation or moisturize the skin. Topical steroid ointments can be used for itchiness.
Eczema	Genetic predisposition to allergies (in infants it may include reaction to milk / dairy products, other foods), stress.	Skin inflammation, redness, vesicles, scales, crusting, pustules.	Visual examination, history.	No true cure-treat symptoms. Eliminate offending food, reduce stress, topical cortiosteroidal creams, skin moisturizers, antihistamines.
Folliculitis	Bacteria (usually staphylococcus).	Small pustules that form around the base of the hair follicle.	Visual examination, site culture.	Proper daily cleansing with an antiseptic cleanser, oral antibiotics (chronic or severe cases).
Herpes	Herpes family of viruses.	Clusters of fluid filled vesicles in patterns specific to the condition, skin inflammation, rash, pain related to the involved sensory nerve. Remains dormant until immunosuppression.	Visual examination, site culture.	Antiviral drugs, usually self-limiting.

8-7 Please go to the DVD for this chapter to view videos on diseases of the skin, such as decubitus ulcers, eczema, and skin cancer.

A Quick Trip Through Some Diseases of the Integumentary System (*Continued*)

DISEASE	ETIOLOGY	SIGNS AND SYMPTOMS	DIAGNOSTIC TEST(S)	TREATMENT
Hives (urticaria)	Allergic reaction to an external agent such as bee stings, plants, temperature extremes, sunlight, or internal agents such as foods, food additives, medication, antibiotics, or specific disease conditions.	Itchy wheals surrounded by a red inflamed area. Can cover most of body.	Visual examination, patient history.	Antihistamines, allergen avoidance.
Keloid	Tissue trauma or surgical incision.	Overproduction of collagen during tissue repair often creating a larger structure than the original scar/traumatized area.	Visual inspection.	Surgical removal, however there is a great potential for keloids to grow back.
Lyme disease	Tick bite containing a specific spirochetal bacterium.	"Bull's eye" macule/papule at the site of the tick bite, flu-like symptoms, stiff neck, swollen lymph node(s), joint aches, fever, headache, persistent sore throat, dry cough. Possible neurologic, cardiac, and arthritic complications if left undiagnosed/untreated.	Visual examination, blood test, patient history.	Vaccine, antibiotics, treat secondary conditions; repeat infection is a possibility.
Malignant melanoma	Occurs in melanocytes, excessive exposure to the sun.	May appear as a brown or black irregular patch which appears suddenly. A color or size change in a preexisting wart or mole may also indicate melanoma. Metastasizes to other areas quickly.	Biopsy.	Surgical removal of the melanoma and the surrounding area and chemotherapy.
Pediculosis	Lice.	Lice and nits (egg deposits).	Visual inspection.	Proper cleansing techniques with medicated soap/shampoo, cleaning of all clothing, bedding, towels, combs, etc. to remove the infestation.

continued

A Quick Trip Through Some Diseases of the Integumentary System (*Continued*)

DISEASE	ETIOLOGY	SIGNS AND SYMPTOMS	DIAGNOSTIC TEST(S)	TREATMENT
Psoriasis	Possible genetic basis with attacks triggered by emotional stress, illness, sunlight, or skin damage.	Red skin with silvery patches, rapid replacement of the epidermal cells, dry cracking skin with crusting, can be painful. Common to have periods of remission then exacerbation. May be an arthritic component.	Visual examination, patient history.	Supportive, skin applications to deal with the symptoms. Medications: steroids, ultraviolet light.
Scabies	Mites.	Elevated, grayish-white lines (burrows), vesicle and pustule formation (due to bite, feces, ova of the offending mite), intense itching.	Visual inspection.	Proper cleaning technique, application of medicated cream, all infected individuals must be treated to prevent reinfection.
Seborrheic keratosis	Unknown agent(s) causing the benign overgrowth of epithelial cells.	A well-defined, warty-scaled lesion that can present in a variety of colors from yellow to brown.	Visual inspection, laboratory examination.	Scraping (curettage), or freezing.
Squamous carcinoma	Abnormal cell growth, arising from the epidermis and occurs most often on the scalp and lower lip.	Abnormal appearance of skin, ulcer, pimple, or mole.	Biopsy.	Surgical removal or radiation.
Tinea Barbae (barber's itch) *Tinea Capitis* (scalp) *Tinea Corporis* (ringworm) *Tinea Cruris* (jock itch) *Tinea Pedis* (athlete's foot) *Tinea Unguium* (nails)	Fungi.	Case dependent. Red ring-shaped patches (mimicking a worm), red inflamed skin, cracked and weeping area(s), itch, discoloration of affected nails.	Visual examination, microscopic examination, site culture.	Maintain clean dry condition of affected area, antifungal medication (topical or systemic).
Warts: Common, Plantar, or Genital	Viruses.	Raised, rubbery, scaly growths of varying sizes and colors.	Visual examination.	Chemical or physical removal.

There are so many common pathological conditions of the integumentary system that a picture truly is worth a thousand words. Please see Figure 8-14 ▧ to view additional photos on types of integumentary conditions.

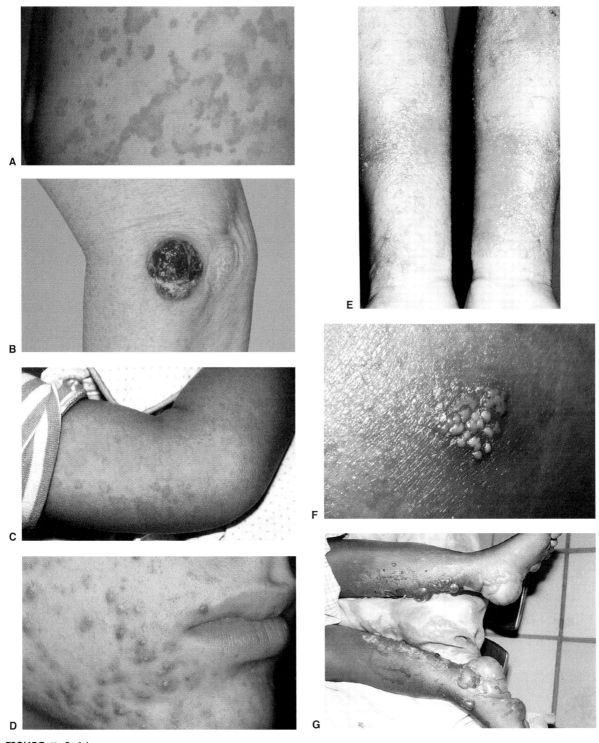

FIGURE ▧ 8-14

Various types of integumentary conditions. **A.** Urticaria (hives). (Courtesy of Jason L. Smith, MD.) **B.** Malignant melanoma. (Source: Biophoto Associates/Photo Researchers, Inc.) **C.** Erythema infectiosum (fifth disease). (Courtesy of Jason L. Smith, MD.) **D.** Acne. (Courtesy of Jason L. Smith, MD.) **E.** Poison ivy (dermatitis). (Courtesy of Jason L. Smith, MD.) **F.** Herpes simplex. (Courtesy of Jason L. Smith, MD.) **G.** Burn, second degree. (Courtesy of Jason L. Smith, MD.)

℞ Pharmacology Corner

Although medications are often delivered orally in the form of pills or injected into the body, they can also be delivered through the skin. Transdermal patches allow medication to be slowly and uniformly absorbed over time. These patches are convenient because they only have to be placed on the skin usually in the morning and then forgotten about until the next morning. Examples of medication delivered in this fashion include heart medication (nitroglycerin), smoking cessation aids (nicotine patches), and birth control. Another advantage of the nicotine patch is that the craving for nicotine is satisfied without all the other thousands of toxic and cancer causing chemicals found in cigarettes.

The treatment of skin irritations and inflammatory reactions may require the use of topical creams. Mild creams can be used to stop the itching associated with many conditions. More powerful creams containing corticosteroids can provide anti-inflammatory relief without the systemic side effects associated with steroids that are fully absorbed into the blood stream.

A variety of medicated shampoos can be used to treat lice and dandruff (excessive dry scalp with sloughing skin.) Anti-fungal preparations can be effective on skin fungal diseases such as ring worn and athlete's foot. Anti-viral creams can be used to treat herpes and other viral skin conditions. Finally, antibiotics can be used to treat bacterial skin infections and also as a prophylactic treatment of cuts to prevent opportunistic infection.

---

www.prenhall.com/colbert

The nursing profession requires much interaction with the integumentary system. Nurses administer medications through all delivery routes, routinely perform wound care, and help prevent skin problems such as pressure or bed sores. This exciting profession has training in and interactions with all the body systems. Please go to the companion website for information and to view videos concerning the nursing profession.

8-8 For a video of topical wound/burn care, please visit your DVD.

SUMMARY

Snapshots from the Journey

→ Your skin is your largest organ.

→ Your skin is an amazing organ that does the following:

a. Acts as a barrier to infection both as a physical shield and through secretion of an antibacterial substance

b. Acts as a physical barrier to injury

c. Helps to keep the body from dehydrating

d. Stores fat (yes, you do need some fat!)

e. Synthesizes and secretes vitamin D with the help of sunshine

f. Regulates body temperature

g. Provides a minor excretory function in the elimination of water, salts, and urea

h. Provides sensory input

→ The skin is composed of three layers: epidermis, dermis, and subcutaneous fascia.

→ Skin is not static; it constantly recreates itself.

→ Various glands in the skin help moisturize, waterproof, and control body temperature as well as excrete some waste products.

→ The severity of burns to the skin is evaluated by the depth of the burn and the area that the burn covers.

→ Nails are protective devices composed of dead material.

→ Hair (also dead material) aids in controlling body temperature.

Case Study

A 27-year-old female presents to her doctor's office with complaints of red, itching, and oozing skin for the past two days. Physical examination and history reveal the following: a well-nourished white female who is in otherwise good health, no known allergies, normal vital signs, pupils normal and reactive, reflexes good, breath sounds normal, liquid-filled vesicles and scabbing on both legs from the top of her sock lines to the bottom of her shorts, new vesicles have formed around her eyes. The patient stated that she returned from a primitive camping and hiking vacation in Virginia two days ago.

1. Based on this information, what do you think is the diagnosis?

   _____

   _____

2. What caused the vesicles to begin to form around her eyes?

   _____

   _____

RAY'S STORY

Family education will play an important role for Ray, our quadriplegic, once he gets home. Although proper ventilator care will be crucial for his survival, why do you think skin care is so important? How will Ray's inability to move potentially cause skin problems? Discuss areas of concern and potential problems involving Ray's integumentary system. What education/training might be important for members of Ray's family?

_____

_____

_____

_____

MARIA'S STORY

Maria, our 35-year-old diabetic, has been developing a series of integumentary system problems over the past several years. Diabetes can, in some ways, be considered a vascular disease. Research the effects of diabetes as it relates to the integumentary system. How and why does diabetes affect wound healing? Why is there a high incidence of toe and leg amputations in the diabetic population? What preventative care measures can be taken to ensure good health of the diabetic's integumentary system?

_____

_____

_____

_____

REVIEW QUESTIONS

Multiple Choice

1. The substance that is mainly responsible for skin color is:
   a   melanin
   b. pigmentin
   c. carrots
   d. luna

2. Whether you have naturally curly or straight hair is dependent on the shape of your:
   a   hair follicle
   b. hair shaft
   c. sebum
   d. melanin

3. The fibrous protein that makes up your hair and nails and fills your epidermal cells is called:
   a. carotene
   b. myelin
   c. keratin
   d. dermasene

4. In a cold environment, in order to maintain a core body temperature, peripheral blood vessels:
   a. vasodilate
   b. venospasm
   c. shiver
   d. vasoconstrict

5. The hair on your scalp can last:
   a   3 to 4 months
   b. until your 40s
   c. 3 to 4 years
   d. until you have kids

6. Athlete's foot is caused by a:
   a. virus
   b. bacterium
   c. fungus
   d. prion

7. Ringworm is also known as:
   a. *Tinea pedis*
   b. *Tinea cruris*
   c. *Tinea corporis*
   d. *Tinea pennia*

Fill in the Blank

1. The three main layers of skin are the _____, _____, and
   _____.

2. The two main types of sweat glands are the _____ and the _____ glands.

3. Sebaceous glands secrete an oily substance called _____.

4. For some individuals, melanin locates in small patches called _____.

5. Yellow jaundice, a condition associated with liver disease, occurs as a result of the build-up of
   _____.

6. _____ is an inflammatory condition of the skin and subcutaneous tissue that produces a
   red, swollen, painful area. It is caused by the staphylococcus bacterium, and if left untreated may lead to
   endocarditis or septicemia.

7. Cold sores, or fever blisters are caused by the virus_____.

8. Tiny mites that burrow into the skin to lay eggs cause an itching condition with vesicles and pustules called
   _____ .

Short Answer

1. Discuss three functions of the integumentary system.

   _____
   _____
   _____
   _____

2. Discuss the functions of sebum.

   _____
   _____
   _____
   _____

3. Why is there an increased production of melanin when there is increased sun exposure?

   _____
   _____
   _____
   _____

4. How does the integumentary system assist in the regulation of body temperature?

_____

_____

_____

_____

5. Discuss the three forms of pediculosis found in this chapter. How is this condition treated?

_____

_____

_____

_____

6. Although Lyme disease produces symptoms that are flu-like in nature as well as having the potential to create neurological and cardiovascular problems, why have we discussed it in the integumentary chapter?

_____

_____

_____

_____

Suggested Activities

1. Contact a local dermatologist to speak to your class on diagnosing and treating skin diseases.

2. Research the variety of medications used to treat acne. Focus on the mode of action, benefits, and side effects.

8-9
Now that you have completed your journey through this chapter, please go to the DVD for interactive games and puzzles concerning the medical terms and concepts contained in this chapter. By playing the games you will reinforce your learning of the integumentary system in a fun way.

THE
NERVOUS
System
The Body's Control Center

To pass the time, it's a good idea to take an interesting novel along when we travel. The nervous system, due to its complexity and importance, is our novel for this journey! Don't be dismayed by its length, because Chapter 10, "The Endocrine System," the related control system, will be a "short story" by comparison. So far on our journey, we have seen infrastructure, the building blocks and support systems of the city. Soon we will visit transportation, protection, and energy delivery systems. Like any efficient city, the body must have a control system, a system to monitor conditions, take corrective action when necessary, and keep everything running smoothly. Imagine what would happen if a traffic light network in a city suddenly failed. The control systems of the body are the nervous and endocrine systems, which receive help from your various senses. Like any control system, they have a large, complex job that is sometimes difficult to understand. They must keep track of everything that is happening in the body. Therefore, the nervous and endocrine systems are perhaps the most complex and vital systems we will visit. In this chapter, we start at the bottom of the control hierarchy, with the cells and the spinal cord. Then we focus on the higher level control at the brain and finally we pull everything together to see how all the pieces of the puzzle form the big picture.

Chapter

9

LEARNING OBJECTIVES

Upon completion of your journey through this chapter, you will be able to:

→ List and describe the components and basic operation of the nervous system

→ Contrast the central and peripheral nervous systems

→ Define the parts and functions of nervous tissue

→ Discuss the anatomy and physiology of the spinal cord

→ Organize the hierarchy of the nervous system

→ Locate and define the external structures of the brain and their corresponding functions

→ Locate and define the internal structures of the brain and their corresponding functions

→ Describe the sensory functions of the brain with related structures

→ Describe the motor functions of the brain with related structures

→ Contrast the parasympathetic and sympathetic branches of the autonomic nervous system

→ Discuss some representative diseases of the nervous system

MULTIMEDIA APPLICATIONS

DVD Interactive Exercises

→ Animation of multiple sclerosis, 9-1

→ Animation of neurochemical synaptic transmission, 9-2

→ Videos of epidural placement, 9-3

→ 3-D animation of brachial and lumbosacral plexus, spinal cord, and cervical spine injuries, 9-4

→ Animation of neuroreflex arc, 9-5

→ Video of carpal tunnel syndrome (CTS), 9-6

→ Positron emission tomography (PET) scan of the brain, 9-7

→ Drag-and-drop exercise: brainstem and subarachnoid space, 9-8

→ Video on Glasgow Coma Scale assessment, 9-9

→ Animation on Coup/Contracoup Injuries, 9-10

→ Video on Alzheimer's disease, 9-11

→ Videos on assessing pain and pain management, 9-12

→ Video on Parkinson's disease, 9-13

→ Videos of related nervous system disorders, 9-14

→ Interactive games and puzzles, 9-15

www.prenhall.com/colbert

→ Professional Profiles:

- Pharmacy

- Electroneurodiagnostician

→ Related Internet Links

→ Additional Review Questions

Pronunciation Guide

 Correct pronunciation is important in any journey so that you and others are completely understood. Here is a "see and say" Pronunciation Guide for the more difficult terms to pronounce in this chapter.

acetylcholine (ah SEET ul KOE leen)

acetylcholinesterase
   (AHS eh till KOH lin ehs tur ace)

amyotrophic (AA mee o TROF ic)

anterior commissure
   (an TEE ree or KAH mih sure)

arachnoid mater (ah RAK noyd MAY ter)

astrocytes (ASS troh SITES)

axon (AK son)

basal nuclei (BAY sal noo KLEE ie)

cerebellum (ser eh BELL um)

cerebrospinal fluid
   (ser eh broh SPY nal FLOO id)

cerebrum (ser EE brum)

commissures (KAH mih sures)

corpus callosum
   (KOR pus kah LOH sum)

corticobulbar tract
   (KOR ti coe BUL bar)

corticospinal tract (KOR ti coe SPY nal)

dendrites (DEN drights)

diencephalon (dye en SEFF ah lon)

dorsal root ganglion
   (DOR sal ROOT GANG lee on)

dura mater (DOO ra MAY ter)

fornix (FOR niks)

ependymal cells (ep PEN deh mall)

epidural space (eh pih DURE all)

ganglia (GANG lee ah)

glial Cells (GLEE all sells)

Guillain Barre (Gey yan bar RAY)

gyri (JIE rie)

hydrocephalus (high droh SEF ah lus)

hypothalamus (high poh THAL ah mus)

limbic system (LIM bick)

medulla oblongata
   (meh DULL lah ob long GA ta)

meninges (men IN jeez)

microglia (mie crow GLEE ah)

myelin (MY eh lin)

neuroglia (glial cells)
   (noo ROH glee ah)

nodes of Ranvier (ron vee AYE)

occipital lobe (ok SIP eh tal)

oligodendrocytes
   (AH li go DEN droe site)

parietal lobe (pah RYE eh tal)

pia mater (PEE ah MAY ter)

pineal body (pih NEE al)

plexus (PLECK sus)

Schwann cells (SHWAN sells)

somatic nervous system (so MAT ick)

spinocerebellar tract
   (SPY no ser eh BELL ar)

spinothalamic tract
   (SPY no thol AH mic)

subarachnoid space
   (SUB ah RACK noyd)

subdural space (sub DOO ral)

sulcus (SULL cuss)

vesicles (VESS ih kulz)

OVERVIEW

We begin this journey with an overview of the entire system to show how all the components are interrelated. Let's start with the basic operations.

The Parts and Basic Operation of the Nervous System

The organization of the nervous system can be compared to a computer. Information is entered into the computer by various "senses:" keyboard, mouse, microphone, Internet connection, and so on. The main components of the computer's "brain" are the hard drive (long-term memory), random access memory (short-term memory), and central processing unit (thinking and decision making). The computer's output—its interaction with the world—exits via ports and cables to printers, displays, speakers, and other devices.

Typically, we refer to the brain and spinal cord as the **central nervous system (CNS)** and everything outside of the brain and spinal cord, which represents the input and output pathways, as the **peripheral nervous system (PNS)**. Figure 9–1 ■ is a schematic of the organization of the branches of the nervous system that will be discussed as we journey through this chapter.

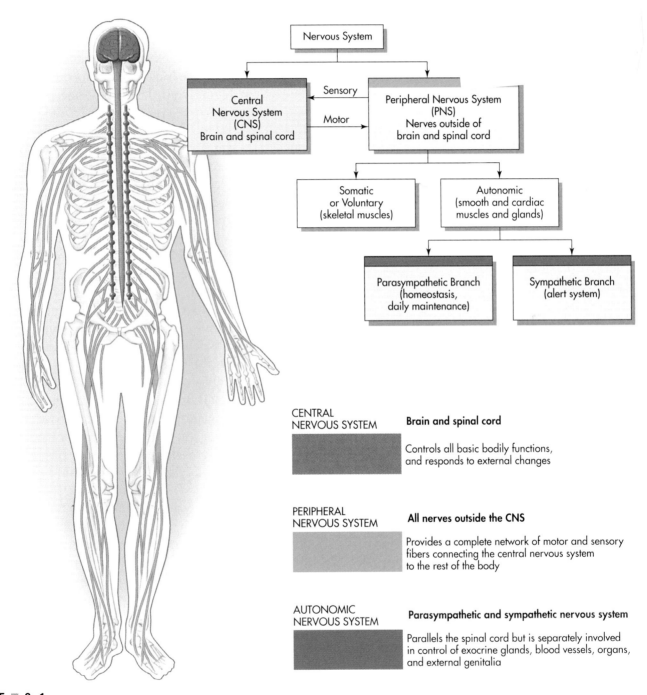

FIGURE ■ 9–1

Organization of the nervous system.

The nervous system's "input devices" comprise the **sensory system.** Your senses sample the environment and bring the information to the nervous system, as explored further in Chapter 11. Your sensory system absorbs and measures information about everything and anything, both within your body and in the outside environment. This sensory information then goes into the nervous system, where it is handled by the brain and spinal cord. The brain and spinal cord combine the input with other kinds of information, compare it to information from past experiences, and make decisions about how to respond to the information.

Once the brain and spinal cord determine an appropriate response to manage the new information, the output side is activated. The output side carries out the orders from the brain and spinal cord. The output side, often called the **motor system,** carries orders to all three types of muscles and to the body's glands, telling them how to respond to the new information. The motor system is divided into two branches: the **somatic nervous system,** which controls skeletal muscle and voluntary movements, and the **autonomic nervous system,** which controls smooth and cardiac muscle in your organs and also several glands. Autonomic output is involuntary and not under conscious control. The autonomic nervous system is further divided into two branches: **parasympathetic** and **sympathetic.** The parasympathetic branch, often called "resting and digesting," deals with normal body functioning, while the sympathetic branch is the body's alert system, commonly known as the "fight-or-flight" response system. Again, please refer to Figure 9-1.

The organization of the nervous system is easier to understand if we look at a real-life event. Imagine that you drive over to your friend's house. As you step onto her walkway, a large dog bounds down the front steps barking and snarling at you. Your sensory system gathers the following information about the new stimulus: large unfriendly dog; distance from your car; lack of immediate help. The information goes into your spinal cord and brain, which makes several decisions. You are in danger; something must be done. To gear up for action, your brain and spinal cord send directions via your autonomic nervous system to your organs. Your heart rate, blood pressure, and respiration rate rise. More blood is delivered to your skeletal muscles and heart in order to get you fully ready to respond. This is all involuntary, meaning you cannot consciously control it. Your nervous system readies your skeletal muscles to get you out of there. This fight-or-flight response is discussed later in further depth. If you can control your fear, you back slowly away from the situation. If you are scared witless, you run from the yard as fast as you can. Either way, you can hopefully escape the danger with your skin and your pride intact.

somatic *(so MAT ick)*
motor = *movement*
para = *near or around*

Amazing Body Facts

FASTER THAN A SPEEDING BULLET

Well, not literally! Your nervous system must respond very quickly to stimuli. Think about how fast you pull your hand away from a hot stove or step on the brake when something runs in front of your car. Nerve impulses can move very quickly. Some neurons have speeds as fast as 100 meters per second. That's in the neighborhood of 200 miles per hour. Bullets, on the other hand, can travel 3,000 miles per hour!

Each part of the nervous system has a separate but unmistakably connected role to play in assessing a situation and responding to it. The nervous system is active 24 hours a day, seven days a week, for your entire life. Some situations, like the unfriendly dog, a sudden drop in blood pressure, or a terrible car accident, may be life-threatening. Others, like an ant crawling across your foot or a pencil rolling under the desk just out of reach, are simply annoying. Everything that happens in your world is monitored and responded to by your nervous system. It's a truly Herculean task!

NERVOUS TISSUE

Like all organs, the components of the nervous system are made up of tissue. But unlike other systems, the nervous system contains no epithelium, connective tissue, or muscle tissue. Nervous tissue is made up of two different types of cells: neuroglia and neurons.

Neuroglia

The **neuroglia** or **glial cells** are specialized cells in nervous tissue that allow it to perform nervous system functions. In the CNS, there are four types of glial cells:

- **Astrocytes** are metabolic and structural support cells that hold the neurons and blood vessels close together.
- **Microglia** can attack microbes and remove debris.
- **Ependymal cells** do the job of epithelial cells, covering surfaces and lining cavities.
- **Oligodendrocytes** hold nerve fibers together and make a lipid insulation called **myelin,** described in more detail a bit later.

In the PNS, there are only two types of glial cells: **Schwann cells,** which make myelin for the PNS, and **satellite cells,** which are support cells. Please see Figure 9–2 ▥, which shows a cutaway view of the spinal cord of the CNS showing the four types of neuroglial cells found there and a peripheral neuron with its glial cells.

Neurons

The glial cells do all the support activities for the nervous system, such as lining and covering cavities and supporting and protecting structures. None of the glial cells, however, are capable of measuring the environment, making decisions, or sending orders. All of the control functions of the nervous system must be carried out by a second group of cells called **neurons.**

neuroglia *(noo ROH glee ah)*
glial cells *(GLEE all sells)*
   glia *= glue*
astrocytes *(ASS troh SITES)*
   astro *= star*
   cyte *= cell*
microglia *(mie crow GLEE ah)*
ependymal cells *(ep PEN deh mall)*
oligodendrocytes
   (AH li go DEN droe sites)
   olig/o *= few*
   dendr/o *= branches*
myelin *(MY eh lin)*
Schwann cells *(SHWAN sells)*

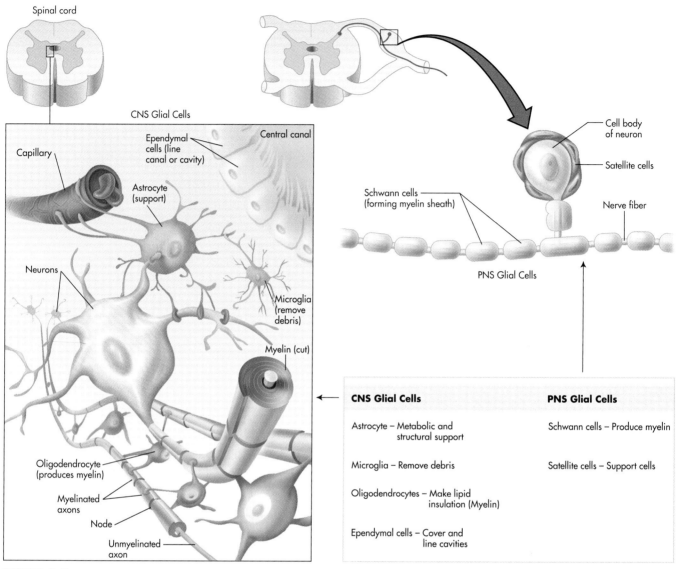

FIGURE 9-2

Glial cells and their functions.

Neurons are rather bizarre-looking cells, often with many branches and what appears to be a tail (see Figure 9–3 ■). The main function of a neuron is to transmit messages from one cell to another, throughout the **body.** Each part of a neuron has a specific function. The neuron cell body's main function is that of cell metabolism. Its **dendrites** receive information from the environment or from other cells and carry that information to the cell body. The **axon** generates and sends signals to other cells. Those signals travel down the axon until they reach the **axon terminal,** which then connects to a receiving cell. This space between the axon terminal and receiving cell is called a **synapse.** If the receiving cell is a skeletal muscle cell, then this particular synapse is called the **neuromuscular synapse** or **junction** (see Chapter 7).

dendrite *(DEN dright) = tree-like or branching*
axon *(AK son)*

synapse *(SIGH naps)*

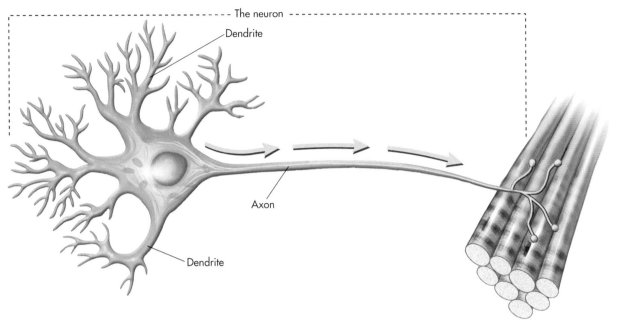

FIGURE ▓ 9–3

A neuron connecting to a skeletal muscle.

Neurons can be classified by how they look (structure) or what they do (function). From a structural point of view, they can be classified as follows:

- neurons that have one axon and one dendrite (bipolar)
- neurons that have one axon and many dendrites or branches (multipolar) as in Figure 9-3
- neurons that have one process that splits into a central and a peripheral projection (unipolar)

Classified by function, input neurons are known as **sensory neurons** (afferent neurons). They carry impulses from the skin and sensory organs to the spinal cord and the brain. Output neurons are known as **motor neurons** (efferent neurons). They carry messages from the brain and spinal cord to muscles and glands. Neurons that carry information between neurons are called **interneurons** (*inter* means between) or **association neurons.**

interneurons
(in ter NOO rons)

 TEST YOUR KNOWLEDGE 9-1

Choose the best answer:

1. Which cells are the support cells in the nervous system?
   a. neurons
   b. neuroglia
   c. epithelium
   d. all of the above

2. The output side of your nervous system is called:
   a. sensory
   b. motor
   c. central
   d. exit

3. The part of the nervous system that integrates and processes information is known as the:
   a. PNS
   b. PBS
   c. CNS
   d. CIA

4. The lipid insulation of nervous tissue is called:
   a. glia
   b. myelin
   c. Schwann cells
   d. astrocytes

HOW NEURONS WORK

Now that we know the basic structure of neurons, let's take a closer look at how they function.

Excitable Cells

A neuron is an **excitable cell**. An excitable cell carries a small electrical charge when stimulated. Each time charged particles flow across a cell membrane, their movement generates a tiny electrical current. (Electricity is just the movement of charges from one place to the other.) All three types of muscle cells are excitable cells, as are many gland cells. Because neurons are excitable cells, it makes sense that the way neurons send and receive signals is via tiny electrical currents.

How, you might ask, can cells carry electricity? It seems hard to believe, but cells are like miniature batteries, able to generate tiny currents simply by changing the permeability of their membranes. Perhaps the plot of the *Matrix* movies in which humans are used by machines as batteries is not so far-fetched.

Action Potentials

A cell that is not stimulated or excited is called a **resting cell** and is said to be **polarized**. It has a difference in charge across its membrane so that it is more negative on the inside than on the outside. When that cell gets stimulated (excited), sodium gates in the cell membrane spring open. When these sodium gates open, they allow sodium ions (Na^+) to travel across the cell membrane. These sodium ions are positively charged, so when they go into the cell, the cell becomes more positive. A cell that is more positive than resting becomes **depolarized.** In less than a millisecond, the gates on the sodium channels shut, just like the automatic doors at the supermarket. Then other gates open for potassium channels. Potassium (K^+), which is also positive, leaves the cell, taking its positive charges with it. The inside of the cell becomes more negative again, eventually returning to rest. This is **repolarization** (see Figure 9–4 ■).

Sometimes a cell overshoots its charge and becomes more negative than when it is at rest. Then the cell is **hyperpolarized.** Eventually the cell will return to resting. A cell is unable to accept another stimulus until it repolarizes (returns to its resting state), and the period during which it cannot accept another stimulus is called the **refractory period.** This whole series of permeability changes within the cell and the resultant changes in charge across the cell membrane is called the **action potential.**

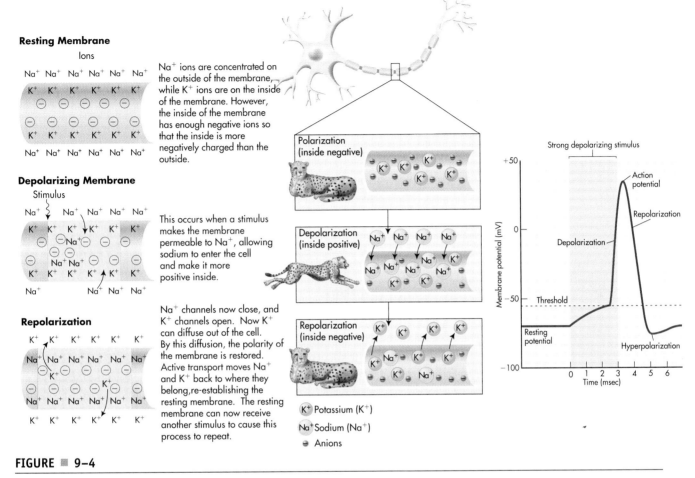

Resting Membrane

Na$^+$ ions are concentrated on the outside of the membrane, while K$^+$ ions are on the inside of the membrane. However, the inside of the membrane has enough negative ions so that the inside is more negatively charged than the outside.

Depolarizing Membrane

This occurs when a stimulus makes the membrane permeable to Na$^+$, allowing sodium to enter the cell and make it more positive inside.

Repolarization

Na$^+$ channels now close, and K$^+$ channels open. Now K$^+$ can diffuse out of the cell. By this diffusion, the polarity of the membrane is restored. Active transport moves Na$^+$ and K$^+$ back to where they belong, re-establishing the resting membrane. The resting membrane can now receive another stimulus to cause this process to repeat.

K$^+$ Potassium (K$^+$)
Na$^+$ Sodium (Na$^+$)
⇒ Anions

FIGURE ■ 9–4

The action potential.

Action potentials are "all-or-none," which means that the action potential, once it starts, will always finish and will always be the same size. Either a cell has one or it doesn't. There are no small or large action potentials.

Local Potentials

Neurons can use their ability to generate electricity to send, receive, and interpret signals. Let's look at an example of how this works. You are hammering a picture hanger into the wall of your newly painted bedroom, and you hit your thumb. The blow from the hammer stimulates the dendrites in your thumb, and sodium gates open. Sodium flows into the dendrites, which

Applied Science

FUGU

The puffer fish, fugu, is considered a dangerous delicacy in Japan because if you eat it, you could die! Fugu can be served only by specially certified chefs trained to prepare the fish so it is safe to eat. Puffer fish contain a poison, tetrodotoxin (TTX), in their tissues that blocks sodium channels, preventing sodium from entering cells. Cells exposed to TTX cannot depolarize. Thus, neurons cannot fire action potentials. People who consume improperly prepared fugu become paralyzed. Symptoms develop as headache, dizziness, vomiting, and chest tightness within 15 minutes to 20 hours post ingestion. Eventually victims develop ascending paralysis. If untreated, death can result in as little as 4 hours, usually from respiratory paralysis.

become depolarized. If you hit your thumb softly, the cell is stimulated only a little, only a few gates open, and the cell does not depolarize very much. (You may not even feel any pain.) If you hit your thumb really hard, more gates open and the cell depolarizes much more. (It hurts a lot more, too!) This phenomenon, known as a **local potential,** is different from the action potential previously discussed. In a local potential, the *size or amount* of the stimulus determines the excitement of the cell. A big stimulus causes a bigger depolarization than a small stimulus. (Local potentials, then, are not all-or-none but vary upon the degree of stimulation.)

Once several local potentials reach a threshold level, an action potential is triggered. The various dendrites carry the depolarization to the sensory neuron cell body. The cell body takes that information and generates an action potential, if the stimulus is big enough. Once an action potential is formed, it travels down the axon from the cell body to the terminal. This movement is called **impulse conduction**.

Impulse Conduction

The speed of impulse conduction along an axon is determined by two characteristics: the presence of a myelin sheath and the diameter of the axon. Myelin is a lipid insulation or sheath formed by the **oligodendrocytes** in the CNS and the **Schwann cells** in the PNS. In preserved brains, the myelinated axons look white. Unmyelinated parts of the CNS, like cell bodies, look gray. Therefore, what we call "white matter" is typically made of axons, and what we call "gray matter" is made of cell bodies. The cell membranes of oligodendrocytes or entire Schwann cells are wrapped around an axon like a bandage. Between adjacent glial cells are tiny bare spots called **nodes of Ranvier** (see Figure 9–5 ■). When an axon is wrapped in myelin, we call the axon myelinated. Myelin is essential for the speedy flow of action potentials down axons. In an unmyelinated axon, the action potential can only flow down the axon by depolarizing each and every millimeter of the axon. Every single sodium channel must open, and every single potassium channel must open. It is a relatively slow process. In a myelinated axon, only the channels at the nodes must open for the action potential to flow down the axon. Myelin prevents ions from passing through channels, because ions are water soluble and myelin is a lipid. (Remember, charged particles cannot get through lipids.) The action potential therefore "skips" down the axon from node to node (nodal transmission) rather than creeping along the entire length of the axon. It works the same way as if you try to walk across the floor heel to toe, never missing a spot and then skip across the floor in large leaps. Which is faster?

The diameter of the axon also affects the speed of action potential flow. Think about moving through two different pipes, say on a playground or obstacle course. One pipe is one-half meter in diameter (less than 2 feet). To move through the pipe, you

oligodendrocytes
(AH li go DEN droe site)

Clinical Application

EPILEPSY

Epilepsy is a disorder in which irregular electrical signals are suddenly discharged from neurons, causing the body to become overloaded with impulses. The body reacts by going into a seizure. A mild seizure, or petit mal seizure, can cause a drop in awareness or muscle coordination. The person experiencing a petit mal seizure may seem to be daydreaming or just staring into space. A severe seizure, or grand mal seizure, can cause convulsions. The person experiencing a grand mal seizure may drop to the floor, shake, and twitch. Epilepsy may be caused by earlier brain damage due to trauma, stroke, exposure to toxins, or lack of oxygen during childbirth. In most cases, the cause of epilepsy is never identified.

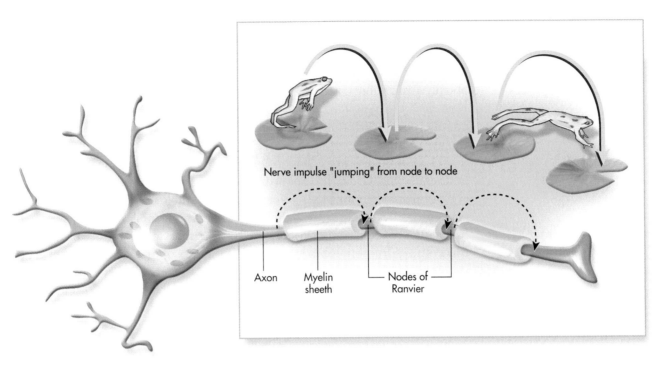

FIGURE ■ 9-5

Impulse conduction via myelinated axon.

must crawl. The other pipe is 2 meters in diameter (more than 6 feet). Most of you can stand in this pipe with room to spare. Which trip would be faster? Obviously, it's the one through the larger diameter pipe. You don't have to crawl, so you don't get hung up on the side. Heck, you could even run through the big pipe! For ions, the axon is essentially a pipe. The wider the axon, the faster the ions flow. The combination of myelination and large diameter makes a huge difference in speed. Small, unmyelinated axons have speeds as low as 0.5 meters per second, while large-diameter, myelinated axons may be as fast as 100 meters per second. That's 200 times faster!

Pathology Connection: Myelin Disorders

MULTIPLE SCLEROSIS

Multiple sclerosis (MS) is a disorder of the myelin in the CNS, which affects 200,000–400,000 people in the U.S. alone. Patients with multiple sclerosis have many areas in their nervous system where the myelin has been destroyed, probably by autoimmune attack. In some patients, axons and cell bodies are also destroyed. The brain, spinal cord, and optic nerves are often involved. In areas without myelin, impulse conduction is slow or impossible. Imagine an electrical wire with many bare spots or with no insulation. These bare or exposed parts can "short out" and prevent the current from passing further down the wire. These areas of damaged myelin often have plaques, or scarred areas, that give MS its name. Symptoms of multiple sclerosis differ from patient to patient depending on where the myelin damage occurs. Patients may have disturbances in vision, balance, speech, or movement.

sclerosis *(skluh ROH sis)* = scars

MS can be grouped into two types based on the pattern of the disease. Relapsing-remitting MS is characterized by flare-ups, with obvious symptoms, and periods in which the patients have no symptoms, called remissions. Chronic progressive MS has no remission periods, and patients become steadily more disabled. Most patients are initially diagnosed with relapsing-remitting MS, but at least 50% of them will progress to the chronic progressive form. MS is more common in women than in men and is diagnosed most often in people under the age of fifty.

There is no definitive diagnosis for MS. Patient history suggesting flare-ups and remissions, visual disturbances along with muscle weakness, and the presence of plaques in an MRI, all can aid in the diagnosis of MS. As with most other progressive neurological disorders, there is no cure for MS. Especially during acute flare-ups, some patients are helped by steroids, **plasma exchange**, or intravenous Immunglobin G. Several immunosuppressant drugs have been shown to decrease the number of relapses and prevent conversion from relapsing-remitting to chronic progressive and may slow progression in some patients. Symptom management, decreasing fatigue, treating depression, etc. is important for many people living with MS.

plasma exchange = *the therapeutic removal of diseased blood and subsequent replacement with fresh healthy blood.*

GUILLAIN-BARRÉ SYNDROME

Guillain-Barré *(Gey yan bar RAY)*

Guillain-Barré syndrome (GBS) is a paralysis caused by an autoimmune attack on peripheral myelin, Schwann cells, or peripheral axons. There are 1-2 cases of GBS/100,000 people in the U.S. each year. Patients develop, over variable periods of time, weakness and usually ascending paralysis of the limbs, face, and diaphragm. The course of the disease can be divided into three phases. The acute phase is the initial onset of the disease, during which the patient gets steadily worse. The plateau phase is a period of days to weeks in which the patient's condition is stable. During the recovery phase, which may last up to 2 years, patients recover function. Some patients may have a mild form of Guillain-Barré syndrome, but many with severe disease must be kept on a ventilator until the paralysis resolves. Patients may recover full function, but a significant portion of those with severe GBS still have measurable disability two years after recovery. The cause of the disease is not known, though many patients develop Guillain-Barré syndrome after a viral infection.

Diagnosis of GBS is difficult because its chief symptom, weakness, may be caused by many different disorders. Patient history is the best way to diagnose the disorder. Rapid-onset ascending paralysis after an infection could indicate GBS. Nerve conduction and EMG (muscle recordings) can be used to help confirm diagnosis. Often there is high protein in the cerebrospinal fluid (CSF) but no white blood cells. There is no effective treatment for GBS except supportive care to treat the symptoms, including ventilation support, prevention of blood clots and bed sores, and pain medication. Some patients improve more rapidly with plasma exchange, which decreases immune function. Fortunately, the disorder is usually temporary. Many patients need rehabilitation after their PNS recovers.

9-1 For more information on multiple sclerosis and to view an animation showing the destruction of the myelin sheath and subsequent pathophysiologic changes, please go to the DVD for this chapter.

How Synapses Work

In order for neurons to communicate, there must be some way for a message to be sent from one neuron to the other. This communication occurs at the synapse. Before synaptic transmission occurs, an action potential is generated and flows toward the axon terminal.

CHEMICAL SYNAPSES

When an impulse arrives at the axon terminal, the terminal depolarizes and calcium gates open. Calcium ions flow into the cell, which causes more depolarization, which causes more calcium to come in, etc. In addition, when calcium flows in, it triggers a change in the terminal. There are tiny sacs in the terminal called **vesicles,** which release their contents from the cell via exocytosis (to expel) when calcium flows in. These vesicles are filled with molecules called **neurotransmitters.**

These neurotransmitters are used to send the signal from the neuron across the synapse to the next cell in line. The neurotransmitter binds to the cell receiving the signal and causes gates to open or close. Some neurotransmitters excite the receiving cell and some calm it down. In the case of the hammered thumb, the neurotransmitter would be released in your spinal cord and would excite a pain neuron. The receiving neuron would be stimulated and take the information about your thumb to your brain, where you would register the pain. You have already learned about a very common chemical synapse: the neuromuscular junction.

The last step in the transfer of information is clean-up. The neurotransmitter must be taken away from the synapse or it will continually bind to the receiving cell. This clean-up is accomplished through an inactivator, usually an enzyme. For example, if the neurotransmitter chemical is **acetylcholine** (ACh), like in the neuromuscular junction, it will be inactivated by acetylcholinesterase, an enzyme that breaks down ACh. This type of information flow, using neurotransmitters, is called a **chemical synapse,** because chemicals (neurotransmitters) are used to carry the information from one cell to another. See Figure 9–6 ■ for the steps in chemical synaptic transmission.

Our understanding of chemical synapses has led to several breakthroughs for treating mental illness. Many current medications are designed to modify synapses. Selective serotonin reuptake inhibitors (SSRI) are good examples. These medications prevent the clean-up of the neurotransmitter serotonin from synapses, thereby increasing the effects of serotonin on the receiving cell. Many antidepressants and anti-anxiety medications are SSRIs. Table 9–1 contains other examples of clinically important neurotransmitters. Remember the neuromuscular junction from Chapter 7? That is perhaps the best understood chemical synapse in the human body. The transmitting neuron is a motor neuron, the receiving cell is a skeletal muscle fiber and the neurotransmitter is acetylcholine.

vesicles *(VESS ih kulz)*

neurotransmitter
(noo roh TRANS mit ter)

acetylcholine
(ah SEET ul KOE leen)
ase ending = *enzyme*

 9-2 To see an animation of neurochemical synaptic transmission, please see your DVD.

 www.prenhall.com/colbert
The electroneurodiagnostician (END) is involved in assessing and testing patients with potential neurological diseases to aid the physician in an accurate diagnosis. To view a video on this health care career, please go to the companion website for this chapter.

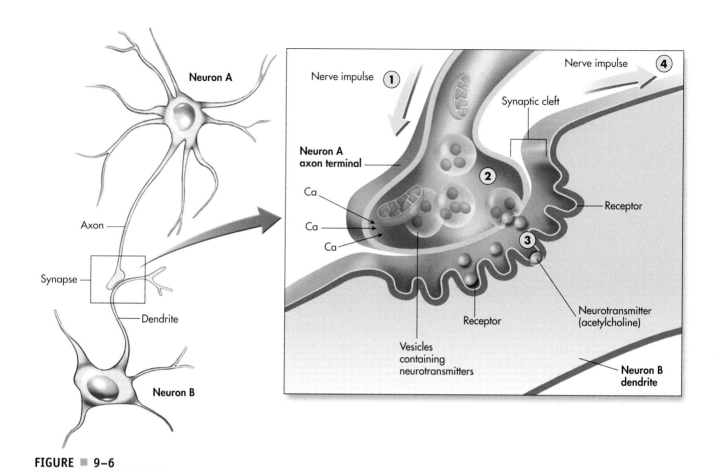

FIGURE ■ 9–6

The chemical synapse. Step 1: the impulse travels down the axon. Step 2: vesicles are stimulated to release neurotransmitter (exocytosis). Step 3: the neurotransmitter travels across the synapse and binds with the receptor site of post synaptic cell. Step 4: the impulse continues down the dendrite.

TABLE 9-1   Selected Common Neurotransmitters

NEUROTRANSMITTER	LOCATION	FUNCTION	COMMENTS
Acetylcholine	CNS* and PNS*	Generally excitatory but is inhibitory to some organs and glands	Found in skeletal neuromuscular junctions and in many ANS* synapses
Norepinephrine	CNS and PNS	May be excitatory or inhibitory depending on the receptors	Found at visceral and cardiac muscle synapses, ANS
Epinephrine	CNS and PNS	May be excitatory or inhibitory depending on the receptors	Found in pathways concerned with behavior and mood
Serotonin	CNS	Generally inhibitory	Found in pathways that regulate temperature, sensory perception, mood, onset of sleep
Endorphins	CNS	Generally inhibitory	Inhibit release of pain neurotransmitters

*CNS = central nervous system; PNS = peripheral nervous system; ANS = autonomic nervous system.

ELECTRICAL SYNAPSES

Some cells do not need chemicals to transmit information from one cell to another. Instead, they use **electrical synapses.** The cells in an electrical synapse can transfer information freely because they have special connections called **gap junctions.** Such connections can exist between many types of excitable cells. They are found, for example, in the intercalated disks between cardiac muscle fibers.

Clinical Application

BIOTERRORISM

If acetylcholinesterase activity is prevented, acetylcholine continually stimulates the muscle, eventually paralyzing it. Some insecticides and the nerve gas Sarin, which killed 12 people and injured hundreds of others during a terrorist attack in a Tokyo subway tunnel in 1995, are acetylcholinesterase inhibitors. They kill by causing paralysis of skeletal muscles, including the diaphragm, the most important respiratory muscle. Victims of Sarin or overexposure to organophosphate insecticides can die from respiratory arrest.

TEST YOUR KNOWLEDGE 9-2

Choose the best answer:

1. Which ions move *into* neurons during the action potential?
   a. potassium
   b. calcium
   c. sodium
   d. acetylcholine

2. Which of the following is all-or-none?
   a. local potential
   b. action potential
   c. chemical synapse
   d. impulse conduction

3. The molecules used to send signals across synapses are called:
   a. hormones
   b. ions
   c. neurotransmitters
   d. messengers

4. What is another name for myelinated axons?
   a. gray matter
   b. dura mater
   c. white matter
   d. duzitmater

5. Multiple sclerosis symptoms are caused by:
   a. destruction of CNS myelin
   b. destruction of PNS myelin
   c. neuron death
   d. damage to the spinal cord

6. Which of the following is true of Guillain-Barré syndrome?
   a. it is genetic
   b. it is always permanently disabling
   c. it causes brain damage
   d. it may be associated with a virus

SPINAL CORD AND SPINAL NERVES

So far we've discussed the nervous system at the cellular and tissue level. In addition, we've described how impulses are transmitted in the nervous system. Now let's focus on the main highway that these impulses travel.

cauda equina
(KAW duh EE kwine ah)
cauda = *tail*
equina = *horse*
cerebrospinal fluid
(ser eh broh SPY nal FLOO id)

dura mater *(DOO ra MAY ter)*
arachnoid mater
(ah RAK noyd MAY ter)
mater = *mother; the three layers formed by the meninges literally mean "hard mother" (dura mater), "soft mother" (pia mater), and "spider mother" (arachnoid mater)*
pia mater *(PEE ah MAY ter)*
epidural space *(eh pih DURE all)*
epi = *on top*
subarachnoid space
(SUB ah RACK noyd)
subdural space *(sub DOO ral)*
sub = *under*

External Anatomy of the Spinal Cord

The **spinal cord** is located in a hollow tube running inside the vertebral column from the foramen magnum to the second lumbar (L2) vertebrae. You can think of your spinal cord as a very sophisticated neural information superhighway that allows nerve impulses to travel to and from the brain. You can also think of the spinal cord as a nerve center that controls many reflexes. The spinal cord is divided into 31 segments, each with a pair of **spinal nerves.** Both the spinal cord segments and the spinal nerves are named for their corresponding vertebrae. The spinal cord ends at L2 in a pointed structure called the **conus medullaris**. Hanging from the conus medullaris is the **cauda equina.** The cauda equina is the bunch of spinal nerves, L2 through the coccygeal (Co) nerves, dangling loosely in a bath of **cerebrospinal fluid (CSF)**, which acts as a shock absorber for both the brain and spinal cord. The spinal cord has two widened areas, the cervical and lumbar enlargements, which contain the neurons for the limbs. See Figure 9–7 ■.

MENINGES

The CNS, both brain and spinal cord, are surrounded by a series of protective membranes called the **meninges.** The purpose of the meninges is to cover the delicate structures of the brain and spinal cord. In essence, they help to set up layers that act as cushions and shock absorbers for the brain and spinal cord. The meninges form three distinct layers. The outer layer of thick, fibrous tissue is called the **dura mater.** The middle layer, a wispy, delicate layer resembling spider webs, is the **arachnoid mater**, which is composed of collagen and elastic fibers. The third, innermost layer, which is fused to the neural tissue of the CNS, is the **pia mater.** This layer contains blood vessels that serve the brain and spinal cord.

A series of spaces are associated with the meninges. Between the dura and the vertebral column is a space filled with fat and blood vessels called the **epidural space.** Between the dura mater and the arachnoid mater, under the dura mater, is the **subdural space,** which is filled with a tiny bit of fluid. Between the arachnoid mater and the pia mater is the large **subarachnoid space,** filled with cerebrospinal fluid, acting as a fluid cushion for the CNS. In addition, CSF can also transport dissolved gases and nutrients as well as chemical messengers and waste products, since there is no blood in the CNS itself. These three membranes and their fluid-filled spaces, together with the bones of the skull and

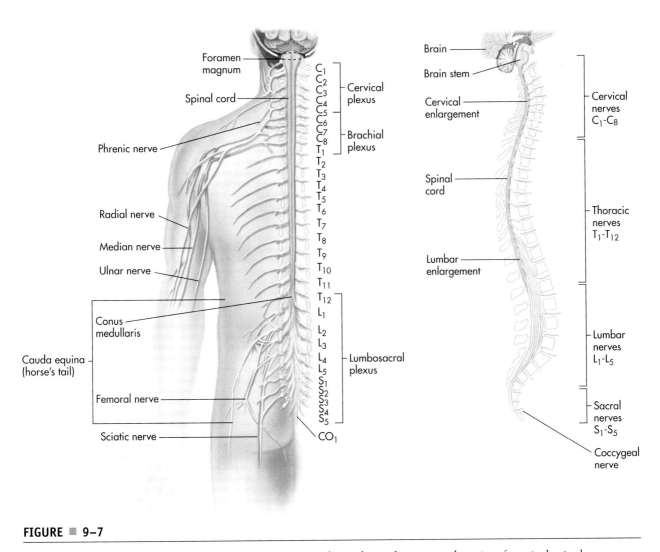

FIGURE ■ 9–7

The spinal cord. Please note that while there are 7 cervical vertebrae, there are eight pairs of cervical spinal nerves.

vertebral column, form a strong protection system against CNS injury. Please see Figure 9–8 ■.

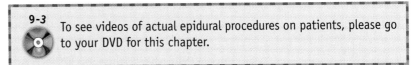

9-3 To see videos of actual epidural procedures on patients, please go to your DVD for this chapter.

Internal Anatomy of the Spinal Cord

The spinal cord is divided by an anterior median fissure and a posterior median sulcus. A **fissure** is a deep groove on the CNS surface; a **sulcus** is a shallow groove on the CNS surface. Please see Figure 9–9 ■. The interior of the spinal cord is then divided into a series of sections of white matter **columns** and gray matter **horns.** The horns are the regions which contain neuron cell bodies. There are three types of horns. The posterior or dorsal horn is involved in sensory functions, the anterior or ventral horn in motor functions, and the lateral horn in autonomic functions.

fissure *(FISH er)*
sulcus *(SULL cuss)*

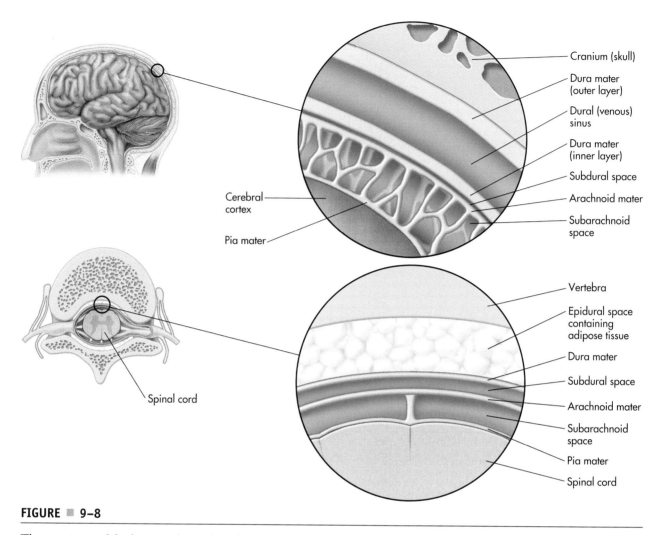

Cranium (skull)

Dura mater (outer layer)

Dural (venous) sinus

Dura mater (inner layer)

Subdural space

Arachnoid mater

Subarachnoid space

Cerebral cortex

Pia mater

Vertebra

Epidural space containing adipose tissue

Dura mater

Subdural space

Arachnoid mater

Subarachnoid space

Pia mater

Spinal cord

Spinal cord

FIGURE ■ 9–8

The meninges of the brain and spinal cord.

spinothalamic tract
(SPY no THAL ah mic)

spinocerebellar tract
(SPY no ser eh BELL ar)

There are also dorsal, lateral and ventral columns. These columns act as nerve tracts, pathways, or axons, running up and down the spinal cord to and from the brain. Three ascending pathways, the **dorsal column tract**, the **spinothalamic tract,** and the **spinocerebellar tract,** carry information from your sense of touch to the spinal cord and then to your brain from all parts of the skin, joints, and tendons. Again see Figure 9-9.

- The dorsal column tract carries fine-touch and vibration information to the brain.

- The spinothalamic tract carries temperature, pain, and crude touch information to the brain.

- The spinocerebellar tract carries information about posture and position to the brain.

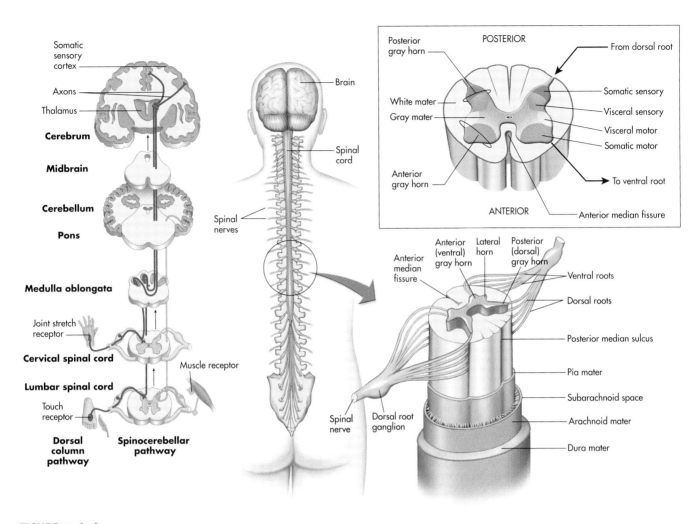

FIGURE ■ 9–9

Internal anatomy of the spinal cord.

Several descending pathways carry motor information (orders for voluntary movements) from the brain to the spinal cord, including the **corticospinal tract**, **corticobulbar tract**, **reticulospinal tract**, and **rubrospinal tract**. The axons from all pathways synapse on motor neurons in the ventral horn.

- The corticospinal tract carries orders from the brain to the motor neurons in the ventral horn of the spinal cord.
- The corticobulbar tract carries orders from the brain to motor neurons in the brain stem (more details later).
- The reticulospinal and rubrospinal tracts (along with several other tracts) carry information from the brain to the brain stem and ventral horn which helps to coordinate movements.

corticospinal tract
 (KOR ti coe SPY nal)
corticobulbar tract
 (KOR ti coe BUL bar)
reticulospinal *(reh TICK you low)*

Whether the tracts are ascending or descending, it helps to think of the columns as individual communication wires, part of a vast network of wires, transporting information to the appropriate parts of the system. This is the neural information superhighway.

In addition, the spinal cord has several other features. The **commissures,** gray and white, connect left and right halves of the cord so the two sides of the CNS can communicate. (See, the right hand *does* usually know what the left hand is doing, even if it doesn't always seem like it!) The central canal is a cavity in the center of the spinal cord that is filled with cerebrospinal fluid. The **spinal roots,** projecting from both sides of the spinal cord in pairs, fuse to form spinal nerves. (The roots are the on-ramps and off-ramps of the neural superhighway.) The dorsal root, with the embedded **dorsal root ganglion,** a collection of sensory neurons, carries sensory information, while the **ventral root** carries motor information. Again, please refer to Figure 9–9.

commissures
(KAH mih sures) = the meeting of structures

ganglion *(GANG lee on)*

Pathology Connection: Polio and Post-Polio Syndrome

Polio is a form of paralysis caused by the poliomyelitis virus. Polio was a rather common virus prior to the start of large scale vaccinations in the 1950s. Today polio has nearly been wiped off the face of the earth. The last naturally acquired case of polio in the U.S. was in 1979. Even before vaccinations, most people (99%) who caught polio only had a mild upper respiratory or digestive illness, lasting for a few days. It would hardly seem worth ridding the world of the virus. However, the other 1% of the polio patients develop the paralytic form. In paralytic polio, the virus somehow gets into and kills the neurons in the ventral horn of the spinal cord. Sensory neurons are spared, so patients with polio can feel, but they cannot move the affected body parts. Of these patients, 25% suffer permanent disability. In 1952 there were 21,000 cases of the paralytic form of polio in the U.S. alone. There is no treatment for polio. Patients are kept alive during the acute phase and, if they survive, undergo extensive rehabilitation (if available). Polio is very rare today. It was eradicated from the Western Hemisphere in 1994 and should be destroyed completely in our lifetime. Clinicians in the U.S. are unlikely to ever see a case of polio.

Polio, however, is coming back to haunt many patients who thought they had put the disease behind them. Post-polio syndrome (PPS) is a progressive weakness appearing several decades after the polio infection. It is estimated that 25–40% of all people who had paralytic polio will develop PPS. The cause of PPS is not known. However, it appears to be directly related to the damage left by the polio virus. In the parts of the spinal cord damaged by polio, neurons are actually destroyed. Patients recover by using the few surviving motor neurons to power the muscles. For example, a normal motor neuron might communicate with 1000 muscle fibers. In a recovered polio patient that same neuron might need to communicate with 5000 or 10,000 muscle fibers. Current thinking is that these surviving neurons become severely overworked and eventually begin to die themselves, causing the muscle weakness. This may seem like a minor disorder, but keep in mind that there are approximately 500,000 paralytic polio survivors in the U.S. today. (We aren't really sure of the number because the past records aren't great.) Worldwide there are millions of paralytic polio patients who have recovered and are at risk for PPS. As you might imagine, diagnosis is difficult and consists mainly of ruling out other causes for progressive muscle weakness in polio survivors. There are no treatments to stop the progression of PPS, but exercise has been shown to improve muscle function in some patients.

SPINAL NERVES

Nerves are the connection between the CNS (brain and spinal cord) and the world outside the CNS. Nerves are therefore part of the PNS. Isn't it amazing that the brain, which is totally encased in darkness, can receive and interpret the nerve messages from the PNS to allow us to see the wonderful world around us? All nerves consist of bundles of axons, blood vessels, and connective tissue. Nerves run between the CNS and organs or tissues, carrying information into and out of the CNS. The nerves connected to the spinal cord are called **spinal nerves**. There are 31 pairs of spinal nerves, each named for the spinal cord segment to which they are attached. Since each spinal nerve is a fusion of dorsal and ventral roots, spinal nerves carry both sensory and motor information. A nerve that carries both types of information is called a **mixed nerve.** All spinal nerves are mixed. When spinal nerves leave the vertebral column, they can go through a number of different pathways to reach the peripheral tissues. Spinal nerves from the thoracic spinal column project directly to the thoracic body wall without branching. Spinal nerves from the cervical, lumbar, and sacral regions of the spinal cord go through complex branching patterns, recombining with nerves from other spinal cord segments before projecting to peripheral structures. These complex branching patterns are called **plexuses.** See Figure 9–10 ▪.

Clinical Application

SPINAL BLOCKS, EPIDURAL ANESTHESIA, AND LUMBAR PUNCTURES (SPINAL TAPS)

The anatomy of the spinal cord, particularly in the lumbar region, makes for easy access to the nervous system for pain relief and testing of CSF.

During labor or in preparation for a Cesarean section, women often receive pain relief by drugs delivered to the spinal cord. Surgeries on body parts below the waist are now being performed under spinal anesthesia instead of general anesthesia, allowing the patient to be awake during the procedure and decreasing complications due to general anesthesia. A **spinal block** is the injection of an anesthetic into the subarachnoid space via a needle. It is a one-time injection that wears off after a couple of hours, but gives rapid, total anesthesia. Patients lose both sensory and motor activity below the injection. An **epidural** is the delivery of local anesthetic into the epidural space by a catheter (a small tube). The catheter allows continuous injection of the anesthetic. Ideally, epidural anesthesia allows a woman to continue to participate actively in the birth without severe labor pains. Epidural injections of steroids are sometimes prescribed for patients with chronic lower back injuries to relieve pain and inflammation.

During a neurological exam, a physician may want to sample CSF. The best place to sample CSF is in the lumbar region because of the loose nerves of the cauda equina. Taking the sample in the lumbar region decreases the likelihood of damage to the spinal cord. Thus, the technique for sampling CSF most often used is the **lumbar puncture**, also known as the **spinal tap**. CSF can be removed by inserting a needle between L3 and L4 or L4 and L5. Then the fluid can be examined for pathogens, chemical abnormalities, or the presence of blood.

9-4 For a 3-D animation of the spinal cord, brachial plexus, lumbosacral plexus, cervical spinal injuries, and spinal immobilization, please go to this chapter on your DVD.

REFLEXES

Reflexes are the simplest form of motor output. Reflexes are generally protective, keeping you from harm without thought. They are involuntary, and usually the response is proportionate to the stimulus. Some familiar reflexes are the **withdrawal reflex**, activated for example, when you pound your thumb with a hammer or touch a hot stove; the **vestibular reflex**, which keeps you vertical; and the **startle reflex**, which causes you to jump at

9-5 For an animation of the neuroreflex arc, please go to your DVD.

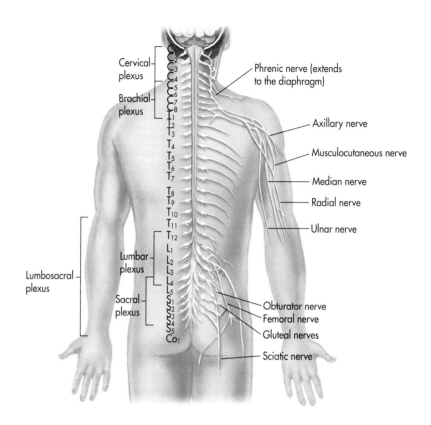

SPINAL NERVE PLEXUSES				
PLEXUS	**LOCATION**	**SPINAL NERVES INVOLVED**	**REGION SUPPLIED**	**MAJOR NERVES LEAVING PLEXUS**
Cervical	Deep in the neck, under the sternocleidomastoid muscle	C_1-C_4	Skin and muscles of neck and shoulder; diaphragm	Phrenic (Diaphragm)
Brachial	Deep to the clavicle, between the neck and the axilla	C_5-C_8, T_1	Skin and muscles of upper extremity	Musculocutaneous Ulnar Median Radial Axillary
Lumbosacral	Lumbar region of the back	T_{12}, L_1-L_5, S_1-S_4	Skin and muscles of lower abdominal wall, lower extremity, buttocks, external genitalia	Obturator Femoral Sciatic Pudendal

FIGURE ◾ 9–10

Spinal cord plexuses.

loud sounds. The amazing thing about reflexes is that they can often occur without your brain being involved. For many reflexes, the spinal cord serves as the reflex center and no impulses need to go to the brain.

Pathology Connection: Peripheral Neuropathy and Spinal Cord Injury

Peripheral neuropathy encompasses a number of disorders involving damage to peripheral nerves. Because peripheral nerves are involved in sensory, motor, and autonomic function, the symptoms of peripheral neuropathy vary greatly among patients. Symptoms include muscle weakness and decreased reflexes, numbness, tingling, paralysis, pain, difficulty controlling blood pressure, abnormal sweating, and digestive abnormalities. Nongenetic causes of neuropathy can be grouped into three broad categories: systemic disease, trauma, and infection or autoimmune disorders. Trauma, such as falls or automobile accidents, causes mechanical injury to nerves, and is the most common cause of peripheral neuropathy. Nerves may be severed, crushed, or bruised. Systemic disorders that cause peripheral neuropathy include kidney disorders, hormonal imbalance, alcoholism, vascular damage, repetitive stress, chronic inflammation, diabetes, toxins, and tumors. The most common of these is diabetes. Infections such as shingles, Epstein-Barr virus, herpes, HIV, Lyme disease, and polio cause peripheral neuropathy. Guillain-Barré syndrome is an acute form of peripheral neuropathy. Some forms of neuropathy, like Charcot Marie Tooth disease, are inherited. The pain that Maria, our diabetes patient, has been feeling may be the early stages of peripheral neuropathy.

Clinical Application

KNEE JERK REFLEX

The most common example of a reflex is the patellar tendon (knee-jerk) reflex. You may have experienced this reflex at the doctor's office. The doctor taps your knee with a hammer and your leg kicks, seemingly against your will. What exactly is going on? When the doctor taps your knee, the hammer gently tugs on a tendon that is connected to your quadriceps muscles. The quads are gently tugged. This causes them to lengthen slightly. This length change is received by a sensory neuron and transmitted back to the spinal cord. The sensory neuron synapses with a motor neuron in the ventral horn. The motor neuron sends a signal to the quadriceps to stop the stretch. How do the quads stop the stretch? They shorten (contract). When a muscle shortens, it causes a movement. The shortening of the quadriceps muscles extends your knee, causing your leg to kick.

This action is totally reflexive (no pun intended). You can't stop yourself from kicking, even if you try. You do not have to think about kicking. The action happens without your brain. Only the spinal cord is necessary. The real purpose of this reflex is obviously not to react to being hit with a rubber hammer. The reflex actually prevents your knees from collapsing while you are standing up. Any time your knees bend, even a little bit, the reflex kicks in (pun intended) and straightens your knees. The reflex is so small that you don't even notice.

Diagnosis of peripheral neuropathy is not easy. History of symptoms, presence of other conditions that cause neuropathy, CT, MRI, electromyogram (EMG), and biopsy can all aid in the diagnosis. Treatment consists of treating the underlying cause of the neuropathy and using medication and therapy to treat the symptoms.

Even though the spinal cord is well protected by the bones and meninges, trauma can cause damage to the delicate neural tissue. The most common cause of spinal cord injuries is car accidents, followed by violence, falls, work injuries, and disease. More than half of all spinal cord injuries occur in people between 16 and 30 years old. The overwhelming majority of spinal cord injuries are in males. There are 10,000 spinal cord injuries per year that leave patients with some paralysis and a quarter of a million spinal cord injury survivors currently living in the U.S. The spinal cord may be partially or completely severed, crushed, or bruised. Bruises to the spinal cord may resolve with time and rehabilitation, but a severed or crushed spinal cord is usually a permanent injury.

The initial injury to the spinal cord is often not the biggest problem. After the injury, the body's response to the injury often does more damage to the already injured tissue. The spinal cord swells, blood flow decreases, the immune system removes tissue and demyelinates the surviving tissue, release of excess neurotransmitter kills cells, and damaged neurons self-destruct.

Spinal cord injury usually results in paralysis and sensory loss below the injury, and the extent of the paralysis is related to the location of the spinal cord injury. Patients with injuries to the cervical spinal cord are **quadriplegics,** paralyzed in all four limbs. Some quadriplegics, with damage very high in the

Clinical Application

CARPAL TUNNEL SYNDROME

A very common form of peripheral neuropathy caused by repetitive motion is **carpal tunnel syndrome (CTS)**, an inflammation and swelling of the tendon sheaths surrounding the flexor tendon of the palm. This is a result of repetitive motion such as typing on a keyboard. As a result of this inflammation and swelling, the median nerve is compressed, producing a tingling sensation or numbness of the palm and first three fingers. Carpal tunnel syndrome is often corrected surgically.

9-6 To view a video on carpal tunnel syndrome, please go to your DVD for this chapter.

cervical spinal cord, have paralyzed diaphragms and cannot breathe on their own. Patients with injuries in the thoracic spinal cord and lower have **paraplegia**. They can move their arms but their legs are paralyzed.

The difference between being able to breathe on your own after a spinal cord injury and being dependent on a ventilator is literally a matter of centimeters. One of the nerves that projects from the cervical plexus is called the **phrenic nerve**. This nerve is the motor nerve for your diaphragm, your main breathing muscle. If the spinal cord is damaged below the cervical plexus, such as in the high thoracic region, the phrenic nerve can receive signals from the brain and send them to the diaphragm and the injured person is able to breathe. On the other hand, if the damage to the spinal cord is between the brain and the cervical plexus, the path from the brain to the phrenic nerve is blocked and the signals can't get to the diaphragm from the brain. The diaphragm is paralyzed, and the person cannot breathe on his or her own. However, even patients with thoracic or low cervical injuries may have difficulty coughing or breathing deeply due to paralysis of abdominal muscles.

In an interesting side note, sexual function is often preserved in patients with spinal cord injuries. Many men with spinal cord injuries can still have an erection. Penile erection is a reflex, so damage to spinal cord pathways doesn't prevent the penis from becoming erect under the right circumstances. Though ejaculation (sperm leaving the penis) is often impaired, sperm are normal and can be used for conception with medical intervention. Even men who are quadriplegic may be able to father a child. In women, the menstrual cycle may be abnormal due to post-injury hormonal changes, but many women with spinal cord injuries remain fertile and may be able to carry a child, with adequate medical supervision.

Sensory information below the injury is also lost. Patients with spinal cord injuries have no sense of touch—including pain, temperature, pressure or position—below their injury. Remember Ray, who was injured in the swimming pool accident? His injury is above C3, so he has no sense of touch below his chin, including heat, cold, or pain.

Diagnosis of a spinal cord injury is accomplished by a neurological exam that tests sensory and motor function and by any of several different types of imaging studies including MRI, X-ray, CT scans, and myelography (a special X-ray using radioactive dye).

Treatment for spinal cord injury in the acute stage, right after injury, consists of attempts to prevent further damage. The injury is immobilized, respiration is aided, low blood pressure or cardiac problems are treated, steroids are given to reduce the damage caused by inflammation and cell death, and the injury is permanently stabilized using a variety of surgical techniques. After the acute phase, the focus of treatment changes. Long-term problems caused by the injury must be treated or prevented, including respiratory difficulties, blood pressure abnormalities, pneumonia, blood clots, organ dysfunction, pressure sores, pain, and bladder and bowel dysfunction. Once stabilized, patients receive lengthy rehabilitation. In the past, rehabilitation focused on teaching patients to live with their disabilities. More recent evidence suggests that extensive and vigorous rehabilitation may allow spinal cord injury patients to recover more function than initially anticipated after their injury. Thus, rehabilitation often includes both coping skills and exercises designed to aid recovery of as much function as possible.

TEST YOUR KNOWLEDGE 9-3

Choose the best answer:

1. The segments of the spinal cord are named for:
   a. the bones of the skull
   b. the vertebrae
   c. their function
   d. none of the above

2. This layer of the meninges is fused to the surface of the CNS:
   a. dura mater
   b. pia mater
   c. arachnoid mater
   d. whatsa mater

3. The ventral root carries _____ information, and the dorsal root carries _____ information.
   a. motor, motor
   b. motor, sensory
   c. sensory, sensory
   d. sensory, motor

4. What is the term for axon pathways carrying information up and down the spinal cord?
   a. horns
   b. roots
   c. columns
   d. ganglia

5. How is spinal cord injury treated initially?
   a. stabilization of injury
   b. drugs to prevent tissue damage
   c. cardiovascular and respiratory support if needed
   d. all of the above

6. A patient complains of progressive weakness in the muscles of her legs. You begin a patient history. During your conversation she relates to you that she had some kind of virus as a kid that left her with a temporary limp. Apparently she was really sick for a while, but she doesn't remember much about it. What disorder should you rule out first?
   a. Multiple sclerosis
   b. Charcot Marie Tooth
   c. Post-polio syndrome
   d. Spinal cord injury

THE BRAIN AND CRANIAL NERVES

The brain lies at the top of the spinal cord, beginning at the level of the foramen magnum and filling the cranial cavity. It acts as the main processor and director of the nervous system. The cranial nerves leave the brain and go to specific body areas where they receive information and send it back to the brain (sensory), and the brain then sends back instructions as to the appropriate response (motor).

Just like a grocery store needs to be organized into sections such as produce, meats, and deli, the brain can be divided into several anatomical and functional sections. We will talk about each section separately and then describe the interactions between the brain parts and the spinal cord.

The Brain's External Anatomy

Let's start our journey looking at the outside of the brain, and then we will zoom in on the internal structures. From the outside, you can see that the brain consists of a **cerebrum, cerebellum,** and **brain stem.** Please see Figure 9–11 ▪.

cerebrum *(ser EE brum)*
cerebellum *(ser eh BELL um)*

CEREBRUM

The cerebrum, the largest part of the brain, is divided into two **hemispheres,** a right and left, by the longitudinal fissure, and divided from the cerebellum ("little brain") by the transverse fissure. The surface of the cerebrum is not smooth, but broken by ridges (**gyri**) and grooves (**sulci**) collectively known as **convolutions**. These convolutions serve a very important purpose by increasing the surface area of the brain, yet allowing it to be "folded" into a smaller space. Most of the sulci are extremely variable in their locations among humans, but a few are in basically the same place in every brain. These less variable sulci are used to divide the brain into four large sections called **lobes,** much like the major departments in a grocery store are separated by aisles.

gyri *(JIE rie)*
sulci *(SULL kie)*

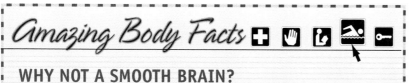

Amazing Body Facts

WHY NOT A SMOOTH BRAIN?

The surface of the brain is folded and rippled into a series of convolutions. These convolutions allow lots of brain surface area to fit into a very small space. If you could lay the convolutions flat, they would be the size of a pillow case. Talk about your big heads! Another fact to keep in "mind" is that the vital brain can die if deprived of oxygen in as little as 4-8 minutes.

The lobes (see upcoming Table 9-2) are named for the skull bones that cover them, and they occur in pairs, one in each hemisphere. The most anterior lobes, separated from the rest of the brain by the central sulci, are the **frontal lobes.** Posterior to the frontal lobe are the **parietal lobes.** Posterior to the parietal lobes are the **occipital lobes.** (There is no obvious dividing line between the parietal and occipital lobes, and these represent the most posterior lobes at the back of the skull.) The most inferior lobes, separated by the lateral fissures, are the **temporal lobes.** Again, please see Figure 9–11 which shows the lobes of the brain and the sulci that separate them. There is a section of the brain, the **insula,** deep inside the temporal lobes, which is often listed as a fifth lobe, but it is not visible on the surface of the cerebrum.

parietal lobe *(pah RYE eh tal)*
occipital lobe *(ok SIP eh tal)*

Much of the information coming into and leaving your brain is **contralateral.** That is, the left side of the body is controlled by the right side of your cerebrum, and the right side of the body is controlled by your left brain.

contra = *against, opposite*
lateral = *side*

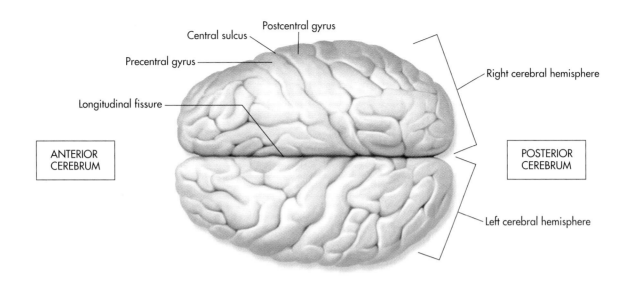

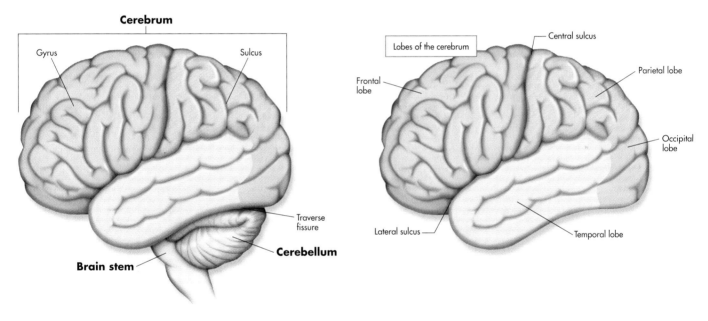

FIGURE ■ 9–11

External brain anatomy and lobes.

CEREBELLUM

The cerebellum is posterior to the brain stem and plays an important role in sensory and motor coordination and balance. Its surface is also convoluted like that of the cerebrum. From its external appearance, it is easy to see why anatomists consider the cerebellum the "little brain."

TABLE 9-2 Cerebral Lobes and Cerebellum

STRUCTURE	MAJOR FUNCTIONS
Cerebral Lobes	
Frontal lobe	Motor function, behavior and emotions, memory storage, thinking, smell
Parietal lobe	Body sense, perception, and speech
Occipital lobe	Vision
Temporal lobe	Hearing, taste, language comprehension, integration of emotions
Insula	Autonomic functions
Cerebellum	Sensory and motor coordination and balance

THE BRAIN STEM

The brain stem (see Table 9-3) is a stalk-like structure inferior to and partially covered by the cerebrum. It is divided into three sections. The **medulla oblongata** is continuous with the spinal cord. The **pons** is just superior to the medulla oblongata and connects the medulla oblongata and the cerebellum with the upper portions of the brain. The **midbrain** is the most superior portion of the brain stem (see Figure 9-12 ■).

medulla oblongata
(meh DULL lah ob long GA ta)

TABLE 9-3 The Brain Stem

STRUCTURE	FUNCTION
Midbrain	Relays sensory and motor information
Pons	Relays sensory and motor information; role in breathing
Medulla oblongata	Regulates vital functions of heart rate, blood pressure, breathing, and reflex center for coughing, sneezing, swallowing and vomiting

The **reticular system** is a diffuse network of neurons in the brain stem that is responsible for "waking up" your cerebral cortex. Reticular system activity is vital for maintaining conscious awareness of your surroundings. When your alarm clock wakes you in the morning, your reticular system is responsible for nudging your cortex out of slumber. General anesthesia inhibits the reticular system, rendering surgery patients unconscious. Injury due to

9-7
The brain can be imaged using positron emission tomography, more commonly known as PET scan. To view a video of this procedure, please refer to your DVD for this chapter.

ischemia, mechanical damage, or drugs can damage the reticular system and lead to coma.

The brain stem receives sensory information and contains control systems for vital functions such as blood pressure, heart rate, and breathing. The brain-stem controls the vital functions of life, and patients with severe brain injuries with an intact brain stem can continue in a vegetative state as long as they are nutritionally supported. This condition is called a **persistent vegetative state** (PVS).

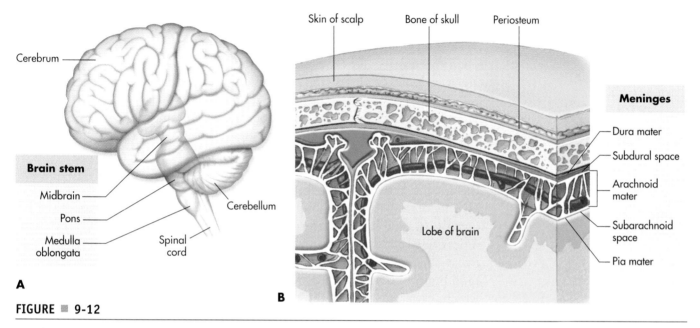

FIGURE ■ 9-12

A. The brain stem and B. meninges

The brain, like the spinal cord, is covered with protective membranes called **meninges**. Again, please see Figure 9–12. The meninges of the brain are continuous with the spinal cord meninges. A potentially fatal condition is an infection of the meninges called **meningitis,** which can rapidly spread and affect the brain and spinal cord through this common covering.

Meningitis causes inflammation of the meninges. Its symptoms include headache, fever, and stiff neck. Most cases of meningitis are caused by viral infection, but very serious cases can be caused by bacterial infection. If untreated, bacterial meningitis can cause seizures, coma, or death.

Meningitis can be diagnosed by a spinal tap or lumbar puncture. In this procedure, the physician collects a sample of cerebrospinal fluid. Fluid from a patient with meningitis generally has low glucose levels and elevated white blood cell and protein levels. The fluid may show signs of specific bacteria. If the patient has viral meningitis, genetic material from viruses will show up in the fluid.

If bacterial meningitis is diagnosed, the patient is immediately given intravenous antibiotics. Fluids may also be drained from around the brain. If viral meningitis is diagnosed, the patient is prescribed bed rest, fluids, and medication to relieve pain and swelling. Viral meningitis usually goes away in about 1-2 weeks.

meningitis (men in JYE tiss)
itis = *inflammation of*

9-8   To see a 3-D view of the brain and to perform an interactive drag-and-drop exercise on the brain and the brain stem, please go to your DVD.

Bacterial meningitis may be difficult to treat due to the *blood-brain barrier*. The blood-brain barrier prevents the entry of various drugs and other substances into brain tissue. This barrier is normally protective of the brain, but it can allow damage to brain tissue when it blocks life-saving medicines. Large concentrations of intravenous drugs are used to deliver drugs to infection around brain tissue.

Pathology Connection: Brain Injury

Traumatic Brain Injury (TBI), or simply head injury, occurs when force is applied to the skull, causing damage to brain tissue. Injury to the brain can occur due to vehicle accidents (the most common cause), falls, violence, and sports injuries, and damage similar to TBI can be caused by lack of oxygen, strokes, or brain hemorrhage. Fifty percent of traumatic brain injuries caused by accidents or violence are alcohol-related. There are 100 cases of TBI per 100,000 people in the U.S. each year and between 2 and 6 million brain injury survivors in the U.S. today. The most likely ages for a traumatic brain injury are under the age of 5, between the ages of 15 and 24 (males), and over the age of 75. The severity of post-injury disability from TBI depends on the severity of the injury, the age and health of the patient, and the location of the injury. Brain injuries can be closed (the skull is not open) or penetrating (the skull is punctured by an object), and can range in severity from mild to moderate to severe. A mild brain injury is more commonly known as a **concussion**. A **contusion**, which is swelling and bleeding at the site of injury, is more serious. In more serious cases, a severe hemorrhage (hematoma) may result.

Symptoms of head injury include dizziness, confusion, headache, blurred vision, ringing ears, sleepiness, and mood or cognitive changes. More serious injuries have more severe symptoms, including vomiting, dilated pupils, convulsions, motor deficits, slurred speech, confusion or agitation and perhaps prolonged unconsciousness. Motor and sensory deficits from head injuries are often confined to one side of the body. (Remember that the cerebrum is contralateral: the right side controls the left body and vice-versa.) Patients with severe injuries may be in a **stupor** (brief period of impaired consciousness or arousal), a **coma** (no consciousness), a **vegetative state** (may have arousal and sleep wake cycles but no response to surroundings) or a **persistent vegetative state** (often a permanent condition) depending on the severity of their injury.

The condition known as a **stroke**, or **cerebrovascular accident (CVA)**, which can cause permanent brain injury, is caused by the disruption of blood flow to a portion of the brain due to either hemorrhage or blood clot. If the oxygen is disrupted for long enough, brain tissue will die. The symptoms of stroke vary depending on the location of the stroke and can include sudden severe headache, dizziness, loss of vision in one eye, aphasia, dysphasia, coma, and possible death. Paralysis and hemiplegia are also possible. According to the Centers for Disease Control and Prevention, stroke is the third leading cause of death in the United States and is the number one cause of serious long-term disability. Clots are the cause of the majority of strokes. Risk factors include smoking, hypertension, heart disease, and a family history of having the disease. Strokes can also rob patients of the ability to speak and can cause blindness and destroy memory. Symptoms of stroke appear suddenly. Some patients have a series of minor strokes (almost like tiny earthquakes before the big one) with minor, temporary symptoms before they have a major stroke. These mini-strokes are called **transient ischemic accidents**, or **TIAs**. Please see Figure 9-13 ■, which shows a cerebral embolism and cross-section of a brain showing a cerebrovascular attack.

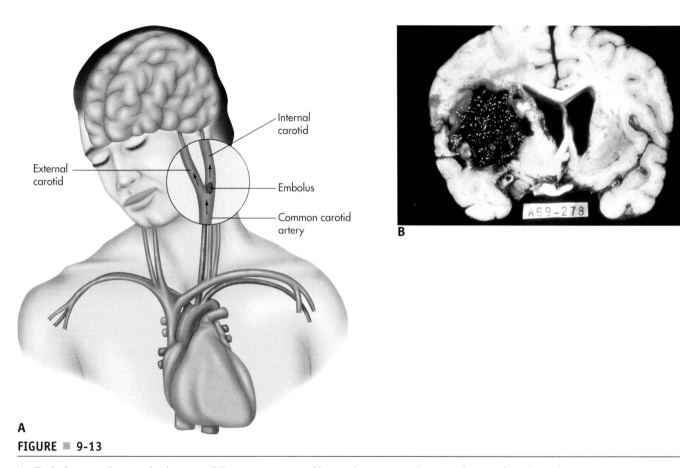

A

FIGURE ■ **9-13**

A. Embolus traveling to the brain and **B.** cross-section of brain showing cerebrovascular accident (CVA).

A **hematoma** is a pool of blood between any of the layers of meninges and the skull. The most common locations for hematomas are **epidural** (between the dura mater and the skull), **subdural** (between the dura mater and the arachnoid mater) and **subarachnoid** (in the subarachnoid space). A blow to the head can rupture tiny blood vessels in the skull, causing them to bleed into the space. A stoke or ruptured aneurysm (weak spot in a blood vessel inside the skull) can also cause a hematoma.

Several techniques are used to diagnose a brain injury. The Glasgow Coma Scale is a scale from 3 to 15 based on the patient's ability to open their eyes on command, respond verbally to questions, and move their limbs when requested. A lower number in-

9-9
To view a video on the actual procedure for assessing the Glasgow Coma scale, please see your DVD.

dicates a more severe injury. Imaging, CT, MRI, and PET scanning are used to pinpoint the location and severity of injury and to monitor any changes.

Treatment of head injury, like treatment of spinal cord injury, involves both prevention of further injury and treatment of existing injury. Remember that the spinal cord self-destructs after an injury, doing damage to itself? Like the spinal cord, the injured brain will self-destruct due to increased swelling and cell death caused by the tissue's attempt to repair the damage. In addition, in closed head injuries if the brain tissue swells or if there is significant bleeding, pressure inside the skull (intracranial pressure) will rise. Since the fluid (blood or tissue fluid) can't be compressed and the skull will not expand to relieve the pressure, more brain tissue will be destroyed by the pressure. In addition, as the brain swells it presses the brain stem downward through the foramen magnum, actually causing a herniation (protrusion). The herniation causes damage to respiratory centers and decreases blood flow to the brain. The combination of abnormal ventilation and decreased blood flow causes more damage to brain tissue.

Acute care of head injury involves immobilization of the head, stabilization of cardiovascular and respiratory functions, monitoring of intracranial pressure, medication to decrease intracranial pressure, and surgery to remove clots, blood, or foreign objects (for example, bullet or bone fragments) from the brain. Post-injury complications must also be treated, including hydrocephalus, infection, pain, bed sores, and multiple organ failure. In addition, many head injury patients, especially those injured in accidents, have other injuries that complicate their treatment and recovery.

After the acute period, many patients with brain injuries continue to have problems. Approximately 40% of brain injured people, even with mild injuries, experience post-concussion syndrome. Several days or weeks after the injury, patients experience dizziness, headache, memory and concentration problems, irritability, disordered sleep, and anxiety and depression. The condition is usually temporary. Many patients with severe injury need lengthy rehabilitation to regain function. Patients with moderate or severe brain injuries are often left with permanent disabilities, including motor, sensory, and speech deficits, memory loss (the most common intellectual deficit), attention or problem solving deficits, vision problems, and persistent pain. For many years, the medical community thought that injured brains could not recover functions, but new data suggests that the human brain is capable of some degree of "re-wiring." This ability to reroute information is probably responsible for the extraordinary recovery sometimes seen after brain injury. However, many people remain permanently disabled.

9-10 To view an animation on a specific brain injury called contracoup injury, please see your DVD.

Internal Anatomy of the Brain

The inside of the brain has, like the spinal cord, white matter and gray matter, and hollow cavities containing cerebrospinal fluid. (See! We really do have holes in our heads.) Unlike the spinal cord, however, the white matter is surrounded by the gray matter in the brain. (Remember, in the spinal cord, the white matter surrounds the gray matter.) The layer of gray matter surrounding the white matter is called the **cortex.** In the cerebrum it is called the cerebral cortex, and in the cerebellum it is called the cerebellar cortex. Throughout the brain there are deep "islands" of gray matter surrounded by white matter. These islands are called **nuclei.** The nuclei in the cerebrum can be part of the **basal nuclei,** which is a motor coordination system, or part of the **limbic system,** which controls emotion, mood, and memory.

limbic system *(LIM bick)*

CEREBRUM

The inside of the cerebrum reflects its external anatomy (see Figure 9-14 ■). The lobes—frontal, temporal, parietal, occipital, and insula—are clearly visible. On either side of the central sulcus are two gyri named for their locations: the **precentral gyrus,** anterior to the central sulcus (frontal lobe), and the **postcentral gyrus,** posterior to the central sulcus (parietal lobe). The structures in the cerebrum are responsible for conscious thought, judgment, reasoning, memory, and will power.

gyrus *(JIE rus)*

Located in the precentral gyrus in the frontal lobe is a specialized area called the **primary motor cortex**. The primary motor cortex has a map of the body with each area dedicated to motor function of a specific body area. Check out the size of the primary motor cortex dedicated to each body part in Figure 9-15 ■. Do the sizes of the areas correspond to the sizes of the body parts? Not really: the hands, lips, and head have very large maps, while the legs and arms have very small maps compared to each body part's size. What do the hand and lips have in common that makes them different from the arms and legs? They are required to perform more finely coordinated movements, such as speaking and handwriting (or typing), and, therefore, have more motor output. If you aren't convinced, think about how much harder it would be to type with your feet than with your hands. This map can be drawn as a homunculus (little man) with huge hands, lips, and head (see Figure 9-15).

Also in the frontal lobe are the premotor and prefrontal areas, which *plan* movements. The plan from these two areas is sent to the primary motor cortex. The premotor and prefrontal areas are anterior to the primary motor cortex. In addition there is a motor speech area in the frontal lobe, called **Broca's area**, that controls movements associated with speech.

The brain must also have a specialized area for sensory input and processing. In the parietal lobe is the **primary somatic sensory cortex**, in the postcentral gyrus. Like the primary motor cortex, primary somatic sensory cortex has a map. Please see Figure 9-16 ■. In this case, however, the sizes of the body maps are proportional to the amount of the sensory input provided. For example, the hands provide much more sensory input due to touch than your neck would, and as a result the area devoted to the neck is much smaller.

There is another area of the parietal lobe that allows *understanding* and *interpretation* of somatic sensory information. It is located in the parietal lobe just posterior to somatic sensory cortex and is known as the **somatic sensory association area**. **Wernicke's area**, also in the parietal lobe, controls understanding of language.

The right and left hemispheres are connected by a collection of white matter surrounding the lateral ventricles, called the **corpus callosum.** This connection allows for cross-communication between the right and left sides of the brain. Many of our day-to-day activities, like walking or driving, require both sides of the body, and therefore both sides of the brain, to be well coordinated. Imagine walking if your legs were acting independently!

corpus callosum
(KOR pus kah LOH sum)

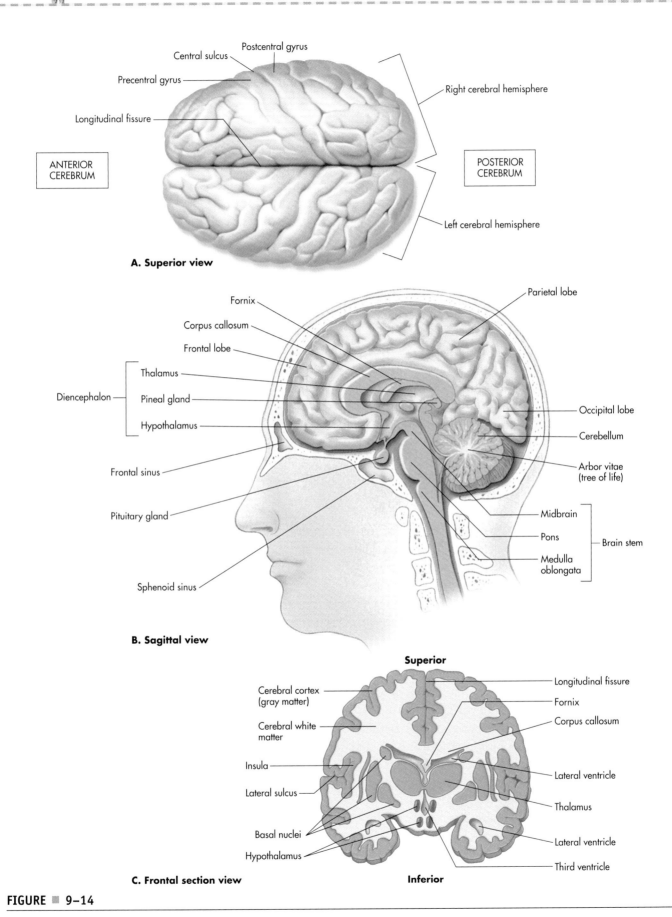

A. Superior view

B. Sagittal view

C. Frontal section view

FIGURE ■ 9–14

A. Superior, **B.** sagittal, and **C.** frontal section view of the brain.

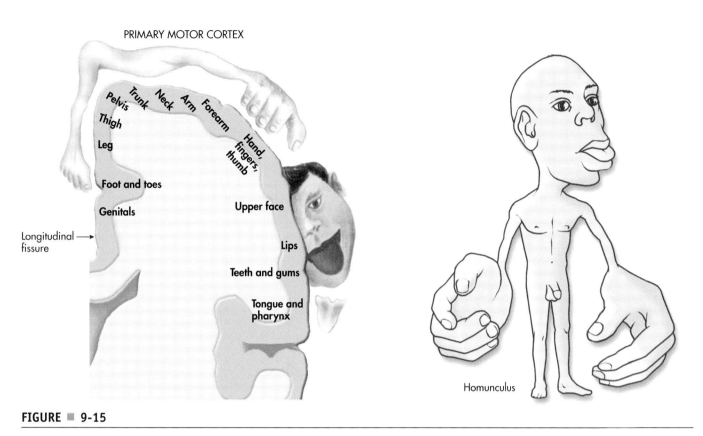

PRIMARY MOTOR CORTEX

Homunculus

FIGURE ■ 9-15

Motor areas of the brain, with homunculus. Notice the body parts that require a lot of motor movements are allotted a much larger region in the primary motor cortex.

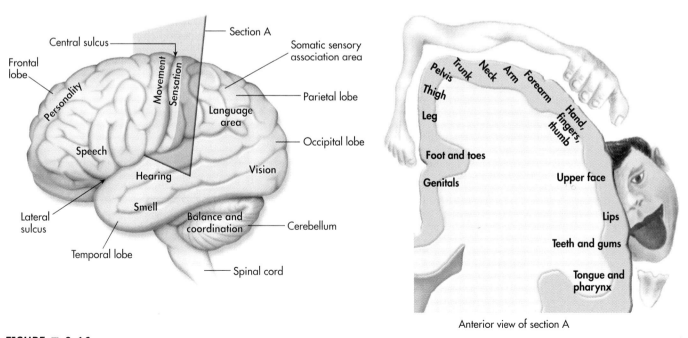

Anterior view of section A

FIGURE ■ 9-16

Primary somatic sensory area. Notice the size of the body parts are proportional to the amount of the sensory input provided. For example, the hands provide much more sensory input due to touch than your neck would and as a result the area devoted to the neck is much smaller.

diencephalon *(dye en SEFF ah lon)*
thalamus *(THAL ah mus)*
hypothalamus
  (high poh THAL ah mus)
pineal body *(pih NEE al)*
basal nuclei *(BAY sal noo KLEE ie)*

DIENCEPHALON

Inferior to the cerebrum is a section of the brain that is not visible from the exterior, called the **diencephalon** (shown in Figure 9-14). The diencephalon can be divided into the **thalamus, hypothalamus, pineal body,** and the **pituitary gland.** See Table 9–4 for diencephalon structures and their corresponding functions. The hypothalamus, pituitary, and pineal represent an interface with the endocrine system (the other control system), covered in Chapter 10. The diencephalon contains a number of nuclei that are part of the basal nuclei and limbic system.

The hypothalamus appears to play a role in emotional state. Experimental stimulus of the hypothalamus can produce extreme responses, such as rage or uncontrolled laughter. Milder emotions such as pleasure and displeasure can also be evoked when certain parts of the hypothalamus are stimulated. It would be a mistake to say that the hypothalamus causes these emotions, however. Emotional state appears to depend on complex interactions between external stimuli, experience, and structures in the brain.

TABLE 9-4 Diencephalon

STRUCTURE	FUNCTION
Thalamus	Relays and processes information going to the cerebrum
Hypothalamus	Regulates hormone levels, temperature, water-balance, thirst, appetite, and some emotions (pleasure and fear); regulates the pituitary gland and controls the endocrine system
Pineal body	Responsible for secretion of melatonin (body clock)
Pituitary gland	Secretes hormones for various functions (explained in Chapter 10)

THE CEREBELLUM

The external similarities between the cerebellum and cerebrum are also obviously internally (again see Figure 9-14). The cerebellum has a gray matter cortex and a white matter center, known as the **arbor vitae** (tree of life). The cerebellum also has nuclei that coordinate motor and sensory activity. Essentially, the cerebellum fine-tunes voluntary skeletal muscle activity and helps in the maintenance of balance, by comparing planned movements to actual.

☤ Pathology Connection: Alzheimer's Disease

Alzheimer's disease is a progressive degenerative disease of the brain and is the most common cause of dementia among people 65 years or older. This disease may progress over 5-18 years and is characterized by memory loss and diminishing cognitive functions. As Alzheimer's disease progresses, tangles of fibers develop in nerve cells and abnormal protein deposits surround the cells. Both the tangles and deposits interfere with brain function. The etiology of this disease is unknown, but age is the most important risk factor.

Alzheimer's symptoms begin gradually with mild forgetfulness, such as difficulty remembering recent events or familiar names. In the first stage of the disease, patients experience confusion, anxiety, some memory loss, and poor judgment. During the second stage, patients experience more memory loss, have difficulty recognizing people, have trouble remembering words, experience sleeplessness and confusion, and start to experience motor problems and a loss of social skills. In the third stage, patients have difficulty speaking, reading and writing, and even maintaining personal hygiene. They become bedridden and develop conditions related to inactivity, such as bed sores and pneumonia. As the condition worsens, patients' personalities change to the point that they become anxious or aggressive and can become so confused that they get lost even in familiar locations. This has led to a classification of symptoms known as the four A's: Anger, Aggression, Anxiety and Apathy.

 9-11 To view a video on Alzheimer's disease, please see your DVD.

There is no known cure for this disease. Certain medications may help to slow the progression of the early and middle stages of this disease. Recent research has focused on the use of preventative medications such as statins which may reduce the build-up of plaque in the brain. Smoking appears to be a contributing factor.

Cerebrospinal Fluid and Ventricles

As in the central canal in the spinal cord, there are fluid-filled cavities in the brain. The cavities in the brain are called **ventricles**, and they are continuous with the central canal of the spinal cord and the **subarachnoid space** of both the brain and the spinal cord. These ventricles, four in total, allow for the circulation of cerebrospinal fluid throughout the brain. The two lateral ventricles are in the cerebrum; the third ventricle is in the diencephalon, a region between the cerebrum and brain stem, and the fourth ventricle is in the inferior part of the brain between the medulla oblongata and the cerebellum (see Figure 9-17 ■).

The ventricles of the brain, the central canal of the spinal cord, and the subarachnoid space, surrounding both parts of the CNS, are all filled with **cerebrospinal fluid (CSF)**. The CSF is filtered from blood in the ventricles by tissue called **choroid plexus**. Produced at the rate of 750 milliliters daily, CSF made in the lateral ventricles flows through a tiny opening into the third ventricle and then through another opening into the fourth ventricle. From the fourth ventricle, CSF flows into the central canal of the spinal cord and the subarachnoid space. CSF is returned to the blood via special "ports" (the **arachnoid villi**) between subarachnoid space and blood spaces in the dura mater (the **dural sinuses**).

ventricles *(VEN trik lz)*
subarachnoid space
(sub ah RAK noyd)

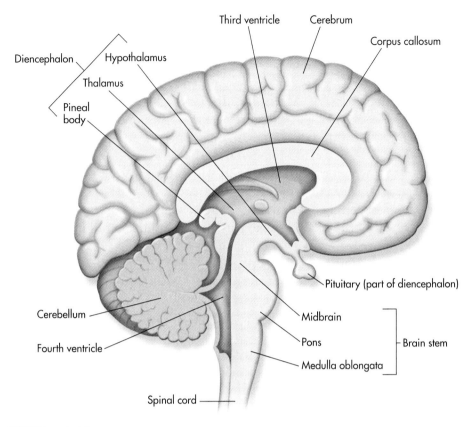

FIGURE ■ 9-17

Internal anatomy of the brain.

 Pathology Connection: Hydrocephalus

The balance of CSF made and CSF reabsorbed by the blood is very important. The brain is a very delicate organ captured between the liquid CSF and the bones of the skull. If there is too much CSF, a condition called **hydrocephalus** (literally "water in the head") occurs, damaging brain tissue. See Figure 9-18 ■. Hydrocephalus can be caused by blockage of the narrow passages between the ventricles due to trauma, birth defects or tumors, or decreased reabsorption of CSF due to subarachnoid bleeding. Hydrocephalus may be accompanied by increased intracranial pressure or may have no apparent rise in pressure (normal pressure hydrocephalus).

Symptoms of hydrocephalus vary depending on the age of the patient and the type of hydrocephalus. In infants, whose skulls have not completely hardened, the skull may expand. Patients may experience vomiting, irritability, and seizures. In older children and adults, the skull does not change size, but the rising pressure causes headache, nausea, blurred vision, balance and coordination problems, sleepiness, and personality changes. Normal pressure hydrocephalus, seen mainly in the elderly, causes dementia and motor problems. It is a chronic progressive disease often accompanied by or caused by other neurological problems.

Hydrocephalus is diagnosed using imaging including CT and MRI (looking for enlarged ventricles) and monitoring of intracranial pressure. Hydrocephalus may be treated by medication, but the most common treatment is insertion of a **shunt**, a tube that drains the extra CSF into the patient's heart or abdominal cavity (again see Figure 9-18). Long-term prognosis for patients with hydrocephalus is mixed. Many have long-term problems.

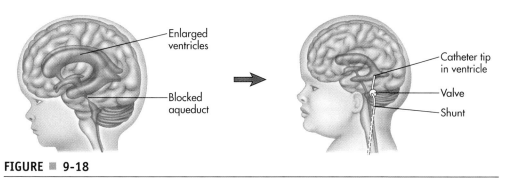

FIGURE ▪ 9-18

Hydrocephalus.

Cranial Nerves

In order for the CNS to function, it must be connected to the outside world via nerves of the PNS. We have already seen that the spinal cord is connected to the outside via spinal nerves. The brain also has nerves to connect it to the outside, aptly named **cranial nerves** (see Figure 9-19 ▪). Cranial nerves are like spinal nerves in that they are the input and output pathways for the brain, just as the spinal nerves are the pathways for the spinal cord. However, that is really where the similarities end. You should remember that there are 31 pairs of spinal nerves and that all of them are mixed nerves: they carry both sensory and motor information because they are formed by a combination of dorsal and ventral roots. There are far fewer cranial nerves—only 12 pairs. All but the first two pairs arise from the brain stem. Cranial nerves are not all mixed. Some cranial nerves are mainly sensory nerves, providing input; some are mainly motor nerves, directing activity; and some are mixed. Cranial nerves are much more specialized than spinal nerves and are named based on their specialty. Some cranial nerves carry sensory and motor information for the head, face, and neck, while others carry visual, auditory, smell, or taste sensation. Table 9-5 lists cranial nerves and their functions.

Peripheral neuropathy, damage to peripheral nerves, is not just about spinal nerves. Cranial nerves are also subject to peripheral neuropathy. The symptoms, however, would be seen on the face and head or in the special senses (vision, hearing, taste, and smell).

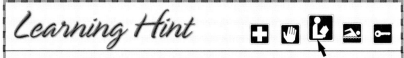

Learning Hint

MNEMONIC DEVICES

A *mnemonic device* is a tool used to help you memorize long lists. It can be very useful in anatomy. To make a mnemonic device, take the first letter of each part of the list you are trying to memorize and make it into a sentence. An example for the cranial nerves is this one: **O**n **O**ld **O**lympus **T**owering **T**ops **A** **F**inn **V**ith **G**erman **V**alked **A**nd **H**opped. Then when you know the nerves in order, their names will often help you remember their functions.

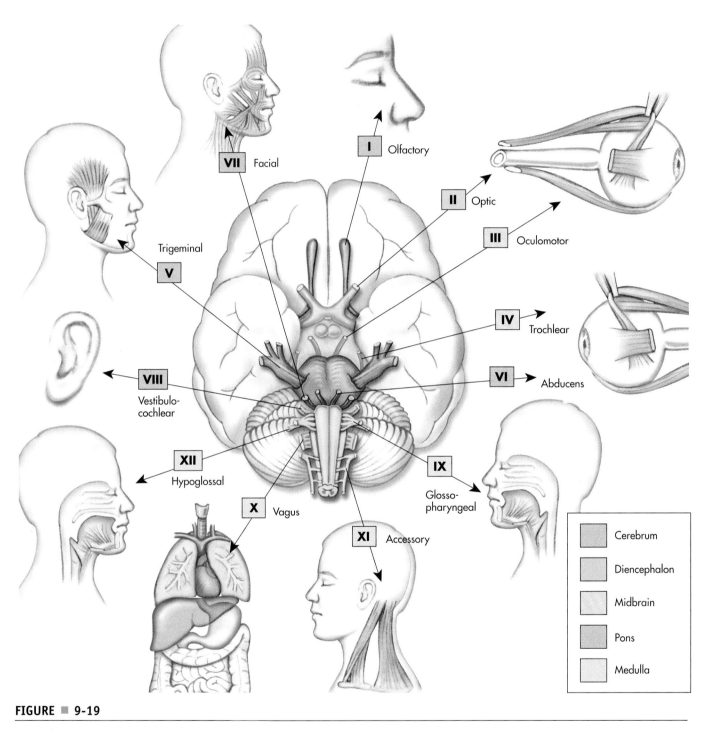

FIGURE ■ 9-19

Cranial nerves.

TABLE 9-5 Cranial Nerves and Functions

NERVE	FUNCTION
Olfactory (I)	Sensory (smell)
Optic (II)	Sensory (vision)
Oculomotor (III)	Mixed, chiefly motor for eye movements
Trochlear (IV)	Mixed, chiefly motor for eye movements
Trigeminal (V)	Mixed, sensory for face, motor for chewing
Abducens (VI)	Mixed, chiefly motor for eye movements
Facial (VII)	Motor for facial expression
Vestibulocochlear (VIII)	Sensory, hearing, and balance
Glossopharyngeal (IX)	Mixed, motor for throat muscles; sensory for taste
Vagus (X)	Mixed, motor for autonomic heart, lungs, viscera; sensory for viscera, taste buds, and so on
Accessory (XI)	Mixed, chiefly motor; motor and sensory for larynx, soft palate, trapezius, and sternocleidomastoid muscles
Hypoglossal (XII)	Chiefly motor for tongue muscles

TEST YOUR KNOWLEDGE 9-4

Choose the best answer:

1. Which of the following is not part of the brainstem?
   a. pons
   b. medulla oblongata
   c. midbrain
   d. diencephalon

2. Deep islands of gray matter are known as:
   a. hemispheres
   b. gyri
   c. nuclei
   d. nerves

3. Which of the following is not found in the diencephalon?
   a. thalamus
   b. pineal
   c. postcentral gyrus
   d. hypothalamus

4. This lobe contains the primary visual cortex:
   a. frontal
   b. temporal
   c. occipital
   d. parietal

5. To prevent further brain injury after TBI, doctors must control:
   a. body temperature
   b. intracranial pressure
   c. demyelination
   d. peripheral neuropathy

THE BIG PICTURE: INTEGRATION OF BRAIN, SPINAL CORD, AND PNS

So how does the brain work with the spinal cord and peripheral nervous system (PNS)? Let's begin to put the pieces together by revisiting the overall organizational chart of the nervous system. Figure 9-20 ■ will serve as our guide to keep our location clear.

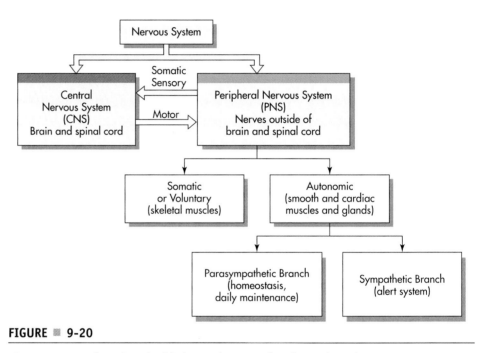

FIGURE ■ **9-20**

Nervous system flow chart highlighting the areas thus far explained.

The Somatic Sensory System

somatic *(soh MAHT ik)*

The **somatic sensory system** provides sensory input for your nervous system. We have already mentioned the parts of your brain that are dedicated to your special senses: vision (occipital lobe), hearing (temporal and parietal lobes), taste (temporal lobe), and smell (frontal lobe). We have not yet talked in detail about the sense of touch, called **somatic sensation.** Somatic sensation allows you to feel the world around you. Somatic sensation includes not just fine touch, which enables you to tell the difference between a peach and a nectarine or a golf ball and a ping pong ball, but also crude touch, vibration, pain, temperature, and body position. While your special senses are all carried on cranial nerves, information for somatic sensation comes into *both* the brain and the spinal cord. Ultimately, for you to attach meaning to the sensation, the sensory information must get to your brain. The pain that Maria, our diabetes patient, feels is due to damage to her peripheral nerves. Her somatic sensory system perceives their malfunction as pain.

Let's start with the spinal cord. When we talked about reflexes, we saw somatic sensory information come into the spinal cord via the dorsal root and synapse with a motor neuron in the ventral horn. Let's go back to hitting your thumb with a hammer. The pain neurons, with bodies in the dorsal root ganglion, are depolarized and send signals to the spinal

9-12 To see videos on pain issues to include assessing pain and pain management, please go to your DVD.

cord via the spinal nerve and dorsal root. The neurons synapse with motor neurons, causing you to pull your hand away from the stimulus. These same pain neurons join the spinothalamic tract in your spinal cord and simultaneously send a pain signal to your brain. The signal goes first to the thalamus and other parts of the diencephalon, causing physiological symptoms like sweating and increased heart rate. The pain then continues along the pathway to your somatic sensory cortex (in the postcentral gyrus). There, the exact location of the pain becomes apparent to you. Now you know *where* the pain is but still may not recognize it as pain. The *understanding* occurs when the information is integrated with sensation in the somatic sensory association area (in the parietal lobe). Now you know that you have hurt yourself! All the receiving and processing of information and associated actions happen almost instantaneously.

Because there are many variations of "touch" information that come in from all parts of the body, your body needs different neural highways to send that information to the brain effectively. See Table 9–6 for a description of sensory spinal cord pathways.

TABLE 9-6 Spinal Cord Pathways for Sensory Information

PATHWAY	INFORMATION	FROM	TO
Spinothalamic		skin	somatic sensory cortex
Lateral	pain; temperature	skin	somatic sensory cortex
Anterior	itch; pressure; tickle	skin	somatic sensory cortex
Dorsal column	fine touch; limb position	skin; joints	somatic sensory cortex; cerebellum
Spinocerebellar	posture	joints; tendons	cerebellum

The sensory information coming into the brain must go to a specific area for processing. As you can see from Table 9–6, the dorsal column and spinothalamic tracts both transport sensory information from your skin to a portion of the cerebrum known as the **primary somatic sensory cortex,** located in the postcentral gyrus of the parietal lobe.

The axons transport information to specific parts of the somatic sensory cortex that correspond to parts of the body as previously discussed in the sensory cortex map (refer to Figure 9-16).

The neurons in the somatic sensory cortex allow conscious sensation. You feel an insect crawling on your arm because the insect stimulates the "arm" neurons in your somatic sensory cortex. The somatic sensory system works on a kind of hierarchy, with the sensory neurons in the spinal cord and brain stem collecting information and passing it to areas in the thalamus, cerebellum, and cerebral cortex for processing. Your *actual* understanding of complex sensory input happens only *after* the information is passed to the somatic sensory cortex and somatic sensory association area.

THE BIG PICTURE: THE MOTOR SYSTEM

The motor system is also a hierarchy, working in parallel with the somatic sensory system. Remember from earlier in the chapter that the output side carries out the orders from the brain and spinal cord. The motor system carries orders to all three types of muscles and to the body's glands, telling them how to respond to the new information.

However, now information is moving in the opposite direction, from brain to spinal cord. See Figure 9-21 ▥ to show our current progress on the nervous system organizational chart.

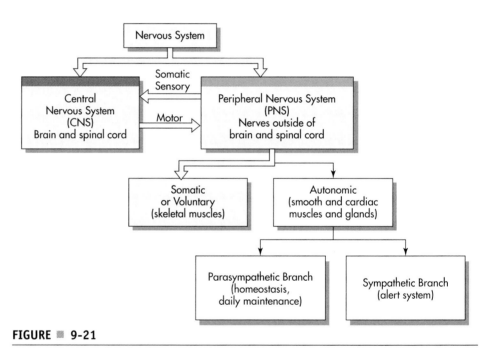

FIGURE ▥ 9-21

Progress thus far on the nervous system flow chart.

Cerebral Cortex

The somatic motor system controls voluntary movements under orders from the cerebral cortex. When you decide to move, the movements are planned in the prefrontal cortex and premotor cortex in the frontal lobe. This plan is then sent to the primary motor cortex, in the precentral gyrus. The orders are then sent to the spinal cord directly and also to a number of coordination centers, including the thalamus, basal nuclei, and cerebellum.

Subcortical Structures

The plan for movement leaves the motor cortex and connects with neurons in the thalamus, which is located in the diencephalon. The thalamus, basal nuclei, and cerebellum are part of a complicated **motor coordination loop**. Here, the movement is fine-tuned, posture and limb positions are taken into account, other movements are turned off, and movement and senses are integrated. This loop is fundamental. Without the coordination loop among the subcortical (under the cortex) structures , movements would be at the very least jerky and inaccurate.

CEREBELLUM

So far we have glossed over the important function of the cerebellum. The cerebellum has both motor and sensory inputs and outputs from the cerebral cortex, the thalamus, the basal nuclei, and the spinal cord. The cerebellum gets information about the *planned* movement and the *actual* movement and compares the plan to the actual. If the plan and the actual do not match, the cerebellum can adjust the actual movement to fit the plan. The function of the cerebellum is subtle and still a bit of a mystery, but without the cerebellum, movements would be inaccurate at best.

SPINAL CORD PATHWAYS

While the movement information plan is being processed by the thalamus, basal nuclei, and cerebellum, it moves direcly to the spinal cord and brain stem via the **corticospinal** and **corticobulbar tracts**. The corticospinal and corticobulbar tracts from your motor cortex are *direct* pathways, carrying orders to the neurons in the brainstem and ventral horn of the spinal cord. The pathways coming from subcortical structures are considered *indirect* pathways, which help coordinate movement.

The spinal cord pathways send orders from the brain to the motor neurons in the spinal cord and brainstem. The motor neurons in your spinal cord connect to skeletal muscles via the cranial nerves (in the brain stem) or spinal nerves (in the spinal cord), sending orders to the skeletal muscles to carry out the planned movement or coordinate ongoing movements. (Remember, these neurons communicate with the muscles via the neuromuscular junction, and therefore release the neurotransmitter acetylcholine across the synapse.) A second function of the motor tracts is to fine-tune reflexes. These tracts inhibit reflexes, making them softer than they would be if they had no influence from the brain.

Let's return to the hitting your thumb with the hammer example. After you hit your thumb, the first motor activity is a withdrawal reflex. You pull your thumb away from the painful stimulus. As we mentioned before, this is not a voluntary movement. It isn't planned. Only your spinal cord motor neurons are necessary for withdrawal. However, after the initial withdrawal, what do you do next? You look at your thumb. That is a planned movement. It requires coordination between your motor system, to uncover your thumb and move it into your visual field, and your visual system to look at your thumb. You walk to the kitchen, open the freezer, get out the ice, and make an ice pack. Then you grab a cold soda from the fridge and walk back into the living room, nursing your wounds. All of this activity requires motor planning and coordination.

You must reach for the freezer door and open it accurately. You must stay upright with respect to gravity. You must open a plastic bag and put ice in it. None of this happens without careful motor planning and coordination.

Pathology Connection: Motor Pathologies

Cerebral Palsy (CP) is a permanent, non-progressive set of motor deficits diagnosed in infants and young children, generally thought to be due to damage to the motor cortex. No single cause has been identified for CP, but many risk factors are associated with the disorder including low birth weight, premature birth, multiple births, infection during pregnancy, developmental abnormalities, brain hemorrhage, **perinatal** brain injury, lack of oxygen, and childhood illness. The symptoms of CP vary from mild foot drag and uncoordination to complete paralysis including the inability to speak. Symptoms of CP include increased muscle tone, overactive reflexes, lack of coordination of voluntary movements, foot drag, drooling, speech difficulties, fine motor problems and tremor or other uncontrollable movements.

Some people with CP are also mentally retarded, autistic, or have seizures, but many CP patients are of normal or above normal intelligence. Diagnosis of CP, especially mild cases, can be difficult. Observing childhood motor skills and developmental milestones, taking images such as CT scans or MRIs, and ruling out other causes of motor deficits are the best ways to diagnose CP. Treatment includes a variety of physical and occupational therapy, assistive devices, and drugs to control the symptoms. Stem cell usage shows promise.

Parkinson's disease (PD) is a chronic progressive motor disorder, common in elderly patients, characterized by resting tremor (rhythmic shaking), slow movement, impaired balance, rigidity, fatigue, and a number of cognitive and emotional disturbances. PD is caused by the disappearance of dopamine neurons in one of the basal nuclei, which later spreads to the cerebral cortex. The cause of PD is unknown, but toxins, mitochondrial malfunctions, or a virus may be involved. Some cases of PD are genetic. Most PD patients are elderly, but some younger people, like the actor Michael J. Fox, have an early onset form of the disease.

Diagnosis of PD is easier than for some other progressive motor disorders. Patient history and a thorough physical exam is often enough to diagnose PD. PD patients have a characteristic shuffling gait and their muscles pop or snap. The rigidity appears almost as if the muscles catch and then release. This type of rigidity is called "cogwheel" rigidity, and it is diagnostic for PD. Imaging studies are useful for ruling out other disorders because, unlike many other disorders, early stage PD patients will have absolutely normal scans. There are several numerical scales that can be used to describe the severity of the disease.

Treatment of PD is generally with the use of dopamine-enhancing drugs. The most common and most effective of these is L-dopa. Unfortunately,

perinatal *(men in JYE tiss)*
  peri = *around*
  natal = *birth*

9-13 For more information on Parkinson's disease, see your DVD for this chapter.

L-dopa, when used long term, has nasty side effects including hallucinations and excessive uncontrollable movement. PD patients on L-dopa also have "on" and "off" periods that are unpredictable. Other medications are used to control PD symptoms of tremor, depression, and blood pressure effects. Deep brain stimulation may also be used to treat the symptoms of PD. For example, deep brain stimulation of the thalamus decreases tremor.

Amyotrophic Lateral Sclerosis (ALS) also known as Lou Gehrig's disease, is a rapidly progressive, fatal degeneration of the motor system. On average, patients with ALS die within 5 years of diagnosis, though many patients, like the physicist Stephen Hawking, live much longer. Symptoms usually appear between the ages of 40 and 60 and begin with muscle weakness, twitching, and cramping, and then progress to complete paralysis including difficulty speaking and swallowing and eventually to paralysis of the diaphragm. At that point, patients will become ventilator dependent. Eye movements and control of the bladder and bowels are usually retained. Death is usually due to respiratory failure.

amyotrophic *(AA mee o TROF ic)*

The cause of ALS is unknown, but we do know that there is an increase in self destruction of motor neurons in the cerebral cortex, brainstem, and spinal cord. There is evidence that toxins, damage from free radicals or mitochondrial problems may be involved. Usually there is excess activity of neuroglia including microglia and astrocytes, and excess production of the neurotransmitter glutamate, all of which increase neuron death.

There is no definitive diagnostic test for ALS, but it has one unique characteristic. ALS is the only motor disorder that exhibits both flaccid and spastic paralysis. Most other disorders affect either upper motor neurons (spastic paralysis) or lower motor neurons (flaccid paralysis). ALS affects both. In ALS, unlike some other forms of paralysis, sensation is normal. EMG, imaging and blood and urine tests can help rule out other disorders. A neural biopsy may also be helpful.

As we have seen with many other neurological disorders, there is no cure for ALS. Only one drug, riluzole, has been approved for treatment. This drug decreases the neurotransmitter glutamate and decreases cell death, slowing the progression of the disease. There are other treatments to reduce symptoms and improve quality of life.

Clinical Application

SPASTIC VERSUS FLACCID PARALYSIS

A good illustration of the relationship between the brain, spinal cord pathways, and ventral horn motor neurons are the two types of paralysis. **Paralysis** is the inability to control voluntary movements. Paralysis can be **spastic** or **flaccid**. Spastic paralysis is characterized by muscle rigidity or increased muscle tone and overactive reflexes. In spastic paralysis, the muscles are rigid and the reflexes are overactive because of decreased communication from the brain to the ventral horn motor neurons in the spinal cord and brainstem motor neurons. Muscles contract randomly, and reflexes do not have any control signals from the brain. Stroke, head injuries, cerebral palsy, multiple sclerosis, basal nuclei disorders (like Parkinson's disease), and spinal cord injuries can cause spastic paralysis.

Flaccid paralysis is characterized by floppy muscles (decreased muscle tone) and decreased reflexes. In flaccid paralysis, the damage is to the spinal nerves or ventral horn (or brainstem) motor neurons. Impulses cannot get to the muscles from the motor neurons. Therefore, the muscles are floppy and reflexes are absent. Flaccid paralysis can be caused by peripheral neuropathy or disorders like polio and Guillain-Barré syndrome. Maria, our diabetes patient, with the beginnings of peripheral neuropathy, may eventually experience localized flaccid paralysis.

TEST YOUR KNOWLEDGE 9-5

Choose the best answer:

1. The size of the map for a particular body part in the precentral gyrus is determined by:
   a. the amount of fine motor control
   b. the size of the body part
   c. the importance of the body part
   d. all of the above

2. The cerebellum compares:
   a. shapes of objects
   b. planned movement to actual movement
   c. textures
   d. position of objects in space

3. Which of the following spinal cord pathways carries pain and temperature information?
   a. dorsal column
   b. spinothalamic
   c. corticospinal
   d. thermostatic

4. The _____ sends direct messages to ventral horn motor neurons, while the _____ are involved in indirect pathways.
   a. thalamus, basal nuclei
   b. primary sensory cortex, thalamus
   c. cerebellum, primary motor cortex
   d. primary motor cortex, basal nuclei

5. This disorder is characterized by slow movements, resting tremor and popping rigidity:
   a. ALS
   b. Huntington's disease
   c. spinal cord injury
   d. none of the above

6. This disorder has both upper and lower motor neurons signs:
   a. ALS
   b. Huntington's disease
   c. spinal cord injury
   d. none of the above

THE AUTONOMIC NERVOUS SYSTEM

The peripheral nervous system is divided into two systems: the somatic system, which controls skeletal muscles, and the autonomic system, which controls involuntary muscles. These muscles are the smooth muscles found in structures such as the blood vessels and airways and cardiac muscle found in the heart. Glands are also controlled by this system. The autonomic system controls physiological characteristics such as blood pressure, heart rate, respiration rate, digestion, and sweating.

The neurons for the autonomic system, like the somatic motor neurons, are located in the spinal cord and brain stem, and release the neurotransmitter acetylcholine. This is where the similarity ends. The autonomic motor neurons are all located in the lateral horn rather than the ventral horns, and unlike the somatic motor neurons, autonomic neurons do not connect directly to muscles. Instead, they make a synapse in a ganglion outside of the CNS. A **ganglion** is group of nerve cell bodies outside of the CNS. You can think of this as a junction box where the signal can be passed on to the next part of the circuit. Then a second motor neuron, called a **postganglionic neuron**, connects to the smooth muscle or gland.

The autonomic nervous system is divided into two subdivisions, or branches: the **sympathetic** and the **parasympathetic.** Please see Figure 9-22 ▓.

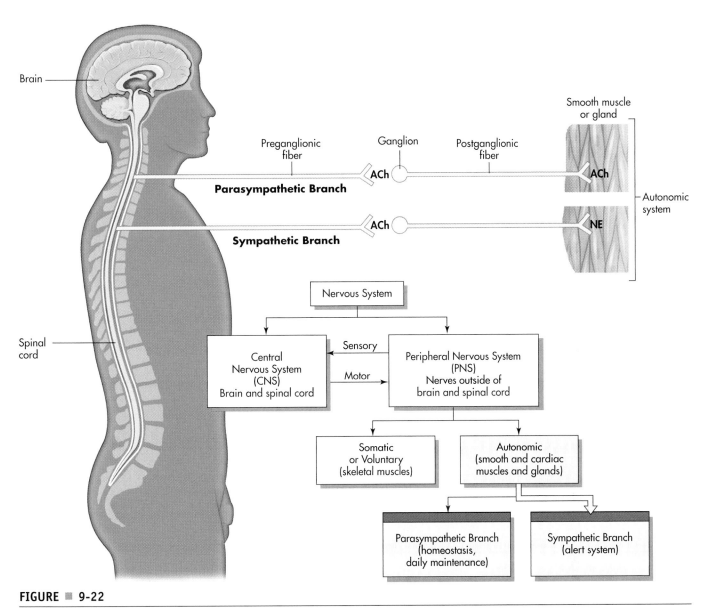

FIGURE ■ 9-22

General representation of autonomic nervous system. ACh = acetylcholine and NE = norepinephrine.

The Sympathetic Branch

The sympathetic branch controls the flight-or-fight response. It is charged with getting your body ready to expend energy. Sympathetic effects include increased heart rate, increased blood pressure, sweating, and dry mouth--all the symptoms of an adrenaline rush. This part of the autonomic system was responsible for your racing heart, rapid breathing, and intense sweating when confronted in our earlier example of the snarling dog. Your heart pumped more blood to your muscles, and your lungs took in more oxygen, both of which got you "up" to either fight or, most likely, flee. Another sympathetic response is dilation of your pupils to help you see the situation at hand much better. Sympathetic output is pretty strong when you hit your thumb with the hammer too.

The **preganglionic neurons** for the sympathetic system are located in the thoracic and first two lumbar segments of the spinal cord. Due to their location, they are classified as **thoracolumbar**. The preganglionic neurons, which secrete acetylcholine, synapse with the **postganglionic neurons** in the sympathetic ganglia. The ganglia for the sympathetic division form a pair of chainlike structures that run parallel to the spinal cord. These are called **paravertebral ganglia**. The postganglionic neurons release the neurotransmitter **norepinephrine**. One of the most important effects of the sympathetic system is its stimulation of the adrenal gland to release the hormone **epinephrine** (adrenalin), the chemical that causes that familiar adrenalin rush by circulating in the blood stream. (Again, please refer to Figure 9-22.)

The Parasympathetic Branch

If you have a gas pedal, you also need a brake. The parasympathetic division is often called "resting and digesting" because it has the opposite effect of the sympathetic division. The parasympathetic division is responsible for maintenance of everyday activities. It also helps bring you back down to normal from a sympathetic response. Parasympathetic effects include decreased heart rate, respiration, and blood pressure and increased digestive activity, including salivation and even stomach rumbling. The parasympathetic system allows you to calm down after you get the ice pack on your thumb or the dog is safely locked inside its cage.

The neurons of the parasympathetic system are in the brain stem and the sacral spinal cord and thus are called **craniosacral**. Like the preganglionic synapses of the sympathetic system, they release acetylcholine. The preganglionic neurons synapse with postganglionic neurons in the parasympathetic ganglia. Parasympathetic ganglia are located near the organs. Postganglionic neurons release the neurotransmitter acetylcholine. (Remember, acetylcholine excites skeletal muscle, but inhibits smooth and cardiac muscle.)

> **9-14** Due to the complexity of the nervous system, there are a host of diseases we could cover. For a representative sampling of diseases not discussed in this chapter but that you may have heard about, we have included additional information on your DVD. To view videos on autism, bipolar disorders, dissociative identity disorders, obsessive compulsive disorder (OCD), schizophrenia, epilepsy, and seizures disorders, go to your DVD for this chapter.

> **www.prenhall.com/colbert**
> Numerous drugs are used to treat nervous system disorders. There are various career opportunities and levels within the pharmaceutical profession. To learn more about it and view a video on the profession of pharmacy, please go to the Web site for this chapter.

A Quick Trip Through Diseases of the Nervous System

DISORDER	ETIOLOGY	SIGNS AND SYMPTOMS	DIAGNOSTIC TEST(S)	TREATMENTS
Multiple sclerosis (MS)	Autoimmune attack on myelin in CNS.	Varies depending on location of plaques, may be sensory, motor, or cognitive.	Patient history, imaging.	Steroids, plasma exchange, immunosuppressant drugs, symptom management.
Guillain-Barré syndrome	Autoimmune destruction of PNS myelin, often after a virus.	Rapid onset of ascending paralysis.	History, EMG, spinal tap.	No real treatment available, supportive care, sometimes plasma exchange.
Charcot Marie Tooth disorder	Genetic destruction of PNS myelin and/or axons.	Ascending muscle weakness and atrophy, decreased sensation in affected limbs.	History, EMG, biopsy, genetic testing.	PT, OT, surgery, pain medication, symptom management, no treatment to stop deterioration.
Myasthenia gravis	Autoimmune attack of acetylcholine receptor at neuromuscular junction.	Progressive fluctuating muscle weakness, often starting with facial or eye muscles.	Blood tests, EMG.	Steroids, immunosuppressant drugs, plasma exchange, acetylcholinesterase inhibitors.
Polio and post-polio syndrome	Polio is caused by destruction of ventral horn motor neurons by poliomyelitis virus; Post-polio syndrome is caused by late onset fatigue of motor neurons originally affected by polio.	Muscle weakness in recovered polio victims.	For polio, sudden onset of paralysis after flu-like symptoms (rare in US); Rule out other causes of weakness for post-polio syndrome.	No treatment for polio infection; supportive care until virus runs its course, therapy to regain strength in damaged muscles. No treatment for post-polio syndrome, although exercise may help.
Spinal cord injury	Usually mechanical injury to spinal cord tissue.	Loss of sensory and motor function, depends on location of injury.	Neurological exam, imaging.	Acute treatment to prevent further injury: immobilization, steroids. Long term care: PT, OT, supportive care and symptom management.
Peripheral neuropathy	Damage to peripheral nerves due to injury or illness.	Motor and sensory abnormalities; including weakness, pain, numbness, and tingling.	Imaging, biopsy, patient history.	Symptom management, treatment of underlying disorder causing nerve damage.

continues

A Quick Trip Through Diseases of the Nervous System (*Continued*)

DISORDER	ETIOLOGY	SIGNS AND SYMPTOMS	DIAGNOSTIC TEST(S)	TREATMENTS
Traumatic brain injury	Damage to brain tissue due to mechanical injury, lack of oxygen, or brain hemorrhage.	Depends on injury severity and location. Ranges from dizziness and nausea to severe cognitive disturbances, memory loss, seizures, and unconsciousness.	Glasgow coma scale, imaging.	Acute treatment to prevent further injury: immobilization, surgery, medication to relieve intracranial pressure. Long term care: PT, OT, supportive care, and symptom management.
Hydrocephalus	Excess CSF in brain due to trauma, birth defects, tumors, etc.	Symptoms vary with age: skull expansion and irritability in babies, irritability, vomiting, seizures, sleepiness, and dementia in older patients.	Imaging, pressure monitoring.	Shunt insertion.
Cerebral palsy (CP)	Risk factors include premature birth, low birth weight, developmental abnormalities, perinatal brain injury.	Non-progressive motor deficits in young children.	Difficult: observation of childhood motor skills, rule out other disorders.	No cure: PT, OT, assistive devices, symptom management.
Parkinson's disease (PD)	Progressive loss of dopamine neurons.	Resting tremor, slow movement, rigidity, cognitive and emotional disturbance.	History, imaging, neurological exam.	No cure: Dopamine enhancing drugs, symptom management, deep brain stimulation.
Huntington's disease	Genetic; progressive loss of neurons from basal nuclei and cerebral cortex.	Mid-life onset of chorea (excessive movement), mood swings and memory loss, progressing to dementia and paralysis.	Family history, imaging, genetic testing.	No cure; medication to control emotional and motor symptoms, no drug treatment for dementia.
Amyotrophic lateral sclerosis (ALS)	Progressive loss of motor neurons in CNS and PNS, cause unknown.	Progressive muscle weakness, twitching, and cramping, eventually leading to complete paralysis. Sensation is normal.	No definitive diagnosis, but presence of both flaccid and spastic paralysis indicates ALS. Rule out other disorders.	No cure; drug treatment may slow progression, symptom management.
Alzheimer's disease	Not fully known; believed to be a buildup of plaque in the brain and/or a defect in the neurotransmitter system in the brain.	Varying degrees of confusion, memory loss, cognitive defects, and personality changes.	History and physical exam, interview of family members, and various cognitive tests.	No cure; drug therapy can prevent or decrease some symptoms in early to middle stages of disease.

Pharmacology Corner

The **blood brain barrier** is an anatomical/physiological protective structure of the brain believed to consist of walls of capillaries in the CNS and surrounding glial membranes. The blood brain barrier functions to prevent or slow down the passage of a variety of chemical compounds and pathogens from the blood and into the central nervous system. Some beneficial drugs cannot pass through this barrier easily and as a result make it difficult to treat brain diseases such as meningitis.

There are dozens of medications that affect the nervous system. Most medications affect chemical synaptic transmission. Nicotine (in tobacco), and opiates (like morphine and heroin), L-dopa (used to treat Parkinson's disease) and benzodiazepines (found in tranquilizers like Valium), bind to the receptors for neurotransmitters, mimicking the effects of the neurotransmitter. Some of these drugs are stimulants and some are depressants, depending on the activity of the neurotransmitter they mimic. Monoamine oxide (MAO) inhibitors (like Selegiline for Parkinson's Disease and Nardil for depression), selective serotonin reuptake inhibitors (SSRIs like Prozac for depression) and acetylcholinesterase inhibitors (like Neostigmine for Myasthenia Gravis or donepezil/Aricept for Alzheimer's disease) stop neurotransmitters from being removed from the synapse, prolonging their activity. Again, their effects depend on the effects of the neurotransmitter. Receptor inhibitors block receptors, preventing the neurotransmitters effects. Examples include Scopolamine (for motion sickness) and Succinylcholine (a muscle paralyzer), which block acetylcholine receptors, and Lopressor and Zobeta, beta blockers that block norepinephrine receptors to treat cardiac problems.

SUMMARY

Snapshots from the Journey

→ The nervous system is the body's computer, its information superhighway. It has a sensory (input) system, an integration center, the CNS, and a motor (output) system. The input and output nerves are in the PNS, and the brain and spinal cord are the CNS.

→ The tissue of the nervous system is made up of two types of cells: neurons, which send, receive, and process information; and neuroglia, which support the neurons.

→ Neurons are excitable cells. They do their jobs by carrying tiny electrical currents caused by changes in cell permeability to certain ions. These tiny electrical currents can be all-or-none (action potentials), can change depending on the size of the stimulus (local potentials), can travel down axons (impulse conduction), or can be used to transmit information from one cell to another (synapses).

→ The CNS is composed of the brain and spinal cord.

→ The CNS is surrounded by a three-layered membrane system: dura mater, arachnoid mater, and pia mater, collectively known as the meninges. Cerebrospinal fluid (CSF) is also contained in the space between the arachnoid and pia maters.

→ The spinal cord has 31 segments, each with a pair of spinal nerves. The spinal nerves are a part of the peripheral nervous system and made of a pair of spinal roots. The ventral root is integral to motor function, and the dorsal root is integral to sensory function. Spinal nerves are mixed: they carry both sensory and motor information.

→ Tracts run up and down the spinal cord to and from the brain. The tracts going toward the brain carry sensory information to the brain. The tracts coming from the brain toward the spinal cord carry motor information from the brain.

→ The brain is a hierarchical organ. It is divided into compartments (lobes), each with very specific functions. The nerves attached to the brain are called cranial nerves, which occur in 12 pairs. They can be sensory, motor, or mixed.

→ The cerebrum controls your conscious movement and sensation. Beneath the cerebrum are the diencephalon, brain stem, and cerebellum. Each part plays an important role in coordinating sensory and motor information for the cerebrum.

→ The cerebral cortex contains motor and sensory maps of the body. Orders for voluntary movements originate in the primary motor cortex, located in the precentral gyrus of the frontal lobe, and travel down the spinal cord via direct spinal cord tracts. Subcortical structures coordinate this information via indirect tracts.

→ The somatic sensory cortex is in the postcentral gyrus of the parietal lobe. Sensory information from the spinal cord tracts eventually ends up in this part of the cortex. When the information arrives there, you become aware of your sense of touch. Other parts of the brain, called association areas, allow you to make connections between different types of sensory information and to compare current experience to memories.

→ Ventricles are the cavities in your brain. The spinal cord cavity is the central canal. These cavities are part of an elaborate protection system for your CNS and are filled with cerebrospinal fluid.

→ The nervous system also controls involuntary movement via a part of the system known as the autonomic nervous system. The sympathetic division controls the flight-or-fight response, and the parasympathetic division controls day-to-day activities.

→ The nervous system can have many disorders. Most are chronic, progressive, and incurable. In all cases, whether the disorder is central or peripheral, genetic, or caused by disease or trauma, the symptoms are often the same. Muscle weakness, loss of sensation and respiratory difficulty often accompany problems in the nervous system. The subtle differences between the symptoms are the key to making these difficult diagnoses and finding the appropriate treatment. Treatment generally consists of symptom management and prevention of further deterioration.

RAY'S STORY

 Let's visit with Ray again. This is the young man who hit his chin on the bottom of the swimming pool doing a back flip. He had struck the bottom of the pool on the point of his chin, jamming and twisting his neck. His neck is broken, the first two vertebrae shattered and his spinal cord irreversibly damaged. He is paralyzed from the neck down and is ventilator dependent.

a. What is the level of Ray's injury? (Cervical, thoracic, lumbar or sacral.)

_____

b. Why can't he breathe? (Be specific.)

_____

_____

c. What do you call the paralysis of all four limbs?

_____

d. What complications might be a problem for Ray?

_____

_____

e. Does Ray have spastic or flaccid paralysis?

_____

REVIEW QUESTIONS

Multiple Choice

1. The input side of your nervous system is known as:
   a. motor
   b. sensory
   c. association
   d. all of the above

2. During depolarization, _____ ions move _____ a neuron.
   a. K^+, out of
   b. K^+, into
   c. Na^+, out of
   d. Na^+, into

3. Bob is stopped across a busy street waiting to take a left turn, when a drunk driver runs the light, T-boning Bob's corvette. When the EMTs arrive, Bob is still pinned in the car and is in respiratory arrest. When he regains consciousness at the hospital, Bob is dismayed to discover that he cannot move or feel his body and that he is on a ventilator. Where is his injury?
   a. C_1
   b. C_8
   c. T_{10}
   d. L_1

4. One of the following brain parts is *not* subcortical. Which one?
   a. hypothalamus
   b. medulla oblongata
   c. precentral gyrus
   d. pineal body

5. The size of the map of each body part in the postcentral gyrus is determined by the:
   a sensitivity of the body part
   b size of the body part
   c. importance of the body part
   d. fine motor control of the body part

6. The sympathetic nervous system:
   a. causes decreased heart rate
   b. has ganglia near the organs
   c has ganglia near the spinal cord
   d. all of the above

7. This part of the brain contains the body's set points and controls most of its physiology, including blood pressure and hunger level:
   a. thalamus
   b hypothalamus
   c. amygdale
   d hippocampus

8. Which of these disorders is caused by loss of neurons from the basal nuclei?
   a. Multiple sclerosis
   b. Amyotrophic lateral sclerosis
   c. Cerebral palsy
   d. Parkinson's disease

9. Amyotrophic lateral sclerosis is different from the many other progressive neurological disorders because patients:
   a. have normal sensation
   b. are easy to diagnose
   c. have symptoms of both flaccid and spastic paralysis
   d. rarely die from the disease

10. Which of the following is *not* progressive?
    a. Amyotrophic lateral sclerosis
    b. Multiple sclerosis
    c. Parkinson's disease
    d. Cerebral palsy

Fill in the Blank

1. The speed of impulse conduction is determined by _____ and _____.

2. _____ fluid is contained in the _____ space between the arachnoid mater and pia mater.

3. Most of the symptoms of MS are caused by destruction of _____ in the brain and spinal cord.

4. The most common cause of spinal cord injuries is _____.

5. Rapid onset, ascending paralysis is a hallmark of this disorder _____.

6. The occipital lobe is responsible for this sensation: _____.

7. The white matter of the spinal cord contains _____ tracts, which are motor, and _____ tracts, which are sensory.

8. Emotion, mood, and memory are controlled by this collection of nuclei: _____.

9. _____ is characterized by tremor, rigidity and slow movement.

10. What is one of the leading killers of patients with progressive neurological disorders? _____

Short Answer

1. Explain the changes in a neuron during an action potential.

_____

_____

_____

_____

2. List the steps in chemical synaptic transmission.

_____

_____

_____

_____

3. List the types of neuroglia and their functions.

_____

_____

_____

_____

4. Explain what happens to spinal cord tissue after injury. Why is it important to treat the injury with steroids?

_____

_____

_____

_____

5. List the differences between the sympathetic and parasympathetic nervous systems.

_____

_____

_____

_____

6. Explain how the cerebral cortex and subcortical structures interact to produce motor output.

_____

_____

_____

_____

7. Explain how the location of a stroke or traumatic brain injury can determines the symptoms.

_____

_____

_____

_____

Suggested Activities

1. With a partner, test your knee-jerk reflexes using reflex hammers. Try to control the reflex. Can you? Try it with your eyes closed. What does that say about reflexes?

2. Grab a friend and explain the action potential and chemical synapse to her or him. Can you do it?

3. Find the parts of the brain on a preserved brain, diagram, or model.

4. Select a partner and play nervous system jeopardy. Can you identify the function of all of the parts of the nervous system?

> **9-15** Now that you have completed your journey through this chapter, please go to the DVD for interactive games and puzzles concerning the medical terms and concepts contained in this chapter. By playing the games you will reinforce your learning of the nervous system in a fun way.

THE ENDOCRINE System

The Body's Other Control Center

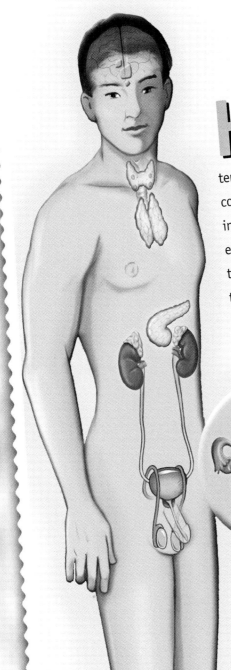

Here we are again, talking about control. We have already visited one of the control systems: the complex structure of cells and connections known as the nervous system. Now we visit yet another control system: the endocrine system. These two control systems may seem like separate systems, but they are interconnected and always monitor each other's activities. The nervous system collects information and sends orders with a speed that is truly mind-boggling. Whereas the endocrine system also collects information and sends orders, it's a slower, more subtle control system. The endocrine system's orders to the body also last much longer than those made by the nervous system. You might think of the endocrine system as sending "standing orders," or orders meant to be obeyed indefinitely unless changed by another set of orders. The orders change subtly on a regular basis, but their intention is constant. The nervous system, on the other hand, issues orders that are to be obeyed instantaneously but are used for short-term situations. The endocrine system demands organs to "carry on" while the nervous system expects them to respond immediately.

On our journey, suppose we stop at an amusement park to ride a roller coaster. Afterward, en route to our next destination, we are forced off the road because of a near miss with a truck. Even after the roller coaster ride is over or the truck is long gone, your legs still shake, your heart continues to race, and your blood pressure remains elevated, even though you are no longer in danger. We call such lingering effects the "adrenaline rush." These lingering effects are not due to continued activity of the nervous system, but rather to endocrine activity deliberately triggered by the autonomic nervous system.

Chapter 10

LEARNING OBJECTIVES

At the completion of your journey through this chapter, you will be able to:

→ Discuss the functions of the various endocrine glands

→ Describe the purpose and effects of hormones within the body

→ Discuss the process of homeostatic control of hormone levels

→ Differentiate between hormonal, humoral control, and neural control

→ Explain common diseases of the endocrine system

MULTIMEDIA APPLICATIONS

DVD Interactive Exercises

→ 3-D animation and drag-and-drop exercise on the endocrine system, 10-1

→ Videos on the effects of aging on the endocrine system, 10-2

→ Animation and video spotlighting the pathology of diabetes and how to monitor blood glucose levels, 10-3

→ Videos on the importance of insulin and monitoring glucose, 10-4

→ Interactive games and puzzles, 10-5

www.prenhall.com/colbert

→ Professional Profiles:
  - Phlebotomist
  - Dietician

Pronunciation Guide

Correct pronunciation is important in any journey so that you and others are completely understood. Here is a "see and say" Pronunciation Guide for the more difficult terms to pronounce in this chapter.

adrenal cortex (ah DREE nal KOR teks)

adrenal medulla
(ah DREE nal meh DULL lah)

endocrine (EHN doh krin)

epinephrine (EP ih NEFF rinn)

homeostasis (hoh mee oh STAY siss)

hypercholesterolemia
(HIGH per koh LESS ter ohl ee me ah)

hyperpituitarism
(HIGH per peh TOO eh tare izm)

hypopituitarism
(HIGH poh peh TOO eh tare izm)

hypothalamus (high poh THAL ah mus)

norepinephrine (NOR ep ih NEFF rinn)

oxytocin (AHK see TOH sin)

parathyroid gland (PAIR ah THIGH royd)

pineal gland (pih NEE al)

pituitary (pih TOO ih tair ee)

prolactin (proh LAK tinn)

testes (TESS teez)

thymus (THIGH muss)

ORGANIZATION OF THE ENDOCRINE SYSTEM

The endocrine system has many organs that secrete a variety of chemical substances. Let's begin by looking at the basic organization of the system.

Endocrine Organs

endocrine *(EHN doh krin)*
endo = *into*
crine = *to secrete*

The **endocrine system** is a series of organs and ductless glands (see Figure 10–1 ▪) in your body that secrete chemical messengers called **hormones** *into* your bloodstream. In contrast, **exocrine** glands and organs such as the pancreas and the sweat, salivary, and lacrimal glands produce secretions that must *exit* that particular gland through a duct.

We have already discussed some of the endocrine glands, such as the hypothalamus, pituitary, and pineal glands, because they are part of the nervous system and provide a link between the two control systems. We will visit some of the other endocrine organs later when we journey through the urinary, reproductive, and digestive systems. Many endocrine glands, like the hypothalamus and pancreas, have multiple functions.

It may seem like an overwhelming task to learn all the endocrine glands and their associated hormones. Therefore, we will begin our discussion with a concise overview to lay a foundation upon which to build. As we journey through this chapter, these concepts are reinforced and expanded. For now, see Table 10–1, which lists the wide variety of functions of endocrine organs.

Amazing Body Facts

LESSER KNOWN ENDOCRINE GLANDS

Did you know that many organs, such as the heart, small intestine, kidneys, stomach, and placenta can also secrete hormones, and therefore have endocrine-like functions? These and many other organs are not listed as endocrine organs because their primary jobs are focused on other tasks, like pumping blood, storing and digesting food, or nourishing an embryo. But the hormones secreted by other organs are still an important part of the body's control systems.

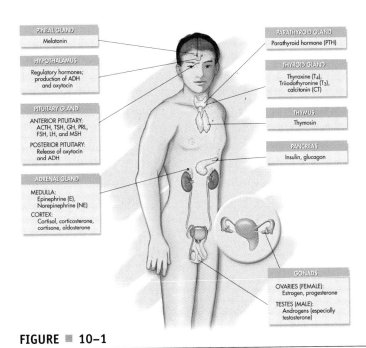

FIGURE ■ 10–1

The endocrine glands and their hormones.

TABLE 10–1 Endocrine Organ Functions

ENDOCRINE ORGAN	HORMONE RELEASED	EFFECT
Hypothalamus	A variety of hormones; see Table 10-3	Controls pituitary hormone levels
Pineal	Melatonin	Believed to regulate sleep
Pituitary	A variety of hormones; see Table 10-3	Controls other endocrine organs
Thyroid	Thyroxine, triiodothyronine	Controls cellular metabolism
	Calcitonin	Decreases blood calcium
Parathyroid glands	Parathyroid hormone	Increases blood calcium
Pancreas	Insulin	Lowers blood sugar
	Glucagon	Raises blood sugar
Adrenal glands	Epinephrine, norepinephrine	Flight-or-fight response
	Adrenocorticosteroids	Many different effects
Ovaries	Estrogen, progesterone	Control sexual reproduction and secondary sexual characteristics, such as pubic hair and axillary hair
Testes	Testosterone	Control secondary sexual characteristics, such as growth of beard and other hair, deepening of voice, increase in musculature, production of sperm

hypothalamus
(high poh THAL ah mus)

Hormones

hormone *(HOR moan)*

The chemical messengers released by endocrine glands are called **hormones.** We have already seen one type of chemical messenger: the neurotransmitter. Neurotransmitters are released by neurons at chemical synapses. They diffuse across the synapse (a very tiny space) to a cell on the other side, and they bind to that cell. They are cleaned up quickly, so their effects are localized and short-lived. Hormones, on the other hand, are released into the bloodstream and travel all over your body. Some hormones can affect millions of cells simultaneously. Their effects last for minutes or even hours or days. Many hormones are secreted constantly, and the amount secreted changes as needed. To help clarify the similarities and differences between a neurotransmitter and hormone, please see Table 10–2.

HOW HORMONES WORK

Like neurotransmitters, hormones work by binding to receptors on target cells. But hormones may bind not only to sites on the outside of the cell, like neurotransmitters, but also to sites inside the cell. If hormones bind to the outside of the cell, they can have several different effects, either changing cellular permeability or sending the target cell a message that changes enzyme activity inside the cell. Thus, the target cell changes what it has been doing, usually by making a new protein or turning off a protein it has been making.

One special class of hormones, **steroids,** is particularly powerful because steroids can bind to sites inside cells. Steroids are lipid molecules that can pass easily through the target cell membrane. These hormones can then interact directly with the cell's DNA, the genetic material, to change cell activity. These hormones are carefully regulated by the body because of their ability, even in very small amounts, to control target cells. See the section "The Adrenal Glands" for further discussion of steroid hormones and the dangers of taking steroids.

10-1 To view a 3-D animation of the endocrine system along with an interactive drag-and-drop exercise, please go to your DVD.

TABLE 10–2 Comparison of Neurotransmitters and Hormones

NEUROTRANSMITTERS	HORMONES
Chemical messengers	Chemical messengers
Bind to receiving cell	Bind to receiving cell
Control cell excitation	Control cell activities
Released by neurons	Released by neurons, glands or organs
Released at chemical synapse	Released into bloodstream
Intended target very close	Travel to distant target
Effects happen quickly (less than a second)	Effects take time (seconds or minutes)
Effects wear off quickly (few seconds)	Effects long lasting (minutes or hours)
Affect single cell	Can affect many cells

TEST YOUR KNOWLEDGE 10-1

Choose the best answer:

1. What chemical, when secreted into the bloodstream, controls the metabolic processes of target cells?
   a. neurotransmitter
   b. secretion
   c. hormone
   d. ligand

2. Steroid hormones are very powerful because they:
   a. are hormones
   b. are medicine
   c. interact directly with DNA
   d. are secreted outside the body

3. Which of the following is true of hormones?
   a. They last a short time.
   b. They are fast acting.
   c. They affect distant targets.
   d. They leave the body.

CONTROL OF ENDOCRINE ACTIVITY

Many endocrine organs are active all the time. The amount of hormones they secrete changes as the situation demands, but unlike neurons, the cells in the endocrine organs often secrete hormones continuously. How is the activity controlled? How do the organs know how much hormone to secrete?

Homeostasis and Negative Feedback

In order to understand how the endocrine system is controlled, we first have to revisit the concept of **homeostasis** discussed at the beginning of our journey in Chapter 1 (see Figure 10–2 ■).

Recall that many of the chemical and physical characteristics of the body have a standard level, or **set point,** that is the ideal level for that particular value. Blood pressure, blood oxygen, heart rate, and blood sugar, for example, all have "normal" ranges. Your control systems, nervous and endocrine, work to keep the levels at or near ideal. There is a way for your body to measure the variable, a place where the "ideal" level is stored, and a way for the body to correct levels that are not near ideal. For example, neurons measure your body temperature. The hypothalamus stores the set point. If your temperature falls below the set point temperature, the hypothalamus causes shivering to produce additional heat. If body temperature rises above the set point, the hypothalamus causes sweating.

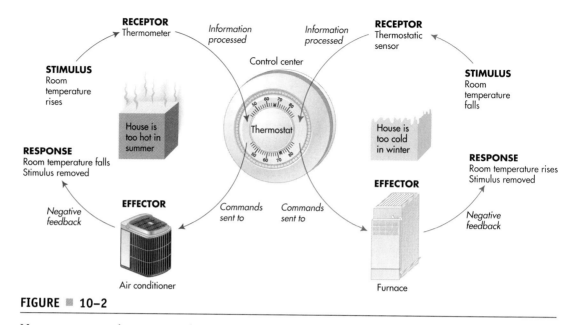

FIGURE 10–2

Homeostasis is analogous to regulation of temperature via a thermostat.

If any of the body's dozens of homeostatic values become seriously disrupted, the control systems work to bring them back to set point. This process is called **negative feedback** (see Figure 10–3). Most of you are familiar with negative feedback in real life. When you pump gas, there is a sensor in the nozzle that turns off the flow of gas when the tank is full. That is negative feedback. The gas is flowing, the tank is filling, and when the goal is reached, the gas stops flowing. In the body, negative feedback counteracts a change. As blood pressure rises, for example, your body works to bring it down to normal. If blood pressure falls, your body works to raise it back to the normal set point. Hormones work the same way. If hormone levels rise, negative feedback turns off the endocrine organ that is secreting the hormone.

The body is also capable of **positive feedback,** which increases the magnitude of a change. The flow of sodium into a neuron during depolarization is a real-life example we have already visited. The more depolarized a neuron becomes, the more sodium flows in, so it becomes more depolarized, so more flows in, and so on. Therefore, positive feedback is not a way to regulate your body because it increases a change away from set point. What if instead of shivering when you got cold to raise your body temperature, you got colder and colder and colder?

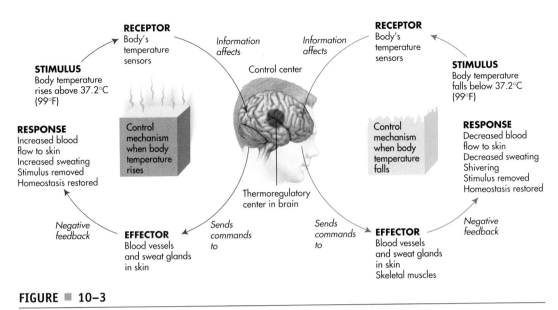

FIGURE ▪ 10–3

Homeostasis and negative feedback as related to control of body temperature.

Sources of Control of Hormone Levels

Hormone levels can be controlled by the nervous system (**neural control**), by other hormones (**hormonal control**), or by body fluids such as the blood (**humoral control**).

NEURAL CONTROL

There are three basic ways in which endocrine organs function to maintain hormone levels and body function. Some hormones are directly controlled by the nervous system. For example, the adrenal glands receive signals from the sympathetic nervous system. When the sympathetic nervous system is active (remember the near-wreck), it sends signals to the adrenal glands to release epinephrine and norepinephrine as hormones, prolonging the effects of sympathetic activity (see Figure 10–4 ▪).

HORMONAL CONTROL

Other hormones are part of a hierarchy of hormonal control in which one gland is controlled by the release of hormones from another gland higher in the chain, which is controlled by another gland's release of hormones yet higher in

---

Clinical Application

CHILD BIRTH AND POSITIVE FEEDBACK

Positive feedback is often harmful if the cycle cannot be broken, but sometimes positive feedback is necessary for a process to run to completion. A good example of necessary positive feedback is the continued contraction of the uterus during childbirth. When a baby is ready to be born, a signal tells the hypothalamus to release the hormone **oxytocin** from the posterior pituitary. Oxytocin increases the intensity of uterine contractions. As the uterus contracts, the pressure inside the uterus caused by the baby moving down the birth canal increases the signal to the hypothalamus: more oxytocin is released and the uterus contracts harder. As pressure gets higher inside the uterus, the hypothalamus is signaled to release more oxytocin, and the uterus contracts even harder. This cycle of ever-increasing uterine contractions due to ever-increasing release of oxytocin from the hypothalamus continues until the pressure inside the uterus decreases—that is, when the baby is born.

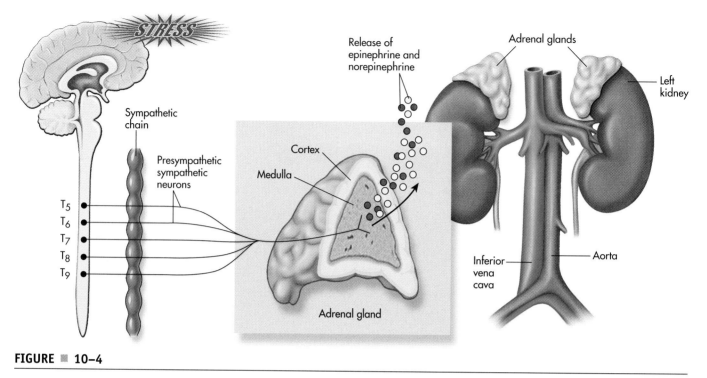

FIGURE ■ 10–4

Sympathetic control of adrenal gland.

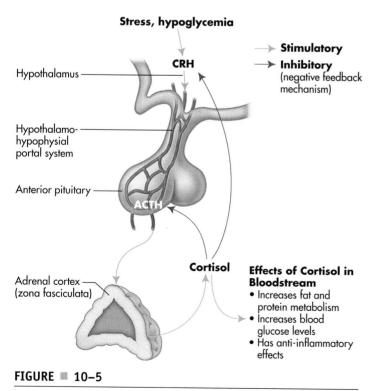

FIGURE ■ 10–5

Hormonal control of adrenal gland. CRH = corticotropin releasing hormone; ACTH = adrenocorticotropic hormone

the chain. Orders are sent from one organ to another. This is very similar to a relay race at a track meet where the baton is smoothly handed from one runner to the next, hopefully in a smooth manner, to send the baton to the finish line. Negative feedback controls the flow of orders via hormones from one part of the "chain of command" to the other. For example, the hypothalamus has control over the pituitary, which has control over the adrenal gland, which secretes the hormone cortisol. Increased cortisol secretion is one way that the body copes with stress, and as cortisol levels rise in the blood, further release of hormones at the hypothalamus is depressed. Please see Figure 10–5 ■.

HUMORAL CONTROL

Still other endocrine organs can directly monitor the body's internal environment by monitoring the body fluids, such as the blood, and respond accordingly. *Humoral* pertains to body fluids or substances, and therefore control through body fluids is called **humoral control**. For example, the pancreas secretes insulin in response to rising blood sugar, as shown in Figure 10–6 ■. In diabetics, like Maria, this response is impaired due to destruction of the pancreas by the immune system.

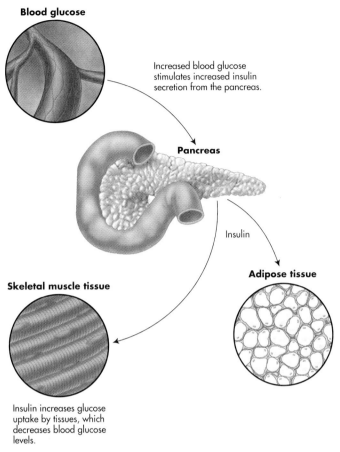

FIGURE ■ 10–6

Humoral control of blood sugar levels.

TEST YOUR KNOWLEDGE 10-2

Choose the best answer:

1. Which of the following is *not* a way that hormone levels are regulated?
   a. negative feedback
   b. chain of command
   c. positive feedback
   d. direct control by nervous system

2. The "ideal" value for a body characteristic is called the:
   a. set point
   b. average
   c. goal
   d. feedback point

3. _____ feedback enhances a change in body chemistry.
   a. negative
   b. positive
   c. regular
   d. cyclic

THE MAJOR ENDOCRINE ORGANS

The endocrine system has several organs, and each has specific tasks within the body. The messages to carry out these tasks are related to the hormones they release.

The Hypothalamus

We already visited the **hypothalamus** when we stopped at the central nervous system. Located in the diencephalon, this gland is an important link between the two control systems: nervous and endocrine (please see Figure 10–7 ■). The hypothalamus controls much of the body's physiology, including hunger, thirst, fluid balance, and body temperature, to name only a few of its functions. The hypothalamus is also, in part, the "commander-in-chief" of the endocrine system, because it controls the pituitary gland and therefore most of the other glands in the endocrine system. Table 10–3 lists the hypothalamic and pituitary hormones.

The Pituitary

pituitary *(pih TOO ih tair ee)*
prolactin *(proh LAK tinn)*
  pro = *for*
  lactin = *milk*

The **pituitary,** also a part of the diencephalon, is a small gland, about the size of a grape, located at the base of the brain. The pituitary is commonly known as the "master gland," indicating its important role in control of other endocrine glands. However, that name is misleading because the pituitary gland rarely acts on its own. The pituitary acts only under orders from the hypothalamus. If the hypothalamus is the "commander-in-chief," the pituitary is a high-ranking soldier who carries out the orders.

THE POSTERIOR PITUITARY (OR NEUROHYPOPHYSIS)

The pituitary is split into two segments: the **posterior pituitary** and the **anterior pituitary**. The posterior pituitary is an extension of the hypothalamus. Hypothalamic neurons, specialized to secrete hormones instead of neurotransmitters, extend their axons through a stalk to the posterior pituitary. Using the posterior pituitary as a sort of launch pad, these neurons secrete two hormones: **antidiuretic hormone (ADH)** (also called vasopressin) and **oxytocin.** Both of these hormones are secreted from the posterior pituitary, but they are made by the hypothalamus.

antidiuretic hormone
  (AN tye dye yoo RET ik)
oxytocin *(AHK see TOH sin)*

ADH does exactly what its name suggests: it is an antidiuretic. A diuretic is a chemical that increases urination, so an antidiuretic *decreases* urination. The effect of ADH, then, is to decrease fluid lost due to urination, increasing body fluid volume. ADH is secreted when the hypothalamus senses decreased

---

Learning Hint

HORMONE NAMES

Most hormones are named according to where they are secreted or what they do. If you learn the meanings of their names, you can usually tell something about the hormone. For example, growth hormone stimulates cells to grow. Prolactin increases milk production. Even a complicated hormone name like adrenocorticotropic hormone can be picked apart fairly easily. *Adreno* refers to the adrenal gland, and *cortico* refers to the cortex. Therefore, adrenocorticotropic hormone is a hormone that changes the activity (in this case increases) of the adrenal cortex. Also keep in mind that most hormones are known by their abbreviations, for obvious reasons. Adrenocorticotropic hormone, for example, is abbreviated ACTH, which is much easier to say and write.

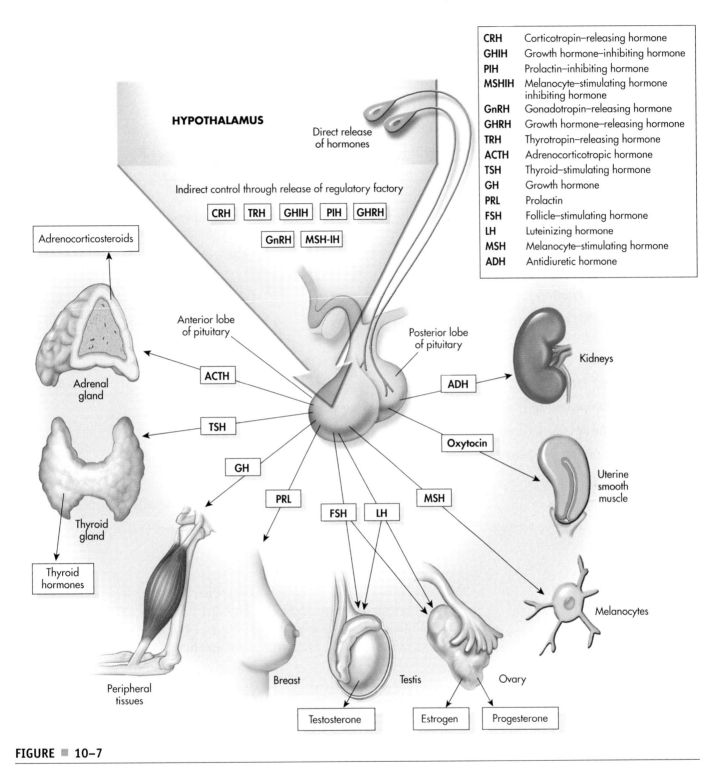

CRH	Corticotropin–releasing hormone
GHIH	Growth hormone–inhibiting hormone
PIH	Prolactin–inhibiting hormone
MSHIH	Melanocyte–stimulating hormone inhibiting hormone
GnRH	Gonadotropin–releasing hormone
GHRH	Growth hormone–releasing hormone
TRH	Thyrotropin–releasing hormone
ACTH	Adrenocorticotropic hormone
TSH	Thyroid–stimulating hormone
GH	Growth hormone
PRL	Prolactin
FSH	Follicle–stimulating hormone
LH	Luteinizing hormone
MSH	Melanocyte–stimulating hormone
ADH	Antidiuretic hormone

FIGURE ■ 10–7

The hypothalamus, anterior and posterior pituitary glands, and their targets and associated hormones.

blood volume or increased blood osmolarity (more solids suspended in blood). ADH circulates through the bloodstream and targets the kidneys specifically, causing them to absorb more water. It is very important in long-term control of blood pressure, especially during dehydration. (For more information on ADH, see Chapter 16, "The Urinary System.")

TABLE 10–3 Selected Hypothalamic and Pituitary Hormones

HORMONE	FUNCTION
Hypothalamus	
growth hormone–releasing hormone (GHRH)	Increases the release of growth hormone from the pituitary gland
growth hormone–inhibiting hormone (GHIH)	Decreases the release of growth hormone from the pituitary gland
corticotropin-releasing hormone (CRH)	Increases the release of adrenocorticotropic hormone from the pituitary gland
gonadotropin-releasing hormone (GRH)	Increases the release of luteinizing hormone and follicle-stimulating hormone from the pituitary gland
Thyrotropin-releasing hormone (TRH)	Increases the release of thyroid-stimulating hormone from pituitary gland
Posterior Pituitary	
Antidiuretic hormone (ADH), also known as vasopressin	Dilutes blood and increases fluid volume by increasing water reabsorption in the kidney
Oxytocin	Increases uterine contractions
Anterior Pituitary	
growth hormone (GH)	Increases tissue growth
thyroid-stimulating hormone (TSH)	Increases secretion of thyroid hormones
adrenocorticotropic hormone (ACTH)	Increases steroid secretion from adrenal gland
Prolactin	Increases milk production
luteinizing hormone (LH)	Stimulates ovaries and testes for ovulation and sperm production
follicle-stimulating hormone (FSH)	Stimulates estrogen secretion and sperm production

As we discussed in the section on positive feedback, the second hypothalamic hormone released from the posterior pituitary is oxytocin. Oxytocin is important in maintaining uterine contractions during labor in women and is involved in milk ejection in nursing mothers. Oxytocin's function in males is unknown.

 Pathology Connection: Posterior Pituitary and Antidiuretic Hormone

Drinking too much alcohol can lead to several unpleasant consequences, not the least of which is the development of a hangover. The symptoms of a hangover are due to many side effects from alcohol consumption, but the most important may be that alcohol turns off ADH. The more alcohol you drink, the less ADH you secrete and the more dehydrated you become. That makes you thirsty, so you drink some more beer, and you secrete even less ADH and become even more dehydrated. The more beer you drink, the more you urinate and the more dehydrated you are. This is part of the reason why consuming too much alcohol on Friday night can make you miserable on Saturday morning. The caffeine in coffee also has a similar but less potent effect in inhibiting ADH. That's why you make frequent trips to the bathroom if you pull an "all-nighter" studying for your A&P tests using coffee (caffeine) to stay awake. Too much caffeine will even make you feel pretty miserable.

DIABETES INSIPIDUS

Diabetes Insipidus (DI) is a condition characterized by the production of large amounts of very dilute urine. It is typically caused by underproduction of ADH due to noncancerous pituitary tumors. Generally, excessive urination is the only symptom of DI because thirst compensates for fluid loss. In severe cases, DI can be treated by taking medications which function like ADH.

THE ANTERIOR PITUITARY (OR ADENOHYPOPHYSIS)

The **anterior pituitary** is also controlled by the hypothalamus, but is an endocrine gland in its own right. The anterior pituitary makes and secretes a number of hormones, under hormonal control of the hypothalamus, that control other endocrine glands. (Growth hormone and prolactin are exceptions to this rule.) Refer back to Table 10–3 and Figure 10–7 for a list of hypothalamic and pituitary hormones. We discussed this relationship previously when we talked about hormonal control. The hypothalamus secretes a hormone that controls hormone secretion by the anterior pituitary, which usually controls the secretion of hormones by another endocrine gland. The hormone levels are controlled by negative feedback to both the pituitary and the hypothalamus.

Pathology Connection: Anterior Pituitary

HYPOPITUITARISM

Hypopituitarism is a decrease in pituitary function caused by tumors, surgery, radiation, or head injury. Hypopituitarism is characterized by loss of any or all of the anterior pituitary hormones including ACTH, GH, LH, and TSH, causing a variety of symptoms. LH and GH are usually the most severely affected.

Hypopituitarism is difficult to diagnose and treat because symptoms are often vague or subtle. In one study, one quarter of patients with traumatic brain injuries had symptoms of hypopituitarism that were untreated one year after injury, even though they had recovered enough to be released from rehabilitation.

If hypopituitarism is caused by a pituitary tumor, the tumor may be removed. No matter the cause, hormone replacement is the treatment of choice. Effective replacement of pituitary hormones is difficult and it often takes months to adjust the medication to appropriate therapeutic levels.

HYPERPITUITARISM

Hyperpituitarism, overproduction of pituitary hormones, is usually cased by benign pituitary tumors. Symptoms include reproductive abnormalities, acromegaly, cardiac dysfunction, sleep apnea, Cushing's Syndrome, and hyperthyroidism.

Like many hormonal disorders, hyperpituitarism is difficult to diagnose due to the vagueness of symptoms. Imaging studies and hormone levels are generally used. Hyperpituitarism is treated by decreasing the size of the tumor or removing it.

CUSHING'S SYNDROME

Cushing's Syndrome is caused by over-secretion of cortisol. Symptoms include upper body obesity, round face, easy bruising, osteoporosis, fatigue, depression, hypertension, and hyperglycemia (high blood glucose). Women may have excess facial hair and irregular periods; men may have decreased fertility and decreased sex drive. Cushing's Syndrome may be a side effect of medical use of steroids, like prednisone, or may be due to pituitary tumors, adrenal tumors, or one of several genetic disorders. New data suggests that some cases of metabolic syndrome (abdominal obesity occurring with high cholesterol, high blood pressure, and hyperglycemia or insulin resistance) may be minor or subtle cases of Cushing's Syndrome.

Diagnosis of Cushing's, particularly telling pituitary causes of Cushing's apart from other causes, is relatively complicated. Techniques include MRI, blood cortisol levels, clinical profile, 24 hour urine cortisol testing, hormone levels, dexamethasone suppression tests, genetic testing, and pituitary sampling.

Treatment depends on the underlying cause of the disorder. Classic Cushing's is due to a pituitary tumor that produces too much ACTH, stimulating the adrenal gland to produce too much cortisol. There are no approved drug therapies for classic Cushing's, though several clinical trials are underway. The tumor must be removed and in the case of a pituitary tumor, damage to the pituitary usually requires hormone replacement for the rest of the patient's life. Other cases of Cushing's are caused by overproduction of cortisol by the adrenal gland independent of what the pituitary and/or hypothalamus are doing. Surgery may be necessary but in many cases the disease can be managed by drug therapy to decrease cortisol secretion.

STATURE DISORDERS

Stature disorders are those disorders that result in well-below-average height (called **dwarfism**) or well-above-normal height (called **giantism** or **gigantism**). Some of these disorders are caused by abnormalities in skeletal development or nutritional deficiencies. However, growth hormone (GH) problems are often implicated. If GH secretion is insufficient during childhood, children do not grow to "standard" height. This type of dwarfism results in stunted adult height. However, if GH deficiency is diagnosed before closure of the growth zones of the long bones, it can be treated with GH injections. Children treated with GH injection may attain full height.

On the other end of the spectrum are those who secrete too much GH. If the oversecretion happens during childhood, people get extremely tall. (The current record in the *Guinness Book of World Records* is more than 8 feet tall.) Gigantism causes many health problems. The body gets so big that it cannot support itself. Treatment is often difficult. Surgery is the treatment of choice for this disorder, but must often be supplemented by radiation and drug therapies.

Oversecretion of GH in adults, after bones have stopped growing, causes **acromegaly**, a painful, often crippling disorder characterized by excess growth of body tissue, especially those of the face and extremities. Bones become deformed and organs may malfunction due to excess growth. Like gigantism, acromegaly is usually caused by noncancerous pituitary tumors. Treatment for acromegaly is the same as for gigantism.

 10-2 The endocrine system, like all systems, is affected by the aging process. To view a video on the effects of aging on the endocrine system, please see your DVD.

TEST YOUR KNOWLEDGE 10-3

Choose the best answer:

1. Oxytocin:
   a. is secreted by the anterior pituitary
   b. decreases uterine contractions
   c. is secreted from the posterior pituitary
   d. is a way to get more oxygen to your toes

2. The _____ is controlled by hormones from the hypothalamus, while the _____ actually secretes hypothalamic hormones.
   a. posterior pituitary, posterior pituitary
   b. anterior pituitary, anterior pituitary
   c. anterior pituitary, posterior pituitary
   d. posterior pituitary, anterior pituitary

3. This gland, under orders from the hypothalamus, releases hormones that control other endocrine glands:
   a. adrenal gland
   b. anterior pituitary
   c. thyroid gland
   d. pancreas

4. Which gland does ACTH control?
   a. adrenal gland
   b. anterior pituitary
   c. thyroid gland
   d. pancreas

5. Cushing's Syndrome is caused by excess secretion of:
   a. cholesterol
   b. epinephrine
   c. growth hormone
   d. cortisol

6. Alcohol inhibits secretion of ADH. This causes:
   a. headache
   b. lots of dilute urine
   c. decreased urination
   d. stupidity

The Thyroid Gland

The **thyroid** gland, located in the anterior portion of your neck on either side of the larynx over the trachea, is a butterfly-shaped (H-shaped) organ. The thyroid gland is responsible for secreting the hormones **thyroxine** (T_4) and **triiodothyronine** (T_3), under orders from the pituitary gland (see Figure 10–8 ▦). Thyroxine is controlled by secretion of TSH. The designations T_4 and T_3 refer to the number of iodine atoms in the hormone.

thyroid *(THIGH royd)*

T_4 and T_3 control cell metabolism and growth. Maintenance of these hormone levels takes place in a negative feedback loop. When blood levels of T_4 or T_3 drop, neurons in the hypothalamus sense the drop and produce thyroid-releasing hormone, or TRH. TRH stimulates the pituitary gland to produce TSH, or thyroid-stimulating hormone. TSH binds to cells on the thyroid gland and stimulates the thyroid to produce T_4 and T_3. Neurons in the hypothalamus sense when T_4 and T_3 blood levels are back to normal, and they stop producing TRH. This shuts down the secretion of TSH in the pituitary gland. Without TSH to trigger the thyroid gland, the production of T_4 and T_3 shuts down. When levels of thyroid hormone get too low, the cycle starts again.

The thyroid gland also secretes a third hormone, **calcitonin**, which decreases blood calcium by stimulating bone-building cells. This prevents hypercalcemia (high blood calcium levels), which can cause conditions such as arrhythmia, kidney stones, and osteoporosis.

T_4 and T_3 are generally referred to as "thyroid hormones" and are of great clinical importance. Overproduction (hyperthyroidism) or underproduction (hypothyroidism) can cause a variety of clinical symptoms, because the level of these hormones is essential in controlling growth and metabolism of body tissues, particularly in the nervous system. The importance of these hormones is so great that in many countries, table salt contains added iodine to ensure that people get enough iodine in their diets to make adequate amounts of thyroid hormones.

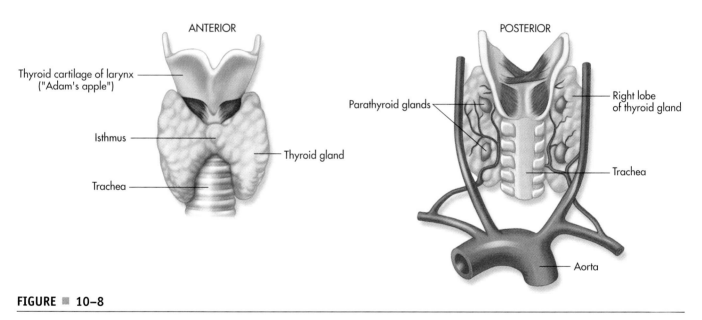

FIGURE ▓ 10–8

The thyroid and parathyroid glands.

Pathology Connection: The Thyroid

HYPOTHYROIDISM

Hypothyroidism occurs when too little thyroid hormone is produced. It can be caused by decreased production of hypothalamic (TRH), pituitary (TSH) or thyroid (T_4 and T_3) hormones. Symptoms of hypothyroidism include fatigue, feeling cold, dry, itchy skin, brittle or thinning hair, constipation, brittle nails, bradycardia, leg cramps, muscle pain, weight gain, hyperlipidemia, hypercholesterolemia, forgetfulness, depression, and sexual dysfunction.

Hashimoto's thyroiditis is the most common cause of hypothyroidism. It is caused by autoimmune attack on the thyroid gland. For unknown reasons, the immune system begins to attack the cells in the thyroid, causing inflammation and damage to the gland. This damage eventually leads to decreased production of thyroid hormones—hypothyroidism. In addition, the thyroid may swell, causing pain and difficulty swallowing. Hashimoto's thyroiditis, like many autoimmune disorders, is most common in women between 30 and 50 years old.

Hashimoto's thyroiditis can be distinguished from other kinds of thyroid disorders by blood tests. The disease is characterized by high TSH levels and low T_4 levels. It can be treated by daily doses of synthetic T_4.

HYPERTHYROIDISM

Hyperthyroidism is the overproduction of thyroid hormone. Symptoms of hyperthyroidism include feeling hot, muscle tremors, sweating, muscle weakness, tachycardia, cardiac arrhythmia, loose bowels, infertility, nervousness, irritability and enlarged, bulging eyes. Patients with hyperthyroidism are often hypersensitive to the effects of the sympathetic hormones norepinephrine and epinephrine. People with hyperthyroidism eat large quantities of food, but continue to lose weight.

Graves' disease, the most common cause of hyperthyroidism, is also an autoimmune disorder. In this case, the immune system attacks TSH receptors, blocking TSH from getting to the thyroid. To compensate, the thyroid overproduces thyroid hormones, causing hyperthyroidism. Like Hashimoto's thyroiditis, Graves' disease is more common in women of childbearing age. Acute Graves' disease can often result in a potentially fatal form of hyperthyroidism called a **thyroid storm**.

Diagnosis of Graves' disease is made by blood tests (low TSH levels with high T_4 levels) and increased levels of radioactive iodine uptake by the thyroid. Patients may also have a condition called exophthalmos, in which swelling of tissue behind the eyes makes the eyeballs bulge, giving the appearance of "bug eyes". There are three treatment options for Graves' disease: antithyroid medications to decrease thyroid activity or beta blockers to decrease the activity of epinephrine and norepinephrine, treatment with radioactive iodine to kill thyroid cells, and removal of the thyroid gland. If the thyroid gland is removed or destroyed, daily doses of thyroid hormones are required.

Goiter is the enlargement of the thyroid and can be the result of either hypothyroidism or hyperthyroidism. A goiter can interfere with swallowing or breathing, and can produce tightness in the throat. The swelling itself may not be painful. Medications or radioactive iodine treatment may shrink a goiter. A goiter may also be removed surgically. A simple goiter is caused by lack of dietary intake of iodine. See Figure 10-9 ■ for examples of goiter and Graves' disease.

PARATHYROID GLANDS

The thyroid gland has two small pairs of glands embedded in its posterior surface. These glands are called the **parathyroid glands,** and they produce **parathyroid hormone (PTH)** which regulates the levels of calcium in the bloodstream. There are four parathyroid glands and each is the size of a grain of rice. If calcium levels get too low, the parathyroid glands are stimulated to release PTH, which stimulates bone-dissolving cells and thereby releases needed calcium in the bloodstream. Again, see Figure 10–8.

parathyroid *(PAIR ah THIGH royd)*

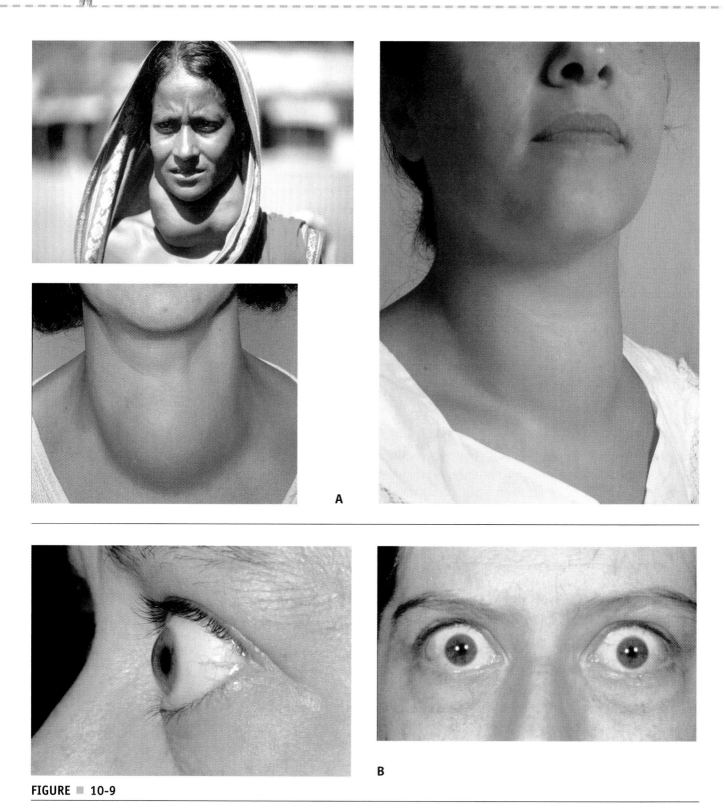

FIGURE ■ 10-9

A. Examples of various goiters **B.** Ocular changes in Grave's disease.

Damage to the parathyroid glands (during surgery or, in rare instances, during radioactive iodine treatment for hyperthyroidism) can reduce the production of parathyroid hormone. Hypoparathyroidism, or insufficient production of parathyroid hormone, can lead to a drop in calcium levels in the blood.

Low blood calcium can interfere with nerve function and cause the condition called **tetany**. Patients with tetany may experience muscle spasms, cramps, uncontrolled twitching, or seizures. Spasms of respiratory muscles, including the larynx, can cut off air supply and cause death. Tetany can be treated with calcium and vitamin D supplements. Parathyroid hormone treatment may also be used. If the tetany is severe, the patient may be treated with intravenous calcium to restore blood calcium levels.

The Thymus Gland

The **thymus gland** is both an endocrine gland and a lymphatic organ. It is located in the upper thorax behind the sternum and plays an important function in the immune system. It produces a hormone called **thymosin**, which helps with the maturation of white blood cells during childhood to fight infections. It begins to disappear during puberty. This gland is further discussed in Chapter 14, "The Lymphatic and Immune System."

thymus *(THIGH muss)*

The Pineal Gland

The tiny **pineal gland** is found in the brain, and its full function still remains unknown. However, it has been shown to produce the hormone **melatonin**, which rises and falls during the waking and sleeping hours. It is believed this hormone is what triggers our sleep by peaking at night and causing drowsiness.

pineal gland *(PIN ee al)*
melatonin *(MELL ah TOH ninn)*

The Pancreas

The **pancreas** is an accessory organ of the digestive system. It is located in the upper abdomen, behind the stomach. Part of the pancreas acts as an exocrine organ. This part produces and secretes digestive enzymes that help to break down starches, fats, and proteins. The other part of the pancreas acts as an endocrine organ. This part produces hormones that regulate blood sugar.

pancreas *(PAN kree ass)*

The pancreas is largely responsible for maintaining blood sugar (glucose) levels at or near a set point. The normal clinical range for blood glucose levels is 70 to 105 mg/dL (milligrams per deciliter). The pancreas can measure blood sugar, and if the blood glucose is high or low, the pancreas releases a hormone to correct the level.

To understand the importance of the pancreas, let's go back in your journey to the chapter on cells. Why does it matter how much glucose is in your blood? Why devote much of an organ, and a pretty big one at that, to controlling blood sugar? There are two reasons why blood glucose is important. Too much glucose floating around in your blood causes many problems with the fluid balance of your cells. Recall that if the concentration of the fluid outside a cell is high in solids, the cell will lose water to the surroundings. If the solids are low outside the cell, the cell will fill with water and eventually can even explode! Obviously, that's a serious problem. It does not matter if the solids are salts or glucose, the result is the same. Blood glucose must therefore be maintained at a certain level for cells to neither gain nor lose water. Why else is glucose important? Glucose is vital for cellular respiration. Cellular respiration is needed to get energy, by making adenosine triphosphate (ATP). Remember from Chapter 3 that the chief way that most cells make ATP, the energy that powers cells, is by breaking down glucose. Therefore, cells need to have enough glucose that they can make sufficient ATP to carry out their daily activities.

The pancreas makes two hormones that control blood glucose: **insulin**, which most of you have heard of before, and **glucagon**. These hormones are produced by specialized groups of cells in pancreatic tissue called the **Islets of Langerhans**. Insulin, the hormone that is missing or ineffective in diabetes mellitus (diabetes), removes glucose from the blood by directing the liver to store excess glucose and by helping glucose to get inside the cells so it can be used to make ATP. (Remember that glucose, a carbohydrate, is big and water soluble, so it cannot get into cells by itself.) When would insulin be secreted by the pancreas: when blood glucose is high or when blood glucose is low? Because insulin removes glucose from the blood, it *lowers* blood sugar, so it is released when blood sugar is high (hyperglycemia), like right after a meal.

Glucagon does the opposite of insulin. Glucagon puts glucose into the bloodstream mainly by directing the liver to release stored glucose in the form of **glycogen**. Glucagon is released typically several hours after a meal to prevent blood glucose from dropping too low (hypoglycemia). These two hormones control blood glucose very tightly in healthy humans (see Figure 10-10 ■). Other hormones, like the adrenal hormone cortisol, also aid in the control of blood sugar.

Pathology Connection: Diabetes Mellitus

Diabetes mellitus is a condition characterized by abnormally high blood glucose (hyperglycemia) because of the decreased secretion of insulin or the body's insensitivity to insulin. Type 1 or juvenile onset is caused by the immune destruction of the insulin-producing cells of the pancreas and is generally diagnosed in people under the age of 30. Patients with type 1 diabetes, such as our patient Maria, do not produce enough insulin. They are always dependent on daily doses of insulin.

Type 2 or late onset is caused by insensitivity of the body's tissues to insulin, and has typically been diagnosed in patients over 50, particularly those with other complicating conditions like obesity. Patients in early stages of type 2 diabetes can often be treated with a carefully controlled diet and weight-loss regimen or anti-diabetic drugs which help maintain glucose levels. To confuse matters further, some type 2 patients become insulin dependent as the disease progresses and there is evidence to suggest that some type 2 patients may actually have delayed onset autoimmune diabetes.

Diagnosis is usually done by blood tests and urinalysis. Some blood substances, such as glucose, will "spill" over into the urine. This is why diabetics, like Maria, sometimes have glucose in their urine. Sugar (Glucose) tests can be used to detect diabetes, confirm the diagnosis of diabetes, or determine the effectiveness of the treatment for diabetes. Note that sugar in urine, known as glucosuria or glycosuria, is not always abnormal. Situations of emotional stress, diet, or a low reabsorption rate by the kidneys with normal blood levels of glucose may cause high readings. However, you will find that in most cases, sugar in the urine of your patient is a result of diabetes mellitus. Normal Urine glucose: < 0.5 mg/dl.

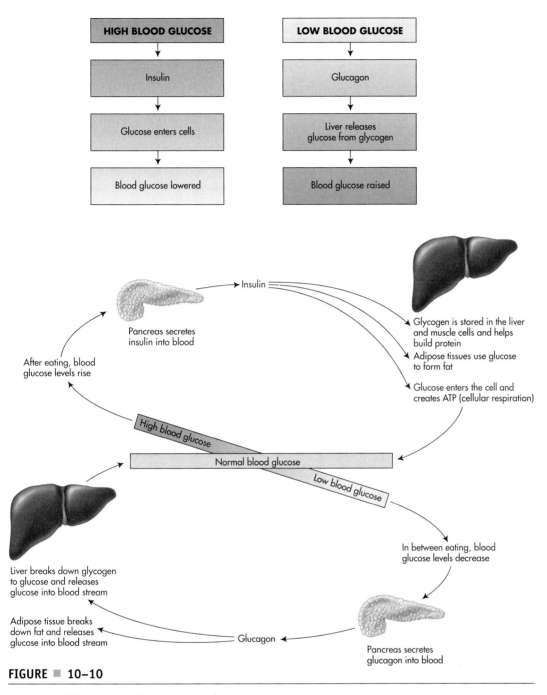

FIGURE 10–10

Control of blood glucose by pancreatic hormones.

In both types of diabetes, abnormally high blood glucose (hyperglycemia) must be resolved. If blood glucose remains high, the kidneys work overtime to excrete the excess sugar. Increased urination and dehydration are the most obvious symptoms. The stress of trying to get rid of the excess blood sugar eventually causes kidney damage. In addition, if insulin is not effective, glucose cannot get into cells. Cells must have glucose in order to make ATP. If cells can't get glucose and can't make ATP, they will look for other sources of energy. Untreated diabetics often lose weight as their body searches for other

TABLE 10-4   Comparison of type 1 and type 2 diabetes

	TYPE 1 DIABETES	TYPE 2 DIABETES
Cause	Autoimmune destruction of pancreas; not enough insulin is produced	Insulin resistance, obesity and sedentary lifestyle; insulin produced by pancreas does not work effectively
Age at onset	Typically before the age of 40	Typically later in life, though some obese and/or sedentary children are being diagnosed
Treatment	Insulin injections, insulin pump, pancreas transplant	Diet and exercise, medications, insulin in later stages
Symptoms	Usually sudden and severe, excess urination, extreme thirst, weight loss	Often more subtle than type 1 symptoms, excess urination, thirst

energy sources. Often, their blood becomes increasingly acidic due to their abnormal metabolism. There is often difficulty in wound healing and permanent damage to the peripheral nervous system. The changes in blood chemistry lead to tissue and organ damage. Left untreated, diabetes mellitus may lead to coma and death. Even diabetics who follow a strict treatment regime often have poorly controlled blood glucose levels due to the complexity of our blood glucose control system. Table 10-4 compares type 1 and type 2 diabetes.

Blood glucose that is too low, hypoglycemia, is actually a more acute problem than hyperglycemia for most diabetics, and is the primary side effect of insulin therapy. The early symptoms of hypoglycemia include hunger, nervousness, dizziness, anxiety, difficulty speaking, and weakness. If not resolved, hypoglycemia can progress to cause mental confusion, seizures, coma, and even death. Too little blood glucose causes fluid balance problems. In addition, too little blood glucose also means that cells, particularly those in the brain, cannot get the glucose they need to produce energy.

Therefore, the body has an elaborate defense mechanism against hypoglycemia. If blood glucose drops too low, the pancreas decreases its insulin secretion, and increases secretion of glucagon. In addition, the adrenal medulla secretes epinephrine. The hypothalamus is involved in that it senses the decrease in blood sugar and triggers the action of the adrenal gland via sympathetic control. (Remember from our discussion of hormonal control that the hypothalamus has some control over the sympathetic nervous system, which controls epinephrine release by the adrenal gland.) The hypothalamus also triggers feelings of hunger during hypoglycemia, so patients will eat food, increasing glucose levels. For some diabetics, hypoglycemia becomes a recurrent problem. It is thought that damage to the autonomic nervous system by repeated hypoglycemic incidents makes it difficult

10-3 For videos and animations spotlighting the pathology of diabetes, please go to your DVD.

www.prenhall.com/colbert
It is critically important to monitor blood glucose levels and maintain an appropriate diet for patients with diabetes. **Phlebotomists** are specially trained allied health professionals who draw and test blood samples. **Dieticians** are intensively involved in the design of specialized diets for diabetes and many other illnesses. To learn more about these important allied health professions, please go to the companion website for this chapter.

for the body to defend against future hypoglycemic episodes and even decreases awareness of symptoms of hypoglycemia. The only way to treat hypoglycemia is to get sugar quickly into the bloodstream. For patients in early hypoglycemia, sugar consumption, sugar-sweetened sodas, juice, or hard candy are among the best ways to increase blood sugar. Severe hypoglycemia requires medical attention. People who are not diabetic may also experience symptoms of hypoglycemia if blood glucose drops too low. A person experiencing hypoglycemic symptoms must get medical treatment immediately.

The Adrenal Glands

The **adrenal** glands are a pair of small glands that sit on top of your kidneys like baseball hats. The adrenal glands are split into two regions: the **adrenal cortex,** an outer layer, and the **adrenal medulla,** the middle of the gland. (Note that it is best to be specific when talking about the *adrenal* cortex, since your cerebrum has a cortex too.)

adrenal *(ah DREE nal)*
adrenal cortex
   (ad REE nal KOR teks)
adrenal medulla
   (ad REE nal meh DULL lah)

THE ADRENAL MEDULLA

The adrenal medulla releases two hormones: **epinephrine** (also known as **adrenalin**) and **norepinephrine.** These hormones increase the duration of the effects of your sympathetic nervous system. Remember your friend's snarling dog? Cells of the adrenal medulla receive the neurotransmitter norepinephrine from the sympathetic nervous system. (It is both a hormone and a neurotransmitter, depending on where it is released.) The neurotransmitter triggers the release of norepinephrine and epinephrine into the bloodstream, increasing heart rate, blood pressure, and respiration rate, and giving you sweaty palms and dry mouth. Again, the effects of the hormones last much longer than effects of the neurotransmitter.

epinephrine *(EP ih NEFF rinn)*
norepinephrine
   (NOR ep ih NEFF rinn)

THE ADRENAL CORTEX

The **adrenal cortex** makes dozens of steroid hormones known collectively as **adrenocorticosteroids** (steroids in the adrenal cortex). The adrenal cortex releases steroid hormones under the direction of the anterior pituitary. Many of these steroid hormones are so important that a decrease in their production could be fatal relatively quickly. Some of these hormones, the **mineralocorticoids**, regulate electrolyte, salt, and fluid balance. The **glucocorticoids** regulate blood sugar. Others produce androgens such as the male sex hormone testosterone, and are responsible for regulation of reproduction and secondary sexual characteristics, and still others control cell metabolism, growth, and immune system function. The sexual hormones are produced in the testes in the male and in the ovaries of the female.

Pathology Connection: Steroid Related Conditions

ADDISON'S DISEASE

Addison's disease is caused by insufficient production of adrenocorticosteroids, particularly cortisol, a glucocorticoid, and aldosterone, a mineralocorticoid. The disease can be fatal if not treated. Symptoms include weight loss, muscle weakness, fatigue, low blood pressure, hypoglycemia, irritability, depression, and excessive skin pigmentation. (Compare these to the symptoms of Cushing's disease, which is characterized by the oversecretion of cortisol.) Many cases of Addison's disease are autoimmune (body attacking itself), but other causes may include infection and cancer. Abnormalities of the hypothalamus and pituitary may also cause Addison's disease. It is diagnosed with blood tests and imaging studies and is treated with hormone replacement.

SIDE EFFECTS OF THERAPEUTIC STEROIDS

Prednisone is clinically important in the treatment of inflammation, organ transplant rejection, and immune disorders, but prescription steroids are a double-edged sword. These medications are so powerful that they can cause dangerous side effects, such as bone density loss, weight gain, fat deposits, and delayed wound healing, much like Cushing's Syndrome. In addition, these drugs, even when taken for a short time, cannot be discontinued suddenly. Patients must be weaned off their medication slowly. Why? When a patient takes a steroid medication, the adrenal gland decreases steroid production in response. (Remember negative feedback: as hormone levels rise, hormone secretion decreases.) Therefore, if the medication is removed suddenly, patients are left with a severe hormone deficiency, which could be fatal. They must be given some time to "gear up" to secrete hormones at the appropriate level.

STEROID ABUSE

It is no surprise, then, that taking steroid medications for the purpose of increasing athletic performance is prohibited by amateur and professional sports organizations. **Anabolic steroids** are a class of steroid molecules that cause large increases in muscle mass when compared to working out without steroids. Some athletes use anabolic steroids to enhance performance or to get big muscles much faster than they would without steroids. Because steroid hormone levels are so tightly controlled by the body, anabolic steroids have a number of side effects. (Think back to the discussion of prednisone at therapeutic levels; abuse levels are much higher.) Men abusing steroids may experience changes in sperm production, enlarged breasts, and shrinking of their testicles. Women may experience deepening of the voice, decreased breast size, and excessive body hair growth. Steroid abuse may lead to cardiovascular diseases and increased cholesterol levels. Many steroids suppress immune function, and because steroid use is illegal, many abusers expose themselves to hepatitis B and HIV when sharing needles. Steroid abuse has also been linked to increased aggressive behavior. All major professional and amateur athletic organizations ban the use of steroids.

CORTISOL AND THE STRESS RESPONSE

The typical cortisol secretion pattern is peak cortisol levels just before waking with cortisol decreasing as the day goes on. During the day, there are peaks and valleys superimposed on the overall levels. Cortisol levels also peak during times of stress.

Your body has a set response to any stressor, whether physiological – such as hypothermia, decreased blood volume, or hyperglycemia – or psychological – such as a big exam or a near collision. When you are exposed to stressors, both the sympathetic nervous system and the adrenal cortex are activated. Epinephrine and norepinephrine are released from the adrenal medulla, raising blood pressure and heart rate, raising respiration rate, increasing blood glucose, and decreasing digestion and other non-essential physiological responses. The adrenal cortex, under orders from the hypothalamus and pituitary, releases cortisol, a glucocorticoid hormone, causing increased blood glucose, and changes in immune response. This initial stress response is appropriate and useful, allowing your body to be ready to expend energy. If the stress continues, this response also continues, entering a phase of adaptation in which there is a balance between stress response and homeostatic mechanisms, and homeostasis is maintained. This is still a useful and healthy response to stress. However, if stress becomes chronic, the secretion of epinephrine, and cortisol in particular, becomes pathological.

Chronic secretion of cortisol has a number of effects on metabolism including increased appetite, changes in the immune system that increase autoimmunity and decrease defense against infection, increased heart rate, hypertension, hyperglycemia, hypercholesterolemia, increased abdominal fat, anxiety, and depression. (Remember the symptoms of Cushing's Syndrome?) So there appears to be a tradeoff at work. A certain amount of stress is productive and useful, sharpening awareness and increasing energy levels, but chronic stress causes detrimental physiological and psychological changes.

The response of the body to chronic stress is also influenced by many other factors, including psychological factors. Increased optimism, greater social support mechanisms, and lower perceived stress are all associated with lower cortisol levels, whereas depression and anxiety are associated with higher levels of cortisol.

In some overweight people, high cortisol levels are associated with increased risk of type 2 diabetes, heart disease, and stroke. However, these are only associations. That cortisol *causes* these conditions has not been shown. It is a "chicken vs. egg" problem. Does high cortisol cause these problems or do these problems cause high cortisol? Results are complicated due to the complexity of hormonal regulation, studies are often conflicting and age, gender and the effects of other conditions complicate the picture. There is much more research to be done before we understand the complex relationship between cortisol levels and disease.

Clinical Application

CORTISOL BUSTERS AND WEIGHT LOSS

It is clear that very high cortisol levels, such as in Cushing's disease, cause weight gain, and that very low cortisol, such as in Addison's disease, causes weight loss. There is evidence that chronic high cortisol is associated with risk factors for metabolic syndrome: abdominal obesity, hypercholesterolemia, hyperglycemia, and hypertension. However, this association is more apparent only in a subset of people who are overweight. Conversely some overweight people have low cortisol levels. Perhaps the combination of high body mass index and high cortisol is the culprit, but the interaction of hormone levels, blood chemistry, and body fat is also complicated by gender, age and, no great surprise, stress levels. At this time there is no clear picture of how cortisol is related to obesity and risk factors for health problems associated with increased body weight, although high cortisol combined with obesity and other risk factors is often a sign of problems to come.

You may have all seen the advertisements for "cortisol busters" which claim to cause significant, rapid weight loss simply by decreasing blood cortisol levels. These products claim that abdominal fat is caused by high cortisol levels, regardless of what you are eating or how much you are exercising. (They do, however, suggest a sensible diet and exercise plan while taking their product.) Given the available scientific information is seems unlikely that simply decreasing cortisol will cause rapid, healthy weight loss. You are better off exercising more and eating more nutritiously if you wish to lose weight, though it's probably not a bad idea to decrease your overall stress levels.

THE GONADS

The chief function of the gonads—the testes in males and the ovaries in females—is to produce and store gametes (reproductive cells), eggs, and sperm. However, the gonads also produce a number of sex hormones that control reproduction in both males and females, including testosterone in males and estrogen and progesterone in females. For more details on the sexual hormones produced by the gonads, see Chapter 17.

Estrogen, produced by the ovary in the female, is responsible for the development of both the reproductive organs and the secondary sex characteristics in females. Progesterone, also produced by the ovary, helps to regulate the menstrual cycle in females. In the male, the testes produces testosterone, which is responsible for development of the reproductive organs and secondary sex characteristics, such as the male's deep voice. For more details on these hormones, see Chapter 17.

PROSTAGLANDINS

Prostaglandins are molecules that act like hormones. Many different body tissues produce prostaglandins. These molecules have a variety of functions, depending on where they were produced. Some prostaglandins help to constrict and dilate smooth muscle—in the blood vessels, digestive tract, or the bronchial tubes of the lungs. Others help with contraction of the uterus during childbirth. Other prostaglandins respond to tissue inflammation. Each prostaglandin generally acts in a small area of the body—on the cells or tissues where it was produced. Prostaglandins have a short-term but powerful effect on these local tissues.

A Quick Trip Through the Diseases of the Endocrine System

DISORDER	ETIOLOGY	SIGNS AND SYMPTOMS	DIAGNOSTIC TEST(S)	TREATMENTS
Diabetes Insipidus	ADH deficiency.	Copious, dilute urine.	Rule out diabetes mellitus, ADH levels.	If caused by a medication, eliminate the medication. If low ADH give synthetic vasopressin.
Cushing's Disease	Excess cortisol usually due to benign pituitary tumor.	Abdominal obesity, easy bruising, hypertension, hyperglycemia, hypercholesterolemia, depression.	Imaging, blood cortisol levels, urine cortisol, CTH levels, dexamethasone suppression test, pituitary biopsy.	Tumor removal followed by hormone replacement. If caused by overuse of steroids, decrease or eliminate use.
Gigantism/ Acromegaly	Excess growth hormone from hyperfunction of pituitary, often due to pituitary tumor.	In children, overgrowth of long bones results in rapid growth to height in great excess of normal height, in adults, excess growth and deformity of body tissues.	Imaging, blood tests for hormone levels.	Removal of pituitary tumor, followed by hormone replacement therapy.
Dwarfism	Growth hormone deficiency from hypofunction of pituitary in childhood.	In children, failure to attain normal height, abnormally slow growth.	Blood tests, rule out other causes of dwarfism.	Early diagnosis, growth hormone injections.
Graves' Disease	Hyperthyroidism due to autoimmune attack on thyroid.	Tremors, sweating, weakness, tachycardia, arrhythmia, irritability.	Blood tests – low TSH levels but high T_4 levels, radioactive iodine uptake.	Removal or destruction of thyroid or anti-thyroid medication, followed by daily thyroid hormone.
Hashimoto's Thyroiditis	Hypothyroidism due to autoimmune attack on thyroid.	Fatigue, thinning hair, bradycardia, hypercholesterolemia, depression.	Blood tests – high TSH with low T_4.	Daily thyroid hormone.
Diabetes Mellitus	Insulin deficiency caused by autoimmune attack on pancreas, insulin resistance.	Excess urination and increased thirst due to high blood glucose.	Urine glucose, blood glucose, glucose tolerance testing.	Diet, exercise, anti-diabetic medications, insulin injections.
Addison's Disease	Deficiency of Adrenocorticosteroids due to autoimmune attack on adrenal cortex.	Weakness, fatigue, hypotension, hypoglycemia, depression.	Imaging and blood tests for corticosteroids.	Hormone replacement.
Chronic Stress	Excess cortisol in response to stressors.	Depression, anxiety, perhaps obesity, hypercholesterolemia, hypertension, increased susceptibility to infection.	No good test. Effects of stress vary widely among individuals.	Decrease stress, treatment of symptoms.

Figure 10–11 ■ shows examples of some common endocrine disorders.

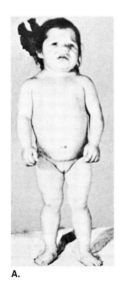
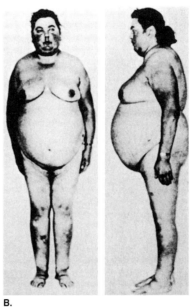

FIGURE ■ **10–11**

Examples of Endocrine Disorders. **A.** A 6-year-old child with congenital hypothyroidism. **B.** a patient with Cushing's syndrome. **C.** a patient with gigantism. (Source: Bettina Cirrone/Photo researchers, Inc.)

℞ Pharmacology Corner

Many hormonal disorders can be treated with hormone replacement therapy. For example Synthroid (a synthetic hormone) is used in the treatment of hypothyroidism. Hydrocortisone, a steroid, is used to replace cortisol in the treatment of Addison's disease. Prednisone, which acts like a corticosteroid, can be used to replace corticosteroids or as an anti-inflammatory drug (in patients with normal corticosteroid levels) and to treat severe allergies, rheumatoid arthritis, or other chronic conditions. Children with growth hormone deficiency are treated by growth hormone injections.

One of the most commonly used medications in the treatment of diabetes is insulin. There are various ways to administer insulin, including a recently introduced aerosol inhaler that replaces the more invasive daily injections. Often in late onset diabetes the insulin is functioning but very poorly. Drugs such as Glucophage help to potentiate (make stronger) the affect of insulin, making it function more efficiently.

Sometimes hormones need to be inhibited rather than replaced. Mitotane, a cortisol inhibitor, can be used to treat some kinds of Cushing's syndrome. Somatostatin analogs can be used to decrease growth hormone production in the treatment of acromegaly. In addition, disorders may be treated by blocking the receptors that receive the hormone at the target cell. For example, scientists are experimenting with specific aldosterone blockers to treat high blood pressure.

10-4 To view videos on the importance of insulin and monitoring and injection techniques, please see your DVD.

SUMMARY

Snapshots from the Journey

→ The endocrine system works together with the nervous system to regulate the activities of all the body systems. The endocrine system secretes hormones that act very slowly on distant targets. Their effects are long lasting.

→ Most hormones act on cells by binding to external receptors, causing changes in enzyme activity inside the target cell. Steroids, however, can enter cells and interact directly with DNA, which makes steroids very powerful.

→ Hormone levels are controlled largely by negative feedback. When hormone levels rise, signals are transmitted to the endocrine organ releasing the hormone, telling the organ to decrease the amount of hormone released. Hormone levels will then decrease. The optimal level of the hormone is called the set point. If the signal brings a hormone back to set point, the action is called negative feedback. If the signal causes the hormone to get further away from set point, the action is positive feedback.

→ Hormone levels can be regulated by three mechanisms: changes in the body's internal environment, control by hormones released by another endocrine gland, and direct control by the nervous system.

→ The hypothalamus, a part of the diencephalon, controls much of the endocrine system by controlling the pituitary gland. The pituitary gland has two parts: the posterior pituitary, which is part of the hypothalamus and actually secretes hypothalamic hormones (ADH and oxytocin), and the anterior pituitary, which secretes several different hormones under the influence of hormones from the hypothalamus. The hormones secreted by the anterior pituitary typically control other endocrine glands. (Growth hormone is an exception.)

→ Non-cancerous pituitary tumors cause a number of different endocrine disorders depending on whether hormones are over-secreted or under-secreted. Hyperpituitarism, acromegaly, gigantism and Cushing's syndrome are caused by over-secretion of pituitary hormones.

→ Several other endocrine glands have important control functions. The thyroid gland secretes the iodine-containing hormones triiodothyronine (T_3) and thyroxine (T_4), which control growth and cellular metabolism.

→ Autoimmune attack on the thyroid can cause serious problems. Hashimoto's thyroiditis, the most common form of hypothyroidism, causes the typical symptoms of hypothyroidism, including bradycardia, thinning hair, leg cramps, fatigue, forgetfulness and depression. Graves' disease, the most common cause of hyperthyroidism, causes tremors, anxiety, irritability, cardiac arrhythmia, and excessive sweating.

→ The pancreas secretes two hormones: insulin, which lowers blood sugar, and glucagon, which raises blood sugar. Diabetes is caused by a decrease in insulin secretion or decreased sensitivity to insulin. Very high blood sugar is the result. Hypoglycemia (low blood sugar) is also a problem for diabetics, and is the chief side effect of insulin therapy. Many diabetics have health problems due to the difficulty of effectively controlling blood glucose.

→ The adrenal glands are split into two parts. The adrenal medulla is an extension of the sympathetic nervous system, releasing epinephrine and norepinephrine as hormones during fight-or-flight response. The adrenal cortex releases many different adrenocorticosteroid hormones, which control reproduction, inflammation, tissue growth, and immunity.

→ One of the most important hormones released by the adrenal cortex is cortisol, a glucocorticoid. The actions of cortisol are somewhat controversial. It is well known that too much cortisol (Cushing's) causes weight gain and that too little cortisol (Addison's) causes weight loss. But these are diseases. Our understanding of how cortisol is involved in obesity and its related conditions that make up metabolic syndrome—hypertension, hypercholesterolemia, and insulin resistance—is not complete.

MARIA'S STORY

Let's revisit Maria's case. She is a 35-year-old insulin-dependent diabetic who currently takes good care of herself. She did not take good care of herself as a teenager. She is on an insulin pump to try to control her blood sugar, but has recently passed out in public several times.

a. What condition causes her to pass out? (hyperglycemia or hypoglycemia)

_____

b. Why does this condition develop?

_____

_____

c. What is the appropriate treatment for the early stages of this condition?

_____

_____

d. Given that she has been a diabetic for so long, why doesn't Maria realize she is in trouble before she passes out?

_____

_____

REVIEW QUESTIONS

Matching

1. Match the hormone or neurotransmitter on the left with the description on the right.

   _____ ADH   A. decreases blood sugar

   _____ insulin   B. increases thyroid hormone secretion

   _____ glucagon   C. regulates cell metabolism

   _____ oxytocin   D. increases steroid release

   _____ epinephrine   E. increases uterine contractions

   _____ thyroxine   F. decreases urination

   _____ prolactin   G. prolongs sympathetic response

   _____ ACTH (adrenocorticotropic   H. stimulates tissue growth
   hormone)   I. increases blood sugar

   _____ TSH   J. increases milk production in females

   _____ growth hormone

2. Match the hormone abnormality on the left with the disease on the right.

   _____ too little ADH   A. gigantism

   _____ too little insulin   B. Diabetes mellitus

   _____ too little cortisol   C. pheochromocytoma

   _____ too little thyroid hormone   D. Diabetes insipidus

   _____ too much epinephrine   E. Graves' disease

   _____ too much thyroxine   F. Addison's disease

   _____ too much cortisol   G. changes in sexual characteristics, immune suppression, rage

   _____ too much growth hormone   H. Hashimoto's thyroiditis

   _____ too many steroids   I. Cushing's disease

Multiple Choice

1. ADH stands for:
   a. vasodilate
   a. antidiuretic hormone
   b. androdoginin hormone
   c. American Department of Health
   d. all-diglyceride hormone

2. The "master gland" is the:
   a. adrenals
   b. pituitary
   c. Graves'
   d. pancreas

3. The thymus gland's main function is for:
   a. reproduction
   b. growth
   c. immunity
   d. RBC levels

4. The pineal gland is located in/on the:
   a. kidneys
   b. brain
   c. thorax
   d. abdomen

5. Glucagon performs the opposite action of:
   a. glucose
   b. insulin
   c. ATP
   d. WBCs

6. Anxiety, rapid heartbeat, and sweating are symptoms of:
   a. Graves' disease
   b. Cushing's syndrome
   c. Hashimoto's thyroiditis
   d. all of the above

7. Glucocorticoid hormones control:
   a. glucose
   b. insulin
   c. ATP
   d. WBCs

8. This hormone may be involved in metabolic syndrome:
   a. glucagon
   b. ADH
   c. TSH
   d. cortisol

Short Answer

1. Compare and contrast neurotransmitters and hormones.

   _____
   _____
   _____

2. List the sources of control of hormone levels.

   _____
   _____
   _____

3. Explain negative feedback and its role in controlling hormone levels.

   _____
   _____
   _____

4. Discuss why the use of anabolic steroids should be outlawed for performance enhancement.

   _____
   _____
   _____

5. What is the difference between neural control and humoral control of endocrine glands?

_____

_____

_____

6. Explain the difference between type 1 and type 2 diabetes.

_____

_____

_____

7. Why are hormone disorders so difficult to diagnose?

_____

_____

_____

8. Explain the difference between "good stress" and "bad stress."

_____

_____

_____

Suggested Activities

1. Review the hormones secreted by each endocrine organ and their functions. Make up 3 × 5 cards with the endocrine gland on front and the hormones it produces on back. Pick a partner and quiz each other. Once you and your partner can match the gland to the hormones, make up another set of cards with the hormone on front and its activity or function on back, and again quiz each other.

2. Review the endocrine diseases listed in this chapter. Make up 3 × 5 cards with the signs and symptoms on the front and the actual disease printed on the back. In a small group or with a partner, try to stump each other in determining the correct disease diagnosis.

10-5 Now that you have completed your journey through this chapter, please go to the DVD for interactive games and puzzles concerning the medical terms and concepts contained in this chapter. By playing the games you will reinforce your learning of the endocrine system in a fun way.

T H E
SENSES

The Sights and Sounds

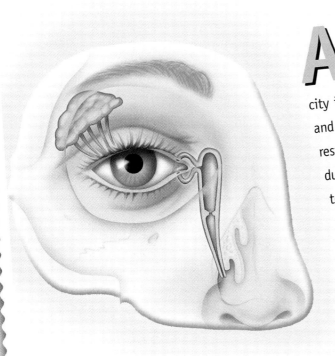

As we stroll through the city on our journey, we take in many sights, sounds, and smells. As you look around you may see many different buildings and types of people. The city is also noisy, full of car horns honking, brakes squealing and people talking. You will smell delicious food wafting out restaurant windows or the stench of garbage over-flowing dumpsters. Our senses receive all of this input and send it to the brain for interpretation so we can understand and appreciate what is happening around us. These senses are highly integrated with the nervous system, enabling us to respond quickly and thereby protecting us from harm. For example, we need to *see* that oncoming speeding car as we step away from the curb and *hear* the blaring horn in order to respond by quickly stepping back to the safe confines of the sidewalk. See how sensory input can determine motor response?

Chapter 11

LEARNING OBJECTIVES

At the end of your journey through this chapter, you will be able to:

→ Differentiate between general and special senses

→ Describe the internal and external anatomy and functions of the eye

→ Describe the internal and external anatomy and functions of the ear

→ Discuss the process involved with the senses of taste, smell, and touch

→ Contrast the types of pain and the pain response

→ Explain several common disorders of the eye, ear, and senses.

MULTIMEDIA APPLICATIONS

DVD Interactive Exercises

→ Interactive drag-and-drop exercise of the eye structures and 3-D animation of the eye, 11-1

→ Pathophysiologic spotlights on eye disorders such as cataracts, macular degeneration, and conjunctivitis, 11-2

→ Animation of the workings of the middle ear, 11-3

→ Video on tympanic membrane thermometer measurements, 11-4

→ Interactive drag-and-drop exercise of the ear structures, 3-D animation of the ear, and animations of the child and adolescent ear, 11-5

→ Videos on ear disorders such as otitis media and hearing tests, 11-6

→ Video on heat and cold therapy procedures, 11-7

→ Video on pain perception and pain scales, 11-8

→ Videos on ophthalmic and otic medications and their delivery, along with ear and eye irrigation, 11-9

→ Interactive games and puzzles, 11-10

www.prenhall.com/colbert

→ Professional Profiles:
  • Ophthalmology and Opticians
  • Audiologist

→ Related Internet Links

→ Additional Review Questions

Pronunciation Guide

Correct pronunciation is important in any journey so that you and others are completely understood. Here is a "see and say" Pronunciation Guide for the more difficult terms to pronounce in this chapter.

amblyopia (am blee OH pee ah)

aqueous humor
   (AY kwee uss HYOO mer)

auricle (AW rih kl)

cataract (KAT ah rakt)

cerumen (seh ROO men)

ceruminous glands (seh ROO men us)

choroid (KOH royd)

ciliary muscles (SILL ee air ee)

cochlea (KOHK lee ah)

conjunctiva (kon JUNK tih vah)

endolymph (EN doe limf)

Eustachian tubes (yoo STAY she ehn)

external auditory meatus
   (AW dih tor ee mee AY tuss)

glaucoma (glaw KOH mah)

gustatory sense (GUSS ta tore ee)

hyperopia (HIGH per OH pee ah)

incus (ING kuss)

labyrinth (LAB ih rinth)

lacrimal apparatus
   (LAK rim al app ah RA tuss)

malleus (MALL ee us)

Ménière's disease (MAIN ee yairs)

myopia (my OH pee ah)

ossicle (AH sih kel)

otitis media (oh TYE tiss MEH dee ah)

perilymph (per ih LIMF)

pinna (PINN ah)

presbyopia (PRESS bee OH pee ah)

sclera (SKLAIR ah)

stapes (STAY peez)

tactile corpuscle (KOR puss el)

tinnitus (tinn EYE tuss)

tympanic membrane (tihm PAN ik)

vestibule chamber (VESS tih byool)

vitreous humor
   (VITT ree uss HYOO mer)

THE SENSES

Our body's senses enable us to experience all aspects of our journey. They are truly remarkable sensors through which we see, hear, smell, taste, and feel the world around us.

Our senses monitor and detect changes in the environment and send this information from the receptor to the brain via sensory (afferent) neurons. The brain interprets the information and in many circumstances makes the appropriate motor (efferent) response.

Traditionally, we are taught that we possess five senses: vision, hearing, smell, taste, and touch. However, there are more areas of sensory input into the brain. What about pain and pressure sensations? How do we "feel" hot and cold? How does our body sense position and balance (equilibrium)? What about feelings of hunger and thirst? These, too, are senses that are very important to our survival.

The senses of sight (eyes), hearing and balance (ears), taste (tongue), and smell (nose) are referred to as our **special senses**. These senses are found in well-defined regions of the head. However, other senses, called **general senses**, are scattered throughout various regions of the body. These include the sensations of heat, cold, pain, nausea, hunger, thirst, and pressure or touch.

The senses can be further broken down. For example, the receptors of the skin are called the **cutaneous senses** and include touch, heat, cold, and pain. The **visceral senses** include nausea, hunger, thirst, and the need to urinate and defecate.

cutane/o = *skin*
visceral = *pertaining to organs*

Before finishing this discussion of the various senses, we should also mention a final, more controversial sense. By "reading your mind," we see you already identified it as extrasensory perception or ESP. The phrase *extrasensory perception* literally means senses outside the normal sensory perceptions. While there is still debate over whether this phenomenon exists, we just "know" this chapter will be an "eye-opening" experience for you. We hope the puns aren't stimulating your visceral senses and making you nauseous.

SENSE OF SIGHT

While on our journey, we see many amazing sights. To record these sights for later viewing, we may take along a camera and film. The eye has many similarities to a camera. The light rays from the images photographed with a camera pass through the small opening (comparable to the pupil) and through the transparent lens (lens of the eye) where the rays are then focused on a photoreceptive film (retina). The shutter of a camera opens and closes at various speeds to adjust the amount of light exposure on the film. The shutter (iris) must allow just the right amount of light to enter and focus properly on the film for a clear image. A camera may be packed away in a suitcase or thrown in a car and therefore needs protection from being dropped or exposed to a harsh environment. You will soon see how the external structures of the eye help protect it from injury, much like a camera case protects a camera. In addition, the camera lens must be kept clean to ensure a clear picture. The lacrimal glands secrete tears to help to perform this function. Now that you have this analogy to give you the big picture, let's explore the specific structures and functions of the eye.

The External Structures of the Eye

The **orbit** (or **orbital cavity**) is a cone-shaped cavity formed by the skull that houses and protects the **eyeball**, a one-inch sphere. This cavity is padded with fatty tissue that cushions and protects the eyeball from injury and has several openings through which nerves and blood vessels can pass. The eyeball is connected to the orbital cavity by six short muscles that provide support and allow rotary movement so you can see in all directions. Also protecting the eye is a pair of movable folds of skin, commonly called **eyelids**, which contain **eyelashes** to help prevent large particles from entering. The eyelashes act as sensors to cause rapid closure as a foreign object approaches the eyeball. The eyelids close over the eye much like the lens cover of a camera to protect the eye from intense light, foreign particles, and impact injuries. The eyelids also contain sebaceous glands that secrete sebum onto the eyelids to keep them soft, pliable, and a little sticky to trap particles.

A protective membrane called the **conjunctiva** covers the exposed surface of the eyeball and protects the exposed eye surface. Each eye has a **lacrimal apparatus** that produces and stores tears. The lacrimal apparatus includes the **lacrimal gland** and its corresponding ducts or passageways that transport the tears. The lacrimal glands (*exocrine* because their secretions of tears go *outside* the body) produce tears, which are needed for constant cleansing and lubrication and which are spread over the eye surface by blinking. Our eyes are

conjunctiva *(kon JUNK tih vah)*
lacrimal apparatus
 (LAK rim al app ah RA tuss)

constantly tearing, but they do not overflow because excess tears drain into the nasal cavity via two small holes in the inner corner of the eye. However, when we cry, the excess runs down our cheeks and more drains into our nose, causing it to run. The tears also act as an **antiseptic** to keep the eyeballs free of germs. Please see Figure 11–1 ▪ for the structures involved with tearing.

The Internal Structures of the Eye

The globe-shaped eyeball is the organ of vision and is separated into two chambers of fluid that help to protect the eye. These "fluids of the eye" are called *humors*. The **aqueous humor** ("watery" humor) bathes the iris, pupil, and lens and fills the anterior and posterior chambers, in the front of the eye. The **vitreous humor**, which maintains eye shape and refracts light rays, is a clear, jellylike fluid that occupies the entire eye cavity behind the lens.

The eyeball has three layers. Please see Figure 11–2 ▪ for the layers and internal structures of the eye. These layers are the **sclera, choroid,** and **retina.** The sclera is the outermost layer and is a tough, fibrous tissue that serves as the protective shield we commonly call the "whites of the eye." The fluids within the eye exert pressure on the sclera, helping maintain the shape of the eye. The extrinsic muscles responsible for moving the eye are attached to the sclera. The sclera contains a specialized portion called the **cornea** (window of the eye),

aqueous humor
    (AY kwee uss HYOO mer)
vitreous humor
    (VITT ree uss HYOO mer)

sclera *(SKLAIR ah)*
choroid *(KOH royd)*
retina *(RETT in ah)*

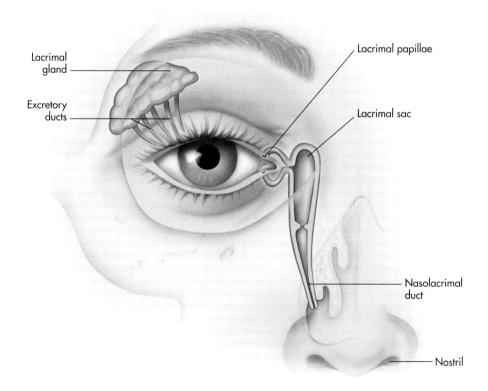

FIGURE ▪ 11-1

Lacrimal structures of the eye.

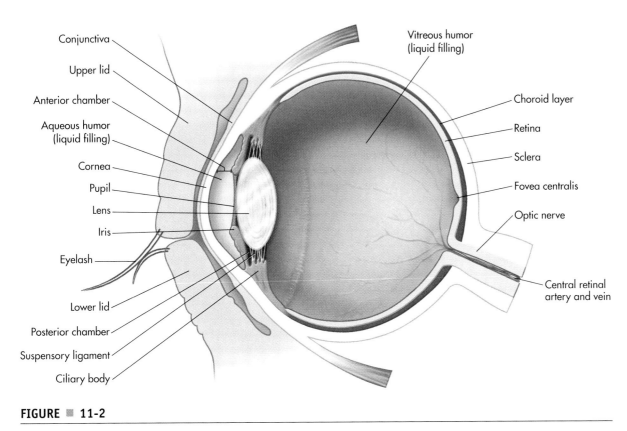

Conjunctiva
Upper lid
Anterior chamber
Aqueous humor (liquid filling)
Cornea
Pupil
Lens
Iris
Eyelash
Lower lid
Posterior chamber
Suspensory ligament
Ciliary body

Vitreous humor (liquid filling)
Choroid layer
Retina
Sclera
Fovea centralis
Optic nerve
Central retinal artery and vein

FIGURE ▦ 11-2

The internal structures of the eye.

which contains no blood vessels and is transparent to allow light rays to pass into the eye. The cornea has a curved surface that enables it to bend the entering light waves to focus them on the surface of the retina.

The middle layer, or choroid (also called choroid coat or choroidea), is a highly vascularized and pigmented region that provides nourishment to the eye. This layer also contains the iris and the pupil. The **iris** is the colored portion of the eye that controls the size of the opening (**pupil**) where light passes into the eye. The iris is a sphincter, which means it has intrinsic muscles that can relax or contract depending on light conditions. In low light, the iris relaxes, causing the pupil to dilate and allowing more light into the eye for a better image. In bright light, the pupil constricts, causing less light into the eye.

The third and innermost layer is the retina. This area contains the nerve endings that receive and interpret the rays of light. Located behind the iris and pupil is an elastic, disc-shaped, biconvex crystalline structure called the **lens**. The lens lies between the anterior and posterior chambers of the eye. The lens is surrounded by **ciliary muscles.** When light enters the eye, it is refracted, or bent, by the lens. Ciliary muscles can alter the shape of the lens, making it thinner or thicker to change the angle of refraction, which allows the focusing of incoming light rays on the retinal area. This process, called **accommodation**, combines changes in the size of the pupil and the lens curvature to make sure the image converges in the same place on the retina and therefore is properly focused.

iris *(EYE riss)*
pupil *(PYOO pill)*

ciliary muscles *(SILL ee air ee)*

In summation, light rays enter the eye and pass through the conjunctival membrane, cornea, aqueous humor, pupil, lens, and vitreous humor, and are focused on the retina. Here the photoreceptors in the retina cause an impulse to be sent to the optic nerve, (cranial nerve II) which carries it to the brain for the interpretation we call vision. See if you can trace the pathway light takes in Figure 11–2. Refer to Table 11–1, which summarizes the major structures and their corresponding functions of the eye.

TABLE 11–1 Structures and Functions of the Eye

ORGAN OR STRUCTURE	PRIMARY FUNCTION
Orbit	Cone-shaped cavity that contains the eyeball; padded with fatty tissues, the orbit has several openings for nerves and blood vessels to pass through
Eye muscles	Six short muscles that provide support and rotary movement
Eyelids	Moveable folds of skin containing eyelashes to protect eye from intense light, foreign particles, and impact injuries
Conjunctiva	Protective membrane that covers the exposed surface of eyeball
Lacrimal apparatus	Includes the lacrimal gland that produces tears that lubricate and cleanse the eye and the corresponding ducts or passageways to transport the tears
Eyeball	Globe-shaped organ of vision
Sclera	Outermost layer of the eyball, known as the "white" of the eye; maintains shape of eye; contains the transparent curved cornea, which bends outside light rays to focus them on the surface of the retina
Choroid	Middle layer of the eyball, containing rich blood vessels and pigmentation to prevent internal reflection of light rays; also contains the iris and pupil
Iris	Colored portion of eye, the part that controls the size of the opening (pupil) where light passes into the eye
Pupil	The opening through which light passes into the eye
Retina	Innermost layer of the eyeball, containing the nerve endings that receive and interpret the rays of light for vision
Optic disk	The area where the neurons of the optic nerve exit the retina
Lens	Located behind the pupil, the lens is controlled by ciliary muscles that shape the lens by thinning or thickening the lens to allow light to focus on the retinal surface; this process, called accommodation, combines changes in pupil size and the lens curvature to insure the light rays focus properly on the retina.

 ## Pathology Connection: Injuries and Disorders of the Eyes

Eye injuries can result in permanent impairment or loss of sight. Injuries often result from exposure to chemicals, physical blows, and the insertion of foreign objects. Some chemicals, such as soaps, cause only mild irritation. Other chemicals, such as acids and cleaning products, can cause permanent damage. Symptoms include redness, swelling, and pain. Chemicals should be flushed from the eye using large amounts of water or saline solution. Medical care is required if exposure is persistent or involves a dangerous chemical. Physical blows may cause injuries such as retinal detachment, in which the retina separates partially or fully from the back of the eye. Symptoms include blurred vision or loss of vision in the injured eye. In addition to damaging the eye, physical blows can cause fractures in the orbital area, so medical care is required. Insertion of foreign objects, such as sand or broken glass, causes corneal abrasions and may puncture the eye. Such injuries are treated by covering the eye and immediately seeking medical help. The object should not be removed, as doing so may cause more damage.

A **stye**, which is also known as a **hordeolum**, is an abscess that forms at the base of an eyelash due to infection of a sebaceous gland. Often the result of bacterial infections, styes are red, swollen, and painful. They resemble acne pimples. Styes are treated by applying warm compresses, which help relieve discomfort and speed the bursting of the stye.

Conjunctivitis is an inflammation of the membrane that lines the eye, and is characterized by red swollen eyes. This condition can either be acute or chronic and can be caused by a variety of irritants such as fumes or onions. Pathogens such as viruses and bacteria can also be a culprit. The acute infective phase is commonly called **pinkeye**; it is a highly contagious form caused by *Staphylococcus aureus* bacteria. Chronic conjunctivitis is called **trachoma** and is caused by *Chlamydia trachomatis*. These cases are often highly contagious and can cause corneal damage and impair vision. They can be treated with antibiotics.

A **cataract** is a condition in which the lens loses its flexibility and transparency and light cannot easily pass through the clouded lens. While the exact cause of cataracts is unknown, it has been shown that increased exposure to sunlight may speed up the development of cataracts. Congenital defects, trauma, and aging are attributed to cataract formation. Untreated, this condition can lead to blindness. It is interesting to note that cataract surgery was one of the earliest recorded surgical procedures, dating back to ancient Greece.

Glaucoma is caused by increased pressure in the fluid of the eye, which interferes with optic nerve functioning, and can also lead to blindness. This

---

Amazing Body Facts ➕ ✋ 🚶 🏊 🔑

WHY WE CAN SEE IN THE DARK

The retina is a delicate membrane that continues posteriorly and narrows to become the **optic nerve**. It contains two types of light-sensing receptors called rods and cones. The **rods** are active in dim light and do not perceive color, whereas the **cones** are active in bright light and do perceive color. The retina of each eye contains about 3 million cones and 100 million rods. Together they enable you to see in a variety of light conditions.

The rods and cones contain **photopigments** that cause a chemical change when light hits them. This chemical change causes an impulse to be sent to the optic nerve, and then to the brain, where the impulse is interpreted and we "see" the object. This interpretation occurs in the visual part of the cerebral cortex, located in the occipital lobe. The area where the neurons of the optic nerve exit the retina is known as the **optic disk**. This area is called the **blind spot** because it lacks rods and cones, so it cannot sense light or color.

---

optic = *pertaining to the eye*
photopigments
  (FOE toe pig ments)

conjunctivitis
  (kon JUNK tih VYE tiss)

---

www.prenhall.com/colbert
An optician specializes in making optical products and accessories. To learn more about this allied health profession and the field of ophthalmology, please visit the companion website. An ophthalmologist is a doctor who specializes in the care of the eyes and in the treatment of eye disease and injury. An ophthalmologist may use an ophthalmoscope, an instrument used to view the inside of the eye. Its use is crucial in determining the health of the retina and the vitreous humor.

---

cataract *(KAT ah rakt)*

glaucoma *(glaw KOH mah)*

increased pressure is due to overproduction of aqueous humor or lack of drainage. The symptoms develop gradually, and include aching and the appearance of a halo around sources of light. Peripheral vision is reduced, which eventually progresses to tunnel vision and then blindness. Glaucoma occurs in 20% of adults over 40 and accounts for 15% of the cases of blindness in America. This is a tragic loss because glaucoma can be readily diagnosed and treated with medication and surgery. Annual visits to an ophthalmologist can lead to early diagnosis and treatment. To identify the onset of glaucoma, an ophthalmologist uses a tool called a tonometer to measure pressure within an eye. The goal is to lower intraocular pressure.

Macular degeneration is the reduction or loss of central vision in which peripheral vision is unaffected. It is estimated to develop in 10% of the elderly. There are two forms of macular degeneration: **atrophic** (dry) and **exudative** (hemorrhagic). There is no cure for the atrophic form, but the exudative form can be improved slightly with argon laser therapy. Irreversible damage to the retina (**retinopathy**) can be caused by hypertension or diabetes mellitus and can result in vision impairment or blindness. Smoking can further aggravate this condition. Maria is at risk for macular degeneration as a result of her diabetes.

The eye can have several defects that impair vision. The eyes may have difficulty focusing on near or far objects because the light rays are not focusing properly on the retina. **Hyperopia** (farsightedness) occurs when the eye cannot focus properly on nearby objects. Hyperopia usually occurs because the eyeball is too short, and the focal point of the lens is beyond the retina. Glasses using convex lenses are used to correct hyperopia, refocusing the focal point forward and on the retina. As the ciliary muscles age, they weaken and pupil size is decreased, reducing the amount of light coming into the retina. **Presbyopia** is farsightedness that occurs with age, usually between 40 and 45 years. The lens becomes stiff and yellowish. Such age-related changes make it difficult for older adults to focus and make them more sensitive to glare that can impair their nighttime driving abilities. Corrective lenses called **bifocals** can be used to treat presbyopia. Bifocal lenses have two points of focus: an upper part for distance vision and a lower part for near vision. **Myopia** (nearsightedness) causes objects at a distance to appear blurred. In myopia, the eyeball is too long, so the focal point of the lens falls in front of the retina. Glasses that use concave lenses correct myopia by refocusing the focal point backward and onto the retina. Students with myopia may have trouble seeing the board at the front of the class. **Amblyopia,** or lazy eye, causes poor vision in one eye because of abnormal dominance of the other eye, which does most of the work. Lazy eye usually occurs in childhood. If not corrected, it can lead to blindness in the lazy eye. The treatment for amblyopia is to cover the dominant eye, forcing the weak eye to work. Corrective lenses and surgery may also be needed.

Normally, the brain perceives the images received by each eye as one image. In **diplopia**, the brain perceives two

hyperopia *(HIGH per OH pee ah)*

presbyopia *(PRESS bee OH pee ah)*
   presby/o = *old*
   opia = *refers to vision*

myopia *(my OH pee ah)*
amblyopia *(am blee OH pee ah)*

Clinical Application

SNELLEN CHARTS

Eye care professionals and others often measure visual acuity using a Snellen chart. You have probably seen one of these if you have taken a vision test for a driver's license. A traditional Snellen chart has 11 lines of block letters, getting progressively smaller toward the bottom. The patient covers one eye and then reads aloud the letters of each row. The smallest row that the patient can read accurately indicates the visual acuity in that eye.

images, so a person affected by diplopia has double vision. Double vision occurs because one eye is not aligned with the other, so the image does not fall on a corresponding point on the retina of each eye. Diplopia may be the result of a non-ocular disease, so the disease needs to be treated. Otherwise, diplopia treatment includes placing a temporary eye patch over the misaligned eye, using special corrective lenses, or realigning the misaligned eye surgically.

In **strabismus**, one eye is misaligned due to an inability in the muscles of the eye to coordinate movement with the other eye. The affected eye turns inward, outward, upward, or downward in relation to the other eye, often giving the affected person a cross-eyed appearance. Children usually suffer from strabismus, which often results from a congenital defect. Acquired strabismus may result from injury or disease. In some cases, strabismus can be treated with exercises or corrective lenses. For many people, strabismus must be treated surgically to modify the affected muscles in an effort to realign the eyes.

Degeneration of the retina can cause difficulty seeing at night or in dim light. **Nyctalopia** is the name for this condition, often caused by a vitamin A deficiency. Vitamin A is a vital ingredient needed to make photopigments in rod cells. Photopigment is a light-sensitive chemical that triggers the photoreceptors along the visual nerve pathway. As a child you might have been told to eat your carrots so you could see well. Of course, carrots are very high in vitamin A.

Retinal disorders can be inherited, such as red-green color blindness due to genes on the X chromosome that cause abnormal photopigments in the cones. People with this condition can see colors but can't distinguish the difference between red and green. Colored figures are used to screen for color blindness to see if patients can differentiate between green, blue and red. Since red-green color blindness is an X linked disorder, more men get the condition. See Figure 11-3 to determine if you have color blindness.

Annual eye exams diagnose many of the problems listed previously, and early diagnosis can improve treatment. Ophthalmologists and optometrists use various tools such as eye charts, tonometers for measuring ocular pressure, and ophthalmoscopes for viewing the interior of the eye to identify and assess eye problems.

Amazing Body Facts

EYE DOMINANCE

Most people have a dominant eye. To find yours, look at a small, distant object with both eyes open. Extend both arms with both palms facing that object. Slowly bring your hands together so there is a small opening or window formed in the space between the thumbs and the index figure. Now locate your distant object in that opening with both eyes open. Now alternately close each eye to determine which eye still sees that object. This is your dominant eye. Generally speaking, right-handed people have right-dominant eyes.

11-2 To view videos on eye disorders such as cataracts, macular degeneration, and conjunctivitis, please see your DVD.

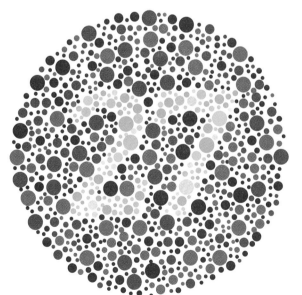

FIGURE ■ 11-3

Ishihara charts to determine color blindness. Can you see the number?

In addition, the eyes can be used to help diagnose a variety of non-visual diseases. For example, a yellow tint to the conjunctiva (jaundice) may indicate a liver disease. A neurological assessment called PERLA, which stands for *pupils equal, reactive to light and accommodation*, can be used to assess brain injury. The rapid eye movement, or REM, stage of sleep is measured during sleep studies and helps to diagnose sleep disorders.

TEST YOUR KNOWLEDGE 11-1

Choose the best answer:

1. Which of the following is *not* a layer of the eye?
   a. photo optic
   b. sclera
   c. choroid
   d. retina

2. Which layer of the eye contains the iris and the pupil?
   a. cornea
   b. sclera
   c. choroid
   d. retina

3. The _____ is the colored portion of the eye that controls the opening, or _____, where light passes through.
   a. retina, pupil
   b. pupil, iris
   c. sclera, pupil
   d. iris, pupil

Complete the following:

4. _____ is a disease caused by increased pressure in the fluid of the eye that can lead to blindness.

5. _____ is a condition in which the eye can't focus on nearby objects.

THE SENSE OF HEARING

On our journey, we hear many interesting sounds. Without our ears, we would miss all the pleasant noises, and the not-so-pleasant ones, that add to the appreciation of the journey. Also, on our journey, we may walk over some rough and uneven terrain. Our ears are also responsible for our sense of balance so we don't fall and get hurt. We can "hear" your heart pounding with anticipation, so let's explore the specific structures and functions of the ear.

Structures of the Ear

The ear is responsible for hearing and for maintaining our equilibrium, or sense of balance. We hear by receiving sound vibrations usually via the air (unless we are under water) and translating them into an interpretable sound via the vestibulo-cochlear nerve. The ear can be separated into three divisions: the **external ear,** the **middle ear** (also called the **tympanic cavity**), and the **inner ear** (also called the **labyrinth.**) See Figure 11–4 ▪ for the structures of the ear.

▱ Applied Science
📋 ✋ 🔊 🏊 🔑

SOUND CONDUCTION

Many people are under the false impression that sound travels best through air because air is the medium in which we are immersed. However, sound is transmitted due to molecular collisions, and is *better* transmitted where the molecules are closer together, such as in a liquid or solid medium. This is why whales can talk to each other up to two miles apart under water. This is also why trapped underground miners tap on the wall instead of shouting for help.

tympanic *(tihm PAN ik)*
labyrinth *(LAB ih rinth)*

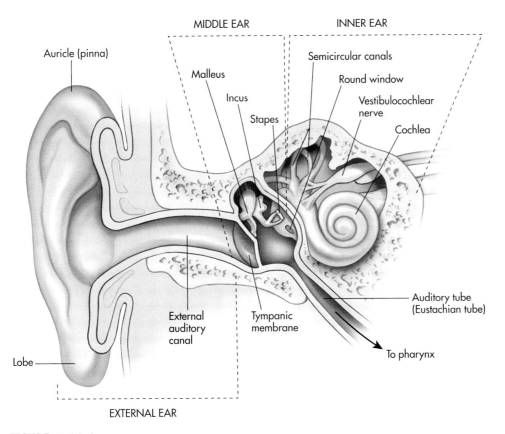

FIGURE ▪ **11-4**

Structures of the ear.

THE EXTERNAL (OUTER) EAR

The external ear is the outer projection, the part we can see. It comes in all shapes and sizes. (Remember Dumbo the Elephant?) It also includes the canal (where people put cotton swabs despite the warning label) leading into the middle ear. The projecting part is called the **pinna** or **auricle,** which collects and directs sound waves into the **auditory canal** or **external auditory meatus.** The canal contains

pinna *(PINN ah)*
auricle *(AW rih kl)*
external auditory meatus
  (AW dih tor ee mee AY tuss)

ceruminous glands
(seh ROO men us)

cerumen *(seh ROO men)*

tympanic membrane *(tihm PAN ik)*

    tympanon = *Greek word for drum*

ossicle *(AH sih kel)*

malleus *(MALL ee us)*
incus *(ING kuss)*
stapes *(STAY peez)*

Eustachian tubes
(yoo STAY she ehn)

cochlea *(KOHK lee ah)* = *Latin for snail shell*

vestibule chamber
  (VESS tih byool)

  vestibule = *a small space or cavity at the beginning of a canal.*

perilymph *(per ih LIMF)*

endolymph *(EN doe limf)*

vestibulocochlear
  (VESS tih byool o KOHK lee are)

earwax, called **cerumen,** which is secreted by the **ceruminous glands** to lubricate and protect the ear. At the end of the canal is the **eardrum,** or **tympanic membrane,** where the external ear ends and the middle ear begins. Don't you wish there were just *one* term everyone agreed upon for these structures?

THE MIDDLE EAR

The middle ear, or tympanic cavity, is basically a space in the temporal bone that contains the three smallest bones of your body. The three bones, or **ossicles,** are joined so they can amplify the sound waves the tympanic membrane receives from the external ear. (Sound travels best through a solid). Once amplified, the sound waves are transmitted to the fluid contained in the internal ear. Once again, this is another example of sound waves being transmitted more efficiently though a liquid medium than in air.

The bones of the ears are named according to their shapes. The first ossicle attached to the tympanic membrane is the **hammer,** or **malleus.** The **anvil,** or **incus,** is attached to the hammer. Finally, the **stirrup,** or **stapes,** connects to a membrane called the **oval window.** The oval window begins the inner ear and carries the amplified vibrations from the tympanic ossicles. During transmission, the sound or vibrations can be amplified as much as 22 times their original level.

Also contained within the middle ear are the **Eustachian tubes** which connect the pharynx to the middle ear. This effectively equalizes the air pressure on either side of the eardrum. The Eustachian tubes act as a passageway to allow for equalization of pressure between the outside atmospheric pressure coming through the nose and throat openings and the inner ear. This allows for equalization of pressure between the atmosphere and the middle ear. This pressure equalization allows the eardrum to vibrate freely with incoming sound waves. Sudden pressure changes, such as those caused by flying in an airplane, can affect this area. This is why, when flying, you are instructed to chew gum or swallow so the inner ear can better sense and adjust to the rapidly changing outside atmosphere via the Eustachian tubes.

11-3 The middle ear is critical for hearing and maintaining proper ear pressure. Go to your DVD to view an animation of the middle ear at work.

THE INNER EAR

The oval window membrane is the portal into the inner ear. This area comprises three separate, hollow, bony spaces that form a complex maze of winding and twisting channels. Since another name for a maze is labyrinth, this area can also be called the bony labyrinth. The three areas are the **cochlea;** the **vestibule chamber,** which houses the inner ear; and the **semicircular canals.**

The cochlea is the bony spiral or snail shell–shaped portion of the internal ear connected to the oval window membrane (see Figure 11–5 ■). The cochlea contains a membranous tube, the cochlear duct, filled with a fluid called **perilymph** that helps to transmit the sound through this area. The sound is then transmitted to the back of the cochlea, which contains another fluid called **endolymph.** Here the sound is carried to tiny hair-like receptors of the **Organ of Corti** that pick up the vibrations of the fluid and conduct the signal to the brain via the **acoustic** (auditory) or **vestibulocochlear nerve** (cranial nerve VIII). Table 11–2 lists the major structures and functions of the ear.

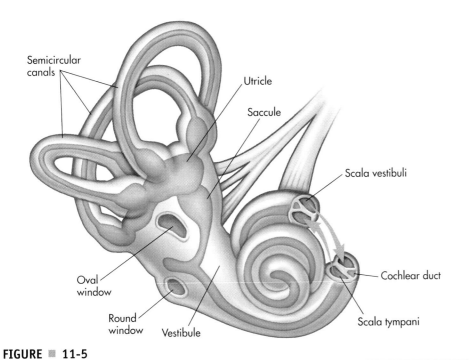

FIGURE ▧ 11-5

The internal structures of the cochlea.

TABLE 11–2 Structures and Functions of the Ear

ORGAN OR STRUCTURE	FUNCTION
External Ear	
Auricle, or pinna	Cartilaginous projection that collects and directs sound waves into the auditory canal much like a satellite dish collects transmissions from space
Auditory canal, or external auditory meatus	Canal that contains earwax, or cerumen, secreted by the ceruminous glands that lubricate and trap foreign particles
Eardrum, or tympanic membrane	Membrane that separates the external and middle ear
Middle Ear	
Ossicles	Three small bones (hammer-, anvil-, and stirrup-shaped) that help amplify and transmit sound
Eustachian tubes	Allow for equalization of external (atmospheric) and internal (within the middle ear) pressure on the tympanic membrane so the eardrum can freely vibrate with incoming sound
Inner Ear, or Labyrinth	
Cochlea	Bony, snail-shaped portion of the internal ear containing perilymph fluid, which helps to transmit sound
Semicircular canals	Three canals containing endolymph fluid, which transmits positional changes to tiny, hair-like receptors that are stimulated and conduct the signal to the brain via the acoustic or vestibulocochlear nerve (eighth cranial nerve) to help maintain balance

11-4 Body temperature can be quickly and accurately measured by the use of a tympanic thermometer. To view a video of this procedure, you can go to your DVD for this chapter.

11-5 To reinforce your knowledge of ear structure and function, go to your DVD to perform an interactive drag-and-drop exercise on ear structures and to view several 3-D animations of the ear.

Sound waves enter the external canal and vibrate the eardrum or tympanic membrane in a process called **sound conduction**. The middle ear then amplifies the sound through the respective ossicles. This process is called **bone conduction** of sound. The last ossicle (stapes) vibrates and causes a gentle pumping against the oval window membrane. This causes cochlear fluid to vibrate small hair-like nerve cells found in the **organ of Corti**. As a result of the vibrating sensory cells (hair-like nerves), a nerve impulse is sent to the temporal lobe of the brain where it is interpreted as sound, a process called **sensorineural conduction**.

Low-intensity sound waves, similar to a clock ticking, send vibrations that cause the sensory cells to move in waves that are interpreted by the brain as that "tick tock" sound. In extreme cases where intense sound waves are produced, such as from a gun blast, it is believed that the vibrations are so great that they may knock over the hair-like cells much like an earthquake knocks over tall trees. Repeated assaults can lead to permanent hearing damage. Therefore, it is a "sound" investment to wear proper hearing protection around loud noises.

The ear is also responsible for your sense of balance or equilibrium. The semicircular canals process sensory input related to equilibrium. They contain nerve endings or receptors in the form of hair cells. The semicircular canals are three loops within the inner ear that help to maintain balance. Like the cochlea, they are filled with endolymphatic fluid, and each canal duct contains a sensory receptor. This fluid moves when you change body position. The movement is picked up by the sensory receptor, which triggers a nerve impulse to travel to the brain stem and the cerebellum via the vestibulo-cochlear nerve. Here the impulse is interpreted as body position to help maintain muscle coordination and body equilibrium.

Pathology Connection: Diseases of the Ear

Deafness can be either partial or complete and is caused by a variety of conditions, ranging from inflammation and scarring of the tympanic membrane to auditory nerve and brain damage. Hearing loss may also be caused by chronic exposure to loud sounds. Conductive hearing loss occurs when sound waves are prevented from reaching the inner ear. Excess earwax and otosclerosis cause conductive hearing loss. When the nerves of the ear are damaged, sensorineural hearing loss occurs.

External otitis, commonly called "swimmer's ear," is an infection caused by bacteria and fungi. Symptoms can include pain, fever, and temporary hearing loss. It is commonly contracted from contaminated swimming pools and beaches, and can be prevented by cleaning and drying the external ear after swimming or wearing ear plugs.

Otitis media is an acute infection of the middle ear, usually caused by a bacteria or virus, and is frequently found in infants and young children because of their immature Eustacian tubes. It is commonly associated with an upper respiratory infection (URI) such as a cold. Symptoms can include pain, edema, and pus. If left untreated, otitis media can perforate the tympanic membrane (eardrum). Otitis media is first treated with antibiotics. If the otis media is recurrent and chronic in nature doctors perform a **myringotomy,** a surgical procedure during which tiny tubes are inserted through the tympanic membrane to relieve pressure. The tubes eventually fall out of the ear as the eardrum heals.

otitis media
(oh TYE tiss MEE dee ah)
ot/o = *ear*
itis = *inflammation*
media = *middle*

By examining the structure of the ear, can you see how a sinus infection (sinusitis) can spread to an ear infection and vice versa? There is a potential for invasion of the nearby honeycombed sinus area called the mastoid process. If colonized it leads to **mastoiditis.** Mastoiditis can potentially lead to brain infections.

Labyrinthitis is an inflammation of the inner ear and is usually due to infection. Labyrinthitis can cause **vertigo,** which is a feeling of dizziness or whirling in space. If you are not sure what vertigo is, watch the Alfred Hitchcock classic movie of the same name. Not only does it clearly demonstrate vertigo, but it is a great mystery movie. **Ménière's disease** is a chronic condition that affects the labyrinth (inner ear) and leads to progressive hearing loss and vertigo.

labyrinthitis *(LAB ih rinn THYE tiss)*

Ménière's disease *(MAIN ee yairs)*

Otosclerosis is a chronic, progressive middle ear disorder that is characterized by excess bone growth in the middle ear. It primarily affects the stapes, or stirrup bone, causing excess growth of spongy bone tissue. The spongy bone tissue hardens and immobilizes the bone, causing deafness. A hereditary disease that eventually affects both ears, otosclerosis is treated by stapedectomy surgery. During a **stapedectomy,** the stapes is removed and replaced with a plastic or wire substitute.

Tinnitus is a symptom of sensorineural hearing loss. It is a ringing sound in the ears, which according to superstition, means someone is talking about you. Clinically, it can occur as a result of chronic exposure to loud noises, Ménière's disease, some medications, wax build-up, or various disturbances to the auditory nerve.

tinnitus *(tinn EYE tuss)*

11-6 To view videos on the ear such as otitis media, please go to your DVD. In addition, you can view videos on an actual ear exam and acuity screening along with otoscopic exam.

TEST YOUR KNOWLEDGE 11-2

Choose the best answer:

1. The structure that marks the end of the external ear and the beginning of the middle ear is called the:

    a. pinna

    b. hammer

    c. tympanic membrane

    d. labyrinth

2. The structures of the ear that are important for balance are the:

    a. ceruminous glands

    b. incus, malleus and stapes

    c. semicircular canals

    d. Eustachian tubes

Provide the synonymous terms for each of the following:

3. auricle _____

4. earwax _____

5. malleus _____

6. anvil _____

7. stirrups _____

Complete the following:

8. _____ is a condition of ringing in the ears.

9. _____ is an inflammation of the inner ear usually due to infection and causes vertigo.

OTHER SENSES

Other senses also help us to interpret the world around us. These include the senses of taste, smell, and touch.

Taste

gustatory sense *(GUSS ta tore ee)*

The sense of taste is referred to as the **gustatory** sense. The tongue is covered with tiny bumps called papillae, shown in Figure 11– 6 ■, each of which contain several taste receptors. These taste receptors are called **taste buds** and can also be found in other parts of the mouth, such as the lips and the back of the throat. Taste buds send signals to the brain by way of three distinct cranial nerves. One nerve detects the anterior two-thirds of the tongue, a second detects the posterior portion of the tongue, and a third nerve detects the throat.

umami = *Japanese for savory or delicious*

Taste buds detect five tastes: *sweet, sour, salty, bitter,* and **umami**. Umami has been recently added because it is the distinct taste of glutamates, which cannot be duplicated by the combination of any of the other four tastes. Quite often, taste preferences change with the body's need, which is why, for example, pregnant women may crave a variety of food throughout their pregnancy. The refinement of food taste is primarily dependent on the sense of smell.

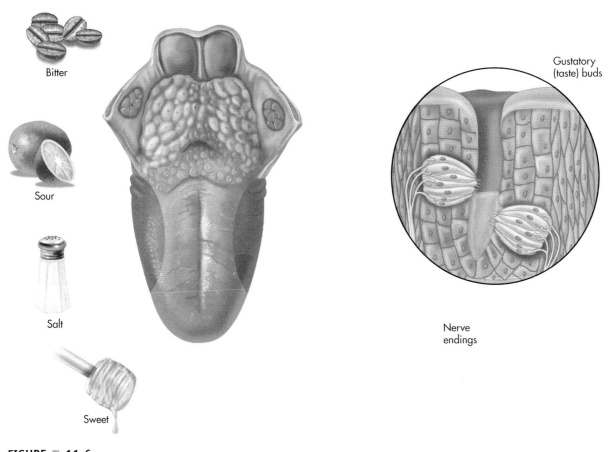

FIGURE ▪ 11-6

The sense of taste.

Smell

The sense of smell arises from the receptors located in the olfactory region or the upper part of the nasal cavity (see Figure 11– 7 ▪). We "sniff" in order to bring smells up to this area, where they can be interpreted by tissues in the nose—olfactory epithelium—which contain specialized nerve cell receptors. Remember that taste and smell are closely related—and in fact smell accounts for 90% of taste—which is why we can't taste foods when we have a severe head cold. Pleasant food odors also initiate digestive enzymes, so when you smell that cinnamon apple pie baking, your mouth really may water in anticipation.

Smell is also closely linked to memory. Have you ever smelled cookies or cinnamon and automatically thought about a special holiday or family celebration? Smell has been shown to trigger memories. This may be useful someday with the treatment of dementia type diseases.

Viral and bacterial infections, allergies, harsh odors, and the use of certain illegal drugs, such as cocaine, all contribute to a condition known as **rhinitis**. This condition causes inflammation of the mucous membranes that line the nasal passage and is characterized by congestion and drainage. Rhinitis develops in response to the body's release of **histamine**, a molecule released during an immune response. Rhinitis treatment includes removing the irritant causing the reaction and the use of antihistamine medications. In cases of chronic rhinitis, surgery may be performed to clear nasal obstructions.

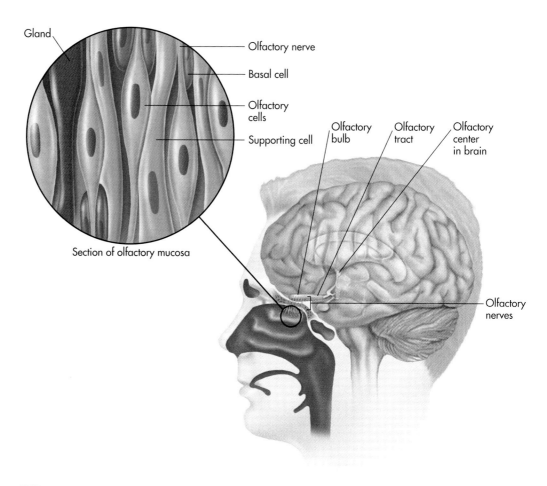

Gland
Olfactory nerve
Basal cell
Olfactory cells
Supporting cell

Olfactory bulb
Olfactory tract
Olfactory center in brain

Section of olfactory mucosa

Olfactory nerves

FIGURE ■ 11-7

The sense of smell.

Touch

tactile corpuscles
(TAK tle KOR puss els)

Touch receptors are small, rounded bodies called **tactile corpuscles** located in skin and especially concentrated in the fingertips. They are also located on the tip of the tongue. It is interesting to note that even when a patient is anesthetized (and awake), there are no pain sensations, but the patient is still conscious of pressure through these deep touch sensors.

Temperature sensors are also found in the skin. The body has separate heat and cold receptors. These receptors may cause an interesting phenomenon called **adaptation** to occur. Continued sensory stimulation causes the sensors to desensitize or adapt. For example, if you are continually exposed to the cold, the receptors adjust so you won't feel so cold. That is why an outside temperature of 50 degrees seems warm after a long cold winter, but that same temperature seems cold after a long hot summer. Another example is when you first enter a hot bath, it may feel extremely hot. After a few seconds, it doesn't feel so hot, yet the actual temperature of the water has not yet changed.

Pain is a very important protective sense. It is the body's way of drawing attention to a particular danger, such as the "hitting your finger with a hammer" example used in the nervous system chapter. Pain is the most widely distributed sense, found in skin, muscle, joints, and internal organs. Pain receptors are merely branches of nerve fibers called **nociceptors (free nerve endings).**

There are even different types of pain. **Referred pain** originates in an internal organ yet is felt in another region of the skin, because internal organs are not mapped on your somatic sensory cortex. For example, liver and gallbladder disease often cause pain in the right shoulder. See Figure 11-8 ▪ to view various sites of referred pain. **Phantom pain** can result from an amputated limb: an individual can feel pain in an arm or leg he or she no longer has.

HEAT AND COLD THERAPY

Heat and cold therapy are used for a variety of injuries. These therapies rely on physiological changes in the body in response to either heat or cold. For example, heat relaxes muscles and dilates blood vessels, thereby bringing more blood flow to the site of injury. Cold therapy constricts blood vessels and minimizes the amount of blood and swelling at a site.

11-7 To view a video on heat and cold therapy procedures, please go to your DVD.

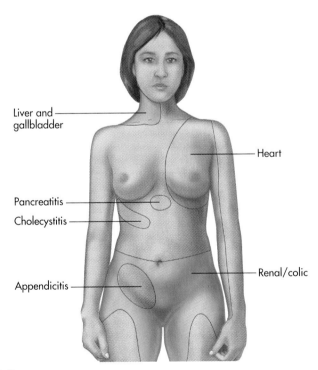

FIGURE ▪ **11-8**

Various sites of referred pain.

Pain receptors do not adapt as heat and cold receptors do. Pain is felt for as long as the stimulus is there that is causing it or unless a person is under **anesthesia.** An interesting debate is whether or not some people have higher or lower thresholds of pain. See Figure 11–9 ■ for the senses of touch.

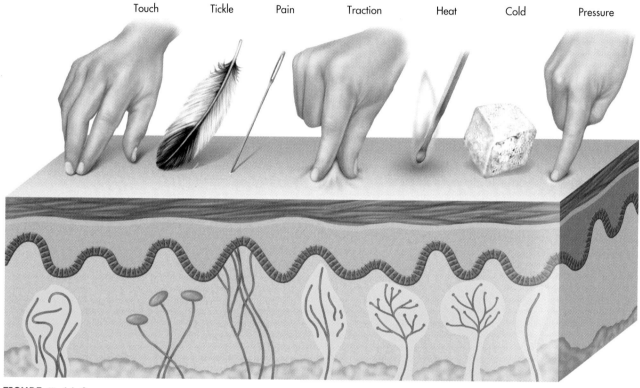

Touch   Tickle   Pain   Traction   Heat   Cold   Pressure

FIGURE ■ 11-9

The sense of touch.

Proprioception is your body orientation sense that allows you for example to locate a body part with your eyes closed or know your arm is raised over your head without seeing it. Proprioreceptors are found in muscles, tendons, joints, and the inner ear, where they help with equilibrium. Motion sickness such as car sickness or sea sickness occurs in response to excessive stimulation of the equilibrium receptors in the ear.

11-8 Videos on perception of pain and selected pain scales are available on your DVD.

A Quick Trip Through Some Diseases of the Eye

DISEASE	ETIOLOGY	SIGNS AND SYMPTOMS	DIAGNOSTIC TEST(S)	TREATMENT
Astigmatism	Alteration in the shape of the cornea in which it becomes more oblong or football shaped than spherical. Usually hereditary but can result from eye injury.	Blurred vision, eye strain and headaches.	Eye exam to view shape of cornea.	Unless extreme, it can be treated with eyeglasses or contact lenses. If extreme, refractive eye surgery.
Amblyopia (lazy eye)	Eye trauma or strong refractive error (nearsightedness or farsightedness); generally develops in young children.	Squinting or completely closing one eye to see, poor vision, eyestrain and headaches.	History and eye exam.	Patching stronger eye to force weaker eye to function better, atropine eye drops, correcting refractive problems or surgery.
Blepharitis	Inflammation of eyelids, usually caused by bacterial infection.	Eye irritation, burning, tearing, dryness and foreign body irritation.	History and eye exam.	Warm compresses, cleansing the eyes, antibiotics, artificial tears, and sometimes steroids.
Cataracts	Clouding of the lens, usually caused by protein clumping together as we age. Cause unknown but excessive UV light exposure suspected. Diabetics are at higher risk. Frequently occurs in people over 70.	Painful, gradual blurring and loss of vision.	Eye exam.	In early stages, glasses or magnification aids may help. Surgical correction of the lens is indicated when vision is seriously impaired.
Color blindness	Not really blindness at all but a genetic disorder that causes a deficiency in the way we see color. Red-green color deficiency is most common form. Caused by retinal cells inability to respond to certain colors. Usually hereditary via the mother, and more common in males. Aging and disease can also cause.	Difficulty distinguishing red from green or blue from yellow.	Ishihara plates (see Figure 11-3).	No cure but needs to be recognized early to prevent developmental issues.

continues

A Quick Trip Through Some Diseases of the Eye (*Continued*)

DISEASE	ETIOLOGY	SIGNS AND SYMPTOMS	DIAGNOSTIC TEST(S)	TREATMENT
Conjunctivitis (pink eye)	Inflammation of the conjunctiva membrane caused by bacteria, viral or allergies.	Pinkish eye, discharge, excessive eye watering, pain, and itching.	History, eye exam and culture and sensitivity.	Avoidance of cause and infection precautions. Warm compresses. Antibiotic eye drops for bacterial infections and antihistamines for allergic forms
Diabetic Retinopathy	High blood sugar, damages the blood vessels in the eyes. Later stages can have new blood vessel growth over the retina, which can cause scar tissue and lead to retinal detachment. If untreated can cause blindness.	Floaters, double vision, fluid leakage, and swelling.	Fluorescein angiography. A contrast dye is injected and the blood flow to the retina is then assessed.	Laser photocoagulation to seal leaking blood vessels and destroy new growth. Some drugs helpful in early stages. May be prevented by good blood sugar control.
Dry eye syndrome	Chronic lack of eye lubrication and moisture due to lack of tear production. Can be part of aging process, side effect of medications, climate related (dry, dusty or windy), long term contact use, lack of blinking or certain disease state. More common in women; smoking increases risk.	Dryness, scratchiness and burning.	History and eye exam.	Avoid irritating cause. When prevention or cure not possible, artificial tears (lubricating eye drops) can be used to alleviate symptoms.

Clinical Application

CATARACT SURGERY

Cataract surgery is the most frequently performed surgery in the United States with a very high success rate. The clouded lens is removed and replaced with a clear plastic intraocular lens (IOL).

Figure 11-10 ■ shows some common eye disorders.

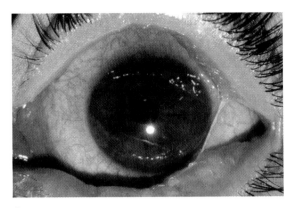

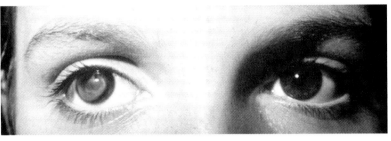

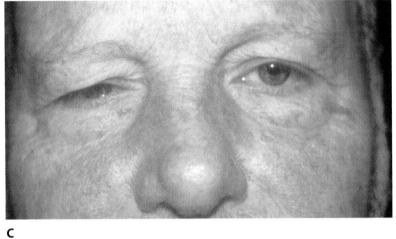

FIGURE ■ 11-10

Some common eye disorders. **A.** conjunctivitis ("pink eye") (Source: Buddy Crofton/Medical Images, Inc. **B.** Cataract of right eye. **C.** blepharitis.

A Quick Trip Through Some Diseases of the Ears

DISEASE	ETIOLOGY	SIGNS AND SYMPTOMS	DIAGNOSTIC TEST(S)	TREATMENT
External otitis, (Swimmer's ear)	Commonly caused by infection from water contaminated by bacteria or fungi.	Pain, fever and temporary hearing loss.	Visual ear exam.	Prevention by cleaning and drying external ear after swimming or by wearing ear plugs.
Otitis media	Acute infection of the middle ear, usually caused by a bacteria or virus. Commonly associated with an upper respiratory infection, or URI, such as a cold.	Pain, edema, pus. If left untreated can perforate the tympanic membrane. There is potential for invasion of nearby mastoid process, leading to mastoiditis.	Otoscopic examination.	Treatment of underlying infection. Use of antibiotics if identified as bacterial infection.

continues

A Quick Trip Through Some Diseases of the Ears (*Continued*)

DISEASE	ETIOLOGY	SIGNS AND SYMPTOMS	DIAGNOSTIC TEST(S)	TREATMENT
Labyrinthitis	Inflammation of the inner ear usually caused by infection.	Vertigo; feeling of dizziness or whirling in space.	Patient history, radiologic studies.	Antivert class drugs, and in many cases Benadryl has been successful.
Ménière's disease	Chronic condition that affects the labyrinth.	Progressive hearing loss tinnitus and vertigo.	Patient history and exam; audiologic and radiologic exams.	Diuretics to "dry" out the labyrinth, decrease caffeine consumption.
tinnitus	Can occur as a result of chronic exposure to loud noises, Ménière's disease, some medications, wax build-up or various disturbances to the auditory nerve.	Ringing in ears.	History.	Eliminate loud noises.

℞ Pharmacology Corner

There are times when the pupils of the eye should be dilated for an eye examination to allow for better visualization of the inner parts of the eye. The class of pupil dilating drugs is called **mydriatic agents**; atropine is an example. Patients are advised to wear sunglasses after an eye exam because the pupil cannot react to bright light until the drug wears off. On the other hand, certain drugs such as narcotics cause the pupils to constrict, sometimes so severely that the pupils look like pin points. The class of pupil constricting agents are called **miotic agents**.

Topical antibiotics (liquid drop and salve form) can be used for eye infections.

Ear medications can also consist of topically applied antibiotics applied for infections. In addition, topical anesthetics can be used for severe ear pain such as results from swimmer's ear. Antivert is a type of medication that affects the neural pathways that originate in the labyrinth and as a result inhibits nausea and vomiting caused by motion sickness.

 11-9 Drugs can be given through the eyes as eye drops or ophthalmic medication. To see a video on this topic and procedure, please go to your DVD. Otic drops for the ear and both eye and ear irrigation videos are also available.

SUMMARY

Snapshots from the Journey

→ The senses of sight (eyes), sound and equilibrium (ears), taste (tongue), and smell (nose) are called special senses. The body feels other sensations, such as touch, heat, cold, and pain, which are called general senses.

→ The eye is very similar to a camera, with lens cover (eyelids), opening (pupils), shutter (iris), lens (eye lens), and photoreceptive film (retina).

→ Light rays enter the eye and pass through the conjunctival membrane, cornea, aqueous humor, pupil, lens, and vitreous humor, and are focused on the retina. The photoreceptors in the retina cause a chemical impulse to be sent to the optic nerve, which carries it to the brain for the interpretation we call vision.

→ The ear has three major divisions: the external, middle, and inner ear. The ear is the organ for hearing and maintaining our sense of balance.

→ Sound waves enter the external canal and vibrate the eardrum, or tympanic membrane. The middle ear then amplifies the sound through the respective tiny bones, or ossicles. The last ossicle (stapes) vibrates and causes a gentle pumping against the oval window membrane. This causes cochlear fluid (perilymph) to move and vibrates tiny, hair-like neurons, which transmit an impulse to the hearing centers in the brain where the sound is interpreted.

→ The semicircular canals are responsible for maintaining body balance.

→ Our sense of taste, or gustatory sense, has traditionally been thought to consist of sweet, sour, salty, and bitter. A fifth taste, umami, has recently been distinguished as its own category. The sense of taste originates on taste buds on the tongue and is closely associated with the sense of smell.

→ The sense of smell arises from the olfactory region of the nose.

→ The sense of touch allows perceptions of pain, temperature, pressure, traction, and the sensation of being "tickled."

MARIA'S STORY

Maria has been going through a period of high stress and has been poorly monitoring her diabetes. She has noticed increasing visual problems and often sees specks floating in her visual field. Twice she had double vision episodes.

a. What should Maria's action plan include to get her back on track?

_____

_____

b. What could be the cause of the visual disturbances and what should she do about them?

_____

_____

c. What structures of the eye are affected and why?

_____

_____

d. Ultimately what eye disease(s) can Maria expect if she does not control her diabetes?

_____

_____

REVIEW QUESTIONS

Multiple Choice

1. The part of the eye that allows for varying amounts of light onto the retina is the:
   a. lens
   b. humor
   c. iris
   d. optic nerve

2. The photopigment structures responsible for the ability to see colors are:
   a. cones
   b. rods
   c. iris
   d. pupil

3. The incus is found in the:
   a. inner ear
   b. middle ear
   c. external ear
   d. region of South America

4. What is the correct descending order for the media through which sound travels, with the most efficient conductor of sound listed first:
   a. liquid, air, solid
   b. solid, liquid, air
   c. air, liquid, solid
   d. they are all equal

5. Another word for the sense of taste is:
   a. olfactory
   b. vertigo
   c. mastication
   d. gustatory

6. This highly contagious disease is also known as pink eye:
   a. conjunctivitis
   b. myopia
   c. glaucoma
   d. contactitis

Fill in the Blank

1. The two functions of the auditory system are _____ and _____.

2. The three ossicles of the ear are the _____, _____, and _____.

3. _____ is a chronic disease that affects the middle ear and includes progressive hearing loss, tinnitus, and vertigo.

Short Answer

1. Differentiate between special and general senses.

   _____

   _____

2. What are the five basic tastes?

   _____

   _____

3. Define *adaptation* in relation to temperature sensors.

   _____

   _____

4. How does the body protect the eyes?

   _____

   _____

5. How is an eye like a camera?

   _____

   _____

Suggested Activities

1. Ask a local ophthalmologist to visit your class and explain the latest concepts and procedures in corrective vision surgery. In addition, have him or her bring an eye chart and explain 20/20 vision.

2. The sense of sight is also important to the interpretation of our sense of taste. Blindfold participants and have them identify similar products with varying tastes, such as different types of sodas or jelly beans, and see if they can correctly identify the flavor.

3. Hold this page about 15 inches from you face. Close your left eye and stare at the black dot while bringing the page slowly closer to your face. While staring at the dot, you will notice the X in the periphery disappear. You have just discovered your blind spot.

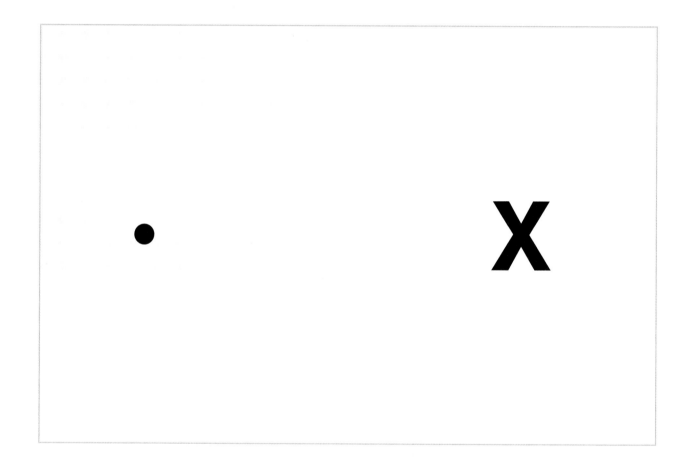

www.prenhall.com/colbert
To learn more about the exciting profession of audiology, please visit the book's companion website.

11-10 Now that you have completed your journey through this chapter, please go to the DVD for interactive games and puzzles concerning the medical terms and concepts contained in this chapter. By playing the games you will reinforce your learning of the senses in a fun way.

T H E
CARDIOVASCULAR
System
Transport and Supply

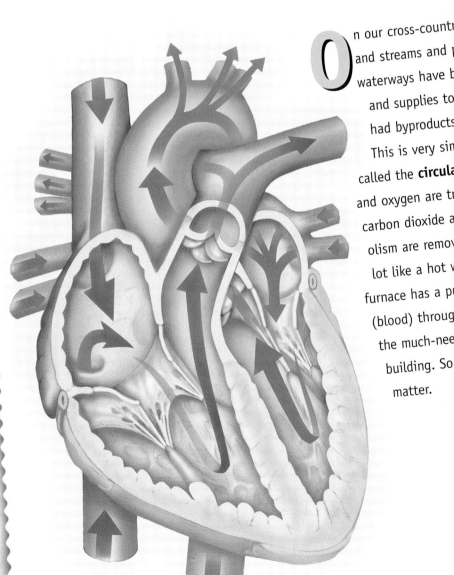

On our cross-country journey, we will see a series of rivers and streams and perhaps some canals. Historically, these waterways have been used as a way of transporting food and supplies to people. These very same waterways also had byproducts of industry dumped back into them. This is very similar to the **cardiovascular system**, also called the **circulatory system**, through which nutrients and oxygen are transported to the cells in the body and carbon dioxide and other waste products of cells' metabolism are removed. The cardiovascular system is also a lot like a hot water heating system in a house. The furnace has a pump (heart) to circulate the hot water (blood) through the piping system (vessels) to deliver the much-needed heat to every room throughout the building. So let's get right to the "heart" of the matter.

Chapter 12

LEARNING OBJECTIVES

Upon completion of your journey through this chapter, you will be able to:

→ Identify structures and functions of the cardiovascular system

→ Trace the blood flow through the vessels and chambers of the heart

→ Explain the coronary circulation

→ Describe the contraction of the heart and the conduction system

→ Differentiate between arteries, veins, and capillaries

→ List the major components of blood and their functions

→ Discuss the importance of blood typing

→ Explain the process of blood clotting

→ Describe various cardiovascular diseases

MULTIMEDIA APPLICATIONS

DVD Interactive Exercises

→ Animation of the chambers of the heart, interactive drag-and-drop exercises labeling various parts of the heart, 12-1

→ Animation of the heart contraction and related blood flow, 12-2

→ Animations and videos on heart attacks and angina, 12-3

→ Videos and animations of CAD, coronary heart disease, and heart catheterization, 12-4

→ Animations and videos on performing ECGs/EKGs, electrode placement, and dysrhythmias, 12-5

→ Videos on Automated External Defibrillator (AED), defibrillation, and cardioversion, 12-6

→ Drag-and-drop exercise on blood types and a video on administering blood, 12-7

→ Videos and animations on sickle cell anemia and leukemia, 12-8

→ 3-D views of the heart and blood vessels of the body and interactive exercise on labeling the circulatory system, 12-9

→ Videos on peripheral artery assessment and carotid artery assessment, 12-10

→ Animations and videos on congenital heart disease and aging of the cardiovascular system, 12-11

→ Animation and videos on shock, bleeding control, administration of digoxin, dopamine, and nitroglycerin, and starting an IV line, 12-12

→ Interactive puzzles and games, 12-13

www.prenhall.com/colbert

→ Professional Profile:
  • Cardiovascular Technology

→ Related Internet Links

→ Additional Review Questions

Pronunciation Guide

Correct pronunciation is important in any journey so that you and others are completely understood. Here is a "see and say" Pronunciation Guide for the more difficult terms to pronounce in this chapter.

agglutinate (ah GLUE tin ate)

albumin (AL byoo men)

anastomosis (ah NASS te MOE sis)

anemia (ah NEE mee ah)

aneurysm (AN yoo rizm)

arterioles (ahr TEE ree ohls)

arteriosclerosis
　(ar TEE ree oh skleh ROH sis)

atherosclerosis
　(ATH er oh skleh ROH sis)

atrioventricular
　(AY tree oh vehn TRIK yoo lahr)

atrium; atria (AY tree um; AY tree ah)

autorhythmicity
　(aw toe rith MIH sih tee)

basophils (BAY soh fills)

bundle of His (HISS)

diastole (dye ASS toe lee)

embolus (EM boh luss)

endocardium (ehn doh KAR dee um)

eosinophils (EE oh SIN oh fillz)

erythrocytes (eh RITH roh sights)

fibrinogen (fye BRINN oh jenn)

hemophilia (HEE moh FILL ee ah)

hemopoiesis (HEE moh poy EE sus)

hemostasis (HEE moh STAY siss)

homeostasis (HOE mee oh STAY siss)

infarct (in FARKT)

inotropism (EYE no TROPE izm)

ischemia (iss KEE mee ah)

leukocytes (LOO koh sights)

lysosomes (LIE so soams)

neutrophils (NOO trow fillz)

phagocytosis (fag oh sigh TOH siss)

polycythemia
　(pall ee sigh THEE mee ah)

prothrombin (pro THROM bin)

sinoatrial (sigh noh AT tree al)

systole (SISS toh lee)

septum (SEHP tum)

thrombocytes (THROM boh sights)

thrombocytopenia
　(THROM boh sigh toh PEE nee ah)

thrombus (THROM buss)

tunica externa (TOO nik ah ex TERN ah)

tunica interna (TOO nik ah in TERN ah)

tunica media (TOO nik ah mee DEE ah)

venules (VEHN yules)

SYSTEM OVERVIEW

cardi/o = heart
vascul/o = vessels

The major components of the **cardiovascular** system include the **heart** (which is the organ that pumps blood through the system), **blood** (a form of connective tissue that has a fluid component called **plasma** and a variety of cells and substances), and **blood vessels** (a network of passageways to transport the blood to and from the body's cells).

Blood vessels can be further classified. Vessels that carry blood away from the heart are called **arteries**. These main vessels branch out into ever smaller vessels called **arterioles,** which eventually become **capillaries**. Capillaries are where the exchange of nutrients, gases, and waste products occurs at the cellular level. Capillaries are also the transition vessels where blood begins its trip back to the heart through ever-merging vessels that form larger **veins**. The tiniest of these veins are called **venules**. To view this transitional region along with the cardiovascular system, see Figure 12–1 ■.

arterioles
　(ahr TEE ree ohls) = small arteries

venules *(VEHN yules) = small veins*

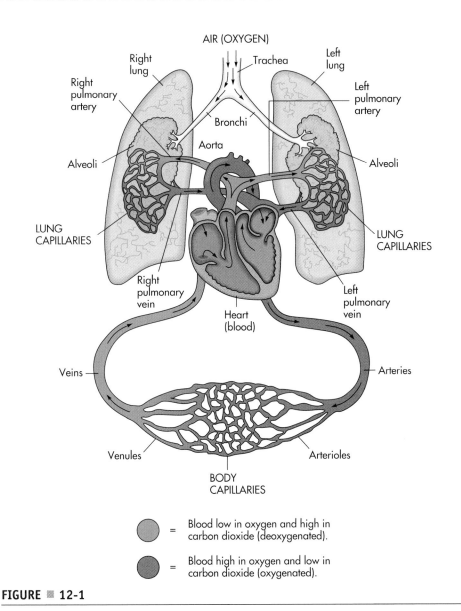

FIGURE ▧ 12-1

Overview of the cardiovascular system.

    The cardiovascular system includes the heart, blood, and blood vessels. The circulatory system includes the cardiovascular system and the lymphatic system, which will be further covered in the immune system chapter. The cardiovascular system has four main functions. First, it functions to pump materials throughout the body. Second, it is designed specifically to transport blood to various parts of the body. Third, this movement of materials facilitates the delivery of vital materials, such as oxygen and nutrients, to the cells of the body and the removal of cellular wastes. Finally, the circulatory system functions to return excess tissue fluid, often in the form of lymph, to general circulation.

In general, veins differ from arteries not only because veins bring blood back to the heart but because the blood now is *deoxygenated* (contains less than the normal arterial amount of oxygen) and has a higher level of carbon dioxide and other waste products of cellular metabolism. Veins also have thinner walls than arteries, are more numerous, and have a larger capacity to hold blood. Arteries also transport blood under higher pressure than do veins, because the blood is being forcefully pushed out of the heart via arteries. In contrast, with veins blood is passively returning to the heart, driven mainly by the pressure of the blood circulation.

INCREDIBLE PUMPS: THE HEART

Let's begin our journey with the main organ of the cardiovascular system: the heart. While it is often describe as the "pump" of the cardiovascular system, you will soon see that it is actually *two* pumps working together.

General Structure and Function

The heart is a specially shaped muscle, about the size of your fist, containing a series of chambers that move blood throughout the body. The heart is surrounded by a serous membrane called the **pericardium**. The outer layer is a tough fibrous pericardium. Inside the fibrous pericardium is the **parietal layer**. The **visceral layer** is fused to the heart surface and there is a potential cavity between the layers called the **pericardial cavity**. The outer layer of the heart wall, the visceral pericardium, is also known as the **epicardium**. The middle layer of the heart wall, the **myocardium**, is made of cardiac muscle. The heart is lined by epithelium, called the **endocardium**. See Figure 12–2 ■.

The heart is located in the thoracic cavity, slightly left of the center of your chest and above your diaphragm. As strange as it may initially appear, the *base* of the heart is proximal to (closer to) your head, while the *apex* of the heart is distal (further away from it).

epicardium *(ep ih KAR dee um)*
myocardium *(my oh KAR dee um)*
endocardium
  (EHN doh KAR dee um)

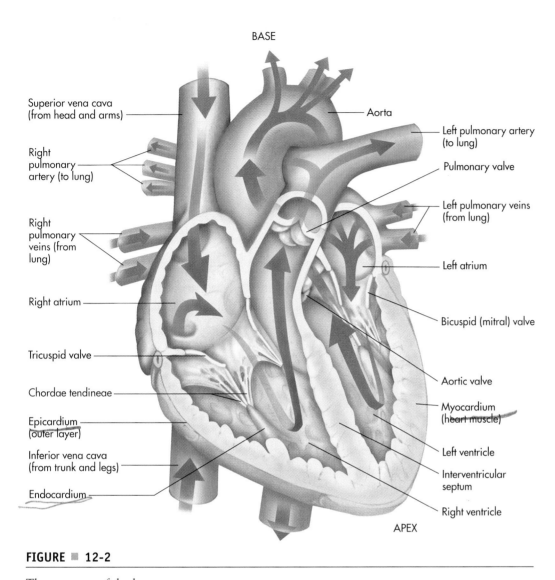

BASE

Superior vena cava (from head and arms)

Right pulmonary artery (to lung)

Right pulmonary veins (from lung)

Right atrium

Tricuspid valve

Chordae tendineae

Epicardium (outer layer)

Inferior vena cava (from trunk and legs)

Endocardium

Aorta

Left pulmonary artery (to lung)

Pulmonary valve

Left pulmonary veins (from lung)

Left atrium

Bicuspid (mitral) valve

Aortic valve

Myocardium (heart muscle)

Left ventricle

Interventricular septum

Right ventricle

APEX

FIGURE ■ 12-2

The anatomy of the heart.

Although the heart is a single organ, it is easier to understand its function if you think of it as two separate pumps working together. The right side of the heart is responsible for collecting blood from the body and sending it to the lungs to pick up oxygen and get rid of carbon dioxide. The left side of the heart collects blood from the lungs and pumps it through the body. Before we explore the heart, it's important to review that blood returning to the heart travels through **veins** (thus, venous blood) and blood traveling away from the heart travels through **arteries** (thus, arterial blood). You have probably already heard of some veins and arteries. For example, the carotid arteries are the large arteries on each side of the neck that deliver blood from the heart to the head. The jugular veins are large veins that take blood from the head to the heart.

Inside the heart are four chambers. The chambers of the right side of the heart are separated from the chambers of the left side, so there is normally no mixing of blood from one side to the other. The wall that separates the two

septum *(SEHP tum)*
  septum = *wall*
atrium *(AY tree um)*
atrioventricular
  (AY tree oh vehn TRIK yoo lahr)
  tri = *three*

smaller chambers is called the **interatrial septum;** the wall between the two larger chambers is called the **interventricular septum.** When looking at Figure 12–2, you will notice the small chamber in the upper left quadrant of the picture. That is the **right atrium.** (Remember, locations are based on the *patient's* perspective). This is a collecting chamber where blood is returned to the heart after its trip through the body. The two large veins that bring the blood to the right atrium are the **superior vena cava** (blood from the head, neck, chest, and upper extremities) and **inferior vena cava** (blood from the trunk, organs, abdomen, pelvic region, and lower extremities). Once the blood is collected, it drains through a one-way valve to a larger chamber in the right side called the **right ventricle.** That one-way valve is called the right **atrioventricular valve,** or **AV valve.** This is also called a **tricuspid** valve because the valve is formed with three cusps. The opening and closing of these valves produces the "lubb dupp" sound characteristic of the heart.

 12-1 To view an animation on the chambers of the heart, and perform interactive drag-and-drop exercises on labeling the various parts of the heart, please go to your DVD for this chapter.

Cardiac Cycle

systole *(siss TOH lee)*
diastole *(dye ASS toe lee)*

The movements of the heart, called the **cardiac cycle,** can be divided into two phases called **systole** and **diastole.** Usually when discussing heart movement we refer to ventricle activity.

When the right ventricle is full of blood, the heart contracts (systole). Because valves are one-way, as the right ventricular pressure increases, the tricuspid valve shuts so blood doesn't squirt back into the right atrium. As the pressure increases, the blood has to go somewhere. The only way for the blood to travel is through the **pulmonary semilunar valve** to the pulmonary trunk, which divides into the left and right **pulmonary arteries.**

Cardiopulmonary circulation is the process of circulating blood through the lungs, picking up oxygen from the lungs and depositing carbon dioxide there to be released from the body during an exhalation. Each pulmonary artery goes to its respective lung and branches down into ever-smaller vessels to the point where they become capillaries that form a network around each air sac in the lungs. This is where the blood gives up carbon dioxide, one of the waste products of metabolism by cells, and picks up a fresh supply of oxygen from the lungs. These capillaries containing freshly oxygenated blood converge into increasingly larger vessels until they form the left and right pulmonary veins. Figure 12–3 ■ illustrates the blood flow through the heart.

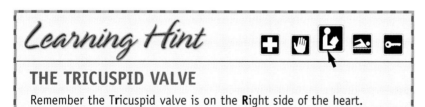

Learning Hint

THE TRICUSPID VALVE

Remember the Tricuspid valve is on the **R**ight side of the heart.

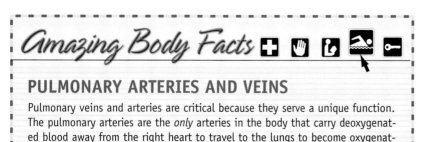

Amazing Body Facts

PULMONARY ARTERIES AND VEINS

Pulmonary veins and arteries are critical because they serve a unique function. The pulmonary arteries are the *only* arteries in the body that carry deoxygenated blood away from the right heart to travel to the lungs to become oxygenated. The pulmonary vein collects this oxygenated blood and returns it to the left side of the heart and therefore is the *only* vein to carry oxygenated blood.

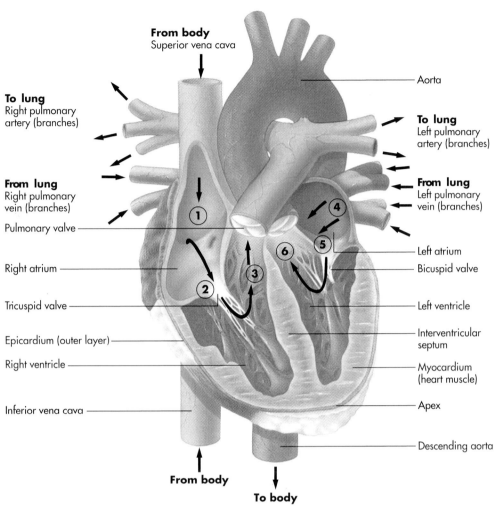

From body
Superior vena cava

To lung
Right pulmonary
artery (branches)

From lung
Right pulmonary
vein (branches)

Pulmonary valve

Right atrium

Tricuspid valve

Epicardium (outer layer)

Right ventricle

Inferior vena cava

Aorta

To lung
Left pulmonary
artery (branches)

From lung
Left pulmonary
vein (branches)

Left atrium

Bicuspid valve

Left ventricle

Interventricular
septum

Myocardium
(heart muscle)

Apex

Descending aorta

From body

To body

RIGHT HEART PUMP

1. Deoxygenated blood returns from the upper and lower body to fill the right atrium of the heart creating a pressure against the atrioventricular (AV) or tricuspid valve.

2. This pressure of the returning blood forces the AV valve open and begins filling the ventricle. The final filling of the ventricle is achieved by the contracting of the right atrium.

3. The right ventricle contracts increasing the internal pressure. This pressure closes the tricuspid valve and forces open the pulmonary valve thus sending blood toward the lung via the pulmonary artery. This blood will become oxygenated as it travels through the capillary beds of the lung and then return to the left side of the heart.

LEFT HEART PUMP

4. Oxygenated blood returns from the lung via the pulmonary vein and fills the left atrium creating a pressure against the bicuspid valve.

5. This pressure of returning blood forces the bicuspid valve open and begins filling the left ventricle. The final filling of the left ventricle is achieved by the contracting of the left atrium.

6. The left ventricle contracts increasing internal pressure. This pressure closes the bicuspid valve and forces open the aortic valve causing oxygenated blood to flow through the aorta to deliver oxygen throughout the body.

FIGURE ■ **12-3**

The functioning of the heart valves and blood flow.

Amazing Body Facts

HOW DOES IT KEEP GOING AND GOING AND GOING?

One of the amazing things about your heart is that it continues to beat day after day without your even thinking about it. Here is a neat experiment. Previously, we said that your heart is about the size of your fist. Let's pretend that your fist is your heart. To mimic the pumping action of your heart, open your fist with your fingers fully extended. Now make a tight fist. Continue this action of fully opening and tightly closing your hand for the next *60 seconds*. Chances are good that your hand will feel like it's ready to fall off of your arm! If this is how your hand feels after 60 seconds, how does your heart constantly beat approximately 100,000 times and move approximately 1,800 gallons of blood each day for decades and decades? Think back to Chapter 7 on muscles. Cardiac muscle cells have specialized connections called **intercalated discs**. These connections, along with associated pores, provide an efficient connection with adjacent muscle cells so electrical impulses, ions, and various small molecules can readily travel throughout the heart, allowing a smooth contraction from one area of the heart to another. The vasculature of the heart takes approximately 5 percent of oxygenated blood from each heartbeat to ensure there is a blood-rich environment so plenty of oxygen and nutrients are available.

The pulmonary veins meet and pour their contents into the **left atrium**. Once the left atrium is filled, blood flows through the left AV valve and into the **left ventricle**. When the left ventricle is full, the heart contracts again. The ventricular pressure increases, forcing the left AV valve shut and ejecting the blood out of the left ventricle through the aortic semilunar valve to the ascending aorta, sending it on its way throughout the body. Two more familiar names for the left AV valve are **bicuspid** (formed by only two leaves or cusps) or **mitral** valves. Now the ventricles and atria can rest (**diastole**) as the atria fill with blood before they squeeze another load of blood into the ventricles. It is important to note that the chambers of the heart fill during diastole, or the relaxation phase, and eject blood during systole, or the contraction phase.

Heart rate can be measured by counting the **pulse**, which is the rhythmic expansion and contraction of arteries due to the opening and closing of the aortic semilunar valve. Pulse can be measured in several places on the body:

- Brachial arteries: Pulse is checked by placing fingers on the anterior side of the elbow, where the arm flexes toward the body.
- Carotid arteries: Pulse is checked by placing fingers on the large arteries on either side of the neck.
- Femoral arteries: Pulse is taken by placing fingers on the anterior midline of the leg where the leg flexes in relation to the trunk.
- Pedal arteries: Pulse is taken by placing fingers on arteries on top of the foot.
- Popliteal arteries: Pulse is taken by placing fingers on arteries in the posterior knee.
- Radial pulse: Pulse is taken by placing fingers on arteries in the anterior wrist on the thumb side.

Some points to remember are that:

- Both atria fill at the same time.
- Both ventricles fill at the same time.
- Both ventricles eject blood at the same time when the heart contracts.

Do you ever get yelled at or get mad at the *other* person for squeezing the toothpaste tube in the middle? What's the big deal? When you eat a freeze pop, do you squeeze it in the middle or from the bottom and work the contents up to your mouth? These two examples are important to visualize because the heart has to contract in a certain way to make sure that all of the blood is squeezed out during each contraction. In order to do this, the contraction begins at the apex and travels upward. As you trace the flow of blood in Figure 12–3, you will see how efficient this is. You should also notice that there is a semilunar valve located at the exit of both ventricles. Since your circulatory system is a pressurized system, these valves are necessary to prevent any ejected blood from leaking back in.

If you further examine the heart illustration, you will notice that the walls of the atria are thinner than the ventricular walls. This is because higher pressures are generated in the ventricles to move blood. You should also note that the walls of the left ventricle are thicker than the walls of the right ventricle. When you think about it, this makes total sense. The right ventricle has to pump blood only a short distance through the vasculature of the lungs and back to the heart. This is called the **pulmonary circulation**. The left ventricle, on the other hand, has to pump all of the blood throughout the body and back to the right atrium! This is referred to as the **systemic circulation**. The resistance of all those blood vessels in the body (systemic) is six times greater than the resistance of the pulmonary vasculature.

In the Amazing Body Facts box, you learned that the heart muscles have a blood-rich environment. In Figure 12–4 ▪, you can see that a portion of the newly oxygen-enriched blood is diverted from the aorta by the right and left **coronary arteries.** These arteries continuously divide into smaller branches, forming a web of interconnections known as **anastomoses,** which enable the heart muscle to constantly and fairly consistently receive a rich supply of blood. It is interesting to note that regular aerobic exercise can increase the density of these blood vessels that supply the heart. The number of anastomoses also increases, as does their number of locations. This is important for people with blockage of a small coronary artery. This may lead to an increased survival rate during a heart attack because blood may have alternate routes to travel, which could help to prevent heart muscle damage.

The right coronary artery provides blood for the right ventricle, posterior portion of the interventricular septum, and inferior parts of the heart. The left coronary artery provides blood to the left lateral and anterior walls of the left ventricle and to portions of the right ventricle and interventricular septum.

anastomosis (ah NASS te MOE sis)

Clinical Application

ANGIOGRAPHY

One way to get a good picture of the blood flow in a person's body (and any problems with it) is to perform an **angiogram**. An angiogram is an X-ray of the body after a **radiocontrast agent** (a dye that makes the blood contrast with its surroundings on an X-ray) has been injected into the blood. The angiographic X-ray shows shadows of the openings within the cardiovascular structures carrying blood; the blood vessels themselves do not appear. The most common use of an angiogram is to visualize the blood flow in the coronary arteries.

 12-2 For a visual animation of the heart contraction and the related blood flow, please go to your DVD for this chapter.

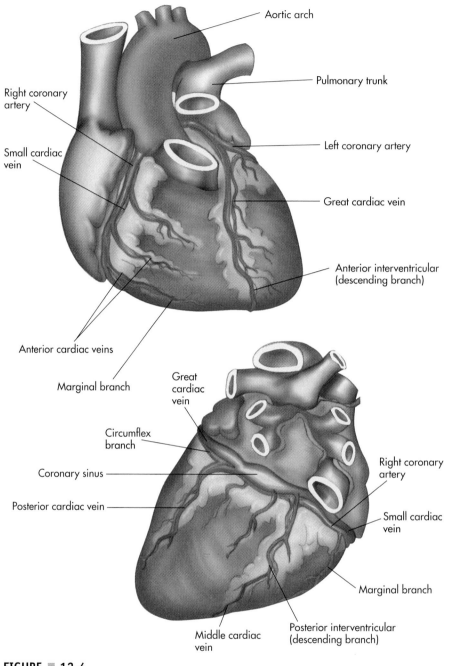

FIGURE ■ 12-4

Coronary circulation.

Pathology Connection: Problems with Cardiac Muscle and Cardiac Vasculature

PUMP PROBLEMS

So far, we have discussed how a healthy heart works. However, certain events can affect the efficiency of the heart's pumping action. Remember that this is a two-pump system. What may damage one pump may not always damage the other pump.

Let's look at the right side of the heart first. Right side heart failure is a potentially serious condition in which the right-side pump can't move blood as efficiently as it should. This is a result of the heart muscle of the right pump *chronically* working harder than normal. As with any muscle that you exercise over time, heart muscle also becomes larger. In this case, the muscles on the right side of the heart become too large and can no longer efficiently pump blood. Remember that the left side is pumping normally, and we are working with a closed system, much like a water pump and cooling system in a car. As the healthy left side pumps blood through your body and back to the right side, the now-inefficient right side cannot take all of the returning blood to pump it to the lungs. As a result, the blood begins to back up. The vessels in the body are flexible and can expand a little to take that extra volume of blood, so extended neck veins can be a sign of a right-side pumping problem. Certain organs can hold more blood than usual, so an engorged liver and spleen can also be a sign of right heart failure. Tissues in the periphery can hold extra fluid, so swollen ankles, feet, and/or hands can also be a sign of right-sided heart failure. Disease conditions such as **polycythemia**, in which the blood is thicker than normal and, therefore, is harder to pump, or blood vessels in the lungs that constrict more than normal, making it harder to push blood through them, can cause the right ventricle to work harder. Because these two conditions are often related to certain lung diseases, it is no surprise that 85% of the patients with chronic obstructive pulmonary disease develop right side heart failure.

polycythemia
(pall ee sigh THEE mee ah)

Heart failure (in the past, often called *congestive* heart failure) is a potentially life-threatening condition that is usually a problem of the left pump of the heart and can be caused by decreased pumping efficiency of damaged muscle fibers of the left ventricle following a heart attack. This is a situation in which the pumping action of the heart cannot overcome an increased systemic blood pressure, or in which the systemic blood pressure is normal but the left ventricle is too weakened to effectively pump blood. As a result of this increasing vascular pressure, fluid begins to leak out of the vessels and into the tissues of the body. Now let's consider a left-sided heart failure scenario. The healthy right pump pushes blood through the vasculature of the lungs on its way to the left-side pump. If the left side can't keep up with the blood being delivered to it, the blood backs up into the lungs, increasing the pressure in those blood vessels. (Remember, this is a closed system!) Once that pressure reaches a certain point, fluid leaks out of the vessels and into the lung tissue and even the air spaces. **Pulmonary edema** is the term for fluid that forms in the lungs and causes difficulty breathing. See Figure 12–5 ■. Heart failure can be treated with cardiotonic agents that increase the rate and force of contraction along with diuretics to rid the system of excess fluid.

Myocardial Infarction (MI) is a condition related to the upcoming "vessel problems" section in which there is too little or no blood flow to the myocardium (the heart muscles). At first, an inadequate flow of blood to the cardiac muscles known as **ischemia** will make the patient feel uncomfortable and anxious and often they will have a pain or discomfort in their chest known as **angina pectoris**, which comes from the lack of O_2 to the heart. The chest pain may radiate to the left shoulder, arm, neck, and jaw; the patient may also experience nausea, diaphoresis, and dyspnea. Application of nitroglycerin commonly reduces or eliminates angina pectoris due to its ability to increase blood flow to the cardiac muscles by dilating coronary blood vessels. There is no permanent

ischemia *(iss KEE mee ah)*

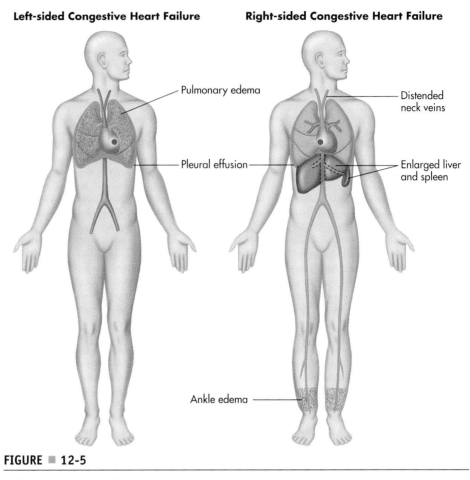

Left-sided Congestive Heart Failure   **Right-sided Congestive Heart Failure**

Pulmonary edema

Distended
neck veins

Pleural effusion

Enlarged liver
and spleen

Ankle edema

FIGURE ■ 12-5

Left-side and right-side heart failure.

damage to cardiac muscle at this point; patients are typically prescribed bed rest and oxygen. If a situation occurs in which there is blockage of coronary blood flow due to a clot (**coronary thrombosis**), a heart attack occurs. If the blood supply is not quickly returned to that region, death (necrosis) of that heart tissue will occur. In this case, there is actual tissue damage and death known as an **infarct**—hence the term **myocardial infarction**. This type of damage is permanent. Remember, cardiac muscle cannot repair itself. Anticoagulant medications may be administered to prevent additional clots from forming. Surgery may be required to break up clots or to bypass a clogged artery.

Sometimes the heart muscle is fine, but there is a problem with one or more of the heart valves that seal off the chambers during contraction. There are two possible types of problems: either the passageway through the valve is too small (**stenotic**) and restricts sufficient blood flow, or the passageway is too large (**valvular insufficiency**) and blood squirts backward into the chamber on contraction.

A **murmur** is a heart sound that provides the clinician a potential clue to the presence of a valve problem. It is caused by the vibrations created by the flow of blood through the heart or proximal great veins. This sound may also be indicative of an aortic aneurysm or aortic stenosis. A problem that can occur in either case is the increased tendency to form clots in the damaged valve area. Such

infarct *(in FARKT)*

clots can detach, flow through the blood vessels, and cause a pulmonary embolus in the lungs or a stroke in the brain.

There are also potential problems with the small, specialized muscles called the **papillary muscles** that attach to the undersides of the cusps and contract when the ventricles contract so the cusps don't flap back up into the atria. The failure of the papillary muscles to properly contract will allow blood to flow backward into the atrium instead of flowing forward. One of the classic causes of valvular damage is the streptococcal organism responsible for rheumatic fever. As the body fights this specific infection, antibodies are created that not only fight the infection, but unfortunately attack the patient's own heart valves, especially the mitral valve. Even though we are now treating infections much more efficiently than in the past, this disease may still occur.

Often the effectiveness of valvular function can be determined clinically through the use of **echocardiography**. This is accomplished by a device that utilizes sound waves (ultrasound which we can't hear) in a way similar to submarine sonar systems or "fish finders" used by sport fishermen. These sound waves bounce off of the various structures of the heart and can give us a real-time view of how well everything is working in the heart.

Clinical Application

CARDIAC ENZYMES

Because there is tissue damage and tissue death as a result of a heart attack, we can look for several enzymes that are released when cardiac muscle cells die. Although this enzyme is commonly found in the cells of skeletal muscles, creatine kinase (CK) in a special form, called CK-MB, is found in heart muscle cells. Once an MI occurs, this special cardiac enzyme is released and can be detected in the patients' blood within 2 – 6 hours. While also found in a variety of body cells including cardiac muscle cells, lactate dehydrogenase levels in the blood will increase in approximately 12 hours following a heart attack. Two proteins, troponin I and II, are also released when there is cardiac tissue death. Their blood levels begin to rise around 4 – 6 hours following an MI. These levels continue to stay elevated for up to ten days. As you can surmise, it is important to monitor the blood levels for several days to note any changes in blood levels of these substances to aid you in determining if an MI did occur.

12-3 To view videos and animations on angina and heart attacks, please see your DVD.

HEART VESSEL PROBLEMS

A common problem with blood vessels occurs to some degree for all of us as we age. **Arteriosclerosis,** also known as hardening of the arteries, is a result of the thickening of the inner layer, which causes the involved vessels to become less flexible or even brittle. Blood vessels in this condition have a tendency to rupture. Since these vessels are less flexible and can't readily accommodate increases in blood volume, the body is more susceptible to high blood pressure, or hypertension. Treatment begins with lifestyle changes that overcome at-risk behaviors such as lack of exercise, improper diet, and smoking. Medications and surgeries are used to treat specific types of arteriosclerosis.

Normally, blood vessels have a smooth inner lining, which promotes efficient blood flow by decreasing resistance. **Atherosclerosis** is a potentially life-threatening condition in which fatty deposits, called **plaque,** build up on the inner lining of blood vessels. As a result, blood flow can become greatly restricted or totally blocked. The fatty material that makes up plaque is composed mostly of **cholesterol.** Interestingly, all blood vessels are susceptible to atherosclerosis, but the aorta and coronary arteries seem particularly susceptible to developing this condition. Cerebral arteries can also be affected (see Figure 12–6 ■).

arteriosclerosis
(ar TEE ree oh skleh ROH sis)
sclerosis = hardening

atherosclerosis
(ATH er oh skleh ROH siss)
ather/o = fatty or porridge-like

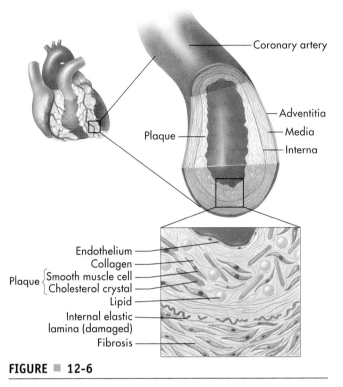

Coronary artery

Adventitia

Plaque

Media

Interna

Endothelium
Collagen
Plaque { Smooth muscle cell
Cholesterol crystal
Lipid
Internal elastic
lamina (damaged)
Fibrosis

FIGURE ▪ 12-6

Atherosclerosis.

As with arteriosclerosis, treatment begins with lifestyle changes. In some cases, the patient may be prescribed medications that control cholesterol, anti-platelet medications, anti-coagulants, and blood pressure medications. Surgeries, such as angioplasty and bypass surgery, can be used to open affected blood vessels.

Heredity seems to be one factor for atherosclerosis, and atherosclerosis is also a common complication of diabetes. It is interesting that many medical professionals feel that diabetes should be classified as a cardiovascular disease. Diet and lifestyle also seem to predispose some individuals to atherosclerosis.

There are several ways to improve blood flow to the heart's tissues. A coronary angioplasty (see Figure 12-7 ▪) is a procedure in which a balloon-tipped catheter is threaded to the coronary artery with the aid of a fluoroscope to where plaque is reducing the blood flow to the heart tissues. Once in the proper position, the balloon is inflated to "open up" the lumen of the artery and thus increase the blood flow. As time goes on, these areas often start to occlude again. To help prevent this reoccurrence, wire **stents** are inserted via the catheter to hold those areas open. See Figure 12-8 ▪ for an example of an intracoronary artery stent.

A more involved surgical procedure to restore proper blood flow to the heart muscle is the coronary artery bypass graft, or CABG. In this situation, healthy

12-4 To view animations and videos of coronary artery disease, coronary heart disease, and catheterizations, please see your DVD.

blood vessels from other regions of the patient's body (such as leg veins or the mammary artery) are used as replacements for the clogged coronary arteries. As with any surgical procedure there are risks, but this procedure has a low mortality rate and with proper rehabilitation and lifestyle changes these patients generally can expect a normal quality of life.

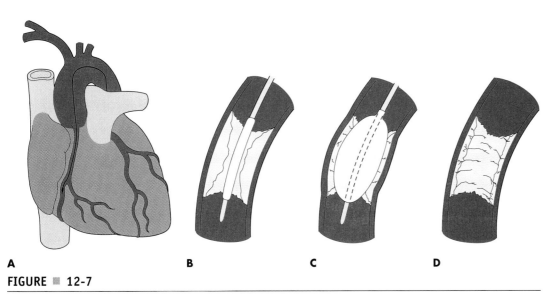

A   **B**   **C**   **D**

FIGURE ▪ 12-7

Balloon angioplasty. **A.** The balloon catheter is threaded into the affected coronary artery. **B.** The balloon is positioned in the area of obstruction. **C.** The balloon is inflated, flattening the plaque against the arterial wall. **D.**

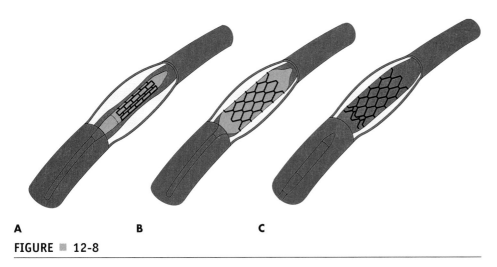

A   **B**   **C**

FIGURE ▪ 12-8

Placement of an intracoronary stent. **A.** The stainless steel stent is fitted over a balloon-tipped catheter. **B.** The stent is positioned at the blockage and expanded. **C.** The balloon is deflated and removed, leaving the stent in place.

TEST YOUR KNOWLEDGE 12-1

Choose the best answer:

1. Which blood vessels carry blood away from the heart?
   a. capillaries
   b. venules
   c. veins
   d. arteries

2. How many chambers are found in the human heart?
   a. one
   b. two
   c. three
   d. four

3. Which blood vessel type is involved in the exchange of oxygen and nutrients with the tissues of the body?
   a. arterioles
   b. sphincters
   c. arteries
   d. capillaries

Complete the following:

4. The chamber responsible for pumping blood to the body's various organs is the _____.

5. Which side of the heart pumps blood to the lungs? _____.

6. Reduced blood flow to cardiac tissue that leads to injury is called _____.

7. _____ means death of cardiac tissue.

THE ELECTRIC PATHWAY

autorhythmicity
(aw toe rith MIH sih tee)

Cardiac muscles don't always rely on nerve impulses or hormones to contract. In fact, they can contract on their own. This unique ability is known as **autorhythmicity.** The problem with this ability is that uncontrolled individual contractions would prohibit the heart from contracting effectively. This potential problem is solved through the use of specialized cardiac cells that create and distribute an electrical current that causes a controlled and directed heart contraction. Think of this as the electric company that supplies electricity to your home and to the fine educational institution you are attending.

Nodal cells (also called **pacemaker cells**) are specialized cells that not only create an electrical impulse but create these impulses at a regular interval. These cells are connected to each other and to the conducting network, which we discuss soon. Nodal cells are divided into two groups. The main pacemaker cells are found in the wall of the right atrium, near the entrance of the superior vena cava. This collection of pacemaker cells forms the **sinoatrial node,** or **SA node.** The SA node generates an electric impulse at approximately 70 to 80 impulses per minute. There is a second collection of pacemaker cells located at the point where the atria and the ventricles meet. This collection forms what is called the **atrioventricular node,** or **AV node.** The cells in the AV node generate an electric impulse at a rate of 40 to 60 beats per minute.

sinoatrial *(sigh noh AY tree al)*

Which one dictates how fast the heart beats? Think of the SA node as the power station and the AV node as a substation that supplies electricity via an electrical grid for a small city. The SA node sends its impulse to the AV node for distribution before the AV node can send its own. However, if the SA node cannot generate an impulse, the AV node takes control and sends out impulses that result in a slower heartbeat, but a heartbeat nonetheless. Figure 12–9 ■ shows the conduction system of the heart.

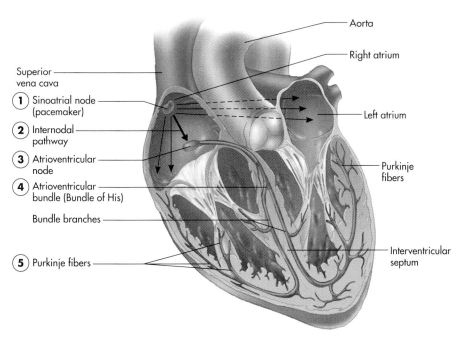

1. The sinoatrial (SA) node fires a stimulus across the walls of both left and right atria causing them to contract.

2. The stimulus arrives at the atrioventricular (AV) node.

3. The stimulus is directed to follow the AV bundle (Bundle of His).

4. The stimulus now travels through the apex of the heart through the bundle branches.

5. The Purkinje fibers distribute the stimulus across both ventricles causing ventricular contraction.

FIGURE ■ 12-9

Conduction system of the heart.

"How come my heart rate isn't always 70–80 beats per minute, like when I play sports or get scared?" you might ask. It's true that the SA node sets the heart rate when the body is at rest, but several influences can increase or decrease heart rate. The **autonomic nervous system** (both the sympathetic and parasympathetic divisions) has direct connections to the SA and AV nodes as well as to the myocardium. The **sympathetic division** can release neurotransmitters that increase heart rate and the force of the contraction (**inotropism**). To counteract this, the **parasympathetic division**, through the **vagus nerves** (cranial nerve X), can release a neurotransmitter that can decrease both the pulse and force of contraction.

inotropism (EYE no TROPE izm)

Clinical Application

ECG/EKG

Since the myocardial contraction is initiated and continues because of an electrical impulse, that charge can actually be detected on the surface of the body. This surface detection of the electric impulse traveling through the heart can be recorded by using an **electrocardiograph**, which records an **electrocardiogram** (ECG or, the German form EKG). See Figure 12–10 ■.

The normal ECG has three distinct waves that represent specific heart activities. The **P wave** is the first wave on the ECG and represents the impulse generated by the SA node and depolarization of the atria right before they contract. The next wave is called the **QRS complex** (a combination of Q, R, and S waves). It represents the depolarization of the ventricles that occurs right before the ventricles contract. The ventricles begin contracting right after the peak of the R wave. Due to the greater muscle mass of the ventricles compared to the atria, this wave is greater in size than the P wave. The final wave is the **T wave**, which represents the repolarization of the ventricles when they are at rest before the next contraction. "Aha," you say, "where is the repolarization of the atria?" It occurs during the QRS complex but is usually overshadowed by the ventricles' activity! In the recording of a healthy heart, there are set ranges for the height, depth, and length of time for each of the waves and wave complexes. Changes in those parameters, or the addition of other abnormal types of waves, known as **cardiac arrhythmias** or **dysrhythmias,** can indicate health problems that involve the heart.

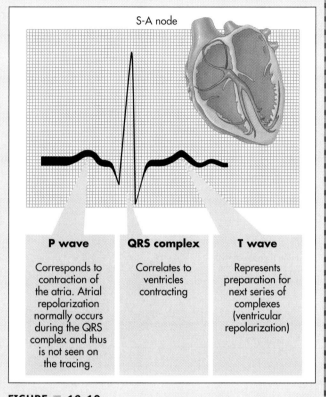

P wave	QRS complex	T wave
Corresponds to contraction of the atria. Atrial repolarization normally occurs during the QRS complex and thus is not seen on the tracing.	Correlates to ventricles contracting	Represents preparation for next series of complexes (ventricular repolarization)

FIGURE ■ 12-10

Typical ECG tracing.

In addition, ions, hormones, and body temperature can alter heart rate. For example, as body temperature increases, so does the rate and force of contractions because of the increased metabolic rate of cardiac muscle cells. Conversely, as the body cools down below normal temperature, the rate and force of cardiac contractions decrease. Epinephrine is a hormonal substance that has the same effects as the sympathetic nerves, so it increases heart rate. Electrolytes are important players, especially when there is an imbalance of too many or too few specific ones. Low sodium or potassium can alter heart activity, as can abnormal levels of calcium. Low potassium can lead to a weak heartbeat, while high levels of calcium can prolong heart muscle contractions to the point where the heart can stop beating. These are just a few examples of what can happen when electrolyte levels are outside of normal range. Age, gender, and a history of exercise or lack thereof can all impact your heart rate. Normally, the resting heart rate for a female is 72 to 80 beats per minute, and the average resting heart rate for males is 64 to 72 beats per minute. That's about 100,000 beats per day.

So far, we have our two generators and regulators of electricity. On our journey we can see how electricity is moved in the form of electric lines along the road. In the heart, the movement of the electric impulses is done by specialized conducting cells. This power grid of electric distribution has to be set up so the following actions can occur:

1. First, the right and left atria contract together and *before* the right and left ventricles.

2. Then, the two ventricles must contract together. But the direction of the wave of ventricular contraction has to be from the *apex* to the *base* of the heart. This ensures that all of the blood is squeezed out of the ventricles. (Remember the tube of toothpaste?)

So let's retrace the electrical "wiring," as was illustrated in Figure 12–9. Once an electric impulse is generated at the SA node, several pathways composed of conducting cells transmit that impulse to the AV node. A slight signal delay allows for the atria to fill with blood before contraction occurs. Once this charge reaches the AV node, it continues its journey through the **AV bundle,** also known as the **bundle of His** (which sounds like "hiss"). Traveling down the interventricular septum, the AV bundle eventually divides into the **right bundle branch** and the **left bundle branch**. These branches spread across the inner surfaces of both ventricles. Finally, another type of specialized cells called **Purkinje fibers** carry the impulse to the contractile muscle cells of the ventricles. And so the contraction begins at the apex, and the wave of contraction smoothly continues up the ventricles, squeezing out all of the blood.

Pathology Connection: Electrical Disturbances

Previously, we discussed problems with the pump and vessels of the heart. Those conditions can lead to a disruption of the normal generation of or the conduction of the electrical impulses in the heart, which can bring about an abnormal rhythm to the heartbeat (arrhythmia or dysrhythmia). For example, infarcted cardiac muscle does not conduct electrical impulses well because scar tissue does not function like normal cardiac muscle. There are also situations in which an area of cardiac tissue is injured and becomes excited to the point where it initiates contractions. Some of those abnormal heartbeats can be very rapid (up to 300 abnormal contractions per minute!) and are known as a **flutter** if those contractions are coordinated. If those rapid contractions are erratic, then they are called **fibrillations**. Both of these types are potentially life threatening, especially if they occur in the ventricular area. Rapid contractions such as those do not allow time for the ventricles to effectively fill with blood so it can be effectively pumped to where it needs to go.

Sometimes the electrical impulse is blocked so a smooth heart contraction can't occur. This is known as a **heart block** and can vary in the amount of electrical impulse that can pass through the heart. The problem with this situation is that the main electrical charge needed to get a smooth and effective heart

Clinical Application ✚ ✋ ♿ 🏊 🔑

CPR

If the patient's heart has stopped beating or the patient has stopped breathing, **cardiopulmonary resuscitation (CPR)** is performed. CPR consists of artificial blood circulation (chest compressions) and artificial respiration (lung ventilation). CPR is generally continued until the patient regains a heart beat or is declared dead. CPR is unlikely to restart the heart; its purpose is to maintain a flow of oxygenated blood to the brain and heart, thereby delaying tissue death and extending the brief window of opportunity for a successful resuscitation without permanent brain damage. Defibrillation is usually needed to restart the heart. The newer CPR guidelines emphasize compressions over ventilation.

contraction is delayed and the heart muscle starts to contract out of synch with the other chambers.

An artificial pacemaker may be needed to be implanted to bring about a normal electrical impulse to the heart. This is accomplished by threading one or more leads into the heart chamber(s) that need a proper electrical impulse. By using dual leads, impulses between the atrium and ventricle can be properly coordinated. The pacemaker can then be set to provide the right impulse for the correct heartbeat. Some pacemakers are set up to provide impulses only when the heart can't do it on its own. For an example of a pacemaker and its leads, please look at Figure 12–11 ▦.

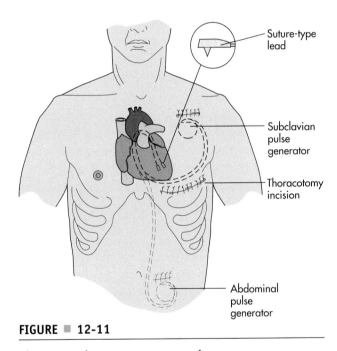

Suture-type lead

Subclavian pulse generator

Thoracotomy incision

Abdominal pulse generator

FIGURE ▦ 12-11

Placement of a permanent pacemaker.

A **defibrillator** (with an Automated External Defibrillator, or AED being one example of this device) is an "extreme pacemaker." A defibrillator is a device that is used to overpower ineffective electrical impulses and arrhythmias (such as ventricular fibrillation) by applying an electrical impulse to all of the cardiac muscle at once so a normal rhythm can begin again.

12-5 To view animations and videos on performing an ECG and various dysrhythmias, please see your DVD.

12-6 To view videos on the proper use of an Automated External Defibrillator (AED) and clinical procedures of defibrillation and cardioversion, please see your DVD.

---

www.prenhall.com/colbert
Cardiovascular technicians perform ECGs and other cardiovascular tests. Please visit the companion website for information and to view a video about this exciting profession.

Clinical Application ✚ ✋ 🚶 🏊 🔑

HEART ATTACKS

You may think that the classic heart attack is when the victim clutches his or her chest in extreme pain and falls over. However, many heart attacks start out slowly with little or no pain (often called a **silent MI**) and may progress over a few hours, days, or even a week, showing only subtle signs.

Symptoms that can be indicative of an MI are centrally located chest pain, chest heaviness, or vague discomfort; pain in the left shoulder or shoulder blade, neck, and jaw (where it mimics a toothache) and radiating down the left arm; nausea; heartburn; generalized weakness; or a clammy, sweaty feeling. Shortness of breath and/or dizziness and anxiety can also be warning signs. Women often exhibit "nontraditional" signs and symptoms, such as pain in the shoulder blade or jaw, and such symptoms are missed as indicators of a heart attack, often with tragic results.

Another big problem is that the victim often goes into denial, trying to explain away the symptoms as the result of some other problem, such as indigestion from eating too much or food that "doesn't agree with me" or thinking that it's a gallbladder problem, or a pulled muscle. This can cost the patient valuable time in the treatment for a heart attack. Indeed, the first hour is when much of the heart damage occurs.

Call 911 *immediately* if you even suspect a heart attack. Research shows that an individual who is experiencing a heart attack should also immediately chew and swallow an aspirin tablet, preferably a plain, non-enteric coated adult aspirin. The anti-coagulating ability of aspirin helps to keep blood flowing through the coronary vasculature, hopefully decreasing the amount of damage to the heart muscle. Most heart attacks are survivable *if* you act quickly and seek treatment immediately.

HEMATOLOGY

Now that we understand the pump, let's talk about what exactly the heart pumps. Blood is a fluid form of connective tissue that is responsible for four very important functions, listed below. **Hematology** is the branch of medicine that is concerned with the study of blood and blood diseases. The body uses blood to:

1. Transport oxygen, nutrients, cellular waste products, and hormones throughout the body
2. Aid in distribution of heat
3. Regulate acid/base balance
4. Protect against infection

Blood transports oxygen from the lungs, nutrients and fat cells from the digestive system, and hormones from endocrine glands to the *trillions* of cells in the body. On the return trip from those cells, blood transports carbon dioxide and other waste products that were formed from metabolic activities of the cells to the kidneys, lungs, and other organs for removal from the body.

Blood helps to regulate body temperature by absorbing heat generated by skeletal muscles, then spreading it throughout the rest of the body. Conversely, blood can radiate excess heat out of the body through the skin. Blood can take in or give up more fluid to help regulate the fluid balance of the body.

Blood helps to regulate a variety of levels in the body to maintain **homeostasis** by ensuring that **pH** (levels of acidity or alkalinity) and **electrolyte** (ion) values are within normal parameters for proper cell functioning.

Finally, but no less important, blood helps to protect us from invasion and infection by pathogens and toxins. This is done by specialized **white blood cells** (often shortened to **WBCs**) and special proteins called **antibodies.**

The amount of blood in the body depends on an individual's size and gender. Normally, the body contains between 4 and 6 liters (8 to 12 pints) of blood, which accounts for 7 to 9 percent of total body weight.

homeostasis
(HOE mee oh STAY siss)

Blood Composition

Although we can classify blood as a connective tissue, it is important to understand all the components that make up blood. When blood is separated by a centrifuge, the major components are **plasma** and **formed elements** (the cells). The centrifuge is a machine that spins a test tube of blood at a very fast rate. Due to the spinning force of the centrifuge, the heavier components, like the formed elements, are forced to the bottom of the tube and the lighter component (plasma) is displaced to the top of the tube (see Figure 12–12 ▪).

Plasma is the yellowish, straw-colored liquid that comprises about 55 percent of the blood's volume and contains about 100 different substances dissolved within the liquid. So, if your total blood volume is 5 liters, you have about 2.6 liters of plasma. While plasma is about 90 percent water; nutrients, salts, and a small amount of oxygen are also dissolved into the plasma for transport to the body's cells. Hormones and other cell activity–regulating substances are found in plasma. **Plasma proteins** are an important group of dissolved substances that include **albumin,** which aids in keeping the correct amount of water in the blood; fibrinogen, which is a substance needed for blood clotting; and globulins, which are antibodies that protect us from infection. Another plasma protein is **prothrombin**, which helps the blood coagulate and is dependent on vitamin K for its production. Blood **serum** is the plasma without fibrinogen and the other blood clotting agents.

albumin *(AL byoo men)*

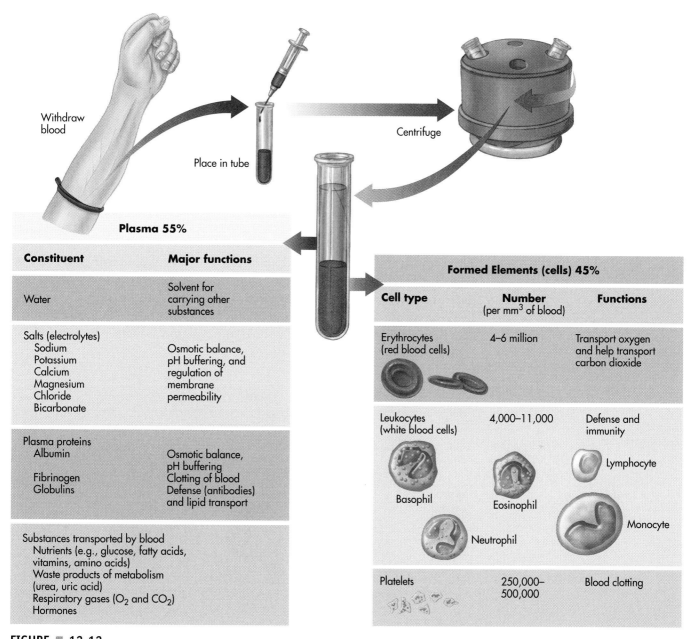

FIGURE 12-12

Composition of blood.

Formed or solid elements include the following:

- **Red blood cells (RBCs)** or **erythrocytes**
- **White blood cells (WBCs),** or **leukocytes,** which can be further classified as neutrophils, basophils, eosinophils, lymphocytes, and monocytes
- **Thrombocytes** (also known as **platelets**), which aid in clotting

erythrocytes *(eh RITH roh sights)*
  erythr/o = *red*
  cytes = *cells*
leukocytes *(LOO koh sights)*
  leuk/o = *white*
thrombocytes *(THROM boh sights)*
  thromb/o = *clotting*

Red Blood Cells

Red blood cells, lacking a nucleus, and therefore unable to divide to form new cells, are created by the red bone marrow through a process called **erythro-poiesis** and are similar in shape to a doughnut. The shape of the red blood cells is also sometimes described as a bi-concave disk. Red blood cells perform two crucial functions. With the aid of an iron-containing red pigment called **hemoglobin** (which comes from the roots *heme*, meaning iron, and *globin*, meaning protein), red blood cells transport oxygen from the lungs to the cells in the body. In addition, they help to transport carbon dioxide, a byproduct of cellular metabolism, from the cells to the lungs for removal from the body.

White Blood Cells

There are several types of white blood cells. **Polymorphonuclear granulocytes** originate from red bone marrow. Also originating from bone marrow but maturing in lymphoid and myeloid tissues are the mononuclear cells. **Leukocytes** are our guardians from invasion and infection. They may be granular, agranular, translucent, or ameboid in shape and are larger than red blood cells (erythrocytes). See Figure 12–13 ▪.

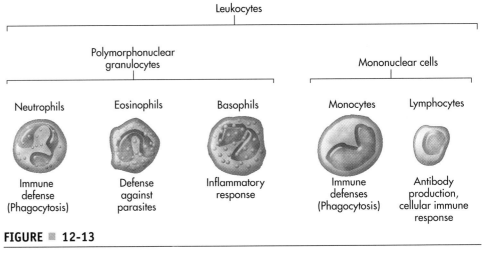

FIGURE ▪ 12-13

Functions of white blood cells.

The types of white blood cells that compose the polymorphonuclear granulocyte group are neutrophils, eosinophils, and basophils. **Neutrophils** are the most aggressive white blood cells in cases where bacteria attempt to destroy tissue. Neutrophils move through capillary walls into surrounding tissue to fight infection and help repair damaged tissue. By degrading the capillary wall in a process called **diapedesis,** neutrophils can escape into the surrounding tissue to perform the work of immune system response. **Phagocytosis** is the process in which neutrophils surround and ingest invader pathogenic, or disease-producing, microorganism and attempt to destroy it by utilizing cell

diapedesis
   dia = *through*
   pedesis = *to leap*
phagocytosis
   (FAG oh sigh TOH siss)
   phag/o = *to eat*
   osis = *process*

particles called **lysosomes** that release powerful enzymes. As an infection occurs, the body produces a higher than normal number of neutrophils. **Eosinophils** are utilized to combat parasitic invasion and a variety of body irritants that lead to allergies. **Basophils** are involved with allergic reactions by enhancing the body's response to irritants that cause allergies. In addition, basophils are important because they secrete the chemical **heparin,** which helps to keep blood from clotting as it courses through blood vessels.

The types of white blood cells that comprise the mononuclear cell group are monocytes and lymphocytes. **Monocytes** are found in higher than normal amounts when a chronic (long-term) infection occurs in the body. Like neutrophils, monocytes destroy invaders through phagocytosis. Even though it takes longer for monocytes to arrive on the scene of infection, their numbers are greater than neutrophils and therefore they destroy more pathogens. **Lymphocytes** are unique cells that protect us from infection. Instead of utilizing phagocytosis, they are involved in a process that produces antibodies that inhibit or directly attack invaders. Fighting infection and foreign invaders is discussed, along with the lymphatic system, in Chapter 14.

lysosomes *(LIE so soams)*
  lys/o = *destruction*
  som(a) = *body*
neutrophils *(NOO trow fillz)*
basophils *(BAY soh fills)*
eosinophils *(EE oh SIN oh fillz)*

Inflammation

In the mad rush to fight infection, injury, or disease, blood cell action can cause several symptoms of the body's natural immune defense system. **Inflammation** is a complex system of responses by vascular tissues to pathogens, damaged cells, or other irritants. Inflammation can cause several visible symptoms, including redness, swelling, pain, and localized heat. These symptoms are due to the increased blood flow at the point of irritation or infection. This excess blood in the tissues is also known as edema.

In some cases, inflammation is accompanied by **pus,** a yellow or whitish-yellow substance produced in the aftermath of phagocytosis. Pus is made up of a fluid that contains the dead neutrophils left over from the immune response, and is therefore an indication of infection. When pus collects in an enclosed body space, it forms an **abscess.** Inflammation is often accompanied by **pyrexia,** or fever, a part of the body's immune response to infection and disease. Fever has been shown to increase the mobility of leukocytes and thereby enhance phagocytosis.

Thrombocytes

Thrombocytes, also known as **blood platelets,** are the smallest of the formed elements and are responsible for the blood's ability to clot. In addition, thrombocytes can release a substance called **serotonin**, which can cause smooth muscle constriction and decreased blood flow.

Thrombocytes are cell fragments produced by budding off other cells called **megakaryocytes,** a type of bone marrow cell. Megakaryocytes circulate in the body for approximately a week, producing between 5,000 and 10,000 platelets, and are then destroyed in the spleen or liver. Thrombocytes are synthesized in the red marrow.

TEST YOUR KNOWLEDGE 12-2

Choose the best answer:

1. Which cell type does not belong in this group?
   a. erythrocyte
   b. lymphocyte
   c. monocyte
   d. eosinophil

2. What substance found in red blood cells is responsible for oxygen transport?
   a. gobuloglobin
   b. hemoglobin
   c. gammaglobulin
   d. cytoplasm

3. Which portion of an ECG tracing represents ventricular depolarization?
   a. T wave
   b. QRS complex
   c. P wave
   d. SA node

Complete the following:

4. This important plasma protein helps to maintain fluid balance of your blood: _____.

5. The main pacemaker of the heart is the _____.

Blood Types and Transfusions

Not all human blood is the same. A person in need of a blood transfusion cannot be given blood from a randomly selected donor. Incompatibility of blood types is due in part to antigens. An **antigen** is a protein on cell surfaces that can stimulate the immune system to produce **antibodies**, which fight foreign invaders. Antigens are often foreign proteins introduced into the body through wounds, blood transfusions, and so on. Because they are not "native" to the body, they are called "non-self" antigens. They differ from our own "self-antigens" that exist on the cell membrane of every cell in the body. The chain of events that occurs between antigens and antibodies is called the antigen–antibody reaction, which is the basis for immune response, as you will see in Chapter 14. Antibodies often react with the antigens that caused them to form, and the antigens stick together, or **agglutinate,** in little clumps. Although there are at least 50 different antigen types found on the surface of a red blood cell, our main focus is on the A, B, and Rh antigens.

agglutinate *(ah GLUE tin ate)*

Everybody has only *one* of the following blood types, inherited from parents:

Type A
Type B
Type AB
Type O

Type A blood is very common. Approximately 41 percent of the American population has this type of blood. A represents a specific type of self-antigen that is found on the cell membrane of each red blood cell in the body of a person with type A blood. Since that person was born with type A blood, no antibodies were created to fight it, so there are no anti–A antibodies in his or her plasma. Type A blood *does* contain anti–B antibodies.

Type B red blood cells possess type B self-antigens and the plasma contains anti–A antibodies. Apparently, these two blood types don't like each other!

Type AB, however, tries to get along with both A and B. Its red blood cells contain *both* of the A and B self-antigens with neither A nor B antibodies in the plasma.

Not to be outdone, type O red blood cells contain *no* A or B antigens, but its plasma contains *both* A and B antibodies.

This information is important to know if there is a need to transfuse blood. If the donor's antigens and the antibodies in the blood recipient's plasma agglutinate, serious harm and even death can occur.

If a *donor* gives blood that contains no A or B antigens, agglutination by anti–A and/or anti–B antibodies in the *recipient's* blood is prevented. Since Type O doesn't have A or B antigens, it can be given to anyone, so a donor with a form of type O blood (O negative) is a **universal donor.** Since type AB doesn't contain plasma anti–A antibodies or anti–B antibodies, it can't clump with any donated blood that contains A or B antigens. Because of this, a type AB person is labeled a **universal recipient.** See Figure 12–14 ■ that shows blood types with matching antigen and antibodies and which recipient's blood would safely match the donor's blood.

Are you with us so far? Good. Now there is one more thing we need to add: the **Rh factor.** Based on a discovery of a special blood antigen first found in the blood of Rhesus monkeys, it was discovered that approximately 85 percent of the white and 88 percent of the African American population of the United States possess the Rh antigen in their blood. Individuals with this antigen in their blood are said to be **Rh-positive.** Conversely, those without this antigen are **Rh-negative.** When typing an individual's blood, the term, *Rh* is eliminated, so an individual would be O-positive or AB-negative, for example. A problem arises when there is an Rh-positive father and an Rh-negative mother. If their first baby inherits the father's Rh-positive blood trait, the mother will develop anti–Rh antibodies (remember the foreign invader scenario?). This baby will be okay, but any future baby born to these parents may be attacked by the anti–Rh antibodies of the mother *IF* that baby has the Rh-positive trait in its blood.

Amazing Body Facts

COCONUT JUICE TRANSFUSIONS AND ARTIFICIAL BLOOD

There are reports that in the Pacific Theater during World War II, coconut juice was used on injured soldiers to bring their blood volumes up when there was a lack of real blood. The juice was supposedly sterile when taken directly from the coconut and transfused into their veins.

Currently, there is much research in the development of artificial blood or blood substitutes. The main objective is to develop a blood form that can be universally used for all humans without the need to match specific blood types. Other objectives include making a substitute that is free of blood-borne diseases, has the ability to be rapidly infused, has increased oxygen-carrying capacity, and has an extended shelf-life. Some artificial blood is made by chemically altering and refining natural blood components, and some artificial blood is synthetic in nature.

 12-7 To perform an interactive drag-and-drop exercise on blood types and view a video on administering blood, please go to your DVD.

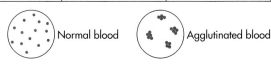

TYPE A	TYPE B	TYPE AB	TYPE O
RBC — Surface antigen A	Surface antigen B	Surface antigens A and B	Neither A nor B surface antigens
PLASMA — Anti-B antibodies	Anti-A antibodies	Neither anti-A nor anti-B antibodies	Anti-A and anti-B antibodies

Recipient's blood		Reactions with donor's blood			
RBC antigens	Plasma antibodies	Donor type O	Donor type A	Donor type B	Donor type AB
None (Type O)	Anti-A Anti-B	• (normal)	(agglutinated)	(agglutinated)	(agglutinated)
A (Type A)	Anti-B	• (normal)	• (normal)	(agglutinated)	(agglutinated)
B (Type B)	Anti-A	• (normal)	(agglutinated)	• (normal)	(agglutinated)
AB (Type AB)	(None)	• (normal)	• (normal)	• (normal)	• (normal)

Normal blood   Agglutinated blood

FIGURE ■ 12-14

Blood types and results of donor and recipient combinations.

Pathology Connection: Assorted Blood Conditions

Secondary polycythemia is a condition in which chronic low levels of oxygen (due to lung disease or living in high altitudes) cause the body to produce more than normal amounts of erythrocytes to transport more efficiently the smaller amounts of available oxygen. **Primary polycythemia** does the same thing but can be caused by bone marrow cancer stimulating over-production.

anemia *(ah NEE mee ah)*

Anemia is a blood condition in which there is a less than normal number of red blood cells or there is abnormal or deficient hemoglobin. Anemia can be a result of bone marrow dysfunction, low levels of iron or vitamins, or the improper formation of red blood cells. Individuals with anemia share these common symptoms:

- pale skin tone (pallor)
- pale mucous membrane and nail beds
- fatigue and muscle weakness
- shortness of breath
- chest pains in some heart patients due to decreased levels of oxygen being supplied to the heart

Specific kinds of anemia are due to particular causes, affect different patient populations, and have different treatments. For example, *iron-deficiency anemia* is caused by a deficiency of dietary iron causing insufficient hemoglobin production. It typically affects women, children, and adolescents and is treated by prescribing iron supplement pills and boosting the amount of green, leafy vegetables in the diet. Another example is *aplastic anemia*, in which the bone marrow does not produce enough blood cells. This type of anemia is typically a side effect of drug or radiation therapy used to treat diseases such as cancer.

Sickle cell anemia is a chronic, inherited condition occurring primarily in African Americans in which red blood cells and hemoglobin molecules do not form properly. The resultant red blood cells are crescent or sickle-shaped and have a tendency to rupture. As they are destroyed, the body is stimulated to produce greater numbers of red blood cells to replace them. Unfortunately, at that high production rate, the blood cells cannot mature fast enough. The ruptured cells also clog up smaller blood vessels. Clogged vessels combined with the increased thickening of the blood from excess red blood cells and cell parts lead to increased clotting and an impaired ability to carry oxygen.

There are several blood problems involving white blood cells. **Leukemia,** usually due to bone marrow cancer (malignancy), is a condition in which a higher than normal number of white blood cells are produced. You might think this would be a good thing; however, the problem is that the white blood cells produced are immature and therefore ineffective in protecting the body from infection. **Leukocytosis** also exhibits as a situation in which there is a higher than normal number of white blood cells. In this case, the cause is often an infection that is being fought. **Leukopenia** is a condition in which the number of white blood cells is lower than normal. This can be a result of drugs that suppress their production, such as corticosteroids and anti-cancer agents. Chronic infections can also wear the body down to the point that it cannot produce the necessary numbers of white blood cells.

Clinical Application

UMBILICAL CORD BLOOD AND LEUKEMIA

Leukemia is most commonly treated via bone marrow transplant, but when a suitable donor is not available, umbilical cord blood, which is rich in stem cells, has been used as an alternative treatment. Special cord blood banks have been established, using umbilical cords that would otherwise be discarded after birth. Cord blood transplants are considered a "last resort" for leukemia patients because of the cost and the complications, but the procedure has saved many people who otherwise would have died.

12-8 To view videos and animations on sickle cell anemia and leukemia, please go to your DVD for this chapter.

BLOOD VESSELS: THE VASCULAR SYSTEM

So far, we have a pump and some fluid. We now need a way to transport the blood away from and then back to the heart. The arteries and veins we discussed previously now come into play.

Structure and Function

Initially, blood leaves the heart through the aorta, which branches into large vessels called **arteries**. Arteries divide into smaller and smaller ones as they spread out through the body. The smallest forms of arteries are **arterioles**. Arterioles feed the capillaries that form the capillary beds in your body's tissues. Here is where fresh oxygen and nutrients are supplied to the cells of your body and carbon dioxide, along with other waste products, are picked up by the blood for removal.

Blood from the capillary beds begins its return trip to the heart by draining into tiny veins called **venules**. Venules combine into veins, which eventually combine into the great veins (superior and inferior vena cavae) that empty back into your heart.

For most blood vessels, the walls are composed of three layers, often called coats or tunics. See Figure 12–15 ■. The **tunica interna** is the innermost layer and is composed of a thin, tightly packed layer of squamous epithelial cells over a layer of loose connective tissue. The compacting of the epithelial cells provides a smooth surface so blood can easily pass through. The next layer is thicker and is composed mainly of smooth muscle and elastic tissue and collagen. This middle layer is called the **tunica media.** By contracting or relaxing those

arterioles *(ahr TEE ree ohls)*

tunica interna
   (TOO nik ah in TERN ah)
tunica media
   (TOO nik ah mee DEE ah)

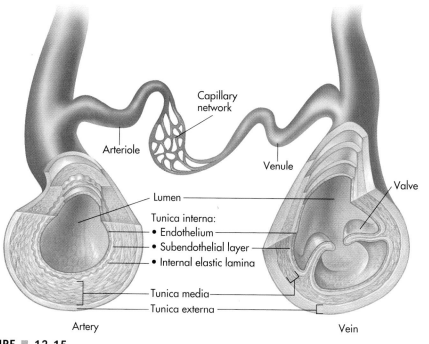

FIGURE ■ 12-15

Blood vessels and the capillary connection.

muscles, this layer actually controls the diameter of the vessels to meet certain blood flow needs of the body at a given time. Your sympathetic nervous system determines these needs and signals the vessels to **vasoconstrict** (decrease the inner diameter or lumen) or **vasodilate** (increase the inner diameter or lumen) as needed. This change in diameter changes blood pressure (**BP**); as the vessels dilate, BP decreases. Constriction leads to BP increases. The outermost, or external, coat is the **tunica externa.** Its job is to provide vessel support and protection, so it is composed of mostly fibrous tissue.

 12-9 To see 3-D views of the heart and blood vessels and perform an interactive labeling of the circulatory system, please go to your DVD for this chapter.

tunica externa
(TOO nik ah ex TERN ah)

TABLE 12–1

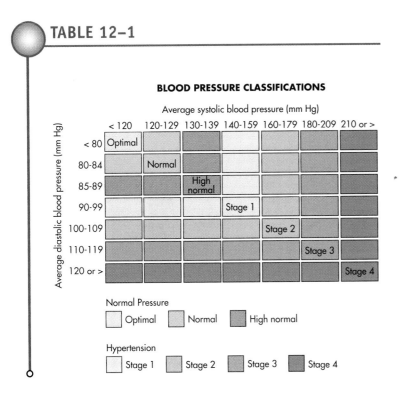

BLOOD PRESSURE CLASSIFICATIONS

Clinical Application

REGULATION OF BLOOD PRESSURE

Blood pressure is controlled by both blood vessel diameter (peripheral resistance) and the volume of blood pumped by the heart (cardiac output). Cardiac output is a function of heart rate and the amount of blood pumped with each contraction (stroke volume). Stroke volume is influenced mainly by blood volume. For example, increased fluid volume, increased heart rate, and increased peripheral resistance would lead to an increased blood pressure. See Table 12-1.

Capillaries and Veins

The structure of a blood vessel varies depending on its job. Arteries possess much thicker walls than veins. As we said earlier, this makes sense because arteries are closer to the heart and have to deal with higher pressures. In fact, larger arteries contain complete sheets of elastic tissue, **elastic laminae**, in their middle walls to help deal with increased pressure. Veins can possess thinner walls because the pressure on them is lower than in arteries, but the inside opening, known as the **lumen,** is larger than in arteries. In addition, the larger veins of the body, especially in the legs, contain valves that prevent blood

Clinical Application

TAKING A BLOOD PRESSURE

An important diagnostic test is the determination of arterial blood pressure (BP). This is done with a stethoscope and a sphygmomanometer. As shown in Figure 12-16 ■, an inflatable cuff is placed around the arm, above the elbow, so when the cuff is inflated, it squeezes the brachial artery shut. Your stethoscope is placed over the brachial artery in the proximity of the patient's elbow. The cuff can then be inflated by repeatedly squeezing the bulb while listening with the stethoscope. As you listen while you inflate the cuff, continue squeezing the bulb until you raise the pressure to about 30 millimeters of mercury (mm Hg) *beyond* the point that the pulse is no longer heard.

Once the pressure is 30 mm Hg above the point where pulse sound is lost, you open the release valve *slightly* so the cuff slowly deflates as you listen to the brachial artery. As the cuff pressure decreases to slightly below systolic pressure (pressure when the heart contracts), the sound of blood being pushed through the artery by the heart can be heard. That is the peak systolic pressure, or top number of the blood pressure (BP) reading.

As the cuff pressure decreases, the sound of the pulse decreases and then disappears. That point is where the cuff pressure is equal to the arterial pressure when the heart is at rest (diastole). This last muffled sound heard is the bottom number in the BP reading. One-twenty over eighty (120/80) has traditionally been accepted as a normal BP for healthy young adults. Recently, lower values are being considered as more in line with a healthy value. Time and continued study will tell which will be the accepted value.

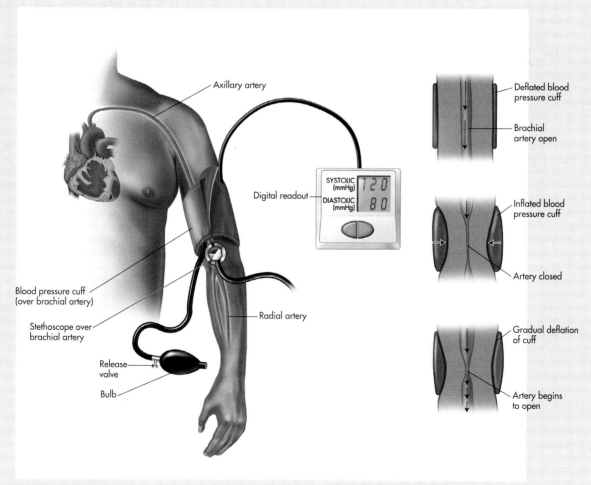

FIGURE ■ 12-16

Blood pressure measurement.

from flowing backward. Remember, the venous side is lower in pressure. Another means the body has to help move venous blood toward the heart is through the use of skeletal muscle. The relaxation and contraction of skeletal muscles that surround veins help to "milk" the blood toward the heart.

Finally, we have the capillaries, which are composed of only the tunic interna. With a diameter of only 0.008 millimeters (slightly larger than the diameter of single red blood cell), this microscopic wall is only one cell in thickness so oxygen and nutrients can easily move into the cells while carbon dioxide and wastes can move into the blood. This is important because even at rest, metabolizing cells require approximately 250 milliliters of oxygen while producing almost 200 milliliters of carbon dioxide *every minute!* Dozens of capillaries form a web, or network, of vessels called a **capillary bed.** As you can see in Figure 12–17 ■, capillary beds are composed of two types of blood vessels: a **vascular shunt,** which is a main road connecting the arteriole to the venule, and **true capillaries,** which make the actual exchanges with cells. True capillaries can be considered the on-ramps and off-ramps to and from the vascular shunt. In Figure 12–17, you will notice a group of structures called **precapillary sphincters.** These structures are composed of smooth muscle and act as toll booths either allowing blood to flow through or stopping blood flow when they contract. If the blood flows through, it travels through the true capillaries and to cells of the tissue. If the blood is stopped at the precapillary sphincters, then the blood travels through the vascular shunt.

Blood Clotting

As we have discussed, the cardiovascular system is a closed and pressurized system. Imagine what could happen if a leak or a break in the system occurred. You have probably seen cars along the road with blown radiator hoses. The steam shoots out and the car goes nowhere. A similar condition can occur in your body; no steam shoots out of your blood vessels, of course, but if enough blood is lost, you won't go anywhere and will die. Thanks to several substances in your blood, some leaks or breaks that occur can be stopped. **Hemostasis** (the stoppage of blood) is accomplished through a chain of events shown in Figure 12–18 ■.

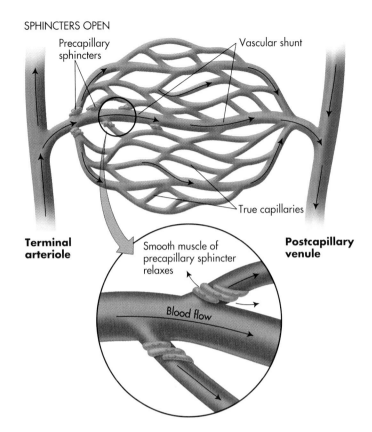

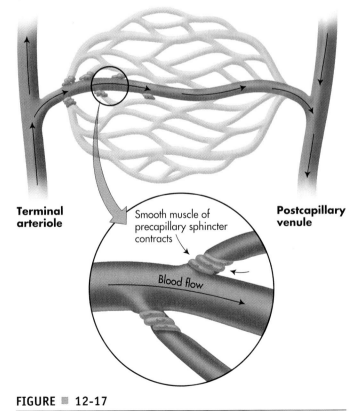

FIGURE ■ **12-17**

Capillary beds and sphincters.

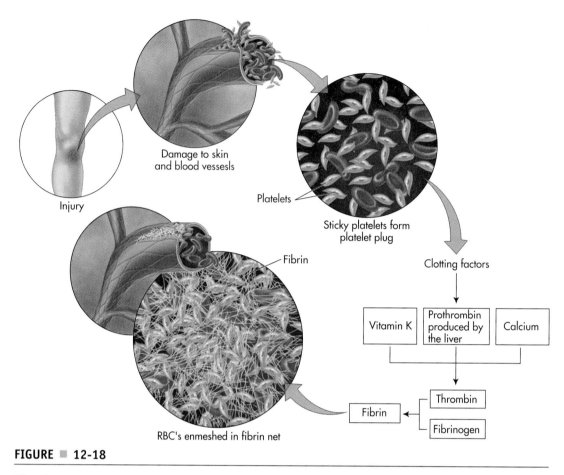

Damage to skin
and blood vessesls

Injury

Platelets

Sticky platelets form
platelet plug

Clotting factors

Fibrin

| Vitamin K | Prothrombin produced by the liver | Calcium |

Thrombin

Fibrin

Fibrinogen

RBC's enmeshed in fibrin net

FIGURE ■ 12-18

The clotting process.

hemostasis (HEE moh STAY siss)
prothrombin (pro THROM bin)
thrombin (THROM bin)
fibrinogen (fye BRINN oh jenn)

Damage to the innermost wall of a vessel exposes the underlying collagen fibers. Platelets that are floating around in the blood begin attaching to the rough, damaged site. The attached platelets release several chemicals that draw more platelets to that site, creating a platelet plug. These platelets also release serotonin, which causes blood vessels to spasm, thereby decreasing blood flow to that area. Within approximately 15 seconds from the time of the initial injury, the actual coagulation (clotting) of blood begins. With the help of calcium ions and 11 different plasma proteins (clotting factors), a chain reaction starts. One of the clotting proteins, **prothrombin,** which is produced by the liver with the help of vitamin K, converts to **thrombin.** Thrombin transforms **fibrinogen,** which is dissolved in the blood, into its insoluble, hair-like form called **fibrin.** Fibrin forms a netlike patch at the injury site and snags more blood cells and platelets, and within 3 to 6 minutes, a clot is created. Once the clot is formed, it eventually begins to retract, and as a result, pulls the damaged edges of the blood vessel together. This allows the edges to regenerate the necessary epithelial tissue to make a permanent repair over time. Once the clot has outlived its usefulness, it is dissolved. Again, please see Figure 12–18.

Pathology Connection: Clotting

Sometimes, there is a problem with the inability of blood to clot properly. **Hemophilia** is a general term used to describe inherited blood conditions that prohibit or slow down the blood's ability to clot. Hemophilia is a hereditary condition transmitted from mother to son, that occurs because of a missing clotting factor. **Thrombocytopenia** is a type of hemophilia in which clotting is impaired because there are fewer than normal circulating platelets. If the platelet count is low enough, even normal movement can lead to bleeding. This condition can be caused by liver dysfunction, vitamin K deficiency, radiation exposure, or bone marrow cancer.

The chain reaction that causes a clot must be stopped when it has accomplished its purpose, or else clotting would continue throughout the vascular system. However, there are times when unwanted clotting occurs. A rough surface on an otherwise smooth lumen of a blood vessel may allow platelets to begin "sticking" there, forming a type of clot called a **thrombus** (or **thrombosis**). A thrombus that forms in the vascular system of the heart can partially or totally block blood flow to a portion of the heart, resulting in a coronary thrombosis, which can cause a heart attack. The degree of blockage along with the heart area affected determines the severity of the attack. If allowed to increase in size as more blood cells attach to it, total blockage of blood flow can happen.

Another scenario is the potential for a portion, or several portions, of the thrombus to break off and flow through the circulatory system like an iceberg at sea. This floating thrombus, called an **embolus,** is not a problem until it travels down too narrow a blood vessel and becomes lodged, partially or totally blocking downstream blood flow. A cerebral embolus would affect blood flow to the brain, causing a stroke; a pulmonary embolus would lodge in the lung region and affect your ability to get oxygen into your blood, as you will see Chapter 13.

When blood does not travel through the vessels at a fast enough rate, unwanted clots can form. People who are bedridden, on long plane flights or bus or car rides, or are immobile for extended periods of time are susceptible to thrombus formation. This will be a continuing problem for Ray. It also appears that women who smoke and use oral contraceptives and individuals on some types of chemotherapy are at a higher risk for clot formation.

Substances that decrease the blood's ability to coagulate, such as aspirin or heparin, help prevent unwanted clotting. Once a clot forms, "clot busters" such as the drug streptokinase, are given to regain proper blood flow. The body's natural way of stopping bleeding is clot formation, but when that doesn't work, health care workers can stop bleeding by applying pressure to several pressure points. These are illustrated in Figure 12-19 ▪.

hemophilia *(HEE moh FILL ee ah)*
thrombocytopenia
(THROM boh sigh toh PEE nee ah)

thrombus *(THROM buss)*

embolus *(EM boh luss)*

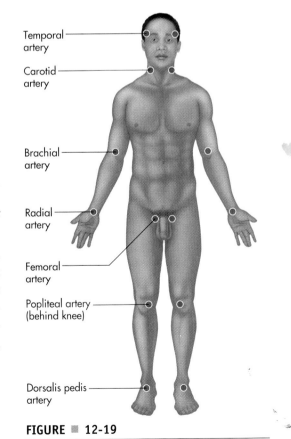

FIGURE ▪ **12-19**

Pressure points.

TEST YOUR KNOWLEDGE 12-3

Choose the best answer:

1. The universal donor blood type is:
   a. O-negative
   b. AB-positive
   c. Rh-positive
   d. B-positive

2. The smallest form of arteries are:
   a. arterules
   b. capillaries
   c. arterioles
   d. vessicles

3. A type of unwanted blood clot is a:
   a. bolus
   b. thrombus
   c. omnibus
   d. skoolbus

Complete the following:

4. List the three layers commonly found in a blood vessel:

   _____

   _____

   _____

Pathology Connection: Problems Associated With Systemic Blood Vessels

Arteries are designed to handle the increased pressures generated by a beating heart. In some individuals, arterial walls may not always be able to maintain their integrity. An **aneurysm** is a localized weakened area of blood vessel wall that may have been caused by a congenital defect, disease, or injury. This weakened area of the blood vessel may balloon. There also appears to be a familial tendency for abdominal aneurysms, which often present no warning symptoms. The danger is when the aneurysm expands to the point that it ruptures, causing hemorrhaging. If it is a major artery, an individual can bleed out in a matter of minutes. If the aneurysm is detected on an X-ray or ultrasound, surgical intervention can remedy the situation. Early detection is important. Aneurysms require surgical treatment. In some cases, the damaged portion of the vessel is removed and replaced with an artificial blood vessel. In other cases, a stent is inserted to support the damaged blood vessel.

Visible **hematomas** are swollen areas that appear as purple, yellow, or greenish splotches in the tissue just under the skin. Hematomas are localized collections of blood in body tissues, internal organs, or body spaces caused by an injury that ruptures a blood vessel. You may know hematomas by their common name: a bruise. They can gradually move or diffuse to nearby tissue—for example, a sprained thumb may initially appear purple around the affected joint, but eventually the entire thumb may turn purple due to flow of the collected blood.

Peripheral vascular disease (PVD) is a condition that is caused by a plaque build-up (atherosclerosis) in the leg arteries. In this situation, there is often a sufficient blood flow to the legs during rest and minimal exertion; however, as activity increases, blood supply to the legs is less than the demand. As a

aneurysm *(AN yoo rizm)*

result, cramping occurs that is relieved, upon rest. Treatment can be either cleaning out the plaque in the affected arteries or vessel replacement.

If smaller arteries are involved, the blood supply to the toes and feet may be insufficient. This may lead to their inability to heal if injured, tissue necrosis, and gangrene. This situation may require surgical amputation of the affected area.

 12-10 To view videos on peripheral artery assessment and carotid artery assessment, please visit your DVD.

A Quick Trip through Some Diseases of the Cardiovascular System

PUMP PROBLEMS

DISEASE	ETIOLOGY	SIGNS AND SYMPTOMS	DIAGNOSTIC TEST(S)	TREATMENT
Cardiac Tamponade (pericarditis)	Inflammatory process that produces a pericardial effusion, trauma, rupture of heart or aorta into pericardium.	Pain possible, tightness/fullness in chest, possible frank dyspnea, loss of consciousness if extreme.	Patient examination/history, radiologic studies.	Medical treatment, removal of the offending fluid if no rapid improvement.
Endocarditis	Inflammatory process/infection.	Anorexia, back pain, fever, murmur, night sweats, weight loss.	Echocardiography.	Antibiotics.
Heart failure A general term for either right or left side heart failure. Formerly known as congestive heart failure. See Figure 12-20 ■	Pulmonary hypertension, high blood pressure, polycythemia, increased work of the heart, MI, valvular disease, cardiac infections.	(Right) peripheral edema, enlarged spleen, distended neck veins, (left) pulmonary edema, shortness of breath.	Patient history, physical examination, radiologic studies.	Treat causative condition, oxygen, medication to improve heart function, diuretics. Decrease the work of the heart.
Murmur	A result of blood leaking backward due to a faulty heart valve or a narrowed valve opening.	Abnormal sound of the blood flow in the heart.	Auscultation, with ultrasonic confirmation.	Correction of the condition creating the abnormal murmur.
Prolapse (valve)	Various connective disease disorders, sometimes coronary artery disease.	Possibly asymptomatic, murmur, fatigue, palpitations, dyspnea, increased chance for clot development.	Auscultation, echocardiography, radiologic studies, cardiac catheterization, stress testing.	Possible valve reconstruction/replacement if severe, medication.

continues

 12-11 Heart disease affects young and old alike. To view animations and videos on congenital heart defects and the effects of aging on the cardiovascular system, please visit your DVD.

A Quick Trip through Some Diseases of the Cardiovascular System (*Continued*)

PUMP PROBLEMS (*continued*)

DISEASE	ETIOLOGY	SIGNS AND SYMPTOMS	DIAGNOSTIC TEST(S)	TREATMENT
Stenosis (valve)	Rheumatic fever, congenital origin, aging process.	Murmur, possible fatigue, chest pain, dyspnea, lower PaO_2, pulmonary edema, syncope.	Auscultation, cardiac catheterization, echocardiography, radiologic studies, stress test.	Possible valve replacement, medication.

BLOOD VESSEL PROBLEMS

DISEASE	ETIOLOGY	SIGNS AND SYMPTOMS	DIAGNOSTIC TEST(S)	TREATMENT
Aneurysm	Genetic predisposition, hypertension, trauma, atherosclerosis, tertiary syphilis (rare).	Pain, differences in right & left femoral pulses.	Radiologic studies.	Observation and , if severe or threatening enough, surgical repair.
Angina	Ischemia of the myocardium, as a result of poor blood perfusion of heart tissue (see *Atherosclerosis*).	Chest discomfort/ pain.	Patient history, blood work to rule out acute MI.	Nitroglycerin, oxygen, procedures to reopen cardiac vasculature/ perfusion.
Arteriosclerosis	"Hardening of the arteries" due to age, materials deposited in the inner walls of arteries, or other degenerative conditions.	See atherosclerosis.	See atherosclerosis.	See atherosclerosis.
Atherosclerosis (a peripheral vascular disease, or PPD)	The most common form of arteriosclerosis ("hardening of the arteries") and is caused by the buildup of a yellow, fat-like substance called plaque. Caused by heredity, improper diet, smoking, sedentary lifestyle, stress.	Often no signs or symptoms for many years, then increased blood pressure, angina, transient ischemic accident (TIA).	Angiogram, blood tests, stress testing, MRI, ultrasound imaging.	Improved diet, exercise, smoking cessation, stress reduction, balloon angioplasty, surgical repair.
Coronary artery disease (CAD)	An insufficient blood flow to heart tissue usually due to atherosclerosis but may be due to clot(s) in coronary artery.	The classic symptom is angina but may include: nausea, sweating, fainting, clammy hands, pulmonary edema, various locations of pain.	See atherosclerosis.	See atherosclerosis.

Figure 12-20 ■ shows the multisymptom effects of heart failure.

continues

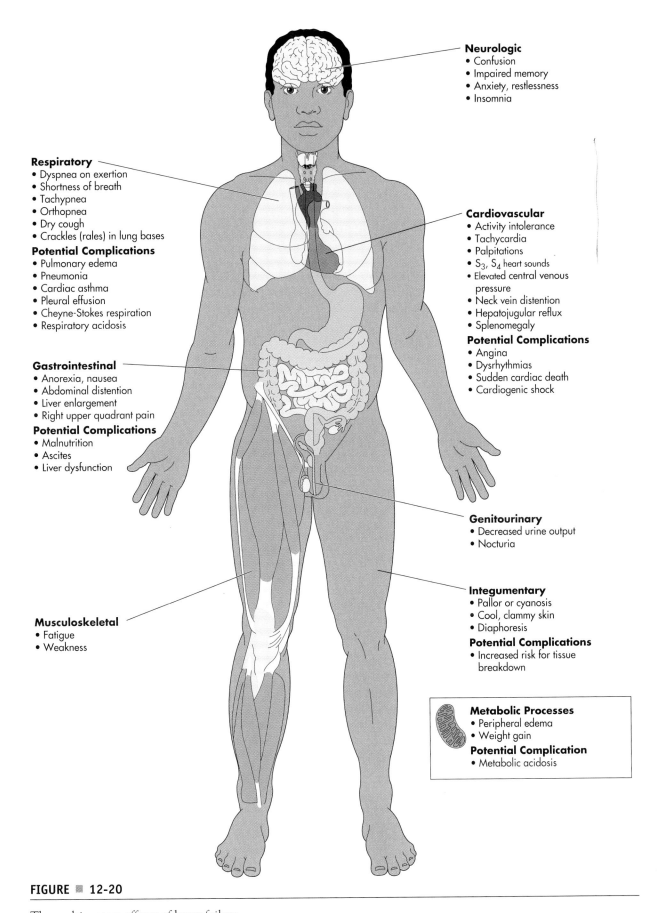

Neurologic
• Confusion
• Impaired memory
• Anxiety, restlessness
• Insomnia

Respiratory
• Dyspnea on exertion
• Shortness of breath
• Tachypnea
• Orthopnea
• Dry cough
• Crackles (rales) in lung bases
Potential Complications
• Pulmonary edema
• Pneumonia
• Cardiac asthma
• Pleural effusion
• Cheyne-Stokes respiration
• Respiratory acidosis

Cardiovascular
• Activity intolerance
• Tachycardia
• Palpitations
• S_3, S_4 heart sounds
• Elevated central venous pressure
• Neck vein distention
• Hepatojugular reflux
• Splenomegaly
Potential Complications
• Angina
• Dysrhythmias
• Sudden cardiac death
• Cardiogenic shock

Gastrointestinal
• Anorexia, nausea
• Abdominal distention
• Liver enlargement
• Right upper quadrant pain
Potential Complications
• Malnutrition
• Ascites
• Liver dysfunction

Genitourinary
• Decreased urine output
• Nocturia

Integumentary
• Pallor or cyanosis
• Cool, clammy skin
• Diaphoresis
Potential Complications
• Increased risk for tissue breakdown

Musculoskeletal
• Fatigue
• Weakness

Metabolic Processes
• Peripheral edema
• Weight gain
Potential Complication
• Metabolic acidosis

FIGURE ▪ 12-20

The multi-system effects of heart failure.

A Quick Trip through Some Diseases of the Cardiovascular System (*Continued*)

BLOOD VESSEL PROBLEMS (*continued*)

DISEASE	ETIOLOGY	SIGNS AND SYMPTOMS	DIAGNOSTIC TEST(S)	TREATMENT
Embolism (traveling blood clot)	Thrombophlebitis, venous stasis, inflammation, smoking, hyper-coagulation disorders, contraceptives, varicose veins, foreign particles, trauma, tumor, pregnancy.	Dependent upon location, size, and number of embolus.	Perfusion scan.	"Clot busting" medication, preventative measures that include: anti-coagulant medications, physical activity, treating thrombophlebitis, smoking cessation, surgical removal.
Hypertension (high blood pressure)	Heredity, diet (high fat and salt intake), age (increases the older you are), obesity, smoking, stress, decreasing flexibility of the blood vessels.	Often not noticed until a heart attack or stroke occurs, easily determined with a blood pressure cuff, increased mass of left ventricle, ringing in ears, not "feeling right," agitated/short temper. May lead to strokes, heart attacks, and kidney failure.	Blood pressure cuff measurement.	Changes in diet, weight loss, exercise, relaxation techniques, decreased alcohol intake, smoking cessation, medications (that either are diuretics, promote increased blood flow to the periphery, or affect heart function).
Myocardial infarct (heart attack)	Conditions/situations that lead to a decreased blood flow to the heart tissue that leads to ischemia and necrosis.	Indigestion, heavy/crushing/squeezing discomfort or pain in the center of the chest, radiating pain to left shoulder/arm, jaw, shoulder blades, anxiety, inability to sleep, dyspnea, nausea/vomiting, sweating.	ECG, physical examination, patient history.	CPR, oxygen, cardiac drugs, cardioversion or pacemaker, surgery.
Peripheral vascular disease (PVD)	Atherosclerosis, plaque buildup, in arteries with a progressive narrowing of the lumen. Diabetes can be a causative factor.	Pain, cramping, decreased blood flow to affected areas, decreased ability to heal, sensitivity to cold temperatures, necrosis, gangrene.	Patient history, physical examination, vascular studies.	Treatment of underlying cause, surgical intervention (cleaning out plaque or grafting on a healthy artery).

A Quick Trip through Some Diseases of the Cardiovascular System (*Continued*)

BLOOD VESSEL PROBLEMS (*continued*)

DISEASE	ETIOLOGY	SIGNS AND SYMPTOMS	DIAGNOSTIC TEST(S)	TREATMENT
Varicose veins	Swollen and distended blood veins normally found superficially in the legs but can be found in the rectal area (hemorrhoids). Caused by excessive blood pressure. Contributing factors: heredity, obesity, sitting or standing still for long periods of time, physical exertion, age, and pregnancy.	Symptoms vary in severity from none to extreme, including night cramps, dull aching, pain in feet and ankles, easily fatigued. Pain, itching and even bleeding for hemorrhoids.	Physical examination.	Weight loss, physical movement to prevent venous stasis (daily walking), support hose, elevation of legs, surgery, laser usage, sclerotherapy (saline injections). Stool softeners, surgery for hemorrhoids.

ELECTRICAL CONDUCTION PROBLEMS

DISEASE	ETIOLOGY	SIGNS AND SYMPTOMS	DIAGNOSTIC TEST(S)	TREATMENT
Bradycardia	Heart damage, increased vagal stimulation, certain drugs, may be idiopathic.	Slower than normal heart rate, usually a rate less than 60 beats per minute or less than 50 beats per minute at rest or during sleep, lethargy.	ECG.	Possible pacemaker.
Fibrillation (atrial or ventricular)	Acute MI, electrolyte imbalances, hypertrophy, electrical shock.	If ventricular, decrease of blood flow – collapse with possible impending cardiac arrest. If atrial – palpitations.	ECG.	CPR, defibrillation, medication.
Flutter	Intrinsic heart disease, or extrinsic influences such as alcoholism.	An unstable rhythm that can revert to a sinus rhythm or to a fibrillation.	ECG.	Cardioversion, medication.
Heart block	Scar tissue in heart muscle may affect the smooth flow of the electrical impulse.	Irregular beat, palpitations, slower pulse.	ECG.	Pacemaker.
Palpitation	Structural abnormalities, irritation/ inflammation/trauma to heart tissue, drugs, stress, various diseases of other body systems.	Irregular / abnormal heart rhythm.	ECG.	Correction of causative agent, medications.

continues

A Quick Trip through Some Diseases of the Cardiovascular System (*Continued*)

ELECTRICAL CONDUCTION PROBLEMS (*continued*)

DISEASE	ETIOLOGY	SIGNS AND SYMPTOMS	DIAGNOSTIC TEST(S)	TREATMENT
Premature contraction	In healthy people: may be due to caffeine, nicotine, or stress. Often common in patients with heart disease.	Irregular rhythm.	ECG.	Eliminate the causative agent.
Tachycardia	Initial response to a decrease in oxygen to the tissues, a result of stress or physical demands, heart disease. May also be idiopathic or congenital.	A more rapid than normal heart rate (more than 100 beats per minute) in absence of exercise / stimulation.	ECG.	Medication, certain types may respond to carotid massage, cardioversion, vagal maneuvers.

HEMATOLOGY RELATED PROBLEMS

DISEASE	ETIOLOGY	SIGNS AND SYMPTOMS	DIAGNOSTIC TEST(S)	TREATMENT
Anemia	Cancer, radiation, or chemotherapy which alters bone marrow; too little amino acids, folic acid, iron, or vitamins B_6 or B_{12}; production of defective erythrocytes; excessive blood loss.	Pallor, fatigue, tissue hypoxia, muscle weakness, shortness of breath, palpitations, angina.	Blood tests, patient history, physical examination.	Treat condition creating anemia, stop bleeding, replace erythrocytes, correct dietary deficiencies.
Erythrocytopenia	Decreased RBC production, increased RBC lysis, hemorrhage, dietary insufficiency, Hodgkin's disease, multiple myeloma, leukemia, Lupus erythematosus, Addison's disease, subacute endocarditis pregnancy, drugs.	Decreased number of blood cells for a given volume of blood, headache, fatigue, pallor, shortness of breath, or may be a life-threatening situation if severe.	Blood tests.	Treat causative agent, medications, blood transfusion.
Erythrocytosis	Polycythemia (vera & secondary), severe diarrhea, dehydration, pulmonary fibrosis, during and immediately after hemorrhaging, acute poisoning, high altitude living, drugs such as gentamicin.	Increased number of red blood cells for a given volume of blood, reddened skin tone, increased blood pressure, increased cardiac workload, blood-shot eyes, splenomegaly, increased risk of clots.	Patient history, physical examination, blood tests, radiologic studies.	Correction of causative condition.

A Quick Trip through Some Diseases of the Cardiovascular System (*Continued*)

HEMATOLOGY RELATED PROBLEMS (*continued*)

DISEASE	ETIOLOGY	SIGNS AND SYMPTOMS	DIAGNOSTIC TEST(S)	TREATMENT
Hemophilia	Inherited bleeding disorder, although ⅓ of the cases exhibit no familial history of hemophilia.	Bleeding episodes often into joint and muscle areas. Pain often accompanies these episodes. Internal bleeding always a possibility.	Blood tests.	Infusions of clotting factor, development of a synthetic clotting factor a consideration.
Hyperlipidemia	An increasing problem in which there is an excessive amount of fatty substances in the blood which include cholesterol, tri-glycerides and fatty acids.	High levels of low density lipoprotiens (LDL) and trigylercerides, lower than expected levels of high density lipoproteins (HDL), fat deposits in arteries, Hypertension and coronary artery disease.	Patient history, physical examination, blood work.	Dietary changes, increased activity level, stress reduction, statin medications, niacin.
Leukemia	Unknown; possibly, inherited factors, radiation, viral causes.	Headache, fatigue, dyspnea, sore throat, fever, swollen lymph nodes, bleeding of mucosal membranes of the mouth and GI tract, bone and/or joint pain, enlarged, lymph nodes/liver/ spleen, repeat infections, anemia.	Physical examination, blood work, radiologic exam.	Chemotherapy, bone marrow transplant.
Leukocytopenia	Viral infection, chronic infections, bone marrow depression, diabetes, alcoholism, acute leukemia, radiation, cancer chemotherapy.	Lower than normal white blood cell count for a given volume, decreased immunity.	Blood test.	Treat causative agent, medication.
Leukocytosis	Infection, leukemia, malignancy (esp. liver, bone, GI tract, metastasis), hemorrhage, trauma, tissue necrosis, certain drugs, toxins.	An increase in the number of white blood cells for a given blood volume.	Blood test.	Treat causative agent.

continues

A Quick Trip through Some Diseases of the Cardiovascular System (*Continued*)

HEMATOLOGY RELATED PROBLEMS (*continued*)

DISEASE	ETIOLOGY	SIGNS AND SYMPTOMS	DIAGNOSTIC TEST(S)	TREATMENT
Polycythemia	Primary result of myeloid leukemia. Often secondary to chronic hypoxia, high altitude living, congestive congenital heart disease. Can be a result of renal disease, liver cancer.	A higher than normal number of circulating red blood cells in relation to the normal blood volume – not to be confused with an increased red blood concentration due to a decrease in plasma volume such as in a burn patient.	Blood test, bone marrow biopsy.	Elimination of the causative agent.
Thrombocytopenia	Less than normal level of platelets, may be due to leukemia, bone marrow aplasia, hypersplenism, certain drugs, or following viral infection.	Spontaneous bleeding into tissue (including G.I. and urinary tracts), petechiae/purpura.	Blood tests, patient history.	Treat causative agent, platelet transfusions.

Pharmacology Corner

As previously stated, quick response time is necessary for the treatment of an MI. The main object is to rapidly get blood flow back to the heart muscle. Nitroglycerin increases blood flow to cardiac muscle through vasodilation of the coronary vasculature. Ischemic areas affected by coronary artery clots must have the blood vessels opened up to return the blood flow to that area. This is accomplished by administering some type of thrombolytic agent that actually dissolves the clot such as streptokinase. To prevent any future clot formations, anti-coagulants such as heparin and coumadin can be administered.

Oxygen is provided to decrease the work of the heart. Even if the blood supply to the heart is still reduced, the blood that gets there will be enriched with oxygen which is needed to sustain life. Other drugs may be given to improve heart function. Surgical intervention such as the previously discussed balloon catheter or CABG may be necessary.

Many groups are affected by high blood pressure, but post-menopausal women and African Americans are most frequently affected. Blood pressure medications can lower blood pressure in several ways: decrease blood volume (diuretics), vasodilate peripheral vessels (ACE, angiotensin-converting enzyme, inhibitors), or slow down the heart rate (beta-blockers or calcium channel blockers).

Although exercise and diet are critical for a healthy cardiovascular system, sometimes it is not enough. Drugs such as the statin class (Zocor, Lipitor) are prescribed to decrease total cholesterol levels in the blood to help prevent clogged arteries.

Extremely low blood pressure as in shock can affect perfusion to vital areas and lead to tissue damage and death. Drugs to treat shock would include agents that stimulate the heart to increase rate and force of contraction such as epinephrine and dopamine. In cases of blood loss, blood replacement and/or IV volume replacement therapy is needed.

 12-12 To view videos and animations on shock, bleeding control and shock management, starting an IV, and the administration of nitroglycerin, digoxin and dopamine, please see your DVD.

SUMMARY

Snapshots from the Journey

→ Like a system of rivers and canals, the cardiovascular system is responsible for transportation of oxygen, hormones, and nutrients to the tissues of the body and for removing the byproducts of metabolism by the cells.

→ The cardiovascular system is a closed, pressurized system much like the engine cooling system of a car.

→ The cardiovascular system also helps maintain proper fluid balance of the body and assists in the control of body temperature.

→ The cardiovascular system is a major player in the body's defense from infection.

→ The heart is an organ that is actually two pumps working together to move blood.

→ The heart's right pump moves blood collected from the body to the lungs, where oxygen is loaded and carbon dioxide is removed to be exhaled by the lungs.

→ The heart's left pump takes the freshly oxygenated blood and pushes it through the body so cells can be kept healthy.

→ Arteries carry blood away from the heart.

→ Veins carry blood to the heart.

→ Capillaries are blood vessels with walls the thickness of only one cell, which readily allows for the transfer of oxygen and nutrients to the tissues in the body.

→ This thinness of capillary walls also allows for waste products of the cells' metabolism to be picked up by the blood for removal.

→ The major components of blood are plasma, erythrocytes (red blood cells, the main transporter of oxygen), leukocytes (white blood cells, protectors from infection), and platelets (aid in the clotting of blood).

MARIA'S STORY

Maria (our diabetic) may develop complications that may impair her wound healing process. She is getting concerned as she gets older that this may become more of a problem and is aware of preventative measures. She is not exactly sure why the physicians are so worried about her cardiovascular status.

a. Explain how the cardiovascular system may be a contributing factor.

_____

_____

_____

b. What preventative measures can Maria take?

_____

_____

_____

c. Which specialist(s) would you consult to prevent these potential problems?

_____

_____

REVIEW QUESTIONS

Multiple Choice

1. A condition in which one side or the other of the heart cannot pump efficiently enough to overcome vascular resistance, resulting in leakage from the vessels into tissues, is:
   a. heart failure
   b. cardiac tamponade
   c. vesiculitis
   d. atherosclerosis

2. Plaque deposits in blood vessels are composed mostly of:
   a. platelets
   b. cholesterol
   c. fibrin
   d. heme

3. A localized weakness in the walls of a blood vessel is called a(n):
   a. aneurysm
   b. altruism
   c. embolism
   d. stint

4. Which of the following symptoms would *not* normally be related to anemia?
   a. shortness of breath
   b. fatigue
   c. increased urine output
   d. pallor

5. A general term to describe an inherited blood-clotting disorder is:
   a. hemorahgeia
   b. hemophilia
   c. hemoglobin
   d. polycythemia

Fill in the Blank

1. List the three ways that blood pressure can be controlled pharmacologically:

_____

_____

_____

2. Decreased blood flow to cardiac muscle that only *injures* the tissue creates a condition known as _____.

3. The structures composed of smooth muscle that direct the flow through capillary beds are called _____.

4. _____ is an important vitamin that is needed for the proper clotting of blood.

5. _____ is a term used for the dividing wall between the ventricles.

Matching

1. _____ Angina

2. _____ Murmur

3. _____ Varicose veins

4. _____ Anemia

5. _____ Aneurysm

6. _____ Atherosclerosis

7. _____ Embolism

a. The sound resulting from blood leaking backward due to a faulty heart valve or a narrowed valve opening

b. Distortion of the walls of a blood vessel due to genetics, hypertension, or trauma that can potentially rupture

c. Chest pain due to ischemia of the myocardium as a result of poor blood perfusion

d. Most common form of arteriosclerosis, caused by the build-up of plaque in the blood vessels

e. A blockage of blood flow due to a traveling clot caused by trauma, venous stasis, smoking, contraceptive use or inflammation

f. Dilated and distorted blood veins normally found superficially in the legs

g. Iron- poor blood due to radiation, chemotherapy, bone marrow disease or dietary deficiencies

Short Answer

1. Why is the direction of the wave of contraction so important in the heart?

_____

_____

_____

_____

2. Describe the flow of blood beginning at the right atrium and ending at the aorta.

_____

_____

_____

_____

3. Provide one reason the proper amount of iron is so important in your diet.

_____

_____

_____

_____

4. Why can polycythemia potentially cause a heart problem?

_____

_____

_____

_____

Suggested Activities

1. Make arrangements for a tour of a local blood bank to better understand the screening process for donors, how blood is typed, how it is stored, and how it is broken down into useful components.

2. Conventional wisdom dictates that a low-fat diet rich in fruits, vegetables, and grains keep you "heart healthy," yet low-carbohydrate, high-fat diets are popular for weight loss (which increases heart health) and for reducing problems associated with diabetes. Research and discuss both sides of the argument.

3. Research and present to the class your findings on all of the useful components of a donated unit of blood.

4. Find a partner and take each other's blood pressure and pulse while sitting in a chair, while lying down, and right after walking up a flight of stairs. Compare your results with your classmates'. Discuss any similarities or differences.

12-13 Now that you have completed your journey through this chapter, please go to the DVD for interactive games and puzzles concerning the medical terms and concepts contained in this chapter. By playing the games you will reinforce your learning of the cardiovascular system in a fun way.

THE RESPIRATORY System

It's a Gas

Without fuel, we wouldn't get very far on our trip. Our car wouldn't start without gasoline, our plane would be grounded without jet fuel, and our bodies would die without the fuel necessary for metabolism. The respiratory system's primary function is to transport the vital fuel, oxygen, from the atmosphere to the bloodstream to be utilized by cells, tissues, and organs for the processes necessary to sustain life. This amazing system is often taken for granted. We don't even consciously realize that the respiratory system is moving 12,000 quarts of air a day in and out of our lungs. On our journey, the fuel for our means of transportation also produces waste by-products or exhaust. For example, our car produces waste gases that it eliminates through its exhaust system. The body also produces a waste gas, carbon dioxide, during metabolism that needs to be eliminated via the respiratory system so it does not build up in toxic levels. In this chapter, we explore the journey oxygen molecules must take from the outside atmosphere to our cells and tissues. In addition, we travel with carbon dioxide as it leaves the respiratory system and is placed back into the atmosphere. Hopefully, your journey through the respiratory system will be a "breath-taking" experience.

Chapter 13

LEARNING OBJECTIVES

At the end of the journey through this chapter, you will be able to:

→ List and state the basic functions of the components of the respiratory system

→ Differentiate between respiration and ventilation

→ Explain how the respiratory system warms and humidifies inhaled air

→ State the purpose and function of the mucociliary escalator

→ Discuss the process of gas exchange at the alveolar level

→ Describe the various skeletal structures related to the respiratory system

→ Explain the actual process of breathing

→ Discuss several common respiratory system diseases

MULTIMEDIA APPLICATIONS

DVD Interactive Exercises

→ 3-D animation of the respiratory system and an interactive drag-and-drop labeling exercise, 13-1

→ Animation on gas exchange, 13-2

→ Video on allergic rhinitis, 13-3

→ Videos on cystic fibrosis, 13-4

→ Videos on intubation and on artificial airways and their care, 13-5

→ Videos on various oxygen therapy and delivery devices, 13-6

→ Interactive drag-and-drop labeling exercise concerning the alveolar area, 13-7

→ Videos on various pulmonary function testing, spirometry, and peak flow monitoring, 13-8

→ Animations and videos on COPD, asthma, and tuberculosis, 13-9

→ Videos on smoking, passive smoke, and smoking cessation, 13-10

→ Videos on assessing the thorax, 13-11

→ Videos on respiratory medications and their delivery systems, 13--12

→ Interactive games and puzzles, 13-13

www.prenhall.com/colbert

→ Professional Profiles:
  • Respiratory Therapist
  • Perfusionist

→ Related Internet Links

→ Additional Review Questions

Pronunciation Guide

 Correct pronunciation is important in any journey so that you and others are completely understood. Here is a "see and say" Pronunciation Guide for the more difficult terms to pronounce in this chapter.

adenoid (AD eh noid)

alveoli (al VEE oh lye)

atelectasis (at eh LEK tah sis)

bronchi (BRONG kye)

bronchioles (BRONG kee ohlz)

carina (kuh RINE uh)

cilia (SIL ee ah)

conchae (KONG Kay)

diaphragm (DIE ah fram)

emphysema (em fih SEE mah)

empyema (em pye EE mah)

epiglottis (ep ih GLOT is)

erythropoiesis (eh RITH roh poy EE sis)

hilum (HIGH lim)

laryngitis (lar in JIGH tis)

laryngopharynx (lahr IN goh FAIR inks)

larynx (LAIR inks)

lingula (LING gu lah)

mediastinum (mee dee ah STY num)

nasopharynx (nay zoh FAIR inks)

parietal pleura
    (pah RYE eh tal PLOO rah)

pharyngitis (far in JIGH tis)

pharynx (FAIR inks)

oropharynx (OR oh FAIR inks)

tuberculosis (too ber kew LOH sis)

trachea (TRAY kee ah)

SYSTEM OVERVIEW

Along your journey, you will undoubtedly have plenty of exercise walking to see the sights. Have you ever wondered why you feel out of breath or why breathing faster and deeper helps you to recover from strenuous exercise? Our body uses energy from the food we eat, but cells can obtain the energy from food only with the help of the vital gas **oxygen** (O_2), which allows for cellular **respiration**. Luckily, oxygen is found in relative abundance in the atmosphere and therefore in the air we breathe.

When the cells use oxygen, they produce the gaseous waste **carbon dioxide**, (CO_2). If allowed to build up in the body, carbon dioxide would become toxic, so the bloodstream carries the carbon dioxide to the lungs to be exhaled and eliminated from the body. The **respiratory system's** primary role, therefore, is to bring oxygen from the atmosphere into the bloodstream and to remove the gaseous waste by-product, carbon dioxide. The respiratory system, also called the **pulmonary system**, is closely connected with the heart and circulatory system. Due to their close relationship, these two systems are often grouped together in medicine to form the **cardiopulmonary system.**

The respiratory system consists of the following major components:

- Two lungs, the vital organs of the respiratory system
- Upper and lower airways that conduct gas in and out of the system

Amazing Body Facts

AUTOCONTROL OF CARDIOPULMONARY SYSTEM

The cardiovascular and pulmonary systems function without any conscious effort on your part. You probably didn't realize it, but as you read the previous paragraph and these last two sentences, your heart beat approximately 70 times and pumped approximately 5 liters of blood around your body. During that same time, you breathed approximately 12 times, moving over 6,000 milliliters of air.

cardi/o = *heart*

pulmon/o = *lungs*

- Terminal air sacs called alveoli surrounded by a network of capillaries that provide for gas exchange
- A thoracic cage that houses, protects, and facilitates function for the system
- Muscles of breathing that include the main muscle, the diaphragm, and accessory muscles

Please take a few minutes to look at Figure 13–1 ■. We will explore each of these components as we travel through the respiratory system.

Ventilation vs. Respiration

Before beginning our journey, it is important to pave the way with a solid understanding of some commonly confused concepts. The air we breathe is a mixture of several gases, listed in Table 13–1. The predominant gas is nitrogen (N_2), but this

13-1 To view a 3-D representation of the respiratory system and to perform an interactive drag-and-drop labeling exercise of the lungs and diaphragm, please go to your DVD for this chapter.

is an inert gas, which means it does not interact in the body. Even though nitrogen travels into the respiratory system and comes out virtually unchanged, it is vitally important as a support gas that keeps the lungs open

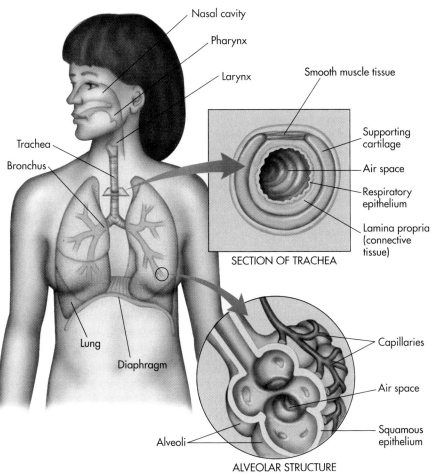

FIGURE ■ 13-1

The various components of the respiratory system.

with its constant volume and pressure. The next greatest concentrated gas is oxygen, and it is very physiologically active within our bodies. You'll notice that carbon dioxide is in low concentration in the air we inhale; it is in much higher concentration in the air we exhale.

TABLE 13-1 Gases in the Atmosphere

GAS	% OF ATMOSPHERE
nitrogen (N_2)	78.08
oxygen (O_2)	20.95
carbon dioxide (CO_2)	.03
Argon	.93

Note: The atmosphere also contains trace gases such as neon and krypton.

ventilation *(ven tih LAY shun)*
respiration *(ress pih RAY shun)*

The respiratory system contains a very intricate set of tubes that conduct gas from the atmosphere to deep within the lungs. This movement of gas is accomplished by breathing. However, a more precise look at the process of breathing shows that it is actually two separate processes. The first is **ventilation,** which is the bulk movement of the air in and out of the lungs. Ventilation moves fresh air down into the terminal end of the lungs, where the gas exchange will take place. The second is **respiration**, the gas exchange process, in which oxygen is added to the blood and carbon dioxide is removed. When an individual breathes, inspiration (breathing in) and expiration (breathing out) occurs, otherwise known as the ventilation process. The measurement of these ventilations per minute is typically referred to as the respiration rate and is approximately 14 to 20 respirations per minute. However, keep in mind respiration is the process of gas exchange.

Since the gas exchange in the lungs occurs between the blood and the air in the external atmosphere, it is more precisely called **external respiration**. The oxygenated blood is transported internally via the cardiovascular system to the cells and tissues, where gas exchange is now termed **internal respiration**, and oxygen moves into the cells as carbon dioxide is removed. Ventilation, therefore, is not the same thing as respiration. When you watch a television show that says place the patient on a respirator, it is incorrect and the person should say "a ventilator" because the

Applied Science

GAS EXCHANGE IN PLANTS

Fortunately for the earth's ecosystem, plant physiology of gas exchange is exactly opposite that of humans. Plants take in the CO_2 in the atmosphere and utilize it to help create energy, and then they release oxygen into our atmosphere. The earth's largest source of oxygen released is the Amazon rainforest, which is unfortunately being destroyed at a high rate every day. We truly need a green earth to survive, so thank the next plant you see.

www.prenhall.com/colbert
The respiratory therapist's main job is to ensure proper ventilation of the respiratory system through a variety of treatment modalities. A perfusionist, on the other hand, runs a machine that actually performs gas exchange. For example, during a heart or lung transplant, the blood supply is temporarily rerouted through a respirator until the new heart or lung is transplanted. Because this procedure requires a lot of time, the rerouted blood must be oxygenated and depleted of carbon dioxide via a special mechanical membrane. To learn more about these two related and exciting professions, please visit the companion website.

machine is only moving the gas mixture (ventilating), not causing gas exchange. See Figure 13–2 ▪, which contrasts these important processes.

13-2 To view an animation on gas exchange, please go to the DVD for this chapter.

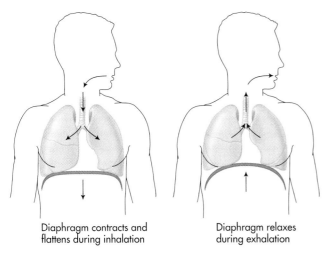

Diaphragm contracts and flattens during inhalation

Diaphragm relaxes during exhalation

VENTILATION

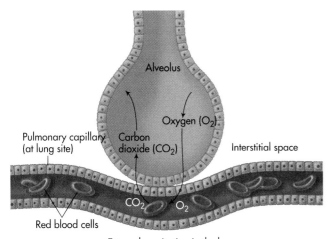

Alveolus

Oxygen (O₂)

Pulmonary capillary (at lung site)

Carbon dioxide (CO₂)

Interstitial space

CO₂   O₂

Red blood cells

External respiration in the lungs

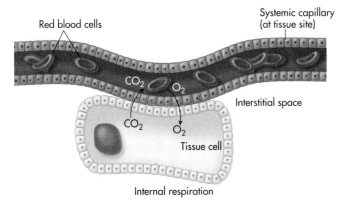

Red blood cells

Systemic capillary (at tissue site)

CO₂   O₂

Interstitial space

CO₂

O₂

Tissue cell

Internal respiration

RESPIRATION

FIGURE ▪ 13-2

Contrast of ventilation and external and internal respiration.

THE RESPIRATORY SYSTEM

The respiratory system is responsible for providing all of the body's oxygen needs and removing carbon dioxide. Unlike cars that have large gas tanks for fuel storage, the human body has a very small reserve of oxygen. In fact, its oxygen reserve lasts only about 4 to 6 minutes. If that reserve is used up and additional oxygen is unavailable, then death is the obvious outcome. Therefore, the body must continually replenish its oxygen by bringing in the oxygen molecules from the atmosphere. Again, this process is called ventilation. We now begin our breathtaking journey through the internal structures of the respiratory system by following the path oxygen molecules take.

The Airways and the Lungs

bronchi *(BRONG kye)*
bronchioles *(BRONG kee ohlz)*
alveoli *(al VEE oh lye)*

In a general sense, the respiratory system is a series of branching tubes called **bronchi** and **bronchioles** that transport the atmospheric gas deep within our lungs to the small air sacs called **alveoli,** which represent the terminal end of the respiratory system. The tissue of the lung is light, porous, and spongy. It floats if placed in water, because of the air in the alveoli.

To better visualize this system, consider the shape of a stalk of broccoli held upside down. The stalk and its branchings represent the airways, and the green bumpy stuff on the end is like the terminal alveoli. See, not only is broccoli good for you, but it can also be a learning tool.

capillaries *(KAP ih LAIR eez)*

Each alveolus is surrounded by a network of small blood vessels called **capillaries.** This combination of the alveolus and the capillary is called the **alveolar-capillary membrane** (respiratory membrane) and represents the connection between the respiratory and cardiovascular systems. This is where the vital process of gas exchange takes place. Before getting in depth into this process, let's trace the journey that oxygen molecules must take in order to arrive at the alveolar-capillary membrane.

The Upper Airways of the Respiratory Tract

The upper airways, which start at the nose, are responsible for initially conditioning the inhaled air. They also perform several other important functions.

UPPER AIRWAY FUNCTIONS

The upper airways begin at the two openings of the nose, called **nostrils** or **nares,** and end at the **vocal cords** (see Figure 13–3 ■). The functions of the upper airway include:

- Heating or cooling inspired (inhaled) gases to body temperature (37 degrees Celsius)
- Filtering particles from the inspired gases
- Humidifying inspired gases to a relative humidity of 100%
- Providing for the sense of smell, or **olfaction**
- Producing sounds, or **phonation**
- Ventilating, or conducting, the gas down to the lower airways

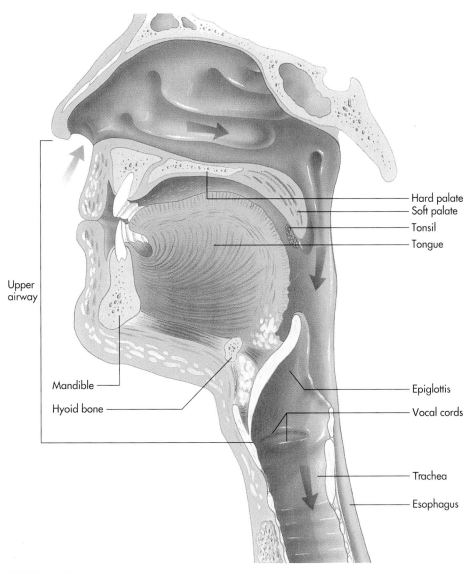

Upper airway

Hard palate
Soft palate
Tonsil
Tongue

Mandible
Hyoid bone

Epiglottis
Vocal cords

Trachea
Esophagus

FIGURE ▪ 13-3

The upper airway.

THE NOSE

While some people do breathe in through their mouths, under normal circumstances we were meant to breathe in through our noses, for reasons that will become clear. The nose is a rigid structure made of cartilage and bone. The **nasal cavity** is the space behind the nose, and consists of three main regions: vestibular, olfactory, and respiratory. The two nasal cavities are separated by a wall called the **septal cartilage**, or **nasal septum**, which divides the nose into right and left sides. Please see Figure 13–4 ▪.

The **vestibular region** is located inside the nostrils and contains the coarse nasal hairs that act as the first line of defense for the respiratory system. These hairs, called **vibrissae**, are covered with sebum, a greasy substance secreted by the sebaceous glands of the nose. Sebum helps to trap large particles and also keeps the nose hairs soft and pliable. Those unwanted nose hairs really do

nasal cavity *(NAY zl CAV ih tee)*

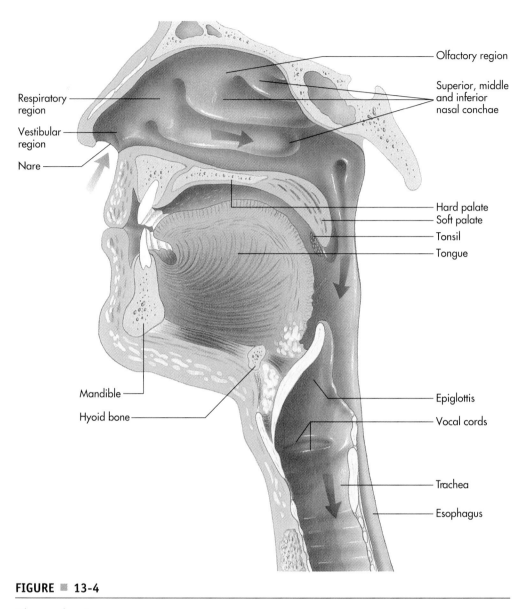

FIGURE ■ 13-4

The nasal regions.

have a role as a gross particle filter. "Gross" in this case means large. The vestibular region helps to filter out large particles so they do not enter the lungs, where they could irritate and clog the airways. You may also sneeze after inhaling particles, such as dust or allergens. Sneezing forces air quickly through the nose, to clear the upper respiratory tract. They try and keep those foreign particles out of your nose. The **olfactory region** is strategically placed on the roof of the nasal cavity. The advantage to this is that sniffing inspired gas into this region keeps it there and does not allow the gas to reach deeper into your lungs. This is a relatively safe way to sample a potentially noxious or dangerous gas without taking a deep breath into the lungs, where it could cause severe damage. Only four molecules of a substance are required for the olfactory senses to detect a gas! It is interesting to note that your ability to taste is related to your sense of smell. If you ever had a nasty head cold and could not taste your food as well, now you know why.

The air from the atmosphere must be warmed to body temperature and must also be moistened so the airways and the lungs do not dry out. This is a job for the **respiratory region**. Even though it is called the respiratory region, no respiration or gas exchange takes place here. Keep in mind that the respiratory region resides in the nasal cavity, which is lined with mucous membranes that are richly supplied with blood. The respiratory region possesses three scroll-like bones known as **turbinates,** or **conchae** (again see Figure 13–4). These split up the gas into three channels, thereby providing more surface area for incoming air to make contact with the nasal mucosa. While the respiratory region has a small volume of 20 milliliters, if you could unfold the turbinates, you would have a surface area of 106 square centimeters (about the size of a post card)!

conchae *(KON kay)*

Because the nasal cavity is not a straight passageway, the air current becomes turbulent so more air makes contact with these richly vascularized mucous membranes that transmit heat and moisture to the inspired gas. Incredibly, these moist mucous membranes add 650 to 1000 milliliters of water *each day* to moisten inspired air to 80% relative humidity within the respiratory region of the nose. In such a short distance, this is a pretty impressive humidification process. When the furnace in your house turns on in cold weather, this may dry the inspired air significantly and make it harder for your respiratory region to work. Therefore, humidifying this dry gas with water, such as from a room humidifier, may keep added stress off of your body's natural humidification system.

Amazing Body Facts

WHY DO WE BREATHE THROUGH OUR MOUTH?

You may have seen your favorite football player wearing an odd-looking strip of plastic across the nose. This is to help make breathing easier by increasing the diameter of the nostrils. The nose is responsible for one half to two thirds of the total airway resistance in breathing. Airway resistance represents the work required to move the gas down the tube. The larger the tube, the less resistance and thus less work involved in breathing. Therefore, mouth breathing predominates during stress and exercise because it is easier for the gas (less resistant) to travel through the larger oral opening. Of course, when you get a head cold and your nasal passages become blocked by secretions, it becomes necessary to breathe through the easier or open route of the mouth. On your journey, you must always be ready for detours.

Pathology Connection: The Nose

Often the nose can indicate the body's sensitivity to certain airborne allergens such as seasonal pollens from grass or ragweed. **Allergic rhinitis** occurs when allergens trigger the nasal mucosa to secrete excessive mucus, causing the infamous runny nose. Because the release of histamine is often the culprit, antihistamines can be used to treat many cases. Some individuals can also take allergy injections to desensitize their reaction to the specific allergen. (For more on allergies, see Chapter 14, "The Immune System.")

Nasal polyps are non-cancerous growths within the nasal cavity. The exact cause is unknown but it is believed that they occur due to chronic inflammation. If they become large enough, they can obstruct the nasal passageway and make breathing difficult. In these cases, surgical removal is necessary.

13-3 To view a video on allergic rhinitis, please see your DVD.

TEST YOUR KNOWLEDGE 13-1

Choose the best answer:

1. Which gas is found in the atmosphere?
   a. oxygen
   b. nitrogen
   c. carbon dioxide
   d. all the above

2. The process of moving gas in to and out of the respiratory system is:
   a. ventilation
   b. external respiration
   c. internal respiration
   d. diffusion

3. The process of gas exchange at the tissue sites is called:
   a. ventilation
   b. external respiration
   c. internal respiration
   d. osmosis

4. Gas exchange takes place across the:
   a. bronchi
   b. bronchioles
   c. alveolar-capillary membrane
   d. heart

5. The bones in the respiratory region of the nose are called:
   a. sinuses
   b. turbinates
   c. tetonic plates
   d. bronchioles

Complete the following:

6. Non-cancerous nasal growths are called _____ and seasonal _____ can occur with the release of airborne pollens.

---

pseudostratified
   pseud/o = *false*
   stratified = *layers*
cilia *(SIL ee ah)*

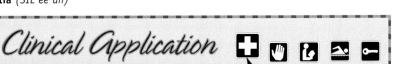

EXCESSIVE MUCUS PRODUCTION

Some diseases, such as chronic bronchitis and cystic fibrosis, cause excess mucus to be produced, which must be expectorated or swallowed. The color of the expectorant is clinically useful. In general, yellow and green sputum indicate an infection, but also indicate that the immune system is helping to fight an infection, because the mucus becomes green when enzymes (from the defense cells) that fight bacteria are activated. Red or brown mucus indicates that bleeding has occurred, which is seen in lung cancer or advanced tuberculosis. Cystic fibrosis causes the exocrine glands to secrete very thick mucus. This blocks the airways and increases the susceptibility to bacterial infections.

GOING TO RIDE THE MUCOCILIARY ESCALATOR

The epithelial lining of the respiratory region of the nose plays a very important role in keeping this region of the nose clean and free of debris buildup. Cells in the epithelial layer (or respiratory mucosa) are called **pseudostratified ciliated columnar epithelium,** and are found not only in the respiratory region of the nose but throughout most of the airways. See Figure 13–5 ■. The **epithelium** is a single layer of tall, column-like cells with nuclei located at different heights, giving the false appearance of two or more layers of cells when in fact there is only one—hence the term *pseudostratified columnar.* Each columnar cell has 200 to 250 **cilia** on its surface.

Goblet cells and submucosal glands are interspersed in the respiratory mucosa and produce about 100 milliliters (approximately 1/4 pint) of mucus per day. The mucus actually forms in two layers. The cilia reside in the **sol layer,** which contains thin, watery fluid that allows them to beat freely. The **gel layer** is on top of the sol layer; as its name suggests, it is more viscous or gelatinous in nature. This sticky gel layer traps small particles, such as dust or pathogens, on the mucus blanket, much like fly paper. Once debris is trapped on the mucus blanket, it must be removed from the lung.

So how does this mucus layer actually work? The microscopic, hair-like cilia act as tiny "oars," and in Figure 13–5, you can see that these oars rest in the watery sol layer. They beat at an incredible rate of 1,000 to 1,500 times per minute and propel the gel layer and its trapped debris onward and upward about 1 inch per minute to be expelled from the body. When this process occurs in the nose, the debris-laden secretions are pushed toward the front of the nasal cavity to be expelled through the nose. The pseudostratified ciliated columnar epithelium, located in the airways of the lungs, propels the gel layer toward the oral cavity to be either expectorated with a cough or swallowed into the stomach. Some texts refer to this epithelial layer as the *mucociliary escalator,* which gives a better picture of what it does. This escalator works 24/7—that is, unless something paralyzes it, such as smoking.

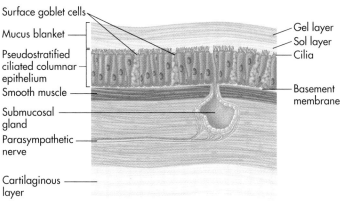

FIGURE ■ 13-5

The mucociliary escalator.

13-4 To view a video on cystic fibrosis, please see your DVD.

The Sinuses

Have you ever heard of someone being called an airhead? Technically we are *all* airheads because the skull contains air-filled cavities (commonly called **sinuses**) that connect with the nasal cavity via small ducts (passageways). Because these are located around the nose, they are called **paranasal sinuses**. They are lined with respiratory mucosa (mucous membrane) that continually drain their secretions into the nasal cavity, warming and moistening the air passing through the nasal cavity. The sinuses are named for the specific facial bones in which they are located. (See Figure 13–6 ■.)

These cavities of air in the bones of the cranium are believed to help prolong and intensify sound produced by the voice. If you ever shouted inside a cave, you noticed a more resonant quality to the sound. In addition, it is theorized that the air-filled sinuses help to lighten your head. Imagine how heavy your head would be if those facial bones were solid!

Sinuses do not exist at birth, but develop as you grow. Facial changes are influenced by the sinuses as you mature. Sinuses also provide further warming and moisturizing of inhaled air.

paranasal sinuses
  para = *around*
  nasal = *referring to nose*
  sinuses = *hollow cavity*

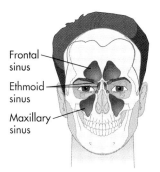
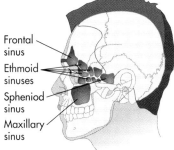

FIGURE ■ 13-6

The paranasal sinuses.

The Pharynx

The **pharynx** (the **throat**) is a hollow, muscular structure, about 2 1/2" long, lined with epithelial tissues. It serves as a common passageway for both air and food. The pharynx begins behind the nasal cavities and is divided into the following three sections, as shown in Figure 13–7 ■.

- nasopharynx
- oropharynx
- laryngopharynx

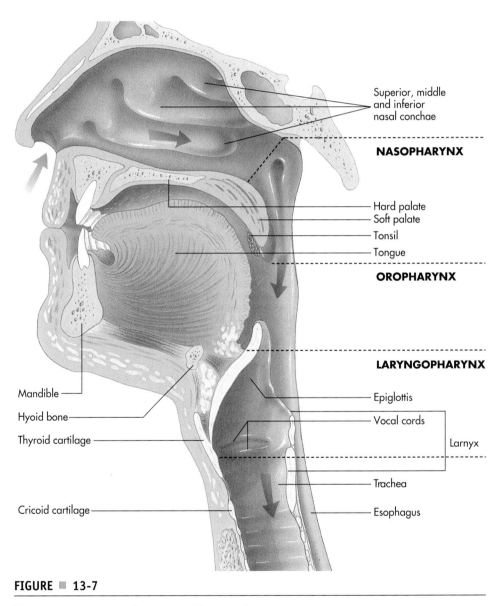

FIGURE ■ 13-7

The nasopharynx, oropharynx, and laryngopharynx.

The **nasopharynx** is the uppermost section of the pharynx and begins right behind the nasal cavities. Air that is breathed through the nose passes through the nasopharynx. This section also contains lymphatic tissue of the immune system, called the **adenoids,** and passageways to the middle ear called **Eustachian tubes.** You can understand how an infection located in the nasal cavities can lead to an ear infection, and vice versa.

The **oropharynx** is the next structure, located right behind the oral or buccal cavity. The oropharynx conducts not only atmospheric gas but also food and liquid. Air breathed through both the nose and the mouth passes through here, as does anything that is swallowed.,

The oral entrance is a strategic area to place "guardians" for the immune system because this is where pathogens can easily enter the body. Lymphoid tissue such as the **palatine tonsils** are located in this area. Another set of tonsils, the **lingual tonsils**, are found at the back of the tongue. During the process of swallowing, the uvula and soft palate move in a posterior and superior position to prevent the nasopharynx and nasal cavity from allowing food or liquid to enter the respiratory system. Swallow and feel this happening within your oral cavity. This protection can be overcome by forceful laughter, and that is why when you laugh with liquid or food in your mouth, the food or liquid can sometimes come up through your nose.

The **laryngopharynx** is the lowermost portion of the pharynx; an older term for it was the hypopharynx because of its position. Air that is breathed and anything that is swallowed passes through the laryngopharynx. Swallowed materials pass through the **esophagus** to get to the stomach, and air travels through the larynx (the next part of the respiratory system) and then the trachea on its way to the lungs. What directs the flow of "traffic" (air to the lungs and food and liquid to the stomach)? Is it a tiny highway worker directing traffic? No, it is directed by the swallowing reflex to be discussed soon.

The Larynx

The **larynx**, located in the neck, is a triangular chamber below the laryngopharynx that houses the important structures needed for speech. Commonly known as the voice box, the larynx is a semi-rigid structure composed of cartilage fibrous plates connected by muscles and ligaments that provide for movement of the vocal cords to control speech. The "Adam's apple" is the largest of the cartilages found in the larynx and is more prominent in males than females. This cartilage is also

adenoid *(AD eh noid)*
Eustachian tubes
(yoo STAY she ehn)

palatine
(PAL ah tine)

KEEPING THE VITAL AIRWAY OPEN

Just like any vital highway, the airway needs to remain open to a flow of traffic, or the oxygen molecules will cease to flow into the alveoli and therefore not get into the bloodstream to supply the tissues. While a traffic jam can last for hours, oxygen flow can be disrupted for only a few minutes without tragic results. For example, if the upper airway swells shut from a severe allergic reaction to a bee sting, an emergency airway must be established. Referring back to Figure 13-7, you see a space between the thyroid and cricoid cartilages and this is where an emergency cricoidthyroidotomy is performed. This space has few blood vessels or nerves, which makes it ideal in an emergency situation to place a temporary breathing tube.

Sometimes a longer-term breathing tube must be inserted into the lungs via a technique called **intubation**. This tube passes through the vocal cords and sits about 2.5 cm above the juncture (carina) between the right and left lung. A machine called a ventilator can move air into and out of the damaged lungs at this juncture. The tube has an inflatable cuff to seal the airway once it is in place. Knowledge that the vocal cords open and close during breathing becomes clinically significant. Adduction during expiration seals the vocal cords, whereas abduction during inspiration opens the cords, increasing the size of the glottic opening. If the patient is breathing, it is better to pass the tube with the deflated cuff through the narrow opening of the vocal cords during an inhalation, when the cords are open. Conversely, when removing the tube (**extubation**), the health care professional must always remember to deflate the cuff so it doesn't damage the cords as the tube is pulled back through them. It should go without saying (no pun intended) that a patient cannot talk when intubated because the vocal cords cannot function properly. If you ever see a TV soap opera or movie where someone is talking with a tube going down his or her mouth into the lungs, you will know it is impossible. More permanent airways, called **tracheostomy tubes**, can be placed in the anterior portion of the neck.

esophagus *(eh SOFF ah guss)*

anatomically known as the **thyroid cartilage**, beneath which is the **cricoid cartilage**. Both cartilages in the exposed areas of airways found in the neck are necessary to provide structure and support for airways so they do not collapse and block the flow of air in and out of the lungs.

The space between the vocal cords is called the **rima glottis**, or simply the **glottis**. The glottis is the opening that leads into the larynx and eventually the lungs. There is a leaf-shaped, flap-like fibrocartilage, located above the glottis, called the **epiglottis.** The epiglottis closes over the opening to the larynx when you swallow and opens up when you breathe, as part of swallowing reflex. This selective closure is called the **glottic** or **sphincter** mechanism. It facilitates the closing of the epiglottis over the glottic opening, sealing it so food does not enter the lungs.

The lungs are closed to traffic when swallowing. As you swallow, the food and liquid travel down the only open tube or route, which is the esophagus, leading into the stomach. When you breathe in, the gas preferentially travels into the lungs through a process that actually draws it into the lungs because of pressure differences. More on this soon.

The **vocal cords** are the area of division between the upper and lower airways, representing the point of transition to the lower airways. The lower airways start below the vocal cords, and we soon continue our journey down the lower airways all the way down to the end point, or alveoli.

glottis *(GLOT is)*

epiglottis *(ep ih GLOT is)*
   epi = *above*

13-5 To view videos on intubation, artificial airways, and their care, please go to your DVD.

Pathology Connection: The Upper Airways

The major pathologies of the upper airways are infections or allergies of the sinuses, nose, or throat. The **common cold** is familiar to everyone and is caused by over 200 different strains of viruses. A cold is a contagious disease with an acute inflammation of the mucous membranes of the upper airway which causes swelling and congestion. You may become more susceptible to the cold virus if your immune system has been weakened by lack of sleep, poor nutrition, stress, or fatigue. Exposure to cigarette smoke can also make you more susceptible. There is no cure for the common cold, but the symptoms can be treated. The best way to prevent the spread of a cold is to wash your hands and cover your mouth when you sneeze or cough. Often colds, allergies, and the flu are confused, and can lead to misdiagnosis and therefore mistreatment. See Table 13-2, which compares and contrasts the differences.

Sinusitis is an infection of the sinus cavities by viruses or bacteria with subsequent inflammation of the mucous membrane linings. Allergies can also affect the sinuses and cause inflammation. This inflammation causes pressure, pain, and often a headache. The tonsils, pharynx, and larynx can also be infected with viruses and bacteria and even other microorganisms such as yeast. Again, the resulting inflammation can lead to swelling, redness and pain. **Tonsillitis** causes the tonsils to swell and become painful, especially when swallowing. If severe, a tonsillectomy may be needed. **Pharyngitis** (sore throat) can cause discomfort, especially when swallowing. **Strep throat** is caused by the streptococcus bacteria, and causes a red, purulent (forming pus), and painful throat. **Laryngitis,** or an inflamed voice box, is characterized by

TABLE 13-2 Comparison of Allergy, Cold, and Influenza

SYMPTON/SIGN	ALLERGY	COLD	FLU
Onset	Response to allergen	Slow	Fast
Duration	> 1+ Weeks	1 week	1-3 weeks
Season	Spring/Summer/ Fall	Fall/Winter	Fall/Winter
Fever	None	Rare, <100	102-104; lasts 3-4 days
Fatigue, weakness	Mild; varies	Quite mild	May last 2-3 weeks
Extreme exhaustion	Rare	Never	Early and prominent
General aches, pains	Varies	Slight	Usual; often severe
Headache	Usual	Rare	Prominent
Sneezing	Usual	Usual	Sometimes
Sore throat	Sometimes	Common	Sometimes
Stuffy nose	Common	Common	Sometimes
Nasal discharge	Clear	Yellow/greenish	Varies
Chest discomfort	Sometimes	Mild to moderate	Common; can be severe
Cough	None; dry	Yes	Yes; dry hacking
Vomit and/or diarrhea	Rare; varies	Rare	Common

hoarseness and loss of speech. Laryngitis can also be caused by excessive use of the voice such as in prolonged singing. Please see Figure 13-8 ■ for examples of locations of these infections. If bacteria are known to be the cause of any of the above infections, an antibiotic can be used. However, if no bacterial infection is involved, the symptoms should be treated until the infection or allergy symptoms run their course.

Croup was the former confusing term used for both LTB and acute epiglottitis, but it is clinically important to differentiate the two because of the life-threatening nature of acute epiglottitis.

Acute epiglottitis is an airway emergency in which the bacteria *Haemophilus influenzae type B* (often abbreviated *H. influenzae type B* or *Hib*) causes acute swelling of the epiglottis and airway obstruction. Its onset is fast, and fever and sore throat are usually the first symptoms. Acute epiglottitis is most prevalent in children ages 2 to 6 and requires rapid recognition and treatment. Since the introduction of the universal Hib vaccine, the incidences have been decreasing. Airway maintenance is the mainstay of treatment, with antibiotic therapy selective against *H. influenzae type B*, although other pathogens can still cause the disease.

Laryngotracheobronchitis (LTB) is different from epiglottitis but also results from an infection of the laryngeal area. It too can cause airway

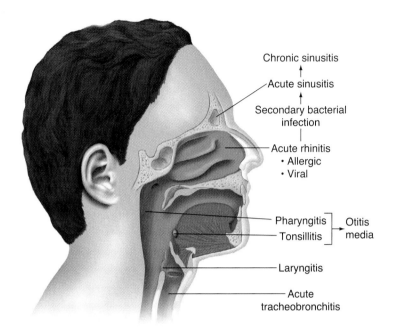

FIGURE ■ **13-8**

The upper airway and related infections.

obstruction and is characterized by noisy breathing, especially on inspiration. LTB can be viral or bacterial and the clinical presentation is that of a barking cough and inspiratory stridor (a high pitched harsh sound often heard without a stethoscope).

Sleep apnea is a condition in which breathing stops during sleep. Sleep apnea is most commonly caused by soft tissue at the back of the throat relaxing and blocking the airway. A more rare form of sleep apnea is caused by failure of the brain to communicate with muscles involved in breathing. When a sleeping person stops breathing, the brain briefly wakes him or her up to trigger restored breathing. A person with sleep apnea may awaken briefly many times during the night but not remember waking up. This interrupted sleep can cause fatigue during the day. Long-term, undiagnosed sleep apnea can cause other health problems, such as high blood pressure, weight gain, and headaches. Sleep apnea can be diagnosed during a sleep study, where the patient is monitored while sleeping. Mild sleep apnea may be treated with special pillows that help to keep the airway open. Weight loss may reduce the severity of sleep apnea. More severe sleep apnea may be treated with oral appliances, breathing devices, or surgery.

Learning Hint

EPIGLOTTITIS

Respiratory distress, drooling, dysphagia, and dysphonia are the four D's that are signs of epiglottitis.

TEST YOUR KNOWLEDGE 13-2

Choose the best answer:

1. The hair-like structures that propel mucus in the airways are:
   a. the sol layer
   b. the gel layer
   c. pathogens
   d. cilia

2. Which of the following is *not* true about the sinuses?
   a. they are air-filled cavities
   b. they are located in the skull and around the nose
   c. they help to lighten the head
   d. gas exchange occurs there

3. Food is prevented from entering the _____ when eating by the closure of the _____.
   a. esophagus, glottis
   b. esophagus, epiglottis
   c. trachea, epiglottis
   d. epiglottis, glottis

4. The vocal cords are found in the:
   a. nasal cavity
   b. nasopharynx
   c. nasal cavity
   d. larynx

5. Match the disease

   1 _____ Common cold   a. sore throat will cause discomfort especially when swallowing.

   2 _____ Sinusitis   b. characterized by hoarseness and loss of speech.

   3 _____ Pharyngitis   c. caused by over 200 different strains of viruses.

   4 _____ Laryngitis   d. airway emergency distress, drooling, dysphagia, and dysphonia

   5 _____ Acute Epiglottitis   e. infection causing pressure, pain and often a headache.

   6 _____ LTB   f. noisy breathing, barking cough and inspiratory stridor

THE LOWER RESPIRATORY TRACT

The airway that leads to the lungs and then branches out into the various lung segments resembles an upside-down tree and is sometimes called the tracheobronchial tree. See Figure 13–9 ■. Upon leaving the vocal cords in the larynx, the inspired air enters the **trachea,** also known as the windpipe, a 4 1/2″ long tube lined with ciliated mucous membrane. Any dust or mucus that also enters the trachea can be removed by expectoration. During expectoration, deep breaths followed by forceful air expulsion (coughing) moves substances out of the lower respiratory tract and into the mouth, where they can be spit out or swallowed. The trachea extends from the cricoid cartilage of the larynx to the sixth cervical vertebrae (approximately to the midpoint of the chest). The cartilage found in the trachea is in the form of C-shaped structures in the anterior portion of the trachea to provide rigidity and protection for the exposed airway in the neck. This C shape also serves another important purpose: the esophagus lies in the area where the C opens up posteriorly. Without the cartilage, there is some "give" in the posterior aspect of the larynx and trachea, so there is room for the esophagus to expand when you swallow larger chunks of food.

trachea *(TRAY kee ah)*

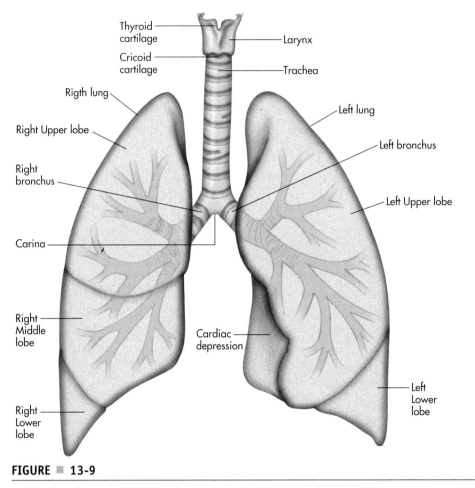

FIGURE ■ 13-9

The tracheobronchial tree.

The trachea is the largest bronchus and can be thought of as the trunk of the tracheobronchial tree. Once the trachea reaches the center of the chest, it begins its first branching, or bifurcation, into two bronchi (*bronchus* is the singular form), the **right mainstem** and the **left mainstem**. The site of bifurcation is called the **carina** (again, see Figure 13–9). One bronchus goes to the right lung and the other bronchus goes to the left lung. The mainstem bronchi are sometimes also referred to as the **primary bronchi**. Now the bronchi must branch into the five lobar bronchi that correspond to the five lobes of the lungs.

Each lung lobe is further divided into specific segments, and the next branching of bronchi are called the **segmental bronchi**. At the point from the trachea down to the segmental bronchi, the tissue layers of the bronchi are all the same, only smaller, as they branch downward. The first layer is the epithelial layer, which contains the mucociliary escalator—or pseudostratified ciliated columnar cells—that keeps the area clean of debris. A middle lamina propria layer contains smooth muscle, lymph, and nerve tracts. The third layer is the protective and supportive cartilaginous layer (see Figure 13–10 ■).

The branching becomes more numerous with tiny **subsegmental bronchi** that branch deep within each lung segment. The diameter of subsegmental bronchi ranges from 1 to 6 millimeters. Cartilaginous rings are now irregular

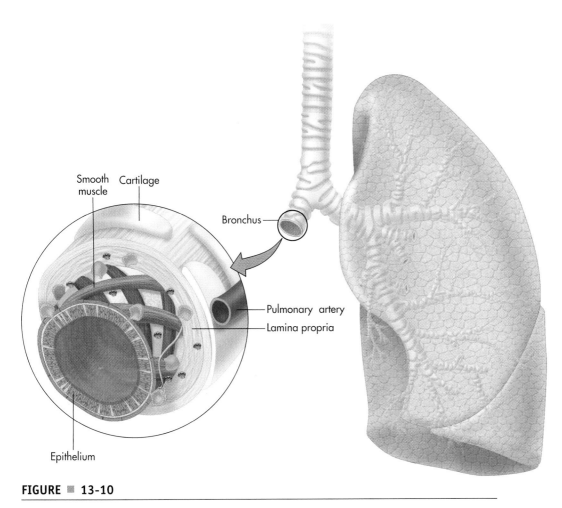

FIGURE ▧ 13-10

Tissues layers in the bronchi.

pieces of cartilage and will soon fade away completely. Notice as we move toward the gas exchange regions that the airways simplify to make it easier for gas molecules to pass through. Next are the very tiny airways called **bronchioles** that average only 1 millimeter in diameter. They have no cartilage layer, and the epithelial lining becomes ciliated cuboidal cells (short squat cells as opposed to large columns). The cilia, goblet cells, and submucosal gland are almost all gone by this point. There is no gas exchange yet, just simple conduction of the gas mixture containing the oxygen molecules down the tree. The **terminal bronchioles,** which have an average diameter of .5 millimeters, no goblet cells, no cartilage, no

Clinical Application ✚ ✋ 🧍 🏊 🗝

THE ANGLE MAKES A DIFFERENCE

The angle of branching is not the same for both sides of the tracheobronchial tree. The right mainstem branches off at a 20 to 30 degree angle from the midline of the chest. The left mainstem branches off at a more pronounced 40 to 60 degree angle. This is important because the lesser angle of the right mainstem branching allows foreign bodies that are accidentally breathed in to more often lodge in the right lung. This is nice to know if a child has aspirated (taken into the lung) an object and the physician must enter the lung with a bronchoscope to remove the object. Time may be critical, and it may make a difference if the search is begun immediately in the right lung because its anatomic structure increases the probability that the object has lodged there. In addition, an endotracheal tube (breathing tube) may be placed too far into the lung and instead of sitting above the carina so both lungs are ventilated, the tube will most likely pass into and ventilate only the right lung. This is why an X-ray for proper tube placement is so important.

Clinical Application

THERAPEUTIC OXYGEN

Often, a distressed respiratory system and sometimes the cardiac system needs supplemental oxygen to assist its function and meet its needs. There are many ways to deliver an enriched oxygen supply to the lungs, including various oxygen masks, nasal cannula (tube), and specialized devices to deliver both oxygen and extra humidity to the lungs to assist their function.

cilia, and no submucosal glands, mark the end of the conducting areas. Finally we journey into the gas exchange or respiratory zone of the lung.

The next airway beyond the terminal bronchiole is called the **respiratory bronchiole** because a small portion of gas exchange takes place here. The epithelial lining consists of simple cuboidal cells interspersed with actual alveoli-type cells, which are flat, pancake-like cells called **simple squamous pneumocytes**. Alveolar ducts originate from the respiratory bronchioles, wherein the walls of the alveolar ducts are completely made up of simple squamous cells arranged in a tubular configuration. The alveolar ducts give way to the grape bunch–like structures of several connected alveoli, better known as the **alveolar sacs**. See Figure 13–11 ■.

The Alveolar Capillary Membrane: Where the Action Is

The alveoli are the terminal air sacs that are surrounded by numerous pulmonary capillaries and together make up the functional unit of the lung known as the **alveolar capillary membrane**. The average number of alveoli in an adult lung ranges from 300 million to 600 million. This gives a total 80 square meters (m^2) surface area for the oxygen molecule to diffuse across (about the size of a tennis court!) into the surrounding pulmonary capillaries, which have about the same cross-sectional area (70 m^2). The blood entering the pulmonary capillaries comes from the right side of the heart and is low in oxygen and high in carbon dioxide because it just came from the body tissues. Gas exchange or external respiration takes place, and the blood leaving the pulmonary capillaries is high in oxygen and travels to the left side of the heart to be pumped around to the tissues. Conversely, carbon dioxide molecules are in high concentration in the blood in the pulmonary capillaries, and very low in the lung (remember, there is little CO_2 in the atmosphere), so CO_2 leaves the blood and enters the lung to be exhaled.

Upon closer inspection of the alveolar capillary membrane, you will see four distinct components. The first layer is the liquid **surfactant** layer that lines the alveoli. This phospholipid helps lower the surface tension in these very tiny spheres (alveoli) that would otherwise collapse due to the high surface tension.

The second component is the actual tissue layer, or alveolar epithelium, comprised of simple squamous cells of two types. The main type (95 percent) comprising the alveolar surface is a flat, pancake-like cell called a **squamous pneumocyte** or Type I cell. This is where the gas molecules can easily pass through in the process of gas exchange. The alveoli also need to produce the valuable surfactant, and this is where the plump Type II, or **granular pneumocytes**, come in.

surfactant *(sir FAC tent)*
pneum/o = *air or lung*
cyte = *cell*

 13-6 To view videos on oxygen therapy, oxygen delivery devices, and assessing oxygen levels with a pulse oximeter, please see your DVD.

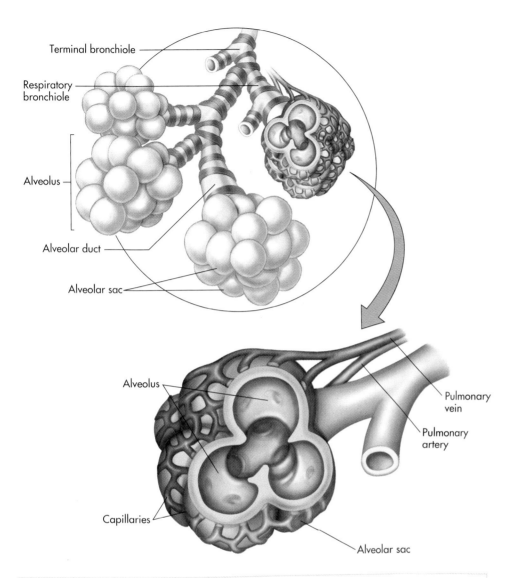

STRUCTURES OF THE LUNGS		GENERATIONS	
Conducting zone	Trachea	0	Cartilaginous airways
	Main stem bronchi	1	
	Lobar bronchi	2	
	Segmental bronchi	3	
	Subsegmental bronchi	4-9	
	Bronchioles	10-15	Noncartilaginous airways
	Terminal bronchioles	16-19	
Respiratory zone	Respiratory bronchioles	20-23	Gas exchange region
	Alveolar ducts	24-27	
	Alveolar sacs	28	

FIGURE ■ 13-11

Conduction and gas exchange structures and functions.

These highly metabolic cells not only produce surfactant but aid in cellular repair responsibilities.

In addition, this area needs to be free of debris that would act as barriers to the vital process of gas exchange. The "clean-up" cells called Type III cells or

Clinical Application

WHAT CAN GO WRONG WITH GAS EXCHANGE?

The membrane between the alveoli and the capillaries is quite thin. In fact, it is only 0.004 millimeters thick! The thinness of this membrane aids in the diffusion of the gases between the lungs and the blood. Anything that would act as a barrier to oxygen molecules getting to or through this barrier would decrease the amount of oxygen that gets into the blood. For example, excessive secretions and fluid such as in pneumonia act as a barrier and reduce the oxygen levels in the blood, which can be measured by sampling arterial blood and analyzing the amount of each gas dissolved in it. This is called an arterial blood gas, or ABG. In the case of severe pneumonia, the level of oxygen known as the PaO_2 goes down because less oxygen can get into the blood, and the $PaCO_2$ in the arterial blood goes up because less CO_2 crosses into the lungs to be exhaled.

Changes in the blood stream can also affect blood gases. Red blood cells, or **erythrocytes,** are responsible for the bulk of the transportation of oxygen and carbon dioxide in the blood via a protein-and iron-containing molecule called **hemoglobin,** which performs the actual transportation. It is estimated that there are about 280 million hemoglobin molecules found in *each* erythrocyte!

In general, if the hemoglobin is carrying large amounts of oxygen, the blood will be bright red. If there is less oxygen and more carbon dioxide being carried, then the blood will be darker in color or may have a "bluish" tint. An obvious example of this can be seen when you look at the veins in your arm. Venous blood has lower levels of oxygen and higher levels of carbon dioxide. As a result, venous blood has a dark red tint. Low levels of red blood cells or anemia would limit the number of hemoglobin molecules that could transport oxygen and thus greatly reduce the amount in the blood available for the tissues. Therefore, the number of red blood cells and the amount of hemoglobin in your blood (both of which can be measured) are important in oxygen delivery to your tissues.

Your body can attempt to respond to low hemoglobin levels by producing more red blood cells by a process called **erythropoiesis.** This process begins once the kidneys detect low levels of oxygen coming to them from the blood. The kidneys release into the bloodstream a hormone called **erythropoietin.** This substance travels through the blood and eventually reaches specialized cells found in the red bone marrow. Once stimulated, these specialized cells begin to increase their production of erythrocytes until demand is met. Having too little iron in the body can also affect oxygen delivery because the iron in the hemoglobin is what holds onto the oxygen molecules. The terms *iron-poor* blood and *tired* blood come from the fact that a patient with low levels of iron tires easily due to low oxygen levels.

erythrocytes *(eh RITH roh sights)*

hemoglobin *(HEE moh GLOH binn)*

erythropoiesis
   (eh RITH roh poy EE suss)
   erythr/o = *red*
   poiesis = *to make*

erythropoietin *(eh RITH roh poy EH tin)*

wandering macrophages, ingest foreign particles as the macrophages wander throughout the alveoli. There are even small holes between the alveoli called **pores of Kohn** that allow for gases and the macrophages to move from one alveolus to another.

The third component of the alveolar capillary membrane is the **interstitial space.** This is the area that separates the basement membrane of alveolar epithelium from the basement membrane of the capillary endothelium and contains interstitial fluid. This space is so small that the membranes of the alveoli and capillary appear fused. However, if too much fluid gets into this space (interstitial edema), the membranes separate, which makes it harder for gas exchange to occur because the gas has to travel a greater distance and through a congested, fluid-filled space.

The fourth component is the **capillary endothelium** (also known as the **simple squamous epithelium**) that forms the wall of the capillary. The capillary contains the blood with the red blood cells that carry the precious gas cargo to its destination.

Pulmonary Function Testing

Lung function can be measured in terms of volumes and flows using pulmonary function testing (PFTs). First, the various volumes can be measured by having the patient breathe normally and then take a maximum deep breath followed by a maximum exhalation. The tracing is recorded as in Figure 13-12 ■. For example, tidal volume (V_T) is the amount of air moved into or out of the lungs at rest during a single breath. The normal tidal volume is 500 ml, although there is considerable variation depending on age, sex, height, and general fitness. Your inspiratory reserve volume is what you can breathe in beyond a normal inspiration. Likewise, your expiratory reserve volume is what you can exhale beyond a normal exhalation. You can never totally exhale all of the air out of your lungs; your residual volume (RV) prevents total lung collapse. We can combine these volumes to get various lung capacities as shown in Figure 13-12:

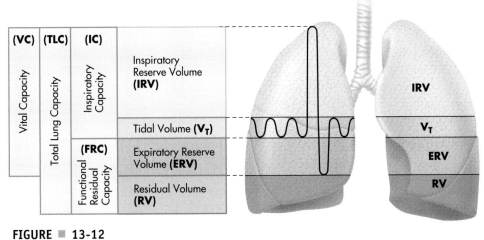

FIGURE ■ 13-12

Normal lung volumes and capacities.

- Function residual capacity (FRC) – the volume of air remaining in the lungs at the end of a normal expiration

- Inspiratory reserve volume (IRV) – the amount of air that can be forcefully inhaled after a normal inspiration

- Expiratory reserve volume (ERV) – the amount of air that can be forcefully exhaled after a normal expiration

- Residual volume (RV) – the volume of air remaining in the lungs after a maximum expiration

- Vital capacity (VC) – the maximum amount of air that can be moved into and out of the respiratory system in a single respiratory cycle

Besides volumes and capacities, we can also measure the flow rates coming out of the lung at various points during a forced (maximum patient effort) vital capacity (FVC). For example we can measure the forced expiratory volume in one second (FEV$_1$) and peak expiratory flow rate (PEFR). Normally one can exhale 75-85% of their FVC in one second. However, someone with an obstructive disease would take longer to exhale and get less than 70% of their total FVC out in one second and therefore have a reduced FEV$_1$.

Applied Science

THE AMAZING SURFACTANT

When the alveoli are small (during end-expiration), the surfactant layer lowers the surface tension, thereby preventing alveoli collapse. However, when you take a deep breath (end-inspiration), your alveoli get larger and the surfactant layer thins and becomes less effective, its surface tension increasing because of its thinning. This prevents overexpansion or rupture of the alveoli. Lack of surfactant can cause "stiff" lungs that resist expansion. Surfactant develops late in fetal development, and premature babies therefore may not have sufficient levels. Without immediate intervention, their tiny lungs would collapse (**atelectasis**) and thus prevent the movement of air in and out of the lung and vital gas exchange. If they are given too much volume to re-expand their stiff lungs, the alveoli may rupture, again because surfactant is not there to prevent overexpansion. Surfactant also has an antibacterial property that helps to fight harmful pathogens. Fortunately, medical science has developed surfactant replacement therapy that can instill surfactant into the lungs to maintain their function until babies have matured and can produce it on their own.

 13-7 To perform an interactive drag-and-drop labeling exercise concerning the alveolar region, please go to your DVD for this chapter.

 13-8 To view videos on various pulmonary function testing, including spirometry and peak flow measurements, please see your DVD.

Another test that helps to establish whether the airways have become 'narrower' than normal (as would be seen in an asthma episode) is the Peak Expiratory Flow Rate or PEFR, which is the maximum flow rate or speed of air a person can rapidly expel after taking the deepest possible breath. PEFR is measured in liters per minute, and should be within a predicted range. This a good test to reflect how the larger airways are functioning and monitor diseases such as asthma that affect these airways.

Pathology Connection: The Lower Airways

Respiratory disease is one of the most common group of diseases seen in health care settings. The following sections review some of these.

ATELECTASIS AND PNEUMONIA

atelectasis *(at eh LEK tah sis)*

Atelectasis, commonly found in the hospital setting, is a condition in which the air sacs of the lungs are either partially or totally collapsed. Atelectasis can occur in patients who cannot or will not take deep breaths to fully expand the lungs, stimulate surfactant production, and keep the air passageways open. Surgery, pain, or an injury of the **thoracic** cage (such as broken ribs) often makes deep breathing painful. Taking periodic deep breaths is important not only to expand the lungs but also to stimulate the production of surfactant, which helps to keep the small alveolar sacs open between breaths.

thorac/o = chest

Patients with large amounts of secretions who cannot cough them up are also at risk for atelectasis because the secretions block airways and lead to areas of collapse. Quite often, if atelectasis is not corrected and secretions are retained, **pneumonia** can develop within 72 hours. Pneumonia is a lung infection that can be caused by a virus, fungus, bacterium, aspiration, or chemical inhalation. Inflammation occurs in the infected areas, with an accumulation of cell debris and thick fluid in the alveoli. In certain pneumonias, lung tissue is destroyed. Pneumonias, if severe enough, can lead to death. Pneumonias can also be classified according to their location as illustrated in Figure 13-13 ▪.

TUBERCULOSIS

tuberculosis *(too ber kyoo LOH siss)*

Tuberculosis (TB) is a bacterial infection that has seen a recent rise in occurrence. Tuberculosis thrives in areas of the body that have high oxygen content such as in the lungs. A healthy immune system can destroy tuberculosis bacteria, or encase the bacteria in lump-like structures called **tubercles.** Tubercles (lesions) form in the lungs. Within a tubercle, the bacteria remain inactive. Tuberculosis can lay dormant in the body for years. However, if the immune system weakens, tubercles can burst and the bacteria can multiply. If the bacteria's spread is unchecked, vast lung damage can occur. There has been recent concern about a form of tuberculosis that is very resistant to the drugs normally used to treat TB, and has a high mortality rate. Someone with tuberculosis will have a cough, low grade fever in the afternoon, weight loss, and night sweats.

CHRONIC OBSTRUCTIVE PULMONARY DISEASE (COPD)

dyspnea *(DIS pee nah)*
dys = difficult
pnea = breathing

Obstructive pulmonary disease is a general term used to describe abnormal pulmonary conditions associated with cough, sputum production, **dyspnea,**

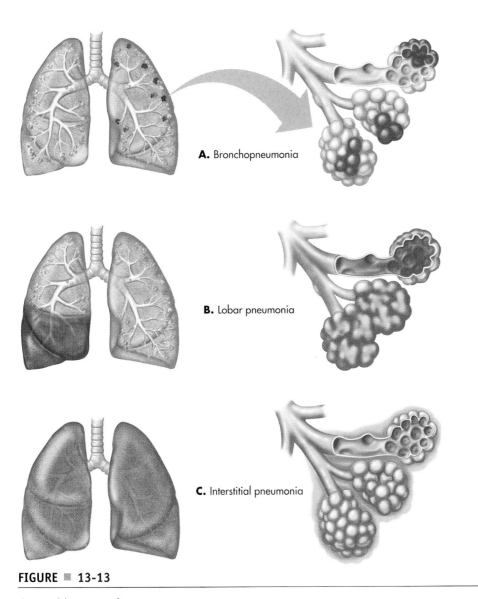

A. Bronchopneumonia

B. Lobar pneumonia

C. Interstitial pneumonia

FIGURE ■ 13-13

General locations for pneumonias.

airflow obstruction, and impaired gas exchange. It is the fourth leading cause of death in the United States. **Chronic obstructive pulmonary disease (COPD)** is a group of diseases in which patients have difficulty getting all the air out of their lungs and often have large amounts of secretions and lung damage. COPD refers to one of or a combination of asthma, emphysema, and chronic bronchitis. **Asthma** is distinguished by having reversible airway narrowing and airway hyperreactivity; and is most commonly characterized as an inflammatory process. **Emphysema** is characterized anatomically as the permanent, abnormal enlargement of distal airway spaces and destruction of the alveolar walls. **Chronic bronchitis** is associated with a productive cough, enlargement of mucous glands, and hypertrophy of the airway smooth muscle. Acute bronchitis is a temporary and very common lung condition that can affect people of any age. It differs from chronic bronchitis in that it is reversible and there are no permanent structural changes. All these illnesses share features of airway obstruction, and therefore share some similar treatments.

asthma *(AZ mah)*

emphysema *(em fih SEE mah)*

chronic bronchitis *(brong KYE tiss)*

However, asthma is typically treated as a separate disease from COPD because of its greater reversibility. Table 13-3 summarizes each of the COPD-associated diseases, each of which is discussed in greater detail later in this chapter.

The primary etiology for COPD is exposure to tobacco smoke. Smokers have higher death rates from both chronic bronchitis and emphysema. Smokers also have more lung-function abnormalities, show more respiratory symptoms, and experience all forms of COPD at a much higher rate than nonsmokers. Age of starting, total pack-years, puff volume, and current smoking status are predictive of COPD mortality. Passive smoking (the exposure of nonsmokers to cigarette smoke) also seems to increase the risk of COPD-related disease. Children of parents who smoke have a higher prevalence of respiratory symptoms and ear infections than children of nonsmokers. Exposure to smoke at an early age may impair lung maturation and the attainment of maximal lung function in adult life. Air pollution, occupational exposure, asthma, and nonspecific airway hyper-responsiveness may all play a role in the development of COPD.

Amazing Body Facts

SMOKING KILLS

The major preventable cause of many of the respiratory diseases is smoking. It has been estimated that smoking kills more people than road accidents, suicides and AIDS combined. This is the equivalent of one person dying every 15 minutes from a smoking-related illness.

TABLE 13-3 COPD Diseases

DISEASE	DESCRIPTION
Asthma	A chronic inflammatory disorder of the airways in which many cells and cellular elements play a role. In susceptible individuals, this inflammation causes recurrent episodes of wheezing, breathlessness, chest tightness, and cough, particularly at night and in the early morning. These episodes are usually associated with widespread but variable airflow obstruction that is often reversible either spontaneously or with treatment.
Chronic Bronchitis	Usually defined in clinical terms as the presence of productive cough during 3 months of the year for 2 consecutive years, provided that other causes of chronic sputum production, such as tuberculosis, are excluded. Airway hyperreactivity may be present, but airflow limitation is not fully reversible.
Emphysema	A pathologic diagnosis marked by destruction of alveolar walls, with resultant loss of elastic recoil in the lung. Dyspnea on exertion is the predominant clinical feature, and airway hyperreactivity may also be present.

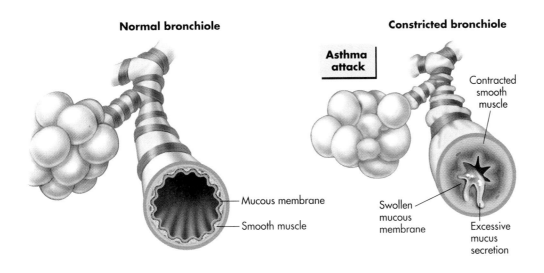

Normal bronchiole

Constricted bronchiole

Asthma attack

Contracted smooth muscle

Mucous membrane

Smooth muscle

Swollen mucous membrane

Excessive mucus secretion

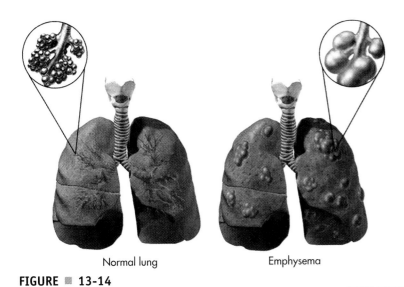

Normal lung

Emphysema

FIGURE ■ 13-14

Asthma and emphysema.

ASTHMA

Asthma is a chronic inflammatory illness of the airways affecting nearly 20 million people in the United States (that's about 3.5% of the population), including 6.2 million children. Asthma is the most common chronic disease of childhood and younger adults, with about 80% of cases developing before the age of 45. It is a potentially life-threatening lung condition in which the airways of the lungs constrict (**bronchospasm**), often in reaction to an allergy. See Figure 13–14 ■. It is difficult to get air in and even more difficult to get air out of the lungs. The

13-9 To view animations and videos on tuberculosis, COPD, and asthma, see your DVD.

Clinical Application

FAMILY HISTORY AND ASTHMA

The patient's history can be very helpful, as many asthmatic patients will have a family history of allergies such as asthma, eczema, or hay fever. Usually, the patient or their parents will be able to identify exposures or circumstances that trigger the patient's symptoms.

inability to get air out of the lungs is known as **gas trapping**. As a result of gas trapping, fresh air cannot get into the lungs, so the victim breathes the same air over and over. This lowers the amount of oxygen in the blood and increases the blood levels of carbon dioxide. Asthma can be controlled with the use of medication. The most common symptoms of asthma include episodic wheezing, shortness of breath, cough, and chest tightness. These symptoms are often worse at nighttime or in the early morning due to diurnal variations in muscle tone of the airways. (For more information about the relationship between asthma and allergies see Chapter 14, "The Immune System.")

Persons with asthma develop bronchospasm on exposure to specific sensitizing substances usually described as "triggers." Common triggers include allergens, inhalants, viruses, cold air, and exercise. Table 13-4 provides a more complete list of asthma triggers. Asthma is a chronic disease, even though certain disease triggers may wax and wane leading to episodic attacks. Therefore, patients and their caregivers need to understand the chronic nature of asthma and the underlying inflammatory process. It is important that patients be able to identify and control their exposure to environmental allergens or other types of triggers.

EMPHYSEMA

Emphysema is a nonreversible lung condition in which the alveolar air sacs are destroyed and the lung itself becomes "floppy" (again see Figure 13–14), much like a balloon that has been inflated and deflated many times. As the alveoli are destroyed, it becomes more difficult for gases to diffuse between the lungs and the blood. The lung tissue becomes fragile and can easily rupture (again, much like a worn tire), causing air to escape into the thoracic cavity and further inhibit gas exchange.

The work of breathing for the severely emphysemic patient expends energy similar to the energy spent jogging. They must continually work to maintain airway pressure in their abnormally large airspaces and must maintain a rapid respiratory rate. Weight loss is typical with severe emphysema. Think how much weight you'd lose if you ran a marathon each day!

About two million persons in the United States are estimated to have emphysema. Of these, 60,000 to 100,000 have a genetic deficiency of alpha$_1$-antitrypsin (α_1-AT) as their underlying etiology. α_1-AT is a glycoprotein that is essential in protecting the lungs against naturally occurring proteases that have the ability to destroy lung tissue. Patients with this deficiency tend to develop emphysema at a younger age (in their 40s or 50s) than do patients with emphysema from other causes. Most persons with α_1-AT deficiency are Caucasians of northern European descent.

Smoking further accelerates the process. Manifestations of severe α_1-AT deficiency involve the lungs, the liver, and the skin, and the major clinical feature is emphysema. The disorder is detected by a decreased level of α_1-AT to below a so-called protective value of 80 mg/dL. This is in comparison to normal serum levels of 150 to 350 mg/dL.

TABLE 13-4 Triggers for Asthmatic Attacks

ALLERGENS

Animal dander (pets with fur or feathers)	House dust mites (in mattresses, pillows, upholstered furniture, carpets)
Pollen (grass, trees, weeds)	Mold
	Cockroaches

INHALED IRRITANTS

Tobacco smoke	Strong odors or spays (perfumes, paint fumes, pesticides, hair sprays, cleaning agents)
Wood smoke	
Sulfur dioxide	Occupational inhalants
Air pollution	

VIRAL RESPIRATORY INFECTIONS (RHINOVIRUS, INFLUENZA, PARAINFLUENZA, CORONA VIRUS, RSV)

COLD AIR

EXERCISE

STRONG EMOTIONS

MENSES

DRUGS

Aspirin	Methacholine (used to provoke bronchoconstriction during diagnostic testing)
NSAIDs	
Beta-adrenergic blockers (oral or ophthalmic)	Histamine (alternative agent to provoke bronchoconstriction during testing)
Preservatives (sulfites and benzalkonium chloride)	

OTHER FACTORS

Allergic rhinitis	Gastroesophageal reflux (GERD)
Rhinosinusitis	

CHRONIC BRONCHITIS

Chronic bronchitis is lung disease in which there are inflamed airways and large amounts of sputum produced. As inflammation occurs, the airways swell and the inner diameter of the airways get smaller. As they get smaller, it becomes difficult to move air in and out, which increases the work of breathing. Because of this increased work level, more oxygen is used and more carbon dioxide is produced.

Chronic bronchitis is more common than emphysema, with a prevalence of over 9 million persons in the U.S. Cigarette smoking is the major causative factor in up to 90% of cases. Patients with chronic bronchitis have an increase in size and number of the mucus-secreting glands, narrowing and inflammation of the small airways, obstruction of airways caused by narrowing and mucus hypersecretion, and bacterial colonization of the airways. Acute episodes are usually brought on by a respiratory tract infection. The usual clinical presentation of chronic bronchitis begins with morning cough productive of sputum. The patient may report a decline in exercise tolerance, although he or she may not have appreciated this decline until questioned. Wheezes may be present, and an increase in the anterior-posterior diameter of the chest (the classic "barrel chest") may be present in both emphysema and chronic bronchitis as a result of excessive accessory muscle use. The patient who has predominantly chronic bronchitis symptoms may undergo repeated episodes of respiratory failure and frequently develop right-sided heart failure.

Table 13-5 lists the differential diagnostic markers for COPD and asthma. No one marker is conclusive, so the entire clinical picture must be assessed for correct diagnosis and treatment.

> **13-10** To view videos on passive smoking and smoking cessation, please visit your DVD.

The Housing of the Lungs and Related Structures

The lungs reside in the thoracic cavity and are separated by a region called the **mediastinum**, which contains the esophagus, heart, great vessels (superior and inferior vena cava and aorta), and trachea (see Figure 13–15 ■).

Breathing in and out causes the lungs to move within the thoracic cavity. Over time, an irritation could occur as the lungs rub the inside of the thoracic cage. To prevent such damage, each lung is wrapped in a sac, or serous membrane (see Chapter 4), called the **visceral pleura.** The thoracic cavity and the upper side of the diaphragm are lined with a continuation of this membrane called the **parietal pleura.** Between these two pleural layers is an intrapleural space (pleural cavity) that contains a slippery liquid called **pleural fluid**. This fluid greatly reduces the friction as an individual breathes. Pleura covers the outer surface of the lungs and lines the inner surface of the rib cage.

mediastinum
(mee dee ah STY num)

visceral pleura
(VISS er al PLOO rah)

parietal pleura
(pah RYE eh tal PLOO rah)

TABLE 13-5 Diagnostic Markers to Differentiate COPD and Asthma

DIAGNOSTIC MARKERS	COPD	ASTHMA
Age of patient when diagnosed	Typically over 40	Often child or young adult
Smoking history	Smokers and ex-smokers	No direct correlation between smoking and asthma, although smoke will trigger attack
Dyspnea	Shortness of breath, especially upon exertion	Episodic attacks, especially upon exposure to allergen/irritant/exercise
Cough	Productive cough, typically in the morning	Cough typically precedes an episode
Triggers (allergens, exercise, temperature, humidity, etc.)	None usually identified for attacks	Exposure leads to attacks
Spirometry	FEV_1/FVC ratio < 70%	FEV_1/FVC ratio low during attacks only
Daily variation in peak expiratory flow rate (PEFR)	Little	Morning dip and day-to-day variability
Effect of corticosteroid trial	Inconclusive (<20% of patients are successful)	Improvement
Eosinophilia (increased number of eosinophils)	No	In allergic reaction forms of asthma
Chest X-ray	Overinflation	Overinflation during attacks

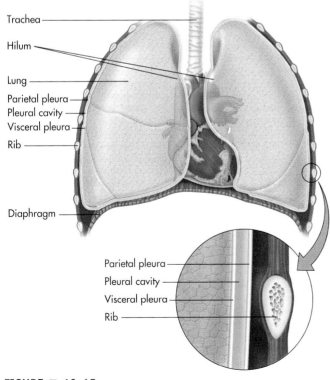

FIGURE ▪ 13-15

Structures of the thoracic cavity.

TEST YOUR KNOWLEDGE 13-3

Choose the best answer:

1. The largest bronchus or trunk of the tracheobronchial tree is the:
   a. right mainstem bronchus
   b. left mainstem bronchus
   c. bronchiole
   d. trachea

2. The site of bifurcation of the right and left lungs is called the:
   a. alveoli
   b. carina
   c. trachea
   d. capillary

3. If an object is aspirated into the airways, it is most likely to go to:
   a. the right lung
   b. the left lung
   c. the stomach
   d. the oropharynx

4. The first portion of the airway where gas exchange begins is the:
   a. terminal bronchiole
   b. trachea
   c. mainstem bronchus
   d. respiratory bronchiole

5. The alveolar layer that lowers surface tension to keep the alveoli expanded is the:
   a. surfactant layer
   b. capillary layer
   c. epithelium layer
   d. macrophage layer

6. The alveolar cell that allows for gas exchange is the:
   a. squamous cell
   b. granular cell
   c. macrophage cell
   d. Kohn cell

Pathology Connection: Pleural Space Problems

pneumothorax
(NOO moh THOH raks)

Trauma or certain disease states can interfere with the structure and therefore the function of the thoracic cavity. A **pneumothorax** is a condition in which there is air inside the thoracic cavity and outside of the lungs. Air can enter the thoracic cavity from two directions. A stab wound or gunshot wound to the chest would allow air to rush into the thoracic cavity from the outside. Alternately, the lung might develop a leak as a result of either a structural deformity or a disease process (such as in emphysema). In this situation, air would enter the thoracic cavity from the lung as air is breathed in. In either case, if the gas cannot escape, it will continue to fill a space in the thoracic cavity and provide less space for the lung or lungs to expand when breathing. If the lungs are too greatly restricted to expand, and begin to collapse, a life-threatening situation may occur.

A **pleural effusion** is a condition in which there is an excessive build-up of fluid in the pleural space between the parietal and the visceral pleura. This fluid may be pus (in which case it is known as **empyema**), serum from the blood (called a **hydrothorax**), or blood (called a **hemothorax**). Because fluids are affected by gravity, pleural effusions tend to move to the lowest point in the pleural space. If a pleural effusion is large enough, it can have the same effect as a large pneumothorax. It can restrict the amount of expansion of a lung or lungs and lead to lung collapse. Since less air can flow in and out of the lungs, the patient has to work harder by breathing in and out more rapidly to meet the body's demands for more oxygen and the removal of carbon dioxide. This additional work of breathing may exhaust an individual to the point that he or she can no longer breathe without intervention. This intervention often includes a chest tube inserted into the pleura space to allow drainage of fluid. This will allow for the lung to re-expand. See Figure 13-16 ▪ for illustrations of a sucking chest wound and how to perform a thoracocentesis to remove pleural fluid.

pleural effusion
(PLOO ral eh FYOO zhun)

empyema *(em pye EE mah)*
hydrothorax *(HIGH dro THOH raks)*
hemothorax *(HEEM oh THOH raks)*

The Lungs

The right and left lungs are conical-shaped organs; the rounded peak is called the **apex** of the lung. The apices of the lung extend 1 to 2 inches above the clavicle. The bases of the lungs rest on the right and left hemidiaphragm. The right lung base is a little higher than the left to accommodate the large liver lying underneath. The medial surface of the lung has a deep, concave cavity that contains the heart and therefore is called the **cardiac impression**; it is deeper on the left side. The **hilum** is the area where the root of each lung is attached. Each root contains the mainstem bronchus, pulmonary artery and vein, nerve tracts, and lymph vessels.

hilum *(HIGH lim)*

The right lung has three lobes—the upper, middle, and lower lobes—that are divided by the horizontal and oblique fissures. The left lung has only one fissure--the oblique fissure--and therefore only two lobes, called the left upper and left lower lobes. Why only two lobes in the left, you may ask? Remember that the heart is located in the left anterior area of the chest and therefore takes up some space of the left lung. The right lung is larger and about 60 percent of gas exchange occurs there. The term **lingula** refers to the area of the left lung that corresponds with the right middle lobe. The lobes are even further divided into specific segments related to their anatomical position. For example, the apical segment of the right upper lobe is the top portion or tip of the right upper lobe.

lingula *(LING gu lah)*

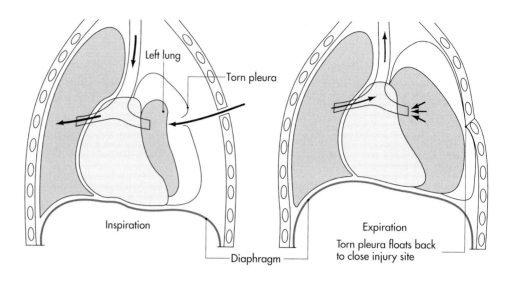

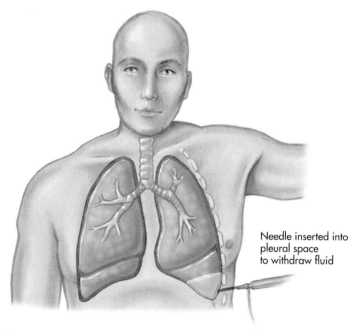

FIGURE ■ 13-16

Pneumothorax (sucking chest wound) and technique for performing thoracocentesis.

The Protective Bony Thorax

The lungs, heart, and great vessels are all protected by the **bony thorax**. This bony and cartilaginous frame provides protection and also movement of the thoracic cage to accommodate breathing. The bony thorax includes the rib cage, the sternum or breastbone, and the corresponding thoracic vertebrae to which the ribs attach (see Figure 13–17 ■).

   The **sternum,** or breastbone, is centrally located at the anterior portion of the thoracic cage and is comprised of the manubrium, body, and xiphoid process. This anatomical landmark is very important for proper hand placement in CPR. The hand is placed over the body of the sternum where com-

sternum *(STER num)*

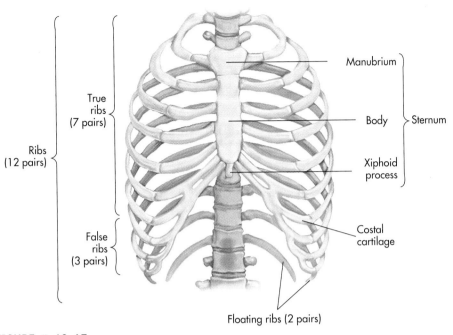

FIGURE ▪ 13-17

The thoracic cage.

pressions squeeze the heart between the body of the sternum and the thoracic vertebrae. If the hand placement is too low on the xiphoid process, it can break off and lacerate the internal organs.

The **thoracic cage** consists of 12 pairs of elastic arches of bone called ribs. The ribs are attached by cartilage to allow for their movement while breathing. The true ribs are pairs 1 through 7 and are called **vertebrosternal** because they connect anteriorly to the sternum and posteriorly to the thoracic vertebrae of the spinal column. Pairs 8, 9, and 10 are called the false ribs, or **vertebrocostal**, because they connect to the costal cartilage of the superior rib and again posterior to the thoracic vertebrae. Rib pairs 11 and 12 are called the floating ribs because they have no anterior attachment.

thoracic *(tho RASS ik)*

 13-11 To view videos on the topic of assessing the thorax, please visit your DVD.

How We Breathe

The control center that tells us to breathe is located in the brain in an area known as the **medulla oblongata.** Inspiration is an active process of ventilation in which the main breathing muscle, the **diaphragm** (a dome-shaped muscle when at rest), is sent a signal via the phrenic nerve and contracts and flattens downward, thereby increasing the space in the thoracic cavity (see Figure 13–18 ▪). The increase in volume in the thoracic cavity causes a decrease in pressure because volume and pressure are related. This creates a lower than atmospheric pressure in the lungs, allowing air to rush into the

medulla oblongata
(meh DULL lah ob lon GOT ah)
diaphragm *(DIE ah fram)*

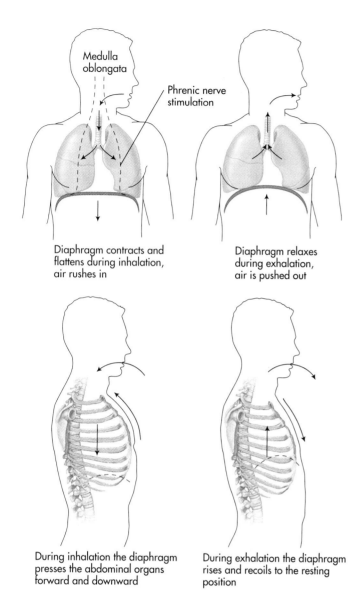

Diaphragm contracts and flattens during inhalation, air rushes in

Diaphragm relaxes during exhalation, air is pushed out

During inhalation the diaphragm presses the abdominal organs forward and downward

During exhalation the diaphragm rises and recoils to the resting position

FIGURE ■ 13-18

How we breathe.

lungs. The external intercostal muscles also assist by moving the sternum and ribs up and outward during inspiration to increase the total volume in the thoracic cavity. By the way, a hiccup happens when your diaphragm muscle has an involuntary spasm that causes the glottis to close suddenly. As you probably know, these spasms can last for awhile and it might be hard to figure out how to stop them.

To achieve changes in volume and pressure, the diaphragm flattens and thereby increases the volume of the thoracic cavity. At the same time, the ribs move upward and outward via the external intercostal muscles, which also increases the thoracic volume. Because the lungs adhere to the pleural membranes, they are moved upwards and outwards. This increases the volume of the lungs and decreases their internal pressure. Air then flows from the relative high pressure of the atmosphere, down the trachea, and into the low pressure of the alveoli. Because air is a gas, it will fill all the available alveoli, and therefore facilitate gas exchange.

The ease by which ventilation occurs is referred to as **compliance**. Low compliance means that it is more difficult to expand the lungs, whereas high compliance means that less effort is required to expand the lungs. This becomes clinically significant when assessing people who have asthma, emphysema, chronic bronchitis, or any other lung disease. In emphysema, the bronchioles become damaged and wider, allowing air into the lungs, increasing compliance but making it much more difficult to breathe out.

Exhalation, on the other hand, is usually a passive act. As the diaphragm relaxes, it forms a dome shape, which decreases the amount of space in the thoracic cavity. As a result, pressure in the lungs becomes greater than the atmospheric pressure and the air is pushed out of the lungs. The fact that the lungs are elastic tissue that is stretched during inspiration also aids in expiration because this elastic tissue now wants to return to rest, much like a stretched rubber band that is released.

What makes the brain tell the lungs how quickly or slowly to breathe? Although we can consciously speed up or slow down our breathing, our breathing rate is normally controlled by the level of

Applied Science

BOYLE'S LAW

The process of ventilation takes place because there is an inverse relationship between pressure and volume, which means that if there is a decrease in the volume of a gas, the pressure will rise; if you increase the volume of a gas, its pressure will fall. This was first recognised by Robert Boyle in the 1600s, and is now called Boyle's law.

carbon dioxide in our blood. If carbon dioxide levels rise, it means that not enough CO_2 is being ventilated. When this occurs, chemoreceptors in the medulla oblongata send signals to the respiratory muscles to increase the rate and depth of breathing. Chemoreceptors in the carotid arteries and aortic arch are also sensitive to high CO_2 levels and low blood pH, and can trigger increased ventilation.

Other factors can change breathing rate as well. Our breathing rate changes as we grow and mature into adults. A developing newborn has a breathing rate of 40-60 breaths/minute. This slows to 14-20 breaths/minute in adulthood. When we sleep, our breathing rate slows down because body cells are using less oxygen and producing less carbon dioxide. Breath rate increases with exercise, when body cells are rapidly producing carbon dioxide. It can also increase when we feel strong emotions. We may yawn when we are tired and need more oxygen. Yawning is a deep breath that fills the lungs and brings oxygen to the blood.

Sometimes the body needs help to breathe beyond resting or normal breathing. For example, during increased physical activity or in disease states in which more oxygen is required, **accessory muscles** are used to help pull up your rib cage to make an even larger space in the thoracic cavity. The accessory muscles used are the scalene muscles in the neck, the sternocleidomastoid, and pectoralis major and pectoralis minor muscles of the chest.

Applied Science

PH AND VENTILATION

CO_2 levels are directly related to hydrogen ion concentration. When CO_2 levels rise, so does hydrogen ion concentration. Consequently, pH falls (that is, blood becomes more acidic). Because of this relationship, your body interprets any decrease in blood pH, no matter the cause, as if it were due to increased CO_2, and ventilation increases. For example, when Maria's diabetes is poorly controlled, she might have lower blood pH (increased hydrogen ion concentration) due to metabolic acidosis. In an attempt to bring her blood pH back to normal, her ventilation rate would increase.

Amazing Body Facts

EXHALED CO_2 AND MOSQUITOES

Since mosquitoes are too small to carry flashlights, how do they find you in the dark? They do it by using carbon dioxide sensors to locate increased concentrations of CO_2 emitted by--you guessed it--your exhaled breath. Once they detect your exhaled CO_2, they use their heat sensors to find an area of skin on which to begin their banquet!

Another amazing carbon dioxide fact is that many people believe swimmers hyperventilate to get more oxygen in their systems before a sustained underwater dive. In reality, they are "blowing off" CO_2 to get their levels low. Since higher levels of CO_2 cause us to want to breathe, the body is fooled into believing it doesn't need to for a while longer, so swimmers can remain underwater longer. Unfortunately, the body still uses up its oxygen at a regular rate, and sometimes uses enough of the reserve that swimmers risk losing consciousness.

COPD patients over time adjust to having a higher than normal level of CO_2 in their blood. As a result these individuals rely on levels of oxygen in their blood to determine when to breathe. When these patients have an exacerbation and are erroneously placed on too much oxygen, the higher levels of oxygen in their blood can actually trigger a slow down or stop in their breathing. Thus, a thorough patient history is imperative.

Although exhalation is a passive process, there are times, especially with certain disease states, when exhalation may need to be assisted. Again, the body has accessory muscles of exhalation that assist in a more forceful and active exhalation by increasing abdominal pressure. The main accessory muscles of exhalation are the various abdominal muscles that push up the diaphragm or the back muscles that pull down and thus compress the thoracic cage. See Figure 13–19 ■ for the specific accessory muscles of exhalation.

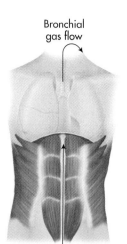
Bronchial
gas flow

Diaphragmatic
pressure

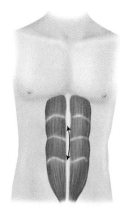
Rectus
abdominis

External
oblique

Internal
oblique

Transversus
abdominis

FIGURE ■ 13-19

The accessory muscles of exhalation.

TEST YOUR KNOWLEDGE 13-4

Choose the best answer:

1. The _____ pleura lines the thoracic cavity.
   a. visceral
   b. parietal
   c. mediastinum
   d. chest

2. The portion of the sternum where CPR is performed is the:
   a. xiphoid process
   b. ribs
   c. manubrium
   d. body

3. The _____ nerve innervates the main breathing muscle, called the _____.
   a. thoracic; internal intercostals
   b. thoracic; diaphragm
   c. phrenic; external intercostals
   d. phrenic; diaphragm

4. Rib pairs 1 to 7 are attached to:
   a. the vertebral column only
   b. the vertebral column and the sternum
   c. the sternum only
   d. are free floating

5. Fill in the blank for the following diseases:
   a. _____ is a condition in which air leaks out of the lung into the thoracic cavity.
   b. _____ is a disease in which there is excessive airway secretions often as the result of smoking.
   c. _____ is a condition that leads to wheezing and shortness of breath, often allergic related.

A Quick Trip Through the Diseases of the Respiratory System

DISEASE	ETIOLOGY	SIGNS AND SYMPTOMS	DIAGNOSTIC TEST(S)	TREATMENTS
Bronchogenic Carcinoma (lung cancer)	Cause not known but linked to smoking and inhalation of carcinogens.	Obstruction of airways interfering with ventilation, weight loss, weakness, cough or change in cough	Bronchoscopy, imaging studies, biopsy, sputum exam, patient exam and history.	Radiation, chemotherapy, surgery.
Infections:				
Acute Bronchitis	Viral or bacterial.	Inflamed mucous membranes of trachea and bronchi; expectorating or dry cough, shortness of breath, fever, rales (raspy sound).	Physical exam.	Antibiotics if bacterial.
Common Cold	Viral.	Upper airway congestion, cough, sore throat.	History and physical exam.	Treat symptoms, pain meds, bed rest, drink fluids, proper nutrition.
Pharyngitis	Viral or bacterial.	Red, sore swollen throat, pus.	History and physical exam, throat culture.	If bacterial use antibiotics, antiseptic gargle.
Laryngitis	Viral or bacterial, allergies, over-use of voice.	Dysphonia, sore throat, trouble swallowing.	History and physical exam.	Rest voice.
Tonsillitis	Viral or bacterial.	Sore throat, swollen tonsils and dysphagia.	History and physical exam, culture.	Antibiotics for bacterial, surgery if needed.
Influenza	Viral.	Fever, cough, body and headaches.	History and physical exam.	Rest, fluids, pain meds. And treat symptoms.
Pneumonia	Viral, bacterial, or fungal.	Productive cough chest pain, weakness, malaise, dyspnea.	Imaging, blood work, sputum culture.	Antibiotics if confirmed bacterial infection. Antifungal drugs if confirmed fungal infection.
Pulmonary tuberculosis	Bacterial.	Primary may be asymptomatic, Secondary , cough (may be blood tinged), fever (night sweats), weight loss.	Imaging, TB skin test, sputum test.	Antibiotic agents.
Inflammations and allergies:				
Seasonal allergic rhinitis (hay fever)	Allergic agents.	Upper airway congestion, watery nose and eyes, sneezing.	History and physical exam, allergy testing.	Antihistamines, preventative allergy shots.
Asthma	Many triggers such as allergens, food, exercise, cold air, inhaled irritants, smoking.	Dyspnea, wheezing, productive cough, hypoxia.	History and physical exam, lung function tests.	Bronchodilators, steroids, and anti-asthmatic agents. Oxygen if needed.

continued

A Quick Trip Through the Diseases of the Respiratory System (*continued*)

DISEASE	ETIOLOGY	SIGNS AND SYMPTOMS	DIAGNOSTIC TEST(S)	TREATMENTS
COPD:				
Chronic Bronchitis	Cigarette smoking and long term exposure to air pollutants, middle or old age.	Dyspnea, wheezing, productive cough, hypoxia.	History and physical exam, lung function tests.	Antibiotics if bacterial, bronchodilators, oxygen if needed.
Emphysema	Cause not fully known but associated with smoking and one genetic form from alpha 1-antityrpsin deficiency.	Dyspnea, tachypnea, wheezing, productive cough, hypoxia.	History and physical exam, lung function tests.	Oxygen therapy, bronchodilators, alpha1 anti-trypsin replacement.
Cystic Fibrosis	Hereditary disease transmitted via a recessive gene.	Excessive thick mucus secretion, repeated infections, large salt losses, difficult digestion.	Sweat test, genetic testing.	Respiratory hygiene therapy, mucus thinning agents, antibiotics, pancreatic enzyme supplements.

Pharmacology Corner

Although oxygen is often not thought of as a drug, it is the major drug used in treating respiratory diseases as well as cardiac disease. The basic purpose is to reduce the work of breathing, increase the oxygen content of the blood, and reduce the work of the heart. Oxygen therapy comes in various forms of delivery, which are demonstrated on your DVD.

Problems with the airway are usually due to inflammatory processes that narrow the airways, making it difficult to breathe. For rapid relief of an acute situation of airway narrowing, aerosolized bronchodilators can be given. For long-term treatment of chronic airway inflammation, inhaled steroids have been used because they have less systemic effects than oral or injectable steroids.

The only method known to prevent or slow the progression of COPD is to stop smoking or eliminate the occupational source. All patients who smoke need to be regularly encouraged to stop. Cigarette smoking kills nearly 450,000 Americans each year and debilitates nearly one half of all long-term smokers. Tobacco dependence is a powerful addiction and one that is extremely difficult to break, often requiring 4-6 attempts. Even after quitting, patients sometimes have a life-long craving that does decrease with time. Some patients will be successful on their own, but most will require behavioral counseling and encouragement in addition to pharmacological therapy. This treatment usually consists of nicotine replacement therapy in the form of gum, skin patches, or inhaled forms.

In many cases premature infants have underdeveloped lungs that do not have adequate surfactant maturation to breathe on their own. Natural and synthetic surfactant forms have been developed that can be instilled into the lungs. This can buy time until the infant develops adequate amounts on his or her own.

Bacterial lung infections can be treated with systemic antibiotics. Although not related to lung disease, new research has allowed for the development of aerosolized insulin that can be inhaled for the treatment of diabetes. See Figure 13-20 ■ for an illustration of drugs used to treat the respiratory system.

 13-12 To view videos on various respiratory medications and their delivery systems, please see your DVD.

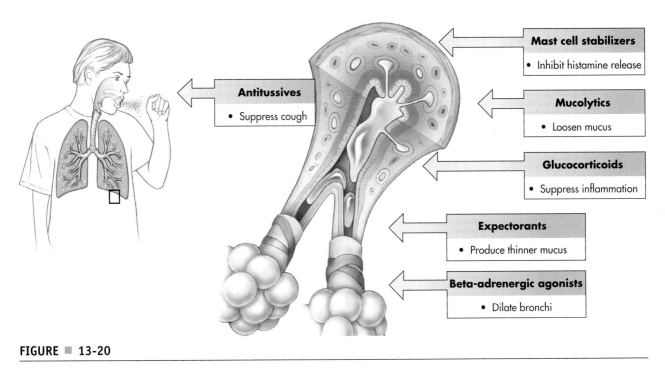

FIGURE ■ 13-20

Classes of drugs used to treat respiratory disorders.

SUMMARY

Snapshots from the Journey

→ Moving approximately 12,000 quarts of air each day, the respiratory system is responsible for providing oxygen for the blood to take to the body's tissues and removing carbon dioxide, one of the waste products of cellular metabolism.

→ Ventilation is the movement of gases in and out of the lungs; during respiration oxygen is added to the blood and carbon dioxide is removed.

→ The lungs contain continually branching airways called bronchi and bronchioles.

→ At the end of bronchioles are alveolar sacs.

→ Each alveolar sac is surrounded by a capillary network where gas exchange occurs with the blood.

→ The purpose of the upper airways is to filter, warm, and moisten inhaled air for its journey to the lungs.

→ In addition, the upper airways provides for *olfaction* (sense of smell) and *phonation* (speech).

→ The mucociliary escalator captures foreign particles, and the hair-like cilia constantly move a layer of mucus up to the upper airways to be swallowed or expelled.

→ Adenoids and tonsils aid in preventing pathogens from entering the body.

→ Since activities of breathing and swallowing share a common pathway, the epiglottis protects the airway to the lungs from accidental aspiration of food and liquids.

→ Vocal cords are the gateway between the upper and lower airways.

→ The tracheobronchial tree is like an upside down tree with ever-branching airways, where the trunk of the tree is represented by the trachea and the leaves by the alveoli.

→ The alveolar capillary membrane is where external respiration or gas exchange occurs.

→ The bony thorax provides support and protection for the respiratory system.

→ The main muscle of breathing is the diaphragm, and accessory muscles assist in times of need such as exercise and disease.

→ The medulla oblongata in the brain is the control center for breathing and sends impulses via the phrenic nerve to the diaphragm.

RAY'S STORY

Ray was brought into the emergency room with thick secretions and a fever. His heart rate, and blood pressure are increased and his lungs do not sound clear upon examination. The carbon dioxide levels in his blood are greatly increased and the oxygen levels are well below normal, even given his dependence on a ventilator.

a. What possible respiratory diseases or conditions could he have?

_____

_____

b. What are some recommended treatments for this patient?

_____

_____

REVIEW QUESTIONS

Multiple Choice

1. The process of gas exchange between the alveolar area and capillary is:
   a. external ventilation
   b. internal ventilation
   c. internal respiration
   d. external respiration

2. The bulk movement of gas into and out of the lung is called:
   a. internal respiration
   b. ventilation
   c. diffusion
   d. gas exchange

3. Which of the following is *not* a function of the upper airway?
   a. humidification
   b. gas exchange
   c. filtration
   d. heating or cooling gases

4. The largest cartilage in the upper airway is the:
   a. cricoid
   b. eustachian
   c. mega cartilage
   d. thyroid

5. Which structure controls the opening to the trachea?
   a. esophagus
   b. hypoglottis
   c. epiglottis
   d. hyperglottis

Fill in the Blank

1. Small bronchi are called _____.

2. The sense of smell is termed _____, and the act of speech is called _____.

3. The hair-like projections called _____ beat within the _____ layer and propel the _____ layer toward the oral cavity to be expectorated.

4. The _____ are thought to lighten the head and provide resonance for the voice.

5. The _____ and the _____ are part of the immune system and are found in the nasopharynx and oral pharynx.

Matching

1 _____ Asthma

2 _____ Chronic Bronchitis

3 _____ Laryngitis

4 _____ Emphysema

5 _____ Cystic Fibrosis

a. An inherited disease transmitted by a recessive gene, causing excessive mucus production

b. Causes permanent destruction of alveolar walls and surrounding capillaries

c. Has many triggers that include pet dander, food allergies and cold air

d. Caused by bacteria or viruses and will affect your voice

e. Cigarette smoking causes large amounts of secretions and inflamed airways

Short Answer

1. Describe the tissue layers in the bronchi.

_____

_____

_____

2. Explain how gas exchange takes place in the lungs.

_____

_____

_____

3. Discuss the importance of surfactant.

_____

_____

_____

4. Describe the process of normal breathing, beginning with the brain.

_____

_____

_____

_____

Suggested Activities

1. Research the effects of second-hand smoking and smokeless tobacco, and share results with the class or develop posters on these subjects.

2. Working as a group, research occupational causes of lung diseases and see how long a list you can develop.

> **13-13** Now that you have completed your journey through this chapter, please go to the DVD for interactive games and puzzles concerning the medical terms and concepts contained in this chapter. By playing the games you will reinforce your learning of the respiratory system in a fun way.

THE
LYMPHATIC
AND IMMUNE
Systems

Your Defense Systems

So far in our travels, we have visited control systems, transport systems, and infrastructure, to name just a few. We have seen how each separate system works together to allow the body to function as an integrated unit. However, like all cities, the body must be protected. Cities have police and fire departments. On a larger scale, countries have armies and bases. Similarly, your body has the immune and lymphatic systems with a variety of protective mechanisms and cells each performing specific duties. These systems help to protect the body from pathogens that can produce disease. Without your immune and lymphatic systems, your journey would be a very short one—the first exposure to a potential pathogenic organism would wreak havoc in your body and literally stop the journey before it began.

Chapter 14

LEARNING OBJECTIVES

Upon completion of your journey through this chapter, you will be able to:

→ List and describe the major components of the lymphatic and immune system and their functions

→ Explain the antigen–antibody relationship

→ Name and describe the functions of the blood cells responsible for protecting the body from invasion

→ Discuss how inflammatory responses and fevers relate to infection

→ Compare innate immunity to adaptive immunity

→ Describe the function of lymphocytes and helper T cells in the immune response

→ List and describe several common diseases of the immune system

MULTIMEDIA APPLICATIONS

DVD Interactive Exercises

→ Video on lymphatic drainage massage therapy, 14-1

→ 3-D illustration of the lymphatic system and interactive drag-and-drop exercise of the lymphatic system, 14-2

→ Video on the topic of skin cancer, 14-3

→ Video on proper hand-washing technique, 14-4

→ Video on the topic of leukemia, 14-5

→ Animation showing T cell destruction by HIV, 14-6

→ Animation and video on the topic of HIV/AIDS, 14-7

→ Animation of the treatment of severe allergic reactions with an EpiPen and a video on the topic of allergic rhinitis, 14-8

→ Interactive puzzles and games, 14-9

www.prenhall.com/colbert

→ Professional Profiles:
  • Nuclear Medicine
  • Pharmacy

→ Related Internet Links

→ Additional Review Questions

Pronunciation Guide

Correct pronunciation is important in any journey so that you and others are completely understood. Here is a "see and say" Pronunciation Guide for the more difficult terms to pronounce in this chapter.

basophils (BAY soh fills)

cytokines (SIGH tow kines)

cytotoxic T cells (sigh tow TOX ick)

dendritic cells (den DRID ick)

eosinophils (ee oh SIN oh fillz)

histamine (HISS tah meen)

interferon (in ter FIR on)

interleukins (in ter LOO kins)

leukocytes (LOO koh sights)

lymph nodes (LIMF nohdz)

macrophages (MAC reh fage ez)

neutrophils (NOO troh filz)

T lymphocytes (T LIMF oh sights)

thoracic duct (thoh RASS ik)

thymus (THIGH muss)

tumor necrosis factor (neh KROH sis)

THE DEFENSE ZONE

Although war may seem a harsh analogy to use, it is the reality of what happens when a potentially dangerous threat invades the body. Suppose that a nasty army of pathogens attempts to invade your body. First, it must get past your barriers. Many invaders will be repelled simply by your intact skin or the secretions of your mucous membranes. If the invader does get inside your body, it is recognized as *not* belonging in your body. This recognition stimulates a series of responses to neutralize the foreign invader. Weapons in the form of specialized cells are engaged by the immune and lymphatic systems to fend off the pathogens. In addition, the immune and lymphatic systems release powerful chemicals to help to fight off the invaders.

These chemicals also stimulate the inflammatory response and leave a chemical mess to clean up. The "war" also leaves behind an area of neutralized pathogens and excessive debris and fluid that has collected around the battlefield and must be cleaned up. This again is accomplished by the combined and integrated efforts of the immune and lymphatic systems until the danger is over and your body can return to normal functioning. Let's begin the discussion with the lymphatic system, then bring in the immune system, and finally show how the two work in concert to keep your body healthy.

THE LYMPHATIC SYSTEM

lymphatic *(lim FAT ik)*

The **lymphatic** system is both the transport system and barracks of your immune system. It is a second circulatory system parallel to the cardiovascular system. As you will soon see, it works closely with the cardiovascular system, and their close proximity is needed for mutual benefit.

The lymphatic system has the following four functions:

- Recycling fluids lost from the cardiovascular system

pathogen *(PATH oh jenn)*
  path/o = *disease*
  gen = *to create*

- Transporting **pathogens** to the lymph nodes where they can be destroyed

- Storing and maturing some types of white blood cells

- Absorbing glycerol and fatty acids from food (see Chapter 15, "The Gastrointestinal System," for details)

Given that the lymphatic system is intimately connected to the function of the cardiovascular system, it should come as no surprise that the smallest pipes of the lymphatic system, called **lymph capillaries,** run parallel to blood capillaries. Lymph capillaries, tubes that form a network between the cells of connective tissues, are so named because they are structurally similar to blood capillaries, but unlike blood capillaries, lymph capillaries are open ended. Proteins and fluids are lost from cardiovascular capillaries and enter the interstitial space. This interstitial fluid lies between the cells and is returned to the blood by way of the lymphatic system. Once this fluid enters the lymph capillaries, it is known as **lymphatic fluid,** or simply **lymph.** Lymph is a straw-colored, clear fluid that is similar to blood plasma. The primary component of lymph is water. It also contains digested nutrients, such as fats, gases such as oxygen, proteins, and white blood cells called lymphocytes and macrophages. You will learn more about these white blood cells later in the chapter. For a map of the lymphatic system, see Figure 14–1 ▪.

lymph capillaries
(LIMF KAP ih lair eez)

lymphatic fluid *(lim FAT ik FLOO id)*
lymph *(LIMF)*

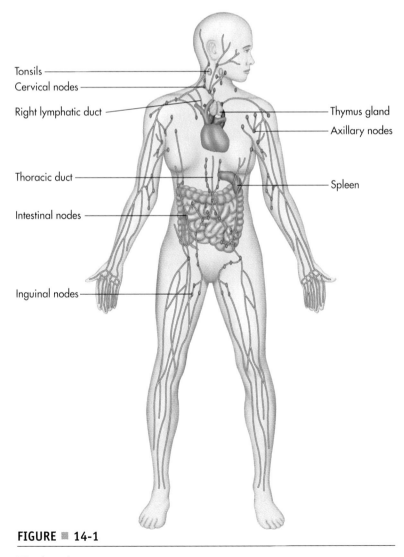

FIGURE ▪ 14-1

The lymphatic system.

lymphatic vessels *(lim FAT ik)*
lymph nodes *(LIMF nohdz)*

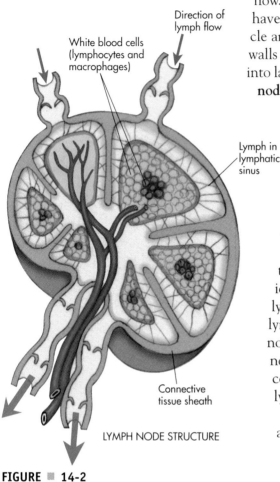

Direction of lymph flow

White blood cells (lymphocytes and macrophages)

Lymph in lymphatic sinus

Connective tissue sheath

LYMPH NODE STRUCTURE

FIGURE ■ 14-2

The lymph node structure.

Several lymph capillary networks empty into **lymphatic vessels,** which are structurally similar to veins, including having valves that prevent backward flow. Lymphatic vessels are located in almost all tissues and organs that have blood vessels. Body movement, e.g., contraction of skeletal muscle around the vessels, and contraction of smooth muscle in the vessel walls propel lymph through the system. Small lymphatic vessels empty into larger lymphatic vessels. Larger lymphatic vessels empty into **lymph nodes.** Ranging in size from a pin head to an olive, and existing either individually or in groups, the lymph nodes can be thought of as filters strategically placed all along the pathways or vessels of the lymphatic system. Think about a water filtration system for a city where a system of pipes (vessels) bring the dirty water into a plant with a filtration system (node) that then recycles clean water back into the system. Lymphatic fluid carries digested food, oxygen, and hormones to cells, and then carries waste back to capillaries for excretion.

Lymph nodes are small, encapsulated bodies divided into sections (see Figure 14–2 ■). Inside the nodes are sections of lymphatic tissue containing white blood cells known as **lymphocytes.** The lymphatic tissue is surrounded by **lymphatic sinuses** filled with lymph fluid. A number of lymphatic vessels enter or exit each lymph node. This ensures that all lymph fluid must pass through a lymph node for filtration and destruction of pathogens by the white blood cells, and it also ensures that the flow slows enough to allow the lymphocytes and macrophages to destroy pathogens.

Lymph nodes are concentrated in several areas around the body and are identified by their regional location: cervical, axillary, inguinal, pelvic, abdominal, thoracic, and supratrochlear. Notice that the lymph nodes are concentrated to catch pathogens where they are most likely to enter the body, such as in the lungs, digestive system, and reproductive system.

14-1 The lymph system is a very low-pressure system and sometimes the nodes or filters can become clogged and circulation impaired. Lymphatic drainage massage is a technique to optimize lymphatic circulation, and you can go to your DVD to view a short video on this important therapy.

lymphatic trunks *(lim FAT ik)*

thoracic duct *(thoh RASS ik)*
right lymphatic duct
   (lim FAT ik duct)

Tonsils are masses of lymphocyte-producing lymphatic tissue located in the rear of the pharynx, or throat. The three sets of tonsils include the **palatine tonsils,** located on either side of the oral pharynx (the rear of the mouth); the **adenoids,** located at the rear of the nasal pharynx (the upper part of the throat); and the **lingual tonsils,** located at the base of the tongue.

Lymphatic vessels exiting lymph nodes empty into one of several **lymphatic trunks.** These trunks, named for their location, are the lumbar, intestinal, intercostal, bronchomediastinal, subclavian, and jugular. Lymphatic trunks empty into one of two collecting ducts. The lumbar, intestinal, and intercostal trunks all empty into the **thoracic duct,** which is the largest lymph vessel. The thoracic duct runs from the abdomen up through the diaphragm and into the left subclavian vein. More than two thirds of the lymphatic system drains into the thoracic duct. The bronchomediastinal, subclavian, and jugular trunks empty into the **right lymphatic duct,** a smaller duct within the right thorax that empties into the right subclavian vein (see Figure 14–3 ■).

Lymphatic fluid flows in only one direction: from the body tissues and organs to the heart. The circulation of lymphatic fluid follows this path: from blood to tissue to lymphatic capillaries to lymphatic vessels to lymph nodes to lymphatic vessels (exiting nodes) to lymphatic trunks to collecting ducts to subclavian veins and then back to the blood. The fluid that moves from blood capillaries into tissues is filtered through the white blood cells in the lymph nodes, where pathogens are removed and destroyed and recycled back to the cardiovascular system. If a substance cannot be destroyed, the node becomes enlarged and inflamed.

Pathology Connection: Tonsillitis

Tonsillitis is an inflammation of the tonsils, usually caused by a bacterial or viral infection. The symptoms of tonsillitis include sore throat, difficulty swallowing, swollen tonsils, fever, upper respiratory symptoms, swollen lymph glands, visible coating or spots on the tonsils, and even upset stomach. Tonsillitis is diagnosed by a physical examination noting inflamed tonsils. It is important, however, to distinguish between viral and bacterial causes of tonsillitis. Viral tonsillitis is generally characterized by a low-grade fever and the presence of upper respiratory symptoms resembling the common cold. Bacterial tonsillitis, generally caused by infection with *Streptococcus* bacteria (which also causes strep throat), is characterized by higher fever and lack of upper respiratory cold-like symptoms. Bacterial tonsillitis can be diagnosed, or ruled out, by the use

14-2 Go to your DVD for this chapter for a 3-D illustration on the lymphatic system and a drag-and-drop exercise on the lymphatic system.

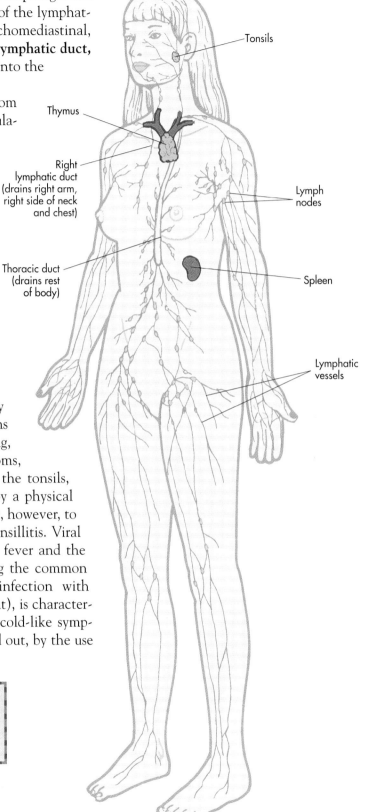

FIGURE ■ 14-3

Spleen, thymus, tonsils, and lymphatic vessels.

of a throat culture that tests for the presence of strep throat. A negative throat culture indicates the absence of a bacterial infection. Bacterial tonsillitis must be treated with antibiotics to combat the bacterial infection. Viral tonsillitis, like all viral infections, does not respond to antibiotics and will usually go away by itself. Treatment of symptoms, rest, fluids, and an analgesic like Tylenol is recommended until symptoms subside. In the past, many children had their tonsils removed after even a single bout of tonsillitis. Today tonsillectomies are performed only if the tonsillitis has become chronic or the tonsils are so enlarged that they obstruct the patient's airway.

Lymph Organs

As previously mentioned, there are two larger collections of lymphatic tissue, known as lymph organs. These lymph organs are the thymus and the spleen. While they are not strictly lymph nodes and are not part of lymph circulation, they are so similar to lymph nodes that they are classified as part of the lymph system.

The **spleen** (again see Figure 14–3) is a spongy sac-like mass of lymphatic tissue in the upper left quadrant of the abdomen. It is structurally similar to lymph nodes but instead of having lymphatic sinuses, the spleen has blood sinuses. The blood sinuses surround islands of white pulp containing lymphocytes and islands of red pulp containing both red blood cells and white blood cells. One of the functions of the spleen is to remove and destroy old, damaged, or fragile red blood cells. Can you guess the second function of the spleen, given its anatomy? Since it is similar to a lymph node, the spleen also filters pathogens from the blood stream and destroys them in the same way that lymph nodes filter pathogens from lymph.

Even though the spleen is a very important organ, it is not vital, and in some cases, such as trauma, it may need to be surgically removed. Because of the spleen's rich supply of blood vessels, injury to the spleen can often cause internal bleeding. While its removal in children can severely compromise their ability to ward off disease, the spleen's removal in an adult has much less effect. As we age, the body becomes better at fighting off infections that it has seen in the past, but new invaders present more of a challenge when we are first exposed to them as children.

thymus *(THIGH muss)*

The **thymus** is a soft organ located between the aortic arch and the sternum (again, see Figure 14–3). The thymus is very large in children because it must defend against many new infections. It gets smaller or even disappears in adults as the immune system fully matures in its ability to fight infection. The thymus produces lymphocytes that mature into a type of white blood cell called a **T lymphocyte**. The thymus also secretes a hormone that stimulates the maturation of T lymphocytes in lymph nodes. Because it secretes a hormone, the thymus is also considered an endocrine gland.

Pathology Connection: Lymphatic Disorders

Lymphadenitis is a disorder of the lymphatic system in which lymph nodes swell. Lymphadenitis may be localized, affecting only a few lymph nodes in a small area, or generalized, affecting several lymph nodes over a larger area. Lymphadenitis is usually the result of bacterial infection, but viruses, parasites, and fungi can cause lymphadenitis.

Initial symptoms of lymphadenitis include swelling of lymph nodes due to fluid buildup and increased production of white blood cells. More advanced symptoms include fever, chills, excessive sweating, rapid pulse, and weakness. Treatment includes infection-specific medication, such as antibacterial or antiviral medication, and rest of the affected area. Hot moist compresses can also be used.

Mononucleosis is a viral infection that typically affects children and young adults. Caused by the Epstein-Barr virus, mononucleosis symptoms are usually milder in younger children than they are in teens and young adults. The disease is often referred to as the "kissing disease" because the virus can be transmitted through oral contact and exchange of saliva. The effects of the disease continue from two or three weeks to two months after infection. Symptoms include fatigue, sore throat, fever, lymphadenitis, and an increased number of leukocytes. Mononucleosis does not have specific treatment. The patient is told to get plenty of bed rest. Antibiotics may be prescribed to treat secondary infections.

Hodgkin's disease, or **Hodgkin's lymphoma**, is a rare cancer that affects the lymphatic system. Early symptoms include swollen lymph nodes, unexplained fevers, night sweats, and fatigue. The disease may also be characterized by weight loss and itchy skin. The cause of Hodgkin's disease is unknown, but factors such as Epstein-Barr infection, being male, and being a young or older adult can contribute to the likelihood of developing the disease. If caught early, the disease can be treated successfully by chemotherapy, radiation, or a combination of both.

Pathology Connection: Cancer Stages and the Lymphatic System

When patients are diagnosed with cancer, they are often told they have a certain "stage" of cancer. Cancer is staged and prognosis is determined by the amount of metastasis (spread), type of tumor, and lymph node involvement. The lymphatic system is amazingly efficient at capturing and transporting pathogens from connective tissue to lymph nodes for destruction. Unfortunately, this ability can also be a liability when cancerous cells develop in one part of the body

14-3 Go to your DVD for a video on the topic of skin cancer that has the potential to be spread via the lymphatic system.

and use the lymphatic system to hitch a ride to distant areas of the body. Once the cancerous cells make their way from their point of origin into lymphatic capillaries, and if the lymphocytes in the lymph nodes do not overpower the tumor cells, cancer cells can easily move around the body, invading many different areas simultaneously.

www.prenhall.com/colbert
Nuclear medicine deals with the treatment of various types of cancer and attempts to prevent the spread of the cancer through the lymphatic system. Please go to the companion website to learn more about nuclear medicine and the related professional profiles of opportunities that exist in this field.

Cancers that are diagnosed after they have already spread are much more likely to be fatal than cancers that are treated before cells have a chance to spread. This is why screening for certain types of cancer is so important and often why lymph nodes are removed for study around surgical sites of cancer. The earlier a patient is diagnosed, the better his or her chances of beating the disease. Though most types of cancer have specific staging criteria, cancer stages generally follow this pattern:

- Stage 1, no spread from origin
- Stage 2, spread to nearby tissues
- Stage 3, spread to nearby lymph nodes
- Stage 4, spread to distant tissues and organs

Stage 4 cancers are often terminal.

TEST YOUR KNOWLEDGE 14-1

Choose the best answer:

1. Any organism that invades your body and causes disease is known as a:
   a. bacteria
   b. fungus
   c. virus
   d. pathogen

2. Which of the following areas do *not* have large number of lymph nodes?
   a. cervical
   b. axillary
   c. abdominal
   d. adrenal

3. Cancer that has entered the lymph nodes is in this stage:
   a. first
   b. second
   c. third
   d. center

4. Which cells are housed in lymph nodes?
   a. red blood cells
   b. white blood cells
   c. platelets
   d. lymph nodules

5. The thoracic duct of the lymphatic system empties into this blood vessel:
   a. right subclavian vein
   b. left subclavian vein
   c. aorta
   d. hepatic portal vein

6. The function of the thymus is to:
   a. remove pathogens from blood
   b. destroy damaged red blood cells
   c. both a & b
   d. none of the above

7. The recommended treatment for uncomplicated viral tonsillitis is:
   a. antibiotics
   b. fluids and pain relievers
   c. tonsillectomy
   d. all of the above

THE IMMUNE SYSTEM

If the lymphatic system can be considered a transport and storage system for the body's defense systems, then the components of the **immune system** are the weapons and the actual troops. Much like the National Guard, the immune system can be called in to protect a city in times of extreme need. The immune system is a series of cells, chemicals, and barriers that protect the body from invasion by pathogens. Some of the weapons are active and some are passive; some are inborn and some change with experience. Together they form a system that is remarkably good at keeping the body free of infection.

Antigens and Antibodies

As you should remember from our discussion of blood types in Chapter 12, cells have molecules on the outer surface of their membranes to distinguish whether they are friend or foe. These molecules are called **antigens.** Each human being has his or her own unique cell surface antigens, as do all other living things, including bacteria, viruses, animals, and plants. The presence of these unique markers or antigens allows the immune system to distinguish between cells that are naturally yours and cells that are not. This ability, called **self-recognition** and **non-self recognition,** is at the heart of immune system function.

antigen *(AN tih jenn)*

Antigens are like the identity codes sent out by airplanes. Air traffic controllers and fighter jets on patrol over a no-fly zone depend on the identity codes to tell which aircraft are friendly. Antigens do the same for your immune system. A well-functioning immune system ignores the body's own antigens (self) and attacks other antigens (non-self). We discuss this in more depth later in this chapter.

As part of its defense system, the body can make proteins that bind to antigens, eventually leading to their destruction. Can you remember these proteins from the discussion of blood types? You got it! They are **antibodies,** one of the most potent weapons in the body's defensive arsenal. Antibodies are called into action when a foreign antigen invades the body.

Innate Versus Adaptive Immunity

The immune system defends the body on two fronts: by innate immunity and adaptive immunity. **Innate (natural) immunity** is the first line of defense against invasion. Innate immunity, as the name suggests, is the body's inborn ability to fight infection. It is inherited from parents, and is permanent. Innate immunity prevents invasion, or if pathogens do get inside, innate immunity recognizes the invasion and takes steps to stop the infection from spreading. However, innate immunity can only recognize that something is not native to the body; it can't identify the invaders. Innate immunity cannot improve with experience, and because it does not recognize specific pathogens, it cannot "remember" an infection that the body has encountered before.

Innate immunity consists of a collection of relatively crude mechanisms for defending the body from infection, sort of like building a wall around a city or having metal detectors at the airport. Keep in mind that metal detectors can't tell car keys from a pocket knife. Other parts of innate immunity are like weapons of mass destruction, indiscriminately killing pathogens and healthy tissue alike.

Innate immunity is backed up by a platoon of mechanisms that specifically target invaders, can remember invaders from previous encounters (and therefore prepare for future invasions), and can improve responses with experience. These mechanisms are known as **adaptive immunity,** or **acquired immunity,** because the mechanisms "learn" and change each time they are engaged. The components of adaptive immunity can be trained as an elite fighting force for particular pathogens. Their goals are "surgical strikes" targeting particular invaders and sparing as much of the healthy body tissue as possible.

It is tempting to think of these two parts of immunity as separate entities. However, the two work closely together. Innate immunity prepares the way for adaptive immunity, weakening some pathogens and stimulating components of adaptive immunity. Adaptive immunity in turn further stimulates innate immunity. It is through the mutual cooperation of both innate and adaptive immunity attacking the pathogen on two fronts that invaders can be removed from the body.

TEST YOUR KNOWLEDGE 14-2

Choose the best answer:

1. Cell surface molecules that can be used to identify cells are called:
    a. antibodies
    b. antigens
    c. antihistamines
    d. antibiotics

2. Proteins that bind to antigens are called:
    a. binding proteins
    b. receptors
    c. hormones
    d. antibodies

3. This type of immunity has no memory and is not specific:
    a. adaptive
    b. acquired
    c. innate
    d. nonspecific

Components of the Immune system

We often think the immune system begins in the blood with the white blood cells. However, physical barriers exist to act as a first line of defense to attempt to stop the infective agents from getting into the body in the first place. This is much like concrete traffic barriers you see in front of public buildings.

BARRIERS

Anything that prevents invaders from getting inside your body prevents infection. Therefore, your body has many barriers located in the places where invaders are most likely to gain entrance. Physical barriers include skin and the mucous membranes of the eyes, digestive system, respiratory system, and reproductive system. Not only are these surfaces difficult to penetrate, but they are packed with white blood cells and lymph capillaries to trap any invaders that might get through. The fluids associated with these physical barriers contain chemicals that act as chemical barriers. These chemicals are contained in tears, saliva, urine, mucous secretions, and sweat. One example is the antibacterial properties of the oil secreted by the sebaceous glands of your integumentary system. These barriers, both chemical and physical, prevent some invaders from ever getting inside the body. They are the "fortress" of the body, part of your innate immunity. After reading this information, can you see why wounds are frequent sources of infection?

 Proper hand-washing technique is the number one way of preventing the spread of disease in a hospital and also helps to protect your immune system from invasion. To view a video on proper hand-washing technique, please go to the DVD for this chapter.

CELLS

If an invader has an opportunity to enter the body, white blood cells (**leukocytes**) are responsible for defending the body against invaders. You learned about red blood cells and platelets in Chapter 12. Red blood cells are responsible for carrying oxygen throughout the body, and platelets are responsible for blood's ability to clot. White blood cells, on the other hand, are the mobile units of the immune system. White blood cells, which form in the bone marrow like red blood cells and platelets, move to other parts of the body to grow and mature until they are needed during an invasion. They are generally not released into the bloodstream in large numbers unless an infection is present. Figure 14-4 ▦ shows the major white blood cells found in the plasma.

leukocytes *(LOO koh sights)*
  leuk/o = *white*
  cyte = *cells*

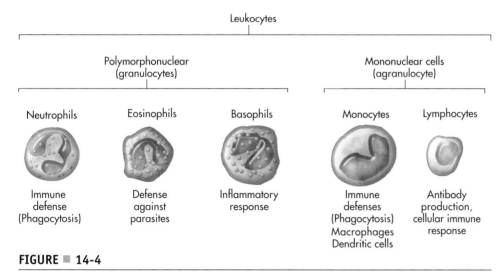

FIGURE ▦ 14-4

Major leukocytes.

TYPES OF WHITE BLOOD CELLS

In a police department of a large city, several types of jobs are needed for the entire department to function: traffic control officers, homicide detectives, forensic crime scene technicians, evidence control officers, internal affairs investigators, and more. Each has a specific duty and is called upon as needed, but in essence they all have the same goal—fighting crime. Similarly, the body has various types of white blood cells that are required to protect the body in varying circumstances. Table 14-1 lists the major white blood cells in the blood and the additional specialized white blood cells of the lymphatic system.

TABLE 14-1 White Blood Cells Involved In Immune Response

CELL TYPE	FUNCTION	INNATE OR ADAPTIVE
Neutrophils	Phagocytosis early in infection	Innate
Macrophages	Phagocytosis later in infection, stimulate immune system, antigen displaying cell	Innate but stimulate adaptive immunity
Basophils and mast cells	Release inflammatory chemicals. Basophils are mobile; mast cells are found in connective tissue	Innate
Eosinophils	Counteract basophils	Innate
Dendritic cells	Antigen displaying cells, weakly phagocytic	Innate but stimulate adaptive immunity
Natural killer cells	Kill cells displaying foreign antigens	Innate
T cells (lymphocytes)	Kill pathogens directly, regulate immune response, activate other lymphocytes, remember past infections	Adaptive
B cells (lymphocytes)	Release antibodies to foreign antigens, remember past infections	Adaptive

Pathology Connection: Leukemia

Leukemia is a cancer of the bone marrow and blood, characterized by the overproduction of white blood cells. In some forms of leukemia, the cells are immature or non-functional. There are several different types of leukemia, differentiated by the type of white blood cells involved and the nature of the symptoms. They include acute **myelogenous leukemia**, chronic myelogenous leukemia, acute lymphocytic leukemia, and chronic lymphocytic leukemia. (There are also some rarer types of leukemia that will not be discussed here.)

In myelogenous leukemia, blood stem cells divide out of control. In **lymphocytic leukemia**, lymphocytes are the culprit. Both myelogenous and lymphocytic leukemia can be either acute or chronic. In acute leukemias, there is a rapid onset of symptoms caused by the overproduction of immature, non-functional cells. In chronic leukemias, symptoms develop slowly and patients may even be symptom-free in the early stages of the disease, because the white blood cells, though overproduced, are functional. The causes of leukemia are not known, though some types are more common in children and/or older adults. Chronic myelogenous leukemia is known to be caused by a particular chromosomal abnormality (the Philadelphia chromosome), and Down syndrome patients are known to be a higher risk for acute myelogenous leukemia.

Symptoms of leukemia are caused by decreased *function* of white blood cells (if cells are immature) and decreased *numbers* of red blood cells and platelets, which are crowded out by the excessive numbers of white blood cells. Thus, symptoms may include anemia, shortness of breath, fatigue, repeated infections, enlarged lymph nodes, weight loss, muscle aches, excessive bruising, and bleeding. Leukemia is diagnosed mainly by laboratory tests

myelogenous *(my eh LOJ gen nuss)*

including complete blood count, bone marrow biopsy, and genetic testing.

Treatment for each type of leukemia is different as are the success rates. Chemotherapy (to kill the abnormal cells), antibody treatments, and stem cell or bone marrow transplants may be useful in treating the disease. In some cases, such as with chronic lymphocytic leukemia, there is no cure. The disease and its symptoms may be managed for years. Acute lymphocytic leukemia, a common childhood leukemia, has a very high rate of successful treatment.

14-5 Go to your DVD to view a video on the topic of leukemia.

CHEMICALS

Not only do blood cells fight invaders, but chemicals found in the body can also assist in neutralizing and destroying invaders. **Cytokines** are proteins produced by damaged tissues and white blood cells that stimulate immune response in a variety of ways, including increasing inflammation, stimulating lymphocytes, and enhancing phagocytosis. Cytokines are involved in both innate and adaptive immunity. **Interferon** is a cytokine produced by cells infected by a virus. Interferon binds to neighboring, uninfected cells and stimulates them to produce chemicals that may protect these cells from viruses. Interferon has also had some success as an anticancer drug, but it is still considered experimental. **Tumor necrosis factor** (TNF), another cytokine, stimulates macrophages and also causes cell death in cancer cells. Many cytokines are types of molecules called **interleukins.** There are at least 10 different interleukins. They are involved in nearly every aspect of innate and adaptive immunity. Interleukins also have been used with moderate success in treating some forms of cancer.

cytokines *(SIGH tow kines)*

interferon *(in ter FIR on)*

tumor necrosis factor *(neh KROH siss)*

interleukins *(in ter LOO kins)*

Complement cascade is a complex series of reactions that activate 20 proteins that are usually inactive in the blood unless activated by a pathogen invasion. When these proteins are activated, they have a variety of effects, including **lysis** of bacterial cell membranes, stimulation of phagocytosis, attraction of white blood cells to the site of infection, clumping of cells with foreign antigens, and alteration of the structure of viruses. Complement cascade is part of both innate and adaptive immunity.

lysis *(LYE siss)* = to breakdown or destroy

INFLAMMATION

Inflammation, or the inflammatory response, is one of the most familiar weapons in the body's arsenal. Nearly everyone has experienced the swelling, pain, heat, and redness associated with inflammation at one time or another. Think back several chapters to the example of hitting your thumb with a hammer. What would happen to your thumb within a few minutes of injury? It would swell, turn red, get hot to the touch, and hurt for some time after the injury. What about an infected cut? Or a sore throat when you have a cold or the flu? Again, the symptoms are the same: redness, heat, swelling, and pain. This reaction is a deliberate action of your body in response to tissue damage, whether a mechanical injury, like hitting your thumb with a hammer, or damage due to the invasion of a pathogen, as in a wound infection or a strep throat. Part of this response helps to wall off the infected area to prevent further spread and allow the battle to focus at this site. This process is called **margination** and is an attempt to isolate the problem.

histamine *(HISS tah meen)*

When tissue is damaged, the cells send out chemicals such as **histamine**, another cytokine, which have several effects. These chemicals attract white blood cells to the site of injury, increase the permeability of capillaries, and cause local vasodilation. Extra fluid moves from the capillaries into the damaged tissue, causing swelling. More blood comes to the site, increasing the temperature of the tissue. White blood cells enter the area, destroying pathogens and clearing away dead and dying cells. The increase in fluid and cells coming to the area increases the pressure and is part of the reason the area remains painful even as the damage is being repaired. Inflammation is an innate immune mechanism, but it is also an important player in adaptive immunity. There are several causes of an inflammatory response. Please see Figure 14–5 ■.

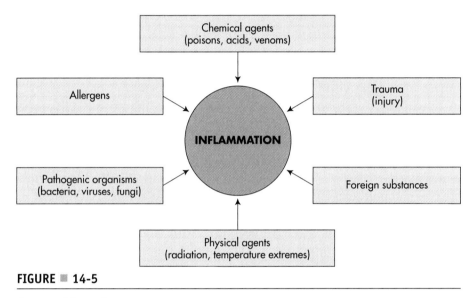

FIGURE ■ **14-5**

Causes of the inflammatory response.

 ## Pathology Connection: Inflammation, a Two-Edged Sword

Inflammation, like many of the body's weapons, has a positive feedback loop. Once inflammation starts, it continues until turned off. This kind of runaway positive feedback can cause problems if localized swelling increases pressure, causing more tissue damage. One of the reasons you "ice" a sprained ankle is to decrease inflammation to prevent further damage. Inflammation is a particular problem in enclosed or small spaces like the brain, spinal cord, respiratory system, and extremities, where a small buildup of pressure can cause serious damage. Remember Ray? Some of the damage to his spinal cord was due to post-injury inflammation. Even more dangerous is inflammation that becomes systemic, spreading throughout the whole body. This type of inflammation, called **anaphylaxis**, often causes blood pressure to plummet due to widespread vasodilation. Some people who are allergic to insect stings, nuts, or shellfish may experience this kind of inflammation. Anaphylaxis can be fatal unless treated by a medical professional immediately. People prone to anaphylaxis usually carry injectable epinephrine (adrenalin) to treat acute reactions. They should also wear medic alert bracelets to inform medical personnel of their allergies in the event of an emergency. Inflammation can be treated with several medications, including non-steroidal anti-inflammatory drugs (NSAIDs) like ibuprofen and naproxen sodium, steroids, and antihistamines (especially for allergic reactions).

FEVER

During an infection, tissues and components of the immune system release a number of cytokines that promote inflammation and immune responses. These cytokines circulate through the bloodstream and often reach distant targets, including the brain. One of the cytokine targets in the brain is the hypothalamus, which is responsible for setting and maintaining body temperature. Under the stimulation of cytokines, the hypothalamus raises the body's temperature set point. You feel cold and put on more clothes, huddle under more blankets, and shiver. Eventually, body temperature rises to the new set point and a fever results. While unpleasant to experience, this rise in body temperature is a deliberate attempt by the immune system to destroy the pathogens that have invaded the body. Like the rest of the innate response system, fever is a crude weapon against invaders. It might help fight off the infection, but it also causes overall discomfort.

TEST YOUR KNOWLEDGE 14-3

Choose the best answer:

1. Which of the following is *not* a function of complement cascade?
   a. lysis of bacterial cell membrane
   b. stimulation of macrophages
   c. chemotaxis
   d. swelling

2. These chemicals may protect the body against viruses and some cancers:
   a. complement
   b. histamine
   c. interferons
   d. immunoglobulins

3. B cells (lymphocytes) are directly responsible for:
   a. cell-mediated immunity
   b. inflammation
   c. complement cascade
   d. none of the above

4. Neutrophils and macrophages aid the innate immune system by doing this:
   a. secreting cytokines
   b. phagocytosis
   c. stimulating immune response
   d. all of the above

5. Redness, heat, swelling, and pain are all symptoms of:
   a. complement
   b. fever
   c. inflammation
   d. infection

6. _____ is a rapidly growing cancer of blood stem cells.
   a. acute myelogenous leukemia
   b. acute lymphocytic leukemia
   c. chronic myelogenous leukemia
   d. chronic lymphocyte leukemia

HOW THE IMMUNE SYSTEM WORKS

Think back to when you were in grade school and remember all the colds and sore throats you and your friends suffered through, not to mention chicken pox and measles for you older students. Compare that to the frequency of colds and sore throats you get as an adult. Chances are the number is much lower. What has happened?

Innate Immunity

As we have seen, for a pathogen to successfully invade your body, it must first get past your physical and chemical barriers. (Think of your body as a castle and the barriers of innate immunity as the alligator-filled moat protecting the castle from the marauding hoards.) Most of the millions of pathogens you encounter each day are kept out by these barriers. However, some pathogens—influenza or the cold virus, for example—are very good at getting past barriers.

neutrophil *(NOO troh fill)*

When a pathogen does get past the barriers, the more active portions of innate immunity are activated. The presence of a foreign antigen is detected by **neutrophils**. Neutrophils ingest the foreign antigen, destroying it, and release chemicals (cytokines, for example) that attract other white blood cells to the site of infection and stimulate inflammation. (Continuing our castle analogy, the neutrophils are the guards who greet any marauders who manage to survive the alligators.)

macrophages *(MAC reh fage ez)*

The release of cytokines and stimulation of inflammation attract **macrophages** and **natural killer (NK)** cells to the infection site. Macrophages destroy more infected cells by phagocytosis. NK cells use chemicals to destroy infected cells. Both cells release chemicals that further stimulate inflammation, activate more immune cells, and trigger the complement cascade. (The castle guards sound the alarm that the castle has been breached, summoning more troops to fight the invaders.)

At this point, the infected cells, or the pathogens themselves, are under attack on several fronts: phagocytosis, noxious chemicals, membrane rupture, clumping, and even alteration of their molecular structure. Chemicals have signaled your hypothalamus to raise your body temperature, and you run a fever. There is no question in your mind that you are ill. (Even though the guards are protecting the castle, the castle will be damaged.)

You would think this would be enough to fight off most pathogens. But keep in mind that this is crude warfare. Innate immunity simply destroys anything non-self. It does not use surgical strikes or specific weapons. Innate immunity lays waste to the infected area with almost indiscriminate attacks. Defending the castle is warfare in the crudest sense—no specialized weapons, just desperate attempts to defeat the invaders. Uninfected cells can be destroyed in the process.

The activities of innate immunity stimulate adaptive immunity. Chemicals released by NK cells, neutrophils, and other cells help activate adaptive immunity. When phagocytic cells ingest pathogens, they display the foreign antigen on their cell membranes. This ability to display foreign antigens without being infected is absolutely necessary for activation of B and T cells.

Adaptive Immunity

Fighting specific pathogens is the job of adaptive immunity. This part of the immune system has memory, "learns" with experience, and recognizes specific pathogens. It is because of adaptive immunity that people get the chicken pox only once. (Thank goodness for that!) It is because of adaptive immunity that immunizations are able to prevent illness. When was the last time you heard of somebody in the United States getting polio? Just 60 years ago, polio was an epidemic in the U.S. Keep in mind that innate and adaptive immunity work hand in hand. One cannot do its job without the other.

LYMPHOCYTE SELECTION

In order to function, lymphocytes must be able to recognize pathogens and ignore the body's own tissues. Think of these cells as members of the local fire company. They have a very specific job—putting out fires. Like any professional firefighter, lymphocytes must be **selected**. During **positive selection**, lymphocytes that recognize and bind to antigens are allowed to survive. Lymphocytes that fail to do their job do not survive, much like a firefighter who cannot carry a hose or climb a ladder will not become part of the company. Unfortunately, some of the selected lymphocytes recognize and bind to *your* antigens. These lymphocytes must be deactivated or they will attack and destroy your own tissues. The destruction of self-recognizing lymphocytes is known as **negative selection**. Both positive and negative selection must work in order for your immune system to function appropriately. You must have lymphocytes that attack invaders but that don't attack you!

 ## Pathology Connection: Autoimmune Disorders

Autoimmune disorders occur when the immune system attacks some part of the body. For some reason, the body fails to recognize "self" and destroys its own tissue as if it were an invader. There are literally hundreds of autoimmune disorders. Any part of the body can come under attack by mistake. Some of the more common disorders (and what they attack) are as follows:

- Rheumatoid arthritis (joint linings)
- Multiple sclerosis (myelin sheath in central nervous system) (See Chapter 9)
- Lupus erythematosis (every tissue, perhaps DNA)
- Type 1 diabetes (beta cells in pancreas) (Discussed several times)
- Myasthenia gravis (acetylcholine receptors in skeletal muscle) (See Chapter 7)
- Graves' disease (thyroid gland) (See Chapter 10)
- Addison's disease (adrenal gland) (See Chapter 10)

Just this short list illustrates how devastating an autoimmune disorder can be. Imagine the full power of your immune system turning against your thyroid gland or myelin sheath. The effects are devastating. Most autoimmune disorders can be treated with immunosuppressant drugs, but treatment may not be spectacularly successful and side effects are often severe.

RHEUMATOID ARTHRITIS (RA)

Rheumatoid arthritis (RA) is an autoimmune disorder in which the immune system attacks the synovial membrane of a patient's joints. The damage to the synovial membrane leads to destruction of bone and cartilage, leading to weakening of the soft tissue supporting the joints. Joints eventually collapse, bones fuse, and tendons shorten, causing disfigurement particularly of the hands and feet. The bone damage happens very early in untreated disease. The destruction of joint tissue itself causes increased inflammation, which causes more destruction in a massive positive feedback loop. Symptoms of RA include joint stiffness, particularly in the morning, symmetrical joint damage, fatigue, and fever. Since RA is a systemic disorder, patients may also have anemia; dry eyes; osteoporosis; lung, pericardial, and blood vessel inflammation; and increased risk of heart attack. Like in any chronic disease, many patients also have problems with depression or anxiety. RA may be moderate or severe, chronic or flaring. Many patients show a relapsing and remitting pattern. There are both juvenile and adult forms of the disease. Interestingly, pregnancy may improve RA symptoms in some patients. The cause of RA is unknown but seems to be caused by a combination of genetics and some environmental trigger, such as a virus, a bacterium, or hormonal changes. There are approximately 2 million RA patients in the US and women are 2 to 3 times more likely than men to have RA.

There is no definitive test for RA. Medical history and physical examination can be used to diagnose the disorder, as can imaging and lab tests including leukocyte counts, erythrocyte sedimentation rates, and a test for the presence of rheumatoid factor. Treatment is long-term and often difficult. Drugs to treat RA include **disease modifying anti-rheumatic drugs (DMARDs)** such as injected gold, methotrexate (a cancer drug), or plaquinil (a malaria drug). For patients who have used DMARDs unsuccessfully or for whom DMARDs have stopped working, there is a newer class of drugs called **biological response modifiers (BRMs)** that includes Humira, Enbrel, and Remicade. These medications suppress the immune system by decreasing immune-enhancing chemicals. In the past, **non-steroidal anti-inflammatory drugs (NSAIDs)** like ibuprofen and naproxen were initially prescribed for RA, but NSAIDs are ineffective in halting the progression of joint damage. Patients are also counseled to make lifestyle changes including reducing stress, increasing moderate exercise, and eating healthy. Assistive devices or surgery are often necessary for severe cases of RA.

SYSTEMIC LUPUS ERYTHEMATOSIS

Systemic lupus erythematosis (SLE) is an autoimmune disorder in which the immune system attacks a variety of structures in the body. The immune system appears to attack connective tissue in general. Nearly every part of the body may be attacked, including joints, skin, kidney, heart, lungs, blood vessels, and the brain. Symptoms include fatigue, joint pain and stiffness, fever, rashes (particularly butterfly-shaped on the face), and kidney disease. Pregnancy may worsen lupus symptoms or even bring on a flare up. Patients may experience chest pain, hair loss, swollen glands, anemia, heart problems, atherosclerosis, and depression or anxiety. Lupus may be moderate to severe, and may be chronic or show a relapsing and remitting pattern. The cause of lupus in unknown but both genetic predisposition and an environmental trigger is probably necessary for the onset of the disease.

Diagnosis is difficult because symptoms vary so much from patient to patient. There is no definitive test, but some patients test positive for antinuclear antibodies. Medical history, tissue biopsies, blood counts, urinalysis, erythrocyte sedimentation rate, complement levels, and imaging studies may be used to diagnose the disorder. Treatment is complicated because of the nature of the disease. Drug treatments include NSAIDs, DMARDs, steroids, and immunosuppressants. Several BRMs that have been successful in treating other autoimmune disorders are being investigated as treatments for lupus.

LYMPHOCYTE ACTIVATION

Lymphocytes develop and mature when you are a baby or a very young child. Just like you, they begin as an undifferentiated cell, meaning they have the potential to become anything. They must undergo a maturation process to become **differentiated**—in other words, to grow up to be a specialized cell

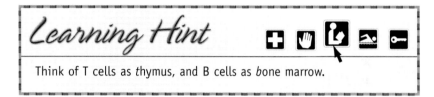

Think of T cells as *thymus*, and B cells as *bone* marrow.

with a specialized function. While undifferentiated lymphocytes are produced in the bone marrow, some migrate to the thymus and are destined to become T cells. Others stay, develop, and mature in the bone marrow to become B cells.

After they are specialized, the lymphocytes hang out in the lymph nodes waiting for a pathogen to come along that they recognize. The lymphocytes can go into suspended animation for the duration. In order for the lymphocytes to fight off the pathogen, they must have a wake-up call. Picture the fire company sleeping in the middle of the night and the alarm going off. This wake-up call is called **lymphocyte activation.** It causes lymphocytes to circulate continuously in the bloodstream and activates the lymph system to combat the pathogen.

Let's go back to innate immune response for a minute. When a pathogen invades your tissues, your innate immunity mounts a response to the pathogen. One part of innate immunity is the phagocytosis of infected cells or bits of pathogen. When these cells—macrophages, **dendritic cells**, and others—eat the pathogen, the pathogen's antigens are displayed on the outside of the phagocytic cells, kind of like little wanted posters. These cells then prowl the lymph nodes, displaying these tiny bits of pathogen, searching for the lymphocytes that can recognize the pathogen bits. When the right lymphocytes meet the right antigen display, the lymphocytes are activated (their wake-up call for battle). This activation is the beginning of adaptive immunity. Remember, water doesn't come out of a fire hose until the hose is turned on. See Figure 14–6 ■ to see how the immune system activates and differentiates lymphocytes.

dendritic cells (den DRID ick)

LYMPHOCYTE PROLIFERATION

The body has only a few lymphocytes that recognize each invader to which it has been exposed. But in order to fight off an infection, hundreds of thousands of lymphocytes are needed to attack the infection. Simple activation of lymphocytes is not enough. The activated lymphocytes must make thousands of copies of themselves in order to fight off the thousands of pathogens reproducing in the body. This reproduction of lymphocytes is called **lymphocyte proliferation** (see Figure 14–7 ■).

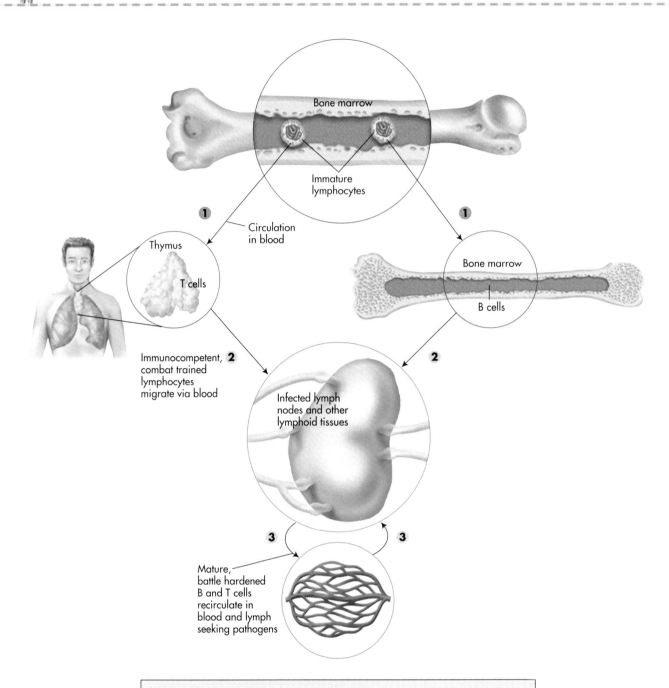

Bone marrow

Immature
lymphocytes

1   Circulation
in blood

1

Thymus

T cells

Bone marrow

B cells

Immunocompetent, **2**
combat trained
lymphocytes
migrate via blood

2

Infected lymph
nodes and other
lymphoid tissues

3   **3**

Mature,
battle hardened
B and T cells
recirculate in
blood and lymph
seeking pathogens

KEY:

1 Site of lymphocyte origin: Some lymphocytes are sent to the thymus where they become T cells, while others remain in the bone marrow to become B cells. These cells are now much like soldiers in boot camp that have learned their combat training but have yet to face the real enemy.

2 Sites of development of immunocompetence as B or T cells: An infection occurs and these cells are now given their marching orders to travel to the site of need. This is usually in the lymphoid tissue where the antigen challenge (the enemy has arrived) and the lymphocytes get placed into actual battle.

3 Site of antigen challenge and final differentiation to mature B and T cells: Now the battle hardened lymphocytes can competently patrol the rest of the body through the blood and lymph system, seeking out the enemy.

FIGURE ■ 14-6

Lymphocyte differentiation and activation.

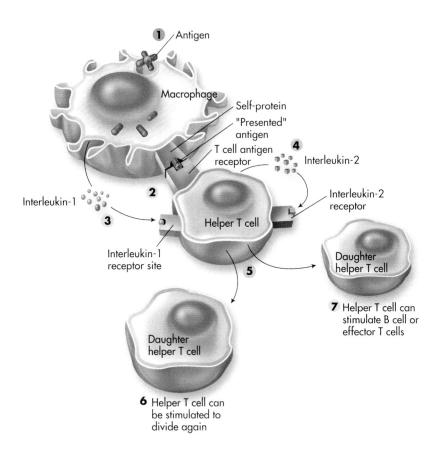

KEY:

1 Antigen processing cell such as a macrophage ingests a pathogen and displays its antigen on the macrophage's cell membrane.

2 Presented antigen is recognized by helper T cell.

3 The macrophage secretes a cytokine called interleukin-1.

4 Interleukin-1 stimulates the helper T cell to secrete interleukin-2 which combines with interleukin receptors.

5 When interleukin-2 binds to the receptor site the helper T cell divides.

6 The daughter cell produced can divide again if exposed to the same antigen. This greatly increases the number of helper T cells which can even further divide proportionate to the antigen exposure.

7 As helper T cells continue to divide they can stimulate B cell activation and produce effector T cells.

FIGURE ■ 14-7

Activation and proliferation of helper T cells.

Just like one platoon would not be enough to repel a full-scale invasion of a country, or one fire company enough to save a city block from burning down, a few lymphocytes are not enough to defend the entire body.

When you first met lymphocytes, earlier in our tour of the troops, you discovered that B cells and T cells have different roles in the defense of the body. Whereas activation and selection is, for our purposes, the same for both B and T cells, proliferation is different depending on the type of lymphocyte that is being reproduced.

There are two types of proliferation: proliferation of helper T cells and proliferation of all other types of lymphocytes. The job of helper T cells is to help other lymphocytes. There must be lots of helper T cells before any other lymphocytes can be activated. Helper T cells are stimulated to divide by binding to antigen-displaying cells (from innate immunity) and by stimulation by cytokines (some secreted by cells from the innate immune response and inflammation). The helper T cells continue to divide, producing more helper T cells, and then help (hence their name) in the proliferation of B cells and other types of T cells (again, see Figure 14–7).

14-6 Go to your DVD to view an animation showing T cell destruction by HIV.

Helper T cells are absolutely necessary for the reproduction of B cells and other types of T cells. This is why HIV, the virus that causes AIDS, is so devastating to the immune system. HIV targets helper T cells specifically. To understand the role of the helper T cells, imagine a huge game of tag in which no lymphocytes can be activated or reproduced until they have been tagged by a helper T cell.

Pathology Connection: HIV and AIDS

Acquired immune deficiency syndrome (AIDS) is an immune deficiency disorder caused by infection with the **human immunodeficiency virus (HIV)**. It is characterized by severe decrease in immune function, particularly helper T cells (a group called CD4). HIV causes AIDS by directly killing helper T cells, by destroying lymph nodes through their chronic infection, and by decreasing the amount of immune-enhancing chemicals in the body. HIV is most commonly contracted during sexual contact with an infected person or by sharing IV needles with an infected drug user. HIV can also be passed via use of blood products (rare in US today), from mother to fetus *in utero* and in breast milk, and via accidental exposure to infected body fluids (rare). It cannot be spread by casual contact, coughing, sneezing, shaking hands, or sharing eating utensils.

Symptoms of HIV infection may appear approximately four weeks after infection and may include flu-like symptoms due to the body's attempt to fight off the infection. In some cases the immune system fights the disease for up to 10-12 years after infection before symptoms appear. Some people who are infected remain asymptomatic. Others may have **AIDS-related complex (ARC)** or full-blown AIDS. Symptoms of ARC include enlarged lymph nodes, fatigue, night sweats, weight loss, and diarrhea. Symptoms of full-blown AIDS include the symptoms of ARC plus opportunistic infections (such as Kaposi's sarcoma, *Pneumocystis jiroveci*, previously known as *Pneumocystis carinii* pneumonia, and toxoplasmosis), CNS dysfunction, and some forms of cancer. Patients with AIDS catch infections that are virtually unknown in humans or become very ill from viruses, fungi, and bacteria that are typically mild in patients with normal immune systems. HIV entered the US in the late 1970s and by the end of 2002 more than 800,000 patients had AIDS and 500,000 had died from the disease.

HIV is diagnosed by medical history and by a blood test that detects very low helper T cell count of less than 200 cells/mm^3. Treatment is complicated. Patients positive for HIV or in full-blown AIDS are treated with one of a number of drug "cocktails" consisting of combinations of drugs. These drugs may block viral reproduction, keep the virus from assembling new viruses, or block the entrance of the virus into cells. These cocktails, while effective are complicated and have severe side effects. Infections are treated appropriately. Advances in treatment in the last 10 years have greatly improved the prognosis of patients who are HIV positive. Many patients have survived for years post-infection without developing full blown AIDS. HIV infection is not necessarily a death sentence. Education is one the of the most important methods of preventing HIV/AIDS.

 14-7 Go to your DVD to view an animation and video on the topic of HIV/AIDS.

TEST YOUR KNOWLEDGE 14-4

Choose the best answer:

1. In order to be activated, B and T cells must bind with:
   a. an antigen displaying cell
   b. a pathogen
   c. a damaged cell
   d. all of the above

2. Selection for immune-competent cells is called:
   a. negative selection
   b. positive selection
   c. immune selection
   d. make a selection

3. After a lymphocyte is activated, what must it do before it can fight off a pathogen?
   a. die
   b. agglutinate
   c. proliferate
   d. congregate

4. The primary function of these cells is the activation of other lymphocytes:
   a. cytotoxic T cells
   b. helper T cells
   c. memory T cells
   d. regulatory T cells

5. HIV targets these cells:
   a. cytotoxic T cells
   b. helper T cells
   c. memory T cells
   d. B cells

6. The best treatment for HIV infection today is:
   a. wait for full blown AIDS to treat patient
   b. use cocktail of drugs to prevent AIDS from developing
   c. practice safe sex
   d. there is no treatment

B and T Cell Action

So, let's get back to the body's response to invasion. The innate immune system has been attacking the pathogen or cells infected with the pathogen on a number of fronts, using phagocytic cells, NK cells, fever, and a variety of noxious chemicals. A number of the body's own cells have been destroyed in the process, but the pathogen has not been defeated. However, while innate immunity has been holding down the fort, **Antigen Displaying Cells (ADCs),** also known as Antigen Presenting Cells (APCs) have sent out a signal calling on the weapons of adaptive immunity, B cells and T cells.

B CELLS

B cells are responsible for a type of adaptive immunity known as **antibody mediated immunity.** B cells fight pathogens by making and releasing antibodies to attack a specific pathogen. B cells develop into **plasma cells** and **memory B cells.** Plasma cells make antibodies and release them into the bloodstream. Antibodies bind to the antigens of infected cells or the antigens on the surface of freely floating pathogens. They are missiles programmed to home in on a specific target. Antibodies destroy pathogens using several methods, including inactivating the antigen (neutralization), causing antigens to clump together (agglutination), activating complement cascade, causing the release of chemicals to stimulate the immune system, and enhancing phagocytosis. This response to the pathogen is called the **primary response.**

All of these antibody-mediated mechanisms not only destroy pathogens specifically, but they also further stimulate both adaptive and innate immune response, continually increasing response to the pathogen. Remember that immune response is a positive feedback loop that must be deliberately turned off. It will not stop on its own. This makes sense if you think in terms of protection. Your protective systems should not give up until the danger is past. Smoke alarms keep wailing until there is no more smoke.

Other B cells, memory B cells, are stored in lymph nodes until they are needed at some future date. If the body is exposed to the same pathogen in the future, memory cells allow it to mount a much faster response to the invasion. This response is known as **secondary response** and is responsible for the ability of adaptive immunity to improve with experience (see Figure 14–8 ■).

plasma cells (PLAZ mah)

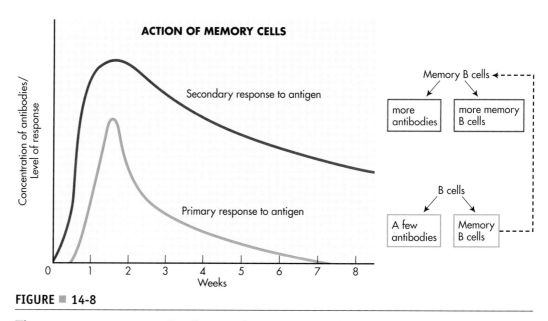

FIGURE ■ 14-8

The primary response causes B cells to produce memory B cells and a few antibodies. The second exposure causes the secondary response to produce more memory B cells and even more antibodies to fight the invaders. Now that the body has antibodies and more memory B cells, the secondary response begins more rapidly after exposure, produces more antibodies, and lasts a longer time.

T CELLS

There are at least four types of T cells: helper T cells, cytotoxic T cells, regulatory T cells (formerly known as suppressor T cells), and memory T cells. We have already seen the action of helper T cells. Helper T cells are responsible for activation of B and T cells.

Cytotoxic T cells are responsible for a part of adaptive immune response known as **cell-mediated immunity,** so called because the cytotoxic T cells are directly responsible for the death of pathogens or pathogen-infected cells. Cytotoxic T cells release a cytokine called **perforin,** which causes infected cells to develop holes in their membranes and die. Cytotoxic T cells also release other cytokines that stimulate both innate and adaptive immunity, especially attracting macrophages to the site of infection to dispose of cellular debris. The response of cytotoxic T cells is the primary response of cell-mediated immunity.

Regulatory T cells are the off-switch for the immune system. Immunity is controlled largely by positive feedback. Tissue damage causes the release of stimulatory chemicals that cause inflammation and increased immune response. Increased immune response results in the release of more chemicals, which causes more stimulation, which causes more chemicals to be released, and so on. Something must actively work to turn off the response when the threat is over or the immune response could become rampant and out of control and thus cause damage. This is the job of the regulatory T cells (along with the **eosinophils** of innate immunity).

Some T cells, rather than becoming cytotoxic T cells, give rise to **memory T cells**. Like memory B cells, memory T cells are responsible for secondary response, storing the recognition of the pathogen until the next encounter (see Figure 14–9 ■).

You can acquire immunity in several different ways. **Natural active immunity** is acquired in the course of daily life. When you catch a virus or a bacterium, your immune system fights it off, and memory cells are created for the next meeting. Anybody old enough to have had the chicken pox as a child (before the vaccine became available) is usually protected from a second round of chicken pox. **Artificial active immunity** is acquired during vaccinations. Getting a measles shot exposes you to small amounts of weakened virus, not enough to make you sick, but enough for your immune system to create memory cells. If you meet the virus later in life, you will be able to fight it off. Babies acquire **natural passive immunity** to many pathogens via antibodies passed across the placenta or during breast feeding. These antibodies, which babies can't yet make, protect them from infection for several months after birth. **Artificial passive immunity** is acquired when antibodies from one person are

cytotoxic *(sigh tow TOX ick)* = *cell death*

eosinophil *(ee oh SIN oh fill)*

Clinical Application

HOW YOU ACQUIRE IMMUNITY TO PATHOGENS

Your adaptive immune system is able to acquire immunity to new pathogens by creating memory cells each time you meet a pathogen. This process, called **immunization**, whether natural or done by medical procedures such as vaccinations, trains your immune system by creating memory cells for a pathogen. When you are exposed to the pathogen a second time, you do not get ill. Your immune system fights off the pathogen very quickly. In active acquired immunity, *you make* antibodies to fight the pathogens. In passive acquired immunity, antibodies *are introduced* to your body and therefore you don't make your own.

injected into another to help fight infection. An example is the HRIG shots that people get right after they are exposed to rabies. These shots are given before the actual rabies vaccine, which confers artificial active immunity. See Figure 14–10 ▪. Passive acquired immunity lasts only a few weeks; active acquired immunity lasts longer.

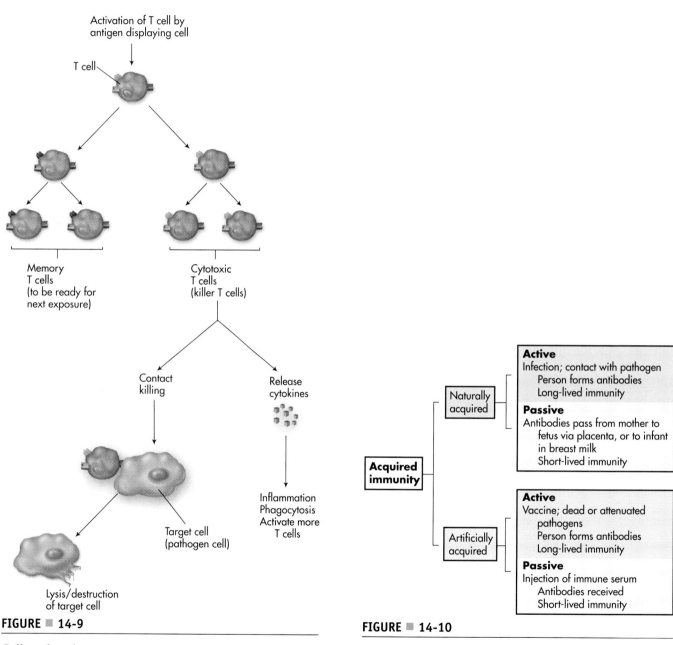

FIGURE ▪ 14-9

Cell-mediated immunity, primary, and secondary response.

FIGURE ▪ 14-10

Types of immunities.

TEST YOUR KNOWLEDGE 14-5

Choose the best answer:

1. B cells, once activated, can become two different kinds of cells: memory B cells and:
   a. cytotoxic T cells
   b. natural killers
   c. plasma cells
   d. macrophages

2. Secondary response is mediated by:
   a. cytotoxic T cells
   b. macrophages
   c. memory cells
   d. all of the above

3. _____ cells are responsible for antibody mediated immunity while cell mediated immunity is performed by _____ cells.
   a. B cells, cytotoxic T cells
   b. B cells, RBCs
   c. B cells, plasma cells
   d. plasma cells, memory cells

4. These cells, which were only discovered recently, are part of the "off-switch" for the immune system:
   a. cytotoxic T cells
   b. memory T cells
   c. regulatory T Cells
   d. helper T cells

5. As a new EMT, Cindy is required to have the Hepatitis B vaccine series. She will gain this type of immunity:
   a. natural active immunity
   b. artificial active immunity
   c. natural passive immunity
   d. artificial passive immunity

THE BIG PICTURE

At this point we have talked about the lymphatic system, innate immunity, and adaptive immunity as separate means to the same end: ridding the body of invading pathogens. We have mentioned repeatedly that the divisions are *not* separate but are intimately connected. Now that we have inspected them separately, it is time to put the entire defense system back together as a single, integrated fighting force (see Figure 14–11 ■).

A nasty army of pathogens attempts to invade your body. First, they must get past your barriers. Many invaders will be repelled simply by your intact skin or the secretions of your mucous membranes. If the invader gets inside your body, a series of weapons are stimulated by the introduction of a non-self antigen. Cells (neutrophils, macrophages, basophils, etc.) are stimulated. Chemicals (cytokines) are also released that stimulate inflammation and phagocytosis.

Macrophages and other cells, having ingested some of the invaders and now wearing the foreign antigens, move to the lymphatic system and search the lymph nodes, looking for the T and B cells that will recognize the intruder. Helper T cells activate and cause the proliferation of B cells and cytotoxic T cells, as well as release chemicals that further stimulate the phagocytic cells and inflammation. B cells produce antibodies that destroy the invaders and

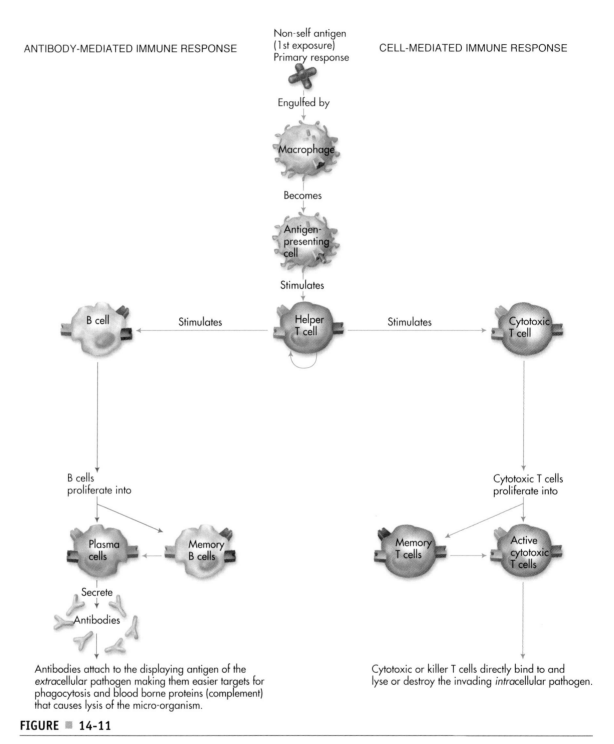

ANTIBODY-MEDIATED IMMUNE RESPONSE

Non-self antigen
(1st exposure)
Primary response

CELL-MEDIATED IMMUNE RESPONSE

Engulfed by

Macrophage

Becomes

Antigen-
presenting
cell

Stimulates

B cell ← Stimulates ← Helper T cell → Stimulates → Cytotoxic T cell

B cells
proliferate into

Plasma cells ← Memory B cells

Cytotoxic T cells
proliferate into

Memory T cells → Active cytotoxic T cells

Secrete

Antibodies

Antibodies attach to the displaying antigen of the
*extra*cellular pathogen making them easier targets for
phagocytosis and blood borne proteins (complement)
that causes lysis of the micro-organism.

Cytotoxic or killer T cells directly bind to and
lyse or destroy the invading *intra*cellular pathogen.

FIGURE ■ 14-11

The battle plan of the body's defenses.

further stimulate immune response. Cytotoxic T cells destroy invaders directly and release chemicals that further stimulate immune response.

Immune response, both innate and adaptive, will continue to be stimulated until the feedback loop is stopped, at least in part by regulatory T cells.
Memory B cells and T cells will be stored in the lymph nodes for later use, in

case another army of those same type of pathogens attempts another invasion. Macrophages and other phagocytic cells will clean up the debris left by the warfare waged by your immune system. Once the danger has passed, your body will return to normal.

Pathology Connection: Hypersensitivity Reactions

During a hypersensitivity reaction, more commonly known as an allergy, the immune system mounts a hyperactive response to a foreign antigen, often treating a harmless antigen, like grass or mold or insect bite, as an invading pathogen. Local hypersensitivity reactions, like hay fever (allergic rhinitis), hives, skin rashes, and asthma, are generally mild and not life-threatening. (Asthma is an obvious exception because it interferes with ventilation.) Millions of people in the U.S. have asthma, hay fever, or both. In contrast, systemic hypersensitivity reactions, known as **anaphylaxis**, are life-threatening. During anaphylaxis, mast cells and basophils release immune-stimulating chemicals throughout the body. The chemicals cause widespread vasodilation, which leads to dangerously low blood pressure and heart failure. Hives and asthma may also accompany an anaphylactic reaction. Figure 14–12 ▦ shows stimulation of the mast cells within the nose due to an allergen, which causes allergic rhinitis (runny nose). Note that mast cells are found throughout the body. If overstimulated in the eyes, mast cells cause red and runny eyes; if overstimulated in the lungs, they cause allergic asthma. In addition to mast cells, other immune cells also get involved in allergic reactions, including basophils, B cells, eosinophils, and macrophages. The activation of these cells lead to increased inflammation and release of immune stimulating chemicals. For many people, one type of allergy may be complicated by another or a mild allergy may progress to a more severe form. For example, many patients with asthma also have hay fever. Other patients may experience what is called **atopic march**. In these patients, early exposure to an allergen may cause a skin reaction, for example, while later exposures cause allergic rhinitis and later exposures cause asthma or even anaphylaxis. That's why a bee sting may cause a minor reaction in a child and a much more severe reaction when that person is stung again as an adult. For many people the worsening of their allergies is predictable.

Allergies are often difficult to diagnose. Blood tests may show increased immune activity but pinpointing the allergen is often difficult. Exposing the patient's skin or a blood sample to suspected allergens may help find the allergen but these are often inconclusive.

The best way to treat an allergy is to avoid coming in contact with whatever causes the problem. That, however is almost impossible if a patient is allergic to grass, mold, or pollen. If the culprit is unavoidable, several treatments are possible and will be elaborated upon in the upcoming Pharmacology Corner.

14-8 Go to your DVD to see an animation of the treatment of anaphylactic shock with an EpiPen (a device that delivers epinephrine, a.k.a. adrenalin). In addition, you can view a video on a nonfatal but fairly common hypersensitivity reaction called allergic rhinitis or, in simple lay terms, the runny nose.

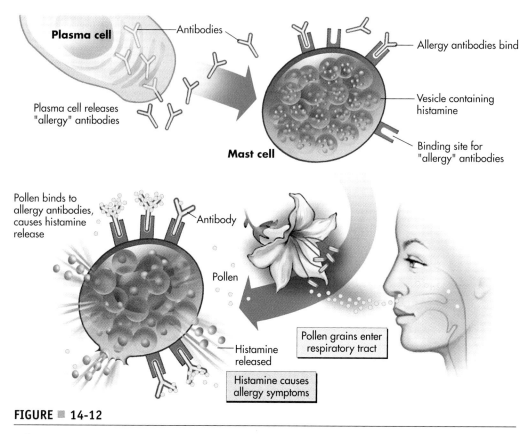

FIGURE ▦ 14-12

Allergic rhinitis.

℞ Pharmacology Corner

Most drugs that are given to control the immune system are immune suppressants of some sort. They are prescribed to decrease the activity of an overactive immune system. The first class of drugs is the anti-inflammatory drugs. Anti-inflammatory drugs specifically decrease inflammation though many have other effects. Most of you are familiar with the non-steroidal anti-inflammatory drugs (NSAIDs), including aspirin, naproxen (Aleve), ibuprofen (Advil), celecoxib (Celebrex), and many others. NSAIDs relieve pain and inflammation by inhibiting a group of hormones called **prostaglandins**. NSAIDs are also notorious for other, less desirable effects, like the formation of stomach ulcers. All NSAIDs inhibit prostaglandins, but they each work in a different way, so one NSAID may work well for one person but not for another.

Another class of anti-inflammatory drugs is the steroids, like prednisone and cortisone. These drugs are powerful anti-inflammatories, but when given systematically (orally or by injection) they may have dangerous side effects when used for long periods. Topical steroids, applied to the skin or nasal passages, are quite effective for treating skin inflammation and upper respiratory symptoms due to allergies. Inhaled steroids are used to control asthma symptoms. Topical and inhaled steroids do not have the severe side effects associated with oral or injected steroids.

Allergies are treated with specialized drugs, like antihistamines. Traditional antihistamines, which decrease mast cell activation, have long been prescribed or available over the counter. They work well but cause severe drowsiness. Newer antihistamines that do not cause drowsiness have proven to be safe and effective, but do not decrease congestion or inflammation. More specific drugs that block cytokines are currently available, as are drugs that block the antibodies that mediate allergies. Experimental medications that block B cells, TNF, and complement are currently being tested. Steroids may often be effective in treating asthma and severe allergies, but long-term use of oral steroids can produce nasty side effects. Topical steroids, either inhaled or administered as nasal sprays, work well for many people. Patients with anaphylactic reactions to common allergens may be required to carry an EpiPen for emergencies. A shot of epinephrine constricts blood vessels, stimulates the heart, and opens air passages, decreasing the effects of anaphylactic reactions and giving a patient time to get to medical care.

Autoimmune disorders like RA and lupus used to be treated with NSAIDs and steroids, but those drugs proved ineffective for many disorders. Today, autoimmune disorders are more likely to be treated with **disease modifying anti-rheumatic drugs (DMARDs)**. DMARDs are confusing pharmacologically. It is not well understood how they work and they are highly toxic. DMARDs include immunosuppressants (Cyclosporine, an anti-rejection drug used after organ transplants), gold shots, methotrexate (a chemotherapy drug) and anti-malarial medications.

For patients who have used DMARDS unsuccessfully or for whom DMARDS have stopped working, there is a newer class of drugs called **biological response modifiers (BRMs)**. Three BRM's on the market—etanercept (Enbrel), infliximab (Remicade), and adalimumab (Humira)—inhibit tumor necrosis factor (TNF), a cytokine. Another, anakinra (Kineret), inhibits interleukin-1, another cytokine, and still another, rituximab (Rituxan), inhibits B cells. These medications suppress the immune system, because they decrease immune enhancing chemicals or inhibit the proliferation of cells. Side effects include increased risk of infection and some forms of cancer, but the drugs are less toxic than most DMARDs. In addition, BRMs seem to do a better job at decreasing fatigue in RA patients than the DMARDs.

> **www.prenhall.com/colbert**
> There are several types of drugs dispensed by a pharmacy to treat autoimmune and allergic disorders. To learn more about the professional opportunities within a pharmacy, please visit the companion website.

A Quick Trip Through Some Diseases of the Immune System

DISEASE	ETIOLOGY	SIGNS AND SYMPTOMS	DIAGNOSTIC TEST(S)	TREATMENT
Lymphoma	Uncontrolled proliferation of lymphocytes (30 different types).	Enlarged lymph nodes, fever, weight loss, other vague symptoms.	Biopsy, imaging, bone marrow tests.	Chemotherapy, radiation, antibody therapy, stem cell or bone marrow transplant.
Tonsillitis	Inflammation of the tonsils due to bacterial or viral infection.	Sore throat, swollen tonsils, fever, upper respiratory symptoms, swollen lymph glands, visible coating or spots on the tonsils, upset stomach.	Examination showing inflamed tonsils, rule out bacterial infection.	If bacterial, treat with antibiotics. If viral, rest and fluids until virus runs its course. If swollen tonsils impair breathing, tonsillectomy may be necessary.
Leukemia	Excess production of white blood cells, causing decreased function of WBCs, RBCs and platelets.	Increased infection, anemia, bleeding, bruising, fatigue, weight loss.	CBC, genetic testing, bone marrow biopsy.	Depending on the type of leukemia, chemotherapy, stem cells, bone marrow transplant.
Hodgkin's Disease	Cancer of lymph nodes, effecting white blood cells—most common in people 5-35 years of age.	Painless swelling of lymph node is early symptom (often in the neck), fever, night sweats, weight loss, weakness, fatigue, and itching. Respiratory symptons may occur with chest involvement.	Biopsy, various types of x-rays, ultrasounds and lymphangiograms.	Possibly chemotherapy and radiation depending upon age and stage.
Rheumatoid arthritis (RA)	Destruction of joints by immune system.	Swollen painful joints, fatigue, anemia, cardiovascular abnormalities.	Blood tests, imaging studies.	DMARDs, BRMs, and lifestyle changes. Difficult to treat effectively.
Lupus	Destruction of many different tissues by immune system.	Swollen painful joints, fatigue, anemia, cardiovascular abnormalities.	Blood tests, imaging studies.	DMARDs, BRMs, and lifestyle changes. Difficult to treat effectively.
HIV/AIDS	Infection of helper T cells by HIV causing decreased numbers of T cells, eventually leading to severe immune deficiency.	Flu-like symptoms in early stages, opportunistic infections and some forms of cancer in full-blown AIDS.	Medical history, very low helper T cell counts. AIDS diagnosis made when T cell counts are less than 200 cells/mm^3.	Drug cocktails designed to prevent infection from progressing to full-blown AIDS, treatment of opportunistic infections when they develop.
Severe combined immune deficiency (SCID)	Severe decreased T cell production (and sometimes other cells) caused by a number of different genetic defects.	Repeated opportunistic or severe infections in babies, fatal if untreated.	Genetic testing.	Depends on the underlying genetic problem: enzyme replacement, complete sterile environment, bone marrow transplant, stem cells, and gene therapy. Most types difficult to treat.
Allergies	Overreaction of the immune system to typically harmless antigens, such as mold, pollen, or animal fur.	Depend on severity: upper respiratory symptoms, asthma, skin rashes, anaphylaxis.	Medical history, elimination of suspected culprits, skin tests, blood tests.	Avoidance, antihistamines, allergy shots, steroids, epinephrine.

SUMMARY

Snapshots from the Journey

→ The lymphatic system is the transport system for the immune system and houses the lymphocytes. It consists of lymph capillaries, vessels, trunks, and ducts containing lymphatic fluid and lymph nodes, which house white blood cells.

→ Lymph nodes are concentrated in several regions of the body: cervical, axillary, inguinal, pelvic, abdominal, thoracic, and supratrochlear. These patches are in areas where pathogens are most likely to enter. Tonsils, adenoids, spleen, and thymus all contain lymphatic tissue.

→ Fluid leaking from blood capillaries enters tissue fluid and flows into lymph capillaries. The fluid is carried through lymph capillaries to lymph vessels to lymph nodes. In the nodes, any pathogens are destroyed by white blood cells. Fluid then flows from nodes to vessels to lymphatic trunks to collecting ducts and into either the right or left subclavian vein, returning the fluid to the bloodstream. Cancer cells may use this "highway" to spread to other parts of the body.

→ The thymus and spleen are lymphatic organs. The spleen contains blood sinuses and removes dead and dying red blood cells as well as pathogens. The thymus is the birthplace of T lymphocytes (T cells).

→ Immune system function is based on its ability to recognize cell surface molecules called antigens. The immune system must ignore self-antigens (the body's antigens) and respond to non-self antigens (foreign cells).

→ The immune system is divided into two separate but extremely interdependent parts, innate immunity and adaptive immunity. Innate immunity is nonspecific, has no memory, and cannot improve performance with experience. Adaptive immunity is specific, has memory, and can improve performance with experience.

→ Barriers prevent pathogens from getting into the body. Barriers can be either physical or chemical. Skin and tears are examples of barriers.

→ The immune system uses a dozen or more different types of cells to combat pathogens. All of these cells are leukocytes or modified leukocytes. Some are part of innate immunity and some are part of adaptive immunity. Their functions range from phagocytosis, chemical stimulation of other cells, and antigen display to antibody secretion and direct destruction of pathogens. Cancer of these cells, which leads to excessive production in the bone marrow, is known as leukemia.

→ Immune response is stimulated by a variety of chemicals, including the cytokines, histamine, and complement.

→ Inflammation, the familiar redness, heat, swelling, and pain associated with infection, is a powerful tool in the immune system's arsenal. During inflammation, white blood cells are stimulated and attracted to the site of infection to destroy pathogens and clean up cellular debris. Inflammation is, like much of immune response, a two-edged sword. Too much inflammation may be more damaging than the infection itself.

→ Fever, like inflammation, is a deliberate attempt by the body to destroy a pathogen. Chemicals trick the hypothalamus into raising the temperature set point in an attempt to make body temperature too hot for pathogens.

→ Innate immune mechanisms are triggered by the presence of foreign antigens in the body. These mechanisms hold off the infection and stimulate adaptive immune mechanisms.

→ Adaptive immunity uses T and B lymphocytes to fight specific pathogens. Lymphocytes must be selected during development to recognize foreign antigens but to ignore the body's own antigens. In order to fight a pathogen, lymphocytes must be activated by binding with antigen displaying cells. Once activated, lymphocytes must proliferate, making thousands of copies of themselves. Helper T cells are required for activation of most types of lymphocytes. If lymphocytes continue to recognize body tissues, an autoimmune disorder may result.

→ Rheumatoid arthritis (RA) is an autoimmune disorder in which the immune system attacks the lining of synovial joints. RA is a systemic illness that involves many joints at once and may also cause damage to the heart, eyes and lungs. Systemic lupus erythematosis (SLE) is a more generalized autoimmune disorder in which the immune system attacks connective tissue, and even DNA. Lupus is also a systemic illness causing joint, heart, and kidney damage.

→ HIV, the virus which causes AIDS, infects and destroys helper T cells. When HIV infection progresses to full-blown AIDS, helper T cell numbers drop so low that activation and proliferation of T and B cells is negatively affected. Adaptive immunity ceases to function and patients are at increased risk of rare infections or may become seriously ill from infections that are not usually dangerous in people with healthy immune systems.

→ B cells are mediators of antibody-mediated immunity. B cells are activated by binding to antigen-presenting cells and helper T cells. Once B cells begin to proliferate, they become either plasma cells or memory B cells. Plasma cells secrete antibodies during primary response. Antibodies help destroy pathogens by binding to antigens on infected cells. Memory B cells are stored for the next time the pathogen is encountered. They mediate secondary response.

→ Cytotoxic T cells mediate cell-mediated immunity by directly killing infected cells. This is the primary T cell response. Like B cells, cytotoxic T cells are activated by helper T cells. Some T cells become memory T cells and mediate secondary response if the pathogen is encountered again.

→ Regulatory T cells are one of the off-switches for the immune system. We don't know very much about how these cells work because they were only discovered in the last 10 years.

→ Keep in mind that innate and adaptive immunity do not work separately. They work together. One stimulates the other in a huge positive feedback loop. If either innate or adaptive immunity stops working, the whole system breaks down.

RAY'S STORY

Remember Ray, the daredevil who broke his neck in the pool and injured his spinal cord? He was treated with steroids as soon as he arrived at the hospital.

a. Why?

_____

_____

_____

b. What were the doctors hoping to prevent?

_____

_____

c. Since the time Ray was injured, doctors have begun experimenting with treating injuries like Ray's by inducing hypothermia. How might making Ray's body very cold have helped treat his injury?

_____

_____

MARIA'S STORY

Maria was diagnosed with diabetes mellitus when she was 12-years-old. After not taking care of herself in her teen years, she had been doing much better until recently. Though she is following all of her doctor's advice, her diabetes is currently not well controlled.

a. What is the cause of her diabetes?

_____

_____

b. How is the immune system involved?

_____

_____

c. It has been suggested that Maria have a pancreas transplant to relieve her symptoms. If she accepts this treatment option, how might her immune system continue to be a problem?

_____

_____

d. What kind of drugs may she have to take after the transplant to prevent problems?

_____

_____

REVIEW QUESTIONS

Multiple Choice

1. Lymphocytes are selected for their ability to:
   a. recognize antigens
   b. ignore self antigens
   c. produce T-cells
   d. both a and b

2. Mounting an excessive immune response to a harmless antigen is called a(n):
   a. autoimmune disorder
   b. allergy
   c. immunodeficiency
   d. AIDS infection

3. Which of the following is an innate immune cell?
   a. neutrophil
   b. memory cell
   c. helper T cell
   d. plasma cell

4. Lymphocytes are activated by binding with antigens on:
   a. bacteria
   b. antigen displaying cells
   c. viruses
   d. lymph nodes

5. Innate immunity is not stimulated by foreign antigens.
   a. true
   b. false
   c. it depends
   d. all of the above

6. Fever and inflammation are both part of what kind of immunity?
   a. auto
   b. adaptive
   c. acquired
   d. innate

7. Which of the following is an autoimmune disorder?
   a. leukemia
   b. AIDS
   c. lupus
   d. allergies

8. Which of the following drugs are not useful in treating rheumatoid arthritis even though they are anti-inflammatory?
   a. DMARDs
   b. BRMs
   c. NSAIDs
   d. chemotherapy

9. Which of the following is not a symptom of leukemia?
   a. bruising
   b. anemia
   c. weight gain
   d. increased infections

Short Answer

1. List the regions of the body containing many lymph nodes.

   _____
   _____
   _____
   _____

2. Trace the circulation of lymph from the cardiovascular system through the lymphatic system and back to the cardiovascular system.

   _____
   _____
   _____
   _____

3. List four types of cells and their functions in immunity.

   _____
   _____
   _____
   _____

4. Explain the differences between innate and adaptive immunity.

_____

_____

_____

_____

5. What is atopic march? Why is it so dangerous?

_____

_____

_____

_____

6. HIV is not the same as AIDS. Explain.

_____

_____

_____

_____

7. Explain the categories of leukemia.

_____

_____

_____

_____

Suggested Activities

1. Draw a flow chart of innate and adaptive immunity and their interactions.

2. Form a group of students and role-play parts of the immune system. Assign some to be macrophages, others to be NK cells, and others to be B and T lymphocytes. Describe the process of fighting off a pathogen as per your role.

3. Using the Internet, do a search of autoimmune disorders. How many can you find? What tissues can be attacked by your immune system?

14-9 Now that you have completed your journey through this chapter, please go to the DVD for interactive games and puzzles concerning the medical terms and concepts contained in this chapter. By playing the games you will reinforce your learning of the lymphatic and immune systems in a fun way.

THE GASTROINTESTINAL *System*

Fuel for the Trip

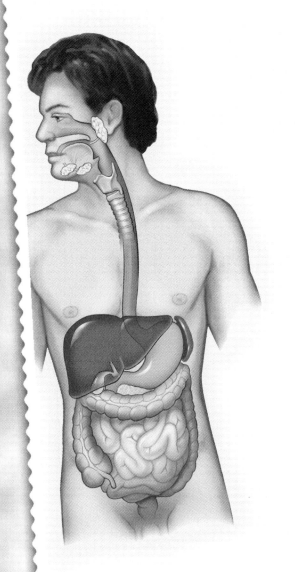

So far, we have discussed a variety of body systems and learned how they are put together and how they function. All of these systems need to function together to create a smooth-running machine. But just like a Ferrari, no matter how well it is designed nor how precisely it is put together, the body cannot function without fuel! This chapter focuses on the digestive, or **gastrointestinal** (GI), system and how it:

→ takes in (ingests) raw materials,

→ breaks them down (digests) both physically and chemically to usable elements,

→ absorbs those elements, and

→ eliminates what isn't usable.

These processes are accomplished through an amazing array of main and accessory organs and substances. Here is a concept to ponder as we journey through the digestive system: the food that enters your mouth, travels through your digestive system, and is eventually eliminated is never once *inside* your body! It remains in a tube-like "highway," with materials contained within the food exiting the highway ramps in specific areas!

Chapter 15

LEARNING OBJECTIVES

Upon completion of your journey through this chapter, you will be able to:

→ Locate and describe the functions of the organs of the digestive system

→ Locate and describe the functions of the accessory organs for digestion

→ Differentiate between ingestion and digestion and between chemical and mechanical processing of food

→ Describe the journey of a bolus of food from the mouth to the anus

→ Discuss the structure of the tooth

→ Describe the various enzymes and chemicals needed for digestion

→ Describe common disorders of the gastrointestinal system

MULTIMEDIA APPLICATIONS

DVD Interactive Exercises

→ 3-D animation of the digestive system, 15-1

→ Animation and videos concerning oral care and assessment along with cavity prevention, 15-2

→ Video and interactive drag-and-drop exercise of the digestive system and intestinal wall, 15-3

→ Animations of carbohydrates, lipids, and proteins, 15-4

→ Videos on nutrition and enteral feedings, 15-5

→ Animation of GERD, 15-6

→ Videos on anorexia, bulimia, and obesity, 15-7

→ Interactive games and puzzles, 15-8

www.prenhall.com/colbert

→ Professional Profiles:
  • Dental Assistants and Hygienists
  • Medical Profession: Dietician

→ Related Internet Links

→ Additional Review Questions

Pronunciation Guide

Correct pronunciation is important in any journey so that you and others are completely understood. Here is a "see and say" Pronunciation Guide for the more difficult terms to pronounce in this chapter.

adventitia (ADD ven TISH ee ah)

alimentary tract (AL ah MEN tair ee)

appendicitis (ah PEN dih SIGH tiss)

appendix (ah PEN dicks)

cecum (SEE kum)

cementum (si MEN tum)

cheilitis (ky LY tis)

cholecystitis (KOH lee siss TYE tiss)

cholelithiasis (KOH lee lih THY ah siss)

chyle (KILE)

chyme (KIME)

defecation (deh fih CAY shun)

duodenum (doo ODD eh num)

emulsification
   (ee MULL sih fih KAY shun)

epiglottis (ep ih GLAH tiss)

esophagus (eh SOFF ah guss)

frenulum (FREN you lum)

fundus (FUN duss)

gingivae (JIN jih vah)

hepatic duct (hep PA tic duct)

ilium (ILL ee um)

jejunum (jee JOO num)

labia (LAY bee ah)

mastication (MASS tih CAY shun)

mesentery (MEZ in tare ee)

pancreatitis (PAN kree ah TYE tiss)

peristalsis (pair ih STALL siss)

pharynx (FAIR inks)

plicae circulares
   (PLY kay sir cue LAIR es)

pyloric sphincter (pye LOR ik SFINK ter)

pylorus (pye LOR us)

rugae (ROO gay)

serosa (seh ROSE ah)

villi (VILL eye)

SYSTEM OVERVIEW

alimentary tract *(AL ah MEN tair ee)*

anus *(AY nuss)*

The digestive tract (often called the **alimentary tract, alimentary canal,** or **GI tract**), is an approximately 30 foot muscular tube or tunnel-like structure that contains the organs of the digestive system. This tube begins at the mouth and ends at the **anus.** Between these two points are the pharynx, esophagus, stomach, and small and large intestines. In addition, accessory organs (such as teeth, tongue, salivary glands, liver, pancreas, and gallbladder) are necessary for processing materials into usable substances. Refer to Figure 15–1 ■ as we journey through the digestive system.

15-1 To view a 3-D animation of the digestive system, please go to your DVD for this chapter.

The components of the digestive system work together to perform the following general steps:

1. ingestion
2. mastication
3. digestion
4. secretion
5. absorption
6. excretion (defecation)

Food first enters the mouth, an activity called **ingestion**. Once food is ingested, the tongue and teeth work together to mechanically process the food by physically breaking it down. The chewing action is called **mastication**.

mastication *(MASS tih CAY shun)*

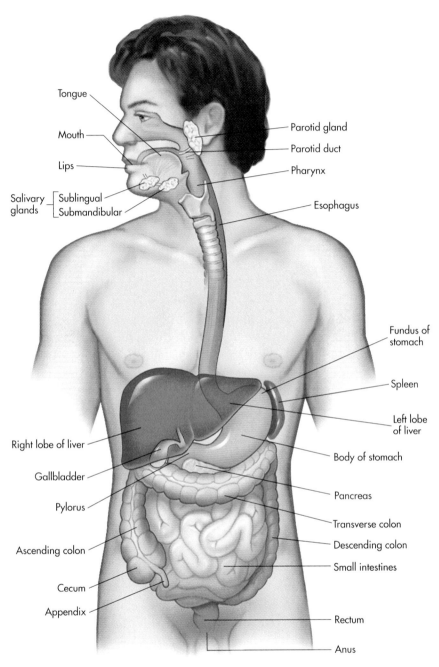

FIGURE ▪ 15-1

The digestive system.

This mechanical mixing process also continues in the muscular motions of the digestive tract, as you will soon see. **Digestion** is the chemical process of breaking down food into small molecules. This is necessary so nutrients can be absorbed by the lining of the digestive tract.

The *secretion* of acids, buffers, enzymes, and water aids in the breakdown of food. Once the food is broken down both physically and chemically in the intestine, it is ready for absorption into the blood and lymph. Finally, waste products and unusable materials are prepared for *excretion* and are eliminated by the body through **defecation.**

Now that you have a general idea of what's going on, let's begin the journey through the digestive system from beginning to end as if we were a slice of double cheese, pepperoni pizza!

THE MOUTH AND ORAL CAVITY

Oral Cavity

labia (LAY bee ah)
palates (PAL ahts)
uvula (YEW view lah) = little grape

pharynx (FAIR inks)

Your mouth (oral cavity) is a mucous-lined opening, also called the **buccal cavity**. Your lips, or **labia,** act as a door to this chamber. The **hard** and **soft palates** create the roof of the chamber, while the **tongue** acts as the floor. The tongue's base (area of attachment) and the **uvula**, that punching bag–shaped object dangling down from the soft palate, act as the boundary between the oral cavity and the next part of the digestive system, the **pharynx.** As we discovered in our travels through the respiratory system, the uvula aids in swallowing because it helps direct food toward the pharynx and helps block food from coming out your nose! There is a pair of lingual tonsils back there too. Although they aren't important for digestion, the tonsils help in fighting infection as part of the lymphatic system. The sides of the cavity are created by your cheeks. Your mouth and oral cavity region receives, or **ingests**, food. The food is tasted, mechanically broken down into smaller pieces, and chemically broken down to some degree. Liquid is added to make it easier to swallow. See Figure 15–2 ■ to view the oral cavity.

Tongue

bolus (BOW luss)

frenulum (FREN you lum)

Would you believe your tongue is actually a skeletal muscle that performs many duties? It provides taste stimuli to your brain via the taste buds on its surface, senses temperature and texture (as does the rest of your mouth), manipulates food while chewing, and aids in swallowing. As the tongue moves the food around in the oral cavity, saliva is added to moisten and soften the food, while teeth continue to crush the food until it reaches the right consistency. The tongue pushes the food into a ball-like mass called a **bolus** so it can be passed on to the pharynx. If you can push that bolus into the pharynx with your tongue, why don't you swallow your tongue too? A membrane under your tongue, called the lingual **frenulum,** which you can see when you lift up your tongue, prevents this from happening. Not only is the frenulum important for swallowing, but it also aids in proper speaking. An abnormally short frenulum prevents clear speech—hence the term "tongue-tied!"

Salivary Glands

parotid salivary gland
    (pah RAH tid SAL ih vair ee glands)
sublingual salivary glands
    (sub LIN gwill SAL ih vair ee glands)
    sub = under
    lingual = pertaining to tongue
submandibular salivary glands
    (sub MAN dih bue lar SAL ih vair ee
    glands)

As you can see in Figure 15–3 ■, there are three pairs of salivary glands, all controlled by the autonomic nervous system. A large **parotid salivary gland** is found slightly inferior and anterior to each ear. These are the ones that swell up and make you look like a chipmunk when you get the mumps. The ducts from these glands empty into the upper portion of the oral cavity. The smallest of the salivary glands, the **sublingual salivary glands,** are located under the tongue. The **submandibular salivary glands** are located on both sides along the inner surfaces of the mandible, or lower jaw.

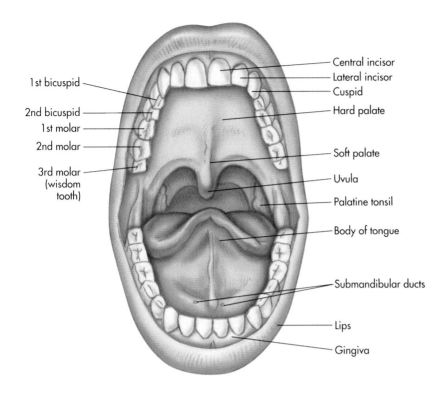

1st bicuspid
2nd bicuspid
1st molar
2nd molar
3rd molar (wisdom tooth)

Central incisor
Lateral incisor
Cuspid
Hard palate
Soft palate
Uvula
Palatine tonsil
Body of tongue
Submandibular ducts
Lips
Gingiva

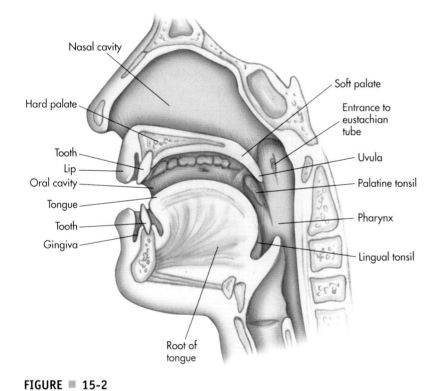

Nasal cavity
Hard palate
Tooth
Lip
Oral cavity
Tongue
Tooth
Gingiva
Root of tongue

Soft palate
Entrance to eustachian tube
Uvula
Palatine tonsil
Pharynx
Lingual tonsil

FIGURE ■ 15-2

The mouth and oral cavity.

On average, these glands collectively produce *1 to 1.5 liters of saliva daily!* Small amounts of saliva are continuously produced to keep your mouth moist, but once you start eating or even think about eating, look out! The flood gates open! Although saliva is almost totally water (99.4%), it also contains some antibodies, buffers, ions, waste products, and **enzymes**. (Remember enzymes from Chapter 3?) Enzymes are formed by cells. They are organic catalysts whose job is to speed up chemical reactions. Salivary **amylase** is a general term for the digestive enzymes that speed up the chemical activity that breaks down carbohydrates, such as starches (carbohydrates), into smaller molecules, such as glucose, that are more easily absorbed by the digestive tract once they get there. So, before you even swallow a bite of pizza, amylase is breaking down the starch in the crust for digestion. **Ptyalin** is a specific salivary amylase that converts starches to simple sugars. Saliva plays an important role even *after* you eat. As the saliva is continuously secreted in small amounts, it cleans the oral surfaces and also aids in reducing the amount of bacteria that grows in your mouth. Just remember, that action is *not* a substitute for brushing your teeth!

ptyalin *(TIE uh lin)*

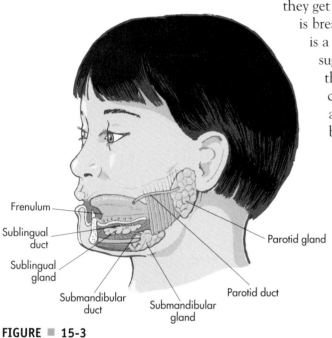

Frenulum

Sublingual duct

Sublingual gland

Submandibular duct

Parotid gland

Parotid duct

Submandibular gland

FIGURE ■ 15-3

The salivary glands.

deciduous *(dee SID you us)*

Teeth

The final important components of the mechanical aspect of digestion in the oral cavity are the teeth. It is unfortunate that we get only two sets of them in a lifetime.

The first set of teeth are called baby teeth, or more properly, **deciduous teeth.** Like the leaves on a deciduous tree, they fall away in time. Beginning at around 6 months of age, they begin to appear, beginning with the lower central incisors, and all 20 are usually in place by 2 1/2 years of age. Between the ages of 6 and 12 years the Tooth Fairy is kept busy as these teeth are pushed out and replaced by the 32 larger **permanent teeth**. The exception to this is the **wisdom teeth**, the four molars in the back, which may not appear until an individual is as old as 21 years.

Not all teeth are the same, as you can see by their shapes and locations. Each has special responsibilities. Look at Figure 15–4 ■ as we discuss the various types of teeth. The first tooth type is the **incisor**, located at the front of your mouth. Incisors are blade-shaped teeth used to cut food. **Canine teeth** are for holding, tearing, or slashing food. Canine teeth are also known as eyeteeth, or **cuspids,** and are located next to the incisors. Next in line are the **bicuspids,** or premolars, which are transitional teeth. **Molars** are the final type of teeth and have flattened tops. Both the bicuspids and molars are responsible for crushing and grinding.

Regardless of its type, the structure of each tooth is pretty much the same. As you can see in Figure 15–4, each tooth has a **crown, neck,** and **root.** The *crown* is the visible part of the tooth. It is covered by the

Clinical Application ✚ ✋ 🚶 🏊 🔑

SUBLINGUAL MEDICATION

Go to a mirror and open your mouth. Lift your tongue and you will notice that there are blood vessels everywhere just barely underneath the surface. This sublingual blood vessel network can readily absorb certain drugs. Rapid absorption into the blood system is vital in some cases, such as for those patients who need the drug nitroglycerin to treat angina quickly. Angina is caused by insufficient blood flow to the heart and consequently not enough oxygen to coronary tissues. Nitroglycerin improves coronary blood flow and in most cases temporarily relieves angina.

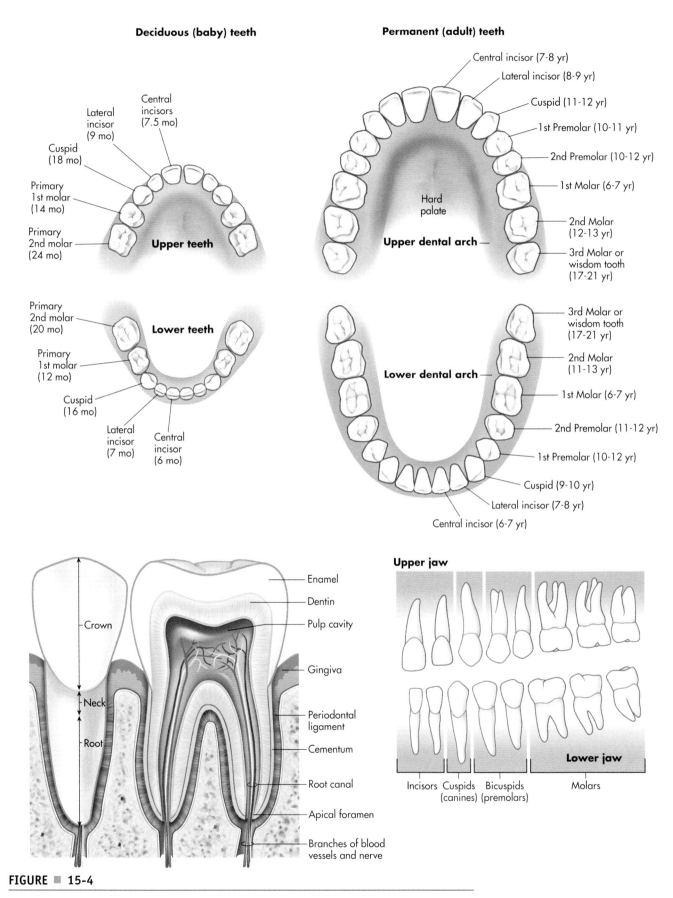

Deciduous (baby) teeth

Central
incisors
(7.5 mo)

Lateral
incisor
(9 mo)

Cuspid
(18 mo)

Primary
1st molar
(14 mo)

Primary
2nd molar
(24 mo)

Upper teeth

Primary
2nd molar
(20 mo)

Primary
1st molar
(12 mo)

Cuspid
(16 mo)

Lateral
incisor
(7 mo)

Central
incisor
(6 mo)

Lower teeth

Permanent (adult) teeth

Central incisor (7-8 yr)

Lateral incisor (8-9 yr)

Cuspid (11-12 yr)

1st Premolar (10-11 yr)

2nd Premolar (10-12 yr)

1st Molar (6-7 yr)

Hard
palate

Upper dental arch

2nd Molar
(12-13 yr)

3rd Molar or
wisdom tooth
(17-21 yr)

3rd Molar or
wisdom tooth
(17-21 yr)

2nd Molar
(11-13 yr)

Lower dental arch

1st Molar (6-7 yr)

2nd Premolar (11-12 yr)

1st Premolar (10-12 yr)

Cuspid (9-10 yr)

Lateral incisor (7-8 yr)

Central incisor (6-7 yr)

Crown

Neck

Root

Enamel

Dentin

Pulp cavity

Gingiva

Periodontal
ligament

Cementum

Root canal

Apical foramen

Branches of blood
vessels and nerve

Upper jaw

Lower jaw

Incisors   Cuspids   Bicuspids   Molars
           (canines) (premolars)

FIGURE ▪ 15-4

Types, location, and structures of teeth.

periodontal *(perr ee oh DON til)*
cementum *(si MEN tum)*
gingiva *(JIN jih VA)*

hardest biologically manufactured substance in the body, **enamel.** The neck is a transitional section that leads to the root.

Internally, most teeth are made up of a mineralized, bonelike substance called **dentin.** The next internal layer is a connective tissue called **pulp,** which is located in the **pulp cavity**. The pulp cavity also contains blood vessels and nerves that provide nutrients and sensations. The nerves and blood vessels get to the pulp cavity via the infamous root canal.

The root is nestled in a bony socket and is held in place by fibers of the **periodontal ligament**. In addition, **cementum** (a soft version of bone) covers the dentin of the root, aiding in securing the periodontal ligament. Healthy gums, or **gingiva,** also help to hold the teeth in place. Epithelial cells form a tight seal around the tooth to prevent bacteria from coming into contact with the tooth's cementum.

Pathology Connection: Oral Disorders

We have all experienced some form of gastrointestinal (GI) disorder at some time in our lives. In fact, GI problems are some of the most common problems in the health care field. That is because the causes of these disorders are so varied. These conditions may have etiologies based on inherited traits, behavioral issues, structural abnormalities, eating habits, lifestyles, or even prescribed medications used to treat other problems! Let's begin at the entrance of the GI system, your mouth.

Dental caries (cavities) are a result of microorganisms attacking tooth enamel. Bacteria, diets rich in carbohydrates, poor dental hygiene, and the lack of regular visits to the dentist all work together to create a soft sticky substance called **dental plaque** which is a great hideout for the bacteria that create the acid that attacks the surface of teeth. Once a cavity forms, it must be cleaned out and filled to protect the tooth. If the cavity is ignored, further destruction of the tooth will occur with potentially major problems that may require tooth removal, surgical intervention, produce systemic infection, or even damage heart valves! Proper dental care and fluoride (found in most municipal water systems and toothpaste) aid tremendously in cavity prevention.

Periodontal disease is a wider problem attributed to the same causative agents as dental caries and sometimes to an inadequate diet. Not only can the teeth be affected, but gums and the supportive structures for your teeth can also be involved. Periodontal disease is the major cause of tooth loss in the adult population. As the plaque builds up at the gum line, the gums become inflamed, causing a condition called **gingivitis** that can cause bleeding at the affected site(s).

Your lips and oral region are also susceptible to cancer. The main causative agents are tobacco products and excessive sun exposure. Consumption of alcohol with tobacco products appears to raise the chances of the development of cancer, not

15-2   For animations and videos concerning oral care and assessment along with cavity prevention, please see your DVD.

only in the oral region but the throat, stomach, and bladder as well. It is believed that alcohol acts as a solvent to dissolve the carcinogenic substances in tobacco, which are more readily absorbed by the highly vascularized tissues in the mouth. **Leukoplakia** is a precancerous condition usually attributed to the use of chewing tobacco or snuff. The affected areas exhibit a white patch of tissue. The use of chewing tobacco can also increase the incidence of cavities due to the amount of sugar normally used in the processing of chewing tobacco. Not all sores or ulcers in the mouth are cancerous. Poor fitting dentures can lead to a condition known as **stomatitis** that inflames areas of oral mucosa. **Canker sores (apthous stomatitis)** are small ulcers in your mouth and are sometimes caused by eating certain foods, although their cause is often unknown. **Cheilitis**, which is cracking and associated inflammation on the lips and corners of the mouth, are not caused by cancer, but rather can be traced to infections, allergies, or even to nutritional deficiency.

leukoplakia *(loo coe PLAY kee ah)*
apthous stomatitis
   (AP thahs stow meh TIGHT es)
cheilitis *(ky LY tis)*
epiglottis *(ep ih GLAH tiss)*
esophagus *(eh SOFF ah guss)*
pharyngoesophageal sphincter
   (fair IN go es SOFF uh jeel)

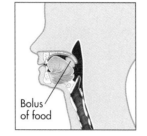

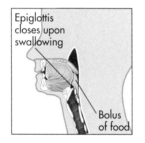

PHARYNX

The pharynx brings us to the next part of the journey. It is a common passageway not only for food but also for water and air. It has three parts: the nasopharynx, the oropharynx, and the laryngopharynx. The **nasopharynx** is primarily part of the respiratory system. The **oropharynx** and the **laryngopharynx** serve double duty as a passageway for food and water and for air. During swallowing, passageways to the nasal and respiratory regions are protected from the accidental introduction of food and liquids. The nasopharynx is blocked by the soft palate, and a flap of tissue called the **epiglottis**, a part of the larynx, covers the airway to the lungs as the larynx rises during swallowing. These actions force the food to enter the only possible route, the esophagus.

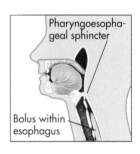

ESOPHAGUS

The **esophagus** is approximately 10 inches (25 centimeters) long and is responsible for transporting food from the pharynx to the stomach. It extends from the pharynx, through the thoracic cavity and diaphragm, to the stomach, which is located in the peritoneal cavity (see Figure 15–5 ■).

The esophagus is normally a collapsed tube, much like a deflated balloon, until a bolus of food is swallowed. As this bolus moves to the esophagus, a muscular ring at the beginning of this structure, known as the **pharyngoesophageal sphincter,** relaxes. This is like opening the door to the esophagus so food can enter. We can't rely on gravity to move food through the esophagus, so the muscles of the esophagus begin rhythmic contractions that work the food down to the stomach. This rhythmic wavelike muscular contraction is known

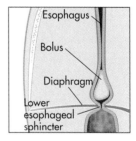

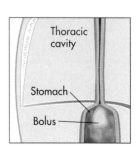

FIGURE ■ 15-5

The movement of a bolus of food from the mouth to the stomach via the esophagus.

peristalsis *(pair ih STALL siss)*

as **peristalsis.** Once the bolus reaches the end of the esophagus, a second door must be opened to allow entry to the stomach. This is the **lower esophageal sphincter,** or **LES,** also known as the **cardiac sphincter.** It relaxes to let food into the stomach and then closes to prevent acidic gastric juices from squirting up into the esophagus. Heartburn (**pyrosis**) occurs if that door inadvertently opens.

The esophagus also helps move the bolus by excreting mucus so its walls are slippery. The esophageal walls are lined with stratified squamous epithelium, which makes the esophagus resistant to abrasion, temperature extremes, and irritation by chemicals.

The whole process of swallowing food takes about 9 seconds on average. Dry or "sticky" food may take longer with repeated attempts to work it down. (Ever try to swallow too large a bite of a peanut butter sandwich and have it stick, pounding on the LES door?) Fluids take only seconds to get to the stomach.

THE WALLS OF THE ALIMENTARY CANAL

It is interesting to note that the same four basic tissue types form the wall of the entire alimentary canal from the esophagus onward (see Figure 15–6 ■). The innermost layer that lines the lumen of the canal is the **mucosa** (just like the respiratory system). This layer is composed mostly of stratified squamous epithelium with some connective tissue and a thin smooth muscle layer surrounding it. The mucosa also possesses cells that secrete digestive enzymes to break down foodstuffs and **goblet cells** that secrete mucus for lubrication.

The **submucosa** is the next layer and is composed of connective tissue. This layer contains blood and lymph vessels, lymph nodes (called **Peyer patches,** similar to tonsils), and nerve endings. The next layer is the **muscularis externa** and is composed of two layers of smooth muscle. The innermost layer encircles the canal, while the outer layer is longitudinal in nature so it lies in the direction of the canal. There is an additional third layer of oblique smooth muscle, but it is found only surrounding the stomach.

The outermost layer is the **serosa,** composed of a single thin layer of flat, serous, fluid-producing cells supported by connective tissue. For most of the canal, the serosa is the **visceral peritoneum.** The peritoneum is a serous membrane in the abdominopelvic cavity. Like all serous membranes, it has two layers. The visceral peritoneum covers the organs, and the parietal peritoneum lines the wall of the abdominopelvic cavity. Between the layers is a fluid-filled potential space called the **peritoneal cavity.** This fluid is important both for keeping the outer surface of the organs moist and allowing friction-free movement of the digestive organs against the abdomenopelvic cavity. Some abdominopelvic organs such as the urinary bladder and the duodenum are not surrounded by peritoneum and are called **retroperitoneal organs.** The esophagus differs in that it possesses only a loose layer of connective tissue called the **adventitia.** (This shouldn't come as any great shock since the esophagus is not in the abdominopelvic cavity!)

15-3 To view a video and perform an interactive drag-and-drop exercise of the GI system and the intestinal wall, please see your DVD.

adventitia *(add ven TISH ah)*

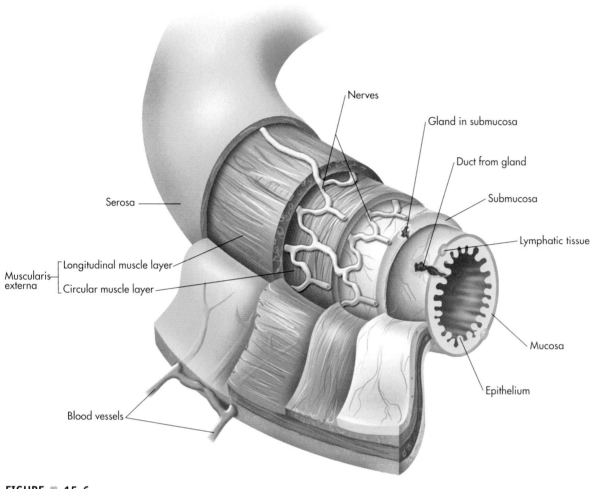

FIGURE ▪ 15-6

Basic tissue types of the alimentary canal.

TEST YOUR KNOWLEDGE 15-1

Choose the best answer:

1. The digestive tract is also called the:
   a. elementary canal
   b. integumentary canal
   c. alimentary canal
   d. panama canal

2. A ball-like mass of food is called a:
   a. bolus
   b. wad
   c. chyme
   d. mastion

3. What is the punching bag–shaped object dangling from the soft palate?
   a. tonsil
   b. uvula
   c. labia
   d. incisor

4. Which of the following is *not* a salivary gland?
   a. parotid
   b. sublingual
   c. submandibular
   d. substernal

◆ TEST YOUR KNOWLEDGE 15-1 *(continued)*

Complete the following:

5. Externally, the three main structural parts of a tooth are the _____ , _____ , and _____.

6. _____ _____ are a result of microorganisms attacking tooth enamel.

7. The consumption of _____ and _____ products appears to increase the chances of oral, throat, and stomach cancers.

STOMACH

The stomach is located in the left side of the abdominal cavity under the diaphragm and is covered almost completely by the liver. This organ is approximately 10 inches (25 centimeters) long with a diameter that varies, depending on how much you eat at any given time. Although the stomach can hold up to 4 liters (about a gallon) when totally filled, it can expand or decrease in diameter thanks to deep, mucous-coated folds in the stomach wall called **rugae** that allow for these size changes. As the stomach receives food from the esophagus, it performs several functions:

rugae (ROO gay)

- Acts as a temporary holding area for the received food
- Secretes gastric acid and enzymes, which it mixes with the food, causing chemical digestion
- Regulates the rate at which the now partially digested food (a thick, heavy, creamlike liquid--semisolid mixture of food and gastric juices—called **chyme**) enters the small intestine
- Absorbs small amounts of water and substances on a very limited basis (although the stomach does absorb alcohol)

chyme (KIME)

15-4 For animations concerning carbohydrates, lipids, and proteins, please see your DVD.

It takes about four hours for the stomach to empty after a meal. Liquids and carbohydrates pass through fairly quickly. Protein takes a little more time, and fats take even longer, usually from 4 to 6 hours.

The stomach is divided into four regions. Located near the heart, the **cardiac region** surrounds the lower esophageal sphincter, sometimes referred to as the cardiac sphincter. Due to the location, that is the reason why indigestion is sometimes referred to as heartburn (see Figure 15–7 ■). The **fundus,** which is lateral and slightly superior to the cardiac region, temporarily holds the food as it first enters the stomach. The **body** is the midportion and largest region of the stomach. The funnel-shaped, terminal end of the stomach is called the **pylorus.** Most of the digestive work of the stomach is performed in the pyloric region. This is also the region where chyme must pass through another door, the **pyloric sphincter,** in order to travel on to the small intestine. Two other points of interest are the concave curve called the **lesser curvature** and the larger convex curve to the left of the lesser curve called the **greater curvature**.

fundus (FUN duss)

pylorus (pye LOR us)

pyloric sphincter
(pye LOR ik SFINK ter)

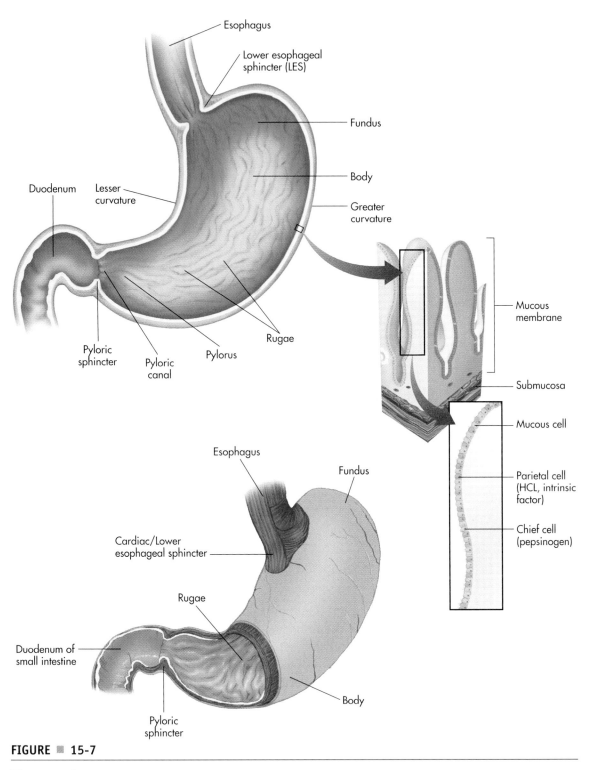

FIGURE ■ 15-7

The stomach.

The muscular action of the stomach works much like a cement mixer and is achieved by the three layers of muscle found in its walls. One layer is longitudinal, one is circular, and the third is oblique in orientation. This arrangement of muscles enables the stomach to churn food as it mixes with gastric juices excreted by **gastric glands** from gastric pits in the columnar epithelial lining of the stomach as well as to work the food toward the pyloric sphincter through the muscle activity, peristalsis. With the combined efforts of muscle and gastric juices, both physical and chemical digestion occur in the stomach.

Gastric juice is a general term for a combination of hydrochloric acid (HCl), pepsinogen, and mucus. The gastric glands produce about 1,500 milliliters of gastric juice each day. Pepsinogen is secreted by the **chief cells**, and HCl is secreted by **parietal cells**. These two cell types have a special relationship in that once pepsinogen makes contact with HCl, the chief digestive enzyme, **pepsin,** is formed. Pepsin is needed to break down proteins (like the ones found in pepperoni). Even though it doesn't actually digest food by itself, HCl does break down connective tissue in meat. The HCl must be very strongly acidic to work. Normally, it has a pH of 1.5 to 2. This highly acidic environment plays another important role: killing off most pathogens that enter the stomach.

You may wonder why the stomach doesn't digest itself. A healthy stomach is protected by mucous cells that generate a thick layer of mucus to shield the stomach lining from the effects of HCl! Other specialized cells secrete what is known as **intrinsic factor**, which is needed for absorption of vitamin B_{12}. Table 15–1 describes gastric glands and their functions.

TABLE 15–1: Gastric Glands and Their Functions

DIGESTIVE CELLS	SECRETION TYPE	FUNCTION
Chief cells	Pepsinogen	Begins digestion of protein
Parietal cells	HCl	Kills pathogens, activates pepsinogen, breaks down connective tissue in meat
Mucous cells	Alkaline mucus	Protects stomach lining
Endocrine cells	The hormone gastrin	Stimulates gastric gland secretion

The stomach's activity is controlled by the parasympathetic nervous system, particularly the **vagus nerve**. Once the vagus nerve is stimulated, the stomach's motility (churning action) increases, as does the secretory rates of the gastric glands.

The three phases of gastric juice production are illustrated in Figure 15–8 . The **cephalic phase** occurs as a result of sensory stimulation such as the sight or smell of food. This sensory input stimulates the parasympathetic nervous system (via the medulla oblongata) and the release of the hormone **gastrin** is increased. Once gastrin travels through the blood and reaches the stomach, gastric gland activity is increased. So, the sight or smell of that pizza literally does get your gastric juices flowing.

This leads us to the **gastric phase**, in which over two thirds of the gastric juices are secreted as the food moves into the stomach. As the food moves in, the stomach begins to distend. As the stomach distends, it sends signals back to the brain, which fires a reply to the gastric glands to step up their work. As chyme is formed, peristalsis pushes it through the pyloric sphincter to the first part of the small intestine, the **duodenum.**

gastrin *(GAS trin)*

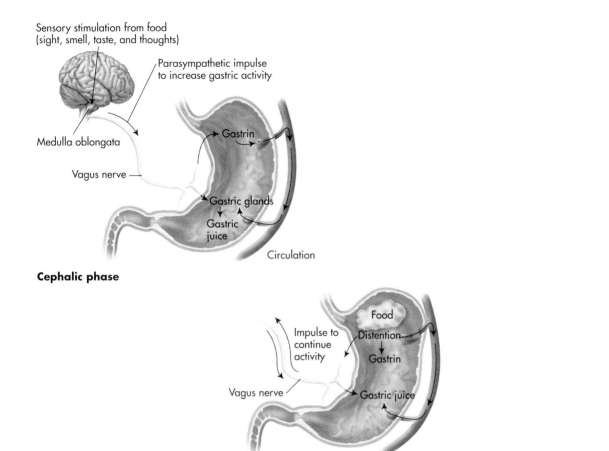

Cephalic phase

Gastric phase

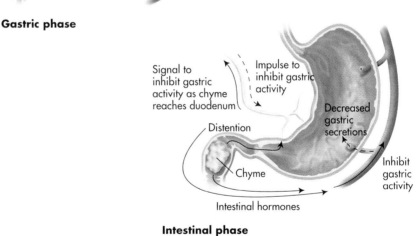

Intestinal phase

FIGURE ▪ 15-8

Phases of gastric secretions.

The entry of the food into the **duodenum** begins the **intestinal phase** of gastric juice regulation. As the duodenum distends and senses the acidity of chyme, intestinal hormones are released that cause the gastric glands in the stomach to decrease gastric juice production. The brain is also signaled and sends a message to inhibit gastric juice secretion because it is no longer needed now that the food bolus (now called chyme) has left the stomach. Once the chyme begins its movement through the duodenum and on to the rest of

duodenum *(doo ODD eh num)*

Clinical Application

PEG TUBE

Patients who are unable to consume nutrients via the normal oral route often have a permanent feeding tube implanted called a PEG tube (percutaneous endoscopic gastrostomy). This tube is attached at the abdominal wall and the stomach wall and allows nutrients to be directly injected to the stomach.

15-5 To view a video on the topic of nutrition and enteral (tube) feedings, please see your DVD.

the small intestine, those inhibitory responses are halted so gastric juice production can continue once again when a new bolus of food enters the stomach.

The rate of the movement of chyme is very important. If it moves too slowly through the stomach, the rate of nutrient digestion and absorption is decreased and the acidity of the chyme may cause erosion of the stomach lining. If chyme passes too quickly through the stomach, the food particles may not be sufficiently mixed with gastric juices, leading to insufficient digestion. Chyme that is not given time to be neutralized may cause acidic erosion of the intestinal lining.

 ## Pathology Connection: Stomach Acid Disorders

Of growing concern over the past several years is the condition known as **gastroesophageal reflux disease** (GERD). The lay term for this is heartburn. GERD occurs when acidic stomach contents "squirt" back (acid reflux) through the lower esophageal sphincter and into the esophagus. Since the esophagus does not have the protective mucus barrier like the stomach, repeated episodes of acid reflux can cause inflammation and eventually ulceration of the affected area. Scar tissue can eventually form, causing a narrowing of the esophagus that may impede digestion. If left untreated, GERD can lead to esophageal cancer. The reason that GERD has become a major concern recently is due to our society's diet and exercise habits. As a nation, we overeat consistently, making sure that we get our money's worth at every buffet. Instead of walking after stuffing ourselves, we flop down on the sofa or easy chair complaining about, "how stuffed" we are. Smoking also may aggravate GERD. So how do we deal with GERD? The quick fix for the burning sensation is the use of antacids and medications that decrease acid production. More importantly, lifestyle changes may prevent GERD. Changes in diet, including limiting fats, alcohol, caffeine, and chocolate may be helpful. Lying down after eating should be avoided for approximately 4 hours. This will give the stomach time to empty. Sleeping with your head elevated and avoiding sleeping on your right side also helps to decrease GERD. If you are obese, lose weight. The pressure from abdominal fat helps to push on stomach contents, increasing the chance of acid reflux occurring. Another rapidly spreading, obesity-related disease called metabolic syndrome is covered with diabetes in the Chapter 10, "The Endocrine System."

Peptic ulcers, also called *duodenal ulcers,* are the result of a breakdown of the mucosal membrane found in the esophagus, stomach, or small intestine. About 80% of the peptic ulcers are located in the duodenum and the bulk of the remaining peptic ulcers are located in the stomach, called *gastric ulcers.* It was once thought that a high-stress lifestyle was the main cause of ulcer development. Stress is a factor in some cases, but the major offender appears to be the *Helicobacter pylori* (H. pylori.) bacterium, which causes the initial erosion of the mucosal lining and leads to the inflammation called **gastritis**. See Figure 15-9 ▪.

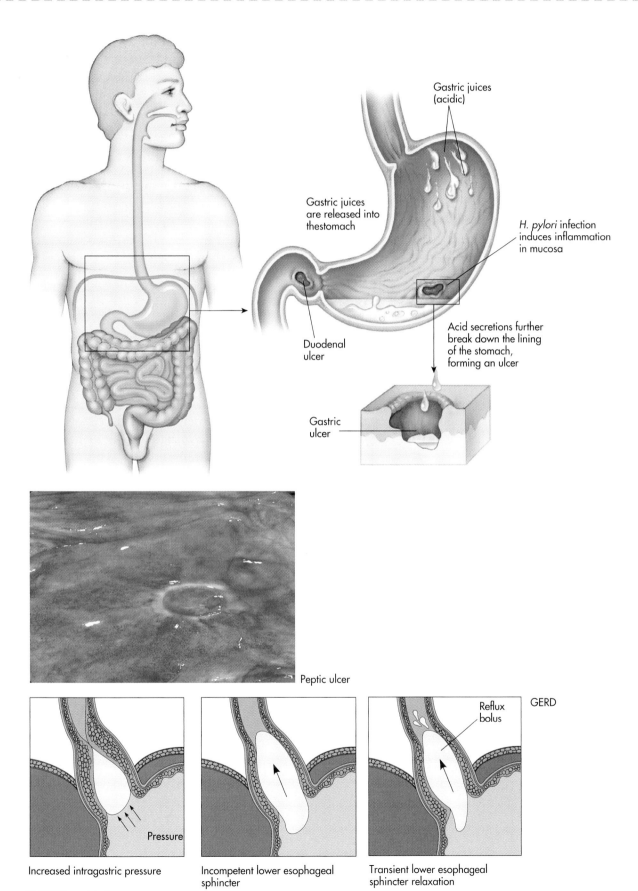

Gastric juices
(acidic)

Gastric juices
are released into
thestomach

H. pylori infection
induces inflammation
in mucosa

Duodenal
ulcer

Acid secretions further
break down the lining
of the stomach,
forming an ulcer

Gastric
ulcer

Peptic ulcer

GERD

Reflux
bolus

Pressure

Increased intragastric pressure

Incompetent lower esophageal
sphincter

Transient lower esophageal
sphincter relaxation

FIGURE ■ 15-9

Mechanisms of peptic ulcers and GERD.

15-6 Gastroesophageal reflux disease (GERD) is a fairly common disorder in which the lower esophageal sphincter does not properly seal off and allows the acidic stomach contents to back up (reflux) into the esophagus, causing a burning sensation. If left untreated, it can cause serious damage to the esophagus. To see an animation of GERD, please go to your DVD.

Other contributing agents can include smoking, heavy and chronic consumption of alcohol, aspirin and other NSAIDs, caffeine, corticosteroids, and other medications. Duodenal ulcers may also be associated with *H. pylori*, but patients with duodenal ulcers appear to overproduce stomach acid and pepsin. In addition, they seem to respond more to the stimuli of acid secretion as well as have a more rapid rate of gastric emptying. This speeded up process may expose the duodenum to greater acidity. Interestingly, patients with gastric (stomach area) ulcers have lower than normal gastric acid. It appears that stomach ulcers are either a result of defective gastric mucosa or of a weakened pyloric sphincter that allows bile to squirt back into the stomach.

Please see Figure 15-9, which shows the pathophysiologic causes of GERD and peptic ulcer disease (PUD).

SMALL INTESTINE

The small intestine, located in the central and lower abdominal cavity between the stomach and large intestine is *the* major organ of digestion. It is where most food is digested (see Figure 15–10 ■). The small intestine is small in diameter, but not in length. Beginning at the pyloric sphincter, the small intestine is the longest section of the alimentary canal, with a length up to 20 feet (up to 6 meters) and a diameter ranging from 4 centimeters where it connects with the stomach to 2.5 centimeters where it meets the large intestine.

The walls of the small intestine secrete several digestive enzymes important for the final stages of chemical digestion and two hormones that stimulate the pancreas and gallbladder to act and that control stomach activity.

In the small intestine, almost 80% of the absorption of usable nutrients takes place when chyme comes in contact with the mucosal walls. Amino acids, fatty acids, ions, simple sugars, vitamins, and water are all absorbed from here into the bloodstream and then on to the body's cells. Some of the remaining 20% was already absorbed by the stomach, with the rest being absorbed by the large intestine. Any residue that cannot be utilized is sent on to the large intestine for removal from the body.

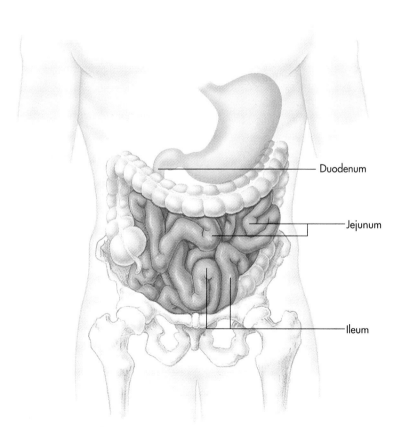

Duodenum

Jejunum

Ileum

FIGURE ■ 15-10

The small intestine.

There are three regions of the small intestine. The **duodenum** is approximately 25 to 30 centimeters (10" to 12") long and is located near the head of the **pancreas.** The duodenum gets its name from *duo* (two), and *denum* (ten), which equal 12, the number of finger widths long that this organ is (10 inches)!

pancreas *(PAN kree ass)*

The **jejunum** is the middle section, and is approximately 2.5 meters (8 feet) long.

jejunum *(jee JOO num)*

The terminal end of the small intestine is the **ileum.** This section (6 – 12 feet long) attaches to the large intestine at the **ileocecal valve**. (Here is one place where spelling counts. The *ileum* is part of your small intestine. The *ilium* is part of your pelvic girdle. You don't want to get them mixed up!)

ileum *(ILL ee um)*

The pyloric valve is important in allowing small portions of chyme to enter the first part of the small intestine (duodenum) because the small intestine can process only small amounts of food at a time. At the duodenum, additional secretions are added from the pancreas and gallbladder. The pancreas provides pancreatic juices. The **gallbladder**, a small green organ located on the inferior surface of the liver, provides bile. **Bile**, produced by the liver and stored in the gallbladder, is yellow-brown or yellow-green and contains acids, mucus, fluids, and two pigments, **biliverdin** (verd/o = green) and **bilirubin** (yellow, rubin/o = red). Whoa! If rubin/o = red, why is the pigment yellow? "Rubin" refers to the fact that bilirubin is formed from the breakdown of old, worn out red blood cells. Bile emulsifies fat—that is, it makes fat able to disperse in water—making the fat found in the cheese of the pizza easier to break down. Pancreatic juice contains enzymes and sodium bicarbonate, which neutralizes the acidic chyme. As chyme enters the duodenum, your gallbladder contracts, sending bile down the cystic duct and into the common bile duct to the duodenum. Gallbladder activity is caused by the hormone **cholecystokinin,** also known as **CCK,** which is also produced by the small intestine. The pancreas is stimulated to secrete as a result of the hormone **secretin,** which is produced by the small intestine. Two types of muscular action occur in the small intestine. **Segmentation** is the muscle action that mixes chyme and digestive juices, working much like a cement mixer. Peristalsis also occurs, moving undigested food remains toward the large intestine. See Table 15–2 for the hormones active in the digestive process.

cholecystokinin *(KOH lee SISS toh KIE nen)*

TABLE 15–2: Hormones in the Digestive Process

HORMONE	SECRETING ORGAN	ACTION
Gastrin	Stomach	Stimulates release of gastric juice
Secretin	Duodenum	Stimulates release of bicarbonate and water from pancreas and bile from liver; slows stomach activity
Cholecystokinin (CCK)	Duodenum	Stimulates digestive enzyme release from pancreas and bile release from gallbladder; slows stomach activity

As previously stated, the small intestine also produces digestive enzymes that are needed to complete chemical digestion. These enzymes (and mucus) are produced by exocrine cells. **Lactase**, **maltase**, and **sucrase** are needed for

Learning Hint

EMULSIFIERS

Think about what a bottle of Italian salad dressing is like before you shake it. Emulsifiers allow the oil, water, and vinegar to blend together.

Learning Hint

THE ENDING ASE

Enzymes end in *ase* and break substances down.

Clinical Application

LACTOSE INTOLERANCE

This unfortunate condition is the inability to digest the sugar (lactose) found in milk and dairy products such as cheese and ice cream. A person with lactose intolerance has a deficiency of lactase, an intestinal enzyme. As a result, lactose is not sufficiently digested. Normal bacteria found in the intestine utilize those undigested sugars with gas production as a byproduct. This is what causes that "bloated feeling" you see on TV advertisements. In addition, the undigested lactose prevents normal water absorption by the small intestine, so diarrhea is formed. So the cheese pizza in our example would not be the meal of choice for these individuals who are lactose intolerant.

Interestingly, it seems that there is a genetic basis for this condition. In some populations, lactase production continues throughout their entire lives. Approximately 15% of the Caucasian population develops lactose intolerance, while 80 to 90% of the African American and Asian populations develop this condition to some degree. To avoid this situation, individuals must either avoid milk and other dairy products or take an oral form of the enzyme lactase before consuming such products.

villi *(VILL eye)*
lacteal *(LACK tee al)*
chyle *(KILE)*

the digestion of double sugars called **disaccharides** that are contained in starches (such as in the pizza crust). **Peptidase** is needed to digest portions of the protein structure called peptides. **Lipase** is needed for digestion of certain fats. It is interesting to note that the secretion of these substances is mainly due to the presence of chyme in the small intestine! Because of the acidity of chyme, both chemical and mechanical irritation occur. This irritation plus the distention of the intestinal wall causes the localized reflex action that results in the release of the enzymes and the two hormones.

The structure of the wall of the small intestine is rather interesting. The wall possesses circular folds called **plicae circulares** and fingerlike protrusions into the lumen called **villi** (see Figure 15–11 ■). The villi also have outer layers of columnar epithelial cells, which possess microscopic extensions known as **microvilli**. These villi are tightly packed, giving a velvety texture and appearance. The purpose of the microvilli, villi, and circular folds is to provide an incredible increase in the surface area of the small intestine. This area, almost the surface area of a tennis court, increases the efficiency of the absorption of nutrients.

Each villus (singular form of villi) contains a network of capillaries and a lymphatic capillary called a **lacteal.** Intestinal glands are located between villi. The capillaries absorb and transport sugars (the result of carbohydrate digestion) and amino acids (the result of protein digestion) to the liver for further processing before they are sent throughout the body. Glycerol and fatty acids (obtained from the digestion of fat) are absorbed by the villi and converted into a lipoprotein that travels on to the lacteal, where it is now a white, milky substance called **chyle.** Chyle goes directly into the lymphatic system for distribution throughout the body.

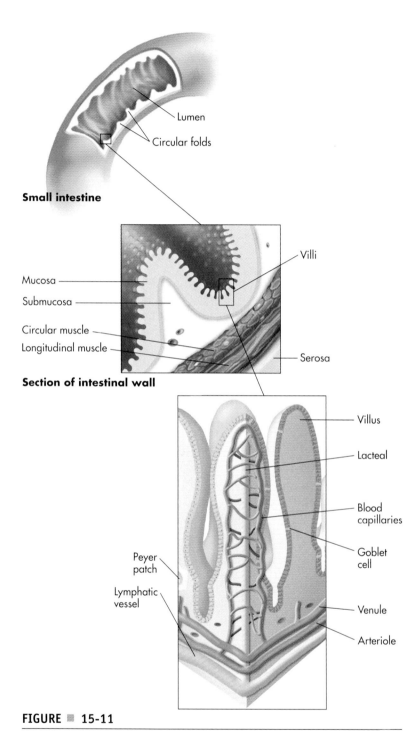

Small intestine

Lumen

Circular folds

Section of intestinal wall

Mucosa

Submucosa

Circular muscle

Longitudinal muscle

Villi

Serosa

Villus

Lacteal

Blood capillaries

Goblet cell

Peyer patch

Lymphatic vessel

Venule

Arteriole

FIGURE ▪ 15-11

Villi.

TEST YOUR KNOWLEDGE 15-2

Choose the best answer:

1. The deep folds of the stomach wall that allow for size changes of the stomach are called:
   a. rugby
   b. sphincter
   c. rugae
   d. glottal folds

2. The final "door" of the stomach that needs to open for chyme to travel to the small intestine is located at the end of the:
   a. fundus
   b. pylorus
   c. epiglottis
   d. adventitia

3. This chief digestive enzyme is needed to break down protein:
   a. guaifenesin
   b. pepsin
   c. pylorin
   d. rugelin

4. A full two-thirds of the gastric juices secreted in the stomach happen as food passes through this gastric juice production phase:
   a. cephalic phase
   b. intestinal phase
   c. gastric phase
   d. pharyngeal phase

Complete the following:

5. _____ _____ is a genetic condition in which an individual cannot digest a sugar found in milk and dairy products.

LARGE INTESTINE

Beginning at the junction with the end of the small intestine (the **ileocecal orifice**) and extending to the **anus,** the **large intestine** almost totally borders the small intestine (see Figure 15–12 ■). The large intestine is responsible for:

- Water reabsorption
- Absorption of vitamins produced by normal bacteria in the large intestine
- Packaging and compacting waste products for elimination from the body

Since there are no villi in the walls of the large intestine, little nutrient absorption occurs here.

Approximately 1.5 meters (5 feet) long and 7.5 centimeters (2.5 inches) in diameter, the large intestine is divided into three main regions: the **cecum, colon,** and **rectum.** The large intestine is large in diameter, but not in length.

cecum *(SEE kum)*

A pouch-shaped structure, the **cecum** receives any undigested food (such as cellulose) and water from the ileum of the small intestine. The infamous **appendix** is attached to the cecum. About 9 centimeters (3 inches) long, the appendix is a slender, hollow, dead-end tube lined with lymphatic tissue. Since

vermiform = *worm like*

it is wormlike in appearance, it is often called the **vermiform** appendix. There is no current reason why we have an appendix. It is considered a *vestigial organ*—an organ whose size and function seem to have been reduced as

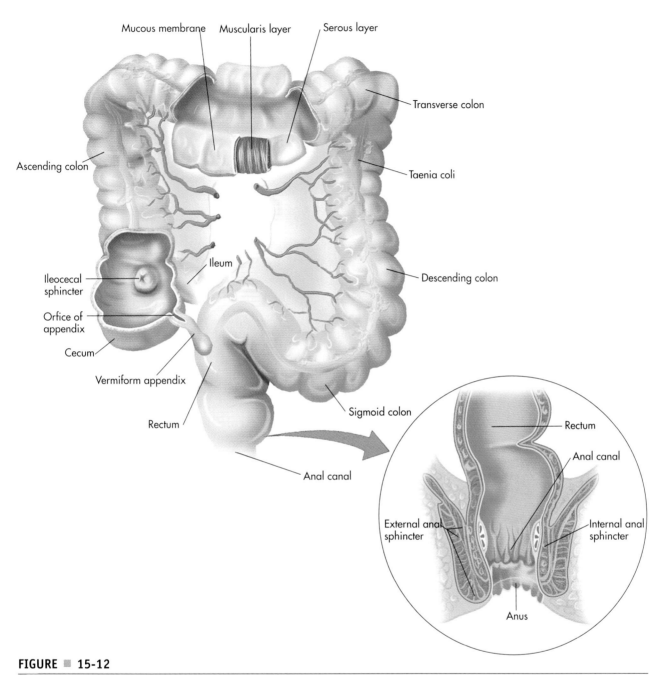

FIGURE ■ 15-12

The large intestine.

humans evolved. Researchers feel that because it possesses lymphatic tissue, it somehow fights infection. Ironically, if the appendix becomes blocked, inflammation can occur, causing **appendicitis.** Treatment for appendicitis is either antibiotics or the surgical removal of the appendix (an **appendectomy**).

Some of the water (used in digestion) and electrolytes are reabsorbed by the cecum and the ascending colon, which we will discuss momentarily. Although this is a relatively small amount of water reabsorption, it is crucial in maintaining the proper fluid balance in the body.

The colon can further be divided into four sections: ascending, transverse, descending, and sigmoid. The **ascending** colon travels up the right side of the

appendicitis
(ah PEN dih SIGH tiss)

body to the level of the liver. The **transverse** colon travels across the abdomen just below the liver and the stomach. Bending downward near the spleen, the **descending** colon goes down the left side, where it becomes the **sigmoid** colon. The sigmoid ("S" shaped) colon extends to the rectum. The rectum opens to the anal canal, which leads to the anus that relaxes and opens to allow the passage of solid waste (feces).

Peristalsis also occurs in the large intestine, but at a slower rate. As these slower, intermittent waves move fecal matter toward the rectum, water is removed, turning feces from a watery soup to a semisolid mass. As the rectum fills with feces, a defecation reflex occurs, which causes rectal muscles to contract and the anal sphincters to relax. If fecal material moves through the large intestine too rapidly, not enough water is removed and diarrhea occurs. Conversely, if fecal matter remains too long in the large intestine, too much water is removed and constipation occurs.

Bacteria found in the large intestine play two important roles. They help to further break down indigestible materials, and they produce B complex vitamins, as well as most of the vitamin K that we need for proper blood clotting. Another by-product of bacterial action in the large intestine is **flatulence**, or the production of gas.

Here is a case where bacteria in the right place keep us healthy. If those same bacteria left the intestinal wall and entered the blood stream, it could be fatal.

Clinical Application

COLOSTOMY

Sometimes a portion of the colon must be bypassed because of disease to allow for healing and/or surgical repair. A new opening needs to be made, and this procedure is called a colostomy. A colostomy can be temporary or permanent depending on the condition. The sites where this procedure is formed are shown in Figure 15–13 ▪.

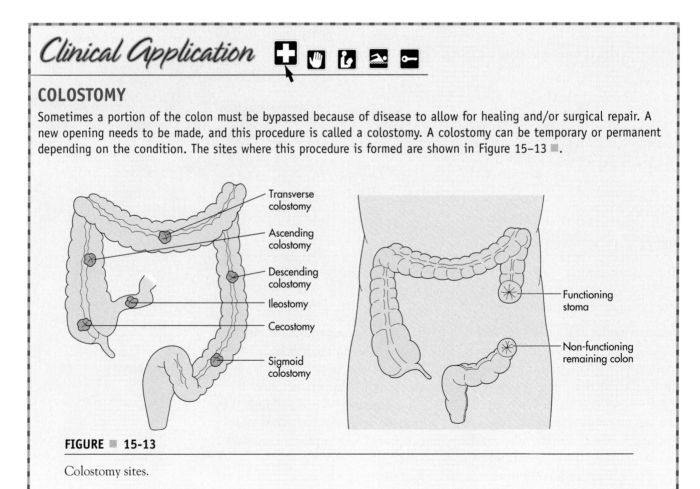

FIGURE ▪ 15-13

Colostomy sites.

Pathology Connection: Disorders of the lower GI Tract

What exactly are hemorrhoids? Interestingly, they are varicose veins caused by increased hydrostatic pressure that weakens the fibromuscular attachments of the blood vessels. Some people have a genetic predisposition for them, while others develop them due to chronic straining during bowel movements (either due to constipation or diarrhea), or from the pressures caused by pregnancy. Hemorrhoids can be either internal (located above the anorectal line) or external (which can actually be seen but are covered by modified anal skin). Bleeding from internal hemorrhoids can produce bright red blood that can merely streak the stool or can reach significant volumes that can occur even when not defecating. Hemorrhoids are often treated with medication, but surgical intervention may be necessary if they become severe.

A very common and potentially lethal condition, **colorectal cancer**, claims over 65,000 lives in the U.S. annually but is readily cured if detected early enough. Although the development of this disease has some genetic predisposition, it appears to be affected more by environmental influences such as:

- A diet rich in animal fat
- A diet lacking inappropriate amounts of fiber and calcium
- Tobacco usage and excessive alcohol consumption
- Higher than normal levels of serum cholesterol (the "bad" form)
- A sedentary lifestyle

A simple colonoscopy is the main diagnostic modality of choice. A barium enema with imaging studies is a less accurate alternative.

Diverticulitis is an infection and inflammation of diverticula or sacs in the intestinal tract, especially in the colon. This is caused by lodged stool and may cause swelling and rupture of the diverticula. **Diverticulosis** is a condition in which there are diverticula in the intestinal tract, especially the colon, that are not inflamed or produce any symptoms. See Figure 15-14 ▪ which shows the pathophysiologic cause of hemorrhoids and diverticulitis.

As noted earlier, appendicitis is an inflammation of the appendix that's caused by a blockage. Symptoms of appendicitis include abdominal pain in the right lower quadrant, nausea, vomiting, and rebound tenderness. Additional symptoms include increased white blood cells, fever, diarrhea or constipation, loss of appetite, and abdominal swelling. In some cases, the appendix ruptures, causing **peritonitis** or abscesses. The most common sign of rupture is initial pain relief followed by extremely high fever and pain. Peritonitis is an infection that occurs when bacteria and the contents of the appendix enter the peritoneal (abdominal) cavity.

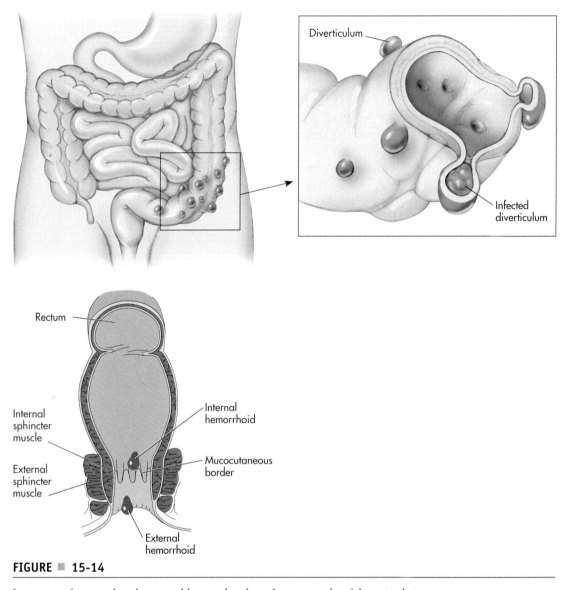

FIGURE ■ 15-14

Location of internal and external hemorrhoids and an example of diverticulitis.

ACCESSORY ORGANS

In addition to the salivary glands of the mouth, other accessory organs are necessary for digestion: the liver, gallbladder, and pancreas.

Liver

Weighing in at approximately 1.5 kilograms (3.3 pounds) and located below the diaphragm in the upper-right quadrant, the liver is the largest glandular organ in the body *and* the largest organ in the abdominopelvic cavity. This organ performs many functions that are vital for survival. As you can see in Figure 15–15 ■, the liver is divided into a larger right side and a smaller left lobe. (Remember, this is from the patient's perspective.) The right side has an anterior lobe and two smaller, inferior lobes.

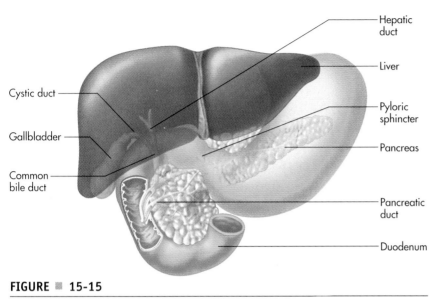

FIGURE ■ 15-15

The liver.

The liver receives about 1.5 quarts of blood *every minute* from the **hepatic** portal vein (carrying blood full of the end products of digestion) and hepatic artery (providing oxygen-rich blood).

Although this chapter is about the digestive system and the liver plays a central role in regulating the metabolism of the body, it is important to understand *all* of the functions that this amazing organ performs. Here is a list of what the liver does:

- Detoxifies (removes poisons from) the body, removing harmful substances such as certain drugs and alcohol
- Creates body heat
- Destroys old blood cells and recycles their usable parts while eliminating unneeded parts such as bilirubin. Bilirubin is eliminated in bile and gives feces its distinctive color.
- Forms blood plasma proteins such as albumin and globulins
- Produces the clotting factors fibrinogen and prothrombin
- Creates the anticoagulant heparin
- Manufactures bile, which is needed for the digestion of fats
- Stores and modifies fats for more efficient usage by the body's cells
- Synthesizes urea, a byproduct of protein metabolism, so it can be eliminated by the body
- Stores the simple sugar glucose as glycogen. When the blood sugar level falls below normal, the liver reconverts glycogen to glucose and releases enough of it into the bloodstream to bring blood sugar back to an acceptable concentration. (Remember the job of the hormones glucagon and insulin from Chapter 10?)
- Stores iron and vitamins A, B_{12}, D, E, and K
- Produces cholesterol

hepatic = *related to the liver*

Stimulated into action by the duodenum's secretion of the hormone **secretin**, bile production is a critical liver digestive function. The salts found in bile act like a detergent to break up fat into tiny droplets that make the work of digestive enzymes easier. Once again, this action of breaking up fat into smaller particles is called **emulsification**, and provides more surface area for the enzymes to do their job of chemically digesting fat.

In addition, bile helps in the absorption of fat from the small intestine and transports bilirubin and excess cholesterol to the intestines for elimination. Once produced by the liver's cells, bile leaves the liver via the **hepatic duct** and travels through the cystic duct to the gallbladder, where it is stored until needed by the small intestine.

emulsification
(ee MULL sih fih KAY shun)

Gallbladder

The **gallbladder** is a sac-shaped organ approximately 7.5 to 10 centimeters (3 to 4 inches) long, located right under the liver's right lobe. Again, please refer to Figure 15–15. While it is storing the bile, your gallbladder also concentrates it by reabsorbing much of its water content. This makes the bile 6 to 10 times more concentrated than it was in the liver. This is a bit of a balancing act: if too much water is reabsorbed and the bile is constantly too concentrated, bile salts may solidify into gall stones.

When fatty foods enter the duodenum, the duodenum releases the hormone CCK (cholecystokinin). This release causes the smooth muscle walls of the gallbladder to contract and squeeze bile into the cystic duct and on through the common bile duct and then into the duodenum.

Pancreas

Although discussed in the endocrine chapter, your pancreas also plays an extremely important role in digestion. This 15-centimeter (6 to 9 inch) organ is located posterior to the stomach and extends laterally from the duodenum to the spleen (see Figure 15–16 ■). The exocrine portion of this organ secretes buffers and digestive enzymes through the pancreatic duct to the duodenum. The buffers are needed to neutralize the acidity of the chyme in the small intestine. With a pH ranging from 7.5 to 8.8, the chyme is neutralized, saving the intestinal wall from damage. This secretory action is activated by the release of hormones by the duodenum.

The general digestive enzymes excreted by the pancreas are **carbohydrases** that work on sugars and starches, **lipases** that work on lipids (fats), **proteinases** that break down proteins, and **nucleases** that break down nucleic acids.

 ### Pathology Connection: Cholecystitis, Pancreatitis, and Hepatitis

As we previously discussed, stones can form from substances in the bile while it is stored in the gallbladder. This condition is called **cholelithiasis.** These stones, which are most often formed from cholesterol, can range in size from grains of sand to marble-size and larger. A person can often pass the small stones, but larger ones need to be removed surgically. This condition can worsen

cholelithiasis
(KOH lee lih THY ah siss)
lith = *stone*

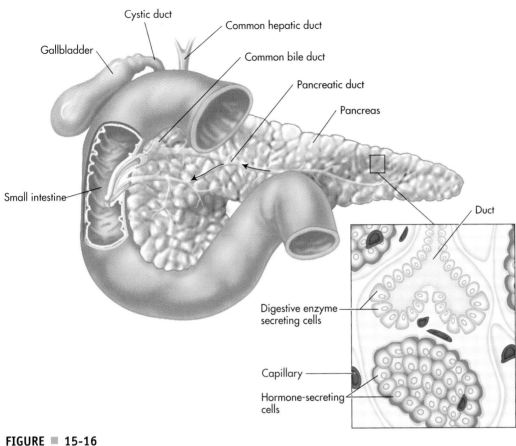

FIGURE ▪ 15-16

The pancreas.

if the stones lodge in the bile ducts, causing extreme pain, which, surprisingly, often radiates to the right shoulder. (This is a good example of referred pain.) If inflammation develops, the condition is called **cholecystitis.** The surgical removal of the gallbladder is called **cholecystectomy**.

If the bile backs up into the liver, the disease **obstructive jaundice** can occur. In this scenario, bilirubin is reabsorbed back into the blood, giving the victim a yellowish tint to the skin and eyes.

The problem must be resolved by unblocking the bile ducts. This can be done by dissolving the stone through medication, using shock waves to smash the stones (**lithotripsy**), or surgically removing them.

Problems can also occur with the pancreas when the bile duct becomes blocked. In some cases, the pancreatic enzymes back up into the pancreas. As a result, those enzymes begin to inflame and destroy the pancreas. This condition is known as **pancreatitis** and can be caused by excessive alcohol consumption, gallbladder disease, or some irritation that causes an abnormally high rate of pancreatic enzyme activation. If this situation is not stopped, death can eventually occur.

Hepatitis is an inflammation of the liver usually caused by an exposure to an infectious agent, a toxin, or certain drugs. Viruses cause infectious hepatitis. There are several types of infectious hepatitis. Hepatitis A virus (HAV) is

cholecystitis
(KOH lee siss TYE tiss)
chole = *gall*
cyst = *bladder*
itis = *inflammation*

the most common type of hepatitis infection worldwide. HAV is spread by food or water that has been contaminated with infected fecal matter. The disease is an acute disease, and the symptoms are typically mild and short lived, requiring only bed rest for treatment.

Infection by the Hepatitis B virus (HBV) causes more serious symptoms than HAV infection causes. HBV is a blood-borne virus that usually causes acute illness, but in a small percentage of cases it develops into a chronic illness that causes permanent liver damage. HBV is transmitted during sexual intercourse, during intravenous drug use, and during organ transplants or blood transfusions. Healthcare workers are especially at risk of HBV infection due to accidental needle sticks and all healthcare workers should be vaccinated against HBV. They should also practice Standard Precautions.

Alcoholism is an example of toxic hepatitis. Some drugs are toxic to the liver in small doses, while others, such as acetaminophen in large doses, can cause liver damage. Chronic liver disease is detected by the elevation of liver enzymes. Acute liver disease can cause jaundice, hepatomegaly (enlarged liver), altered mental status, and possible multi-system failure. Alcohol and drug abuse also contribute to **cirrhosis**, a liver disease in which healthy tissue is replaced by fibrous connective tissue.

COMMON SYMPTOMS OF DIGESTIVE SYSTEM DISORDERS

Symptoms of digestive disorders generally include one or more of the following:

- Vomiting
- Diarrhea
- Constipation
- Pain

Vomiting is a protective means of ridding the digestive tract of an irritant or overload of food. Sensory fibers are stimulated by the irritant or overdistension and send signals to the vomiting center (yes, you have a vomiting center) in the brain. Motor impulses are then sent to the diaphragm and abdominal muscles to contract, which squeezes the sphincter at the esophageal opening, and the contents are regurgitated.

Diarrhea results when the fluid contents in the small intestine are rushed through the large intestine before they can adequately reabsorb the water. Proper absorption of electrolytes and nutrients is also prevented, which can cause serious problems. Diarrhea is characterized by frequent loose and watery

diarrhea

  dia = *through*

  rrhea = *flow; literally diarrhea means "flow through," which is exactly what happens*

bowel movements. It is caused by a number of factors, including infection, poor diet, toxins, food allergies or irritants, and stress.

Constipation is the opposite of diarrhea. There are several possible causative factors:

- The feces may travel so slowly through a colon that too much water gets re-absorbed and the stool becomes hard, dry, and difficult to push through the system.
- There may be an overstimulation of the musculature of the intestine, creating a more narrow passage for the material.
- The musculature has become flaccid or slow, as is often the case in bedridden geriatric patients.

Dietary changes with more liquid and more roughage (such as high-fiber cereals, fruits, and vegetables), reestablishment of regular bowel habits, avoidance of stress/tension, and moderate exercise all can contribute to improving the patient's condition. Chronic use of laxatives and enemas should be avoided, and may in fact aggravate existing conditions (such as hemorrhoids) or create new ones (such as injury to the mucous membrane of the intestinal lining).

GI pain can be sharp, dull, localized, diffuse, radiating, occasional, or constant. The severity of the pain does not always indicate the severity of the disease. Large amounts of flatus (gas) can be extremely painful, but once that wretched cloud is released, all is well (except for the innocent bystanders). Stomach cancer, on the other hand, may initially present with a relatively mild form of indigestion. Any pain from the GI tract should be investigated.

A Quick Trip Through Some Diseases of the Gastrointestinal System

UPPER G.I. REGION

DISEASE	ETIOLOGY	SIGNS AND SYMPTOMS	DIAGNOSTIC TEST(S)	TREATMENT
Anorexia nervosa	Eating disorder characterized by refusal to eat, classified as a psychiatric disorder.	Patients limit the intake of food until starvation occurs but at the same time consider themselves fat.	Physical exam and history.	Counseling, reversal of the malnourished status.
Bulimia nervosa	Eating disorder characterized by recurrent binge eating and purging of the food with laxatives and vomiting; classified as a psychiatric disorder.	Emaciated appearance, erosion of dental enamel, esophageal ulceration.	Physical exam and history.	Counseling, reversal of malnourished status, dental repair.

continues

A Quick Trip Through Some Diseases of the Gastrointestinal System *(continued)*

DISEASE	ETIOLOGY	SIGNS AND SYMPTOMS	DIAGNOSTIC TEST(S)	TREATMENT
Caries	Also known as dental cavities, the gradual decay of teeth; can cause inflamed tissue.	Destruction of tooth enamel, pain, may lead to abscesses and even heart disease.	Visual inspection, X-rays.	Evacuation and filling of cavity, proper dental hygiene, proper diet.
Cholelithiasis	Inflammation of the gallbladder; formation or presence of stones or calculi in the gallbladder or common bile duct. Incidence increases with age. Common in men, women following multiple pregnancies, the obese, diabetics, and patients who have had rapid weight loss.	Asymptomatic/ or mild discomfort to extreme pain, often preceded by ingestion of fatty or greasy foods. Pain is usually steady, lasting from 15–30 minutes up to several hours with a spontaneous resolution. Nausea, vomiting, bloating, flatulence, abdominal tenderness. Acute cholecystitis includes low grade fever, well localized right upper quadrant tenderness.	Physical exam/exam, patient history, radiologic studies, ultrasound, oral cholescintigraphy, blood work with a rise in leukocyte count during acute cholecystitis (although other values usually within normal range).	Changes in diet, observation, surgical removal if deemed severe enough.
Cirrhosis	Chronic inflammatory disease of the liver, leading to irreversible scarring and destruction of active liver tissue. 75% of cases caused by heavy/ chronic alcohol consumption. Other causes include overuse of prescription medications, viral hepatitis, or chronic obstruction of the bile ducts. Cause sometimes is unknown.	Enlargement of the liver (hepatomegaly) with normal tissue being replaced by fibrous tissue. This causes a decrease in its function, along with nausea, vomiting, weakness, jaundice, swollen ankles, loss of weight, loss of body hair, massive hematemesis, coma, and death.	Patient exam and history, blood results.	Cessation of causative agent.
Cleft lip	Congenital anomaly in which the upper lip fails to come together, often seen along with a cleft palate.	Obvious structural deformity making nursing difficult.	Physical examination.	Surgical correction.

15-7 To view videos on anorexia, bulimia, and obesity, please see your DVD.

A Quick Trip Through Some Diseases of the Gastrointestinal System *(continued)*

DISEASE	ETIOLOGY	SIGNS AND SYMPTOMS	DIAGNOSTIC TEST(S)	TREATMENT
Cleft palate	Congenital anomaly in which the roof of the mouth has a split or fissure.	Obvious structural deformity making nursing difficult, allowing milk to be aspirated into the lungs from the nasal cavity.	Physical examination, imaging studies.	Surgical correction.
Crohn's disease	Form of chronic inflammatory bowel disease affecting the ileum and/or colon. Also called regional ileitis/enteritis.	Pain, cramps, diarrhea, bloating, weight loss.	Physical exam and history, radiologic studies.	Anti-inflammatory medications, such as prednisone, surgical intervention if severe.
Esophageal stricture	Narrowing of the esophagus that makes the flow of fluids and food difficult. May be due to GERD.	Difficulty swallowing, indigestion, pain, feeling of fullness in the throat.	Patient history, physical exam, imaging studies.	Surgical repair or dilation.
Gastroesophageal reflux disease (GERD)	Erosion of the esophageal lining by acid reflux, often due to a weakened lower esophageal sphincter. Other causative agents: overeating, reclining too soon after eating, obesity, overconsumption of alcohol, caffeine, or chocolate.	Heartburn, especially following a meal, often while reclining.	Patient history, examination, gastroscopy, barium studies.	Treat causative agents, medications such as antacids and those reducing stomach acid production, changes in the timing of eating and the types of food consumed, such as avoidance of chocolate, peppermint, coffee, citrus, fried or fatty foods, and tomato products. Stop smoking and don't lie down 2–3 hours after eating.
Gingivitis (gum disease)	Inflammation of the gums, leading to tooth loss; generally due to poor dental hygiene.	Sore, bleeding gums, receding gum line.	Physical examination.	Proper diet/ dental hygiene.
Hiatal hernia	Protrusion of the stomach through the diaphragm and extending into the thoracic cavity; reflux esophagitis is a common symptom. More common in women, incidence increases with age.	May be asymptomatic, heartburn within 1–4 hours following eating, esophageal reflux.	Patient history and exam, radiologic studies.	Similar to GERD: Frequent, small meals, loss of excess weight, elevation of head of bed, cough suppressants, smoking cessation, antacids, avoid sleeping on right side.

continues

A Quick Trip Through Some Diseases of the Gastrointestinal System *(continued)*

DISEASE	ETIOLOGY	SIGNS AND SYMPTOMS	DIAGNOSTIC TEST(S)	TREATMENT
Impacted wisdom tooth	Wisdom tooth that is tightly wedged into the jaw bone so that it is unable to erupt.	Pain, discomfort, realignment of normal teeth. May become infected, producing pus.	Oral examination, radiographic study.	Removal.
Leukoplakia	Thick, white patches of tissue of the mucosal membranes in the mouth where there is chronic physical contact with tobacco.	White patches of tissue, which are considered to be pre-cancerous.	Visual examination.	Quit using snuff/chewing tobacco.
Periodontal disease	Disease of the supporting structures of the teeth, including the gums and bones. Approximately 80% of tooth loss in individuals age 45 and older can be attributed to periodontal disease.	Destruction of gum tissue and supporting bone structure, tooth loss.	Visual examination, radiologic studies.	Correct causative agents, improve dental hygiene. Warning—improper or excessive tooth brushing may lead to receding gums
Pyorrhea	Discharge of purulent material from dental tissue.	Pain, offensive odor.	Oral exam, radiographic studies, culture of discharge.	Correct source of the discharge, antibiotics (if a bacterial source), maintain proper oral care.

www.prenhall.com/colbert
Dieticians provide valuable information on nutrition and specialized diets. To view a video and obtain more information, visit the website for this chapter.

A Quick Trip Through Some Diseases of the Gastrointestinal System *(continued)*

MID G.I. REGION

DISEASE	ETIOLOGY	SIGNS AND SYMPTOMS	DIAGNOSTIC TEST(S)	TREATMENT
Cholecystitis	An acute or chronic inflammation due to gallstones (cholelithiasis) blocking ducts in the gallbladder.	Sudden severe pain, fever, nausea, vomiting, tenderness in the right upper quadrant of the abdomen.	Patient history, physical exam, imaging studies.	Avoid fatty foods, surgical removal.
Gastritis	Acute or chronic inflammation of the stomach due to infection, spicy foods, excess acid production, stress, alcohol, aspirin consumption, heavy smoking.	Pain, tenderness, nausea, and vomiting.	Patient history, imaging studies, endoscopy, gastric biopsy.	Antacids, antibiotics (if a bacterial infection), lifestyle changes.
Gastroenteritis	Infection and/or inflammation of the stomach and small intestines due to a virus or bacterium. Source may also be chemical in nature.	Abdominal pain, nausea, vomiting, diarrhea. Dehydration may also result as a complication.	Patient history, imaging studies, identification of offending agent.	Antibiotics (if bacterial infection), antiemetics, antidiarrheal drugs.
Hepatitis, viral	Inflammation of the liver; the most common chronic liver disease. Caused by a virus. Several types of hepatitis (A [food or water-borne], B [blood-borne or sexually transmitted], C, D, E) with varying severity and outcomes.	Jaundice, fever, nausea, anorexia, weakness, arthralgia, enlarged and tender liver, dark urine. Oddly, an aversion to food and cigarette smoking can occur. May lead to chronic hepatitis and cirrhosis.	Patient history and physical examination, blood work, liver biopsy, stool samples, immunodiagnostics.	Preventative vaccine (for hepatitis type B), preventative infection control practices, immunoglobulin injections, antiviral drugs, rest, hydration, safe sex practices. Health care workers at risk should be offered vaccine.
Peptic ulcer	Ulcer occurring in the lower portion of the esophagus, stomach, and duodenum caused by the breakdown of the mucosal membrane by the acid of gastric juices and pepsin. Causative agents can include: *H. pylori* bacterium, stress, excessive consumption of alcohol, tobacco, aspirin/NASIDs, ,caffeine, corticosteroids and other drugs.	Heartburn, indigestion, sharp localized pain, often relieved by eating, only to return again within 2 to 4 hours, possible back pain, possible diarrhea, bleeding a possibility.	Patient history and examination, barium/radiologic studies, biopsy.	Remove causative agent(s), reduce stress, utilize medications such as antacids/antibiotics, or H2 blockers (drugs) that block release of histamine, surgery only in extreme cases.

continues

A Quick Trip Through Some Diseases of the Gastrointestinal System *(continued)*

DISEASE	ETIOLOGY	SIGNS AND SYMPTOMS	DIAGNOSTIC TEST(S)	TREATMENT
Reflux esophagitis	Acid from the stomach backs up into the esophagus, causing inflammation and pain.	Heartburn, continued erosion of the tissue of the esophagus can lead to scar, stricture, difficulty swallowing.	Patient history, radiologic studies, visual examination.	Reduce reflux through smaller meals, stop eating approximately 4 hours before going to bed, smoking cessation, avoid alcohol, caffeine, chocolate, elevate head of bed, avoid sleeping on right side, utilize medications that decrease gastric acidity, surgery only in extreme cases.

LOWER G.I. REGION

DISEASE	ETIOLOGY	SIGNS AND SYMPTOMS	DIAGNOSTIC TEST(S)	TREATMENT
Celiac sprue	A chronic condition in which wheat glutens (or equivalent substances in barley, oats, or rye) cause damage to the mucosa of the small intestine, creating a malabsorption situation. Disease more common in white European groups, especially western Ireland.	Malabsorption, diarrhea, weight loss, flatulence, distended abdomen, anemia, back pain, hypoproteinemia, impotence, osteopenic bone disease and other nutritional deficiency manifestations may occur. Death may occur in severe cases.	Patient history and physical exam, radiologic studies, intestinal biopsy, blood studies for specific antibodies.	Gluten-free diet, anti-inflammatory medication.
Diverticulitis	Infection and inflammation of a diverticulum or sac in the intestinal tract, especially in the colon.	Bleeding, abdominal pain, fever, hyperactive bowel sounds, muscular hypertrophy/ shortening of the colon due to muscular hyperactivity, inflammation, pelvic abscess.	Patient history and physical exam, blood work, radiographic studies, endoscopy.	High fiber diets, stool softeners, antibiotics, surgical intervention.
Hemorrhoids	Varicose veins in the rectum.	Pain, itching, burning sensation, bleeding.	Proctoscopy, stool sample examination.	Dietary changes (more fiber and water), stool softeners, medication to relieve discomfort.
Inguinal hernia	Hernia or outpouching of intestines into the inguinal region of the body.	Visual bulging, pain, discomfort, possible adhesions, incarceration of intestinal loop, strangulation.	Patient history, physical examination.	May be reduced in the early stages, binding, surgical repair.

A Quick Trip Through Some Diseases of the Gastrointestinal System *(continued)*

DISEASE	ETIOLOGY	SIGNS AND SYMPTOMS	DIAGNOSTIC TEST(S)	TREATMENT
Irritable bowel syndrome (IBS)	Disturbance in the functions of the intestine from unknown causes; often seen in young to middle age females. May be stress related; may also be due to laxative abuse or irritating foods.	Symptoms generally include abdominal discomfort/pain, bloating, and an alteration in bowel activity (diarrhea, constipation).	Patient exam and history, radiologic studies, stool sample examination.	Lifestyle alterations, dietary considerations, medication.
Intussusception	Result of the intestine slipping or telescoping into another section of intestine just below it, often in the ileocecal region; more common in children.	Pain.	Radiographic studies.	Surgery.
Malabsorption syndrome	Inadequate absorption of nutrients from the intestinal tract; caused by a variety of diseases and disorders of the pancreas or gallbladder.	Malnourishment, "failure to thrive," failure to absorb fats or other nutrients, abnormal stools (black/tarry, pale/floating), foul odor.	Patient examination and history, stool samples, blood work, radiographic studies.	Treat offending disease, dietary adjustments, supplemental vitamins/minerals.
Peritonitis	An infectious and/or inflammatory process of the peritoneum; may be due to leakage of contents from the gallbladder, appendix, duodenal ulcer, penetrating injuries or the result of a cancerous tumor.	Pain, fever, malaise, shock, abscesses.	Patient history, physical examination, blood work.	Correction of cause, surgical intervention, antibiotics.
Polyposis	Small tumors that contain a pedicle or foot-like attachment in the mucous membranes of the large intestine (colon).	Discomfort, bleeding.	Endoscopic studies.	Surgical removal.
Ulcerative colitis	Ulceration of the mucous membranes of the colon; cause is unknown, but UC tends to run in families; belongs to a group of conditions known as inflammatory bowel disease (IBD).	Inflammation of the region, lesions, decreased flexibility of the bowel, fistulas, abdominal pain, diarrhea, fever, weight loss, bleeding, anemia, malabsorption of nutrients; may lead to carcinoma.	Patient history, radiographic studies, endoscopic studies, colonoscopy, blood work.	Anti-inflammatories, such as prednisone, immunomodulators, biologic therapies, or surgery.

continues

A Quick Trip Through Some Diseases of the Gastrointestinal System *(continued)*

DISEASE	ETIOLOGY	SIGNS AND SYMPTOMS	DIAGNOSTIC TEST(S)	TREATMENT
Volvulus	This condition is a result of a congenital deformity or a foreign body in which the bowel twists upon itself and causes an obstruction. Painful and requires immediate surgery.	Pain. Symptoms generally include abdominal discomfort and an alteration in bowel activity.	Radiologic studies.	Surgery.

Pharmacology Corner

We previously discussed the damage caused by stomach acid in conditions such as GERD and GI ulcers. So how can we decrease the amount or acidity of gastric juices? This can be done in several ways. Antacids directly decrease the acidity of gastric juices by providing an alkaline (base) substance to the stomach. Proton pump inhibitors decrease the production of gastric juices in the stomach.

Diarrhea and vomiting are common symptoms of a variety of GI disorders. Anti-diarrheal agents basically slow the movement of digestive particles through the system to allow for more water absorption back into the body. Sometimes vomiting is something you want to encourage as in the case of certain poisonings. Syrup of Ipecac was commonly used to induce vomiting but now is often replaced with ingestion of activated charcoal and whole bowel irrigation.

The nausea associated with sea or motion sickness can be treated by medications that depress the CNS and parasympathetic nervous system as well as provide mild sedation. Scopolamine is an example in this category and can be used as a transdermal patch or pill to prevent motion sickness. This can also be used to treat vertigo.

SUMMARY

Snapshots from the Journey

→ The digestive tract is a hollow tube extending from the mouth to the anus. It contains a variety of structures that allow the digestion of food and the absorption of nutrients necessary for life.

→ Food is processed mechanically and chemically to efficiently break it down to usable substances.

→ The following are some of the main components of the digestive systems and their functions:

ORGAN	DIGESTIVE ACTIVITY	SUBSTANCE DIGESTED	REQUIRED DIGESTIVE SECRETIONS
Mouth, or oral cavity	Chews food and mixes it with saliva; forms food into a bolus, and swallows; sense of taste	Starch	Salivary amylase and ptyalin
Esophagus	Moves bolus to the stomach through peristalsis	Not applicable	Not applicable
Stomach	Stores food, also churns food while mixing in digestive juices	Proteins	Hydrochloric acid, pepsin
Small intestine	Secretes enzymes, receives secretions from the pancreas and liver, neutralizes the acidity in chyme, absorbs nutrients into the bloodstream and lymphatic system	Carbohydrates, fats, nucleic acid, proteins	Intestinal and pancreatic enzymes, bile from the liver
Large intestine	Creates and absorbs fat-soluble vitamins, reabsorbs water, forms and eliminates feces	Vitamins B_{12} and K	Not applicable

→ Enzymes are formed by special cells and act as a catalyst that speeds up chemical reactions.

→ The bulk of the digestive process and the absorption of most nutrients occur in the small intestine.

→ Although not directly a part of the alimentary canal, the accessory organs (liver, gallbladder, and pancreas) are needed for proper and efficient functioning of the digestive process.

→ The rate of speed that food travels through the gastrointestinal system affects the acidity of the digesting food, the absorption of nutrients, and the quality of the feces.

→ Diseases of the gastrointestinal system can be a result of heredity, the type and amount of food consumed, substance abuse, or emotions.

RAY'S STORY

Eating was always important to Ray as he considered himself somewhat of a chef. However, since is accident the natural act of eating we take for granted has now become a difficult process. Considering Ray's situation as a quadriplegic, research the best way(s) to deliver his food.

a. How effective would a PEG tube be as opposed to a tube utilizing the naso-gastric route?

_____

_____

_____

b. What are the potential hazards that Ray may encounter?

_____

_____

_____

c. Are there any special nutritional considerations for patients in Ray's situation?

_____

_____

_____

REVIEW QUESTIONS

Multiple Choice

1. Which of the following is not a responsibility of the large intestine?
   a. production of vitamin K
   b. absorption of water
   c. digestion of carbohydrates
   d. elimination of feces

2. Starches begin to be digested in the:
   a. oral cavity
   b. esophagus
   c. stomach
   d. large intestine

3. This structure prevents food and liquid from entering the lungs:

a. larynx

b. pharynx

c. epiglottis

d. glottis

4. The liver receives approximately this much blood every minute:

a. 500 milliliters

b. 1 gallon

c. 2 units

d. 1.5 quarts

5. Which of the following is *not* a colon segment?

a. transverse

b. ascending

c. descending

d. absorbing

6. A condition that includes the cracking with associated inflammation of the lips and corners of the mouth is called:

a. chelation

b. cheilitis

c. condrosite

d. leukoplakia

Fill in the Blank

1. _____ is the muscle action that mixes chyme with digestive juices, while _____ is the muscular action that moves food through the digestive system.

2. This vermiform structure is attached to the large intestine and is considered a vestigial organ: _____ .

3. The exocrine portion of this important organ secretes buffers needed to neutralize the acidity of chyme and also secretes several digestive enzymes: _____ .

4. _____ is the breaking up of fat into smaller particles that can more readily be acted upon by digestive enzymes.

5. The end result of fecal matter remaining in the large intestine too long, with too much water being removed from it, is _____ .

6. _____ is a precancerous condition creating white patches in the oral region and is associated with the chronic use of snuff and chewing tobacco.

Short Answer

1. Explain the difference between *chyme* and *chyle*.

_____

_____

_____

_____

_____

2. Explain the importance of bacteria in the large intestine.

_____

_____

_____

_____

3. Discuss the importance of the liver in the digestive process.

_____

_____

_____

_____

4. Could you live without a gallbladder? Defend your answer.

_____

_____

_____

_____

5. Describe GERD and ways to treat / prevent this condition.

_____

_____

_____

_____

Suggested Activities

1. Nutrition is a very important consideration when planning a healthy lifestyle. Research the different food groups and the daily recommended amounts that you should consume. List your daily consumption of food for a week and compare what you eat to what you should eat. List suggestions for changes that you should make in your diet, if any.

2. Although it is felt that most stomach ulcers are a result of bacteria, research ways that you can help to prevent their formation.

> **15-8** Now that you have completed your journey through this chapter, please go to the DVD for interactive games and puzzles concerning the medical terms and concepts contained in this chapter. By playing the games you will reinforce your learning of the GI system in a fun way.

THE URINARY *System*

Filtration and Fluid Balance

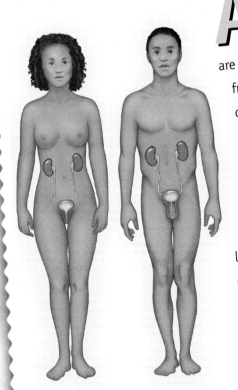

A city must have a safe, reliable water supply, usually obtained by pumping water from a reservoir or diverting a river. Sometimes cities have a series of wells to supply drinkable water. In desert areas, there are even special water treatment plants to remove salt from seawater to make fresh water for human consumption. In any case, there must be a way to clean the water. Water purification plants remove chemicals and debris and disinfect the water before it ever gets to your faucet. You are completely unaware of the activities of your city's water purification plant, but if it stopped working, you would be extremely unhappy. Imagine a glass full of muddy, bacteria-laden water. Yum!

Your body also must have a purification plant for its fluid—blood. Your liver does some of the purification, but your urinary system is responsible for controlling the electrolyte (ion) and fluid balance of your body. The kidneys filter blood, reabsorb and secrete ions, and produce urine. Without them, fluid and ion imbalance, blood pressure irregularities, and nitrogen waste buildup could cause death in a matter of days.

Chapter 16

LEARNING OBJECTIVES

Upon completion of your journey through this chapter, you will be able to:

→ Present an overview of the organs and functions of the urinary system

→ Describe the internal and external anatomy and physiology of the kidneys

→ Discuss the importance of renal blood flow

→ Describe the process of urine formation

→ Trace the pathway of reabsorption or secretion of vital substances

→ List and discuss the importance of hormones for proper kidney function

→ Describe the anatomy and physiology of the bladder and urination

→ Discuss several common disorders of the urinary system

MULTIMEDIA APPLICATIONS

DVD Interactive Exercises

→ 3-D view of the urinary system, 16-1

→ Animation and Interactive labeling exercise of the kidney, 16-2

→ Video on kidney stones, 16-3

→ Videos on urinalysis, including dipstick urinalysis, 16-4

→ Animation of hypovolemic shock, 16-5

→ Video on renal failure, 16-6

→ Videos on basic peritoneal dialysis, 16-7

→ Drag-and-drop exercise on labeling the urinary bladder, 16-8

→ Interactive puzzles and games, 16-9

www.prenhall.com/colbert

→ Professional Profile:
- Ultrasound Technician

→ Related Internet Links

→ Additional Review Questions

Pronunciation Guide

Correct pronunciation is important in any journey so that you and others are completely understood. Here is a "see and say" Pronunciation Guide for the more difficult terms to pronounce in this chapter.

afferent arterioles
    (AFF er ent ahr TEE ree ohlz)

aldosterone (al DOSS ter ohn)

antidiuretic hormone (ADH)
    (AN tye dye yoo RET ik)

atrial natriuretic peptide (AY tree al
    NAY tree your ET ick PEP tide)

calyx, calyces (KAY licks, KAY leh seez)

cortical nephron
    (CORE tih cull NEFF rahn)

efferent arterioles
    (EFF er ent ahr TEE ree ohlz)

external urethral sphincter
    (EKS ter nal yoo REE thral SFINK ter)

glomerulus (gloh MAIR yoo luss)

glomerular capsule
    (gloh MAIR you ler CAP sell)

glomerulonephritis
    (gloh MAIR you loh neh FRY tis)

glomerulosclerosis
    (gloh MAIR you loh sklee ROW sis)

hematuria (HE mah TOO ree ah)

juxtaglomedullary nephron (JUX ta glo
    meh DULL lair ee NEFF rahn)

juxtaglomerular cells
    (JUX ta gla MARE you ler)

lithotripsy (LITH oh TRIP see)

nephropathy (neh FROP ah thee)

renal hilum (REE nal HIGH lum)

renal medulla (REE nal meh DULL lah)

renin-angiotensin-aldosterone
    (REE nen-an gee oh TEN sen-al DOSS
    ter ohn)

urea (you REE ah)

uremia (you REEM ee ah)

ureter (yoo REE ter) or (YOO reh ter)

urethra (yoo REE thrah)

SYSTEM OVERVIEW

ureter *(yoo REE ter)*

urethra *(yoo REE thrah)*

The urinary system (see Figure 16–1 ■) consists of two **kidneys,** bean-shaped organs located in the superior dorsal abdominal cavity between the peritoneum and the back muscles, and several accessory structures. The kidneys filter blood and make urine. A **ureter** is a tube of smooth muscle lined with mucus that carries urine from each kidney to the single **urinary bladder,** located in the inferior ventral pelvic cavity. The urinary bladder is basically an expandable sac that holds the urine. The **urethra** is the tubing that transports urine from the bladder to the outside of the body.

The job of the urinary system is to make urine, thereby controlling the body's fluid and electrolyte (ion) balance, eliminating waste products, and removing nitrogenous wastes, certain salts and excess water from blood through excretion. (The kidney also is instrumental in controlling blood pressure as you will see later in this chapter.) To make urine, three processes are necessary: **filtration, reabsorption,** and **secretion.** Filtration involves filtering the blood to create a filtrate (that which has passed through the filter). This filtrate, which contains various substances, can then be either reabsorbed back into the bloodstream or secreted into the urine and removed from the body as urine. Keep in mind the part of the filtrate that is reabsorbed is conserved and placed back into the bloodstream, while the part that is secreted is eliminated from the body. You'll learn how this process occurs as you journey through this chapter.

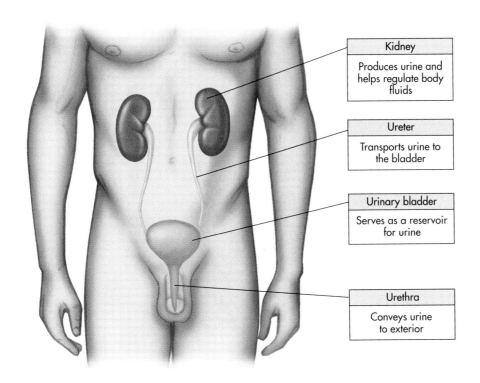

Kidney
Produces urine and helps regulate body fluids

Ureter
Transports urine to the bladder

Urinary bladder
Serves as a reservoir for urine

Urethra
Conveys urine to exterior

FIGURE ▪ 16-1

Anatomy of the urinary system.

THE ANATOMY OF THE KIDNEY

The kidney is a very intricate filtration system. Though you have two kidneys, you can actually function well with only one healthy kidney. That is why someone can donate a kidney while he or she is alive.

External Anatomy

The external anatomy of the kidney is relatively simple. The kidney is covered by a fibrous layer of connective tissue called the **renal capsule.** The indentation that gives the kidney its bean-shaped appearance is called the **renal hilum.** At the hilum, renal arteries bring blood into the kidneys to be filtered. Once filtered, the blood leaves the kidney via the renal vein. The ureter is also attached at the hilum to transport the urine away from the kidney to the bladder (see Figure 16–2 ▪).

renal capsule *(REE nal CAP sell)*
renal hilum *(REE nal HIGH lum)*
  hilum = *root*

Internal Anatomy

The internal anatomy of the kidney (again see Figure 16–2) is considerably more complicated than the external anatomy. The kidney can be divided into three layers. The outer layer is the **renal cortex,** the middle layer is the **renal medulla,** and the innermost layer is the **renal pelvis.** Adding the word *renal* is important here. Remember that the brain and the adrenal gland both have a

renal cortex *(REE nal CORE tex)*
  cortex = *Latin for "rind," or outer layer, like the rind of a watermelon*
renal medulla
  (REE nal meh DULL lah)
  medulla = *inner portion*
renal pelvis *(REE nal PELL vis)*

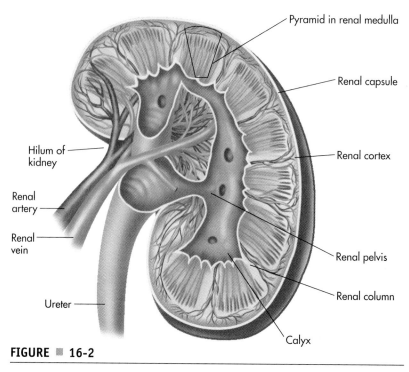

Pyramid in renal medulla

Renal capsule

Renal cortex

Hilum of kidney

Renal pelvis

Renal artery

Renal vein

Renal column

Ureter

Calyx

FIGURE ▪ 16-2

The internal and external anatomy of the kidney.

medulla and a cortex. Also keep in mind that the body also has more than one pelvis, as in the bony pelvis.

The renal cortex is grainy in appearance and has very little obvious structure to the naked eye. The renal cortex is where the blood is actually filtered. The renal medulla contains a number of triangle-shaped striated (striped) areas called **renal pyramids.** The renal pyramids are composed of collecting tubules for the urine that is formed in the kidney. Adjacent pyramids are separated by narrow **renal columns,** which are extensions of the cortical tissue.

The renal pelvis is a funnel. The funnel is divided into two or three large collecting tubes called **major calyces.** Each major **calyx** is divided into several **minor calyces.** The calyces form cup-shaped areas around the tips of the pyramids to collect the urine that continually drains through the pyramids. The kidney is essentially a combination of a filtration and collection system. The blood is filtered by millions of tiny filters in the cortex, the filtered material flows through tiny tubes in the medulla, and the resulting urine is collected in the renal pelvis. The renal pelvis, which is simply the enlarged proximal portion of the ureter, empties into the tube of the ureter. Through peristalsis, the ureter then carries the urine to the urinary bladder, where it is stored and eventually eliminated from the body. (Again, see Figure 16–2.)

calyx *(KAY licks)* = cup or cup-shaped
calyces *(KAY leh seez)*

16-1 To see a 3-D view of the urinary system, please go to your DVD.

BLOOD VESSELS

Because the kidney's job is to filter blood, the blood must reach every part of the kidney. To accomplish this filtration process there is a network of blood vessels throughout the kidney. A single **renal artery** enters each kidney at the hilum (see Figure 16–3 ■). The renal artery then branches into five **segmental arteries**, and the segmental arteries in turn branch into **lobar arteries**. The lobar arteries branch into interlobar arteries, which pass through the renal columns. **Arcuate arteries** originate from the interlobar arteries. The arcuate arteries are so named because they arch around the base of the pyramids in the renal medulla. Many tiny interlobular arteries branch from the arcuate arteries, supplying blood to the renal cortex. These interlobular arteries give rise to numerous **afferent arterioles**.

afferent arterioles
(AFF er ent ahr TEE ree ohlz)

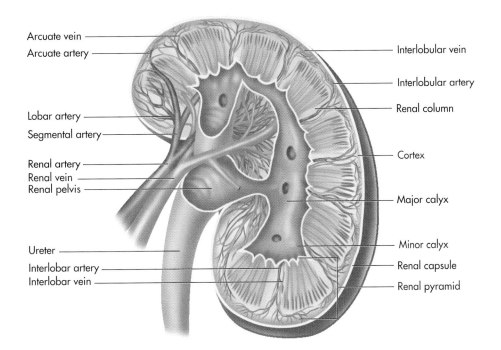

Arcuate vein
Arcuate artery
Lobar artery
Segmental artery
Renal artery
Renal vein
Renal pelvis
Ureter
Interlobar artery
Interlobar vein

Interlobular vein
Interlobular artery
Renal column
Cortex
Major calyx
Minor calyx
Renal capsule
Renal pyramid

Renal artery → Segmental arteries → Lobar arteries → Interlobar arteries → Arcuate arteries → Interlobular arteries → Afferent arterioles → Glomerulus → Efferent arterioles → Peritubular capillaries → Interlobular veins → Arcuate veins → Interlobar veins → Lobar veins → Renal vein

FIGURE ■ **16-3**

Renal blood vessels and the pathway of blood through the renal system.

Each afferent arteriole leads to a ball of capillaries called a **glomerulus.** **Efferent arterioles** then leave from the glomerulus and travel to a specialized series of capillaries called the **peritubular capillaries** and on to the **vasa recta** (straight collecting tubes) that are associated with the **renal nephron,** the functional unit of the kidney. The peritubular capillaries wrap around the tubules of the nephron. You have seen a situation like this in the lungs, where the pulmonary capillaries surround the alveoli. Having blood vessels close to the nephron allows efficient movement of ions between blood and the fluid in the nephron, just as having pulmonary capillaries near the alveoli allows efficient diffusion of respiratory gases between the alveoli and the bloodstream.

glomerulus *(gloh MAIR yoo luss)*
efferent arterioles
(EFF er ent ahr TEE ree ohlz)
peritubular *(per ee TUBE you ler)*
nephron *(NEFF rahn)*

From each set of peritubular capillaries, blood flows out the interlobular veins.
From there, the blood flows out a series of
veins that are the direct reverse pattern of
the arteries, with one exception. There are
no segmental veins. The blood finally
leaves the kidney via the **renal vein.** Please
see Figure 16–3 again for a diagram of the
renal blood vessels.

Microscopic Anatomy of the Kidney: The Nephron

renal corpuscle
(REE nal KOR pus el)

So far we have looked at an overview of the kidneys structure and function.
Now let's take a closer look at what actually happens within the kidneys. The business end of the kidney, the part that performs the real functions of the kidney, consists of millions of microscopic funnels and tubules. These fundamental functional
units of the kidney are called **nephrons** (see Figure16–4 ■). The nephron is divided into two distinct parts: the **renal corpuscle** and the **renal tubule.** The renal
corpuscle is a filter, much like a window screen or coffee filter.

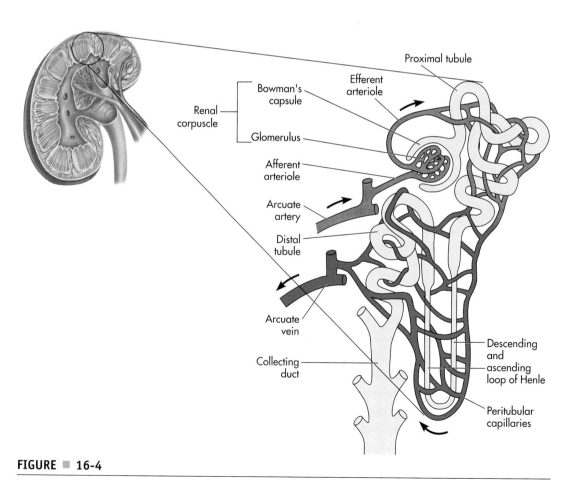

FIGURE ■ 16-4

The nephron.

Blood enters the renal corpuscle via the **glomerulus,** a capillary ball. Surrounding the glomerulus is a double-layered membrane called the **glomerular capsule** (or **Bowman's capsule**). The layers of the glomerular capsule are similar to the layers of a serous membrane like the pleura or pericardium. The inner layer of the glomerular capsule, the **visceral layer**, surrounds the glomerular capillaries. The visceral layer is made of specialized squamous epithelial cells called **podocytes.** The combination of podocytes and the simple squamous epithelium making up the walls of the glomerular capillaries make a very effective filter.

The outer or parietal layer of the glomerular capsule is simple squamous epithelium and completes the container for the filter. Blood flows into the glomerulus and everything *but* blood cells and a few large molecules, mainly proteins, are pushed from the capillaries across the filter and into the glomerular capsule. The material filtered from the blood into the glomerular capsule is called **glomerular filtrate.** Can you see why blood cells or excessive protein found in urine may indicate a kidney filtration problem?

The rest of the nephron is series of tubes known as the **renal tubule**, sort of like a Habitrail system for a gerbil or hamster (see Figure 16–5 ■). Just like the water filtration system in our towns and cities, the water (glomerular filtrate) travels through a network of pipes (tubules) where the impurities remain in the pipes to be discharged while the filtered water is collected (peritubular capillaries) and recycled back into the city's water supply.

Now back to the kidney. Glomerular filtrate flows from the glomerular capsule into the first part of the renal tubule, the **proximal tubule** (sometimes called the proximal *convoluted* tubule). The wall of the proximal tubule is made of cuboidal epithelium with microvilli. From the proximal tubule, glomerular filtrate flows into the **nephron loop** (also called the **loop of Henle**). The nephron loop consists of two segments, the **descending loop,** with a structure similar to the proximal tubule, and the **ascending loop,** with a wall made of simple cuboidal epithelium without microvilli. From the nephron loop, the glomerular filtrate flows into the **distal tubule** (a.k.a. the distal convoluted tubule). The wall of the distal tubule is like that of the ascending branch of the nephron loop.

From the distal tubule, glomerular filtrate flows into one of several **collecting ducts** also made of cuboidal epithelium. The collecting ducts lead to the minor calyces, then to the major calyces, renal pelvis, and ureter. At this point, the glomerular filtrate is urine. Again, if you refer to our previous learning hint, what stays in the tubular system (plastic pipe) eventually gets eliminated as urine.

glomerulus *(gloh MAIR yoo luss)*
glomerular capsule
(gla MARE you ler CAP sell)

glomerular filtrate
(gla MARE you ler FILL trate)

16-2 To view an animation of the kidney and perform a drag-and-drop labeling exercise, please go to your DVD.

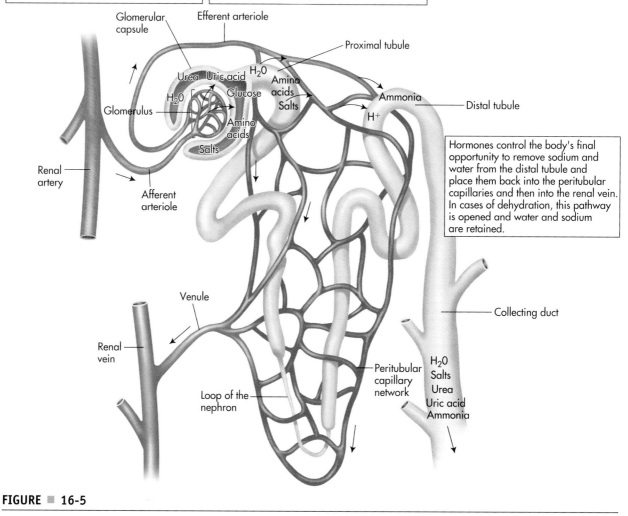

Glomerular Filtration

Blood from the renal artery is filtered in the glomerulus. The filtered product which contains water, salts, nutrients, and waste products is called the glomerular filtrate.

Tubular Reabsorption

Nutrients and salts are actively reabsorbed and transported to the peritubular capillary network and some water is passively reabsorbed into the peritubular capillaries.

Tubular Secretion

Some substances are actively secreted from the peritubular capillaries into the distal tubule for removal from the body.

Hormones control the body's final opportunity to remove sodium and water from the distal tubule and place them back into the peritubular capillaries and then into the renal vein. In cases of dehydration, this pathway is opened and water and sodium are retained.

FIGURE ▬ 16-5

The renal tubule.

Pathology Connection: Kidney Stones and Polycystic Kidney Disease (PKD)

Kidney stones are exactly what their name implies: hard bodies (stones) in the kidney. Kidney stones result when substances in the urine crystallize in the renal tubule, often because the concentration of that substance is higher than normal in the renal tubule. However, sometimes the cause of kidney stones is a mystery. Kidney stones are quite common. As many as 10% of people in North America will have a kidney stone before they turn 70 years old. Stones are more common in men than women. Patients with kidney disease, who have had bariatric (weight control) surgery, or who have bone disease, are at a higher risk of kidney stones.

Stones can be made of many different substances, including calcium (70% of stones) or uric acid (10% of stones), or can be caused by kidney infection. Some individuals appear to be more susceptible to stones than others, and once you have had a kidney stone, you are more susceptible to kidney stones in the future. Many kidney stones pass through the kidney unnoticed. However, larger or irregularly shaped stones may lodge in the kidney tubules, obstructing flow and irritating nearby tissues. Most patients diagnosed with kidney stones are driven to seek treatment because of blood in their urine (hematuria) or excruciating lower back pain. Some patients may have fever, nausea, and urinary urgency. Patients with kidney stones often describe the pain as the worst they have ever felt. Kidney stones are diagnosed using medical exam and patient history, urinalysis, ultrasound, capture and analysis of the stone, and imaging studies. Metabolic analysis may help pinpoint the cause of the stones and thus determine long-term treatment to prevent recurrence.

Treatment for kidney stones depends on the number, size, location and composition of the stones. Many patients are treated for pain and sent home with instructions to drink lots of water and wait for the stone to pass. (Often, they are asked to save the stone as it passes so it can be sent for chemical analysis.) Some patients are prescribed medication (usually corticosteroids and alpha-blockers) to increase the chance of passing the stones. If the stone is large or is obstructing urine flow, several less invasive techniques are available that may enable patients to avoid surgery. **Extracorporeal shock wave lithotripsy**, an outpatient procedure, uses shock waves applied to the outside of the body to break up the stone so it is small enough to pass through the kidney. During **ureteroscopy**, a fiberoptic endoscope is threaded up the urethra, through to bladder, and into the ureter. An instrument attached to the endoscope shatters the stone and captures the pieces. **Percutaneous nephrolithotomy** uses an endoscope and a small incision to remove large stones. See Figure 16-6 ■, which shows three different methods of lithotripsy.

Recurrence is often a problem, with as many as 50% of patients who have had a kidney stone developing another stone. Many patients are advised to adhere to a diet to prevent their particular type of stone, or are prescribed medication to decrease the likelihood that they will

www.prenhall.com/colbert
Ultrasound technicians help to view the kidneys for diagnostic purposes. They also perform diagnostic images on other body organs. To view a video on ultrasound procedures, please go to the website for this chapter.

Clinical Application

TRAUMA, ISCHEMIA, AND KIDNEY DAMAGE

The kidney is obviously very well vascularized. Each nephron is literally surrounded by blood vessels, and the flow of blood around the nephron is controlled by flow through the afferent arteriole. **Ischemia** is a condition of tissue injury resulting from too little oxygen delivery to tissues, usually caused by decreased blood flow. When blood flow to the nephrons decreases for a period of time, oxygen delivery to the nephron decreases and ischemia can result. Decreased blood flow to the kidney results from anything that causes prolonged constriction of the afferent arterioles.

For example, a young boy runs into a storm door, shattering the glass and in the process punctures his femoral artery, which begins to bleed profusely. As his blood volume falls, so does his blood pressure. His body fights desperately to bring his blood pressure back to normal, causing widespread vasoconstriction. The afferent arterioles, under the influence of sympathetic hormones and other vasoconstrictors, get smaller and smaller, greatly decreasing blood supply to the nephrons. If the situation continues long enough, the tissues will become ischemic and eventually begin to die. Even if the boy survives the initial blood loss from the wound, his kidneys may be damaged, resulting in temporary or permanent renal failure. It is not uncommon for trauma patients to survive the initial trauma only to become the victim of organ damage due to ischemia. Kidney failure will be discussed in more detail later in the chapter.

16-3 To view a video on kidney stones, please see your DVD.

develop another kidney stone. Long-term treatment depends on stone composition.

Polycystic kidney disease (PKD) is a genetic disorder in which large cysts (made mainly of epithelial cells) form in the kidneys and liver. It is the most common genetic cause of kidney disease in the US. Patients with PKD are also at higher risk for aneurysms and heart abnormalities. One form of PKD is so serious that patients die in infancy. The more com-

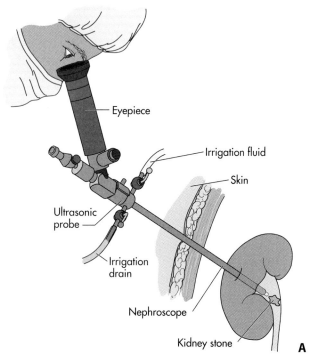

Eyepiece

Irrigation fluid

Skin

Ultrasonic probe

Irrigation drain

Nephroscope

Kidney stone

A

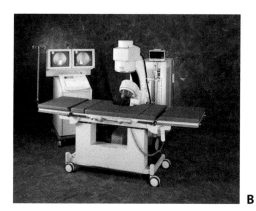

B

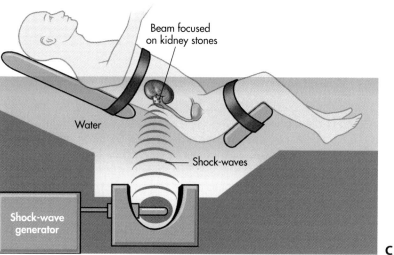

Beam focused on kidney stones

Water

Shock-waves

Shock-wave generator

C

FIGURE ■ 16-6

Three methods of lithotripsy. **A.** Percutaneous ultrasonic lithotripsy. A nephroscope is inserted into the renal pelvis, and ultrasound waves are used to fragment the stones. The fragments are then removed through the nephroscope. **B.** Dornier Compact Delta® system. Acoustic shock waves travel through soft tissue to break the renal stone into fragments, which are then eliminated in the urine (source: Dornier Medical Products, Inc.) and **C.** Water immersion lithotripsy.

mon form (1 in 1000 adults in the U.S.) is an adult-onset disorder character-ized by decreasing kidney function as normal nephrons are displaced by cysts. Few cysts develop in people under the age of 30. As more and more cysts de-velop, kidneys get very large. One polycystic kidney weighed 22 pounds! Symp-toms of PKD are variable but include hypertension, impaired renal functions, urinary tract infections, dilute urine, liver cysts, cerebral hemorrhage, decreased kidney perfusion, pain, hematuria and hormone abnormalities, particularly in-creased erythropoietin and increased activity of the renin-angiotensin-aldos-terone system (RAS). PKD is diagnosed by ultrasound and genetic testing. There is no cure for PKD except kidney transplantation. Patients generally progress to renal failure in their late 40s, 50s or early 60s. Patients are treated for complications of kidney disease, such as hypertension. Several medications are being investigated that may be helpful, including angiotensin converting en-zyme (ACE) inhibitors, statin drugs, and immunosuppressants. For PKD pa-tients who develop end-stage renal disease, either kidney dialysis or kidney transplant is necessary.

TEST YOUR KNOWLEDGE 16-1

Choose the best answer:

1. What carries urine from the kidneys to the bladder?
   a. urethra
   b. ureter
   c. vagina
   d. all of the above

2. The renal _____ is the outer layer of the kidney.
   a. medulla
   b. pelvis
   c. hilum
   d. cortex

3. These vessels carry blood into the glomerulus:
   a. peritubular capillaries
   b. afferent arterioles
   c. segmental arteries
   d. none of the above

4. The fundamental functional unit of the kidney is the:
   a. renal corpuscle
   b. renal pelvis
   c. nephron
   d. pyramid

5. Glomerular filtrate flows from the _____ into the _____.
   a. proximal tubule, distal tubule
   b. ascending loop, descending loop
   c. glomerular capsule, proximal tubule
   d. proximal tubule, collecting duct

6. Kidney stones are most often made of this material:
   a. bacteria
   b. uric acid
   c. calcium
   d. nickel

URINE FORMATION

As the body's water purification system, the job of the kidneys is to control fluid and electrolyte balance by carefully controlling urine volume and composition. The kidneys also remove nitrogen-containing waste, excess water, certain salts,

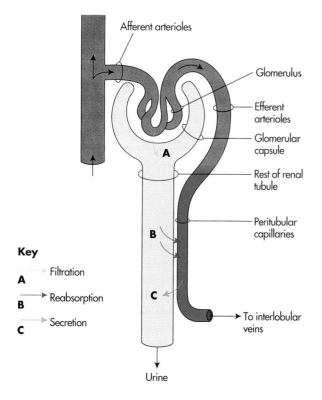

Afferent arterioles

Glomerulus

Efferent arterioles

Glomerular capsule

Rest of renal tubule

Peritubular capillaries

A

B

C

To interlobular veins

Urine

Key

A → Filtration

B → Reabsorption

C → Secretion

Schematic view of the three stages of urine production:
(A) filtration; (B) reabsorption; and (C) secretion.

FIGURE ■ 16-7

The processes involved in urine formation.

and other impurities from blood. In order to form urine, the nephron must perform three processes: **glomerular filtration, tubular reabsorption,** and **tubular secretion.** Please see Figure 16–7 ■ for a diagram of these three processes.

During glomerular filtration, the first step in urine formation, blood from the renal artery enters the glomerulus. Fluid and molecules pass from the glomerular capillaries into the glomerular capsule (also called Bowman's capsule), across a filter composed of the wall of the capillaries and the podocytes of the glomerular capsule. The filtrate flows into the renal tubule, where the composition of the filtrate is controlled by reabsorption and secretion.

Substances are either reabsorbed by the body or secreted. The substances that are reabsorbed pass from the renal tubule into the peritubular capillaries and stay within the body. The body uses secretion and reabsorption to regulate its supply of a variety of chemicals. Substances that are secreted pass from the peritubular capillaries into the renal tubule and eventually leave the body via the urine. For example, Hydrogen ions (H^+) are selectively secreted to maintain the body's acid-base balance. If blood levels of certain substances are high, such as glucose, amino acids, vitamins, or sodium, those substances will not be completely reabsorbed.

The combination of all three processes is necessary for the formation of urine. Filtration moves fluid and chemicals into the nephron from blood, while reabsorption and secretion control the concentration of chemicals and volume of urine. Glomerular filtrate is chemically similar to blood, while urine is chemically very different. Some substances, like glucose, are completely reabsorbed, and other substances, like the metabolic waste products urea and creatinine, are secreted such that urea and creatinine are much more concentrated in urine than in blood. See Table 16–1 for a comparison of plasma, glomerular filtrate, and urine chemistry.

TABLE 16-1 Kidney Fluid Chemistry

SUBSTANCE	PLASMA	GLOMERULAR FILTRATE	URINE
Protein	3,900–5,000 mg/100ml	none	None
Glucose	100 mg/100ml	100 mg/100ml	None
Sodium	142 mg/100ml	142 mg/100ml	128 mg/100ml
Potassium	5 mg/100ml	5 mg/100ml	60 mg/100ml
Urea	26 mg/100ml	26 mg/100ml	1,820 mg/100ml
Creatinine	1.1 mg/100ml	1.1 mg/100ml	140 mg/100ml

Control of Filtration

Think for a moment about filters you are familiar with. What controls these filters? What force drives filtration? What determines whether a substance passes through the filter or stays on one side? The example you are probably most familiar with is a window screen. Why do you have window screens in your windows? To keep out the bugs. Without screens, every insect in the neighborhood would be in your bedroom eating you alive, right? So what determines whether something gets through the screen? The most obvious answer is the size of the mesh in the screen. Dust gets through the screen, but most bugs don't. Imagine a screen with holes twice the size of a typical screen. You would spend all night swatting mosquitoes! The screens would be pretty much useless. Filter size determines what gets through the filter. All filters are *selective*: only some substances pass through, mainly depending on the size of the openings in the filter (see Figure 16–8 ■).

Can you think of a situation where something might get through a window screen? Leaving behind the bugs, let's think about a screened-in porch. During a gentle snow storm, snow doesn't come through the screen. But in a high wind, the floor of the porch becomes covered in snow. The high winds force the snow through the screen. Substances are moved through a filter by differences in pressure across the filter. Pressure pushes stuff through the holes. The higher the pressure on one side of the filter compared to the other side, the faster substances are filtered.

This combination works great in coffee makers. The filters have holes too small to let any but the tiniest coffee grounds pass through. The pressure of the water on top of the filter pushes the water, and the substances in coffee grounds that make drinkable coffee through the filter, as the grounds stay on the other side.

The glomerulus and glomerular capsule work in much the same way. Like most filters, the podocytes and capillary walls of the renal corpuscle create a filter with fixed openings. Plasma and many of the substances dissolved in plasma pass through the filter, but red and white blood cells, platelets, and some large molecules, like proteins, do not pass through the filter in a healthy kidney but

---

16-4 To view a video on urinalysis, including a dipstick urinalysis, please see your DVD.

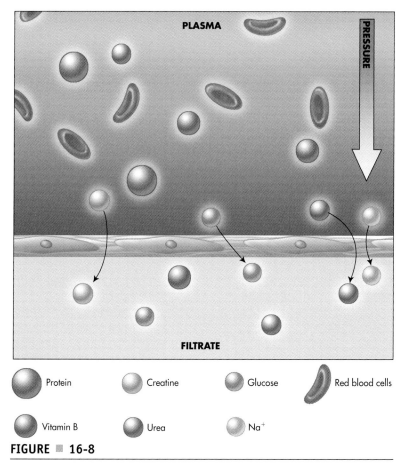

FIGURE ▪ 16-8

Filter selectivity.

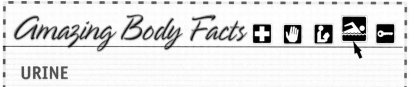

URINE

In a healthy renal system, the urine produced is sterile. The urea found in urine is the same substance that is used to melt ice in the winter. Urea is also a component in plant fertilizer: although nitrogen is a waste product for humans, it is an essential product for plant growth.

remain in the bloodstream. What passes through the filter is predetermined by the size of the openings. This explains why protein in urine is a sign of kidney damage. Under normal circumstances, protein molecules are too large to fit through the glomerular filter, so proteins do not get into the renal tubule or into the urine. Only when the filter is damaged can protein pass into the filtrate and then into the urine (see Figure 16–9 ▪).

FILTRATION RATE

Filtration rate can be controlled by changing the pressure difference across the filter. The most obvious way to control the pressure is to change the pressure of the blood in the glomerular capillaries. Higher pressure in the glomerulus will increase filtration, and lower pressure will decrease filtration. You might think that every minor change in systemic blood pressure would affect glomerular filtration rate. That would be the logical conclusion. However, the glomerulus is protected from minor changes in blood pressure by a mechanism called **autoregulation.** As systemic blood pressure increases, the afferent arterioles leading into the glomeru-

FIGURE ▦ 16-9

Comparison of damaged and healthy kidneys.

lus constrict, decreasing the amount of blood getting into the glomerulus. If less blood gets into the glomerulus, the pressure doesn't rise. Autoregulation protects the delicate filter from repeated rapid changes in blood pressure caused, for example, by walking up steps.

Autoregulation can be overridden in situations when blood pressure must be regulated. Because the kidney controls fluid volume, it is often part of mechanisms, along with the cardiovascular system, that regulate systemic blood pressure. For example, if there is a decrease in systemic blood pressure or volume, such as during severe blood loss, glomerular filtration decreases dramatically in an attempt to conserve fluid volume by producing less urine.

Remember that the flight-or-fight response includes decreased urine production. The sympathetic nervous system and the hormones of the adrenal medulla—epinephrine and norepinephrine—decrease glomerular filtration by causing dramatic vasoconstriction of the afferent arterioles. This prevents blood from flowing to the glomerulus, decreasing glomerular filtration rate. Small changes in systemic blood pressure do not affect glomerular filtration because autoregulation keeps glomerular pressure relatively constant, but during shock, for example, glomerular filtration decreases significantly. This is one of the reasons that urine output is monitored in trauma and surgery patients. As you learned earlier when we discussed renal blood vessels, one serious complication of severe blood loss is permanent kidney damage due to decreased blood flow and subsequent death of kidney tissue.

The reverse is also true: if blood volume is elevated, sympathetic output to the afferent arterioles decreases, the arterioles dilate, and glomerular filtration rate increases, allowing the kidneys to get rid of excess fluid.

GLOMERULAR FILTRATION

The millions of tiny filters in your kidneys perform a truly Herculean task. Average glomerular filtration rate is 110 milliliters per minute, or 160 liters per day. That's more than 40 gallons filtered every 24 hours!

 16-5 To see an animation of hypovolemic shock and how it affects various systems, please go to your DVD.

Pathology Connection: Nephropathy

nephropathy
(neff ROPP ah thee)
nephr/o = *kidneys*
pathy = *disease*

Nephropathy is a general term for kidney disease, particularly damage to the filtering apparatus of the nephrons. Nephropathy is generally progressive, but in the early stages some types of nephropathy can be reversed with treatment. There are many different causes of nephropathy. Symptoms include proteinuria, hematuria, and hypertension, but the onset, type, and severity of symptoms differ depending on the cause and the severity of the nephropathy. Nephropathy is diagnosed using urinalysis and blood tests. Treatments include corticosteroids, immunosuppressants, blood pressure medications, diet, and in cases that progress to end stage renal disease, kidney replacement.

Diabetes mellitus is a condition characterized by abnormally high blood glucose. This hyperglycemia, caused by production of too little insulin or by insensitivity to insulin, wreaks havoc with blood chemistry, including osmotic balance. One of the functions of your kidney is to remove all that extra glucose from your blood. When the blood glucose is high, the kidneys must work that much harder to remove it. Patients with elevated glucose produce a much greater volume of urine than do patients with healthy levels of blood glucose because their kidneys try to excrete the excess blood glucose.

Prolonged high blood glucose causes a type of kidney damage known as **diabetic nephropathy**. Forty percent of diabetic patients will develop diabetic nephropathy. It begins with a thickening of the filter surface of the glomerular capsule and eventually leads to a complete breakdown of that tissue. Once the tissue is damaged, the selectivity of the filter is destroyed. Substances that usually would not pass through the filter and into glomerular filtrate, like proteins and blood cells, begin to appear in urine. The efficiency of filtration is compromised and kidney function begins to deteriorate. Protein and blood cells in the urine are early indicators of renal failure. In the early stages, characterized by increased filtration and even small amounts of protein in the urine (microalbuminuria), diabetic nephropathy can be reversed by strict glycemic control and ACE inhibitors (even if blood pressure is normal). Diabetics can help prevent the onset of kidney damage by keeping blood sugar levels as tightly controlled as possible, preventing high blood pressure, and reducing blood cholesterol to safe levels. Diabetic nephropathy is the leading cause of kidney failure in the United States. One third of patients on dialysis have diabetes mellitus. Death from diabetic nephropathy usually occurs 5-10 years after large amounts of protein begin to show up in the urine (macroalbuminuria).

Many prescription drugs or chemicals can cause nephropathy either by direct toxicity, allergy-induced damage to the kidney, or crystallization in the kidney. Analgesic nephropathy is caused by the use of some pain relievers, such as the nonsteroidal anti-inflammatory drugs (NSAIDs) ibuprofen and naproxyn, particularly in people with renal disease or other risk factors. Even some patients without renal disease have transient decreases in kidney function while on postoperative doses of NSAIDs for pain relief. Antibiotics may also cause nephropathy, mainly due to allergic reactions. Immunosuppresants, like those used after kidney transplant, also increase the risk of drug-induced nephropathy. Contrast dye used during imaging studies causes nephropathy by causing severe vasospasm in the afferent arterioles, and is the third leading cause of kidney failure in hospitalized patients. People with diabetes mellitus, dehydration, or kidney disease, or people who are on diuretics or have congestive heart failure, are at higher risk for kidney damage caused by contrast dyes. As many as 30% of patients are left with permanent kidney damage.

Nephritis is a general term that refers to an inflammation of the kidney. Nephritis may affect the glomerulus, tubules, or interstitial tissue within the kidney. It may cause permanent kidney damage if not diagnosed and treated. Acute nephritis, which comes on suddenly, often occurs as an allergic reaction or hypersensitivity to drugs. Chronic nephritis may be caused by drug hypersensitivity, infection, high blood pressure, urinary blockage, or an autoimmune reaction from the body. Hypercalcemia and other metabolic disorders can also cause chronic nephritis.

Glomerulonephritis is a type of nephritis that results in inflammation of the glomerulus. **Glomerulosclerosis** is scarring of the glomerulus. Both cause damage to the delicate filter apparatus. When the filter is damaged, blood cells and blood proteins enter the filtrate and eventually appear in the urine. Removal of waste products is decreased and electrolyte balance is generally abnormal because of the change in urine chemistry. Glomerulosclerosis and glomerulonephritis may be primary, without any obvious cause or secondary, due to another condition which causes kidney damage. There are many secondary causes of glomerulonephritis and glomerulosclerosis, including bacterial infection, drug toxicity, diabetes mellitus, systemic lupus erythematosus, and genetic disorders such as Alport's syndrome and Goodpasture's syndrome.

glomerulonephritis
(gloh MAIR you loh neh FRY tis)

glomerulosclerosis
(gloh MAIR you loh sklee ROW sis)

CONTROL OF TUBULAR REABSORPTION AND SECRETION

Glomerular filtration controls the rate of urine formation. Tubular reabsorption and secretion control the chemistry and volume of the urine. Substances that are reabsorbed move from the tubule back to the bloodstream via the peritubular capillaries, and stay in the body. Substances that are secreted stay in the tubule and eventually leave the body via the urine. Thus, anything that affects reabsorption and secretion affects urine chemistry.

The first thing that affects tubular reabsorption and secretion is tubule permeability. Each portion of the tubule can reabsorb and secrete different substances. Remember that molecules can move across membranes via several different methods. **Diffusion** is the movement of molecules from high to low concentration. **Osmosis** is the movement of water across a semipermeable membrane. Some molecules can only move across membranes by being carried across by proteins. This type of movement can be either active or passive. Differences in tubule permeability result in dramatic differences in which molecules are reabsorbed or secreted in each part of the tubule. See Table 16–2 for a list of substances reabsorbed or secreted in each part of the tubule.

The second factor that affects tubular reabsorption and secretion is a special type of circulation around the nephron loop, called **countercurrent circulation.** When ions move across cell membranes, they move from areas of high to low concentration. They are said to move "down their concentration gradient." In a solution, if

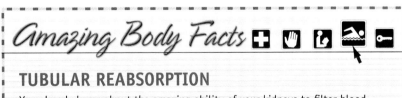

TUBULAR REABSORPTION

You already know about the amazing ability of your kidneys to filter blood. However, the ability of your renal tubule to reabsorb the fluid filtered by the glomerulus is just as incredible. Your kidneys filter 160 liters (40 gallons) per day, yet you produce only about 1500 milliliters (less than 1 gallon) of urine per day!

TABLE 16-2 Individual Tubule Functions

TUBULE	SUBSTANCES REABSORBED OR SECRETED
Proximal tubule	Potassium, chloride, sodium (80% of sodium is normally reabsorbed in the proximal tubule), magnesium, bicarbonate, phosphate, amino acids, glucose, fructose, galactose, lactate, citric acid, water, hydrogen (H⁺), neurotransmitters, bile, uric acid, drugs, toxins, ammonia, urea
Descending loop	Water (90% of the water is normally reabsorbed in the descending loop), urea
Ascending loop	Sodium, potassium, chloride, urea
Distal tubule	Sodium, potassium, chloride, hydrogen (H⁺), water
Collecting duct	Sodium, potassium, chloride, water, urea

there is a lot of solvent (water), there is less solute (dissolved substances), and vice versa, so you would expect that solute and solvent would move in opposite directions down their concentration gradients. Figure 16–10 ■ shows two solutions separated by a diffusible membrane. In solution A, the solvent water is more concentrated and therefore it will move down its concentration gradient into B. The dissolved solutes, in this case ions, are more concentrated in solution B, so they move down their concentration gradient into A. As always in the body, this is an attempt to maintain the homeostatic balance.

Because of the tendency for water and ions to move in opposite directions, as shown in Figure 16–10, the kidneys could not reabsorb both ions and water without the special selective environment around the nephron loop. One would be reabsorbed and one would be secreted if it weren't for selected areas of the nephron being impermeable to one or the other. The characteristics of the nephron that make the countercurrent circulation work include the concentration gradient in the fluid surrounding the nephron, with low ion concentration at the beginning of the descending loop and high concentration at the tip of the loop, and the differences in permeability between the descending loop (water) and ascending loop (ions).

As filtrate flows into the descending loop, *water* is reabsorbed, and the concentration of ions in the loop increases as water leaves the tubule. As the filtrate turns the corner and enters the ascending loop, much of the water has left and the fluid is extremely concentrated. The ascending loop is permeable only to ions, so *ions* are reabsorbed from the ascending loop. Water and ions that leave the renal tubule enter the capillaries and go back to the bloodstream (see Figure 16–11 ■).

The third set of factors that affect reabsorption and secretion are several hormones that regulate blood pressure. It should come as no surprise that these mechanisms affect kidney function, since the

Side A   **Side B**

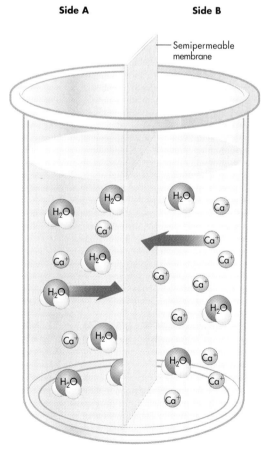

Semipermeable membrane

FIGURE ■ 16-10

Movement of solute (ions) and solvent (water) down their concentration gradient.

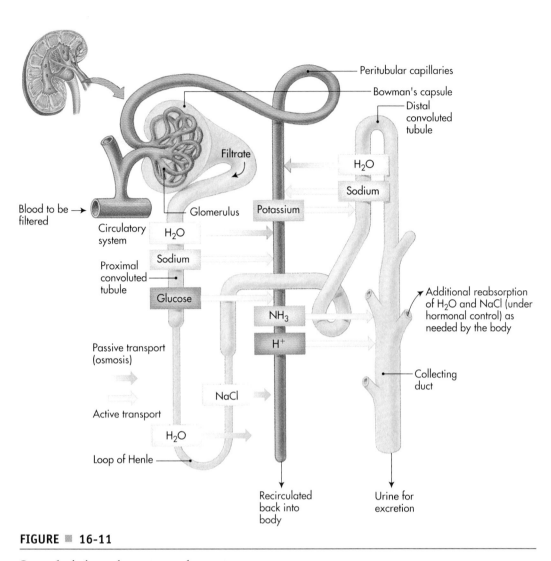

FIGURE ▓ 16-11

Sites of tubular reabsorption and secretion.

kidneys control ion and fluid balance. The hormones that affect the kidneys perform a variety of functions. You met some of these hormones during your visit to the cardiovascular and endocrine systems.

- **Antidiuretic hormone (ADH)** is made by the hypothalamus and secreted from the posterior pituitary when blood pressure decreases or blood ionic concentration increases. ADH increases the permeability of the distal tubule and the collecting duct so that more water is reabsorbed, thereby increasing the blood volume and blood pressure and diluting the ionic concentration. Less urine is produced as more water is reabsorbed, hence the name antidiuretic. Alcohol and caffeine inhibit ADH from working and thus prevent the distal tubules and collecting duct from becoming permeable to water. The water then stays in the collecting tubules and is sent to the bladder, thereby increasing urination. The more beverages containing these substances that you drink, the more dehydrated you become.

- **Aldosterone** is an adrenocorticosteroid, a steroid secreted by the adrenal cortex. Aldosterone is secreted when plasma sodium decreases or plasma potassium increases. It increases the reabsorption of sodium ions (thus bringing more back

antidiuretic hormone
(ADH) (AN tye dye yoo RET ik)

aldosterone *(al DOSS ter ohn)*

into the blood) and secretion of potassium ions (thus decreasing the plasma levels) by the distal tubule and ascending limb of the nephron loop. Because sodium is reabsorbed back into the bloodstream, so is more water, and urine volume therefore decreases under the influence of aldosterone.

atrial natriuretic peptide
(AY tree al nay tree your RET ick PEP tide)

- **Atrial natriuretic peptide (ANP)** is secreted by the atria of the heart when blood volume increases. As blood flow into the right atrium increases, the right atrium stretches and releases ANP. ANP decreases sodium reabsorption and therefore increases urination. Does this make sense? If blood volume has increased dramatically, how would you keep blood pressure constant? Would you get rid of water or keep more water?

renin-angiotensin-aldosterone
(REE nen-an gee oh TEN sen-al DOSS ter ohn)
juxtaglomerular
(JUX ta gla MARE you ler)

- The **renin-angiotensin-aldosterone (RAS) system** is a series of chemical reactions that regulate blood pressure in several different ways. When there is a decrease in blood flow to the kidneys, a special group of cells near the glomerulus called the **juxtaglomerular apparatus** secretes renin into the bloodstream. Renin converts **angiotensinogen**, a protein made by the liver, into angiotensin I. Another enzyme made by the lungs, angiotensin converting enzyme (ACE), converts angiotensin I to angiotensin II. Angiotensin II (active angiotensin) has several effects. It increases thirst, increases ADH secretion, increases aldosterone secretion, and causes vasoconstriction. Blood pressure is therefore increased either by increased fluid volume due to higher water intake or decreased urination caused by increased levels of ADH or aldosterone. Notice how the kidneys, lungs, and liver work together to regulate blood pressure. Many patients with high blood pressure may be given an ACE inhibitor, which inhibits RAS and therefore lowers blood pressure.

Applied Science

ELECTROLYTES AND ACID BASE BALANCE

The kidneys maintain electrolyte balance by selectively excreting or reabsorbing the electrolytes within the tubular system. One very important interaction is the relationship of hydrogen ion (H^+) and bicarbonate ions (HCO_3^-). The relationship between these ions determines the blood's **pH** (level of acidity or alkalinity). This is referred to as the **acid–base relationship**. If too much acid is present in the blood, H^+, which causes increased acidity (pH decrease), will be excreted to a greater level in the urine. At the same time, more bicarbonate ions (which neutralizes acids) will be reabsorbed back into the acidic blood to bring the pH value back toward normal. The respiratory system also plays a role in maintaining the acid–base relationship by increasing ventilation to "blow off" more acid in the form of exhaled carbon dioxide (carbonic acid) if the blood is too acidic.

Pathology Connection: Uremia and Renal Failure

Uremia is a blood condition that accompanies kidney disease but cannot be explained by problems with volume, ion concentration, or hormonal abnormalities. The disorder appears to be due to the accumulation of organic wastes in the blood stream (**azotemia**). The symptoms of uremia include fatigue, seizures, lack of appetite, decreased sense of smell and taste, cramps, mental confusion, metabolic disorders, insulin resistance, itching, hiccups, anemia, clotting abnormalities, and systemic inflammation. Uremia is very difficult to treat. A low-protein diet and kidney replacement therapy (dialysis or transplant) are the only treatments.

A transplant involves acquiring a donor organ from someone with a similar immune system, so finding a compatible transplant donor can take time. In the interim, kidney dialysis serves as a substitute kidney by externally performing the functions of a kidney upon the blood. Dialysis leads to partially treated uremia, prolonging the life of patients but not treating the condition fully. In

2004 more than 400,000 people were on kidney dialysis as a treatment for uremia. The 5 year survival rate in the late 1990s for patients on dialysis was only 35%. Kidney transplant is the only effective treatment. See Figure 16-12 ■ on the location of a transplanted kidney within the body.

Hemolytic uremic syndrome is a disorder caused by an infection with the bacteria *E. coli*, typically from eating undercooked meat. The bacteria infect the digestive tract and release toxins that destroy red blood cells. The damaged red blood cells lodge in the blood vessels in the kidneys, blocking them and preventing blood flow to the nephrons. Without treatment, permanent kidney damage may result.

Renal Failure (also known as **renal disease**, **kidney disease** or **kidney failure**) is a spectrum of disorders including acute renal failure, chronic renal failure and end stage renal failure. There are many risk factors for renal failure, including injury, ischemia, nephrotoxic medications, hypovolemia, hypotension, hypertension, liver disease, diabetes mellitus, lupus, infections, urinary tract obstruction, immune system malfunctions, PKD, and cardiovascular disease. Symptoms of renal failure include decreased urine output, uremia, fluid retention, anorexia, fatigue, hiccups, nausea, mental confusion, clotting disorders, and seizures. In the early stages of renal disease, symptoms are vague and mild. As kidney function worsens, symptoms increase in severity and number.

Acute renal failure is a rapid decrease in kidney function, as measured by blood urea and creatinine levels, that develops over minutes or days. Acute renal failure can be caused by hypoperfusion (decreased blood supply) to the kidney, urinary tract abnormalities, or injury to renal blood vessels or other renal tissues. Decreased blood supply to the kidney due to decreased blood pressure leads to tissue damage, which leads to decreased renal function. It is also possible for local kidney damage to occur even in the range of normal blood pressure. Acute renal failure is exacerbated by the presence of any risk factors for kidney disease, including diabetes mellitus, heart disease, or existing kidney disease. In addition, the use of NSAIDs or contrast dyes increases the likelihood that acute renal failure will develop. Acute renal failure is treated by treating the cause, replacing fluid volume, normalizing BP, decreasing nephrotoxic drugs, and treating any infections. Some patients may require kidney replacement therapy. If no

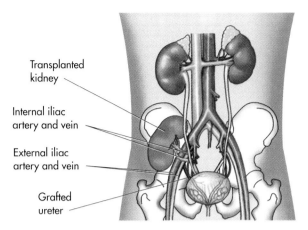

Transplanted kidney

Internal iliac artery and vein

External iliac artery and vein

Grafted ureter

FIGURE ■ 16-12

Placement of transplanted kidney.

Clinical Application

KIDNEY DISEASE AND HYPERTENSION

It should be pretty clear now that blood pressure and the kidneys are intimately connected. One of most important risk factors for kidney disease is hypertension. If blood pressure is high all the time, that pressure damages the delicate blood vessels of the nephron, leading eventually to chronic kidney disease. However, decreased blood flow to the kidneys also causes kidney damage due to ischemia. In addition, the kidneys are an integral part of blood pressure control mechanisms. The connection between kidney function and blood pressure can set up a vicious cycle that exacerbates kidney injuries (positive feedback). For example, in kidneys damaged by nephropathy, blood flow to the kidneys decreases, causing a decrease in glomerular filtration rate. To compensate, the kidneys release renin to raise blood pressure. Systemic blood pressure (BP) rises. However, because the kidneys are damaged, the increased blood pressure does not improve circulation in the kidneys. The kidneys increase blood pressure more, but kidney function does not improve. However, pressure in the renal blood vessels rises with systemic BP, causing more damage to the kidney. If the rise in BP is not stopped, the damage will progress as BP rises. That's why in early nephropathy, ACE inhibitors and other medications that lower BP may reverse or at least slow down the progression of kidney damage.

16-6 To view a video on renal failure, please go to your DVD.

structural changes occur in the kidney, patients can recover from acute renal failure in one or two days if the cause of the failure is treated, however, if the kidneys are permanently damaged, patients may develop chronic renal failure.

Chronic renal failure (chronic kidney disease) is defined as an ongoing, progressive kidney disease lasting three months or more. In the US eight million adults have moderate or severe chronic kidney disease, and that number is on the rise. In the early stages, chronic renal failure is often asymptomatic. In later stages symptoms increase in severity and number as kidney function deteriorates. The most common risk factor for chronic kidney disease is diabetes mellitus (45%), followed by hypertension (20%), cardiovascular disease, age, obesity, ethnicity (African Americans, Latinos and Native Americans are at higher risk), nephrotoxic drugs, polycystic kidney disease, lupus, and structural damage to the kidneys. Often, the progression can be controlled by treating the underlying cause of the damage or controlling blood pressure and blood cholesterol. Screening for and treating risk factors for kidney disease may delay or even prevent the development of the disease. For patients who already have kidney disease, treatment consists of treating the underlying cause, decreasing progression and preparing patients for kidney replacement. Treatments include angiotensin-converting enzyme (ACE) inhibitors, angiotensin receptor blockers, statin drugs, smoking cessation, and good blood sugar control. As the disease progresses, patients may need kidney replacement.

Eventually, chronic kidney failure may lead to **end-stage renal failure**; however, most chronic kidney disease patients do not reach end-stage disease. Instead, most patients with kidney disease die of cardiovascular disease, often in early stages of kidney disease. Thus many of the treatments for chronic kidney disease focus on decreasing cardiovascular risk factors that accompany loss of kidney function. End-stage renal failure is the final stage of renal failure. Nearly half a million patients in the US have end-stage renal failure. Patients in end-stage renal failure can be treated only with kidney dialysis or transplantation.

TEST YOUR KNOWLEDGE 16-2

Choose the best answer:

1. Glomerular filtrate is most similar in composition to:
   a. blood
   b. lymph
   c. urine
   d. none of the above

2. When substances move from the tubule into the bloodstream, this is known as:
   a. secretion
   b. reabsorption
   c. filtration
   d. all of the above

3. Which of the following is usually completely reabsorbed in the proximal tubule?
   a. sodium
   b. urea
   c. glucose
   d. water

4. The descending nephron loop is permeable to _____, while the ascending is permeable to _____.
   a. water, ions
   b. ions, water
   c. water, water
   d. ions, ions

5. If blood pressure increases beyond normal range, what happens to glomerular filtration rate?
   a. it increases
   b. it decreases
   c. it stays the same
   d. none of the above

6. Which of the following is not typically found in urine?
   a. glucose
   b. protein
   c. blood cells
   d. all of the above

7. Renin is released by the kidney when:
   a. blood glucose increases
   b. blood flow to the kidney decreases
   c. blood pressure rises
   d. Angiotensin is in short supply

8. Which of the following is a symptom of renal disease?
   a. high blood pressure
   b. proteinuria
   c. high serum creatinine
   d. all may be symptoms

9. Which of the following is the most common cause of nephropathy and chronic kidney failure?
   a. urinary tract infections
   b. systemic lupus erythematosis
   c. diabetes mellitus
   d. kidney dialysis

Pathology Connection: Kidney Replacement Therapy

For patients in end-stage renal failure, the only treatment is kidney replacement therapy. Kidney replacement therapy consists of either kidney dialysis, in which the patient's blood is filtered by an artificial or biological filter, or kidney transplant, the surgical replacement of the damaged kidney with a donor kidney.

There are two types of kidney dialysis, **hemodialysis** and **peritoneal dialysis**. During hemodialysis, blood is removed from the patient via a needle and IV line and circulated through a dialysis machine. In the machine, blood is passed through a semipermeable membrane that removes fluid, electrolytes, and wastes. The blood is then returned to the patient through another IV line. Hemodialysis must be performed about 2-3 times a week, usually at a dialysis center. Some patients may also do home dialysis. It usually takes about 2-4 hours each time.

During peritoneal dialysis, dialysis fluid is pumped into the peritoneal cavity. Waste from body fluids is filtered across the peritoneum and into the dialysis fluid. After several hours, the dialysis fluid is removed. The peritoneum acts as a biological filter. Peritoneal dialysis is more convenient for patients because it can be done while patients are sleeping and requires less equipment when travelling.

 16-7 To view a video on basic peritoneal dialysis, please go to your DVD.

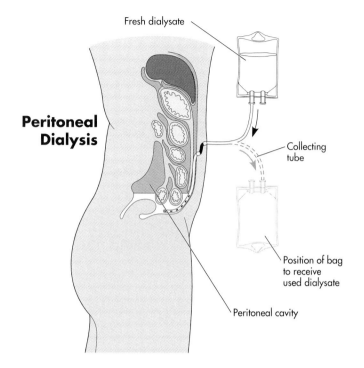

Peritoneal Dialysis

Fresh dialysate

Collecting tube

Position of bag to receive used dialysate

Peritoneal cavity

Hemodialysis

From heparin source

Heparin infusion pump

Blood pump

Arterial blood line (to apparatus)

Dialyzing (bathing) solution

Cellophane membrane (tubing containing blood)

Bubble trap

Venous blood line (from apparatus)

Compressed air

Fresh dialyzing solution

Constant-temperature bath

Used dialyzing solution

FIGURE ▪ 16-13

Peritoneal dialysis and hemodialysis.

Kidney dialysis is only a partial treatment for patients in end-stage renal failure. Dialysis does not work as well as functional kidneys and does a particularly bad job at treating uremia. In addition, there are many side effects of dialysis including low BP, nausea, cramps, headaches, fatigue, infection, fluid overload, and clotting abnormalities. Outpatient dialysis is also inconvenient because patients must be near the dialysis center. See Figure 16-13 ▪ for illustrations of peritoneal and hemodialysis.

Kidney transplant, the replacement of a damaged kidney with a donor kidney, is the only real cure for end-stage renal disease. There are three types of transplants: cadaver transplants, living related-donor transplants, and living unrelated-donor transplants. Kidney transplants are the second most common organ transplant in the U.S. (with corneal transplants being the most common), with 9000 kidneys transplanted each year. Kidney transplants are also quite successful. One year post-transplant, 90% of patients still survive; at 3 years the figure is 80%. Even after 10 years, 50% of kidney transplant recipients still survive. There are several possible complications of kidney transplant, including the typical complications from surgery and the possibility of organ rejection. Kidney transplant recipients must take immunosuppressant drugs after transplant, usually for the rest of their lives. Ironically, some of the best anti-rejection drugs are also nephrotoxic. There are never enough kidneys to go around. Every year thousands of patients die waiting for donor kidneys.

The Urinary Bladder and Urination Reflex

Glomerular filtrate flows out the collecting ducts into the minor calyces and then into the major calyces that form the renal pelvis. Once the glomerular filtrate leaves the collecting ducts, its concentration cannot be changed. At this point, filtrate is urine. Urine collects in the renal pelvis and flows down the ureters to the urinary bladder where it is stored.

The urinary bladder is a small, hollow organ posterior to the pubic symphysis and behind the peritoneum. It is lined with transitional epithelium, the only epithelium stretchy enough to expand as the bladder fills. The stretchiness of the bladder is enhanced by a series of pleats called **rugae** similar to the rugae that allow for stomach expansion. The bladder has a muscular wall consisting of several layers of circular and longitudinal smooth muscle and is covered by connective tissue and parietal peritoneum (see Figure 16–14 ■).

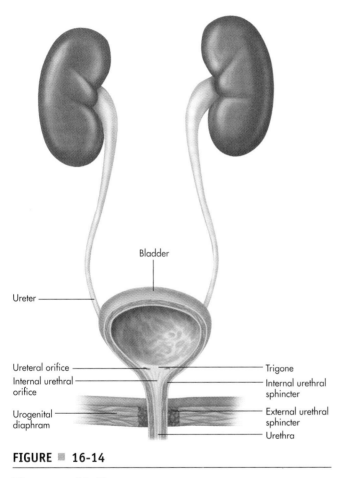

Bladder

Ureter

Ureteral orifice

Internal urethral orifice

Urogenital diaphram

Trigone

Internal urethral sphincter

External urethral sphincter

Urethra

FIGURE ■ 16-14

The urinary bladder.

As urine accumulates, the bladder fills and stretches and can hold approximately 500 ml of fluid. At some point, this stretch triggers urination, or voiding—the emptying of the bladder. For years, urination was thought to be a spinal cord reflex, influenced but not controlled by the brain. New research, however, suggests that the brain actually controls urination. When the bladder is full, signals are sent from the bladder to the spinal cord and then to the pons. The pons sends parasympathetic signals down the spinal cord that cause contraction of the muscular walls of the bladder, and the bladder empties. Emptying urine is involuntary but controlled through the nervous system. There is direct control through nerve impulses on kidney blood vessels and indirect control through stimulation of endocrine glands.

 16-8 For an interactive drag-and-drop labeling exercise of the urinary bladder, please see your DVD.

Urine leaves the bladder via the **urethra**, a thin muscular tube lined with several different types of epithelium along its length. The urine travels from the bladder through the urethra to the final opening called the **urethral meatus**. (For details on the anatomy of the urethra, including sexual differences, see Chapter 17.) Parts of the brain can inhibit urination by controlling the **internal urethral sphincter,** a valve at the junction of the bladder and the urethra, and the **external urethral sphincter,** a valve that is part of the muscles of the pelvic floor. Sympathetic stimulation of these sphincters prevents urine from leaving the body (see Figure 16–15 ■). Fortunately, although you have little control over contractions of the bladder wall, you have very good control over the sphincters.

internal urethral sphincter
(IN ter nal yoo REE thral SFINK ter)

external urethral sphincter
(EKS ter nal yoo REE thral SFINK ter)

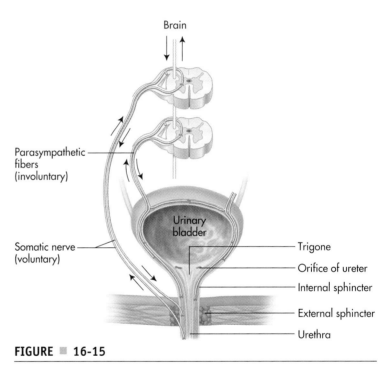

FIGURE ■ 16-15

Control of urination.

Pathology Connection: Incontinence and Urinary Tract Infection (UTI)

Incontinence is the inability to control urination. Often incontinence is accompanied by urinary urgency, urinary frequency and nighttime urination. There are many causes of incontinence including overactive bladder, urinary tract infection, and benign prostatic hypertrophy (BPH). (See Chapter 17 for information about the closely related reproduction system.)

Urinary tract infection (UTI) or cystitis is caused by the movement of fecal bacteria into the urinary tract, infecting the urethra or urinary bladder. Symptoms include frequent, painful urination often with very small volume, nighttime urination, urine cloudiness, foul smell and hematuria. Women are 50 times more likely than men to develop urinary tract infections probably because women's urethras are shorter than men's. Also at risk are elderly hospitalized patients, patients requiring catheters, pregnant women, post-menopausal women, patients with structural abnormalities of the urinary tract, and men with BPH. UTIs are diagnosed via urinalysis (hematuria, bacteria in urine), imaging and bacterial culture of the urine. Urinary tract infections must be treated promptly because they can cause damage to the kidneys if the infection moves from the bladder up the ureters and into the kidneys. Untreated UTIs may develop into pyelonephritis, an infection of the kidneys. Symptoms include pain, fever, chills, nausea, fatigue, mental confusion, and UTI symptoms. In the elderly, mental confusion may be the *only* symptom of kidney infection. Treatment for UTI includes antibiotics and increased fluids.

Many patients with UTI develop more chronic conditions, including recurrent UTI or **interstitial cystitis**. Interstitial cystitis is a disorder with bladder, pelvic, vaginal or abdominal pain, hematuria, and chronic UTI symptoms including urinary frequency, urgency and nighttime urination. These symptoms may overlap with endometriosis. Some patients with interstitial cystitis have a history of UTIs. Treatment for chronic UTI or interstitial cystitis includes preventative antibiotics and adequate hydration. An experimental treatment, the addition of hyaluronic acid into the bladder as a protective barrier, has shown some promise for preventing chronic UTIs and treating interstitial cystitis. The home remedy of consumption of cranberry juice or cranberry extract for treatment of UTI has shown little promise. In studies, the doses necessary to prevent UTI caused unpleasant side effects, including acid reflux and nausea.

Overactive bladder is a common (33 million people in the U.S.) condition with symptoms of urinary urgency, nighttime urination, and urinary frequency with or without incontinence, but without pain or obvious urinary pathology such as hematuria or UTI. Though overactive bladder does not involve infection or progressive degenerative illness as do other renal disorders, patients with untreated overactive bladder experience significant decrease in quality of life, increased stress from dependence on restroom availability, and fatigue from interrupted sleep. Overactive bladder is more common in women than men and much more common in people after age 40. The cause of overactive bladder is unknown, but it is thought to be caused either by abnormalities of the bladder muscles, the nerves innervating the bladder or the stretch receptors in the bladder wall. The cause is probably different in different patients. Diagnosis of overactive bladder is made mainly by ruling out other disorders using medical history, physical exam and urinalysis. Treatment is difficult. The most common and effective non-pharmaceutical treatment is bladder training, designed to decrease the sensitivity of the bladder to fullness and to increase bladder control. The most effective drug treatments are anticholinergic drugs and alpha adrenergic agonists, which block parasympathetic stimulation or increase sympathetic activity. These drugs have side effects including dry mouth and anxiety. These drugs may be dangerous for patients with heart disease or high blood pressure. The combination of training and drugs is more effective than either treatment alone, but nearly 50% of patients are not cured. Many patients make lifestyle changes such as drinking fewer fluids and avoiding situations in which frequent trips to the restroom could be problematic. Medications including antidepressants, nerve blockers and botulinum toxin are also used with varying degrees of effectiveness.

There are a variety of other conditions associated with urination. Some of these conditions are related to production of urine or frequency of urination.

Nocturia is frequent urination at night. It can be caused by a variety of conditions, including diabetes mellitus, urinary tract infection, or congestive heart failure. Nocturia can also result from drinking too much fluid before bedtime. **Polyuria** is the production of large amounts of urine. Polyuria is often a symptom of untreated diabetes mellitus. **Enuresis** is a condition in which the person, usually a child, wets the bed during sleep. Enuresis can be caused by poor muscle control, neurological disease, sleep disorders, or disorders of the urinary tract. **Anuria** is a condition in which *no* urine is produced. Anuria may be caused by an obstruction, such as a kidney stone or tumor, in the ureter or urethra. **Oliguria** is diminished production of urine and can be an early sign of acute renal failure.

Sometimes, conditions related to urination can help a physician to diagnose disease. **Glycosuria** is the secretion of sugar in urine. It is usually a sign of untreated diabetes mellitus. **Pyuria** is the secretion of pus in urine. It can be a sign of a urinary tract infection. **Dysuria** is painful urination. It is also often related to bacterial infection in the urinary tract.

A Quick Trip Through Disorders Of The Urinary System

DISORDER	ETIOLOGY	SIGNS AND SYMPTOMS	DIAGNOSTIC TEST(S)	TREATMENTS
Kidney stones	Increased urine concentration, infection, kidney disease. Stones are composed of calcium and uric acid crystals. They get larger and can block ureters.	Severe pain, hematuria, fever, chills, nausea, urinary urgency.	Symptoms, imaging, urinalysis, patient history.	Pain relief, extracorporeal shock wave lithotripsy, ureteroscopy, percutaneous nephrolithotomy.
Polycystic kidney disease	Genetic disorder causing large cysts to form in kidney.	Enlarged, cystic kidneys, hypertension, UTI, dilute urine, liver cysts, pain, hematuria, aneurysm.	Imaging, genetic tests.	Medication, kidney replacement, no cure.
Ischemic nephropathy	Decreased blood flow to kidneys.	Kidney failure, uremia, hypertension or hypotension, decreased urine output, increased serum creatinine and urea.	Urinalysis, blood tests.	Treat underlying cause and symptoms, kidney replacement if needed.
Diabetic nephropathy	Diabetes mellitus.	Early stages: increased glomerular filtration, protein in urine. Later: uremia, hypertension, pain, and loss of sensation in the extremities.	Blood tests, urinalysis.	Tight glycemic control, blood pressure medications, lipid control, diet, kidney replacement.

A Quick Trip Through Disorders Of The Urinary System (*continued*)

DISORDER	ETIOLOGY	SIGNS AND SYMPTOMS	DIAGNOSTIC TEST(S)	TREATMENTS
Drug induced nephropathy	Drugs toxic to kidney. tissue, especially contrast dye and NSAIDs.	Same as for ischemic nephrothpathy.	Same as for ischemic nephropathy.	Stop drugs, do not use contrast dyes for patients with known risk factors, keep patients well hydrated when using contrast dyes.
Glomerulonephritis and glomerulosclerosis	Inflammation and scarring of the glomerulus.	Same as for ischemic nephrothpathy.	Same as for ischemic nephropathy.	Depends on cause. Treating underlying cause may decrease progression.
Uremia	Build up of organic waste products in blood due to renal insufficiency.	Fatigue, neuropathy, seizures, lack of appetite, decreased smell and taste, mental confusion, insulin resistance, itching, inflammation, clotting.	Blood tests.	Difficult to treat. Dialysis only partial treatment. Transplant only cure.
Diabetes insipidus	ADH deficiency.	Copious, dilute urine.	Rule out diabetes mellitus, ADH levels.	Tumor removal, pre-and post-operative medication
Renal failure (kidney disease)	Either acute or progressive damage to the kidneys from many different causes.	Decreased urine output, uremia, fluid retention, loss of appetite, fatigue, hiccups, nausea, mental confusion, clotting disorders, seizures.	Blood test, urinalysis, imaging.	Blood pressure medication, glucose control, diet, treatment of underlying condition, prevention of cardiovascular disease, kidney replacement when progresses to end stage disease.
Urinary tract infection (UTI)	Bacterial infection of urinary tract, usually bladder.	Painful urination, urinary frequency, night time urination, foul smell, cloudy urine.	Urinalysis.	Antibiotics, prevention of recurrence is difficult.
Overactive bladder	Unknown, may be abnormalities in nervous or muscle control of bladder.	Frequent urination without pain or infection.	Rule out UTI.	Bladder training, sympathetic drugs.
Hemolytic uremic syndrome	Bacterial infection with certain strains of *E. coli*, toxins damage kidneys.	Fever, abdominal pain, pallor, fatigue, bruising, decreased urination, swelling.	Blood tests, history.	Blood transfusion, kidney dialysis.

℞ Pharmacology Corner

High blood pressure is a common condition that can affect the kidneys. In addition, kidney disease can also lead to high blood pressure and therefore treatment of high blood pressure is critical. One way of treating high blood pressure is to decrease blood volume by decreasing body water. This can be accomplished via the use of diuretics, which will increase urine output. Lasix is an older drug that still is used and works by not allowing sodium and thus water to be reabsorbed; therefore more is excreted. Another way of controlling blood pressure is through the use of angiotensin-converting enzyme (ACE) inhibitors. ACE inhibitors decrease blood levels of angiotensin II, which leads to systemic vasodilation. The relaxation of the vessel walls will decrease overall blood pressure.

In cases of severe hypotension (cardiac and septic shock), renal perfusion may be severely decreased which can lead to renal damage. In these cases, dopamine can be administered to increase blood flow to the renal vessels. Dopamine is very dosage-dependent and must be given in low dosages to achieve this affect.

Kidney transplants are becoming more common and successful due to technological advancements. One reason for their long-term success is the use of immunosuppressant drugs that reduce the chance of organ rejection.

Since the kidneys act as a filter for all blood, drugs found in the blood stream may lead to nephrotoxicity. A variety of prescription drugs can cause damage by direct toxicity, allergy-induced damage to the kidney or crystallization in the kidney. The overuse of pain medications, like the nonsteroidal anti-inflammatory drugs (NSAIDs) ibuprofen and naproxyn, can also lead to renal damage. Antibiotics may also cause nephropathy, mainly due to allergic reactions.

UTIs are very common infections that can be treated with antibiotics and increased fluids. Two common antibiotics used are Bactrim and Cipro. The most effective drug treatments for incontinence are anticholinergic drugs which block the parasympathetic stimulation to empty your bladder. These drugs have side effects including dry mouth and anxiety and may be dangerous for patients with heart disease or high blood pressure. Therefore, bladder training may be preferred when possible.

SUMMARY

Snapshots from the Journey

→ The urinary system consists of paired kidneys and paired ureters, which carry urine to the single urinary bladder. The urethra transports urine from the bladder to outside the body. The function of the urinary system is control of fluid and electrolyte balance and elimination of nitrogen-containing waste.

→ A kidney is bean-shaped and covered in a capsule. It has an indentation known as the renal hilum and an interior cavity known as the renal sinus. The kidney can be divided into three layers: renal cortex, renal medulla, and renal pelvis. The renal pelvis is a funnel that is divided into large pipes, the major calyces. Each major calyx is divided into several minor calyces. The renal pelvis empties into the ureter.

→ The kidney is very well vascularized. Blood is supplied to each kidney by a renal artery. The blood vessels split into smaller and smaller branches until there are millions of tiny arterioles, the afferent arterioles. The afferent arterioles supply millions of nephrons, the functional unit of the kidney, with blood. Blood leaves the kidney by a series of veins and ultimately returns to circulation via the renal vein.

→ The nephron is the functional unit of the kidney. There are millions of nephrons in each kidney. The nephron is divided into two parts. The renal corpuscle, consisting of the glomerulus (capillaries) and the glomerular capsule, filters blood and produces glomerular filtrate. The renal tubule, consisting of the proximal tubule, nephron loop, distal tubule, and collecting ducts, controls the concentration and volume of urine by reabsorbing and secreting water, electrolytes, and other molecules. The walls of the nephron are made of epithelium. The type of epithelium changes depending on the specific function of each part of the nephron.

→ Urine is formed by a combination of three processes: glomerular filtration, tubular reabsorption, and tubular secretion. The selectivity of the glomerular filter is determined by the size of the openings in the filter and the difference between the blood pressure of the glomerulus and the pressure in the glomerular capsule. The size of the filter does not change unless the glomerulus is damaged. Protein, for example, cannot pass through the filter. However, the filtration rate will change if the pressure in the glomerulus changes. Most of the time, autoregulation, control of the diameter of the afferent arteriole, keeps glomerular pressure and glomerular filtration rate constant. However, sympathetic stimulation can regulate (in this case decrease) glomerular filtration and urine output due to constriction of afferent arterioles.

→ Tubular reabsorption and secretion is controlled by differences in tubular permeability. The proximal tubule is the most versatile, reabsorbing dozens of different molecules. The nephron loop is part of an elaborate countercurrent mechanism, with the descending loop permeable to water and ascending loop permeable to ions. The distal tubule and collecting ducts reabsorb water. The permeability of the renal tubule can be regulated by a number of hormones that control blood pressure. These hormones, aldosterone, ADH, atrial natriuretic peptide, and others, regulate blood pressure by regulating urine volume and ion secretion. Changes in urine volume change total body fluid volume and thereby change blood pressure.

→ The urinary bladder is a collecting and storage device for urine and is located in the pelvic cavity. It has a muscular wall. Contractions of the muscle result in voiding (urination), emptying the bladder. Urination is a reflex controlled by parasympathetic neurons in the pons. Signals from a full bladder reach the pons. The neurons in the pons then send signals for the bladder to contract. Sympathetic neurons control two valves, the internal and external urethral sphincters, which allow significant conscious control of the urination reflex.

→ The symptoms of renal disease are common no matter what the cause of the kidney dysfunction. They include blood in the urine, decreased removal of wastes from the blood (increased BUN and serum creatinine), and hypertension. Treatment of kidney disease usually involves treating the underlying cause of the disorder, by decreasing blood pressure, tightening blood sugar control, changing life style and decreasing cardiovascular disease risk factors. Most people with kidney disease do not die from uremia caused by end stage renal failure, but from cardiovascular disease caused in part by kidney dysfunction. The only cure for kidney disease is kidney transplant but many patients can be helped by kidney dialysis.

MARIA'S STORY

 During a recent trip to the doctor's office, Maria had a standard urinalysis. The results worry her doctor. In her urine there was a tiny amount of protein. The doctor orders more extensive tests.

a. Why is Maria's doctor concerned? It's only a little protein.

_____

_____

b. What other tests should the doctor order?

_____

_____

c. When the tests come back, the doctor tells Maria she needs to go on an ACE inhibitor even though her blood pressure is normal. What is the diagnosis?

_____

_____

RAY'S STORY

 Ray, paralyzed from the neck down after a swimming pool accident, has been doing pretty well except for one thing. He keeps having urinary tract infections.

a. Why is Ray so susceptible to urinary tract infections?

_____

_____

REVIEW QUESTIONS

Multiple Choice

1. The function of this part of the renal tubule is filtration of blood:
   a. renal calyx
   b. renal corpuscle
   c. renal cortex
   d. renal colums

2. The collecting ducts are found in this part of the kidney:
   a. renal cortex
   b. renal capsule
   c. renal pelvis
   d. renal pyramids

3. This tube leads from the urinary bladder to the outside:
   a. collecting ducts
   b. distal tubule
   c. ureter
   d. none of the above

4. The ion responsible for causing acidic blood is:
   a. Na^+
   b. H^+
   c. K^+
   d. HCO_3^-

5. The renal hormone secreted by the hypothalamus when blood pressure decreases to promote the reabsorption of water is:
   a. aldosterone
   b. atrial natriuretic peptide
   c. antidiuretic hormone
   d. epinephrine

6. The most common genetic disorder of the kidney is:
   a. diabetes mellitus
   b. nephropathy
   c. polycystic kidney disease
   d. overactive bladder

7. BUN is:
   a. blood glucose
   b. nitrogen wastes in blood
   c. blood pressure
   d. where you put a hamburger

8. This disorder appears to be caused by the buildup of organic waste in blood:
   a. nephropathy
   b. diabetes mellitus
   c. uremia
   d. hypertension

9. Why are UTIs more common in women than men?
   a. men have better hygiene
   b. women go through menopause
   c. women have shorter urethras than men
   d. autoimmune disorders are often more common in women

10. The only cure for kidney disease is:
    a. prevention
    b. dialysis
    c. transplant
    d. there is no cure

Fill in the Blank

1. Most substances are reabsorbed or secreted in this part of the renal tubule: _____.

2. This part of the renal tubule has an elaborate countercurrent mechanism for reabsorption of sodium and water: _____.

3. This hormone is released by the heart when fluid volume increases: _____.

4. Blood in the urine is called _____.

5. List two substances that should not be found in urine: _____ and _____.

Short Answer

1. List and explain the activity of three regulators of kidney function.

_____

_____

_____

_____

2. Explain the three processes that are necessary for urine formation. In which part of the nephron are these functions performed?

_____

_____

_____

3. Describe the structure of the wall of the urinary bladder.

_____

_____

_____

4. Explain the control of urination reflex.

_____

_____

_____

5. List the symptoms of kidney failure.

6. Explain the relationship between blood pressure and kidney disease.

7. Explain the difference between overactive bladder and UTI.

Suggested Activities

1. Select a partner and trace the flow of a particular substance from blood through the kidney. What happens to hydrogen ions or vitamin C? Use the Internet to help you find the answers.

2. Quiz yourself. Do you know what happens in each part of the nephron? Do you know the difference between plasma, glomerular filtrate, and urine chemistry?

3. There are many different disorders that affect the kidneys. Use the Internet to research one of these diseases. How does it cause the changes in kidney function?

16-9 Now that you have completed your journey through this chapter, please go to the DVD for interactive games and puzzles concerning the medical terms and concepts contained in this chapter. By playing the games you will reinforce your learning of the urinary system in a fun way.

THE REPRODUCTIVE *System*

Replacement and Repair

In order for a city to run smoothly, it must have an infrastructure, buildings, roads, playgrounds, schools, and more. All of these structures must be repaired on a regular basis, or if damaged beyond repair, they must be replaced. As a city grows along with its population, it needs the resources and means to expand. The same is true of your body. Cells and tissues get damaged or simply wear out. Damaged or worn-out cells and tissues must be repaired or replaced. Asexual reproduction, or mitosis, the process by which cells make exact copies of themselves, is absolutely necessary to maintain a healthy body. Mitosis was discussed in Chapter 3. Ultimately, cellular reproduction leads to the complicated process by which humans produce new humans: sexual reproduction. Without this ability, the human species would die out and the journey would end for the human race. Thankfully, all the splendid diversity of the human race is passed on for generations to come, who will also get to enjoy this amazing journey called "life."

Chapter 17

LEARNING OBJECTIVES

Upon completion of your journey through this chapter, you will be able to:

→ Differentiate mitosis from meiosis

→ Locate and describe the male and female reproductive organs

→ Describe the function of the male and female reproductive organs

→ Discuss the phases of the menstrual cycle

→ Discuss the effects of hormonal control on the male and female reproductive systems

→ Describe the stages of labor and delivery

→ Explain the common disorders of the male and female reproductive systems

MULTIMEDIA APPLICATIONS

DVD Interactive Exercises

→ Animation on cellular division, 17-1

→ Videos on the human genome project 17-2

→ Animation showing fertilization, 17-3

→ 3-D animation of the female reproductive system with a drag-and-drop labeling exercise, 17-4

→ Animation on oogenesis with drag-and-drop labeling exercise, 17-5

→ 3-D animation of the male reproductive system with a drag-and-drop labeling exercise, 17-6

→ Animation on spermatogenesis with drag-and-drop labeling exercise, 17-7

→ Video on erectile dysfunction, 17-8

→ Videos on childbirth, fetal lie, labor, infant delivery, the placenta, and postpartum assessment, 17-9

→ Video on breast cancer, 17-10

→ Interactive games and puzzles, 17-11

www.prenhall.com/colbert

→ Professional Profiles:
  • Doula or Midwife

→ Related Web Sites

→ Additional Review Questions

Pronunciation Guide

Correct pronunciation is important in any journey so that you and others are completely understood. Here is a "see and say" Pronunciation Guide for the more difficult terms to pronounce in this chapter.

bulbourethral gland
   (BUHL boh yoo REE thral)
clitoris (KLIT oh riss)
corpus luteum (KOR pus LOU tee um)
cytokinesis (SIGH toe kih NEE suss)
endometriosis (EHN doh MEE tree oh sis)
endometrium (EHN doh MEE tree um)
epididymis (ep ih DID ih miss)
eukaryotic cell (you care ee AH tic)
fimbria (FIM bree ah)
follicle-stimulating hormone (FALL ih
   kle stim you LAY ting HOR mohn)
gametes (ga MEETS)
genitalia (jen ih TALE ya)
gonadotropin-releasing hormone
   (GO nad oh TROH pin)
human chorionic gonadotropin
   (HUE man core ee AH nik
   GO nad oh TROH pin)

labia majora (LAY bee ah mah JOR ah)
labia minora (LAY bee ah mih NOR ah)
luteal phase (LOO tee al faze)
luteinizing hormone
   (LOO tee en IZE ing)
meiosis (my OH sis)
menses (MEN seez)
menstruation (MEN stroo AY shun)
myometrium (MY oh MEE tree um)
oocyte (OH oh site)
oogenesis (OH oh jenn eh siss)
perimetrium (pair ee MEE tree um)
primordial follicles
   (pry MORE dee all FALL ih kulz)
progesterone (proh JESS ter ohn)
pudendal cleft (PEW den dall cleft)
seminal vesicles
   (SEM ih nal VESS ih kulz)

seminiferous tubules
   (SEM ih NIF er uss TOO byoolz)
sertoli cells (sir TOW lee sells)
spermatocytes (sper MAT oh sites)
spermatids (SPER mat ids)
spermatogonia
   (sper MAT oh GO nee ah)
spermatozoa (sper MAT oh ZOE ah)
testis, testes (TESS tiss, TESS teez)
testicles (TESS tih kulz)
testosterone (tess TOSS ter ohn)
urethra (yoo REE thrah)
uterus (YOO ter uss)
vagina (vah JYE nah)
vas deferens (VAS DEFF er enz)
vulva (VULL vah)
zygote (ZIGH goht)

TISSUE GROWTH AND REPLACEMENT

Although mitosis was discussed in Chapter 3, we will briefly review that form of cellular reproduction in this chapter. Cellular reproduction is the basis for all more complex reproduction, so this is a good starting point.

Mitosis

Cellular reproduction is the process of making a new cell. It is also known as **cell division,** because one cell divides into two cells when it reproduces. Cells can only come from other cells. When cells make *identical* copies of themselves *without the involvement of another cell*, the process is called **asexual reproduction.** Most cells can reproduce themselves asexually, whether they are animal cells, plant cells, or bacteria.

The cells that make up the human body are a type of cell known as an **eukaryotic cell.** Eukaryotic cells have a nucleus, cellular organelles, and usually several chromosomes in the nucleus. (Reminder: the genetic material of the cell, DNA, is bundled into "packages" of chromatin known as chromosomes.)

a = *without*

eukaryotic cell
(you care ee AH tic sell)

Since chromosomes carry all the instructions for the cells, all cells must have a complete set after reproduction. These instructions include how the cell is to function within the body and blueprints for reproduction. No matter whether a cell has one chromosome, like bacteria (not eukaryotic), or 46 chromosomes, like humans, all the chromosomes must be copied before the cell can divide.

Now let's go to the more complex form of cellular reproduction. Eukaryotic cells, like yours, must go through a more complicated set of maneuvers in order to reproduce. Not only do your cells have to duplicate all 46 of their chromosomes, but they have to make sure that each cell gets all of the chromosomes and all of the right organelles. The process of sorting the chromosomes so that each new cell gets the right number of copies of all of the genetic material is called **mitosis**. Mitosis is the only way that eukaryotic cells can reproduce asexually. For a quick review of mitosis, go back to Chapter 3.

mitosis *(MY toe sis)*

MITOSIS IN THE BODY

Mitosis, asexual cellular reproduction, serves many purposes in your body. Any time your cells must be replaced, mitosis is the method used to replace them.

Many of your tissues, such as bone, epithelium, skin, and blood cells, are replaced on a regular basis. Repair and regeneration of damaged tissue is accomplished by mitosis as well. If you cut your hand, the skin is replaced, first by collagen, but eventually by the original tissue. Mitosis increases in cells near the injury so that the damaged or destroyed cells can be replaced. A broken bone is replaced in much the same way.

 17-1 Go to your DVD to view an animation on cellular division.

Growth is also accomplished by mitosis. Lengthening of bones as you grow, increases in muscle mass due to exercise, indeed, most ways that tissue gets bigger, are due to mitosis of cells in the tissues or organs. Without mitosis, your body would not be able to grow or replace old or damaged cells.

SEXUAL REPRODUCTION

Thus far we have discussed cell division for growth and repair, but we must also be able to perpetuate the species. This requires sexual reproduction.

Reduction Division: Meiosis

Many animals reproduce sexually; that is, they produce, with the aid of another individual, offspring that are not exactly identical to themselves. Sexual reproduction involves the union of a cell from one organism with a cell from another organism of the same species. In animals, females produce **eggs** (ova) and males produce **sperm** for the purpose of reproducing sexually. These special cells are known as **gametes.**

Gametes are produced by a specialized type of cell division known as **meiosis,** or **reduction division.** Meiosis is called reduction division because the *daughter* cells produced at the end of meiosis have half as many chromosomes as the original *mother* cell. (Note: the cells are called mother cells and daugh-

gametes *(ga MEETS)*
meiosis *(my OH sis)*

Learning Hint

MITOSIS VERSUS MEIOSIS

These words sound and look alike and are therefore often confused. Remember, meiosis produces gametes or sexual cells, which contain half of the chromosomes because the sexual union of male and female will contribute the other half. Mitosis (I reproduce myself) is asexual and produces exact copies of the cell and the full complement of chromosomes because no union is needed.

Clinical Application

DOWN SYNDROME

Down syndrome is a relatively common birth defect that causes short stature, heart defects, and mental retardation. Down syndrome is caused by the presence of an extra chromosome 21 in a person's cells. Sometime during meiosis, usually in the mother, chromosomes fail to separate, leaving some daughter cells without a chromosome 21 and others with two copies of chromosome 21. If the egg with the extra chromosome is fertilized, the resulting fetus will have three copies of chromosome 21 instead of two.

The possibility of having a baby with Down syndrome increases dramatically in women over 35 years old. Down Syndrome is one of very few disorders resulting from abnormal chromosome numbers, probably because most fertilized eggs with abnormal numbers of chromosomes do not develop normally enough to survive.

Historically babies with Down Syndrome were assumed to be severely intellectually disabled and were often institutionalized. More recently it has been recognized that Down Syndrome patients have a wide variety of intellectual abilities. With early intervention and accommodation to meet their needs many people with Down Syndrome can even live independently as adults.

ter cells even when we are talking about the process in males.) These daughter cells must have half as many chromosomes because they will fuse together during sexual reproduction. In humans, the total number of chromosomes in a cell is 46. If the gametes did not lose half of their chromosomes somewhere along the way, then the cell that resulted from sexual reproduction would have twice as many chromosomes (92 chromosomes) as a normal human cell. It is absolutely necessary to control the number of chromosomes in a cell. Cells with too few or too many chromosomes often die.

The fact that you were produced by the fusion of an egg from your mother and a sperm from your father means that your 46 chromosomes can be thought of as being 23 *pairs* of chromosomes. Each pair of chromosomes consists of one chromosome from your father and one from your mother. They can be thought of as pairs because they can be matched based on size, shape, and the genes they carry. For example, you get a chromosome 1 from your father and a chromosome 1 from your mother, the same for chromosome 2, 3, 4, and up to 22.

The 23rd "pair" of chromosomes is a set of sex chromosomes. These are called sex chromosomes because their identity determines the sex of a baby. XX is female, and XY is male (see Figure 17–1 ■). The female parent always contributes an X, but the male parent can contribute either an X or Y chromosome. The male actually determines the sex of the baby by either contributing an X chromosome for a girl or a Y chromosome for a boy.

17-2 To view a video on the human genome project, please see your DVD.

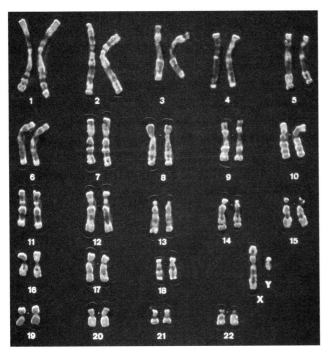

FIGURE ■ 17-1

Photo of human chromosome profile: Both an X and a Y chromosome appear in the image; this is therefore a male sample. (Source: CNRI/SPL, Photo Researchers, Inc.)

THE HUMAN LIFE CYCLE

Both mitosis and meiosis are absolutely necessary parts of human life. Without them, cells could not be replaced, injuries could not be repaired, and new humans could not be produced. The relationship between mitosis and meiosis and their importance can be easily explained by looking at human life as a cycle (see Figure 17–2 ■). Eggs and sperm, with only half as many chromosomes as other cells, are produced by meiosis in specialized organs known as the **gonads.** The male gonads are called **testes.** The female gonads are the **ovaries.** During sexual reproduction, the gametes (egg and sperm) unite and combine their genetic material. This union and combination of genetic material is called **fertilization.** The fertilized egg is known as a **zygote.** Unlike gametes, which have only 23 chromosomes, the zygote has the typical number of chromosomes for a human cell, 46. The zygote undergoes millions of rounds of mitosis and development within the female to change into an **embryo** in 7 days. The embryo then develops into a **fetus** at two months (the infant that is not born yet and resembles the adult form). The rest of this chapter is devoted to describing sexual reproduction in humans.

gonad (GO nad)
testis, testes (TESS tiss, TESS teez)
ovary, ovaries
  (OH vah ree, OH vah reez)
zygote (ZIGH goht)

 17-3 Please go to your DVD to view an animation showing the fertilization of the sperm and egg cell.

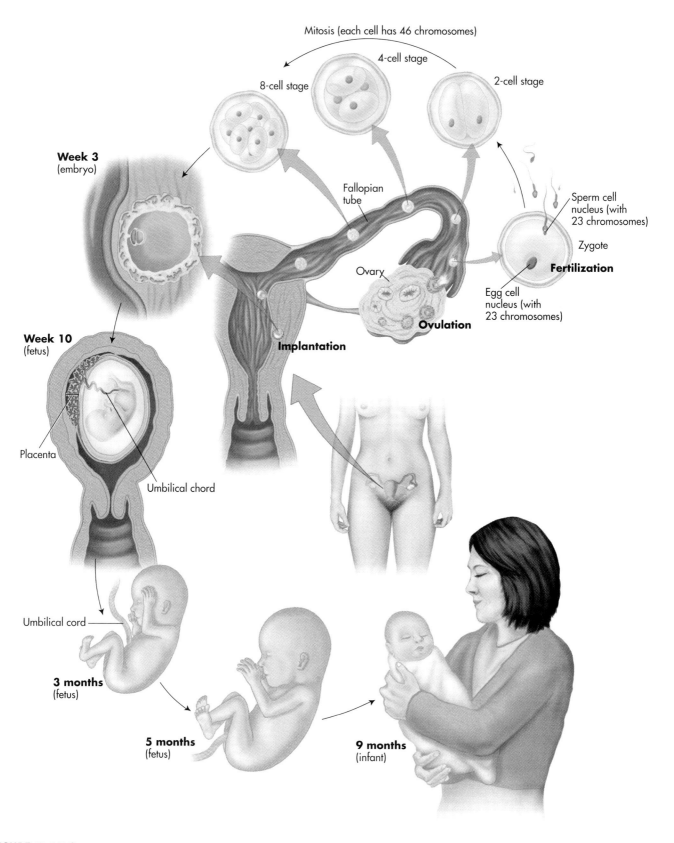

FIGURE ■ 17-2

The early stages of the human life cycle.

TEST YOUR KNOWLEDGE 17-1

Choose the best answer:

1. Cells with half the number of chromosome used for sexual reproduction are known as:

   a. zygotes

   b. daughter cells

   c. gametes

   d. half cells

2. Organs that produce sperm and egg are called:

   a. gametes

   b. gonads

   c. zygotes

   d. chromosomes

3. After _____, the egg is called a _____ and has 46 chromosomes.

   a. reproduction, gamete

   b. meiosis, daughter cell

   c. mitosis, zygote

   d. fertilization, zygote

THE HUMAN REPRODUCTION SYSTEM

Reproductive organs are called **genitalia.** The general term for primary genitalia of both male and female is the gonads, which produce the gametes. The secondary genitalia are all the other structures that aid in the reproductive process. In this section, we will first discuss female anatomy and physiology and then male anatomy and physiology.

genitalia *(jen ih TALE ya)*

Female Anatomy

In females, the primary genitalia are called the ovaries. The secondary genitalia are the **uterine tubes** (also known as **oviducts** or **Fallopian tubes**), the **uterus,** the **vagina,** and the external genitalia, called the **vulva.**

    The ovaries are paired structures, about 3 centimeters long, in the peritoneal cavity (in the lower part of the abdominal cavity). There is one ovary on each side of the uterus. Several ligaments suspend or anchor each ovary. The **mesovarium** suspends the ovary, the **suspensory ligament** attaches the ovary to lateral pelvic wall, and the **ovarian ligament** anchors the ovary to the uterine wall. Blood vessels, the ovarian artery, and the ovarian branch of the uterine artery travel through the mesovarium and suspensory ligament, supplying the ovary with oxygenated blood. Each ovary contains thousands of microscopic, sacs, or follicles. Please see Figure 17–3 ■ for a diagram of internal reproductive anatomy.

uterus *(YOO ter uss)*
vagina *(vah JYE nah)*
vulva *(VULL vah)*

THE OVARY

The ovary is covered by a fibrous capsule called the **tunica albuginea** made of cuboidal epithelium. The interior of the ovary is divided into the cortex, which contains the eggs, and the medulla, which contains blood vessels, nerves, and lymphatic tissue surrounded by loose connective tissue.

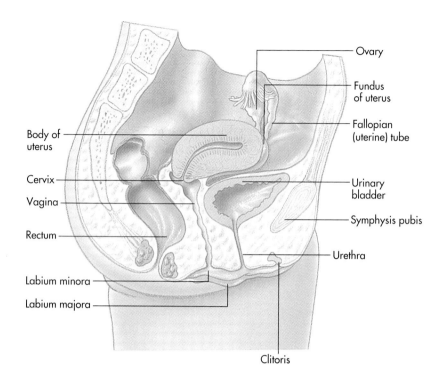

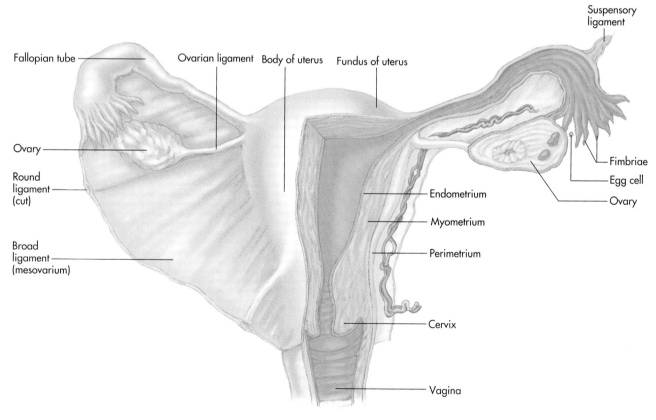

FIGURE ▪ 17-3

Internal female reproductive organs.

The anatomy of the cortex is relatively complicated and will be described during our discussion of physiology. Ovaries produce eggs, or ova (female gametes; singular, ovum), and manufacture the female hormones estrogen and progesterone. Ova production and hormone level fluctuations usually occur as a result of specific reproductive cycles—the menstrual cycle, ovarian cycle, and uterine cycles—which you will learn about later in the chapter.

THE UTERINE TUBES

The uterine tubes, also known as oviducts or Fallopian tubes, are the passageway for the egg to get to the uterus. The uterine tubes, which are about 4 inches long and not attached to the ovaries, begin as a large funnel, the **infundibulum**, surrounded by ciliated projections, the **fimbria.** The infundibulum leads to a widened area, the **ampulla**, followed by a longer, narrower portion known as the **isthmus**. The uterine tubes are connected to the superior portion of the uterus. The tube is constructed of sheets of smooth muscle lined with highly folded, ciliated, simple columnar epithelium that help propel the ova to the uterus. The outside of the tube is covered by the visceral peritoneum and suspended by a mesentery known as the **mesosalpinx**. Again see Figure 17-3.

fimbria *(FIM bree ah)*

THE UTERUS

The thick walled and pear shaped uterus is in the pelvic cavity, posterior and superior to the urinary bladder and anterior to the rectum. The major portion of the uterus is called the **body**. The rounded superior portion between the uterine tubes is the **fundus**, and the narrow inferior part is the **isthmus**. The **cervix** is a valve-like portion of the uterus that protrudes into the vagina, while the **cervical canal** connects with the vagina. Like the ovaries, the uterus is suspended and anchored by a series of ligaments. The **mesometrium** attaches the uterus to the lateral pelvic walls. (The combination of the mesometrium and mesovarium is called the **broad ligament**.) The lateral cervical ligaments attach the cervix and vagina to the lateral pelvic walls. The uterus is anchored to the anterior wall of the pelvic cavity by the **round ligaments**.

cervix *(SER viks)*

Like most of the hollow organs or tubes we have visited, the walls of the uterus consist of three layers. The **perimetrium,** the outermost layer, is also the **visceral peritoneum**, just like the outermost layer of the heart is the visceral pericardium. The middle layer, the **myometrium,** consists of smooth muscle. The **endometrium,** or inner lining, is a mucosa layer of columnar epithelium and secretory cells. The mucosa has two divisions: the basal layer and the functional layer. The **basal layer** is responsible for regenerating the uterine lining each month. The **functional layer** sheds about every 28 days when a woman has her period.

perimetrium *(pair ee MEE tree um)*

myometrium *(MY oh MEE tree um)*
endometrium
(EHN doh MEE tree um)

The endometrium is highly vascular. (This is no big surprise to any woman of child-bearing age.) Blood is supplied by the **uterine arteries** that branch from the internal iliac arteries on each side. The uterine arteries split **into arcuate arteries**, which supply the myometrium, and **radial arteries**, which

Clinical Application

PAP TEST

The cervix is the area where a Pap test (name after George Papanicolaou) is taken from the cervix to examine scrapings from the cervical cells to detect the presence of cancer. Regular Pap tests allow for early cancer detection, which increases the likelihood of successful treatment. The human papilloma virus (HPV) has been directly linked as a cause for cervical cancer. It is also a cause of genital warts, which is a sexually transmitted disease. This has caused a lot of controversy because of Guardisil, the new vaccine for HPV, Guardisil, is recommended to be given to girls ages 12-13 to hopefully prevent some types of cervical cancer.

supply blood to the endometrium. As you might expect, since the endometrium has two separate divisions, there are two different types of radial arteries. *Straight radial arteries* supply the basal layer. **Spiral** *radial arteries* supply the functional layer. The spiral arteries actually decay and regenerate every month, as part of the menstrual cycle, and undergo spasms that contribute to the shedding of the endometrium each month. (For more on control of the menstrual cycle, see the section "Hormonal Control" later in this chapter.) Blood returns to circulation via a network of venous sinuses.

Pathology Connection: Endometriosis

Each month, women of child-bearing age shed and replace the endometrium, the lining of the uterus. In many women, this endometrial tissue escapes the uterus and implants in the abdominal and pelvic cavities. There, the tissue responds to the woman's hormonal cycles, continuing to build up and decay each month. Unfortunately, the continued proliferation and decay and bleeding of endometrial tissue in the abdominal and pelvic cavities can cause scarring and damage to the organs. This condition is called **endometriosis**.

As many as 6 million women in the US alone have endometriosis. Many women who have endometriosis have no noticeable symptoms, while other women experience severe abdominal and back pain around the time of their period, painful intercourse, chronical pelvic pain, painful urination, and painful bowel movements, especially during their periods. Some women even have symptoms that mimic a bladder infection. Untreated endometriosis can cause *adhesions* on the intestines and urinary bladder, which can be extremely painful and must be removed surgically. Endometriosis is among the top three causes of infertility. The scar tissue may block the uterine tubes and therefore block the journey of the egg through the tubes, in addition to scarring the ovaries and uterus. Some women are diagnosed with endometriosis only when they are being treated for infertility.

Endometriosis is diagnosed after a detailed medical history and exam. Ultrasound may be helpful in finding large implants of endometrial tissue outside the uterus. However, the only way to diagnose endometriosis with certainty is to perform a laparoscopic examination of the woman's pelvic cavity. Suspected endometrial implants are removed for biopsy. A biopsy is the only sure way to identify that the tissue is actually endometrial tissue.

The cause of endometriosis is unknown, so treating the disease is often difficult. Many women find relief of symptoms during treatment only to have their symptoms recur after treatement has ended. Endometrial implants often reoccur. Endometrial pain can be treated successfully using NSAIDs, but endometrial implants will continue to progress. In order to halt progression of the

adhesions = holding together of two parts or surfaces as in a wound disease, implants must either be removed surgically or suppressed using hormones. Surgery typically involves removing the implants and scar tissue during laproscopic examination of the pelvic cavity. Severe or intractible cases may require laparotomy (open surgery) or even hysterectomy (including removal of the ovaries). Hormonal treatment includes the use of contraceptives (combined estrogen and progesterone), GnRH agonists, and synthetic androgens (male hormones). Most hormonal treatments have side effects and many cannot be used for longer than a few months, and cannot be used at all in adolescents or in women who are trying to get pregnant. Infertility is treated with a number of fertility drugs. For further discussion of hormone treatment drugs see the Pharmacology Corner at the end of the chapter. See Figure 17-4 ■ for locations of endometriosis outside of the uterus.

THE VAGINA

The **vagina** is a smooth muscle tube with a mucous membrane lining, approximately 10 centimeters long, which runs from the uterus to the outside of the body. Its purpose is to receive the penis during intercourse and to allow for the passage of menstrual fluid out of the uterus. The vagina is also known as the birth canal, since its primary function is to allow the movement of a baby out of the uterus during childbirth. The external opening of the vagina may be covered by a perforated membrane called the **hymen**. A torn hymen was once thought to "prove" that a woman had had intercourse. However, many hymens are highly perforated and easily ruptured by day-to-day activities such as riding a bicycle or jogging. An intact hymen is no longer considered a litmus test for virginity; however, some cultures still hold this erroneous belief to be true.

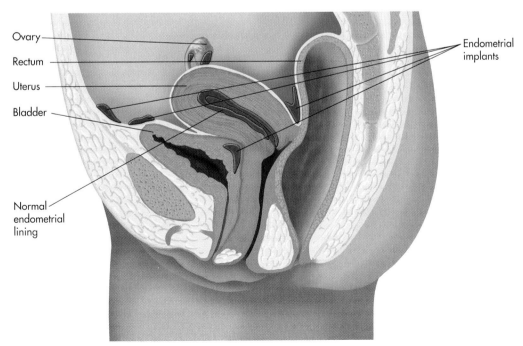

FIGURE ■ **17-4**

Location of endometriosis outside the uterus.

vulva *(VULL vah)*

labia majora
  (LAY bee ah mah JOR ah)

pudendal cleft *(PEW den dall cleft)*

labia minora
  (LAY bee ah mih NOR ah)
prepuce *(PREE pewce)*
clitoris *(KLIT oh riss)*

mammary glands *(MAM ah ree)*

THE EXTERNAL GENITALIA

The external genitalia (Figure 17–5 ▓), collectively known as the **vulva,** while not perhaps as obvious as the male external genitalia, are a complex and important part of reproduction. The **vulva** is surrounded on each side by two prominences called the **labia majora.** The labia majora are rounded fat deposits that surround the entrance to the vagina; they protect the rest of the external genitalia. The labia majora meet anteriorly to form the mons pubis. Both the mons pubis and the labia majora are covered by pubic hair. The **perineum** is the area between the vagina and the rectum.

Between the two halves of the labia majora is an opening known as the **pudendal cleft.** Within the pudendal cleft lies the **vestibule,** a space into which the urethra (anterior) and vagina (posterior) empty. (Unlike the male reproductive system, the female urinary system is completely separate from the reproductive system.) The lateral border of the vestibule is formed by the thin **labia minora,** which meet anteriorly to form the **prepuce.** Several glands surround the vestibule, helping to keep it moist.

Just posterior to the prepuce, and anterior to the vestibule, is the **clitoris.** The clitoris is a small erectile structure, 2 centimeters in diameter. Like the penis (see "Male Anatomy" in this chapter), the clitoris has a shaft, or body, and a glans (tip), and it becomes engorged with blood during sexual arousal. However, the clitoris increases in diameter, not in length.

MAMMARY GLANDS

There is another set of external accessory sexual organs in the female, far removed from the pubic area, the **mammary glands.** The mammary glands are milk production glands housed in the breasts (see Figure 17–6 ▓). In young children, mammary tissue is virtually identical in boys and girls. At puberty, estrogen and progesterone stimulate breast development in girls. In adult females, the breasts consist of 15 to 20 lobes, which are glandular, and lots of adipose tissue. Each lobe is divided into smaller lobules, which house milk-secreting sacs called **alveoli,** when a woman is lactating (producing milk). Milk, made in the alveoli, travels through a series ducts and sinuses, eventually reaching the nipple. The areola is the darkened area that surrounds the nipple. Milk production is controlled by the hormone prolactin.

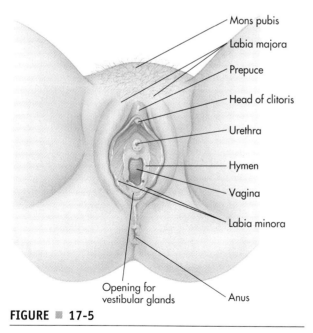

Mons pubis
Labia majora
Prepuce
Head of clitoris
Urethra
Hymen
Vagina
Labia minora
Opening for vestibular glands
Anus

FIGURE ▓ 17-5

The external female genitalia (vulva).

17-4 Please go to your DVD to view a 3-D animation of the female reproductive system along with a drag-and-drop interactive labeling exercise.

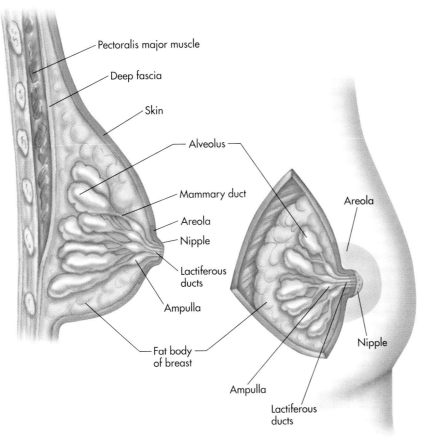

FIGURE ▧ 17-6

The mammary glands.

TEST YOUR KNOWLEDGE 17-2

Choose the best answer:

1. The _____ is also known as the birth canal.
   a. ovary
   b. uterus
   c. uterine tubes
   d. vagina

2. The inner lining of the uterus that is shed each month is the:
   a. perimetrium
   b. endometrium
   c. myometrium
   d. all of the above

3. This part of the female reproductive system is erectile:
   a. ovary
   b. labia majora
   c. clitoris
   d. penis

4. Which of the following is NOT a symptom of endometriosis?
   a. pain
   b. infertility
   c. hot flashes
   d. painful intercourse

REPRODUCTIVE PHYSIOLOGY: FEMALE

The female reproductive physiology is closely tied to a regulated cycle. This cycle is normally regulated by hormonal control.

The Menstrual Cycle

menstrual cycle *(MEN stroo al)*
ovarian cycle *(oh VAIR ee an)*
uterine cycle *(YOO ter in)*

Female reproduction is organized around an approximately 28-day cycle involving both the ovaries and the uterus, collectively known as the **menstrual cycle.** The **ovarian cycle** involves the monthly maturation and release of eggs from the ovary. The **uterine cycle** consists of the monthly build-up, decay, and shedding of the uterine lining. The cycles begin during a woman's teen years at the beginning of puberty, with **menarche**, or the first menstrual cycle, and end usually during her 40s or 50s, with the end of **menopause**, which is the cessation of menstruation and is characterized by symptoms such as hot flashes, dizziness, headaches, and emotional changes. The ultimate goal of the cycle is to release an egg that might be fertilized and to prepare the uterus to receive and nourish the fertilized egg should pregnancy result. If pregnancy does not occur, the uterine lining sheds and the cycle begins again.

The menstrual cycle is divided into four stages, usually over 28 days. During the **follicle stage**, which usually lasts about 10 days, a hormone called **follicle-stimulating hormone (FSH)** is released by the pituitary gland. At the ovary, FSH stimulates the follicle and ovum to mature, resulting in a release of estrogen and preparation of the uterine lining. In the next stage, the **ovulation stage**, the pituitary stops producing FSH and starts producing another hormone, **luteinizing hormone (LH).** At day 14, the follicle ruptures and the ovum is released. During the third stage, the **corpus luteum stage**, the corpus luteum secretes progesterone and continues to do so if the egg is fertilized, preventing further ovulation and maintaining the uterine lining. This stage lasts about 14 days. The final stage is the **menstruation stage**, which occurs if the egg is not fertilized. The corpus luteum diminishes progesterone production and the uterine lining is broken down and discharged over the course of 3 to 6 days. You will learn more about these stages and the structures and hormones involved as you read this chapter.

menses *(MEN seez)*

menstruation *(MEN stroo AY shun)*

The menstrual cycle, occurring approximately every 28 days in sexually mature women who are not pregnant, begins with the first day of **menses.** Menses is the time period when the uterine lining is shed. You are probably more familiar with the term **menstruation.** Menstruation is the actual shedding of the endometrium, while menses is the time during which a woman is menstruating. In more common terms, menses is the time during which a woman is having her "period," and menstruation is the "period" itself. Menses typically lasts 4 to 5 days but can be longer or shorter in different women and can even vary month to month in the same woman.

ovulation *(OV yoo LAY shun)*
oocyte *(OH oh site)*

Once menses is over, the endometrium begins to proliferate, or build up, readying itself for the egg that is about to be released from the ovary, during **ovulation.** From day 1 to day 14, the ovary is also busy. In the ovary, an egg cell, or **oocyte,** is undergoing a number of developmental changes getting ready for ovulation on day 14. Ovulation is the release of a mature egg from the ovary. The egg travels from the ovary to the uterus, by way of the Fallopian tube, which has been getting ready to receive the egg. If the oocyte has been fertilized by a sperm, it will implant in the thickened endometrium. If the egg does not implant within a few days, the endometrium will begin to decay and menstruation will occur about two weeks after ovulation. The time between the end of menses and ovulation is known as the **follicular** or **proliferative phase,** because the endometrium is proliferating and the follicles are maturing in the ovary. **Follicle** is the term used to refer to an egg and associated helper cells. The time between ovulation and menses is known as the **luteal phase** or **secretory phase,** because of the development of a structure called the **corpus luteum** in the ovary and the beginning of secretion in the uterus.

follicular *(fa lik YOU ler)*
proliferative *(pro LIFF er ah TIV)*

luteal *(LOO tee al)*

Sounds simple enough, right? The cycle is deceptively simple when described without details or information about control of the cycle. Now that you understand the cycle itself, let's get down to the nitty gritty. Just how does an egg mature, and how is the cycle controlled?

OOGENESIS, FOLLICLE DEVELOPMENT, AND OVULATION

The process by which eggs are produced is called **oogenesis.** Oogenesis begins with the birth of **oogonia,** or egg stem cells, in the ovary. The oogonia undergo mitosis, producing millions of **primary oocytes.** This happens very early in a woman's life. There are millions of primary oocytes produced in a fetus. That's right: women have all the eggs they will ever have five months before they are born!

oogenesis *(oh oh JENN eh siss)*
oocyte *(OH oh site)*

Primary oocytes, since they are born via mitosis, still have all 46 chromosomes. In order to become gametes, they must undergo meiosis to cut their chromosome number to 23. Remember, gametes must have only 23 chromosomes in order for fertilization (successful combining of sperm and egg) to produce a cell with the normal number or 46 chromosomes. So, primary oocytes begin meiosis. However, they do not fully complete their development. The primary oocytes stay in a kind of suspended animation until puberty when they finish developing. (That's at least 10 years!)

> ### Clinical Application
>
> ### MENSTRUAL DISORDERS
>
> **Amenorrhea** is the absence of menstruation and can be the result of pregnancy, menopause, or other factors such as distress, extreme dieting, or poor health. **Dysmenorrhea** is difficult menstruation and usually results in painful cramping. **Vaginitis** is the inflammation of the vagina that is usually caused by a micro-organism such as a bacteria or yeast. Vaginitis is not always a result of sexually transmitted diseases. For coverage of sexually transmitted diseases, please see Chapter 18.

amenorrhea *(ah MEN oh REE ah)*
 a = *absence*
dysmenorrhea
 (DISS men oh REE ah)
 dys = *difficult*
vaginitis *(vaj ih NYE tiss)*

primordial *(pry MORE dee all)*

These primary oocytes eventually are surrounded by helper cells, called **granulosa cells**. Once surrounded by granulosa cells, the primary oocyte and surrounding cells are known as **primordial follicles.** These primordial follicles stay dormant until puberty. Hormonal signals during puberty cause some primordial follicles to enlarge and increase the number of granulosa cells. These enlarged cells are then called **primary follicles** and represent thousands of microscopic sacs within the ovary.

Once a girl reaches puberty, one primary follicle will become a **secondary follicle.** The secondary follicle will not complete its development unless it is ovulated and fertilized. Just before ovulation, the secondary follicle matures and fills with fluid and moves toward the surface of the ovary, where it forms a visible lump. The mature follicle in its sac is sometimes called a **graafian follicle.**

The fimbria of the uterine tubes brush the surface of the ovary, causing the follicle to rupture. When the follicle ruptures, the egg (now an oocyte) is released into the peritoneal cavity. The fimbria then pulls the egg toward the funnel, drawing it into the uterine tube (see Figure 17–7 ■).

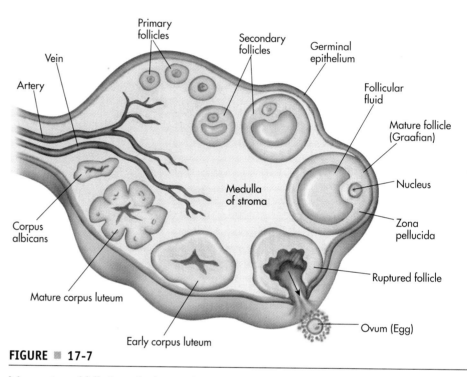

FIGURE ■ 17-7

Maturation of follicle and release of egg.

zygote *(ZIGH goht)*

As the egg travels down the uterine tube (also called the **oviduct** or **Fallopian tube**), it either will or will not be fertilized. If sperm are present in the uterine tube and all the conditions are right, the sperm will penetrate the egg, fertilizing it and triggering the rest of egg development. The successfully fertilized egg has 46 chromosomes and is called a **zygote.** The zygote starts out as a single cell, but the cell divides and subsequent cell divisions multiply the number of cells in the zygote as it travels down the Fallopian tube to implant in the uterus. When the zygote enters the uterus, the newly proliferated endometrium is ready for it. If the zygote successfully implants in the uterus, pregnancy will result and the woman will not menstruate. The ruptured follicle left behind in the ovary during ovulation will become the **corpus luteum** (remember the luteal phase?) and secrete hormones to help maintain the thickened endometrium, which will serve to nourish the growing fetus.

corpus luteum
(KOR pus LOU tee um)

If there is no sperm in the uterine tube, conditions are not right, or something goes wrong after fertilization, the zygote will not implant in the uterus. If there is no implantation within a few days, the uterine lining will begin to degenerate and the woman will have her period. The corpus luteum will become a corpus albicans and eventually disappear (see Figure 17–8 ■).

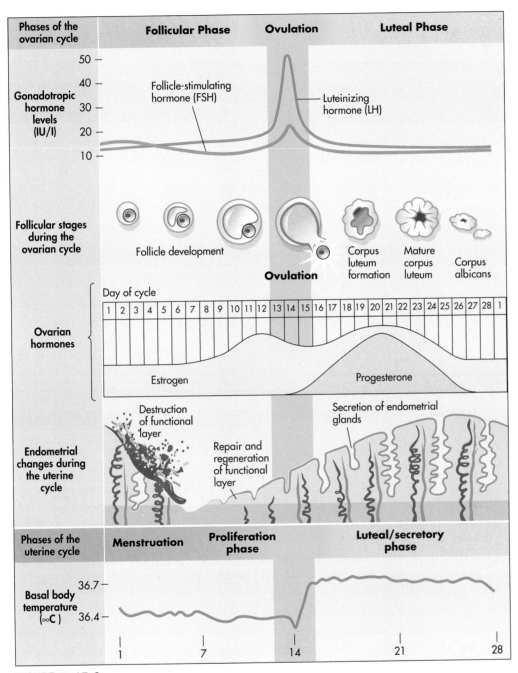

FIGURE ▪ 17-8

The menstrual cycle.

17-5 Please go to your DVD to view an animation on oogenesis and for an interactive drag-and-drop labeling exercise on oogenesis.

TEST YOUR KNOWLEDGE 17-3

Choose the best answer:

1. The time from the end of menses to ovulation is what phase?
   a. luteal phase
   b. follicular phase
   c. mitotic phase
   d. ovarian phase

2. The union of an egg and a sperm is called:
   a. sex
   b. fertilization
   c. oogenesis
   d. ovulation

3. The act of the egg being expelled from the ovary is called:
   a. oogenesis
   b. ovulation
   c. puberty
   d. pregnancy

4. If _____ does not occur, the endometrium will decay and a woman will menstruate.
   a. sex
   b. ovulation
   c. fertilization
   d. oogenesis

HORMONAL CONTROL

It seems obvious that something must control the complex cyclic changes in the female reproductive system. The control system that keeps the uterus and ovaries in synch is the endocrine system, specifically hypothalamic, pituitary, and ovarian hormones.

Remember that hormones are chemical signals released by one organ into the bloodstream to control another organ or tissue some distance away. Hormone levels are generally controlled by negative feedback. As hormone levels rise, the organ releasing the hormone decreases the amount of hormone released. Hormones are often released as part of a hierarchy, with the hypothalamus releasing a hormone that controls the pituitary, which then releases a hormone that controls another organ.

The hormones controlling female reproduction are no exception. The menstrual cycle is controlled by a combination of four hormones: **estrogen** and **progesterone** from the ovary and **luteinizing hormone (LH)** and **follicle-stimulating hormone (FSH)** from the pituitary gland. At the beginning of puberty, estrogen and progesterone secretion from the ovaries increases greatly, increasing the secretion of LH and FSH.

The release of **gonadotropin-releasing hormone** (GnRH) from the hypothalamus causes an increase in the secretion of LH and FSH from the pituitary. FSH initiates the development of primary follicles each month, and LH triggers ovulation. During this part of the cycle (the follicular or proliferative phase), estrogen levels continue to rise as more and more is secreted by the developing follicle. Estrogen exerts a positive influence on the hypothalamus, increasing secretion of GnRH and thus increasing LH and FSH secretion. This positive feedback loop increases the levels of LH and FSH, stimulating follicle development and triggering ovulation. Rising estrogen levels also stimulate proliferation of the uterine lining (endometrium).

progesterone *(proh JESS ter ohn)*

luteinizing hormone
   (LOO tee en IZE ing)

follicle-stimulating hormone
   (FALL ih kle stim you LAY ting HOR mohn)

gonadotropin-releasing hormone
   (GO nad oh TROH pin)

Once ovulation occurs, the feedback loop actually reverses itself! The leftover ruptured follicle, now the corpus luteum, begins to secrete progesterone and secretes a little estrogen. Under the influence of progesterone, estrogen exerts negative feedback on the hypothalamus and pituitary, decreasing GnRH, LH, and FSH secretion. Progesterone also exerts negative feedback on the hypothalamus and pituitary. Thus, during the luteal or secretory phase, LH, FSH, and estrogen levels drop while progesterone levels rise. These hormonal changes prevent another egg from maturing.

For about 10 days after ovulation, progesterone levels remain high as the corpus luteum continues to secrete the hormone. Progesterone's effect on the uterus is to *maintain* the buildup of the endometrium and to decrease uterine contractions. If no pregnancy results, then the corpus luteum will degenerate and stop producing progesterone. Decreased progesterone causes degeneration of the endometrium, followed by menstruation. Decreased progesterone also releases the hypothalamus and pituitary from inhibition. FSH and LH levels rise and the cycle begins again (again see Figure 17-8).

If pregnancy does result, the implanted fertilized egg secretes a hormone called **human chorionic gonadotropin** (HCG). HCG stimulates the corpus luteum, which keeps secreting progesterone and a little estrogen to maintain the uterine lining. At about three months' gestation (pregnancy), the placenta begins to secrete its own progesterone and estrogen, thereby becoming an endocrine organ. For a list of hormones and functions related to pregnancy, please see Table 17–1.

human chorionic gonadotropin
*(HUE man core ee AH nik
GO nad oh TROH pin)*

TABLE 17-1 Hormones Controlling Pregnancy

HORMONE	WHERE SECRETED	CAUSES
Human chorionic gonadotropin	Implanted fertilized egg	Maintains the function of corpus luteum; this is what gives a positive pregnancy test
Estrogen and progesterone	Corpus luteum for first two months of pregnancy; then by the placenta	Both stimulate development of uterine lining and mammary glands; progesterone prohibits uterine contractions during pregnancy; estrogen relaxes the pelvic joints; once labor begins, estrogen negates the effects of progesterone on uterine contractions and makes the myometrium sensitive to oxytocin
Prolactin	Anterior pituitary gland	Stimulates milk production by breasts
Oxytocin	Posterior pituitary gland	Uterine contractions to begin labor; stimulates the release of milk from the breasts (see Figure 17-9 ▨)

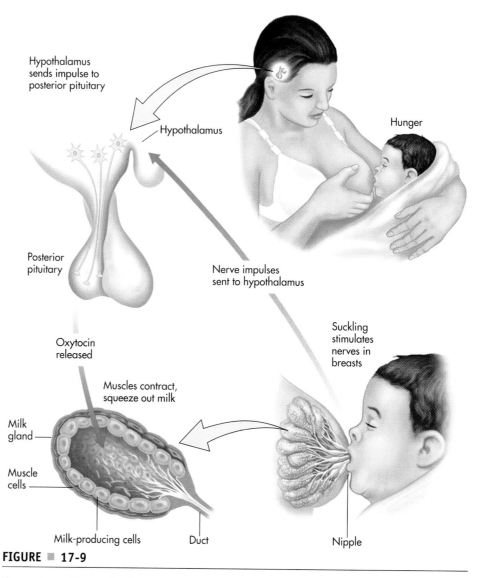

FIGURE ■ 17-9

Oxytocin and breast feeding.

Pathology Connection: Menstrual Disorders

Most women (as high as 90%) have some symptoms a few days before their period; 20-40% say the symptoms are problematic, and 1% of women experience debilitating symptoms that severely interfere with daily activities. When the symptoms are severe enough to effect quality of life, women may be diagnosed with **premenstrual syndrome (PMS)** or **premenstrual dysphoric disorder (PMDD)**.

PMS is characterized by predictable cognitive, physical, emotional, and behavioral symptoms associated with the luteal phase of a women's reproductive cycle. The symptoms begin approximately 5 days before the onset of menses and end within a few days after. There is no definitive criteria for diagnosis because symptoms vary dramatically not just from woman to woman, but even month to month in the same woman. There are no tests that can be taken and to date no differences in hormone levels have been found between women with diagnosed PMS or PMDD and women who do not have the disorders. Currently PMS is diagnosed if a woman has one affective (psychological) and one physical symptom from the following list:

- **Affective/Psychological:** depression, anger, irritability, anxiety, confusion, and withdrawal
- **Physical:** breast tenderness, bloating, swelling extremities, and headache

For a diagnosis of PMS, the symptoms must begin 5 days prior to menses, cease 4 days into menses, and not recur until day 12 of the woman's cycle. This pattern must have been repeated for the last 3 cycles. Other conditions must be ruled out. A woman is diagnosed with PMDD if she has 5 of the criteria, in the same pattern as for PMS, for most of her cycles in the previous year. One of the symptoms must be affective. In addition the woman must not have an underlying condition, such as depression or anxiety. These chronic conditions are usually worsened by hormonal changes during the menstrual cycle, making it impossible to tell whether the patient has PMDD itself, or just increased symptoms of an ongoing chronic condition.

Treatment for PMS and PMDD varies depending on the severity of the disease. Mild/moderate PMS/PMDD is treated by lifestyle changes, including good nutrition, regular exercise, complex carbohydrates, and calcium supplements. NSAIDs or hormonal suppression may be used if more conservative treatments do not work effectively. If affective symptoms are severe, SSRIs or anti-anxiety medications are the next step. If symptoms do not respond to life style changes and mood elevators then GnRH agonists may be used. The cause of PMS/PMDD and relationship between the two disorders remains a mystery.

Toxic Shock Syndrome (TSS) is an inflammatory disease that occurs in menstruating women, often as a result of tampon misuse. It can also be seen in men and children, but not as a result of tampon misuse. TSS is the result of a bacterial infection, usually *Staphylococcus aureus* and *Streptococcus pyogenes* infection. The bacteria produce a toxin that causes symptoms including high fever, headache, vomiting, diarrhea, sore throat, rash, and irritability. In more severe cases, people experience abdominal tenderness, hypotension, shock, respiratory distress, and renal failure. Treatment includes antibiotics and other prescription medications to control symptoms, rest, and hydration.

Polycystic Ovarian Syndrome (PCOS) is a disorder characterized by the formation of multiple cysts on the ovaries, hormonal abnormalities, infertility, obesity and insulin resistance. The ovaries make too much testosterone and estrogen, but too little progesterone. The result is that women may fail to ovulate, and have excess facial hair, male pattern baldness, and irregular menstrual cycles. In addition women with PCOS are more likely to be overweight, insulin resistant (even if they are not overweight), and to have hypertension and hyperlipidemia. Many women with PCOS develop metabolic syndrome and Type II diabetes (regardless of body weight).

PCOS is diagnosed using imaging and blood work. Large ovaries ($>10\ cm^3$) with 12 or more large follicles (2-3 cm diameter) are characteristic of PCOS. Women with PCOS also have increased blood levels of LH, testosterone, prolactin, and insulin.

Because the cause of PCOS is unknown, treatment is mainly symptom relief. First and foremost is weight control. Losing weight causes a decrease in severity of most other symptoms including infertility. Birth control pills can regulate a woman's menstrual cycle. Drugs can treat hyperlipidemia and hyperinsulinemia. Infertility can be treated by fertility drugs and/or by surgery to repair the damaged ovaries.

Surgery for PCOS and other uterine disorders may involve a procedure known as laparoscopy. During laparoscopy, a tube is inserted through an incision in the abdominal wall. The tube is equipped with a scope or a camera and a fiber-optic light source. The tube allows doctors to observe internal structures for defects and to more precisely guide instruments during surgeries. Laparoscopy is a less invasive procedure than traditional surgery, resulting in smaller incisions, shorter recovery time, and less patient discomfort.

TEST YOUR KNOWLEDGE 17-4

Choose the best answer:

1. At puberty, the ovaries begin to secrete:
   a. FSH
   b. LH
   c. estrogen
   d. all of the above

2. These hormones stimulate follicle maturation and ovulation:
   a. estrogen and progesterone
   b. estrogen and GnRH
   c. LH and FSH
   d. FSH and progesterone

3. Before ovulation, estrogen _____ the secretion of FSH and LH.
   a. decreases
   b. increases
   c. stabilizes
   d. there is no relationship between estrogen, FSH, and LH

4. The role of progesterone is:
   a. to promote ovulation
   b. to keep the endometrium from shedding
   c. to enhance fertilization
   d. to increase FSH and LH levels

5. Symptoms of PMS generally end:
   a. just before a woman's period
   b. just after her period begins
   c. at ovulation
   d. there is no pattern

6. PCOS symptoms most closely resemble this endocrine disorder:
   a. hyperthyroidism
   b. Addisons
   c. Cushings
   d. none of the above

Male Anatomy

Like the female reproductive system, the male reproductive system has a pair of primary genitalia, or gonads, called the **testes**. On average, male testes are each about the size of a small chicken egg. The testes produce the male gamete or sperm that must travel along its journey to find the egg. Unlike in the female, the male primary genitalia are external. In addition, the male has many secondary genitalia—the **penis,** an external sperm-delivery organ; several sperm ducts; the **epididymis;** the **vas deferens;** and the **urethra.** The urethra performs dual roles, as a channel for both excreting urine and expelling semen. Males

epididymis *(ep ih DID ih miss)*
vas deferens *(VAS DEFF er enz)*
urethra *(yoo REE thrah)*

also have several accessory glands: the **prostate gland,** the **seminal vesicles,** and the **bulbourethral glands.** Please see Figure 17–10 ▥ for the anatomy of the male reproductive system.

seminal vesicles
(SEM ih nal VESS ih kulz)
bulbourethral gland
(BUHL boh yoo REE thral)

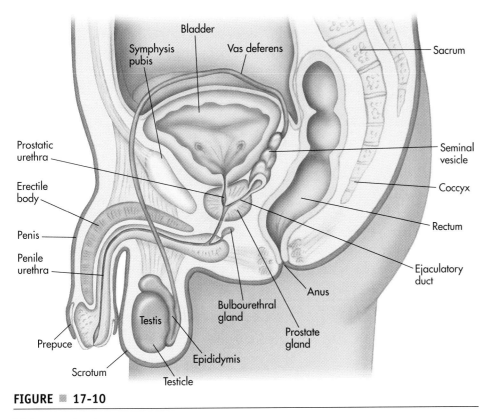

FIGURE ▥ **17-10**

Male reproductive anatomy.

THE TESTES (TESTICLES)

The primary genitalia, the testes (or testicles), are paired glands suspended in a sac called the **scrotum.** The testicles form in the abdomen during the last three months of embryonic development and migrate into the scrotum. The testicles are external genitalia, one hanging on either side of the penis. The failure of the testes to descend into the scrotal sac is termed **cryptorchidism** and may require surgical intervention. If the testes remain undescended, the male may become sterile because of the body heat destroying the sperm. (Viable sperm cannot be made at normal human body temperature.) Another pathologic condition of the testes is **hydrocele,** which is an abnormal collection of fluid within the testes.

cryptorchidism *(kript OR kid izm)*

hydrocele *(HIGH droh seel)*

Each small egg-shaped testis is surrounded by a serous membrane called the **tunica vaginalis,** which originates from the peritoneum. The inside of the testes are divided into 250 to 300 wedges called **lobules,** each of which contain one to four **seminiferous tubules.** The seminiferous tubules, which are made of epithelium and areolar tissue, contain sperm stem cells and sperm helper cells (**sertoli cells** or nurse cells).

seminiferous tubules
(SEM ih NIF er uss TOO byoolz)

sertoli cells *(sir TOW lee sells)*

THE PENIS

The other part of the external genitalia in males is the **penis**. The penis, from the Latin word for "tail," is a sperm-delivery organ that transfers sperm from male

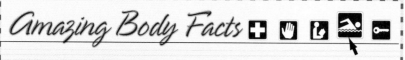

BOXERS OR BRIEFS?

Normal human body temperature is too warm for healthy sperm development. The testes are on the outside of the body because sperm development is extremely temperature sensitive and sperm could not develop normally inside the pelvic cavity. Indeed, there is some speculation that very tight clothing might contribute to male infertility by holding the gonads too close to the body and raising their temperature.

to female. The attached portion of the penis is called the **root**, while the freely moving portion is the **shaft** or **body**. The tip of the penis, the **glans penis**, is covered by a loose section of skin called the **foreskin**, unless a man has been circumcised. Internally, the penis contains the **urethra**, which is a transport passage for both sperm and urine. (But not at the same time!) In addition, the penis has three erectile bodies, or tubes with a sponge-like network of blood sinuses.

THE EPIDIDYMIS

There are several ducts in the male reproductive system. The **epididymis** is a comma-shaped duct on the posterior and lateral part of the testes. (It looks something like a stocking cap for each gonad.) Its highly coiled tubes connect the testes to the vas deferens. If unraveled, the epididymis would be 6 meters long! It is made of pseudostratified ciliated epithelium and smooth muscle. Sperm mature here.

THE VAS DEFERENS

The **vas deferens** is a 45 centimeter long tube. It is lined with ciliated pseudostratified epithelium, like the epididymis, but has a thick smooth muscle layer and is surrounded by a connective tissue layer called the **adventitia**. The vas deferens runs from the scrotum to the penis via a relatively complicated pathway. It runs from the epididymis as a pair of tubes, one on each side, into the abdominal wall (through the inguinal canal) and pelvic cavity, medially over the urethra and along the posterior bladder wall. Posterior to the bladder, the vas deferens joins the seminal vesicles to form the **ejaculatory duct.** The ejaculatory duct then passes through the prostate gland and empties into the urethra. (Remember, the urethra carries both sperm and urine in males.) Between the scrotum and the inguinal canal, the vas deferens runs through a tube, with blood vessels and nerves, collectively called the **spermatic cord.**

ACCESSORY GLANDS

There are three accessory glands in the male reproductive system. The **seminal vesicles** are highly coiled glands posterior to the bladder, made of pseudostratified epithelium, smooth muscle, and connective tissue. The **prostate gland** is a chestnut-sized gland surrounding the urethra just inferior to the bladder. It is a dense mass of connective tissue and smooth muscle with embedded glands. The **bulbourethral glands** are pea-sized glands inferior to the prostate.

Pathology Connection: Male Reproductive Diseases

Benign Prostatic Hyperplasia (BPH), the non-cancerous enlargement of the prostate, is the most common cause of lower urinary tract symptoms in men. It is common in men ages 60 and older, and as many as 70% of men will have some enlargement of their prostate by the age of 70. The cause of BPH is not known but may be due to age-related hormonal changes. Symptoms of BPH, due to the enlargement of the prostate impinging on the urethra and bladder, include urinary urgency and frequency, nighttime urination, weak urine stream, and incomplete voiding. For many men, the problems are serious enough to have significant impact on quality of life. Diagnosis is made by taking a detailed history and performing a physical exam, including examining the prostate. Prostate-specific antigen (PSA) blood levels should be tested to rule out prostate cancer. Other urinary conditions such as UTI, diabetes mellitus, and overactive bladder must also be ruled out. Treatment for BPH has changed in the last twenty years, from primarily surgical intervention (prostatectomy) to primarily hormonal treatment. In mild cases, a "watch and wait" strategy is used with changes in life style to decrease symptoms. Drug treatments, include 5 alpha-reductase inhibitors, which decrease testosterone (and shrink the prostate) and alpha blockers, which cause smooth muscle to relax (improving urine stream). Other interventions include minimally invasive laser or microwave ablation, surgical removal of partial or whole prostate, and the placement of a stent in the urethra. See Figure 17-11 ▦ for an illustration of BPH.

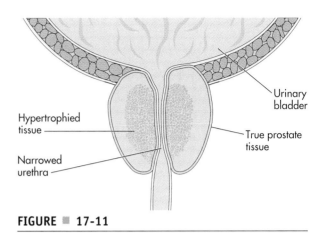

FIGURE ▦ **17-11**

Benign prostatic hyperplasia.

Prostate cancer is the most common non-skin form of cancer that affects American men. In fact, over 32,000 men die each year from prostate cancer, many of whom could have been saved with early detection. A common blood test called the prostate-specific antigen (PSA) can be used to help detect this disease. It is recommended this test be done at age 50 or earlier if there is a family history of prostate cancer.

TEST YOUR KNOWLEDGE 17-5

Choose the best answer:

1. The _____ are the gonads, where sperm are made.
   a. vas deferens
   b. testes
   c. scrotum
   d. penis

2. After passing the seminal vesicles, the vas deferens becomes the:
   a. prostate gland
   b. urethra
   c. penis
   d. none of the above

3. Sperm mature in the:
   a. epidiymis
   b. testes
   c. scrotum
   d. penis

4. _____ has the same symptoms as UTI and overactive bladder.
   a. PCOS
   b. ED
   c. BPH
   d. PMS

REPRODUCTIVE PHYSIOLOGY: MALE

Unlike female reproduction, male reproductive physiology is not organized around a tightly controlled monthly cycle. Let's explore the physiologic processes of the male reproductive system.

Spermatogenesis

Sperm production, in the testes, is a continuous process beginning when a boy reaches puberty and usually continuing until death. As such, the control of spermatogenesis, sperm production, is much less complicated than control of oogenesis. Spermatogenesis occurs in the seminiferous tubules.

spermatogonia
(sper MAT oh GO nee ah)

spermatocytes *(sper MAT oh sites)*

spermatids *(SPER mat ids)*

spermatozoa *(sper MAT oh ZOE ah)*

The **spermatogonia,** sperm stem cells, undergo mitosis to form **primary spermatocytes.** Unlike primary oocytes, the primary spermatocytes do not wait to go through meiosis. Primary spermatocytes form two **secondary spermatocytes.** Secondary spermatocytes complete meiosis to form **spermatids,** and spermatids go through a period of development to form immature **spermatozoa (sperm).** All of this takes place in the testes inside the seminiferous tubules. The distribution of the different stages of sperm development is predictable. Spermatogonia line up against the walls of the tubules, and mature sperm cluster near the lumen of the tubules (see Figure 17–12 ■). Sperm then travel from the seminiferous tubules to the epididymis, where the sperm spend about two weeks maturing and gaining the ability to swim.

17-6 Please go to your DVD to view a 3-D animation of the male reproductive system along with a drag-and-drop labeling exercise.

HORMONAL CONTROL OF MALE REPRODUCTION

testosterone *(tess TOSS ter ohn)*

Testosterone is arguably the most important male sex hormone. Before birth, **human chorionic gonadotropin (HCG)** secreted by the placenta causes the embryonic (still developing) testes to secrete testosterone, masculinizing the fetus. (Fetuses not exposed to testosterone or insensitive to testosterone look female.) After birth, there is little testosterone secreted until puberty. At puberty, two hormonal changes occur that signal the beginning of sexual maturity. First, testosterone secretion by the testes increases dramatically. Second, there is a change in the relationship between testosterone and GnRH, LH, and FSH. (Yes, the same hormones that control female reproduction control male reproduction!) Before puberty, the small amount of testosterone that is secreted by the testes inhibits GnRH secretion. After puberty, testosterone does not inhibit GnRH secretion. So after puberty, GnRH and therefore FSH and LH secretion increase. In addition, FSH and LH also enhance testosterone secretion in a major positive feedback loop, but only after the onset of puberty. Testosterone secretion at puberty is also responsible for the obvious physical changes of the male secondary sexual characteristics, including body, facial, and pubic hair growth; deepening of the voice; and increased muscle and bone mass.

LH and FSH affect males exactly as they do females. They stimulate gamete development. They are controlled in the same way in males as in females. GnRH is released from the hypothalamus, which stimulates LH and FSH secretion from the pituitary.

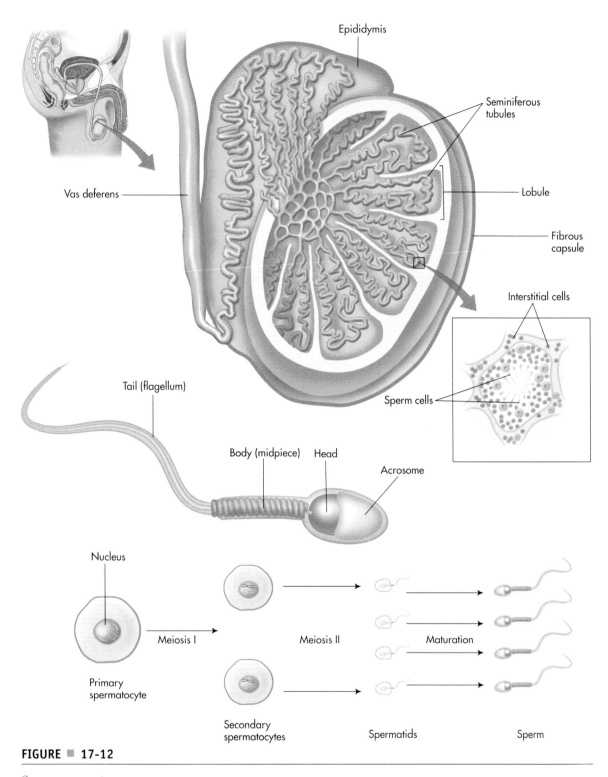

FIGURE ■ 17-12

Spermatogenesis.

ERECTION AND EJACULATION

When a man is sexually aroused, the erectile tissue in the penis becomes engorged with blood, stiffening and expanding the penis. The change in shape of the penis is called an **erection.**

semen *(SEE men)*

In order for sperm to leave the male reproductive system, ejaculation must occur. **Ejaculation** is the expulsion of **semen** (sperm and assorted chemicals) from the urethra. Smooth muscle contracts throughout the ducts and glands of the male reproductive system and propels the sperm from the epididymis into the vas deferens, which then carries the sperm into the pelvic cavity. As the sperm passes the seminal vesicles, sugar and chemicals are added to the sperm. The sperm and chemicals then enter the ejaculatory duct. As the ejaculatory duct passes through the prostate gland, prostatic fluid is added, liquefying the semen. The semen then passes by the bulbourethral glands, which add mucus to the semen and protects sperm from the acid environment of the vagina by the secretion of an alkaline substance. The semen then enters the urethra and is carried outside of the body. A sexual dysfunction called **impotence** often interferes with the process of ejaculation because the penis does not become or will not stay erect. Impotence can result from physical or psychological problems, including alcoholism, endocrine disease, anxiety, depression, stress, poor blood circulation, or prostate disease.

Clinical Application

ANDROGEN INSENSITIVITY SYNDROME

Testosterone secretion *in utero* is important for masculinizing a male fetus. Males who are not sensitive to testosterone will not respond to the masculinizing effects and will not develop normal male reproductive systems. Androgen insensitivity is a genetic disorder that can result in a broad range of malfunctions of the male reproductive system, from complete lack of external or internal male genitalia (testicular feminization—individuals look female though they are genetically male) to patients with ambiguous genitalia to patients who have typical male genitalia but are sterile. Treatment for the disorder depends of the severity of the genital malformation.

17-7 Please go to your DVD to view an animation on spermatogenesis and an interactive drag-and-drop labeling exercise on spermatogenesis.

Amazing Body Facts

HORMONES

Did you know that females have some natural testosterone and males have natural estrogen? It's true. The balance between the hormones—not the presence or absence of the hormones—is what is important.

If a man is engaged in sexual intercourse with a woman and ejaculates, the sperm enter the vagina and make their way to the uterus and into the uterine tubes. The female reproductive system is not particularly hospitable to sperm, and many sperm do not survive the journey to the uterine tubes. If there is an egg waiting to be fertilized in one of the uterine tubes, sperm will find the egg and attempt to penetrate it and fertilize it. New research suggests that the egg is not a passive participant in fertilization, but may actually engulf the sperm and even choose which sperm to allow inside. One sperm and only one sperm will fertilize the egg. If the egg is fertilized, and all other conditions are met, a pregnancy will result.

 ## Pathology Connection: Erectile Dysfunction (ED)

Erectile dysfunction (ED) is the long term (three months or more) inability to maintain an erection. ED increases with age and is relatively common, with 50% of men between the ages of 40 and 70 having some degree of ED. ED has many different causes which can be grouped into organic (caused by an underlying endocrine, neurological, anatomical or vascular condition) or psychogenic (having an underlying emotional cause) or mixed (a combination of organic and psychogenic causes). ED is diagnosed via medical history and specially designed questionnaires. Physical exam and blood work including PSA and testosterone levels as well as blood glucose and cholesterol levels can help determine the underlying cause of ED. Treatment of ED depends on the underlying cause. Improvement of underlying conditions often improves erectile function. In addition there are several drugs which improve the symptoms of ED, including Phosphodiesterase type 5 inhibitors, apomorphine, prostaglandins and androgens. (For more information on medications to treat ED see the pharmacology section at the end of this chapter.) If drugs do not solve the problem or cannot be tolerated, penile prosthetics may restore function.

17-8 To view a video on ED, please go to your DVD for this chapter.

TEST YOUR KNOWLEDGE 17-6

Choose the best answer:

1. The stiffening and expanding of the penis is known as:
   a. ejaculation
   b. erection
   c. emulsification
   d. erectile dysfunction

2. The combination of sperm, sugars, chemicals, and mucus is:
   a. semen
   b. urine
   c. spermatozoa
   d. all of the above

3. Male secondary sexual characteristics, masculinization of the body, and enhancement of FSH and LH levels are caused by:
   a. LH
   b. FSH
   c. estrogen
   d. testosterone

4. ED is always caused by psychological problems.
   a. true
   b. false

PREGNANCY

Pregnancy occurs when an egg is fertilized by the sperm and implants in the female reproductive system. The period of time during which the developing baby grows within the uterus is called the gestational period, approximately 40 weeks. A baby born before 37 weeks gestational period is termed a premature infant. Occasionally, miscarriages occur during pregnancy. A miscarriage is a spontaneous abortion, or expulsion of the fetus, that may be the result of an inherited defect in the fetus, infectious disease, physical injury, or endocrine disorder.

From the time the egg is fertilized by a sperm and implants in the uterine wall until the eighth week, the developing infant is referred to as an embryo. During the embryonic period, the organs and systems are fundamentally formed.

Beyond the eight-week period until the birth, the developing infant is called a fetus. Between the fourth and sixth months of development, the fetus begins to move. See Figure 17-13 ▥ that shows the stages of life of a newborn.

The growing fetus is nourished by a spongy structure called the **placenta**, which is attached to the fetus via the umbilical cord. The fetus is encased in a membranous sac called the **amnion**, which contains the amniotic fluid in which the fetus floats. **Labor** is the actual process whereby the fetus is delivered from the uterus through the vagina and into the outside world.

The amniotic fluid can be tested through a procedure called an **amniocentesis**, which punctures the amniotic sac and withdraws a fluid sample. The fluid can be sampled to test for chromosomal abnormalities leading to birth defects. In addition, biochemical disorders can be detected. See Figure 17-14 ▥, which demonstrates this procedure.

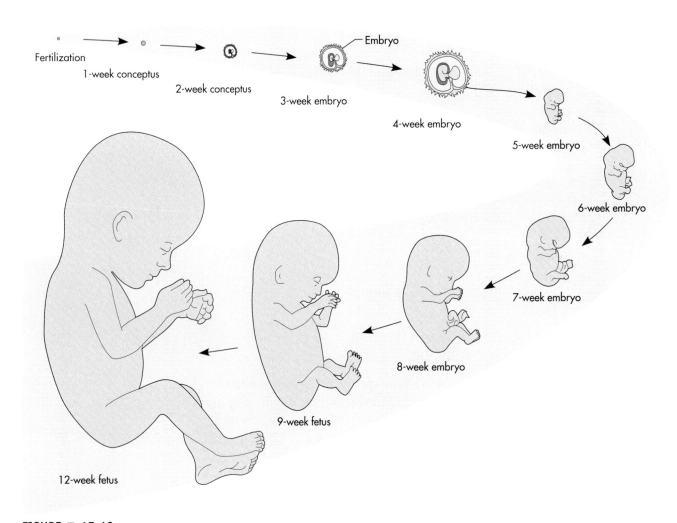

Fertilization

1-week conceptus

2-week conceptus

3-week embryo

Embryo

4-week embryo

5-week embryo

6-week embryo

7-week embryo

8-week embryo

9-week fetus

12-week fetus

FIGURE ▥ 17-13

Actual size of a human from fertilization to the early fetal stage.

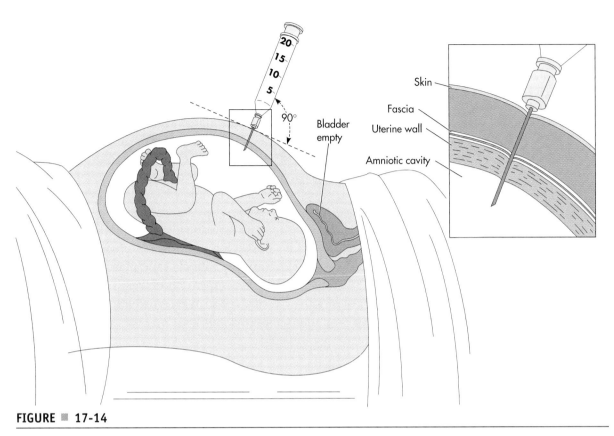

FIGURE ▦ 17-14

Amniocentesis. The woman is usually scanned by ultrasound to determine the placental site and to locate a pocket of fluid. As the needle is inserted, three levels of resistance are felt when the needle penetrates the skin, fascia, and the uterine wall. When the needle is placed within the amniotic cavity, amniotic fluid is withdrawn.

Pathology Connection: Infertility

Infertility, or a failure to conceive offspring, can be caused by numerous factors, some of which are medically treatable. Infertility can be primary or secondary. Couples with primary infertility have never been able to conceive a child together. Couples with secondary infertility have conceived a child in the past but are currently unable to do so.

Some women are infertile because their ovaries do not produce and release eggs. This can sometimes be corrected with injections or oral medications that stimulate follicles in the ovaries or adjust hormone levels (for example, for women who are infertile due to metabolic syndrome or PCOS). Women with Fallopian tube damage may be able to have surgery to repair the damage. In women with thyroid or adrenal abnormalities or diabetes, treating those conditions may help with the ability to conceive.

In males, the primary cause of infertility is insufficient sperm production, which can be caused by hormone imbalance, illness, or injury. Some drugs that treat common problems like high blood pressure and arthritis can cause low sperm count, for example. A fertility specialist can work with a couple to diagnose the root cause of the infertility and suggest medication and lifestyle changes that may improve the odds.

 17-9 For videos on fetal lie, labor, infant delivery, the placenta, and postpartum assessment, please go to your DVD for this chapter.

Clinical Application

CONTRACEPTION

The prevention of pregnancy is termed **contraception** ("against conception") and can be accomplished by a number of means such as intrauterine devices (IUDs), spermicidal agents, birth control pills, or shields such as a condom.

Sterilization of the male can be accomplished through a **vasectomy**, in which part of the vas deferens is removed or blocked to prevent the sperm from traveling out of the penis during sexual intercourse. Females may be sterilized via **tubal ligation**, in which the Fallopian tubes are cut or tied shut to prevent the egg from traveling from the ovary to the uterus.

Labor consists of three stages. In the **dilation stage**, the uterine smooth muscle begins to contract, thereby moving the fetus down the uterus and causing the cervix to begin to dilate. When the cervix is completely dilated (10 centimeters), the **expulsion stage** begins, during which the baby is actually delivered. Generally, the head presents first, which is called **crowning**. At this time the baby's mouth should be suctioned before the baby takes its first breath to prevent aspiration of fluid into the baby's lung. Sometimes, the baby is turned the wrong way and the buttocks appear first in a breech presentation, which makes the delivery difficult. The last stage of labor is the **placental stage**, in which the placenta or afterbirth is delivered due to final uterine contractions (see Figure 17–15 ■).

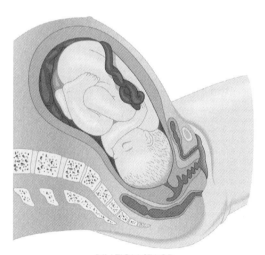

DILATION STAGE:
First uterine contraction to dilation of cervix

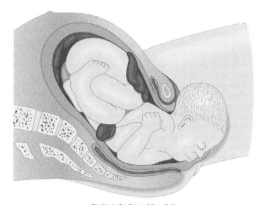

EXPULSION STAGE:
Birth of baby or expulsion

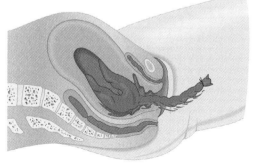

PLACENTAL STAGE:
Delivery of placenta

FIGURE ■ **17-15**

Stages of labor.

Pathology Connection: Breast Health

After birth, the baby is often fed with breast milk. One of the complications is **mastitis**, which is an inflammation of the breast. Mastitis can occur at any age in both males and females, but is usually associated with lactating females. It is caused by bacteria that can enter through a break in the tissue of the nipple, causing localized breast pain, redness, and swelling, with difficulty in getting the milk to flow. Treat-

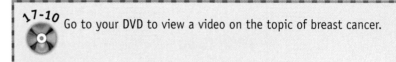

17-10 Go to your DVD to view a video on the topic of breast cancer.

ment is usually local application of heat, analgesics, and—if severe enough—antibiotics and anti-inflammatory drugs.

Breast cancer is one of the leading causes of deaths in woman between the ages of 32 and 52, killing approximately 40,000 women per year. Men can also develop breast cancer, but at a much lower rate.

The earlier breast cancer is detected, the greater the chance for survival. It is often detected by a mammogram, which is an X-ray of the breast. Annual mammograms are typically recommended for women over age 40, or for women under 40 who have a strong family history of breast cancer.

Treatment for breast cancer may include chemotherapy, radiation therapy, or partial or full mastectomy (surgical removal of the breast). Women who have a family history may elect to have a mastectomy before any cancer symptoms occur. This is somewhat controversial. One of the most important procedures to detect cancer at an early, more treatable stage is regular breast self-examination. See Figure 17-16 ■ for how to perform a breast self examination.

Pharmacology Corner

There are many drugs used to treat reproductive disorders. First we will discuss drugs that are prescribed to women and then those that are prescribed to men. **Oral contraceptives**, otherwise known as birth control pills, are used to prevent pregnancy and to regulate the menstrual cycle. These pills, generally combinations of estrogen and progesterone, work by preventing ovulation and decreasing menstruation. Side effects include irregular bleeding, weight gain, increased appetite, depression, and headaches.

Progestins (synthetic progesterone such as Depo-provera), may also be administered to regulate the menstrual cycle, either orally or by injection. The progestins, like the combination estrogen and progesterone medications, prevent ovulation and decrease menstruation. Side effects are the same as for oral contraceptives.

Gonadotropin releasing hormone (GnRH) agonists (such as Lupron) are a newer class of hormones that can be administered by injection or as a nasal spray. These medications inhibit estrogen production and are often used in combination with low-dose estrogen and progesterone. GnRH agonists are the current best hormonal therapy for endometriosis. Side effects are the same as for menopause,

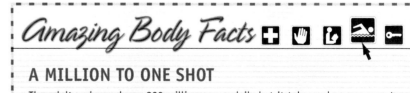

A MILLION TO ONE SHOT

The adult male produces 200 million sperm daily but it takes only one sperm to fertilize the egg.

| WHY DO THE BREAST SELF-EXAM? | There are many good reasons for doing a breast self-exam each month. One reason is that it is easy to do and the more you do it, the better you will get at it. When you get to know how your breasts normally feel, you will quickly be able to feel any change, and early detection is the key to successful treatment and cure.

REMEMBER: A breast self-exam could save your breast – and save your life. Most breast lumps are found by women themselves, but in fact, most lumps in the breast are not cancer. Be safe, be sure. |
|---|---|
| WHEN TO DO BREAST SELF-EXAM | The best time to do breast self-exam is right after your period, when breasts are not tender or swollen. If you do not have regular periods or sometimes skip a month, do it on the same day every month. |

NOW, HOW TO DO BREAST SELF-EXAM

1. Lie down and put a pillow under your right shoulder. Place your right arm behind your head.

2. Use the finger pads of your three middle fingers on your left hand to feel for lumps or thickening. Your finger pads are the top third of each finger.

3. Press firmly enough to know how your breast feels. If you're not sure how hard to press, ask your health care provider. Or try to copy the way your health care provider uses the finger pads during a breast exam. Learn what your breast feels like most of the time. A firm ridge in the lower curve of each breast is normal.

4. Move around the breast in a set way. You can choose either the circle (A), the up and down line (B), or the wedge (C). Do it the same way every time. It will help you to make sure that you've gone over the entire breast area, and to remember how your breast feels.

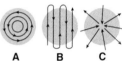

A   B   C

5. Now examine your left breast using your right hand finger pads.

6. If you find any changes, see your doctor right away.

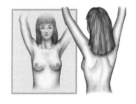

FOR ADDED SAFETY:

You should also check your breasts while standing in front of a mirror right after you do your breast self-exam each month. See if there are any changes in the way your breasts look: dimpling of the skin, changes in the nipple, or redness or swelling.

You might also want to do a breast self-exam while you're in the shower. Your soapy hands will glide over the wet skin, making it easy to check how your breasts feel.

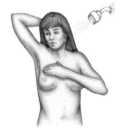

FIGURE ■ 17-16

Breast self-examination.

including hot flashes and bone loss. Prior to the development of GnRH agonists, the synthetic androgen Danazol was the treatment of choice for endometriosis. (It is only rarely prescribed today.) The addition of androgens and the subsequent decrease in natural estrogens stopped ovulation and menstruation which relieved endometriosis symptoms. The side effects, however, were severe and sometimes permanent, including voice deepening, excess facial and body hair growth, acne, hyperlipidemia, decreased sex drive, and weight gain. Many women could not continue treatment because of the side effects.

Drugs prescribed to men are generally used for treatment of BPH or ED. For BPH treatment is via 5-alpha-reductase inhibitors or alpha blockers. 5-alpha-reductase inhibitors, administered orally, decrease testosterone levels, preventing further enlargement of the prostate and even shrinking the enlarged gland. Side effects include decreased sex drive, ED, breast enlargement and decreased sperm count. Alpha blockers cause smooth muscle in the urethra to relax, increasing the urine stream. Side effects include headache, dizziness, decreased blood pressure, and fatigue. Most practitioners suggest combination therapy with 5-alpha-reductase inhibitors and alpha-blockers. To treat ED the most common (and becoming even more common) medications are phosphodiesterase type 5 inhibitors. These medications (for example, Viagra, Levitra, and Cialis) increase blood accumulation in erectile tissue, increasing the likelihood of erection. The side effects for each of these medications are different for each medication but can include visual disturbances, headache, dizziness and flushing. These are the first line medications for treating ED of any cause.

A Quick Trip Through Some Diseases of the Reproductive System

DISEASE	ETIOLOGY	SIGNS AND SYMPTOMS	DIAGNOSTIC TEST(S)	TREATMENTS
Female				
Amenorrhea Dysmenorrhea	Absent or painful menstruation; many different causes.	Lack of menstrual periods; pain during periods.	Physical exam, hormone levels, imaging, exploratory surgery.	Hormones, NSAIDS, surgery, treat underlying conditions.
Premenstrual syndrome (PMS) or premenstrual dysphoric disorder (PMDD)	Exact cause not determined; may be due to irregular levels of estrogen or progesterone.	Constipation, nausea, headache, backache, edema, malaise, irritability, mental confusion, depression.	Currently PMS is diagnosed if a woman has one affective (psychological) and one physical symptom. A woman is diagnosed with PMDD if she has 5 of the criteria as stated in text.	Treat symptoms; proper nutrition, complex carbohydrates, antidepressants, analgesics and hormone therapy.
Endometriosis	Cause unknown.	Hematuria, dysuria, painful intercourse, and excessive menstrual bleeding; increasing painful periods; lower back, vaginal, and abdominal pain.	Patient history and exam, laparoscopic exam.	Early diagnosis is critical. Use of oral contraceptives and other hormones appear to delay onset; analgesics for relief of discomfort; surgery for severe cases.
Polycystic ovarian syndrome	The ovaries produce too much testosterone and estrogen but too little progesterone.	Multiple cysts on the ovaries, hormonal abnormalities, infertility, obesity and insulin resistance; failure to ovulate, excess facial hair, and irregular menstrual cycles.	Imaging and blood work; increased blood levels of LH, testosterone, prolactin, and insulin.	Mainly symptom relief; weight control, birth control pills to regulate a woman's menstrual cycle. Treat hyperlipidemia and hyperinsulinemia if present.

continues

A Quick Trip Through Some Diseases of the Reproductive System (*Continued*)

DISEASE	ETIOLOGY	SIGNS AND SYMPTOMS	DIAGNOSTIC TEST(S)	TREATMENTS
Breast cancer	Most common malignancy of American women and leading cause of death between ages 40–55; uncontrolled growth of breast tissue; familial history.	Bloody, brown, or serous nipple discharge; noticeable lump, changes in breast tissue.	Mammogram, imaging, biopsy.	Lumpectomy, mastectomy, chemotherapy, radiation.
Cervical cancer	Current research shows connection with human human papiloma virus and early and frequent sexually activity; may be genetic predisposition; most common is women between 30–50.	Early stages asymptomatic; watery vaginal discharge; occasional spotting of blood; later stages include foul smelling vaginal discharge.	PAP test with microscopic examination, patient history, and exam.	Cryotherapy (freezing) or laser surgery; hysterectomy (removal of uterus); radiation.
Vaginitis	Inflammation of the vagina (infection, irritation, menopause).	Discharge, itching, pain.	Pelvic exam, lab tests, patient history.	Depending on cause: antibiotics, antifungals, vaginal creams, avoid irritant.
Abruptio placenta	Placenta separates from uterus.	Bleeding, pain, preterm labor.	Ultrasound.	Depends on severity: ranges from nothing to emergency cesarean.
Placenta previa	Placenta grows over cervix; causes unknown.	Bleeding late in pregnancy, pain.	Ultrasound.	Transfusions, medication, cesarean.
Ectopic (tubal) pregnancy (embryo implants in uterine tube)	Scarring, endometriosis.	Bleeding, abdominal pain, nausea, unconsciousness, and shock if tube ruptures.	Ultrasound, exploratory surgery, blood tests, pregnancy test.	Termination of pregnancy, treatment of symptoms.
Postpartum depression	Changes in hormone levels after pregnancy.	Fatigue, anxiety, feelings of worthlessness.	Exam.	Counseling, medication.

Male

DISEASE	ETIOLOGY	SIGNS AND SYMPTOMS	DIAGNOSTIC TEST(S)	TREATMENTS
Erectile dysfunction disorder (ED)	Many underlying physical and/or psychological causes.	Inability to maintain erection (problem lasts 3 months or more).	Exam, blood work.	Medication, counseling, prosthetics.
Benign prostatic hyperplasia (BPH)	Unknown, age related hormone changes perhaps.	Lower urinary tract symptoms similar to UTI or overactive bladder.	Exam, blood work, rule out other causes.	Medication, surgery.
Prostate cancer	Uncontrolled growth of prostate cells, which can spread.	Same as for BPH.	Same as for BPH.	Prostatectomy, chemotherapy.

SUMMARY

Snapshots from the Journey

→ Tissues grow, are replaced, and are repaired by asexual reproduction. Cells make identical copies of themselves. Asexual reproduction takes place all over the body as tissues grow or are repaired. Some tissues, like epidermis, blood, and bone, replace themselves continually, always by asexual reproduction.

→ Asexual reproduction in eukaryotic cells is accomplished by a relatively complex process called mitosis.

→ If an organism is going to reproduce sexually, it must use specialized cells called gametes with only half the typical number of chromosomes for that organism. (The gametes must have half as many chromosomes because, during fertilization, the gametes will fuse and combine their genetic material.)

→ In humans, the gametes are eggs and sperm. Eggs and sperm are produced by a special type of cell division called meiosis, or reduction division. Reduction division produces four daughter cells, each with only 23 chromosomes. These cells are not identical to each other or to the mother cell.

→ Cells undergoing meiosis go through two divisions. The chief difference between mitosis and meiosis is the pairing of homologous chromosomes (alike in size shape and genetic content) during meiosis.

→ Human reproduction can be described as a cycle, known as the human life cycle. Adult humans have specialized organs called gonads (ovaries and testes), which produce gametes via meiosis. Gametes get together during sexual reproduction in a process known as fertilization. The fertilized egg is called a zygote. The zygote undergoes many rounds of mitosis, eventually becoming an embryo, a fetus, a baby, a child, and finally an adult.

→ The female reproductive system consists of several internal genitalia. The gonads (ovaries) produce eggs (gametes). The uterine tubes provide a passageway for the egg to get to the uterus. The uterus is an incubator for the fertilized eggs. The vagina is the birth canal, connecting the uterus with the outside world. The external genitalia include the external opening of the vagina and several protective structures.

→ Female reproductive physiology is relatively complicated, organized around a monthly cycle of changes in both the ovaries and the uterus, collectively known as the menstrual cycle. The cycle begins with menstruation, the shedding of the uterine lining. After the lining, the endometrium, has finished shedding, it begins to build up again in a process called proliferation. As the endometrium is proliferating, an egg (follicle) is maturing in the ovary. Eventually, the follicle will be released from the ovary (ovulation) and travel to the uterus. If fertilization occurs, then the fertilized egg will implant in the thickened endometrium and pregnancy will result. If the egg is not fertilized, the endometrium will degenerate and menstruation will occur.

→ Control of the menstrual cycle is accomplished by four hormones: estrogen, progesterone, follicle-stimulating hormone (FSH), and luteinizing hormone (LH). During puberty, the ovaries begin to secrete larger amounts of estrogen and progesterone, which cause the development of female secondary sexual characteristics and set the cycle in motion.

→ The cycle works this way: The hypothalamus releases gonadotropin-releasing hormone (GnRH) that causes the pituitary to increase secretion of FSH and LH. FSH and LH cause follicles to develop and eventually trigger ovulation. Estrogen levels continue to rise as the developing follicles secrete more and more estrogen. Estrogen causes positive feedback to the hypothalamus and pituitary, which increases the levels of FSH and LH. This positive feedback loop continues right up until ovulation.

→ At ovulation, the ruptured follicle (left behind after ovulation) begins to secrete progesterone and backs off on estrogen secretion. Under the influence of progesterone, the feedback loop between estrogen and the hypothalamus and pituitary reverses itself, becoming a negative feedback loop. Thus, LH and FSH levels drop. Progesterone levels continue to rise, maintaining the thickened endometrium in case fertilization occurs. If fertilization does not occur, progesterone decreases, the endometrium decays, and the cycle begins all over again.

→ The male reproductive system has somewhat more obvious external genitalia, the penis, and the testes. Internal genitalia include a series of ducts, the vas deferens, ejaculatory duct and urethra, and a series of glands, the seminal vesicles, prostate, and bulbourethral glands.

→ Male reproductive physiology is not cyclic and is therefore a bit less complicated than female reproduction. Sperm, like eggs, develop via meiosis under the control of LH and FSH in the testes. Sperm mature in the epididymis. The penis is the organ of copulation. When a man is sexually aroused, the penis becomes engorged with blood, and an erection occurs. If arousal continues, ejaculation, the movement of sperm from the testes to the penis and out of the man's body, may occur. During ejaculation, sperm move from the epididymis through the vas deferens, ejaculatory duct, and out the urethra. Along the journey, the seminal vesicles, prostate, and bulbourethral glands add sugar, chemicals, and mucus to the sperm to form semen.

→ Testosterone is the chief male hormone. It is secreted by the testes and is responsible for masculinizing male fetuses, triggering LH and FSH production, and the development of male secondary sexual characteristics.

MARIA'S STORY

 Maria has always wanted to have children but she has read horror stories about diabetes and pregnancy. Her doctor has told her that if she is healthy, she could try to get pregnant when the time is right. There are significant risks, however, including a high birth weight baby.

a. What characteristic of diabetes mellitus would contribute to increased birth weight?

_____

_____

_____

b. How can the condition be prevented?

_____

_____

_____

c. What other conditions might Maria need to take care to prevent?

_____

_____

_____

REVIEW QUESTIONS

Multiple Choice

1. In each lobule of the testes are several _____, tubes in which the sperm are made and develop.
   a. seminal vesicles
   b. seminiferous tubules
   c. bulbourethral glands
   d. all of the above

2. The female primary genitalia are the:
   a. ovaries
   b. uterus
   c. vagina
   d. testes

3. This division of the endometrium sheds each month:
   a. horny layer
   b. basal layer
   c. menstrual layer
   d. none of the above

4. This hormone stimulates ovulation:
   a. estrogen
   b. progesterone
   c. LH
   d. FSH

5. Which of the following in not a function of asexual reproduction (mitosis)?
   a. making gametes
   b. tissue growth
   c. tissue repair
   d. all are functions of asexual reproduction

6. Danazol, a synthetic testosterone, has which of the following side effects when prescribed to women to treat endometriosis?
   a. facial hair growth
   b. bone loss
   c. increased ovulation
   d. all of the above

7. Which hormone is inhibited during treatment for BPH?
   a. estrogen
   b. progestin
   c. testosterone
   d. GnRH

Fill in the Blank

1. The stiffening of the penis is known as _____, and the actual movement of sperm out of the penis is _____.

2. In the male reproductive system the _____ is posterior to the bladder, and the _____ gland is inferior to the bladder.

3. _____ is the type of cell division that produces egg and sperm.

4. After ovulation, estrogen and progesterone send _____ (positive or negative) feedback to the pituitary and hypothalamus.

5. The union of sperm and egg is called _____.

6. Oral contraceptives are typically a combination of these two hormones _____ and _____.

Short Answer

1. Trace the journey of a sperm from its birth through fertilization.

_____

_____

_____

_____

2. Explain the role of hormones in controlling the female reproductive cycle.

_____

_____

_____

_____

3. Contrast mitosis and meiosis.

_____

_____

_____

_____

4. Distinguish between PMS and PMDD.

_____

_____

_____

_____

5. Explain the treatment strategies for Benign Prostatic Hyperplasia.

_____

_____

_____

_____

Suggested Activities

1. List all the hormones involved in reproduction. Describe what they do.

2. Describe the human life cycle in detail. How much do you know about your own reproductive system?

3. With a partner, list the similarities and differences between male and female reproduction. How many can you list?

> **17-11** Now that you have completed your journey through this chapter, please go to the DVD for interactive games and puzzles concerning the medical terms and concepts contained in this chapter. By playing the games you will reinforce your learning of the reproductive system in a fun way.

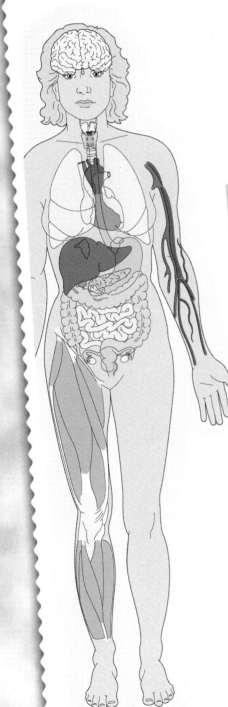

THE JOURNEY'S End

Now What?

For any of you who have ever taken a memorable journey or have lived through a significant event, you know that you have been changed to some degree forever. We hope that this book in some positive way has also changed you.

Your study of anatomy, physiology, and disease has only just begun. The basics that you have learned will act as a foundation of understanding for many things that you will encounter in your life. So go forth and make your mark in the world! The following topics are just a small sampling of how anatomy and physiology and pathology play an important role in many areas. We have added a whole section of Amazing Facts so you can impress friends and family. Enjoy the chapter!

Chapter

18

LEARNING OBJECTIVES

Upon completion of your journey through this chapter, you will be able to:

→ Discuss the relationship between forensic science and anatomy, physiology, and pathology

→ Relate anatomy and physiology changes to the process of aging

→ Describe the concept of wellness and personal choices

→ List and describe wellness concepts for each body system

→ Discuss cancer prevention and treatment

→ Dazzle your friends with amazing anatomy, physiology and pathology facts

MULTIMEDIA APPLICATIONS

DVD Interactive Exercises

→ Animation of the cause and effect of lead poisoning, 18-1

→ Animation of general information on poisons, 18-2

→ Video on geriatric evaluation, multi-system problems, and a variety of geriatric issues, 18-3

→ Video on nutrition and aging, 18-4

→ Video on cancer overview and skin cancer, 18-5

→ Videos on emotional, social, and spiritual well-being, caring, and empathy, 18-6

→ Videos on osteoporosis and carpal tunnel syndrome, 18-7

→ Video on muscle atrophy, 18-8

→ Video on children and obesity, 18-9

→ Videos on smoking, passive smoke, and smoking cessation, 18-10

→ Video on nutrition and eating disorders such as anorexia and bulimia, 18-11

→ Videos on macular degeneration, cataracts, and audiology, 18-12

→ Videos on SIDS and major life transitions of infancy, 18-13

→ Video on AIDS and other sexually transmitted diseases, 18-14

→ Interactive games and puzzles, 18.15

www.prenhall.com/colbert

→ Professional Profiles:
  - Criminalist
  - Mental Health Professionals
  - Physician Assistant

→ Related Web Sites

→ Additional Review Questions

Pronunciation Guide

  Correct pronunciation is important in any journey so that you and others are completely understood. Here is a "see and say" Pronunciation Guide for the more difficult terms to pronounce in this chapter.

anhidrosis (an HIGH droe sis)

bacilli (bah SILL eye)

chlamydia trachomatis
   (klah MID ee ah TRAY koh mah tis)

forensic science (for IN sick)

geriatric (JAIR ee AT rik)

herpes simplex virus 2 (HER peez)

human papilloma virus (pap ih LOW ma)

incontinence (in KAH tih nens)

neisseria gonorrhoeae
   (nye SEE ree ah gon ah REE ah)

spina bifida (SPY nah BIFF ih dah)

thallium (THAL ee um)

treponema pallidum
   (TREP oh NEE mah PAL ih dum)

FORENSIC SCIENCE

forensic science *(for IN sick)*

We've all watched movies or TV shows featuring super-sleuths solving such crimes as murders or disappearances using some amazing scientific technology. Some of the methods used seem pretty far-fetched, as though invented just for entertainment purposes. Actually, much of the **forensic science** we see on TV is very close, if not "dead on," to the actual technologies that are applied to solve crimes. However, it is important to note that this kind of investigation can be extremely tedious, expensive, and take longer than a 46-minute TV show to solve a crime. Most forensic scientists don't look like supermodels or drive $50,000 SUVs either. This section deals with just a sampling of how anatomy and physiology and an understanding of pathology are used to help solve crimes and explain historical mysteries.

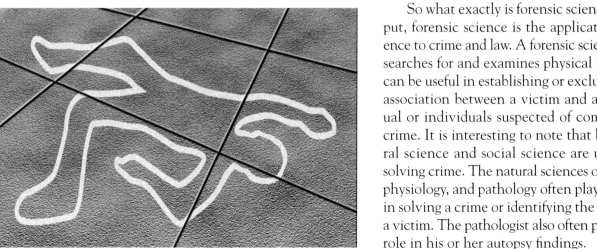

So what exactly is forensic science? Simply put, forensic science is the application of science to crime and law. A forensic scientist often searches for and examines physical traces that can be useful in establishing or excluding some association between a victim and an individual or individuals suspected of committing a crime. It is interesting to note that both natural science and social science are used when solving crime. The natural sciences of anatomy, physiology, and pathology often play a key role in solving a crime or identifying the remains of a victim. The pathologist also often plays a vital role in his or her autopsy findings.

Disease Detection

Forensic science has unlocked ancient mysteries and allowed us to learn something about the health of ancient people. Proof that the ancient Egyptians were stricken with tuberculosis was discovered by examining their skeletons. We normally think of tuberculosis as a lung disease, but tuberculosis thrives in areas of the body that possess high levels of oxygen. These include the lungs, brain, kidneys, and *the ends of long bones!* Since the organs were long ago decomposed, the only thing left to examine were the skeletal remains, particularly the ends of the long bones such as the femur. On examination of these bones, bone scouring that is typically caused by tuberculosis was discovered. Bone scouring is the destruction of the smooth bone surface as a result of colonization by the tuberculosis **bacilli.**

Applied Science

TOMATOES WERE ONCE THOUGHT TO BE POISONOUS

For several hundreds of years, it was believed that tomatoes were poisonous. This erroneous fact occurred because during the Middle Ages, the wealthier people ate from pewter plates that contained high levels of lead. The acidic tomatoes leached out the lead, which was consumed and resulted in lead poisoning. Even in modern times the threat of lead poisoning exists—but not from tomatoes.

18-1 To view an animation on the causes and effects of lead poisoning, please go to your DVD.

bacilli *(bah SILL eye)*

Primitive Surgery

Forensic scientists have discovered evidence of primitive surgery. A skull fragment found in a 400-year-old trash dump at the site of the Jamestown, Virginia settlement provided evidence that skull surgery had been performed on a patient. The circular marks on the fragment indicated that there was an attempt to drill a hole in the skull, probably to relieve pressure on the brain due to a skull fracture that may have been caused by an accident or a tomahawk blow to the head. Scientists analyzing the fragment were able to determine that the patient was a European male on the basis of the fragment's shape and thickness *and* because the bone contained traces of lead that was absorbed by the body from using lead-glazed pottery and/or eating off of pewter plates!

Remains Identification

Nazi Josef Mengele, the "Angel of Death," was responsible for the death of at least 400,000 people during World War II. As the war ended, Mengele escaped to South America and avoided capture. There were numerous reported sightings throughout the years but none produced the war criminal. Then, in 1979, an elderly man named Wolfgang Gerhard drowned and was buried in a Brazilian cemetery. Suspicions that he was Mengele prompted the Brazilian gov-

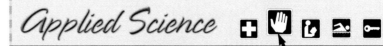

Applied Science

FORENSICS AND HISTORY

DNA fingerprinting techniques are used extensively to clear up mysteries. For example, it is now standard to use this technology to solve rape and murder cases. DNA fingerprinting also helped determine that Thomas Jefferson most likely fathered some children of his slave Sally Hemings. It was also used to identify many of the victims of the World Trade Center, Pentagon, and Flight 93 in Shanksville, Pennsylvania, after 9/11.

Applied Science

IF BONES COULD SPEAK

Skeletal remains can also assist in telling the sex of an individual. The female pelvis is shaped like a basin and is wider than the male pelvis to allow for the larger birth canal. In general, it is broader and lighter than the male pelvis, and the pubic angle is 100 degrees or greater (see Figure 18–1 ▪). The male pelvis is heavier, stronger, and more funnel shaped, with a pubic angle of 90 degrees or less.

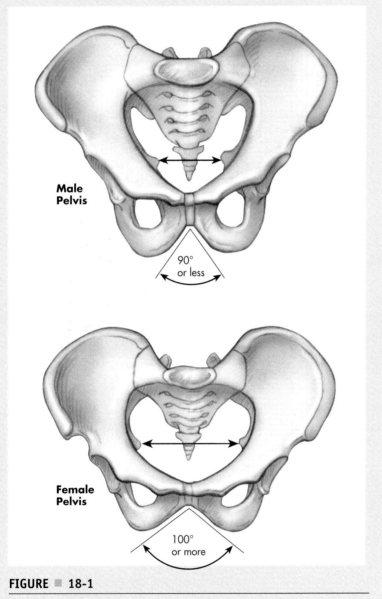

Male Pelvis

90° or less

Female Pelvis

100° or more

FIGURE ▪ 18-1

The male pelvis and female pelvis.

ernment to exhume the body. Analysis of the bones revealed a right-handed male Caucasian between 60 and 70 years old with a height approximated within a half centimeter of Mengele's. Having a pile of bones, a shattered skull, and no current medical information or dental records, the forensic anthropologists turned to a technique called video skull–face superimposition. After reconstructing the shattered skull, the experts marked it with pins at 30 specific structural landmarks. The same thing was done with a picture of Mengele that was scaled to match the size of the skull. Compared side by side, the pins matched up. Photo images of the skull and picture were then superimposed and the points matched up. They had their man. Later, when his old dental records were found, it was again a match, as was a later DNA analysis. More recently, an updated similar procedure was used to determine the actual facial structure of Tutankhamen (King Tut), the Boy Pharaoh of Egypt.

Murder Most Foul!

Although it possesses no vocal chords, your hair speaks volumes about your health and your habits. Hair can reveal the race of an individual, whether the hair was cut or pulled out from the body, or if it was dyed. If the hair shaft has its follicle still attached, genetic information such as DNA or blood type can also be discerned. Hair samples can place an individual at the scene of a crime.

Even though hair is technically dead, just like your fingernails and toenails, it acts like a library (as do your nails), storing information about substances you have been exposed to or ingested. Since hair grows relatively slowly, it can contain a *timeline* of substances that an individual ingested or was exposed to. In an interesting case of thallium poisoning, a man's wife was convicted of murder partially on the basis of hair samples of the victim. The victim's hair was long enough that it provided a record of ingestion of thallium for approximately 330 days. During that period of time, the spikes in concentration of thallium in sections of his hair matched the times the victim was with his wife. Drops in the level of thallium in other sections of his hair coincided with the times he was away from his wife or in the hospital for other illnesses. There was a massive spike in thallium concentration a few days before he died, which was indicative of premeditated poisoning. In a plea bargain, it was revealed that victim's wife had been feeding him rat poison that contained thallium in order to collect on a life insurance policy!

 18-2 To view an animation for more information about poisons, please see your DVD.

www.prenhall.com/colbert
The criminalist is responsible for analyzing the evidence at a crime scene and is one member of the forensic team. To view a video about this profession, please visit the companion website for this chapter.

Fingerprints

Even with all the new advances, fingerprints are still a time honored crime-solving method (see Figure 18–2). Like snowflakes, no two prints are exactly alike. Fingerprints are **friction ridges** that form a pattern on the anterior surface of the hands and also on the plantar surface of the feet. As the name applies, the friction ridges help prevent slippage when grasping or holding objects. They are especially prominent on the skin covering the tips of the fingers and toes. Fingerprints are made inside the womb in response to the pull of the elastic fibers in the dermal papillary layer upon the epidermis. Because fingerprints are not wholly determined by DNA, identical twins do not have identical fingerprints.

Interest in fingerprint analysis has been around for a long time. The uniqueness of fingerprints was discovered in the 1600s. They were described and studied for many years, and then in 1880, Dr. Henry Faulds wrote the first article suggesting that fingerprints may be used to solve crimes. In 1886, he began trying to convince Scotland Yard to adopt a fingerprinting identification system.

DNA Fingerprinting

Forensic scientists may also use DNA, the molecule found in the nucleus of our cells, to identify unknown individuals. DNA can be sampled at a crime scene from body fluids, such as blood or semen, or can be extracted from skeletal or even very tiny soft tissue remains. During the process of DNA fingerprinting (which has nothing to do with actual fingerprints), DNA molecules are split into pieces and separated using electrical currents. These pieces can be compared to pieces of a known DNA sample from a relative of a victim, from hair samples off a comb, or from mouth cells from a toothbrush, or can be compared to a suspect in a criminal case.

FIGURE ■ 18-2

A fingerprint. (Source: Dorling Kindersley Media Library.)

TEST YOUR KNOWLEDGE 18-1

Choose the best answer:

1. This part of your body acts as a library and provides a timeline to record substances that you have ingested or have been exposed to:
   a. heart
   b. lungs
   c. hair
   d. spinal cord

2. This metal has been detected in the bones of medieval skeletons and can be traced to eating off of pewter plates:
   a. silver
   b. gold
   c. kryptonite
   d. lead

3. Proof that ancient Egyptians had tuberculosis was discovered by examining their:
   a. tombs
   b. bones
   c. hair
   d. mommies

4. The difference between male and female skeletal remains can be determined by examining the angle of the:
   a. skull
   b. vertebrae
   c. sternum
   d. pelvis

GERIATRICS

geriatric *(JAIR ee AT rik)*

Many regions in the United States are exhibiting a major change in patient demographics. In general, we are seeing an older population due to safer workplaces, healthier lifestyles, effective vaccines and medications, and opportunities to access health care. As a result, you probably will deal with a high number of **geriatric** patients in the health care profession that you choose. (Even the maternity departments may have 50- to 60-year-old mothers in the future, as witnessed by several recent news reports!)

Geriatric patients differ in many ways from other patient age groups, so it is important to recognize these differences in order to provide the best health care possible for this patient population. Here are some general tidbits to get you started.

The general use of the term *elderly* when describing patients is somewhat misleading. Currently, some experts divide this patient group into the following classifications.

- Age 65 to 75: younger old
- Age 76 to 84: older old
- Age 85 and older: elite old

Using the vague term *aging* as a way to generally describe a patient is a misnomer in that an individual's body does not age uniformly. For example, an individual may look older due to aging skin but may have a cardiovascular system of a person 10 to 15 years younger than his or her chronological age. With that said, there is the general "1% rule" in which we see a 1% decrease in the function of most body systems each year beginning around the age of 30.

So what general characteristics or tendencies do we see in an aging patient? The hallmark sign of aging is a decrease in the ability to maintain homeostasis. These individuals may have normal baselines but begin exhibiting a decrease in the ability to adapt to stressors. Often, disease processes along with aging will accelerate the loss of body reserves that younger patients take for granted, such as recovery time or complications following an accident or surgery. In addition, you may have increased difficulties in evaluating these patients because of decreases in their vision, hearing, and possibly their mental abilities.

Amazing Body Facts

YOU'RE NOT GETTING OLDER, YOU'RE GETTING SMARTER

There are a lot of misconceptions about the brain and aging. In the absence of disease, your brain continues to mature up to the age of 50. Interestingly, we have better overall brain function and decision-making skills at age 60 than at age 30! Even though an older patient may lose up to 1000 neurons a day, we all start out with several billion, so percentage-wise, a loss of 1000 a day is minor. This is why you may not see a decrease in brain function until around the age of 75.

General Body Changes

Let's take a general look at the aging process of the body. There is a decrease in the total body water for both males and females. The clinical significance of this is that older people have a tendency to dehydrate more rapidly, which can potentially affect the excretion rate of medications they are on.

From ages 20 to 70, we see a loss of lean body mass due to an up to 30% loss in the number of muscle cells, atrophy of remaining muscle cells, and a general decrease in muscle strength. Conversely, we see an increase to total body fat, which slowly increases between the ages of 25 and 45. However, this trend can continue up to age 70. The fat accumulated is a "deeper" fat, meaning that it is abdominal fat and found more in the viscera than subcutaneously.

 18-3 To view videos on geriatric evaluation, multi-system problems and a variety of geriatric issues, please see your DVD.

Bone density usually reaches its greatest peak around age 35. Contrary to what you see on TV, loss of bone mass occurs both in women *and* men. For women, during the first 5 years after menopause, 1 to 2% bone loss per year can occur. In general, we see a 1% loss per year between the ages of 55 and 70. After that, it is approximately 0.5% per year.

In general, as an individual ages, he or she loses muscle mass, gains fat, and loses bone density. This varies among individuals and can be affected by lifestyle choices. Later in this chapter, we discuss steps that can be taken to slow (or even reverse) this process.

Gustatory Changes

Sensory changes can also occur in this patient group. The senses of taste and smell begin to deteriorate as a natural process of aging. The number of taste buds decreases by about 50 percent as the body reaches the geriatric stage. Sweet versus bitter tastes becomes less discernable. The acuity of salt and bitter tastes decline. Orange juice may taste metallic. Patients on oxygen via a nasal cannula may have a decrease in their ability to taste. These changes are clinically

18-4 To view a video on nutrition and aging, please see your DVD.

significant because they may make it more difficult to insure a properly balanced diet for good health. This is also a concern coupled with the additional functional and physiologic difficulties affecting grocery shopping and food preparation. It is estimated that 33% of this patient group lives alone. These and other sensory impairments affect activities of daily living (ADLs), and as a result, we see approximately 5 to 15% of this population exhibiting protein and calorie malnutrition.

Additional barriers to proper nutrition include tooth loss, difficulty swallowing, and decreased saliva. In addition, there are decreases in digestive juices acidity and secretion, nutrient absorption, and bowel function.

The Brain and Nervous System

One main problem with the aging process and the nervous system is the slower reaction time that occurs. As a result, there are increased chances for motor vehicle accidents, falls, burns, and so forth.

Pain is something usually associated with the aging process. Even though the authors of this book are young in spirit, vibrant, and good looking, they have occasional creaking joints and stiff, sore muscles! Untreated pain can lead to an overall decrease in the quality of a person's life, including impaired sleep, a decrease in socialization, confusion, depression, malnutrition, polypharmacy (use of multiple medications), and impaired ambulation. If any of these conditions were occurring before the pain was introduced, the pain may cause these conditions to worsen.

Often, geriatric patients are undermedicated for pain. This may be because patients who are debilitated, cognitively impaired, or have a history of substance abuse are sometimes unable to effectively relate how they feel. Sometimes, it is a result of the ignorance of health care professionals who cannot recognize indicators of pain.

Clinically, these are some of the potential behavioral changes related to pain:

- Changes in personality such as becoming agitated, quiet, withdrawn, sad, confused, depressed, grumpy
- Loss of appetite
- Screaming, swearing, name calling, grunting, noisy breathing
- Crying, rocking, fidgeting
- Splinting or rubbing a sore area, wincing
- Cold, clammy, and pale skin

The Cardiovascular System

Within the cardiovascular system, we see changes such as calcification of the heart valves, which decreases their efficiency. We also see a lessening in the flexibility of the blood vessels. This makes vessels less able to deal with blood pressure changes, so increased blood pressure is common. There is also a decrease in cardiac output and an approximate 25% decrease in the maximum heart rate.

Source: Arthur Tilley/Getty Images, Inc.—Taxi.

Genitourinary System

Between ages of 20 and 80, we lose about 50% of our renal function. This is clinically significant when you consider the amount of drugs used by the geriatric population that are excreted by the kidneys. Therefore, a drug taken by someone with impaired renal (and/or liver) function can lead to the drug not being metabolized at the expected rate. This can lead to an accumulation of the drug in the body at harmful or toxic levels. There is also an increase in **incontinence** (loss of bladder control) in this group.

incontinence *(in KAH tih nens)*

Integumentary System

Here we see a loss in skin elasticity, increased skin delicacy, multiple skin lesions, and increases in skin cancer. While many of these problems could have been reduced by limiting sun exposure and not smoking earlier in their lives, the use of medications such as systemic corticosteroids does accelerate many of these conditions.

 18-5 To view videos on cancer overview and skin cancer please see your DVD.

Polypharmacy

Polypharmacy, the administration of many drugs at the same time, is a major concern for this patient group. Contributing factors to this problem include the fact that these patients typically see several specialists for a variety of diseases, often with one doctor not knowing what the other is prescribing. These multiple diseases have competing therapeutic needs, and the combinations of utilized drugs can cause life-threatening situations. Aging also can affect the rates of drug absorption, drug distribution throughout the body, metabolism of the drugs, as well as their excretion from the body.

As a result, here are some clinical considerations. Review the drugs that the individual is taking, looking for possible drug interactions or medications that are no longer needed. Look at your patient's liver and kidney functions. They will affect the rate of removal of medications from the body and resultant drug blood levels. When medicating with opioids, dose by age, not by body weight. Remember, opioids slow the gastrointestinal tract. In general, when considering an *equal* dose given to a geriatric patient compared to a middle-aged patient, analgesics are stronger and last longer in elderly patients. Start low and go slow.

It is important that patients tell all their doctors what medications they are taking. If you work in a physician's office or do home care, you can help your patients relay this information more effectively by reminding them to bring all of the drugs they're taking to each appointment. It is also important to instruct them to keep an updated list of their drugs and dosages in case they have to make a trip to the emergency department.

TEST YOUR KNOWLEDGE 18-2

Choose the best answer:

1. There is a _____ decrease in the function of most body systems each year beginning around the age of 30.
   a. 10%
   b. 20%
   c. 5%
   d. 1%

2. The hallmark sign of aging is the decrease in the ability to maintain:
   a. cardiac enzymes
   b. exercise potential
   c. homeostasis
   d. sense of humor

3. _____ is a term used for the administration of many drugs at the same time.
   a. Addiction
   b. Polypharmacy
   c. Multimeds
   d. Overdose

4. In the absence of disease, your brain continues to mature up to the age of:
   a. 50
   b. 35
   c. 75
   d. 10

WELLNESS

One of the most important personal choices you will ever make will be the decision on what kind of lifestyle you want to live. Choices you make now may have a *profound* effect on your future health and lifespan. Will you eat properly, exercise within reason, smoke, do drugs and/or alcohol, live or work in a dangerous environment, and/or engage in risky behaviors?

Individual accountability and *informed choices* are two key concepts in deciding your lifestyle. It is amazing that individuals have sued fast-food companies for making their food too appealing and, as a result, making the "victim" get fat.

People don't force other people to smoke cigarettes. In each of these cases, the individual made a conscious choice to eat more than necessary and to begin smoking.

Yes, peer pressure is real, but it is too often looked at from the negative side. What about peers and influential people who promote good, healthy lifestyles? Think about it. Do you know people or have friends who do healthy things? Once again, it is a conscious choice to surround yourself with the type of friends you want.

Read and critically analyze what you read. Utilize multiple reliable sources. Even though the Internet is a wonderful source of information, there is a lot of "junk" science that is more opinion than fact.

Let's take a quick review of the body systems that we covered and briefly discuss some ways to improve your health. While there are always some controversial issues in wellness, we have chosen the more commonly held beliefs to discuss here.

Nervous System

Stress is a natural part of life. Stress is also good and necessary: it is a motivator, and it helps you to protect yourself. The problem occurs when stress becomes chronic and you can no longer effectively deal with it. The high cost of stress is that it can affect some or all of your body's various systems to varying degrees. A poor stress response can lead to an assortment of disorders such as eating and digestive disorders, decreased immune response, decreased memory and work capacities, sleep problems, joint and muscle aches, heart problems, and person-

www.prenhall.com/colbert
The mind and body cannot be separated. There are a host of mental health professionals trained in diagnosing and treating mental health issues. For a review of these various professions, please go to our website for this chapter.

18-6 To view videos on emotion, social and spiritual well-being, and caring and empathy, please see your DVD.

ality changes. Some of these problems can be life threatening. For an extensive review of a healthful stress management system, consult the Study Companion Guide found in the back of your textbook.

Often, when dealing with a patient, we focus specifically on the presenting illness and fail to realize we should look at the whole person. The "whole person" includes not only other systems of the body but also the mental and spiritual aspects of the person. Too often in the past, dealing with mental health has had a negative image. We don't hide the fact that we have the flu or a broken arm, so why should we hide the fact that we may be sad for long periods of time (depression) and need help to resolve that condition?

Skeletal System

Diet is extremely important in the growth and protection of your bones. A diet rich in calcium and vitamins helps to maintain good bone growth and development. Weight-bearing exercise has also been shown to be beneficial in maintaining healthy bones over a lifetime.

One common occupational condition related to repetitive motion, such as typing on a keyboard, playing a piano, or hammering, is known as **carpal tunnel**

18-7 To view videos on osteoporosis and carpal tunnel syndrome, please see your DVD.

syndrome. While this syndrome is caused by damage to the median nerve, it is the result of the *skeletal* structure of the wrist being too restrictive during extended periods when the wrist is kept in an upward bent position. See Figure 18–3 ■, which illustrates the structures that affect this repetitive syndrome.

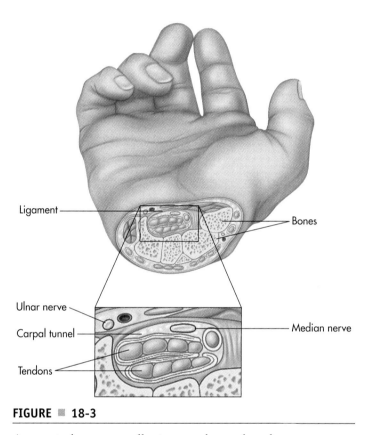

FIGURE ■ 18-3

Anatomical structures affecting carpal tunnel syndrome.

Muscular System

Again, proper exercise and diet will help to develop and maintain properly functioning muscles. While there are many types of muscle-training programs, you need to investigate which type is best for you depending on your needs or desired outcomes. See Figure 18–4 ■, which presents the Activity Pyramid as a possible guide.

18-8 To view a video on muscle atrophy, please see your DVD.

Integumentary System

Proper diet and hydration is important for the functioning of the integumentary system. This is why it is important to drink enough fluids to meet your hydration needs. Other fluids such as caffeinated coffee and alcohol are diuret-

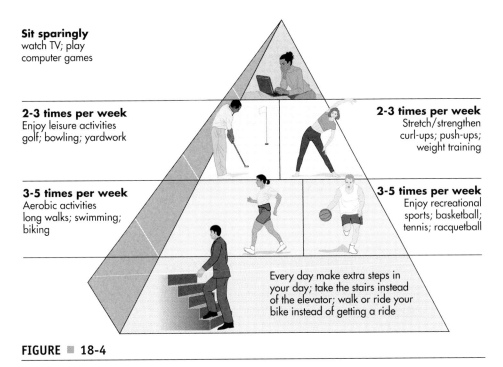

Sit sparingly
watch TV; play
computer games

2-3 times per week
Enjoy leisure activities
golf; bowling; yardwork

2-3 times per week
Stretch/strengthen
curl-ups; push-ups;
weight training

3-5 times per week
Aerobic activities
long walks; swimming;
biking

3-5 times per week
Enjoy recreational
sports; basketball;
tennis; racquetball

Every day make extra steps in
your day; take the stairs instead
of the elevator; walk or ride your
bike instead of getting a ride

FIGURE ▇ 18-4

The Activity Pyramid.

ics and can cause a net water loss. Smoking also affects this system by causing premature aging of the skin.

While sun exposure is important for the production of vitamin D in your body, limiting the amount of sun exposure to prevent skin cancer is equally important. While some forms of skin cancer are treatable, others can be lethal. The effect of the sun is cumulative, and severe sun burns as a child can have severe consequences in adulthood. There are several ways to prevent excessive sun exposure. First, minimize your time in the sun between 10 a.m. and 4 p.m. because this is when the sun's rays are most intense. Wear long-sleeved shirts when possible, brimmed hats, sunglasses (preferably wraparounds), and, of course, sun screen.

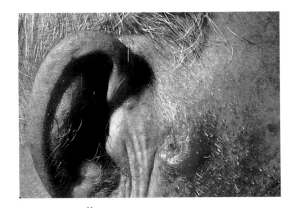

Squamous cell cancer.

Cardiovascular System

A heart-healthy diet low in saturated fats (remember, we all need some fat for proper function) and high in fiber, rich in fruit and vegetables will help maintain an optimal cardiovascular system.

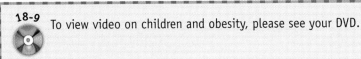

18-9 To view video on children and obesity, please see your DVD.

Diet alone is not sufficient for a healthy heart, however. The proper level and regularity of exercise also helps condition the heart for maximum functioning. The newest guidelines have increased the recommended exercise level to 30-45 minutes of activity, 5-7 days a week. This can be as simple as walking briskly for 30-45 minutes a day, five to seven times a week. Of course, the level of your exercise program depends on many individualized factors. As always, consult your doctor before beginning any exercise program.

Smoking, alcohol, and other drugs can adversely affect the cardiovascular and other systems. For an illustration of the effects of chronic alcoholism on the cardiovascular, digestive, and nervous system, please see Figure 18–5 ▪.

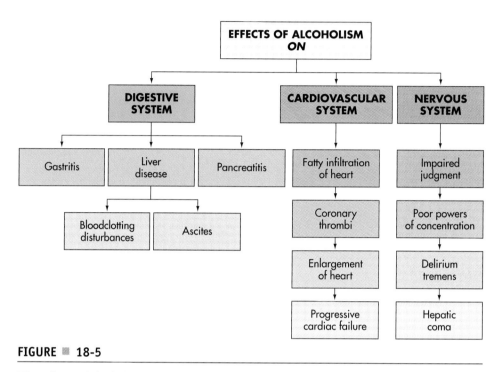

FIGURE ▪ 18-5

The effects of alcoholism on the body systems.

Respiratory System

Smoking is the #1 preventable cause of respiratory diseases. Smoking can lead to damage of lung tissue and chronic diseases such as bronchitis, emphysema, and asthma. In addition, smoking increases the occurrence of lung infections and colds as well as sinus infections. Approximately 80% of all lung cancers can be traced to smoking. Smoking also affects the heart by reducing the availability of oxygen to the heart muscle.

18-10 To view videos on smoking, passive smoke, and smoking cessation, please see your DVD.

Smoking, along with alcohol consumption, leads to an increase in stomach and mouth cancers.

What we breathe in, depending on where we live or work, can also affect the health of the respiratory system. Both outdoor and indoor pollution can lead to a number of respiratory problems.

Occupational hazards can occur when workers are exposed to dust or vapors. For example, coalminers exposed to coal dust without proper protection can develop **black lung disease**, also known as Coalworkers' pneumoconiosis, in which so much coal dust builds up in the lungs that they are unable to expel it. The lung's initial response to an inhaled irritant is to close down or restrict the airway, thereby minimizing the inhalation of the substance. This can lead to severe breathing difficulties. See Figure 18–6 ▪, which shows the constricted airway in an asthma attack.

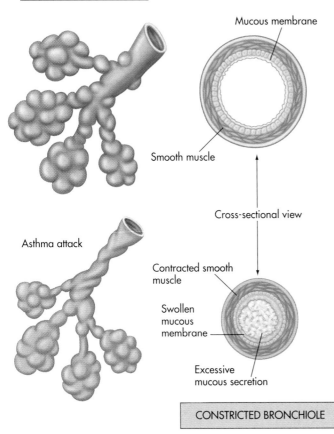

NORMAL BRONCHIOLE

Mucous membrane

Smooth muscle

Cross-sectional view

Asthma attack

Contracted smooth muscle

Swollen mucous membrane

Excessive mucous secretion

CONSTRICTED BRONCHIOLE

FIGURE ■ 18-6

The normal and constricted airway.

Gastrointestinal System

A proper diet is critical for growth, development, and health in general. Lack of a proper diet, which leads to undernourishment, can affect all the systems, as can be seen in Figure 18–7 ■. While most experts agree that the best source of vitamins and minerals comes from natural food sources, the responsible use of supplemental vitamins and minerals can play an important role in health. For example, excessive dosages of fat-soluble vitamins (A, D, E, and K) can actually harm the body because they can build up in toxic levels. See Table 18-1 for more information about vitamins and what they do.

Clinical Application

AGE- AND ACTIVITY-RELATED DIETS AND NUTRITIONAL NEEDS

At some time in our educational careers, we have seen the old food pyramid, which told us what portion of our diet should be dairy, what portion of our diet should be vegetables, and so forth. One of the problems with this pyramid was that it was designed to fit *everyone*. Unfortunately, not everyone is the same sex, height, shape, or age, nor has the same level of activity. Enter MyPyramid.gov. This is the USDA's newest spin on the food pyramid that takes into account your age, sex, and activity level when determining the best diet for YOU.

 18-11 To view videos on nutrition and eating disorders, please see your DVD.

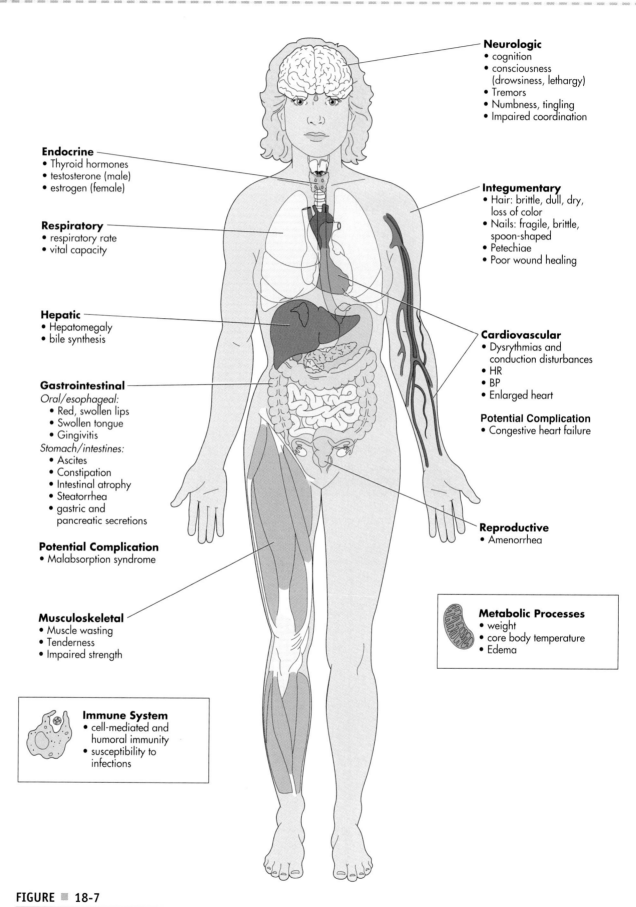

Neurologic
- cognition
- consciousness (drowsiness, lethargy)
- Tremors
- Numbness, tingling
- Impaired coordination

Endocrine
- Thyroid hormones
- testosterone (male)
- estrogen (female)

Respiratory
- respiratory rate
- vital capacity

Hepatic
- Hepatomegaly
- bile synthesis

Gastrointestinal
Oral/esophageal:
- Red, swollen lips
- Swollen tongue
- Gingivitis
Stomach/intestines:
- Ascites
- Constipation
- Intestinal atrophy
- Steatorrhea
- gastric and pancreatic secretions

Potential Complication
- Malabsorption syndrome

Musculoskeletal
- Muscle wasting
- Tenderness
- Impaired strength

Immune System
- cell-mediated and humoral immunity
- susceptibility to infections

Integumentary
- Hair: brittle, dull, dry, loss of color
- Nails: fragile, brittle, spoon-shaped
- Petechiae
- Poor wound healing

Cardiovascular
- Dysrythmias and conduction disturbances
- HR
- BP
- Enlarged heart

Potential Complication
- Congestive heart failure

Reproductive
- Amenorrhea

Metabolic Processes
- weight
- core body temperature
- Edema

FIGURE ▪ 18-7

The effects of undernourishment on the body systems.

TABLE 18-1 Vitamins and the Body Systems

VITAMIN	DESCRIPTION
Vitamin A	Proper night vision; proper development of bones and teeth; mucous membrane and epithelial cell integrity; helps resist infection
Vitamin B	B_2 promotes healthy muscular growth; B_{12} is needed for healthy blood cell development and to treat pernicious anemia; B_1 and B_{12} promote healthy function of nervous tissue; B_1 aids in carbohydrate metabolism, normal digestion, and appetite; niacin is necessary for fat synthesis and cellular respiration
Vitamin C	Aids in absorption of iron; promotes healing of fractures, development of teeth and bone matrix, and wound healing; ensures capillary integrity; bolsters immune system
Vitamin D	Promotes strong bones and teeth, regulates skeletal calcium reabsorption; aids in absorption of calcium and phosphorus from the intestinal tract; D_3 helps regulate release of parathyroid hormone
Vitamin E	Promotes muscle growth; E_1 necessary for hemolytic resistance of red blood cell membranes; helps prevent anemia for proper reproductive system functioning; current research shows that consumption from natural sources may reduce the risk of Parkinson's disease; further investigation is needed
Vitamin K	Needed for proper blood clotting

Endocrine System

Proper diet and exercise also assist the endocrine system. One of the areas of concern, in professional and even high school sports, is the use of performance-enhancement substances. For example, anabolic steroids are used to increase strength and endurance rapidly and to build muscle mass. Anabolic steroids are closely related to the male hormone testosterone. However, these steroids have *serious* side effects that include kidney damage, liver damage, increased risk of heart disease, irritability, and aggressive behavior. Woman taking steroids can develop facial hair and deeper voices. In men, these substances diminish sperm production. Some of these effects can be permanent even after ceasing the use of the drugs. The use of anabolic steroids is banned and is tested for in sports.

Sensory System

Care of the sensory system includes proper diet, wearing hearing and sight protective devices when necessary, and periodic examination of the eyes and ears. Wearing hearing protection during activities that

 18-12 To view videos on macular degeneration, cataracts, and audiology, please see your DVD.

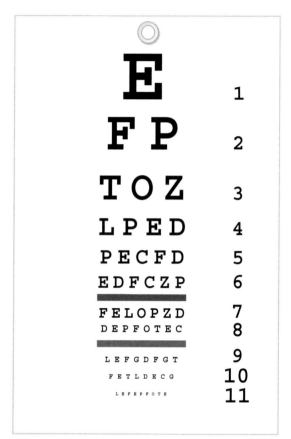

FIGURE ■ 18-8

The Snellen Eye Chart.

produce high levels of noise will greatly extend the functional life of your hearing. Damage to the ear is cumulative, so there is no better time to start than right now. In addition, protective eyewear should be worn any time that the risk of eye injury can occur, such as in certain occupations, hobbies, and sports. Figure 18–8 ■ shows the Snellen Eye chart for determining visual acuity.

Immune System

Proper diet and exercise are needed for optimal functioning of the immune system. In addition, other factors can assist your immune system. One of the simplest and most effective ways to protect not only yourself but your patients is hand-washing. Always wash your hands before and after working with each of your patients. Even though you are a staff member in a hospital or health care institution, and not a patient, you may still be a carrier of infectious organisms. In addition, you could be susceptible to becoming infected by hospital pathogens. Correct washing of your hands goes a long way in stopping the spread of infection.

However, there is also another way to protect yourself and others from the spread of pathogens via body fluids: follow the Standard Precaution Guidelines (refer to Figure 1–6 in Chapter 1). These guidelines are standard precautions that are to be followed for every patient that you treat.

Having current immunizations is important to assist the immune system in being prepared for certain pathogens. Immunization schedules are recommended by the Centers for Disease Control and Prevention (CDC), the American Academy of Pediatrics (AAP), and the American Academy of Family Physicians (AAFP). Many individuals mistakenly believe immunizations only occur in childhood. The influenza vaccine is just one example of an important immunization for the geriatric population.

A big issue is when and when not to take antibiotics. The body handles most infections using its own immune system in a few days, so antibiotics are in most cases not required. Further, the overuse of antibiotics has led to several critical health issues. First, many viral infections are mistakenly treated with antibacterial agents, which do nothing to the virus and cause harm to normal bacteria such as in the intestinal system. Second, overuse does not allow the immune system of a child to properly develop and respond to future infections; this can cause related disorders such as asthma. Finally, many patients do not properly take their antibiotics. For example, they do not take the full dose for the full length of time but discontinue when they "feel better." However, often there are still surviving bacteria that are left that represent a stronger, drug-resistant strain, and they are now free to reproduce stronger drug-resistant offspring. This has led to limited epidemics of drug-resistant infections.

Applied Science

ANTIBIOTICS

The term *antibiotic* means "against life" and technically includes medications that inhibit or destroy any microorganism, including bacteria, viruses, and fungi. However, *antibiotic* in medicine has become associated only with antibacterial agents.

Reproductive System

Don't smoke! Smoking mothers tend to have babies of lower birth weights, tendency toward premature births, and a higher rate of SIDS (sudden infant death syndrome). And while we're talking about babies and children, don't forget about the hazards of second-hand smoke in the home. In homes that have at least one smoking parent, kids have slower than normal lung development and are predisposed to increased incidences of bronchitis, asthma, and ear infections (otitis media).

 18-13 To view videos on SIDS and the major life transition of infancy, please see your DVD.

 www.prenhall.com/colbert
A physician's assistant is trained in the diagnoses and treatment of all the body systems. To learn more about this profession and to view a video, please go to the website for this chapter.

DIET

As the old saying goes, when you are pregnant you are now eating for two! This doesn't mean that mom should pig out at every chance she gets, however. What it means is that diets should be followed that provide important vitamins, minerals, and nutrients for the developing fetus and to maintain the health of the mother. Think about it. If the diet is lacking in calcium, where does the fetus get calcium for bone development? It has to take it from the bones and teeth of the mother, thus decreasing the integrity of her system. The congenital condition of **spina bifida** can be prevented by a dietary supplement of folic acid, a member of the vitamin B complex. The elimination of alcohol during pregnancy is also important to assure the best chances of normal spinal cord and nervous system development.

spina bifida *(SPY nah BIFF ih dah)*

SEXUALLY TRANSMITTED DISEASES (STDS)

STDs are a growing problem and can have serious effects on the reproductive system and lethal effects on the body. There are various types of diseases and organisms that can be transmitted through unprotected sex (including oral sex). Please see Table 18–2 for a list of sexually transmitted diseases.

 18-14 To view videos on AIDS and other sexually transmitted diseases, please see your DVD.

CANCER PREVENTION AND TREATMENT

All the body systems can be ravaged by cancer. Cancer is the runaway reproduction and spreading of abnormal cells and is a very complicated disorder. Each type of cancer, named for the location or the type of cells that are running amok (for example, colon cancer, prostate cancer, squamous cell carcinoma) has its own unique characteristics. However, in the past few years, medical science has learned a number of things about cancer that have resulted in great improvements in cancer prevention and treatment.

Any number of triggers can make a cell cancerous, including genes, radiation, sunlight exposure, smoking, fatty foods, viruses, and chemical exposure.

herpes simplex virus 2 *(HER peez)*

neisseria gonorrhoeae
(nye SEE ree ah gon ah REE ah)

chlamydia trachomatis
(klah MID ee ah TRAY koh mah tis)

treponema pallidum
(TREP oh NEE mah PAL ih dum)

human papilloma virus
(pap ih LOW ma)

TABLE 18-2 Sexually Transmitted Diseases

DISEASE	ORGANISM	SYMPTOMS
Herpes	Herpes simplex virus 2	Male: fluid-filled vesicles on penis
		Female: blisters in and around vagina
Gonorrhea	Neisseria gonorrhoeae	Male: purulent discharge from urethra, dysuria, and urinary frequency
		Female: purulent vaginal discharge, dysuria, urinary frequency, abnormal menstrual bleeding, abdominal tenderness; can lead to sterility
Chlamydia	Chlamydia trachomatis	May be asymptomatic
		Male: mucopurulent discharge from penis, burning and itching in genital area, dysuria, swollen testes; can lead to sterility
		Female: mucopurulent discharge from vagina, inflamed bladder, pelvic pain, inflamed cervix; can lead to sterility
Syphilis	Treponema pallidum	Systemic disease that can lead to lesions, lymph node enlargement, nervous system degradation, chancre sores
Genital Warts	Human papilloma virus (HPV)	Cauliflower-like growths on penis and vagina

Some of these triggers, like genes or some viruses, are difficult to avoid. But others, like smoking, sunlight, radiation exposure, and fatty foods can be pretty easily avoided by eating right, avoiding smoking, and wearing sunscreen. Many types of cancer can be prevented or managed with a healthy diet and exercise. Even genetic susceptibility to cancer does not make cancer unavoidable. Testing, such as mammograms (for breast cancer), colonoscopy (for colon cancer), and Pap tests (for cervical cancer), can improve survival by catching cancers early, before they have spread, or even allowing the removal of abnormal cells before they become cancerous. See Table 18-3 for testing suggestions for various cancers.

Treatments for cancer typically involve surgical removal of the cancerous cells, if possible, and some form of treatment to kill any cells remaining in the body. **Chemotherapy** is the treatment of cancer with chemicals that kill rapidly dividing cells. Radiation uses energy waves to shrink tumors. Biological or immunotherapy targets the cancer by manipulating the immune system to hunt down and kill the cancer cells. New treatments are constantly under development to treat cancers that are difficult to fight.

Let's use the skin cancer, melanoma, as an example. Melanoma is one of the most deadly forms of skin cancer. It is formed by the runaway reproduction of melanocytes, the pigment-forming cells of the skin. People at the highest risk of melanoma are those with fair skin and light eyes or hair who have been exposed to lots of sun during their lifetime. However, new evidence indicates that even people who tan easily may develop melanoma if they get enough sun exposure. Melanoma risk is higher for those living near the equator, but people in the northern parts of the United States are not without risk. Exposure to sunlight, particularly sunburns, even as an adult, is the key risk factor for melanoma. Genetic factors are involved in some cases of melanoma.

Melanoma can be easily prevented by decreasing exposure to UV light. Aside from staying indoors all the time, which isn't very practical, limited exposure and the use of sunscreen is the best way to protect yourself from melanoma. Individuals at risk should have a skin screening on a regular basis.

Standard treatment for melanoma in early stages (stage I, no spread) has been the "watch and wait" approach. The melanoma is removed, and the patient is monitored for several years. Patients with more advanced melanomas often have lymph nodes sampled and removed to prevent further spread. This more extensive surgery was deemed unnecessary for patients in very early-stage disease. However, a 2005 study has shown that even patients with no obvious spread of their cancer benefit from having lymph nodes sampled and removed if they contain cancer cells. Patients who had the procedure, called a sentinel lymph node mapping and biopsy, were 26% less likely to have their cancer return within 5 years than patients who only had the tumor removed. See Figure 18–9 for the possible causes and warning signs of cancer.

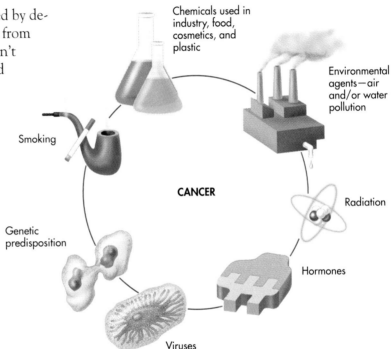

Chemicals used in industry, food, cosmetics, and plastic

Environmental agents—air and/or water pollution

Smoking

CANCER

Radiation

Genetic predisposition

Hormones

Viruses

The American Cancer Society lists several signs of cancer, the first letter of each word spelling the acronym,

CAUTION.

Change in bowel or bladder habits
A sore that does not heal
Unusual bleeding or discharge
Thickening or lump in breast or elsewhere
Indigestion or difficulty in swallowing
Obvious change in wart or mole
Nagging cough or hoarseness

FIGURE ■ 18-9

Possible causes and warning signs of cancer.

TABLE 18-3 Summary of American Cancer Society Recommendations for Early Detection of Cancer in Asymptomatic People

SITE	RECOMMENDATION
Breast	Annual mammography for women over 40
	Annual breast exam
	Monthly self-breast examination
Colon and rectum	Beginning at age 50:
	Fecal occult blood test every year or
	Flexible sigmoidoscopy every 5 years or
	Annual fecal occult blood test and flexible sigmoidoscopy every 5 years
	A double contrast barium enema every 5 to 10 years
	A colonoscopy every 10 years
	Combined testing is preferred over annual tests every 5 years; people at moderate or high risk should talk with doctor about a different testing schedule
Prostate	Prostate-specific antigen test and digital rectal exam should be offered annually beginning at age 50; men at risk should start testing at age 45
Uterus	Cervix: All women who are sexually active or who are 18 and older should have an annual Pap test and pelvic exam
	Endometrium: all women should be informed about the risks and symptoms of endometrial cancer
Cancer-related checkup	A cancer-related check up is recommended every 3 years for people aged 20 to 39 years and every year for people 40 and older; exam should include health counseling and, depending on a person's age, examinations for cancers of the thyroid, oral cavity, skin, lymph nodes, testes, ovaries, as well as for some nonmalignant diseases

Source: American Cancer Society

TEST YOUR KNOWLEDGE 18-3

Choose the best answer:

1. This vitamin is needed for proper blood clotting:
   a. A
   b. B
   c. C
   d. K

2. The abuse of this substance can have side effects such as facial hair and deeper voices in women, kidney and liver damage, and aggressive behavior.
   a. antidepressants
   b. aspirin
   c. anabolic steroids
   d. chocolate

3. The _____ eye chart is used to determine normal vision.

   a. Snellen
   b. Seymour
   c. See Clear
   d. Optical

4. The guidelines that help prevent the spread of disease by contact with body fluids are the:

   a. Barrier reef
   b. Splash guard
   c. Standard precautions
   d. Body shield

MORE AMAZING FACTS

Remember those amazing facts we promised you at the beginning of the chapter? True to our word, here is a list of a few to amaze your friends and family.

- Senior citizens are more prone to food poisoning not only because of decreases in their senses of smell and taste but because their digestive juices are not as acidic as they used to be and so cannot always efficiently destroy all of the food-borne pathogens they ingest.

- Nerve impulses can travel up to 426 feet per second!

- Approximately 450,000 people die in the United States annually because of smoking-related diseases. That is about equivalent to a jumbo jet full of passengers crashing *each day with no survivors*.

- On average, a healthy kidney filters about 180 quarts of fluid every day. What makes this amazing is that the kidney is only about 4 inches long, 2 inches wide, and 1 inch thick!

- Hair grows about a quarter inch each month and grows faster during the day than at night. Hair also grows faster in the summer than in the winter.

- You use a little over a half pint of oxygen each minute when you are at rest.

- Everyone has one nare (nostril) that is larger than the other. If you don't believe it, take a look at people next time you go out to the mall or out to eat.

- Talk about a busy worker! Your heart beats over 36 million times a year!

- You possess over 16,000 miles of capillaries!

- Because viruses are continuously mutating, your immunity to influenza will not last a lifetime.

- More hair facts! You have from 110,000 to 150,000 hairs just on your head! Each strand of hair can support approximately 100 grams of weight, so, at least in theory, a full head of hair could support the weight of two African elephants!

- Research appears to indicate that vitamins and minerals from natural food sources are better utilized than synthetic pills. But the pills are better than nothing!

- The horns of a bull are composed of the same material that makes up your fingernails and toenails.

- You have about a quarter million sweat glands on your feet.

- Based on current research of fibroblasts' doubling ability before they can no longer accurately divide, human bodies have the *potential* to live to 120 years of age.

- Your eyes can see approximately 7 million shades of color.

- The ability to roll your tongue into a tube is inherited; not everyone can do it (or cares to do it).

- Cavities and poor oral hygiene can lead to diabetes and heart attacks. In fact some health experts believe that daily flossing can add 6.4 years to your life! That's because bacteria that grows in the mouth of an individual with poor oral hygiene can escape into the bloodstream and travel throughout the body, causing problems. As a result, in worst-case scenarios, that individual may be at a four times greater risk for stroke and at a 14 times greater risk for heart attack. The risk for diabetes is also increased.

- Current research indicates that stomach cancer, which affects 24,000 Americans annually, may originate from bone marrow cells that enter the stomach to repair damage to the stomach lining.

- Walking uphill or downhill may make a difference in your desired health outcomes. A recent study showed that individuals who walked uphill cleared fats (especially triglycerides) from their blood faster, while downhill hiking reduced blood sugars more readily and improved glucose tolerance. Hiking either way removed LDL, or bad cholesterol. This information may be applicable for exercise regimens for diabetics who may have trouble with aerobic exercises.

anhidrosis *(an HIGH droe sis)*

- CIPA (congenital insensitivity to pain with **anhidrosis**) is a rare genetic disorder that affects the development of the small nerve fibers that transmit the sensations of pain, heat, and cold to the brain. There are only 17 cases known in the United States. Patients with this untreatable condition receive bruises without knowing they are hurt. Since these patients can't sense extreme cold or heat, they don't sweat! Biting through their tongue while eating is a distinct possibility.

SOME THINGS TO PONDER AND SOME THINGS TO KEEP YOU UP AT NIGHT

We hope you enjoyed your journey through our book and that you learned a thing or two! As with any textbook, it is virtually impossible to put everything you want in a book of a given size. Not only that, but with the speed of technologic advancements, what the author or authors consider "the latest" may be old news by the time you, the student, read it in your textbooks. So here are a few things that are developing as we put this book to bed. Some are amusing, some a little scary.

- New research may shed light on the purpose of the appendix! It may be the producer and storage area for the good bacteria your GI tract needs for the proper digestion of food. If a situation occurs where your normal digestive tract bacteria die out or are greatly reduced (such as from amoebic dysentery or cholera), the appendix can "reseed" the GI tract with the normal flora that you previously had!

- Lung cancer, one of our deadliest cancers, can exist and metastasize in a patient for years before it is detected. Patient screening using computed tomography appears to find lung cancer at early, more treatable stages. Does this increase the survivability rate in these patients? Will this become a standard screening routine for patients believed at risk? Time will tell.

- Global warming with brain eating amoebas! The *Naegleria fowleri* amoeba thrives in the bottom sediment of warm, slow-moving waters such as lakes, hot springs, and in rare cases, swimming pools. If water containing this amoeba gets splashed up an individual's nose, the amoeba can attach to the olfactory nerve and eat its way up to the brain, with almost always fatal results. The initial symptoms are headache, stiff neck, and fever that can be attributed to an incredible number of health conditions. Later stages include hallucinations and changes in behavior. The complete clinical scenario from initial exposure to death normally occurs within two weeks.

- A recent problem has been uncovered by doctors that relate to the increase incidents of heart attacks and strokes in people who have received blood transfusions. It appears that stored blood loses a valuable component called **nitric oxide** even after the first day of storage. Nitric oxide is important because it causes blood vessel walls to relax and open up allowing an easier flow of red blood cells, which carry oxygen, to the body's tissues. Current research indicates that adding nitric oxide to stored blood before transfusing it will once again allow the opening up of blood vessels for more effective oxygen delivery and, thus, a decrease in heart attacks and strokes.

- Think that all hunger impulses and cravings are in your head? A small study has found that certain individuals who crave chocolate daily have different bacteria in their GI tract than others! When measuring metabolic byproducts in the blood and urine of the chocolate cravers, higher levels of the amino acid glycine and lower levels of the "bad" cholesterol, LDL were found. This may open up new areas of research where cravings for various foods dictated by the gut and different bacteria may affect weight loss outcomes in patients.

SUMMARY

Snapshots from the Journey

→ Forensic science is the application of science to crime. Natural sciences (including anatomy, physiology and pathology) and social sciences are used when solving crimes.

→ Forensic science has been used to solve not only current mysteries but ancient mysteries as well.

→ The uniqueness of fingerprints was written about as early as the 1600s.

→ DNA fingerprinting is a form of identifying individuals from small samples of body fluids or tissues.

→ The geriatric population is the fastest growing population in the United States.

→ The hallmark sign of aging is the decreased ability to maintain homeostasis.

→ Our bodies don't age evenly, and certain systems age more rapidly than others.

→ The loss of mental capacities is not directly related to aging until around age 75, when it is still minimal unless disease is present.

→ Due to changes in the gastrointestinal, renal, and hepatic systems, geriatric patients respond differently to many medications.

→ Polypharmacy is the use of many drugs at the same time and often is the result of seeing many specialists at the same time.

→ The most important personal choice you will make is a healthy lifestyle.

→ To maintain a healthy lifestyle, it is important to eat properly, exercise, manage stress, and avoid bad habits.

→ Some cancers, such as skin and lung cancer, can be highly preventable. Limiting the amount and intensity of sunlight exposure and not smoking are two ways that can help prevent cancer.

RAY AND MARIA'S STORIES

Now that we have followed Ray and Maria's trials and tribulations throughout this text, let's think about the systemic problems that may befall them as they age.

a. What research can you find that may help or reverse their initial medical conditions (spinal cord injuries and diabetes) in the near future?

_____

_____

_____

_____

b. What can be done proactively to prevent some of the potential problems that come with aging, especially in their individual cases?

_____

_____

_____

_____

REVIEW QUESTIONS

Multiple Choice

1. The pelvic angle of the female is:
   a. less than 90 degrees
   b. 100 degrees or greater
   c. 75 degrees
   d. greater than 180 degrees

2. From ages 20 to 70, there is up to a
   _____ percent loss of lean
   body mass:
   a. 10
   b. 20
   c. 30
   d. 50

3. This vitamin is needed for strong bones and teeth and calcium absorption:
   a. A
   b. B
   c. C
   d. D

4. The overuse of this classification of drugs has caused drug-resistant strains of bacteria:
   a. steroids
   b. antibiotic
   c. diuretics
   d. pain killers

5. The congenital condition of spina bifida can be prevented by the addition of what vitamin during pregnancy?
   a. A
   b. folic acid
   c. niacin
   d. K

Fill in the Blank

1. The new food pyramid takes into account your _____, _____, and _____ when determining the best diet for you.

2. The fat soluble vitamins are _____, _____, _____, and _____.

3. Cauliflower-like growths on the penis and vagina are _____ _____.

4. The test for cervical cancer is called the _____ _____.

Short Answer

1. Discuss various ways to prevent skin cancer.

_____

_____

_____

_____

2. List and discuss way to prevent STDs.

_____

_____

_____

_____

3. Discuss ways to protect yourself and your patient from the spread of infection.

_____

_____

_____

_____

4. Discuss ways that forensic science can be used in solving crimes.

_____

_____

_____

_____

Suggested Activities

1. Discuss the various aging processes you have seen in your parents and grandparents.

2. Be old for a day by changing your senses. This can be experienced by covering your eyes or your glasses with clear plastic wrap. Wear thin cotton gloves or surgical gloves (if you are not allergic to latex). Gently plug your ears with cotton balls (not cotton swabs). During the next class period, discuss the difficulties you had doing simple, everyday activities.

3. List the foods you've eaten for three days. Share the list with your classmates and determine whether or not it is a healthy diet. Discuss ways to improve it.

4. Make poster presentations on one of the following topics or choose your own: STDs, passive smoking, hazards of smoking, healthy lifestyles, forensic science, aging process, proper diet, or healthy pregnancy.

18-15 Now that you have completed your journey through this chapter, please go to the DVD for interactive games and puzzles concerning the medical terms and concepts contained in this chapter. By playing the games you will reinforce your learning of this chapter in a fun way.

THE
STUDY
SUCCESS
Companion
The Key to Your Successful Journey

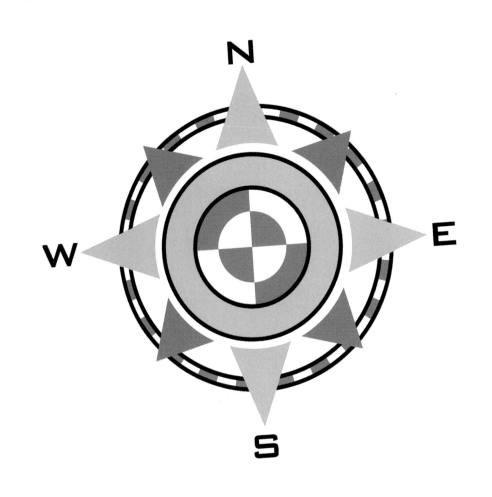

WHY USE THIS STUDY SUCCESS COMPANION?

On any journey, a well-written travel guide can make the experience positive, less stressful, and more productive. This is the main reason for including this guide along with your textbook.

Anatomy, physiology, and pathology can be tough topics to navigate, and it might feel like a very long and arduous journey. It is also a fascinating journey, though, for the human body is the vehicle in which we will all travel through life.

This guide provides effective study strategies, along with a simple and effective stress management system to help you through the rough parts of this journey and even throughout the rest of your life. In addition, this guide helps with "need to know" concepts such as the metric system that are usually not covered in an A&P/Pathology course due to time constraints. Knowing the metric system is critical because it is the mathematical language of anatomy and physiology and of medicine in general.

The *Study Success Companion* covers the following areas:

- Study Skills
- Stress Management
- The Metric System

"The difference between ordinary and extraordinary is that little extra."

Author unknown

STUDY SKILLS

While you may have had courses on study skills, it can never hurt to review the fundamentals. We have chosen the top three general study skills we feel are critical, not just for this course, but for all the current and future courses you will ever take. And the top three choices are:

Clinical Application ✚ ✋ 🚶 🏊 🗝

CLASSICAL CONDITIONING

Throughout the textbook you will notice instances of the mind-body connection. The psychological term *classical conditioning* illustrates this concept and actually has impact upon study skills. The term came from an experiment performed on dogs by the Russian scientist, Ivan Pavlov. In this experiment, a bell was rung and dogs were then fed meat. The meat stimulated the dogs to salivate in anticipation of beginning the digestive process. Repeatedly, experimenters would ring a bell and feed the dogs. The dogs soon learned to connect the ringing of the bell to the arrival of meat. Eventually, the scientists only had to ring a bell and the dogs would salivate like crazy, even if they didn't receive meat.

So what's the purpose of telling you this story other than making you hungry or grossing you out? When you study in bed and then go to sleep repeatedly, you soon connect studying to sleeping. Every time you begin to study (even in mid-afternoon), you may begin to yawn and feel unfocused. You are conditioning yourself to connect studying to sleeping. Simply avoid studying in bed because you will not be as focused and therefore less effective. Besides, studying in bed may interfere with your ability to get a good night's sleep. Classical conditioning can also be used to your advantage in health care. Repeated, positive interactions and therapies with clients will "classically condition" them to feel good each time they see you.

1. Select a Good Time and Place to Study

As with any trip, there is *no* substitute for good planning and preparation. Successful preparation includes developing a daily schedule that includes proper study time. Don't be discouraged if your first few schedules don't work well; just make sure to build in some flexibility. Be sure to schedule relaxation and recreational time, as these are important. Studies show that you learn more in three 30 minute sessions than in one marathon two-hour session. Therefore, studying notes over shorter periods of time at more frequent intervals is more effective than long cramming sessions. We suggest using 30–45 minutes as your *maximum* study time without a break. A good way to check if you are studying efficiently is to periodically ask yourself questions about what you have just read. If you can't answer these questions, you probably are losing interest and need to take a short break and become more focused.

The time of day you choose to study is very important. It may not be the same time for everyone. Some people are "morning people" while others are not. The message for now is to schedule study times when you are most alert and focused.

The place you choose to study is also very important. Ideally, it should be the same place each time so you "connect" this place to studying and consequently become focused. It should have minimal or no distractions and good lighting. The table or desk should have the needed study tools such as pen, paper, calculator, computer, etc.

Which person is getting the most out of their study time?

TEST YOUR KNOWLEDGE 1

Assess your current place of study by answering the following questions. Be critical and honestly circle the response that best answers the question.

1. Do you have a dedicated study area?
   a. Sometimes
   b. Always
   c. Rarely

2. Is it quiet where you study?
   a. Sometimes
   b. Always
   c. Rarely

3. Are the conditions and lighting comfortable?
   a. Sometimes
   b. Always
   c. Rarely

4. Do you have all the tools (pencils, paper, electronic tools, etc.) you need to study at this place?
   a. Sometimes
   b. Always
   c. Rarely

How well did you do? For now, it doesn't matter as long as you were honest. Remember, this is an assessment of where you are now. Your eventual goal should be to have all your responses be "Always." If they are now—great! If not, you need to develop an action plan to make all the responses "Always" in the near future.

2. Use Good Study Habits

Take good, accurate, legible notes. Remember that the purpose of taking notes is to get key points from textbooks and lectures and not to write down every word that is said or written. Instead of high-lighting the chapter, outline your chapters so you can make the connection from your brain to the pencil. This is what you'll need to do on the test. Outlining may *initially* take longer than highlighting, but you will learn the material better. Outlining will actually save you studying time in the long run. Review your lecture and outline notes frequently.

Also, make diagrams and pictures to help visualize concepts within your outline. The more you can "see it" the better you can understand relationships or how it all fits together. Use the Web and DVD visualizations to help reinforce the concepts in your mind's eye.

Anatomy, physiology, and pathology are great subjects to study with a friend or study group. An excellent method to help truly learn the material is to explain concepts to each other. Each time you explain something, the oral recitation reinforces

Applied Science

HOW WE LEARN

A quick generalization of learning theory is as follows:

We learn:
10% of what we read
20% of what we hear
30% of what we see
50% of what we both see and hear
70% of what is discussed with others
80% of what we experience
95% of what we TEACH to someone else

While some may argue the percentages, the general concepts are true to what the research on learning has shown. The more senses you can get involved in the learning process, the more internalized the learning becomes. Therefore, the text illustrations and the DVD and website animations and videos enhance the learning process. Lab experiences and interactive games and exercises will also increase learning. Group or study discussions are highly beneficial and if you have to teach the group your concept, you will really learn what it is all about.

Clinical Application

AIDING YOUR MEMORY

To succeed, it is important to truly "learn" the material and use critical thinking and problem-solving skills. One who knows how to obtain information and understand it, is much better off than someone who merely memorizes for short term storage.

Say, for example, you memorize the steps to cardio-pulmonary resuscitation (CPR) but you truly don't understand *why* you need to establish an open airway, or even *how* to actually do it. You may be able to repeat the steps on a pen and paper test and receive a good grade, but what if, in six months, you are in a situation where you need to perform CPR on an individual in need? You don't want to say, "I had that six months ago and I really didn't learn the material."

While the purpose of education is to encourage thinking skills rather than memorization, memory is vital. Memory is used as an index of success because most of the techniques used to measure learning rely on it. Therefore, a good memory is an asset that you should definitely develop. Try to memorize only when you are well rested, and use memorization techniques such as mnemonics. **Mnemonics** are words, rhymes, or formulas that aid your memory. Acronyms are one type of mnemonic. An **acronym** is a word made from the first letters of other words. For example, the ABCs of CPR remind you that **A** = establish **A**irway, **B** = rescue **B**reathing, and **C** = establish **C**irculation. This helps you to better remember the steps and their proper order in a critical situation.

You can also use rhymes or formulas to assist your memory. For example, "Spring forward, Fall back" helps us to adjust our clocks accordingly for Daylight Savings Time. You can also make up silly stories to help remember facts. In fact, the sillier the story, the easier it often is to remember.

your understanding. *You'll soon learn that there is no better way to learn something than to teach it to someone else.*

Other good study habits include taking personal responsibility for your success. Go to class and read the assigned readings *prior* to class. This will also help you to begin to develop professional responsibility skills that are crucial in health care. Good study habits will greatly reduce the test anxiety you may feel because you will have properly prepared. However, it is normal to have some anxiety about taking an exam no matter how well you have prepared.

Here are some other hints on test taking that may help to further reduce the anxiety and increase your performance. Know what type of test you are taking. With an objective exam (multiple choice, true/false) be sure you understand all directions first. With objective exams, usually your first idea about the answer is your best. If it is an essay exam (short answer), survey the questions, plan your time, and give time to questions in proportion to their value.

Some people develop their own test-taking strategies. For example, they may do all the easy questions first and then return to the more difficult ones. Make sure you mark the questions you skipped or you may forget to return to them. Finally, do not destroy your old exams—keep them and learn from them!

3. Take Care of Yourself

Learning requires a healthy mind, body, and attitude. It is important to exercise your brain to stay mentally fit, but it is also important to stay physically fit. A poor physical condition can distract the mind and minimize your mental focus. Eating right, exercising several times a week, and staying free of drugs will make you feel better and enhance your ability to learn. There may be times you must study when sick or tired. Begin these study sessions with slow rhythmic breathing. This can help you relax and, in turn, your concentration may improve. Remember to get sufficient rest, especially before exams. Learning to manage your stress level is so important to both your mental and physical fitness that the next section is devoted to this topic. Taking time for hobbies, music, or doing things you like to do is important to help refresh your mind.

TEST YOUR KNOWLEDGE 2

A portion of Chapter 18 is devoted to proper exercise, nutrition, and healthy living habits. Circle the answer to the following general questions and make an action plan for each "yes" answer.

1. Do you feel tired during the day when studying?
   a. Yes
   b. No

2. Do you feel stressed out when you start your study session?
   a. Yes
   b. No

3. Do you skip your exercise sessions during the week?
   a. Yes
   b. No

4. Do you eat a diet that is heavy in fats and "junk food?"
   a. Yes
   b. No

STRESS MANAGEMENT

Managing your stress is critical to your academic and personal success. Let's learn more about the concept of stress in order to better manage the stress in a positive manner.

Stress Misconceptions

The major misconception about stress is that all stress is bad for you. This is certainly not true. As you are reading this text, your body is probably in a room that is between 20 and 25 degrees Celsius. This may feel comfortable to you, but it is actually causing stress within your body. In order to survive, the body must react to this stress and make needed physiologic changes to maintain a core temperature of around 36 degrees Celsius and therefore it is continually working at a level we are not consciously aware of to sense and adapt to an externally stressful environment. This response is vital and needed for our survival.

Here is another example of good stress: in Chapter 9, "The Nervous System" you learn about the sympathetic system and the fight or flight response. Picture for a moment the first time you are called upon to do CPR on a cardiac arrest victim. This is going to be a stressful event in your life and may stimulate your sympathetic nervous system. Even though you practiced and trained hard, you are still uncertain of how it will be in a real life-and-death situation. This is normal. Your physical and psychological symptoms may include:

- Increased adrenalin levels for more energy to perform better
- Faster heart rate (*tachycardia*) to supply more oxygen to muscles
- Increased blood pressure to get more blood flow to the brain
- Pupil dilation to bring in more light to see better
- Faster breathing (*tachypnea*) to bring in more oxygen
- Heightened state of awareness to focus on the job at hand
- Mild level of anxiety to keep you sharp and not take the situation too lightly

Applied Science ✚ ✋ 🧗 🏊 🔑

TEST YOURSELF—HOW MANY STRESS SIGNALS DO YOU HAVE?

How many of the following signals do you have on a regular basis (once every week or two)? If you checked two or fewer signals, you're doing pretty good and need minor improvement. Three or more means you need to work hard on how you're handling stress.

_____ Headaches

_____ Shortness of breath

_____ Fast or irregular pulse

_____ Nausea

_____ Insomnia

_____ Difficulty eating

_____ Sadness

_____ Chronic fatigue

_____ Irritability

_____ Diarrhea

_____ Feeling overwhelmed

_____ Difficulty concentrating

_____ Neck or back pain

Clinical Application ✚ ✋ 🧗 🏊 🔑

SADNESS VERSUS DEPRESSION

It's important to differentiate between depression and sadness. If someone is sad following a painful disappointment or the loss of a loved one, this is a normal part of the grieving process. However, if the sadness remains for a prolonged period of time and interferes with the ability to go about your daily business, it becomes depression.

These can all be helpful reactions that enhance your performance. However, they can go too far and lead to a "bad stress" situation where you can't perform at your peak level. Therefore, you can have "good stress" and "bad stress." The key, again, is balance or moderation. A little stress will get you "up" for the task at hand. However, if you let stress get out of hand and panic, you have bad stress and your anxiety level rises to the point where you perform poorly or maybe even not at all.

What Causes Physical and Emotional Stress?

To be able to manage and harness stress to your advantage, you must first understand what stress is all about. First, it is important to realize that no situation or event by itself causes us stress. Rather it is how we _perceive_ a situation that causes stress. It is our learned, internal response to external stimuli.

For example, two individuals can volunteer to give blood. They will both undergo the same procedure with the same technician in the same environment. Yet one individual may not feel any stress or anxiety and the other may be highly stressed at the thought of giving blood. Stress occurs as a result of how we _interpret_ and _react_ to a situation or event. It can be either positive or negative depending on our reaction.

Stress can have many definitions. For now, we will define the _stress reaction_ as "how our mind and body react to an environment that is largely shaped by our perceptions of an event or situation." Notice the mind/body connection. Have you ever heard any of the following sayings?

I lost my breath.

My heart was pounding.

My stomach was twisted in knots.

Stress can also relate to time. We all have temporary stressors in our lives. However, when a stressor becomes constant and negative it has serious effects upon our body and mind. Continual or constant negative stress can lead to:

* High blood pressure, heart attack, or stroke
* Stomach ulcers

- Lack of sleep, or insomnia
- Decreased immune system functioning
- Depression and personality changes
- Poor academic and job performance

However, good or positive stress can actually help us to perform better. Indeed, any time you try something new or meet a new challenge, stress can be a powerful friend. This is how we grow and develop. Continual uncontrolled negative stress will exact a price on our minds and bodies. So the key is learning how to maintain good stress in your life and avoid bad stress.

A Stress Management System

The first step in treating a patient is good assessment. This is also the first step in a good personal stress management system. First, recognizing your own stressors and the symptoms they cause can help you determine when your stress is out of balance or, in other words, when you have entered your bad stress zone. These signals can be valuable to your good health and positive attitude. They represent a "wake up call" that says you need to cope with what's going on in your life before it overtakes you.

As already stated, a certain amount of stress is normal. We need it to develop and grow. This good stress can make you feel energized, focused, and "up" for the event: so you can ace your course exams, for example. However, going beyond your good stress zone and "losing it" by entering your bad stress zone can be harmful. You need to determine when you are losing balance. The best way is to look for physical and emotional signs or indicators that the stress is too much. See Table 1 for some physical and emotional changes that signal too much stress.

Looking at the list of stress signs should paint a pretty good visual picture of what high levels of stress can cause. It's no wonder that individuals who can't handle stress have more accidents, poorer attendance, and are unable to study and learn.

TABLE 1 Physical and Emotional Signs of Stress

PHYSICAL CHANGES	EMOTIONAL CHANGES
headaches	lack of concentration
shortness of breath	irritability
increased pulse rate (tachycardia)	hopeless feelings
nausea	mood swings
insomnia	overreaction
fatigue	depression
neck or back pain	eating disorders
dermatological problems (acne)	anxiety
chronic constipation/diarrhea	low self image

Effective Coping Strategies

Remember, the most important aspect of stress is that it is individually determined. Its meaning lies within us. Therefore, we are the ones to determine what is stressful and whether we are going to use stress to our advantage or let it use us. It should logically follow, if *we* determine the level of stress, *we* should be able to control it.

Someone once said you don't get ulcers from what you eat, but rather from what's eating you. It's important to cope with stress before you suffer from stress overload or burnout. There are two basic ways you can cope with stress. The first is to effectively cope with the emotional side of stress. The second way is to deal with its physical side. Keep in mind the key is to recognize when you are stressed out and intervene as early as possible.

Clinical Application

PREVENTIVE MEDICINE AND EARLY INTERVENTION

In recent years preventive medicine has gained much focus versus the traditional disease model in which we *waited* until the individual got sick and then treated them. The signals in Table 1 are all late signals that mean you have been in your bad stress zone for quite some time. If you can pick up earlier signals and perform a healthy intervention, you can prevent many of these from happening. One hint is that nervous habits such as biting fingernails, pulling your hair, shaking your leg, clicking your pen, etc. are early signs you have just entered your bad stress zone. If you intervene then and there you can prevent more serious problems. This awareness is often difficult because many of these habits are so automatic that we aren't aware of them. So, if you catch yourself shaking your leg and intervene, you can prevent the subsequent muscle tightness, upset stomach, and headache from occurring.

Effective Physical Strategies

You can't separate the mind and body. If we mentally feel bad, it affects our physical well-being. If we don't take care of our physical bodies, we lack energy and focus, can't sleep, and do not reach our full intellectual potential. Therefore, you must balance emotional and physical strategies. Indeed, you'll see that some of these techniques help both our physical and emotional health.

Let's discuss some of the major points you should consider in reducing physical stress. Handling stress "physically" can be broken down into the following four areas:

- Rest and leisure
- Exercise
- Nutrition
- Relaxation techniques

REST AND LEISURE

Adequate sleep is a must for us to function at our peak and handle stress. Research has shown that lack of sleep (sleep deprivation) makes you more susceptible to illness. Of course, lack of sleep also makes you more irritable and less able to focus. Experts recommend that most adults get between seven and nine hours of sleep a night.

Taking leisure time for yourself can also help restore your ability to deal with stress. Even if that leisure time is only 15 minutes, it can help greatly. Sometimes the more you focus on a major problem, the more stress it causes. When this happens it is good to take a break and get away from the problem by doing something else. In many cases, the solution will then just come to you as if by magic. It's not magic, just your subconscious mind working for you.

EXERCISE

Physical exercise and sports are a great way to work off the tensions of everyday life. If done properly they help you to gain both a physical and mental focus. Any type of aerobic activity that gets your heart beating will exercise your muscles and relieve mental tension. Vigorous walking, jogging, running, bicycling, and lifting weights are all things you can do by yourself. Of course, you can find a good workout partner or play team sports also. Vigorous exercise also releases a group of hormones known as *endorphins*. These are our body's natural painkillers. They also are mood elevating chemicals that give us a healthy, natural high.

Remember to properly warm up and stretch before any vigorous exercise. This not only helps prevent injuries, but stretching helps to relieve muscular tension and lower blood pressure. Never rush through the stretches by bouncing or using fast, jerky movements that could strain or tear muscles. Stretch slowly until you feel mild tension and then hold the position for 10 to 30 seconds.

NUTRITION

Good nutrition is a must for growth and development. It also aids in fighting off stress and disease. In addition, it is a good idea to drink plenty of water. Water makes up the majority of the body. Water aids in digestion, absorption of nutrients, and removal of waste products.

Cut down on caffeine, which is found in coffee, tea, and many sodas. Caffeine is a potent central nervous system stimulant. Large amounts can make you anxious, nervous, and unable to get a good night's sleep. Eating a well balanced diet in moderation is important in maintaining nutritional health.

RELAXATION TECHNIQUES

Practicing relaxation techniques will help to clear your mind and make you sharper. Most people will find a million excuses why they can't take the time to relax. Do you see the illogical thinking? If they are *that busy,* then they need to take the time to relax and restore the body and mind. This time also allows you to listen to your body. Two types of relaxation techniques that are very effective are breathing relaxation and meditation.

Slow, deep breathing serves several purposes. First, it increases oxygen to your brain and your body. It also slows down your thinking to help clear your head and relax your muscles.

Finally, don't forget to have a good laugh. Laughter helps bring you back in to perspective and besides, it feels good. Laughter also helps you physically by lowering blood pressure, releasing endorphins, and stimulating the pleasure centers of the brain. Table 2 lists some do's and don'ts to follow when stress is getting the better of you.

TABLE 2 The Do's and Don'ts of Stress Management

DO	DON'T
appropriately confront a problem	think it will resolve itself
discuss calmly	fight or yell
exercise	be a couch potato
accept responsibility	blame others
use relaxation techniques	use alcohol or drugs
accept/learn from your mistakes	be perfectionistic
follow good nutrition	over- or under-eat
be concerned and take action	worry and do nothing
live in the present	agonize over past or future
help others	avoid people

THE MATHEMATICAL LANGUAGE OF MEDICINE

Why Learn the Metric System?

Whereas medical terminology represents the written and spoken language for understanding anatomy, physiology, and pathology, the metric system is the "mathematical language" of medicine. For example, blood pressure is measured in millimeters of mercury (mm Hg) and organ size is usually measured in centimeters (cm). Medications and fluids are given in milliliters (ml) or cubic centimeters (cc) and weight is often measured in kilograms (kg). What exactly does it mean when you are taught that normal cardiac output is 6 liters per minute? You can now see why one must be familiar with the metric system in order to truly understand A&P and medicine.

In addition, you must be able to perform calculations to properly treat patients. For example, a particular drug may be ordered to be given at 5 milligrams/kilogram of body weight. In order to find the right amount to administer, you must first be able to convert the patient's body weight in pounds to kilograms. While it may seem complicated, it really isn't if you have a basic understanding of exponential powers of 10 and systems of measurement, particularly the metric system.

METRIC vs. ENGLISH

Try to visualize the physical comparison between the metric and English system. For example, a meter is a little more than a yard and a liter is a little more than a quart. So a normal cardiac output of 6 liters per minute means that the heart is pumping out approximately 6 quarts of blood every minute.

Exponential Powers of 10

Understanding the powers of 10 gives us a thorough knowledge of the base upon which the metric system is built.

To understand the powers of 10, we need to review some terminology. Consider the expression b^n, where b is called the *base* and n the *exponent*. The **n** represents the number of times that **b** is multiplied by itself. See Figure 1 ■.

If we use 10 as the base, we can develop an exponential representation of the powers of 10 as follows:

10^0 = 1 (mathematically, any number that has an exponent of 0 = 1)

10^1 = 10

$10^2 = 10 \times 10 = 100$

$10^3 = 10 \times 10 \times 10 = 1,000$

$10^4 = 10 \times 10 \times 10 \times 10 = 10,000$

$10^5 = 10 \times 10 \times 10 \times 10 \times 10 = 100,000$

$10^6 = 10 \times 10 \times 10 \times 10 \times 10 \times 10 = 1,000,000$

Thus far, we have discussed positive exponents which result in numbers equal to or greater than 1. However, numbers that are less than one can also be represented in exponential notation, using negative exponents. A negative exponent can be thought of as a fraction. For example,

$10^-1 = 1/10 = .1$

$10^-2 = 1/10 \times 1/10 = .01$

$10^-3 = 1/10 \times 1/10 \times 1/10 = .001$

$10^-4 = 1/10 \times 1/10 \times 1/10 \times 1/10 = .0001$

$10^-5 = 1/10 \times 1/10 \times 1/10 \times 1/10 \times 1/10 = .00001$

$10^-6 = 1/10 \times 1/10 \times 1/10 \times 1/10 \times 1/10 \times 1/10 = .000001$

Systems of Measurement

There are two major systems of measurement in our world today. The United States Customary System (USCS) is used in the United States and Myanmar (formerly Burma), and the System International (SI) is used everywhere else and especially in health care. The SI system is also known as the International or **Metric system.** The metric system is also the system used by drug manufacturers.

The USCS system is based on the British Imperial System and uses several different designations for the basic units of length, weight, and volume. We commonly call this the **English System.** For example, in the English system volumes can be expressed as ounces, pints, quarts, gallons, pecks, bushels, or cubic feet. Distance can be expressed in inches, feet, yards, and miles. Weights are measured in ounces, pounds and tons. This may be the system

(exponent)

b^n $3^2 = 3 \times 3 = 9$

(base)   (base 3)

$10^3 = 10 \times 10 \times 10$

(base 10)

FIGURE ■ **1**

The Exponential Expression

Applied Science

SCIENTIFIC NOTATION—THIS IS NOT THE SCRIBBLINGS OF ALBERT EINSTEIN

In medicine we often use extremely large numbers (for example, there are about 25,000,000,000 blood cells circulating in an adults' body) and extremely small numbers (for example, 0.0000005 meters is the size of some microscopic organisms). It is often useful to write these numbers in more convenient (or shorthand) form based on their power of 10. This abbreviated form is known as *scientific notation*. The basic rule is to move the decimal point in order to have one integer to the left of the decimal point and multiply by the appropriate power of 10 (the number of places you moved the decimal point). So 25,000,000,000 becomes 2.5×10^{10}. The number 0.0000005 would become 5.0×10^{-7}. Notice since this number is less than one, the exponent must be a negative number. In simpler terms, movements of the decimal point to the right are negative values and movements to the left are positive values.

Applied Science

THE APOTHECARY SYSTEM

The Apothecary System developed in the 1700's still has measurements used today. For example, the pint, quart, and gallon were derived from this system. Apothecary measurements for calculating liquid doses of drugs include the minim and fluid dram. Solids are measured in grams, scruples, drams, ounces, and pounds. Two unique features of the apothecary system are the use of Roman numerals and placement of the unit of measure before the Roman numeral. However, the metric system is now used to calculate drug dosages, since the apothecary system is less precise.

you are most familiar with, but it is not the system of choice used throughout the world nor within the medical profession. The reason is the English system is very cumbersome to use because it has no common base. It is very difficult to know the relationship between each of these units because they are not based, in an orderly fashion, according to the powers of 10 as in the metric system. For example, how many pecks are in a gallon? Just what is a peck? How many inches are in a mile? These all require extensive calculations and memorization of certain equivalent values, whereas in the metric system you simply move the decimal point the appropriate power of 10.

Most scientific and medical measurements utilize what is commonly referred to as the metric system. The metric system utilizes three basic units of measure for lengths, volume, and mass and these are the *meter, liter,* and *gram* respectively. While the term "mass" is commonly used for weight, weight is actually the force exerted on a body by gravity. In space, or zero gravity, all objects have mass but are indeed weightless. Since current health care is confined to earthly gravitational forces, we will use the term "weight." Table 3 gives you the metric designation for the three basic units of measure, along with an approximate English system comparison.

Again, notice that there are only three basic types of measures (meter, liter, and gram) and the metric system has only one base unit per measure. Because the metric system is a base 10 system, prefixes are used to indicate different powers of 10. Conversion within the metric system is done by simply moving the decimal point in the appropriate direction by the power of ten according to the prefix before the unit of measure. For example, the prefix kilo means 1000× or 10^3. Therefore, one kilogram is equal to 1000 grams. See Table 4 for the common prefixes and their respective powers of 10.

TABLE 3 Metric and English System Comparison

TYPE	UNIT	ENGLISH SYSTEM COMPARISON (APPROXIMATE SIZE)
length	meter	slightly more than a yard
volume	liter	slightly more than 1 quart
mass/weight	gram	about 1/40 of an ounce

TABLE 4 Common Prefixes of the Metric System

THOUSANDS	HUNDREDS	TENS	BASE UNITS	TENTH	HUNDREDTH	THOUSANDTH
kilo	hecto	deca	liter, meter, or gram	deci	centi	milli
(K)	(H)	(Da)	(l) (m) (g)	(d)	(c)	(m)
10^3	10^2	10^1	10^0 or 1	10^{-1}	10^{-2}	10^{-3}

It can be seen from Table 4 that a kilometer would be 1000 or 10^3 meters. A centigram would be (.01) one-hundredth or 10^{-2} of a gram. Working with the metric system is easy because to change from one prefix to another you simply move the decimal point to the correct place. In other words, to convert within the system, simply move the decimal point for each power of 10 as indicated in the prefix.

EXAMPLE CALCULATION 1

In drug dosage calculations you often need to convert between grams and milligrams and liters and milliliters. A common conversion requirement might be something like: 500 milliliters is equal to how many liters? We know from Table 2 that 500 milliliters (ml) would be equal to 0.5 liters because you would simply move the decimal point 3 spaces (powers of 10) to the left for the equivalent value since you are starting with milliliters and going to the base unit of liters.

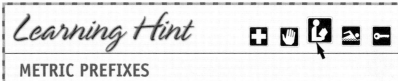

Learning Hint

METRIC PREFIXES

Deci is associated with decade meaning ten years; centi is associated with cents or a hundred cents in a dollar; and milli is associated with a millipede with a thousand legs. Biological note: A millipede doesn't actually have a thousand legs, it just looks like it.

EXAMPLE CALCULATION 2

How many grams are equal to 50.0 kilograms? Start at kilograms on Table 4 and move to the unit you want to convert to, in this case grams. You would need to move the decimal point three places (powers of ten) to the right to give an equivalent answer of 50,000 grams.

This knowledge of the metric system will prove invaluable to you as you work within the medical profession and even if you travel outside the United States. That is, of course, unless you go to Myanmar.

One final note before leaving this part of the discussion on the metric system: it has been determined that one cubic centimeter (cc) would hold the approximate volume of one milliliter (ml). Therefore, 1 cc = 1 ml. See Figure 2 ■. You may hear someone say you have 500 cc of an intravenous (IV) solution on hand, while someone else may say you have 500 ml of solution; either way they are both saying the same thing. Efforts are being made to standardize between cubic centimeters and milliliters, making milliliters the preferred choice. However, you will see and hear both used in health care settings.

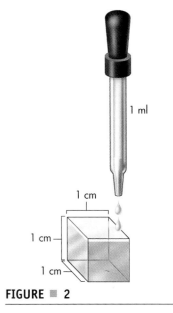

FIGURE ■ 2

1 cc = 1 ml

Learning Hint

CHECK YOUR WORK

Always check your answer to see if it makes sense. For example, a common mistake is moving the decimal point in the wrong direction. If you had done that in example calculation 1, you may have erroneously said that 500 milliliters is equal to 500,000 liters. If you visualize this you would know that 500 comparatively very small units (milliliters) in no way can equal 500,000 comparatively larger units (liters).

You will also note in the healthcare setting that decimals are always preceded by a zero if they are less than one. For example, if you need one-half of a milligram, then it is written 0.5mg. On the other hand zeros are never used after whole numbers. For example, if you need five milligrams, then it is written 5mg. This is part of the effort to reduce the number of medication errors.

Being able to convert temperatures from Fahrenheit to Celsius and from Celsius to Fahrenheit is another important conversion to know. Below is the formula to use.

Fahrenheit to Celsius
$(F-32) \times 5/9$
Celsius to Fahrenheit
$(C \times 9/5) + 32$

TEST YOUR KNOWLEDGE 3

Choose the best answer:

1. The metric system is based on the exponential power of:
   a. 100
   b. 10
   c. 2
   d. 15

2. Which of the following is *not* a basic unit of measure in the metric system?
   a. liter
   b. gram
   c. pound
   d. meter

Complete the following:

3. A cubic centimeter (cc) is equal to:

4. 500 grams is equal to how many kilograms?

5. 200 meters is equal to how many centimeters?

Conversion of Units

You should now be able to work comfortably within the metric system, but what if you need to take an English unit and convert it to a metric unit? For example, in the introduction of this Study Success Guide, we said that a certain drug order read to give a patient 5 milligrams per kilogram of body weight. You would need to know the relationship between pounds in the English system and kilograms in the metric system to properly treat this patient.

The following is a method for changing units or converting between the English and metric system. This method is sometimes referred to as the *Factor-Label Method* or *Fraction Method*. This method allows your starting units to cancel or divide out until you reach your desired unit. There are two basic steps:

Step One: Write down your starting value with units over one. This places it in the form of a fraction, but since the number one is in the denominator it does not change the numerical value.

Step Two: Place the starting unit in the denominator of the next fraction to divide or cancel out and place the desired unit in the numerator along with the corresponding equivalent values. Since the values are equivalent, this is the same as multiplication by 1, which does not change the value of the quantity. This allows you to treat the units as in the multiplication of fractions and "cancel" the units. Notice that by carefully placing the units so that canceling is possible, the units can be converted.

The following section shows an example of this conversion.

EXAMPLE CALCULATION 3

How many inches are there in one mile?
First, put down what value is given as a fraction over 1.

$$\frac{1 \text{ mile}}{1}$$

Next, put miles in the denominator and the desired unit in the numerator with equivalent values. You know that 1 mile = 5,280 feet so:

$$1 \ \cancel{\text{mile}} \times \frac{5,280 \text{ ft}}{1 \ \cancel{\text{mile}}}$$

You've cancelled out miles, but need to go to inches. Just carry the process out until you reach your desired unit.

$$1 \ \cancel{\text{mile}} \times \frac{5,280 \ \cancel{\text{ft}}}{1 \ \cancel{\text{mile}}} \times \frac{12 \text{ inches}}{1 \ \cancel{\text{ft}}} = 63,360 \text{ inches}$$

EXAMPLE CALCULATION 4

How many seconds are there in 8 hours?

Step One: $\dfrac{8 \text{ hours}}{1}$

Step Two: $\dfrac{8 \ \cancel{\text{hours}}}{1} \times \dfrac{60 \ \cancel{\text{minutes}}}{1 \ \cancel{\text{hour}}} \times \dfrac{60 \text{ seconds}}{1 \ \cancel{\text{minute}}} \times 28,800 \text{ seconds}$

TEST YOUR KNOWLEDGE 4

Complete the following:

1.   How many days in 3 years? _____

2.   How many inches in 4.5 yards? _____

3.   How many quarts in 10.5 gallons?

_____

Clinical Application

BODY SURFACE AREA

A six foot man who weighs 240 lbs. may require a different dosage than a six foot man weighing 150 lbs. This is especially true with highly toxic agents such as those used in cancer chemotherapy. A method to determine the total body surface area (BSA) combines both height and weight in a single measurement to determine the true overall body size. Comparisons like this are called nomograms. See Figure 3 ■ for a nomogram used in determining BSA. Simply mark the patient's corresponding height and weight on the respective scale and either draw a straight line or use a ruler to find the intersection point to get the body surface area. Due to America's poor eating habits, the importance of BSA in relationship to the development of a form of diabetes is very important. Often that form of diabetes may improve with simple weight loss.

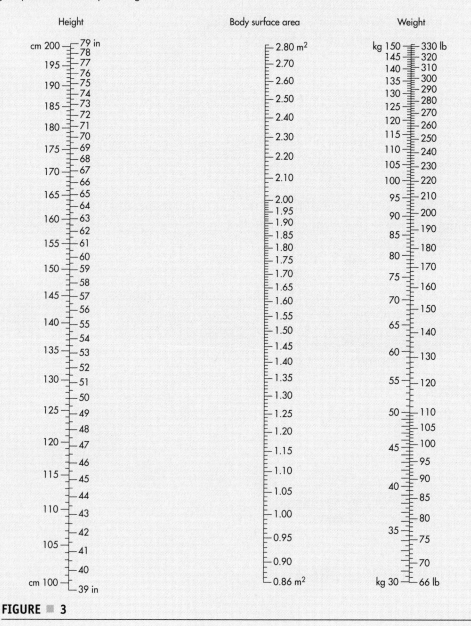

FIGURE ■ 3

Nomogram for Determining Body Surface Area (BSA).

Using the Factor Label Method for Conversion Between Systems

One can attempt to memorize the hundreds of conversions between the English and metric systems, but that would be nearly impossible. All that is needed is to memorize one conversion in each of the three units of measure. This will allow a "bridging of the systems." These conversions are:

1 inch = 2.54 cm	used for units of lengths
2.2 lbs = 1 kg	used for units of mass or weights
1.06 qt = 1 L	used for units of volume

EXAMPLE CALCULATION 5

If an individual weighs 150 pounds, how many kilograms does the patient weigh? First, you must change pounds to kilograms; therefore, write the given weight as a fraction over 1. Then place the unit you want to cancel (pounds) in the denominator and the unit you want to convert to (kilograms) in the numerator of the next fraction.

$$\frac{150 \text{ pounds}}{1} \times \frac{1 \text{ kilogram}}{2.2 \text{ pounds}} = 68.18 \text{ kilograms}$$

EXAMPLE CALCULATION 6

The conversion will not always be as direct as in the previous example, but it is no problem if you simply follow the system. For example, one foot is equal to how many centimeters? There is an equivalency somewhere for feet and centimeters, but you don't need to know that using the Factor Label Method and the conversion for distance.

Now, answer the question of how many centimeters are there in one foot. Remember, your unit conversion for length is 1 inch = 2.54 cm.

$$\frac{1 \text{ foot}}{1} \times \frac{12 \text{ inches}}{1 \text{ foot}} \times \frac{2.54 \text{ cm}}{1 \text{ inch}} = 30.48 \text{ cm}$$

TEST YOUR KNOWLEDGE 5

Choose the best answer:

1. The Body Surface nomogram compares what two units of measure?
   a. weight and sex
   b. height and sex
   c. surface area and length
   d. height and weight

Complete the following:

2. A quart of blood is equal to how many milliliters (ml)?

3. If a patient voids (passes) 3.2 liters of urine in a day, what is the amount in milliliters?

4. Convert 175 lbs to kilograms.

APPENDICES

APPENDIX A: Medical Terminology, Word Parts, and Singular and Plural Endings

Word Parts Arranged Alphabetically and Defined

The word parts that have been presented in this textbook, and additional terms you may encounter, are summarized with their definitions for quick reference.

PREFIX	DEFINITION	PREFIX	DEFINITION
a-	without absence of	im-	not
ab-	away from	inter-	between
ad-	toward	intra-	within
an-	without, absence of	later-	side
ana-	up, toward, apart	lipo-	fat
ante-	before	macro-	large
anti-	against, opposing	mal-	bad
aut-	self, same	mega-	large
bi-	two	meta-	after, change
bin-	two	micro-	small
brady-	slow	mono-	single
circum-	around, about	multi-	many
contra-	in contrast; against	my-	muscle
di-	two	myo-	muscle
dia-	around, passing through	neo-	new
diplo-	double	nulli-	none
dys-	bad, abnormal	ortho-	straight, normal
ecto-	on outer side, situated on	pan-	all, entire
en-	within, upon, on, over	para-	near, alongside; departure from normal
endo-	within, inner, absorbing	per-	through
ep-	upon, on, over	peri-	around, about, surrounding
epi-	upon	poly-	many
eu-	normal, good	post-	after
ex-	outside, away from	pre-	before
exo-	outside, away from	pro-	forward, preceding
hemi-	one-half	pseudo-	false
homo-	same	quad-	four
hydro-	water	re-	back
hyper-	excessive	retro-	backward, located behind
hypo-	under, below normal	semi-	half

PREFIX	DEFINITION
sub-	beneath
sym-	together, joined
syn-	together, joined
tachy-	rapid, fast
tele-	distant
tetra-	four
trans-	through, across, beyond
ultra-	beyond, in excess
uni-	one

COMBINING FORM	DEFINITION
abdomin/o	abdomen, abdominal cavity
acou/o	hearing
acoust/o	hearing
acr/o	extremity, extreme
aden/o	gland
adren/o	adrenal gland
aer/o	air or gas
albumin/o	albumin
alveol/o	alveolus (air sac)
amni/o	amnion, amniotic fluid
amnion/o	amnion, amniotic fluid
an/o	anus
andr/o	male
angi/o	blood vessel
ankyl/o	crooked
anter/o	front
aort/o	aorta
appendic/o	to hang onto, appendix
aque/o	water
arche/o	first, beginning
arter/o	artery
arteri/o	artery
arthr/o	joint
astheni/o	weakness
atel/o	imperfect, incomplete

COMBINING FORM	DEFINITION
ather/o	fat
atri/o	atrium
aud/o	hearing
audi/o	hearing
aur/i	ear
aut/o	self
azot/o	urea, nitrogen
bacter/o	bacteria
balan/o	glans penis
bi/o	life
bil/i	bile
blast/o	germ, bud
blephar/o	eyelid
bronch/i	bronchus (airway)
bronch/o	bronchus (airway)
burs/o	purse or sac; bursa
calc/i	calcium
cancer/o	cancer
carcin/o	cancer
card/o	heart
cardi/o	heart
carp/o	wrist
cartil/o	gristle, cartilage
caud/o	tail
cec/o	blind intestine, cecum
cel/o	hernia, protrusion
celi/o	abdomen, abdominal cavity
cephal/o	head
cerebell/o	cerebellum (little brain)
cerebr/o	cerebrum (brain)
cerumin/o	wax
cervic/o	cervix, neck
cheil/o	lip
chem	chemistry, drug
chir/o	hand
chol/e	bile, gall

COMBINING FORM	DEFINITION	COMBINING FORM	DEFINITION
choledoch/o	common bile duct	duoden/o	twelve, duodenum
chondr/o	cartilage	dur/o	hard
chori/o	membrane, chorion	ech/o	sound
chromat/o	color	electr/o	electricity
clon/o	spasm	embol/o	plug
col/o	colon	embry/o	embryo
colp/o	vagina	encephal/o	brain
conjunctiv/o	to bind together, conjunctiva	endocrin/o	endocrine
cor/o	pupil	enter/o	small intestine
core/o	pupil	epididym/o	epididymis
corne/o	horny, cornea	epiglott/o	epiglottis
coron/o	crown, circle	episi/o	vulva
cortic/o	tree bark, outer covering, cortex	eryth/o	red
cost/o	rib	erythr/o	red
cran/o	skull, cranium	esophag/o	gullet, esophagus
crani/o	skull, cranium	esthesi/o	sensation
crin/o	to secrete	eti/o	cause (of disease)
crypt/o	hidden	fasci/o	fascia
cutane/o	skin	femor/o	thigh
cyan/o	blue	fet/i	fetus
cyes/i	pregnancy	fet/o	fetus
cyes/o	pregnancy	fibr/o	fiber
cyst/o	bladder, sac	fibul/o	clasp of buckle, fibula
cyt/o	cell	fovea/o	small pit
dacry/o	tear	gangli/o	ganglion
dent/i	teeth	gastr/o	stomach
derm/o	skin	gen/o	formation, cause, produce
dermat/o	skin	ger/o	old age
diaphragmat/o	diaphragm	geront/o	old age
dipl/o	double	gingiv/o	gums
dips/o	thirst	gli/o	glue, neuroglia
dist/o	away	glomerul/o	little ball, glomerulus
diverticul/o	small blind pouch, diverticulum	gloss/o	tongue
dors/o	back	gluc/o	glucose, sugar
duct/o	lead, move	glut/o	buttock

COMBINING FORM	DEFINITION
glyc/o	glycogen, sugar
glycos/o	sugar
gravid/o	pregnancy
gravidar/o	pregnancy
gyn/o	woman
gynec/o	woman
halat/o	breath
hem/o	blood
hemat/o	blood
hepat/o	liver
hern/o	protrusion, hernia
herni/o	protrusion, hernia
heter/o	other
hidr/o	sweat
hist/o	tissue
hom/o	sameness, unchanging
hormon/o	to set in motion
hydr/o	water
hymen/o	hymen
hyster/o	uterus
iatr/o	physician
idi/o	person, self
ile/o	ileum of small intestine, to roll
ili/o	flank, groin, ilium of the pelvis
immun/o	exempt; immunity
infect/o	to enter, invade
infer/o	below
inguin/o	groin
ir/o	rainbow, iris
irid/o	rainbow, iris
isch/o	to hold back, deficiency, blockage
ischi/o	haunch, hip joint, ischium
jejun/o	empty, jejunum
kal/i	potassium
kerat/o	hard, horny; cornea
ket/o	ketone bodies

COMBINING FORM	DEFINITION
keton/o	ketone bodies
kinesi/o	motion
kyph/o	hump
labyrinth/o	labyrinth, internal ear
lacrim/o	tear
lact/o	milk
lamin/o	thin, lamina
lapar/o	abdomen, abdominal cavity
laryng/o	larynx
later/o	side
lei/o	smooth
leuk/o	white
lingu/o	tongue
lip/o	fat, lipid
lith/o	stone
lob/o	lobe
lord/o	bent forward
lumb/o	loin, lower back
lymph/o	clear water or fluid
lys/o	dissolution
mal/o	bad
mamm/o	breast
mast/o	breast
meat/o	opening
medi/o	middle
megal/o	abnormally large
melan/o	dark, black
men/o	month, menstruation
menstru/o	month, menstruation
mening/o	meninges, membrane
menisc/o	crescent-shaped moon, meniscus membrane
ment/o	mind
metr/o	uterus
mon/o	one
muc/o	mucus

COMBINING FORM	DEFINITION
my/o	muscle
myc/o	fungus
myel/o	bone marrow; spinal cord; medulla; myelin
myelon/o	bone marrow; spinal cord; medulla; myelin
myos/o	muscle
myring/o	membrane; eardrum
myx/o	mucus
nas/o	nose
nat/o	birth
natr/o	sodium
necr/o	death
nephr/o	kidney
neur/o	sinew or cord, nerve, fascia
noct/i	night
nucl/o	kernel, nucleus
nyct/o	night, nocturnal
nyctal/o	night, nocturnal
obstetr/o	midwife, prenatal development
ocul/o	eye
odont	of or pertaining to teeth
olig/o	few in number
omphal/o	umbilicus (navel)
onc/o	tumor
onych/o	nail
oophor/o	ovary
opt/o	eye, vision
opthalm/o	eye
or/o	mouth
orch/o	testis or testicle
orchi/o	testis or testicle
orchid/o	testis or testicle
organ/o	tool
orth/o	straight
oste/o	bone
ot/o	ear

COMBINING FORM	DEFINITION
ov/i	egg
ov/o	egg
ovari/o	ovary
ox/o	oxygen
pachy/o	thick
palat/o	roof of mouth, palate
pancreat/o	pancreas
par/o	parturition or labor
parathyroid/o	parathyroid
pariet/o	wall
part/o	parturition or labor
patell/o	small pan, patella
path/o	disease
pector/o	chest
ped/o	child
pediatr/o	child
pelv/l	basin, pelvis
pen/o	penis
peps/o	digestion
perine/o	perineum
peritone/o	to stretch over, peritoneum
petr/o	stone
phac/o	lens
phag/o	eat, swallow
phak/o	lens
phalang/o	row of soldiers
pharyng/o	pharynx (throat)
phas/o	speech
phleb/o	vein
phot/o	light
physi/o	nature
physis/o	growth
plegi/o	paralysis
pleur/o	pleura
pneum/o	lung or air
pneumat/o	lung or air

COMBINING FORM	DEFINITION
pod/o	foot
poikil/o	irregular
polyp/o	polyp
poster/o	back
presby/o	old age
prim/i	first
proct/o	anus
prostat/o	prostate gland
proxim/o	near
pseud/o	false
psych/o	mind
pub/o	grown up
puerper/o	childbirth
pulmon/o	lung
py/o	pus
pyel/o	pelvis (renal)
pylor/o	pylorus
quad/o	four
quadr/i	four
rachi/o	spine
radi/o	spoke of a wheel, radius
radic/o	nerve root
radicul/o	nerve root
rect/o	straight, erect, rectum
ren/o	kidney
retin/o	net, retina
rhabd/o	rod
rhin/o	nose
rhytid/o	wrinkles
sacr/o	sacred, sacrum
salping/o	tube: fallopian tube
sarc/o	flesh, muscle
scler/o	thick, hard; sclera
scoli/o	curved
seb/o	sebum, oil
semin/o	seed

COMBINING FORM	DEFINITION
sept/o	wall, partition
sial/o	saliva, salivary gland
sigm/o	sigmoid colon , the letter "s"
sinus/o	cavity
somat/o	body
somn/o	sleep
son/o	sound
sperm/o	seed
spermat/o	seed
sphygm/o	pulse
spin/o	spine or thorn
spir/o	breathe
splen/o	spleen
spondyl/o	vertebra
staped/o	stapes
staphyl/o	grape-like clusters (bacterium)
stasis/o	standing still
steat/o	fat
sten/o	narrowness, constriction
stern/o	chest, sternum
steth/o	chest
stigmat/o	point
stomat/o	mouth
strept/o	twisted or gnarled (bacterium)
super/o	above
synov/o	synovial fluid
synovi/o	synovial fluid
syndesm/o	binding together
tars/o	flat surface
taxi/o	reaction to a stimulus
ten/o	to stretch out; tendon
tend/o	to stretch out; tendon
tendin/o	to stretch out; tendon
test/o	testis, testicle
testicul/o	small testis, testicle
thalam/o	thalamus

COMBINING FORM	DEFINITION
thel/o	nipple
therm/o	heat
thorac/o	thorax (chest)
thromb/o	clot
thym/o	wart-like, thymus gland
thyr/o	shield, thyroid
toc/o	birth, labor
tom/o	cut, section
ton/o	tone, tension, pressure
toxic/o	poison
trache/o	trachea
trachel/o	neck, cervix
trich/o	hair
tubercul/o	little mass of swelling
tympan/o	eardrum
umbilic/o	navel
ur/o	urine
ureter/o	ureter
urethr/o	urethra
urin/o	urine
uter/o	womb, uterus
uvul/o	grape, uvula
vagin/o	sheath, vagina
valvul/o	little valve
varic/o	dilated vein
vas/o	blood vessel; duct
vascul/o	little blood vessel
ven/o	vein
ventr/o	front, belly
ventricul/o	little belly or cavity, ventricle
vers/o	turn
vertebr/o	joint, vertebra
vesic/o	bladder, sac
vesicul/o	vesicle (seminal vesicle)
vitr/o	glassy
vitre/o	glassy

COMBINING FORM	DEFINITION
vulv/o	vulva
xanth/o	yellow
xer/o	dry

SUFFIX	DEFINITION
-a	singular
-ac	pertaining to
-acusis	hearing condition
-ad	toward
-al	pertaining to
-algesia	pain
-algia	pain
-apheresis	removal
-ar	pertaining to
-ary	pertaining to
-ase	enzyme
-asthenia	weakness
-atresia	closure; absence of a normal body opening
-capnia	carbon dioxide
-cele	hernia, swelling, protrusion
-cente	100
-centesis	surgical puncture
-centre	center
-cide	causing death or destruction
-clasia	break apart
-clasis	break apart
-clast	break apart
-crit	separate
-cusis	hearing condition
-cyte	cell
-desis	surgical fixation, fusion
-dipsia	thirst
-drome	run, running
-dynia	pain
-eal	pertaining to
-ectasis	expansion, dilation

Suffix	Definition
-ectomy	excision
-elle	small
-emesis	vomiting
-emetic	vomiting
-emia	blood (condition of)
-esthesia	sensation
-gen	formation, cause, produce
-genesis	origin, cause
-genic	pertaining to formation, causing, producing
-gram	recording
-graph	instrument for recording
-graphy	recording process
-hemia	blood (condition of)
-ia	diseased state (condition of)
-ial	pertaining to
-iasis	condition of
-iatry	treatment, specialty
-ic	pertaining to
-ion	pertaining to
-ior	pertaining to
-is	pertaining to
-ism	condition or disease
-ist	one who practices
-itis	inflammation
-lepsy	seizure
-logist	one who studies
-logy	study of
-lytic, -lysis	to loosen, dissolve
-malacia	softening
-meter	measuring instrument
-metry	measurement
-oid	resemblance to
-oma	abnormal swelling, tumor
-opia	vision
-opsy	view of
-orrhagia	bursting forth

Suffix	Definition
-orrhea	flow
-osis	process or condition that is usually abnormal
-otomy	cutting into, excision
-ous	pertaining to
-oxia	oxygen
-paresis	paralysis (minor)
-pathy	disease
-penia	abnormal reduction in number, deficiency
-pepsia	digestion
-pexy	surgical fixation, suspension
-phagia	eating or swallowing
-phasia	speaking
-phil	loving, affinity for
-philia	loving, affinity for
-phobia	fear
-phonia	sound or voice
-phylaxis	protection
-physis	growth
-plasia	shape, formation
-plasm	something shaped
-plasty	surgical repair
-plegia	paralysis (major)
-pnea	breathing
-poiesis	formation
-practic	one who practices
-ptosis	falling downward (condition of)
-ptysis	spit out a fluid
-rrhagia	bleeding, hemorrhage
-rrhaphy	suturing
-rrhea	excessive discharge
-rrhexis	rupture
-salpinx	trumpet, fallopian tube
-sarcoma	malignant tumor
-schisis	split, fissure
-sclerosis	hardening

Suffix	Definition
-scope	viewing instrument
-scopy	process of viewing
-sis	state of
-some	body
-spasm	sudden, involuntary muscle contraction
-stasis	standing still
-stenosis	narrowing, constriction
-stomy	surgical creation of an opening
-tic	pertaining to

Suffix	Definition
-tocia	birth, labor
-tome	cutting instrument
-tomy	incision
-toxic	toxin, poison
-tripsy	surgical crushing
-trophy	nourishment, development
-um	pertaining to
-uria	urine, urination
-y	process of

Endings in Medical Terminology

Plural endings. The following list provides a summary of plural endings that are in common use with medical terms. Examples are provided to demonstrate how these endings are applied.

ENDINGS SINGULAR	PLURAL
-a	-ae
-ax	-aces
-ex	-ices
-is	-es
-ix	-ices
-ma	-mata
-on	-a
-um	-a
-us	-i
-y	-ies

EXAMPLES SINGULAR	PLURAL
fistula	fistulae
hemothorax	hemothoraces
cortex	cortices
mastoiditis	mastoidites
cicatrix	cicatrices
fibroma	fibromata
contusion	contusia
bacterium	bacteria
fungus	fungi
episiotomy	episiotomies

Adjective endings. The list below provides a summary of suffixes that mean "pertaining to" and form an adjective (a description of a noun) when combined with a root.

ENDING	EXAMPLE	DEFINITION
-ac	cardiac	pertaining to the heart
-al	endotracheal	pertaining to within the trachea
-ar	submandibular	pertaining to below the mandible
-ary	pulmonary	pertaining to a lung
-eal	esophageal	pertaining to the esophagus
-ic	leukemic	pertaining to leukemia
-ous	fibrous	pertaining to fiber
-tic	cyanotic	pertaining to cyanotic (blue)

Diminutive endings. The endings listed below provide the meaning of "small" to the word of origin.

ENDING	EXAMPLE	DEFINITION
-icle	ossicle	small bone
-ole	bronchiole	small bronchus (airway)
-ula	macula	small macule (spot)
-ule	pustule	small pimple

Diagnostic endings. The endings in this list summarize the suffixes that are in common use to indicate measurements, treatments, and procedures.

ENDING	MEANING	EXAMPLE	DEFINITION
-gram	record	bronchogram	recording of bronchus image
-graph	recording instrument	sonograph	ultrasound instrument
-graphy	process of recording	echocardiography	procedure of heart recording
-iatrics	treatment	pediatrics	treatment of children
-iatry	treatment	psychiatry	treatment of the mind
-logy	study of	oncology	study of cancer
-logist	one who studies	audiologist	one who studies hearing
-ist	one who specializes	optometrist	specialist in eye measurement
-meter	instrument of measure	spirometer	instrument measuring breathing
-metry	process of measuring	spirometry	process of measuring breathing
-scope	instrument for exam	endoscope	instrument for examination within
-scopy	examination	endoscopy	examination within

APPENDIX B: Clinical Abbreviations and "Do Not Use" List

Listed below are the currently acceptable abbreviations to use; however, different health care facilities may use others as well. The Joint Commission, which accredits and certifies over 15,000 health care organizations and programs in the United States, has created a list of abbreviations which should not be used because in handwritten documents or forms they may cause confusion or problems. Those are also included below. It is important that health care employees follow their facility policy regarding the use of medical abbreviations.

ABBREVIATION	MEANING	ABBREVIATION	MEANING
ABGs	arterial blood gases	ante	before
ac	before meals	AP	anteroposterior
ACAT	automated computerized axial tomography	APAP	acetaminophen (Tylenol)
		aq	aqueous (water)
Acc	accommodation	ARC	AIDS-related complex
ACL	anterior cruciate ligament	ARD	acute respiratory disease
ACTH	adrenocorticotropic hormone	ARDS	adult respiratory distress syndrome
AD	right ear, Alzheimer's disease		
ad lib	as desired	ARF	acute respiratory failure, acute renal failure
ADD	attention deficit disorder		
ADH	antidiuretic hormone	ARMD	age-related macular degeneration
ADHD	attention-deficit-hyperactivity disorder	AROM	active range of motion
		AS	aortic stenosis, arteriosclerosis, left ear
ADL	activities of daily living		
AE	above elbow	ASA	aspirin
AF	atrial fibrillation	ASCVD	arteriosclerotic cardiovascular disease
AGN	acute glomerulonephritis		
AHF	antihemophilic factor	ASD	atrial septal defect
AI	artificial insemination	ASHD	arteriosclerotic heart disease
AIDS	acquired immunodeficiency syndrome	ASL	American Sign Language
		AST	aspartate transaminase
AK	above knee	Astigm.	astigmatism
ALL	acute lymphocytic leukemia	ATN	acute tubular necrosis
ALS	amyotrophic lateral sclerosis	AU	both ears
ALT	alanine transaminase	AuD	doctor of audiology
am, AM	morning	AV, A-V	atrioventricular
AMI	acute myocardial infarction	Ba	Barium
AML	acute myelogenous leukemia	BaE	barium enema
amt	amount	basos	basophil
Angio	angiography	BBB	bundle branch block (L for left; R for right)
ANS	autonomic nervous system	BC	bone conduction

ABBREVIATION	MEANING
BCC	basal cell carcinoma
BDT	bone density testing
BE	barium enema, below elbow
bid	twice a day
BK	below knee
BM	bowel movement
BMR	basal metabolic rate
BMT	bone marrow transplant
BNO	bladder neck obstruction
BP	blood pressure
BPD	bipolar disorder
BPH	benign prostatic hypertrophy
bpm	beats per minute
Bronch	bronchoscopy
BS	bowel sounds
BSE	breast self-examination
BSN	bachelor of science in nursing
BUN	blood urea nitrogen
BX, bx	biopsy
$\bar{c}$	with
C	100
C1, C2, etc.	first cervical vertebra, second cervical vertebra, etc.
Ca^{2+}	calcium
CA	cancer, chronological age
CABG	coronary artery bypass graft
CAD	coronary artery disease
cap(s)	capsule(s)
CAPD	continuous ambulatory peritoneal dialysis
CAT	computerized axial tomography
cath	catheterization
CBC	complete blood count
CBD	common bile duct
CC	clean catch urine specimen, cardiac catheterization, chief complaint
CCS	certified coding specialist

ABBREVIATION	MEANING
CCU	cardiac care unit, coronary care unit
CD4	protein on T-cell helper lymphocyte
CDH	congenital dislocation of the hip
c.gl.	correction with glasses
CGL	chronic granulocytic leukemia
chemo	chemotherapy
CHF	congestive heart failure
chol	cholesterol
Ci	Curie
CIS	carcinoma in situ
Cl^-	chloride
CLL	chronic lymphocytic leukemia
CLS	clinical laboratory scientist
CLT	clinical laboratory technician
CMA	certified medical assistant
CML	chronic myelogenous leukemia
CNA	certified nurse aide
CNIM	certification in neurophysiologic intraoperative monitoring
CNS	central nervous system
CO_2	carbon dioxide
CoA	coarctation of the aorta
COLD	chronic obstructive lung disease
COPD	chronic obstructive pulmonary disease
COTA	certified occupational therapy assistant
CP	cerebral palsy, chest pain
CPD	cephalopelvic disproportion
CPK	creatine phosphokinase
CPR	cardiopulmonary resuscitation
CRF	chronic renal failure
crit	hematocrit
CRT	certified respiratory therapist
C & S	culture and sensitivity test
CS, CS-section	Cesarean section

ABBREVIATION	MEANING
CSD	congenital septal defect
CSF	cerebrospinal fluid
CT	computerized tomography, cytotechnologist
CTA	clear to auscultation
CTS	carpal tunnel syndrome
CUC	chronic ulcerative colitis
CV	cardiovascular
CVA	cerebrovascular accident
CVD	cerebrovascular disease
CVS	chorionic villus biopsy
Cx	cervix
CXR	chest X-ray
cyl lens	cylindrical lens
cysto	cystoscopic exam
D	diopter (lens strength)
D & C	dilation and curettage
D/C, d/c	discontinue
dB	decibel
DC	doctor of chiropractic
DDM	doctor of dental medicine
DDS	doctor of dental surgery
DEA	Drug Enforcement Agency
decub	lying down, decubitus ulcer
Derm, derm	dermatology
DI	diabetes insipidus, diagnostic imaging
diff	differential
dil	dilute
disc	discontinue
disp	dispense
DJD	degenerative joint disease
DM	diabetes mellitus
DO	doctor of osteopathy
DOB	date of birth
DOE	dyspnea on exertion

ABBREVIATION	MEANING
DPT	diphtheria, pertussis, tetanus; doctor of physical therapy
dr	dram
DRE	digital rectal exam
DSA	digital subtraction angiography
DSM-IV	*Diagnostic and Statistical Manual for Mental Disorders, Fourth Edition*
DTR	deep tendon reflex; dietetic technician, registered
DUB	dysfunctional uterine bleeding
DVA	distance visual acuity
DVT	deep vein thrombosis
Dx	diagnosis
E. coli	*Escherichia coli*
EAU	exam under anesthesia
EBV	Epstein-Barr virus
ECC	endocervical curettage, extracorporeal circulation
ECCE	extracapsular cataract extraction
ECG	electrocardiogram
Echo	echocardiogram
ECT	electroconvulsive therapy
ED	erectile dysfunction
EDC	estimated date of confinement
EEG	electroencephalogram, electroencephalography
EENT	eyes, ears, nose, throat
EGD	esophagogastroduodenoscopy
EKG	electrocardiogram
ELISA	enzyme-linked immunosorbent assay
elix	elixir
EM	emmetropia (normal vision)
EMB	endometrial biopsy
EMG	electromyogram
EMT-B	emergency medical technician-basic

ABBREVIATION	MEANING
EMT-I	emergency medical technician-intermediate
EMT-P	emergency medical technician-paramedic
emul	emulsion
Endo	endoscopy
ENT	ear, nose, throat
EOM	extraocular movement
eosins, eos	eosinophil
ER	emergency room
ERCP	endoscopic retrograde cholangiopancreatography
ERT	estrogen replacement therapy
ERV	expiratory reserve volume
ESR	erythrocyte sedimentation rate
ESRD	end-stage renal disease
e-stim	electrical stimulation
ESWL	extracorporeal shock-wave lithotripsy
et	and
ET	endotracheal
FBS	fasting blood sugar
FDA	Federal Drug Administration
Fe	iron
FEF	forced expiratory flow
FEKG	fetal electrocardiogram
FEV	forced expiratory volume
FHR	fetal heart rate
FHT	fetal heart tone
fl	fluid
FOBT	fecal occult blood test
FRC	functional residual capacity
FS	frozen section
FSH	follicle-stimulating hormone
FTND	full-term normal delivery
FVC	forced vital capacity
Fx, FX	fracture
G_1	first pregnancy

ABBREVIATION	MEANING
Ga	gallium
GA	general anesthesia
GB	gallbladder
GC	gonorrhea
GERD	gastroesophageal reflux disease
GH	growth hormone
GI	gastrointestinal
gm	gram
GOT	glutamic oxaloacetic transaminase
gr	grain
grav I	first pregnancy
gt	drop
gtt	drops
GTT	glucose tolerance test
GU	genitourinary
GVHD	graft vs. host disease
GYN, gyn	gynecology
H_2O	water
HA	headache
HAV	hepatitis A virus
Hb	hemoglobin
HBOT	hyperbaric oxygen therapy
HBV	hepatitis B virus
HCG, hCG	human chorionic gonadotropin
HCO_3^-	bicarbonate
HCT, Hct	hematocrit
HCV	hepatitis C virus
HD	Hodgkin's disease
HDL	high-density lipoproteins
HDN	hemolytic disease of the newborn
HEENT	head, ears, eyes, nose, throat
HF	heart failure
Hgb, HGB	hemoglobin
HIV	human immunodeficiency virus
HMD	hyaline membrane disease
HNP	herniated nucleus pulposus

ABBREVIATION	MEANING
HPV	human papilloma virus
HRT	hormone replacement therapy
hs	hour of sleep
HSG	hysterosalpingography
HSV	*Herpes simplex* virus
HTN	Hypertension
HZ	hertz
IBD	inflammatory bowel disease
IBS	irritable bowel syndrome
IC	inspiratory capacity
ICCE	intracapsular cataract cryoextraction
ICP	intracranial pressure
ICU	intensive care unit
I & D	incision and drainage
ID	Intradermal
IDDM	insulin-dependent diabetes mellitus
Ig	immunoglobins (IgA, IgD, IgE, IgG, IgM)
i	one
ii	two
iii	three
IM	intramuscular
inj	injection
I & O	intake and output
IOL	intraocular lens
IOP	intraocular pressure
IPD	intermittent peritoneal dialysis
IPPB	intermittent positive pressure breathing
IRDS	infant respiratory distress syndrome
IRV	inspiratory reserve volume
IUD	intrauterine device
IV	intravenous
IVC	intravenous cholangiogram
IVF	*in vitro* fertilization

ABBREVIATION	MEANING
IVP	intravenous pyelogram
JRA	juvenile rheumatoid arthritis
JVP	jugular venous pulse
K^+	potassium
kg	kilogram
KS	Kaposi's sarcoma
KUB	kidney, ureter, bladder
L	left, liter
L1, L2, etc.	first lumbar vertebra, second lumbar vertebra, etc.
LASIK	laser-assisted in-situ keratomileusis
LAT, lat	lateral
LAVH	laparoscopic-assisted vaginal hysterectomy
LBW	low birth weight
LDH	lactate dehydrogenase
LDL	low-density lipoproteins
LE	lower extremity
LGI	lower gastrointestinal series
LH	luteinizing hormone
liq	liquid
LL	left lateral
LLE	left lower extremity
LLL	left lower lobe
LLQ	left lower quadrant
LMP	last menstrual period
LP	lumbar puncture
LPN	licensed practical nurse
LUE	left upper extremity
LUL	left upper lobe
LUQ	left upper quadrant
LVAD	left ventricular assist device
LVH	left ventricular hypertrophy
lymphs	lymphocyte
LVN	licensed vocational nurse
mA	milliampere

ABBREVIATION	MEANING
MA	mental age
MAO	monoamine oxidase
mcg	microgram
mCi	millicurie
MCV	mean corpuscular volume
MD	doctor of medicine, muscular dystrophy
mEq	milliequivalent
mets	metastases
mg	milligram
MH	marital history
MI	myocardial infarction, mitral insufficiency
mL	milliliter
MLT	medical laboratory technician
mm	millimeter
MM	malignant melanoma
mm Hg	millimeters of mercury
MMPI	Minnesota Multiphasic Personality Inventory
Mono	mononucleosis
monos	monocyte
MR	mitral regurgitation
MRA	magnetic resonance angiography
MRI	magnetic resonance imaging
MRSA	Methicillin-Resistant *Staphylococcus Aureus*
MS	mitral stenosis, multiple sclerosis, musculoskeletal
MSII	melanocyte-stimulating hormone
MSN	master of science in nursing
MT	medical technologist
MTX	methotrexate
MUA	manipulation under anesthesia
MV	minute volume
MVP	mitral valve prolapse
n & v	nausea and vomiting
Na$^+$	sodium

ABBREVIATION	MEANING
NB	newborn
NG	nasogastric (tube)
NGU	nongonococcal urethritis
NHL	non-Hodgkin's lymphoma
NIDDM	non-insulin-dependent diabetes mellitus
NK	natural killer cells
NMR	nuclear magnetic resonance
no sub	no substitute
noc	night
non rep	do not repeat
NP	nurse practitioner
NPDL	modular, poorly differentiated lymphocytes
NPH	neutral protamine Hagedorn (insulin)
NPO	nothing by mouth
NS	normal saline
NSAID	nonsteroidal anti-inflammatory drug
NSR	normal sinus rhythm
O$_2$	oxygen
OA	osteoarthritis
OB	obstetrics
OCD	obsessive-compulsive disorder
OCPs	oral contraceptive pills
OD	overdose, right eye, doctor of optometry
oint.	ointment
OM	otitis media
O & P	ova and parasites
Ophth.	ophthalmology
OR	operating room
ORIF	open reduction-internal fixation
Orth, ortho	orthopedics
OS	left eye
OT	occupational therapy
OTC	over the counter

ABBREVIATION	MEANING
Oto	otology
OTR	occupational therapist
OU	each eye
oz	ounce
$\bar{p}$	after
P	pulse
PI	first delivery
PA	posteroanterior, physician assistant, pernicious anemia
PAC	premature atrial contraction
PAP	Papanicolaou test, pulmonary arterial pressure
para I	first delivery
PARR	postanesthetic recovery room
PBI	protein-bound iodine
pc	after meals
PCA	patient-controlled administration
PCP	*Pneumocystis carinii* pneumonia
PCV	packed cell volume
PDA	patent ductus arteriosus
PDR	*Physician's Desk Reference*
PE tube	polyethylene tube placed in the eardrum
PEG	pneumoencephalogram, percutaneous endoscopic gastrostomy
per	with
PERRLA	pupils equal, round, react to light and accommodation
PET	positron emission tomography
PFT	pulmonary function test
pH	acidity or alkalinity
PharmD	doctor of pharmacy
PID	pelvic inflammatory disease
PKU	phenylketonuria
PM, pm	evening
PMN	polymorphonuclear neutrophil
PMP	previous menstrual period

ABBREVIATION	MEANING
PMS	premenstrual syndrome
PND	paroxysmal nocturnal dyspnea, postnasal drip
PNS	peripheral nervous system
PO, po	phone order, by mouth
polys	polymorphonuclear neutrophil
PORP	partial ossicular replacement prosthesis
pp	postprandial (after meals)
PPD	purified protein derivative (tuberculin test)
preop, pre-op	preoperative
prep	preparation, prepared
PRK	photo refractive keratectomy
PRL	prolactin
prn	as needed
Pro-time	prothrombin time
PROM	passive range of motion
prot	protocol
PSA	prostate specific antigen
pt	pint
PT	prothrombin time, physical therapy, physical therapist
PTA	physical therapy assistant
PTC	percutaneous transhepatic cholangiography
PTCA	percutaneous transluminal coronary angioplasty
PTH	parathyroid hormone
PUD	peptic ulcer disease
PVC	premature ventricular contraction
$\bar{q}$	every
qam	every morning
qh	every hour
qhs	every night
qs	quantity sufficient
R	respiration, right, roentgen
Ra	radium

ABBREVIATION	MEANING	ABBREVIATION	MEANING
RA	rheumatoid arthritis	RUE	right upper extremity
rad	radiation absorbed dose	RUL	right upper lobe
RAI	radioactive iodine	RUQ	right upper quadrant
RAIU	radioactive iodine uptake	RV	reserve volume
RBC	red blood cell	Rx	take
RD	respiratory disease, registered dietitian	s̄	without
RDA	recommended daily allowance	S1	first heart sound
RDH	registered dental hygienist	S2	second heart sound
RDS	respiratory distress syndrome	SA, S-A	sinoatrial
REEGT	registered electroencephalography technologist	SAD	seasonal affective disorder
		SAH	subarachnoid hemorrhage
REM	rapid eye movement	SBFT	small bowel follow-through
REPT	registered evoked potential technologist	SC, sc	subcutaneous
		SCC	squamous cell carcinoma
Rh-	Rh-negative	SCI	spinal cord injury
Rh+	Rh-positive	SCIDS	severe combined immunodeficiency syndrome
RHIA	registered health information administrator	Sed-rate	erythrocyte sedimentation rate
RHIT	registered health information technician	SEE-2	Signing Exact English
		SG	skin graft, specific gravity
RIA	radioimmunoassay	s.gl.	without correction or glasses
RL	right lateral	SGOT	serum glutamic oxaloacetic transaminase
RLE	right lower extremity		
RLL	right lower lobe	SIDS	sudden infant death syndrome
RLQ	right lower quadrant	Sig	label as follows/directions
RML	right mediolateral, right middle lobe	SK	streptokinase
		sl	under the tongue
RN	registered nurse	SLE	systemic lupus erythematosus
ROM	range of motion	SMAC	sequential multiple analyzer computer
RP	retrograde pyelogram		
RPh	registered pharmacist	SMD	senile macular degeneration
RPR	rapid plasma reagin (test for syphilis)	SOB	shortness of breath
		sol	solution
RPSGT	registered polysomnographic technologist	SOM	serous otitis media
		SPP	suprapubic prostatectomy
RRT	registered radiologic technologist, registered respiratory therapist	SR	erythrocyte sedimentation rate
		ss	one-half

ABBREVIATION	MEANING
st	stage
ST	skin test, esotropia
stat, STAT	at once, immediately
STD	skin test done, sexually transmitted disease
STSG	split-thickness skin graft
subcu	subcutaneous
subq	subcutaneous
supp.	suppository
suppos	suppository
susp	suspension
syr	syrup
T	tablespoon
t	teaspoon
T & A	tonsillectomy and adenoidectomy
T1, T2, etc.	first thoracic vertebra, second thoracic vertebra, etc.
T_3	triiodothyronine
T_4	thyroxine
T_7	free thyroxine index
tab	tablet
TAH-BSO	total abdominal hysterectomy-bilateral salpingo-oophorectomy
TB	tuberculosis
tbsp	tablespoon
TENS	transcutaneous electrical nerve stimulation
TFT	thyroid function test
THA	total hip arthroplasty
THR	total hip replacement
TIA	transient ischemic attack
tid	three times a day
tinc	tincture
TKA	total knee arthroplasty
TKR	total knee replacement
TLC	total lung capacity
TMJ	temporomandibular joint
TNM	tumor, nodes, metastases

ABBREVIATION	MEANING
TO	telephone order
top	apply topically
TORP	total ossicular replacement prothesis
tPA	tissue-type plasminogen activator
TPN	total parenteral nutrition
TPR	temperature, pulse, and respiration
tr	tincture
TSH	thyroid-stimulating hormone
tsp	teaspoon
TSS	toxic shock syndrome
TUR	transurethral resection
TURP	transurethral resection of prostate
TV	tidal volume
TX, Tx	traction, treatment
U/A, UA	urinalysis
UC	uterine contractions, urine culture
UE	upper extremity
UGI	upper gastrointestinal, upper gastrointestinal series
ung	ointment
URI	upper respiratory infection
US	ultrasound
UTI	urinary tract infection
UV	ultraviolet
VA	visual acuity
VC	vital capacity
VCUG	voiding cystourethrography
VD	venereal disease
VF	visual field
VFib	ventricular fibrillation
VLDL	very low density lipoproteins
VO	verbal order
VRE	Vancomycin-resistant enterococcus

ABBREVIATION	MEANING	ABBREVIATION	MEANING
VS	vital signs	WBC	white blood cell
VSD	ventricular septal defect	wt	weight
VT	ventricular tachycardia	x	times

Official "Do Not Use" List[1]

DO NOT USE	POTENTIAL PROBLEM	USE INSTEAD
U (unit)	Mistaken for "0" (zero), the number "4" (four) or "cc"	Write "unit"
IU (International Unit)	Mistaken for IV (intravenous) or the number 10 (ten)	Write "International Unit"
Q.D., QD, q.d., qd (daily)	Mistaken for each other	Write "daily"
Q.O.D., QOD, q.o.d, qod (every other day)	Period after the Q mistaken for "I" and the "O" mistaken for "I"	Write "every other day"
Trailing zero (X.0 mg)*	Decimal point is missed	Write X mg
Lack of leading zero (.X mg)	Decimal point is missed	Write 0.X mg
MS	Can mean morphine sulfate or magnesium sulfate	Write "morphine sulfate"
MSO4 and MgSO4	Confused for one another	Write "magnesium sulfate"

1 Applies to all orders and all medication-related documentation that is handwritten (including free-text computer entry) or on pre-printed forms.
Exception: A "trailing zero" may be used only where required to demonstrate the level of precision of the value being reported, such as for laboratory results, imaging studies that report size of lesions, or catheter/tube sizes. It may not be used in medication orders or other medication-related documentation.

Additional Abbreviations, Acronyms, and Symbols
(FOR POSSIBLE FUTURE INCLUSION IN THE OFFICIAL "DO NOT USE" LIST)

DO NOT USE	POTENTIAL PROBLEM	USE INSTEAD
> (greater than)	Misinterpreted as the number "7" (seven) or the letter "L"	Write "greater than"
< (less than)	Confused for one another	Write "less than"
Abbreviations for drug names	Misinterpreted due to similar abbreviations for multiple drugs	Write drug names in full
Apothecary units	Unfamiliar to many practitioners Confused with metric units	Use metric units
@	Mistaken for the number "2" (two)	Write "at"
cc	Mistaken for U (units) when poorly written	Write "ml" or "milliliters"
μg	Mistaken for mg (milligrams) resulting in one thousand-fold overdose	Write "mcg" or "micrograms"

© The Joint Commission, 2008. Reprinted with permission.

APPENDIX C: Infectious Diseases and Their Pathogens

Viral Conditions

DISEASE	VIRUS
Acquired immunodeficiency syndrome (AIDS)	Human immunodeficiency virus (HIV)
Chickenpox (varicella) and shingles (herpes zoster)	Varicella zoster virus (VZV)
Common cold and upper respiratory infections (URIs)	Rhinoviruses
Fever blisters and herpes	Herpes simplex 1 and 2
Hantavirus pulmonary syndrome	Hantavirus
Hepatitis (infectious)	Hepatitis A virus
Hepatitis (serum)	Hepatitis B virus
Hepatitis (non A; non B)	Hepatitis C
Infectious mononucleosis	Epstein-Barr virus (EBV)
Influenza	Influenza A, B, C, etc.
Measles	Rubeola virus
Mumps	Paramyxovirus
Poliomyelitis	Poliovirus 1, 2, and 3
Rabies	Rabies virus
Rubella (German measles)	Rubella virus
Viral encephalitis	(many)
Warts, genital warts, and cervical cancer	Human papilloma viruses (HPV)

Bacterial Conditions

DISEASE	ORGANISM
Acute bacterial conjunctivitis	Staphylococcus, Haemophilus, Proteus, and other organisms
Anthrax	Bacillus anthracis
Botulism	Clostridium botulinum (bacillus)
Brucellosis	Brucella species (bacilli)
Cholera	Vibrio cholerae (curved)
Dental caries	Streptococcus mutans (coccus) and other organisms
Diarrhea	Campylobacter jejuni
Diphtheria	Corynebacterium diphtheriae (bacillus)
Epiglottitis	Haemophilus influenzae
External otitis (swimmer's ear)	Pseudomonas aeruginosa, Staphylococcus aureus, Streptococcus pyogenes, etc.

Bacterial Conditions *(continued)*

DISEASE	ORGANISM
Gonorrhea	*Neisseria gonorrhoeae* (coccus)
Legionnaires' disease	*Legionella pneumophila* (bacillus)
Lyme disease	*Borrelia burgdorferi* (spirochete)
Meningitis	*Streptococcus pneumoniae, Neisseria meningitidis, Haemophilus influenzae,* and other organisms
Parrot fever (psittacosis)	*Chlamydia psittaci*
Pelvic inflammatory disease (PID)	*Neisseria gonorrhoeae* (coccus), *Mycoplasma hominis,* and other organisms
Peptic ulcer	*Helicobacter pylori*
Pertussis (whooping cough)	*Bordetella pertussis* (bacillus)
Pharyngitis (strep throat)	*Streptococcus pyogenes*
Pneumonia	*Streptococcus pneumoniae* (coccus), *Pseudomonas aeruginosa,* and other organisms
Rheumatic fever	*Group A beta-hemolytic streptococci* (cocci)
Rocky Mountain spotted fever (RMSF)	*Rickettsia rickettsii* (small)
Salmonellosis	*Salmonella* species (bacilli)
Shigellosis (Shigella dysentery and bacillary dysentery)	*Shigella* species (bacilli)
Skin infections	*Staphylococcus* species (cocci)
Syphilis	*Treponema pallidum* (spirochete)
Tetanus	*Clostridium tetani* (bacillus)
Toxic shock syndrome (TSS)	*Staphylococcus aureus strains* (cocci)
Trachoma (chlamydial conjunctivitis)	*Chlamydia trachomatis*
Tuberculosis	*Mycobacterium tuberculosis*
Typhoid fever	*Salmonella typhi* (bacillus)
Urinary tract infections	*Pseudomonas aeruginosa, E. coli*

Note: Although the organisms listed are the most common causes of these diseases, other organisms may also be causative agents.

Mycotic (Fungal) Conditions

DISEASE	ORGANISM
Aspergillosis	*Aspergillus* species
Blastomycosis	*Blastomyces dermatitidis*
Candidiasis (thrush)	*Candida albicans* and other species
Coccidioidomycosis (San Joaquin fever)	*Coccidioides immitis*
Histoplasmosis	*Histoplasma capsulatum*
Tinea (includes athlete's foot and jock itch)	*Epidermophyton, Microsporum,* and *Trichophyton* species

Conditions Caused by Protozoa

DISEASE	ORGANISM
Amebiasis and amebic dysentery	*Entamoeba histolytica*, *Entamoeba polecki*, and other organisms (ameba)
Giardiasis (traveler's diarrhea)	*Giardia lamblia* (flagellate)
Isosporiasis	*Isospora belli* (sporozoan)
Malaria	*Plasmodium* species (sporozoa)
Pneumonia associated with AIDS	*Pneumocystis jiroveci*
Toxoplasmosis	*Toxoplasma gondii* (sporozoan)
Trichomoniasis	*Trichomonas vaginalis* (flagellate)

APPENDIX D: Laboratory Reference Values

Abbreviations Used in Reporting Laboratory Values

ABBREVIATION	MEANING	ABBREVIATION	MEANING
cm^3	cubic centimeter	mg/dL	milligram per deciliter
cu μ	cubic microns	mm	millimeter
dL	deciliter	mmol	millimole
fL	femtoliter	mm^3	cubic millimeter
g	gram	mm Hg	millimeter of mercury
g/dL	grams per deciliter	ng	nanogram
kg	kilogram	ng/dL	nanogram per deciliter
L	liter	ng/mL	nanogram per milliliter
mol (M)	mole	pg	picogram
mEq	milliequivalent	U/L	units per liter
mg	milligram	uIU/mL	units International Unit per milliliter

Hematology Tests: Normal Ranges*

Erythrocytes–Red blood cells (RBC)	Females 4.2–5.4 million/mm^3 Males 4.6–6.2 million/mm^3 Children 4.5–5.1 million/mm^3
Hemoglobin (HGB, Hgb)	Females 12.0–14.0 g/dL Males 14.0–16.0 g/dL
Hematocrit (HCT)	37.0–54 % Females 37–47 % Males 40–54 %
Leukocytes–White Blood Cells	4500–11,000/mm^3 blood cells (WBC)
Differential (WBCs)	Neutrophils 54–62 % Lymphocytes 20–40 % Monocytes 2–10 % Eosinophils 1–2 % Basophils 0–1 %
Thrombocytes– Platelets	200,000–400,000/mm^3

*These Values May Vary Between Hospital Laboratories

Hematology Tests: Coagulation Tests

Bleeding time	2.75–8.0 min
Prothrombin time (PT)	12–14 sec
Partial Thromboplastin Time (PTT)	30–45 sec

Blood Chemistries: Normal Ranges

A/G ratio	0.7–2.0 g/dL
Alanine aminotransferase	5–30 U/L (ALT, SGPT)
Albumin	3.5–5.5 g/dL
Alkaline phosphatase (ALP)	20–90 U/L
Anion gap	10–17 mEq/L
Aspartate aminotransferase	10–30 U/L (AST, SGOT)
Bilirubin	0.3–1.1 mg/dL
Blood urea nitrogen (BUN)	8–20 mg/dL
Calcium (Ca)	9.0–11.0 mg/dL
Chloride (Cl)	100–108 mmol/L
Cholesterol	<200 mg/dL
High density lipoprotein (HDL)	>60 mg/dL
Low density lipoprotein (LDL)	<100 mg/dL
Creatine phosphokinase (CPK)	Females 30–135 U/L Males 55–170 U/L
Creatinine	0.9–1.5 mg/dL
Free T_4	0.8–1.8 ng/dL
Globulin	1.4–4.8 g/dL
Glucose (fasting)	70–115 mg/dL
Lactate dehydrogenase (LDH)	100–190 U/L
Phosphate (PO_4)	3.0–4.5 mg/dL
Potassium (K)	3.5–5.0 mEq/L
Prostate specific antigen	0.0–4.0 ng/mL (PSA) Male
Sodium (Na)	136–145 mEq/L
Testosterone	241–827 ng/dL
Thyroid stimulating	0.5–6.0 ulU/mL hormone (TSH)
Thyroxine (T_4)	4.4–9.9 µg/dL
Triglycerides	<150 mg/dL
Uric acid	Females 1.5–7.0 mg/dL Males 2.5–8.0 mg/dL

Urinalysis: Normal Ranges

Color	yellow to amber
Turbidity (appearance)	Clear to slightly hazy
Specific gravity	1.003–1.030
Reaction (pH)	5.0–7.0
Odor	Faintly aromatic
Output	1–1.5 liters per day
Protein	Negative
Glucose	Negative
Ketones	Negative
Bilirubin	Negative
Blood	Negative
Urobilinogen	0.1–1.0
Nitrite	Negative
Leukocytes	Negative

Normal Values of Arterial Blood Gases*

pH	7.35–7.45
PaO_2	80–100 mm Hg
$PaCO_2$	35–45 mm Hg
HCO_3^-	22–26 mEq/L
O_2 saturation	95–98%

*Some normal values will vary according to the kind of test carried out in the laboratory.

Common Blood Chemistries and Examples of Disorders They Indicate

TEST	Abbreviation	Normal Range	EXAMPLES OF POSSIBLE DIAGNOSIS	
			Results Increased	Results Decreased
Alkaline phosphate	ALP	30–115 mU/mL	Liver disease, bone disease, mononucleosis	Malnutrition hypothyroidism, chronic nephritis
Blood urea nitrogen	BUN	8–25 mg/dL	Kidney disease, dehydration, GI bleeding	Liver failure, malnutrition
Calcium	Ca	8.5–10.5 mg/dL	Hypercalcemia, bone metastases, Hodgkin's disease	Hypocalcemia, renal failure, pancreatitis

continued

Common Blood Chemistries and Examples of Disorders They Indicate (*continued*)

TEST	Abbreviation	Normal Range	Results Increased	Results Decreased
Chloride	Cl	96–11 mEq/L	Dehydration, eclampsia, anemia	Ulcerative colitis, burns, heat exhaustion
Cholesterol	CHOL	120–200 mg/dL	Atherosclerosis, nephrosis, obstructive jaundice	Malabsorption, liver disease, hyperthyroidism
Creatinine	Creat	0.4–1.5 mg/dL	Chronic nephritis, muscle disease, obstruction of urinary tract	Muscular dystrophy
Globulin	Glob	1.0–3.5 g/dL	Brucellosis, rheumatoid arthritis, hepatic carcinoma	Severe burns
Glucose fasting blood sugar	FBS	70–110 mg/100 mL	Diabetes mellitus	Excess insulin
Two-hour postprandial	2-hr PPBS	140 mg/dL	Cushing syndrome, brain damage	Addison's disease, CA of pancreas
Lactic acid	LDH	100–225 mU/mL	Acute MI, acute leukemia, hepatic disease	
Potassium	K	3.5–5.5 mEq/L	Renal failure, acidosis, cell damage	Malabsorption, severe burn, diarrhea
Serum glutamic-oxaloacetic	SGOT	0–41 mU/mL	MI, liver disease, pancreatitis	Uncontrolled diabetes mellitus with acidosis
Serum glutamic pyruvic transaminase	SGPT	0–45 mU/mL	Active cirrhosis, pancreatitis, obstructive jaundice	
Sodium	Na	135–145 mEq/L	Diabetes insipidus, coma, Cushing syndrome	Severe diarrhea, severe nephritis, vomiting
Free thyroxine	T4	1–2.3 mg/dL	Thyroiditis, hyperthyroidism, Graves disease	Goiter, myxedema, hypothyroidism
Total bilirubin	TB	0.1–1.2 mg/dL	Liver disease, hemolytic anemia, lupus erythremia	
Triglycerides	TRIG	40–170 mg/dL	Liver disease, atherosclerosis, pancreatitis	Malnutrition
Uric acid	UA	2.2–9.0 mg/dL	Renal failure, gout, leukemia, eclampsia	

Diabetes Mellitus

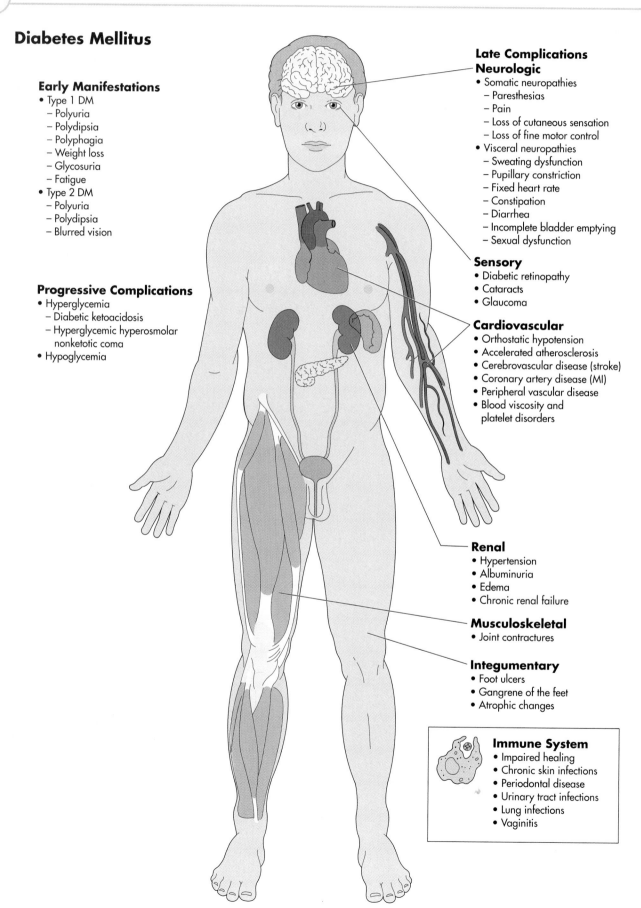

Early Manifestations
- Type 1 DM
  - Polyuria
  - Polydipsia
  - Polyphagia
  - Weight loss
  - Glycosuria
  - Fatigue
- Type 2 DM
  - Polyuria
  - Polydipsia
  - Blurred vision

Progressive Complications
- Hyperglycemia
  - Diabetic ketoacidosis
  - Hyperglycemic hyperosmolar nonketotic coma
- Hypoglycemia

Late Complications
Neurologic
- Somatic neuropathies
  - Paresthesias
  - Pain
  - Loss of cutaneous sensation
  - Loss of fine motor control
- Visceral neuropathies
  - Sweating dysfunction
  - Pupillary constriction
  - Fixed heart rate
  - Constipation
  - Diarrhea
  - Incomplete bladder emptying
  - Sexual dysfunction

Sensory
- Diabetic retinopathy
- Cataracts
- Glaucoma

Cardiovascular
- Orthostatic hypotension
- Accelerated atherosclerosis
- Cerebrovascular disease (stroke)
- Coronary artery disease (MI)
- Peripheral vascular disease
- Blood viscosity and platelet disorders

Renal
- Hypertension
- Albuminuria
- Edema
- Chronic renal failure

Musculoskeletal
- Joint contractures

Integumentary
- Foot ulcers
- Gangrene of the feet
- Atrophic changes

Immune System
- Impaired healing
- Chronic skin infections
- Periodontal disease
- Urinary tract infections
- Lung infections
- Vaginitis

Heart Failure

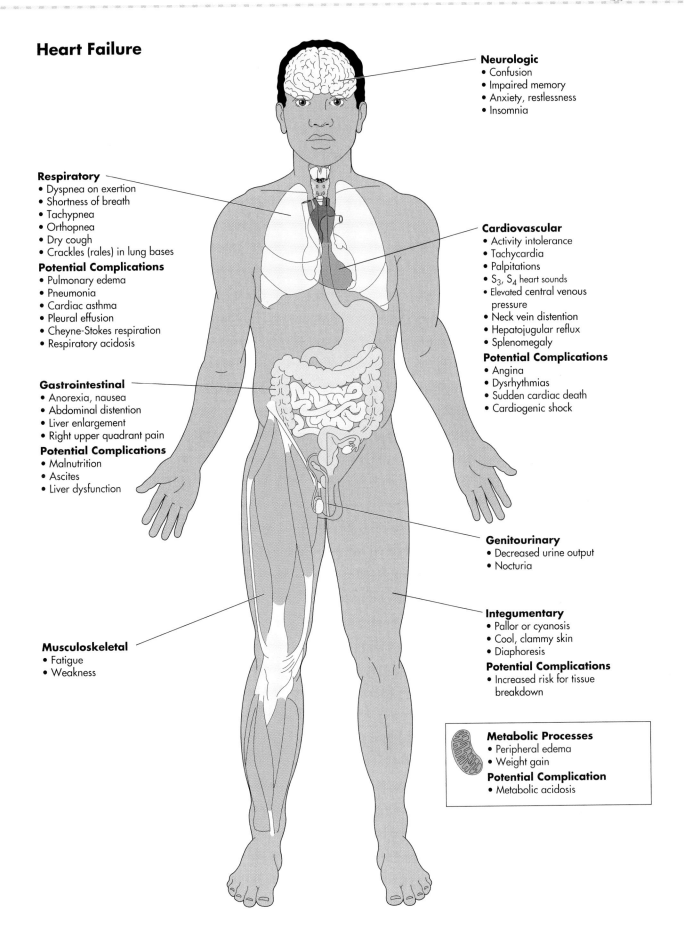

Neurologic
- Confusion
- Impaired memory
- Anxiety, restlessness
- Insomnia

Respiratory
- Dyspnea on exertion
- Shortness of breath
- Tachypnea
- Orthopnea
- Dry cough
- Crackles (rales) in lung bases

Potential Complications
- Pulmonary edema
- Pneumonia
- Cardiac asthma
- Pleural effusion
- Cheyne-Stokes respiration
- Respiratory acidosis

Cardiovascular
- Activity intolerance
- Tachycardia
- Palpitations
- S_3, S_4 heart sounds
- Elevated central venous pressure
- Neck vein distention
- Hepatojugular reflux
- Splenomegaly

Potential Complications
- Angina
- Dysrhythmias
- Sudden cardiac death
- Cardiogenic shock

Gastrointestinal
- Anorexia, nausea
- Abdominal distention
- Liver enlargement
- Right upper quadrant pain

Potential Complications
- Malnutrition
- Ascites
- Liver dysfunction

Genitourinary
- Decreased urine output
- Nocturia

Musculoskeletal
- Fatigue
- Weakness

Integumentary
- Pallor or cyanosis
- Cool, clammy skin
- Diaphoresis

Potential Complications
- Increased risk for tissue breakdown

Metabolic Processes
- Peripheral edema
- Weight gain

Potential Complication
- Metabolic acidosis

Hyperthyroidism

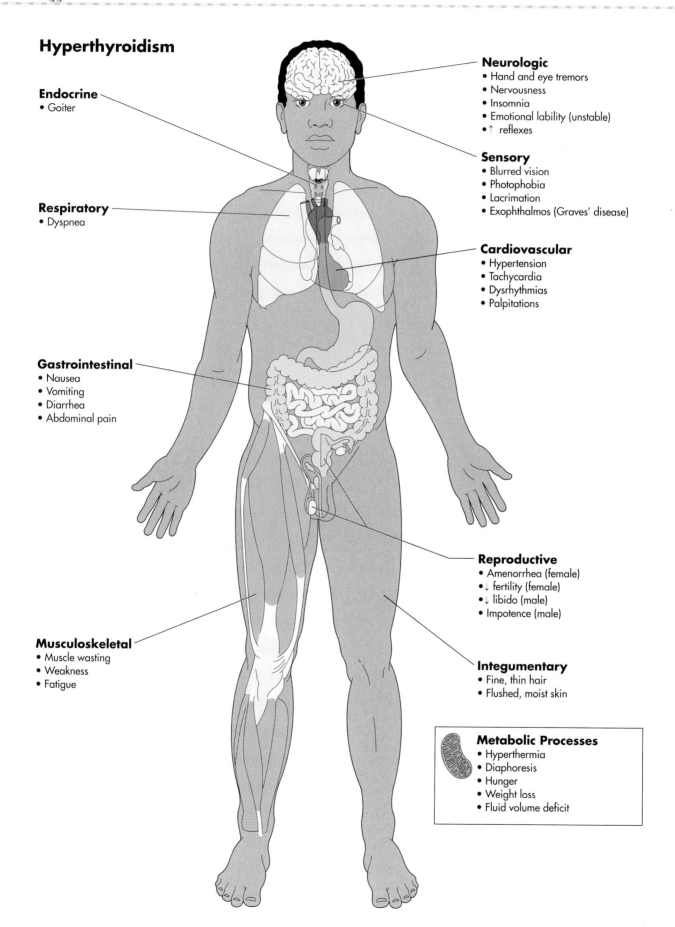

Endocrine
• Goiter

Respiratory
• Dyspnea

Gastrointestinal
• Nausea
• Vomiting
• Diarrhea
• Abdominal pain

Musculoskeletal
• Muscle wasting
• Weakness
• Fatigue

Neurologic
• Hand and eye tremors
• Nervousness
• Insomnia
• Emotional lability (unstable)
• ↑ reflexes

Sensory
• Blurred vision
• Photophobia
• Lacrimation
• Exophthalmos (Graves' disease)

Cardiovascular
• Hypertension
• Tachycardia
• Dysrhythmias
• Palpitations

Reproductive
• Amenorrhea (female)
• ↓ fertility (female)
• ↓ libido (male)
• Impotence (male)

Integumentary
• Fine, thin hair
• Flushed, moist skin

Metabolic Processes
• Hyperthermia
• Diaphoresis
• Hunger
• Weight loss
• Fluid volume deficit

Hypothyroidism

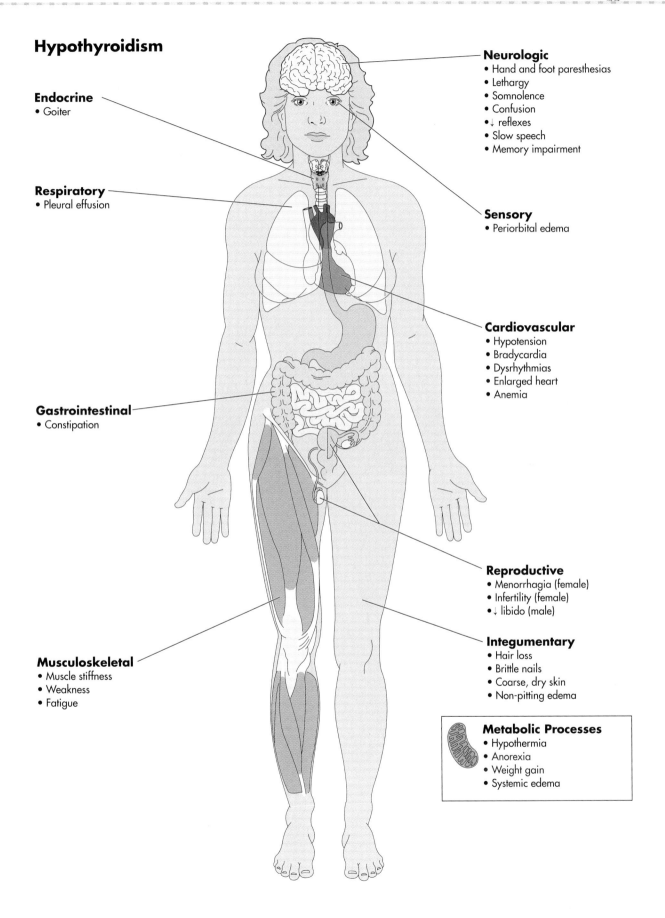

Endocrine
• Goiter

Respiratory
• Pleural effusion

Gastrointestinal
• Constipation

Musculoskeletal
• Muscle stiffness
• Weakness
• Fatigue

Neurologic
• Hand and foot paresthesias
• Lethargy
• Somnolence
• Confusion
• ↓ reflexes
• Slow speech
• Memory impairment

Sensory
• Periorbital edema

Cardiovascular
• Hypotension
• Bradycardia
• Dysrhythmias
• Enlarged heart
• Anemia

Reproductive
• Menorrhagia (female)
• Infertility (female)
• ↓ libido (male)

Integumentary
• Hair loss
• Brittle nails
• Coarse, dry skin
• Non-pitting edema

Metabolic Processes
• Hypothermia
• Anorexia
• Weight gain
• Systemic edema

Leukemia

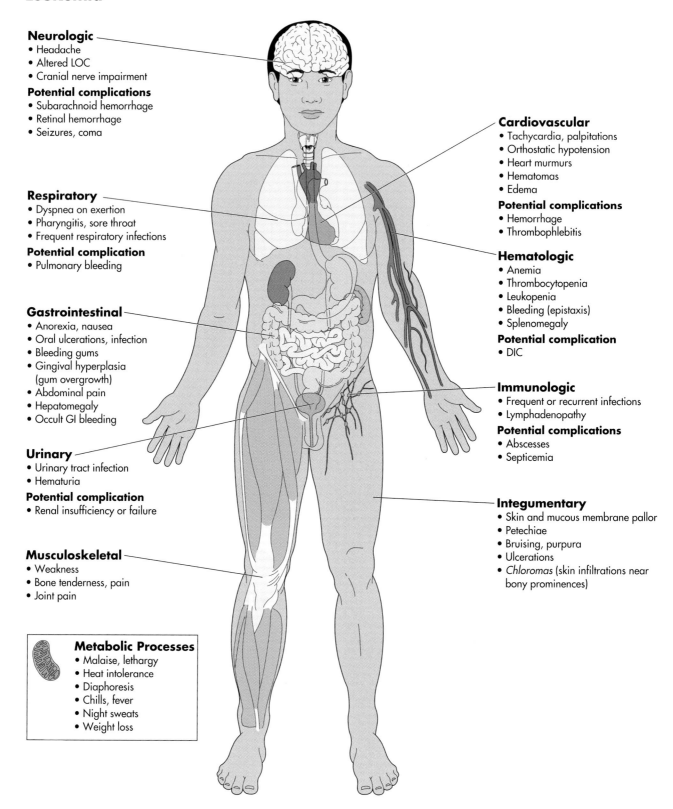

Neurologic
- Headache
- Altered LOC
- Cranial nerve impairment

Potential complications
- Subarachnoid hemorrhage
- Retinal hemorrhage
- Seizures, coma

Respiratory
- Dyspnea on exertion
- Pharyngitis, sore throat
- Frequent respiratory infections

Potential complication
- Pulmonary bleeding

Gastrointestinal
- Anorexia, nausea
- Oral ulcerations, infection
- Bleeding gums
- Gingival hyperplasia
  (gum overgrowth)
- Abdominal pain
- Hepatomegaly
- Occult GI bleeding

Urinary
- Urinary tract infection
- Hematuria

Potential complication
- Renal insufficiency or failure

Musculoskeletal
- Weakness
- Bone tenderness, pain
- Joint pain

Cardiovascular
- Tachycardia, palpitations
- Orthostatic hypotension
- Heart murmurs
- Hematomas
- Edema

Potential complications
- Hemorrhage
- Thrombophlebitis

Hematologic
- Anemia
- Thrombocytopenia
- Leukopenia
- Bleeding (epistaxis)
- Splenomegaly

Potential complication
- DIC

Immunologic
- Frequent or recurrent infections
- Lymphadenopathy

Potential complications
- Abscesses
- Septicemia

Integumentary
- Skin and mucous membrane pallor
- Petechiae
- Bruising, purpura
- Ulcerations
- *Chloromas* (skin infiltrations near
  bony prominences)

Metabolic Processes
- Malaise, lethargy
- Heat intolerance
- Diaphoresis
- Chills, fever
- Night sweats
- Weight loss

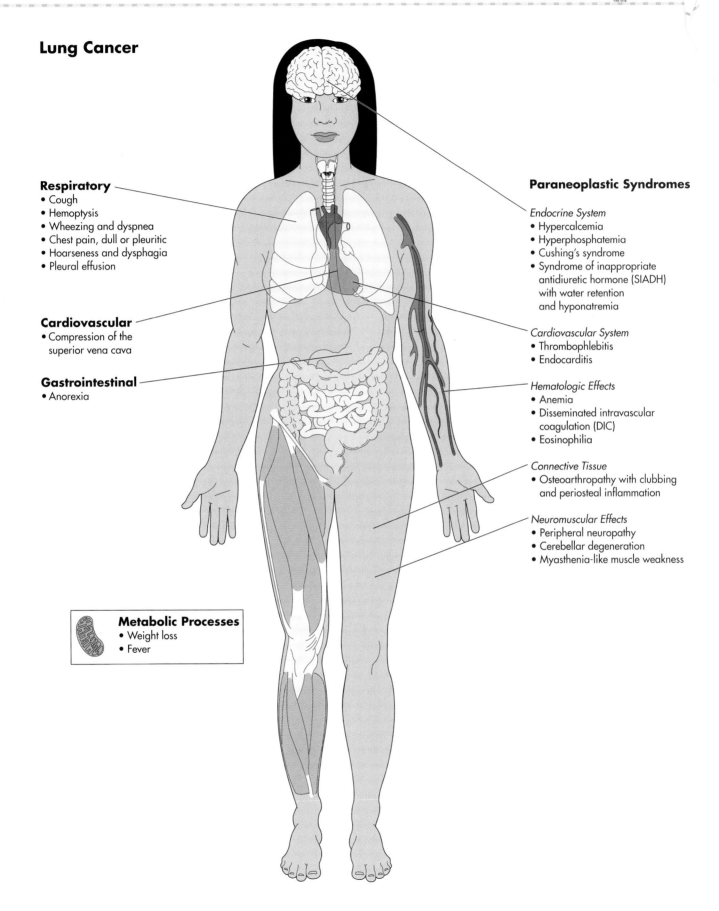

Lung Cancer

Respiratory
- Cough
- Hemoptysis
- Wheezing and dyspnea
- Chest pain, dull or pleuritic
- Hoarseness and dysphagia
- Pleural effusion

Cardiovascular
- Compression of the superior vena cava

Gastrointestinal
- Anorexia

Metabolic Processes
- Weight loss
- Fever

Paraneoplastic Syndromes

Endocrine System
- Hypercalcemia
- Hyperphosphatemia
- Cushing's syndrome
- Syndrome of inappropriate antidiuretic hormone (SIADH) with water retention and hyponatremia

Cardiovascular System
- Thrombophlebitis
- Endocarditis

Hematologic Effects
- Anemia
- Disseminated intravascular coagulation (DIC)
- Eosinophilia

Connective Tissue
- Osteoarthropathy with clubbing and periosteal inflammation

Neuromuscular Effects
- Peripheral neuropathy
- Cerebellar degeneration
- Myasthenia-like muscle weakness

Multiple Sclerosis

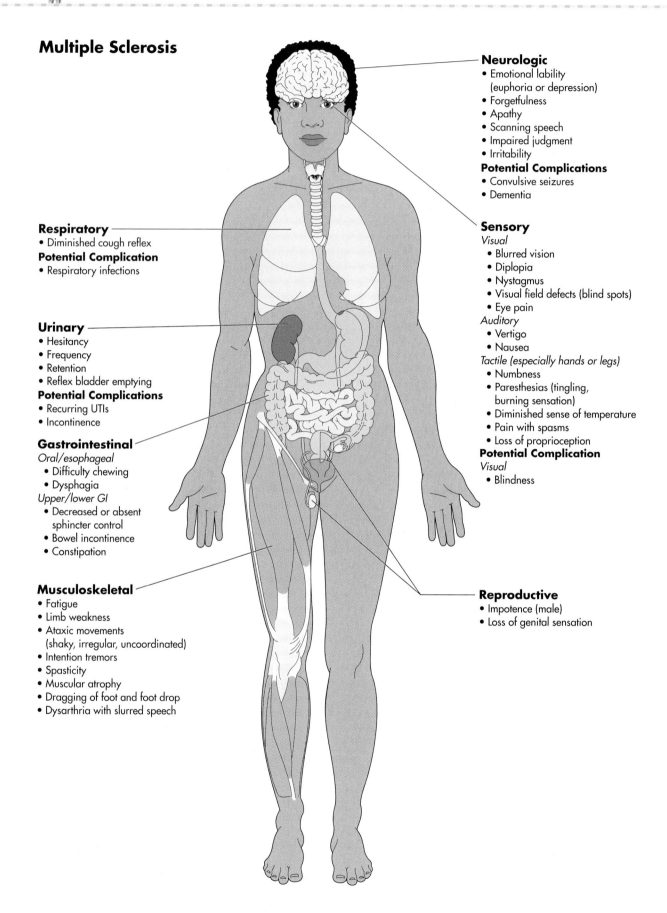

Neurologic
- Emotional lability
  (euphoria or depression)
- Forgetfulness
- Apathy
- Scanning speech
- Impaired judgment
- Irritability

Potential Complications
- Convulsive seizures
- Dementia

Sensory

Visual
- Blurred vision
- Diplopia
- Nystagmus
- Visual field defects (blind spots)
- Eye pain

Auditory
- Vertigo
- Nausea

Tactile (especially hands or legs)
- Numbness
- Paresthesias (tingling,
  burning sensation)
- Diminished sense of temperature
- Pain with spasms
- Loss of proprioception

Potential Complication

Visual
- Blindness

Respiratory
- Diminished cough reflex

Potential Complication
- Respiratory infections

Urinary
- Hesitancy
- Frequency
- Retention
- Reflex bladder emptying

Potential Complications
- Recurring UTIs
- Incontinence

Gastrointestinal

Oral/esophageal
- Difficulty chewing
- Dysphagia

Upper/lower GI
- Decreased or absent
  sphincter control
- Bowel incontinence
- Constipation

Musculoskeletal
- Fatigue
- Limb weakness
- Ataxic movements
  (shaky, irregular, uncoordinated)
- Intention tremors
- Spasticity
- Muscular atrophy
- Dragging of foot and foot drop
- Dysarthria with slurred speech

Reproductive
- Impotence (male)
- Loss of genital sensation

Premenstrual Syndrome

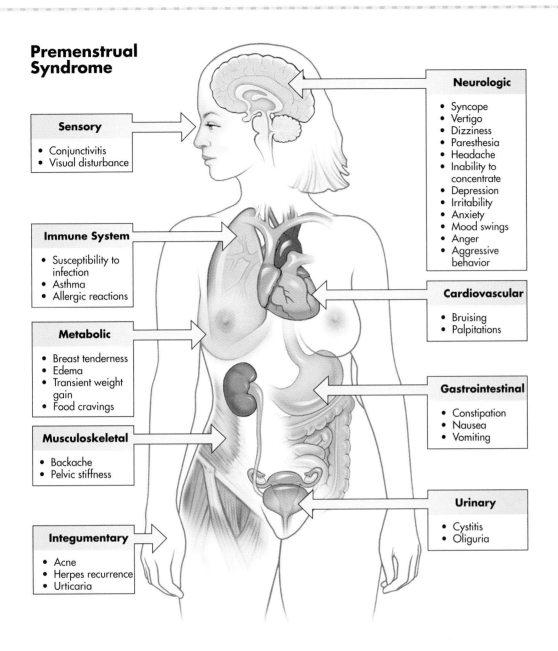

Sensory
- Conjunctivitis
- Visual disturbance

Immune System
- Susceptibility to infection
- Asthma
- Allergic reactions

Metabolic
- Breast tenderness
- Edema
- Transient weight gain
- Food cravings

Musculoskeletal
- Backache
- Pelvic stiffness

Integumentary
- Acne
- Herpes recurrence
- Urticaria

Neurologic
- Syncope
- Vertigo
- Dizziness
- Paresthesia
- Headache
- Inability to concentrate
- Depression
- Irritability
- Anxiety
- Mood swings
- Anger
- Aggressive behavior

Cardiovascular
- Bruising
- Palpitations

Gastrointestinal
- Constipation
- Nausea
- Vomiting

Urinary
- Cystitis
- Oliguria

Rheumatoid Arthritis

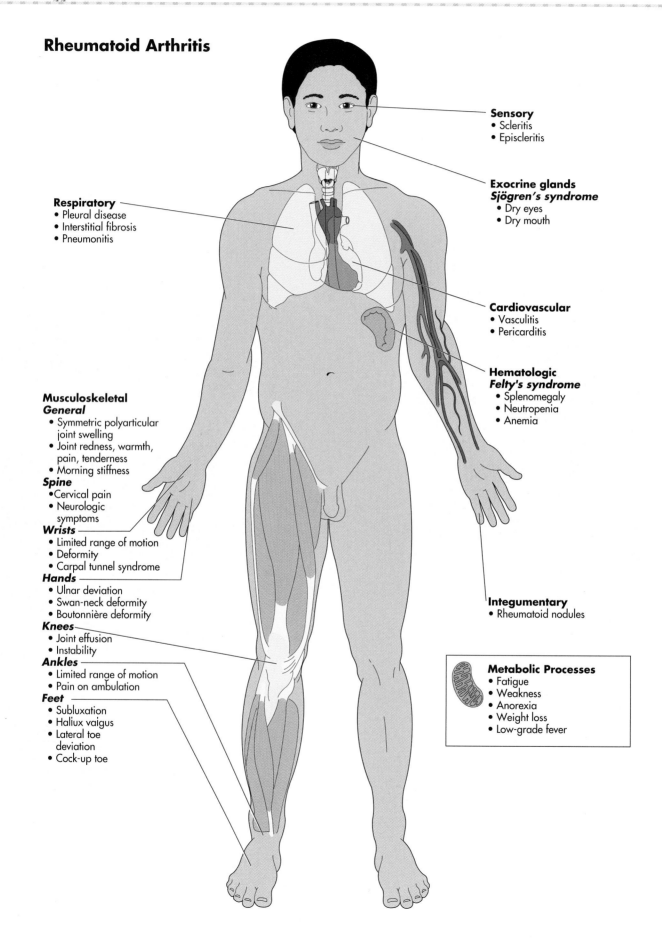

Sensory
- Scleritis
- Episcleritis

Exocrine glands
Sjögren's syndrome
- Dry eyes
- Dry mouth

Respiratory
- Pleural disease
- Interstitial fibrosis
- Pneumonitis

Cardiovascular
- Vasculitis
- Pericarditis

Hematologic
Felty's syndrome
- Splenomegaly
- Neutropenia
- Anemia

Musculoskeletal
General
- Symmetric polyarticular joint swelling
- Joint redness, warmth, pain, tenderness
- Morning stiffness

Spine
- Cervical pain
- Neurologic symptoms

Wrists
- Limited range of motion
- Deformity
- Carpal tunnel syndrome

Hands
- Ulnar deviation
- Swan-neck deformity
- Boutonnière deformity

Knees
- Joint effusion
- Instability

Ankles
- Limited range of motion
- Pain on ambulation

Feet
- Subluxation
- Haliux vaigus
- Lateral toe deviation
- Cock-up toe

Integumentary
- Rheumatoid nodules

Metabolic Processes
- Fatigue
- Weakness
- Anorexia
- Weight loss
- Low-grade fever

Shock

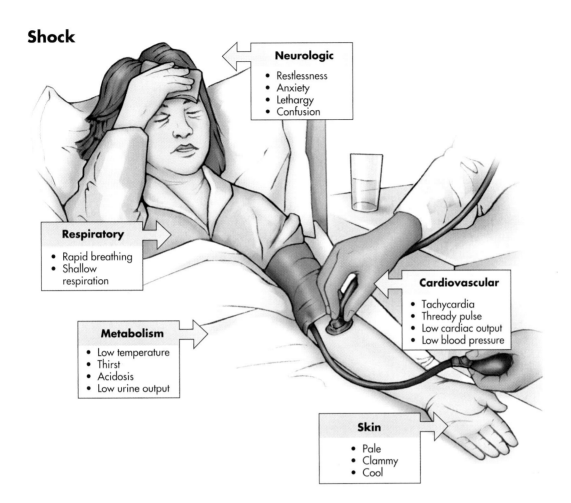

Neurologic
- Restlessness
- Anxiety
- Lethargy
- Confusion

Respiratory
- Rapid breathing
- Shallow respiration

Metabolism
- Low temperature
- Thirst
- Acidosis
- Low urine output

Cardiovascular
- Tachycardia
- Thready pulse
- Low cardiac output
- Low blood pressure

Skin
- Pale
- Clammy
- Cool

Uremia

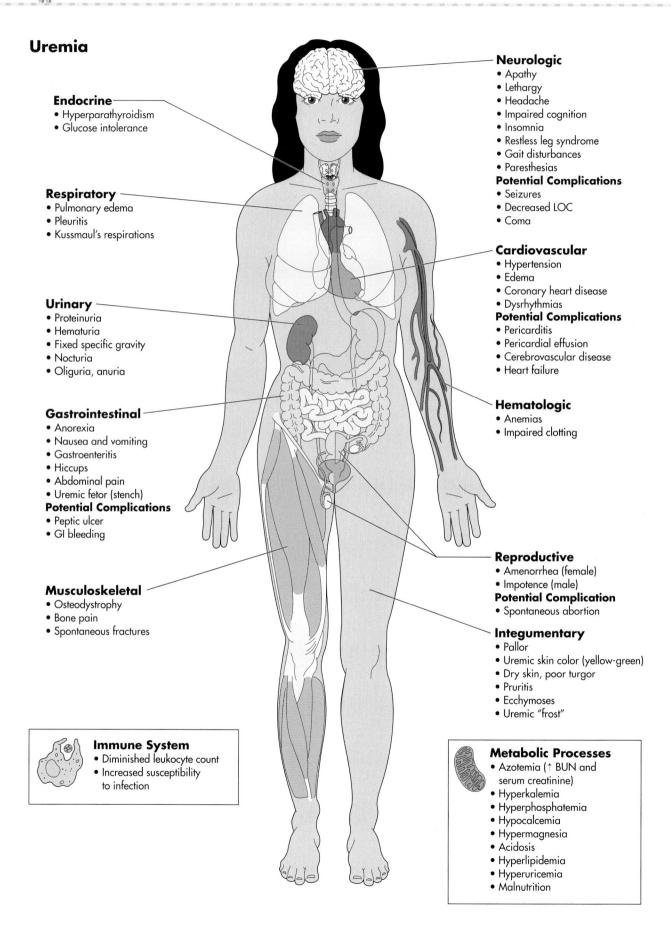

Endocrine
- Hyperparathyroidism
- Glucose intolerance

Respiratory
- Pulmonary edema
- Pleuritis
- Kussmaul's respirations

Urinary
- Proteinuria
- Hematuria
- Fixed specific gravity
- Nocturia
- Oliguria, anuria

Gastrointestinal
- Anorexia
- Nausea and vomiting
- Gastroenteritis
- Hiccups
- Abdominal pain
- Uremic fetor (stench)

Potential Complications
- Peptic ulcer
- GI bleeding

Musculoskeletal
- Osteodystrophy
- Bone pain
- Spontaneous fractures

Immune System
- Diminished leukocyte count
- Increased susceptibility to infection

Neurologic
- Apathy
- Lethargy
- Headache
- Impaired cognition
- Insomnia
- Restless leg syndrome
- Gait disturbances
- Paresthesias

Potential Complications
- Seizures
- Decreased LOC
- Coma

Cardiovascular
- Hypertension
- Edema
- Coronary heart disease
- Dysrhythmias

Potential Complications
- Pericarditis
- Pericardial effusion
- Cerebrovascular disease
- Heart failure

Hematologic
- Anemias
- Impaired clotting

Reproductive
- Amenorrhea (female)
- Impotence (male)

Potential Complication
- Spontaneous abortion

Integumentary
- Pallor
- Uremic skin color (yellow-green)
- Dry skin, poor turgor
- Pruritis
- Ecchymoses
- Uremic "frost"

Metabolic Processes
- Azotemia (↑ BUN and serum creatinine)
- Hyperkalemia
- Hyperphosphatemia
- Hypocalcemia
- Hypermagnesia
- Acidosis
- Hyperlipidemia
- Hyperuricemia
- Malnutrition

APPENDIX F: Vitamins and Minerals: Sources, Function, and Effects of Deficiency and Toxicity

NUTRIENT	PRINCIPAL SOURCES	FUNCTIONS	EFFECTS OF DEFICIENCY AND TOXICITY
Vitamin A (retinol)	Fish liver oils, liver, egg yolk, butter, cream, vitamin A-fortified margarine; as carotenoids: dark green, leafy vegetables; yellow fruits; red palm oil	Photoreceptor mechanism of retina, integrity of skin	**Deficiency:** night blindness, keratomalacia, increased morbidity and mortality in young children **Toxicity:** headache, peeling of skin
Vitamin D (cholecalciferol, ergocalciferol)	Ultraviolet irradiation of the skin (main source); fortified milk (main dietary source), fish liver oils, butter, egg yolk, liver	Calcium and phosphorus absorption; resorption, mineralization, and maturation of bone	**Deficiency:** rickets (sometimes with tetany), osteomalacia **Toxicity:** anorexia, renal failure
Vitamin E group (α-tocopherol and other tocopherols)	Vegetable oil, wheat germ, leafy vegetables, egg yolk, margarine, legumes	Intracellular antioxidant, scavenger of free radicals in biologic membranes	**Deficiency:** RBC hemolysis or rupture of red blood cells, neurologic damage **Toxicity:** interference with enzymes
Vitamin K group (phylloquinone and menaquinones)	Leafy vegetables, pork, liver, vegetable oils, intestinal flora after newborn period	Formation of coagulation factors, and bone proteins	**Deficiency:** hemorrhage from deficiency of coagulation factors
Thiamine (vitamin B$_1$)	Dried yeast, whole grains, meat (especially pork, liver), enriched cereal products, nuts, legumes, potatoes	Carbohydrate metabolism, central and peripheral nerve cell function, heart function	**Deficiency:** beriberi (peripheral nerve degeneration, heart failure, Wernicke Korsakoff syndrome
Riboflavin (vitamin B$_2$)	Milk, cheese, liver, meat, eggs, enriched cereal products	Many aspects of energy and protein metabolism, integrity of mucous membranes	**Deficiency:** dry lips, mouth sores, dermatitis
Niacin (nicotinic acid, niacinamide)	Dried yeast, liver, meat, fish, legumes, whole-grain enriched cereal products	Carbohydrate metabolism	**Deficiency:** pellagra (dermatitis, swelling of the tongue, gastrointestinal and nervous system disturbance)
Vitamin B6 group (pyridoxine, pyridoxal, pyridoxamine)	Dried yeast, liver, organ meats, whole-grain cereals, fish, legumes	Many aspects of protein metabolism	**Deficiency:** convulsions, nervous system disturbance **Toxicity:** peripheral degeneration of nervous tissue
Folic acid	Fresh green leafy vegetables, fruits, organ meats, liver, dried yeast	Maturation of red blood cells	**Deficiency:** anemia, neural tube defects
Vitamin B$_{12}$ (cobalamins)	Liver, meats (especially beef, pork, organ meats), eggs, milk and milk products	Maturation of red blood cells, neural function, DNA synthesis	**Deficiency:** pernicious anemia, megaloblastic anemia

NUTRIENT	PRINCIPAL SOURCES	FUNCTIONS	EFFECTS OF DEFICIENCY AND TOXICITY
Biotin	Liver, kidney, egg yolk, yeast, cauliflower, nuts, legumes	Amino acid and fatty acid metabolism	**Deficiency:** dermatitis, swelling of the tongue
Vitamin C (ascorbic acid)	Citrus fruits, tomatoes, potatoes, cabbage, green peppers	Essential to bone tissue, collagen formation, vascular function, tissue oxygenation, and wound healing	**Deficiency:** scurvy (hemorrhages, loose teeth, gingivitis, bone disease)
Sodium	Wide distribution—beef, pork, sardines, cheese, green olives, corn bread, potato chips, sauerkraut	Acid-base balance, fluid balance, blood acidity, muscle contractility, nerve transmission	**Deficiency:** confusion, coma **Toxicity:** confusion, coma
Chloride	Wide distribution—mainly animal products but some vegetables; similar to sodium	Acid-base balance, fluid balance, blood acidity, kidney function	**Deficiency:** failure to thrive in infants **Toxicity:** increase in extracellular volume, hypertension
Potassium	Wide distribution—whole and skim milk, bananas, prunes, raisins, meats	Muscle activity, nerve transmission, intracellular acid-base balance and water retention	**Deficiency:** paralysis, cardiac disturbances **Toxicity:** paralysis, cardiac disturbances
Calcium	Milk and milk products, meat, fish, eggs, cereal products, beans, fruits, vegetables	Bone and tooth formation, blood coagulation, neuro-muscular irritability, muscle contractility, myocardial conduction	**Deficiency:** long term: osteoporosis tetany, neuromuscular hyperexcitability **Toxicity:** gastrointestinal paralysis, kidney failure, psychosis
Phosphorus	Milk, cheese, meat, poultry, fish, cereals, nuts, legumes	Bone and tooth formation, acid-base balance, component of nucleic acids, energy production	**Deficiency:** irritability, weakness, blood cell disorders, gastrointestinal tract and kidney dysfunction **Toxicity:** accumulation in kidney disease
Magnesium	Green leaves, nuts, cereal grains, seafood	Bone and tooth formation, nerve conduction, muscle contraction, enzyme activation	**Deficiency:** neuromuscular irritability **Toxicity:** hypotension, respiratory failure, cardiac disturbances
Iron	Wide distribution (except dairy products)—soybean flour, beef, kidney, liver, beans, clams, peaches Heme iron in meat well absorbed (10–30%); non-heme iron in vegetables poorly absorbed (1–10%)	Hemoglobin and myoglobin formation, cytochrome enzymes, iron-sulfur proteins	**Deficiency:** anemia, decreased work performance, impaired learning ability **Toxicity:** nausea, vomiting, diarrhea, gastrointestinal damage; fatal in children

NUTRIENT	PRINCIPAL SOURCES	FUNCTIONS	EFFECTS OF DEFICIENCY AND TOXICITY
Iodine	Seafood, iodized salt, eggs, dairy products, drinking water in varying amounts	Synthesis of thyroid hormones	**Deficiency:** hypothyroid, impaired fetal growth and brain development **Toxicity:** hyperthyroidism
Fluorine	Seafood, vegetables, grains, tea, coffee, fluoridated water (sodium fluoride 1.0–2.0 ppm)	Bone and tooth formation	**Deficiency:** predisposition to dental caries, osteoporosis **Toxicity:** mottling and pitting of permanent teeth
Zinc	Meat, liver, eggs, oysters, peanuts, whole grains	Component of enzymes; skin integrity, wound healing, growth	**Deficiency:** growth retardation, small reproductive gland (testes) acrodermatitis enteropathica cause zinc deficiency
Copper	Organ meats, oysters, nuts, dried legumes, whole-grain cereals	Enzyme component, red blood cell synthesis, bone formation	**Deficiency:** anemia in malnourished children **Toxicity:** nausea, vomiting, diarrhea, brain damage
Chromium	Brewer's yeast, liver, processed meats, whole-grain cereals, spices	Promotion of glucose tolerance	**Deficiency:** impaired glucose tolerance in malnourished children, some diabetics, and some elderly persons
Selenium	Wide distribution—meats and other animal products	Functions as an antioxidant with vitamin E	**Deficiency:** rare; muscle weakness **Toxicity:** loss of hair and nails, nausea, dermatitis, polyneuritis
Manganese	Whole-grain cereals, green leafy vegetables, nuts, tea	Component of enzymes	**Primary Deficiency:** rare **Toxicity:** Rare due to occupational exposure
Molybdenum	Milk, beans, breads, cereals	Component of coenzyme	**Deficiency:** tachycardia, headache, nausea, disorientation

Source: Modified from Merck Manual, 1999

APPENDIX G: STANDARD AND TRANSMISSION-BASED PRECAUTIONS

Transmission of infectious agents requires three elements: 1) a source of infectious agents; 2) a susceptible host with a source of entry receptive to the agent; and 3) a mode of transmission for the agent. All blood, body fluids, secretions, excretions (except sweat), non-intact skin, and mucous membranes may contain contagious infectious agents. As discussed in Chapter 1, because of the risk of transmitting these infectious agents, a set of Standard and Transmission-Based Precautions were developed. The guidelines, *The Guideline for Isolation Precautions: Preventing Transmission of Infectious Agents in Healthcare Settings 2007*, were updated in 2007 by the Centers for Disease Control. Go to http://www.cdc.gov for the complete text of the 2007 CDC Standard Precautions as well as any updates to the guidelines.

Standard Precautions serves as the basis for preventing the transmission of infectious agents in all health care settings. Transmission-Based Precautions should be implemented until the origin of the infection is known. These guidelines include practices that apply to all health care providers, patients, visitors, and family members. Hand hygiene is the single-most effective way to prevent the spread of infections and is a critical element in Standard Precautions. Other safe practices may involve the use of gloves, goggles, face shield, and/or a gown when there is a risk of splashing or spray of blood or body fluids. Safe handling of soiled linens also reduces the risks of spreading infectious agents. Listed below are procedures to follow to minimize the risk of spreading infectious agents.

Hand Hygiene

Hand hygiene includes effective hand washing with plain or antiseptic-containing soap and water or the use of alcohol-based products that do not require the use of water. You should wash your hands:

- Between patient contacts.
- After removing protective gloves.
- After contact with body fluids, even if gloves have been worn.

HAND-WASHING PROCEDURE

1. Stand at sink. Avoid contact of your uniform with the sink. Roll a paper towel out to have ready to use after washing your hands.
2. Turn on water and adjust water temperature.
3. Wet the hands and wrist thoroughly. Keep hands lower than elbows.
4. Apply skin cleanser or soap to hands.
5. Using friction, wash the palms, backs of the hands, fingers, between the fingers, knuckles, wrists and forearms. Clean nails. If no nail brush is available, rub nails across palms of hands.
6. Continue washing for at least 15 seconds.
7. Rinse thoroughly all surfaces of wrists, hands and fingers, keeping hands lower than the elbows and fingertips down
8. Dry hands thoroughly using a clean, dry paper towel.
9. Use a clean, dry paper towel to turn off faucets. Do not touch inside of sink at any time
10. Dispose of used paper towel(s) in wastebasket immediately after shutting off faucet.

Respiratory Hygiene/Cough Etiquette

It is essential that sneezes and coughs are covered so that the air is not contaminated with respiratory secretions. This may include placing a mask on a coughing patient. Respiratory hygiene/cough etiquette procedures should be used for all patients with respiratory symptoms (e.g., coughing and sneezing). The elements of cough etiquette are listed below:

1. Education of healthcare facility staff, patients, and visitors
2. Posted signs, in language(s) appropriate to the population served, with instructions to patients and accompanying family members or friends
3. Source control measures (e.g.,covering the mouth/nose with a tissue when coughing and prompt disposal of used tissues, using surgical masks on the coughing person when tolerated and appropriate)
4. Hand hygiene after contact with respiratory secretions
5. Spatial separation, ideally greater than 3 feet, of persons with respiratory infections in common waiting areas when possible

Personal Protective Equipment

Personal protective equipment (PPE) includes a variety of barriers and respirators used alone or in combination to protect mucous membranes, airways, skin, and clothing from contact with infectious agents. It can include gowns, masks, gloves, and so on. The PPE items chosen should be based on how you will be interacting with the patient. First, determine the expected level of exposure. Then refer to the CDC guidelines for the use of Personal Protective Equipment to determine the appropriate PPE equipment to use. Always practice safe work practices, which include:

- Keep hands away from face.
- Work from clean to dirty.
- Limit surfaces touched.
- Change when torn or heavily contaminated.
- Perform hand hygiene.

PROCEDURES

Note - When using Personal Protective Equipment, don in the order the procedures are given.

Prior to donning protective equipment:

1. Remove your watch or push it well up your arm.
2. Wash your hands.

GOWN

Isolation gowns are usually worn to protect the employee's arms and exposed body areas and clothing from blood and body fluids and are used in combination with gloves and other PPE as necessary.

GOWN PROCEDURES

1. Pick up gown and unfold.
2. Facing the back opening of the gown, place arms through each sleeve.
3. Fasten the neck opening.
4. Secure gown at waist making sure that back of clothing is covered by gown (as much as possible).

5. Put on gloves (see gloving procedure). Cuffs of gloves overlap cuffs of gown.

6. Prior to removing gown, with one gloved hand, grasp the other glove at the palm, and pull glove off.

7. Slip fingers from ungloved hand underneath cuff of remaining glove at wrist, and remove glove, turning it inside out as it is removed.

8. Dispose of gloves into designated waste container/waste container without contaminating self.

9. Unfasten gown at neck after removing gloves.

10. Remove gown without touching outside of gown.

11. While removing gown, hold gown away from body, turn gown inward and keep it inside out, and dispose of in appropriate receptacle.

MASK

Surgical masks are used to cover the nose and mouth and provide protection from contact with large infectious droplets (over 5 mm in size). For protection from inhalation of small particles or droplet nuclei, particulate respirators are recommended. The healthcare worker must be fitted for these masks prior to use in order to maintain appropriate seal and protection. The Infection Control Department staff will do fit testing for the employee during the employee's orientation period. Masks may be used with goggles in order to protect the mouth, nose, and eyes.

MASK PROCEDURES

1. Determine the appropriate type of face mask to be used.

2. Place the mask snugly over your nose and mouth. Secure the mask by tying the strings behind your head or placing the loops around your ears.

3. If the mask has a metal strip, squeeze it to fit your nose firmly.

4. If you wear eyeglasses, tuck the mask under their lower edge.

5. Front of mask is contaminated – DO NOT TOUCH!

6. To remove, grasp ONLY bottom then top ties/elastics and remove.

7. Discard in waste container.

8. Wash hands.

GOGGLES/FACE SHIELD

Goggles or face shield should be worn when there is a risk of contaminating the eyes. The eye protection chosen for specific work situations (e.g., goggles or face shield) depends upon the circumstances of exposure, other PPE used, and personal vision needs. Note that personal eyeglasses and contact lenses are not considered appropriate eye protection. Safe removal of a face shield, goggles, and mask should be performed after gloves have been removed and hand hygiene performed. The front of a mask, goggles, and face shield are considered contaminated. The ties, ear pieces, and/or headband used to secure the equipment to the head are considered clean to touch with bare hands.

GOGGLES/FACE SHIELD PROCEDURES

1. Determine whether to use face shield or goggles.

2. Put on face and adjust to fit.

3. Once used, the outside of goggles or face shield are contaminated.

4. To remove, handle by "clean" head band or ear pieces.

5. Place in designated receptacle for reprocessing or in waste container.

6. Wash hands.

GLOVES

Gloves should be worn when there is risk of touching blood or body fluids. Gloves are worn only once and are discarded according to agency policy. Some care activities for an individual patient may require changing gloves more than once. Hands should be thoroughly washed after gloves are removed.

GLOVES PROCEDURES

1. Use clean, disposable gloves.
2. Select a glove that provides appropriate fit.
3. Place the gloves on the hands so that the gloves extend to cover the wrist of the isolation gown.
   Note: Once used, the outside of gloves are contaminated.
4. When removing, grasp outside of glove with opposite gloved hand; peel off.
5. Hold removed glove in gloved hand.
6. Slide fingers of ungloved hand under remaining glove at wrist.
7. Dispose of in appropriate waste receptacle.
8. Wash hands.

REMOVING PERSONAL PROTECTIVE EQUIPMENT

The order or sequence used for removing personal protective equipment is very important. Remember that the outside surfaces of your Personal Protective Equipment are contaminated. The CDC recommends removing PPE at the doorway to the patient's room, or anteroom (if applicable). Practice hand hygiene immediately after removing the PPE. The following procedure summarizes the correct removal of gown, gloves, and mask/eye protection.

REMOVING PERSONAL PROTECTIVE EQUIPMENT PROCEDURES

1. While wearing gloves, untie the gown's waist strings.
2. With your gloved left hand, remove the right glove by pulling on the cuff, turning the glove inside out as you pull.
3. Remove the left glove by placing two fingers in the glove and pulling it off, turning it inside out as you remove it. Discard. Wash your hands.
4. Untie the neck strings of your gown. Grasp the outside of the gown at the back of the shoulders and pull the gown down over your arms, turning it inside out as you remove it.
5. Holding the gown well away from your uniform, fold it inside out. Discard it in the laundry hamper.
6. Wash your hands. Turn off the faucet using a paper towel and discard the towel in trash container.
7. Remove the mask/goggles to avoid contaminating your face or hair in the process. Untie your mask and/or remove your goggles by holding only the strings/strap. Discard.
8. Wash your hands and forearms with soap or antiseptic after leaving the room.

Mouthpieces, Resuscitation Bags, and Other Ventilation Devices

Use a ventilation device, which provides protection for the caregiver from oral contact, as an alternative to mouth to mouth resuscitation. Using a ventilation device prevents the health care provider from being exposed to oral and respiratory fluids during the procedure.

Caring for Soiled Linens

The hospital laundry uses water temperature and cleaning products capable of destroying any pathogens that might be present in patient's bed linens. Before you care for soiled linens, wash your hands and don disposable gloves.

CARING FOR SOILED LINENS PROCEDURES

1. Fold the soiled bed linens inward upon themselves when removing them from the bed.
2. Hold the linens away from your uniform when removing from room.
3. Place in the nearest hamper.
4. Should the outside surface of the linen hamper bag become soiled, it should be placed in a clean outer bag prior to pickup by the laundry staff.

Disposing of Sharps

Because of the danger inherent in handling sharp objects or contaminated needles, the health care industry has moved towards needleless systems whenever possible. However, when a needleless system cannot be provided, needles will have a protective sheath used to cover before and after use. There is no longer any need to recap used needles. Be sure to select the safest method of using sharps, using a needleless system whenever possible. Before you handle the sharp items, be sure to wash your hands and don disposable gloves. Locate the closest sharps container to the area in which you will be working.

Note: Gloves are not meant to protect the caregiver from needle stick or sharps injuries.

DISPOSING OF SHARPS PROCEDURES

1. After the use of any needle, place the protective sheath on the needle before leaving the patient area.
2. If sharps are used during a procedure, carefully separate them from other items on the tray and cover them with the procedure tray cover.
3. Place any disposable sharps items in the closest sharps container.
4. Place non-disposable sharps items in the designated area of the closest dirty utility room.
5. Remove and discard disposable gloves.
6. Wash hands.

Communicable Diseases

A communicable disease, also called an infectious disease, is any type of disease that can be transmitted from one person to another. Transmission of the disease can occur through one or more ways, including body fluids, airborne inhalation, food, or physical contact with an infected individual.

THE AIDS PATIENT

The first cases of AIDS (Acquired Immune Deficiency Syndrome) were reported in the 1980s. AIDS is a chronic life-threatening condition caused by the Human Immunodeficiency Virus (HIV). HIV destroys the cells of your immune system, which affects your body's ability to fight bacteria, fungi, and viruses. This makes the body more susceptible to certain types of cancers, and to opportunistic infections your body would normally resist, such as pneumonia, Kaposi's sarcoma, and tuberculosis. When AIDS was first discovered, there was rampant fear of anyone with HIV or AIDS. People avoided contact with anyone diagnosed with HIV or AIDS. Even some healthcare workers devoted to the care of the ill refused to care for these patients.

On a positive note, the recognition of this disease was instrumental in the creation of Standard Precautions and the continuing revision of *Precautions for Occupational Health and Bloodborne Pathogens*. Education about the causes and treatment of HIV and AIDS has progressed to the point that healthcare workers and the general public are more accepting and much less fearful of those infected.

THINGS TO DO WHEN TREATING HIV-POSITIVE OR AIDS PATIENTS

- Educate yourself and the patient.
- Help the patient understand options for treatment.
- Be supportive.
- Promote the patient's role as an individual and a member of the community.

THINGS NOT TO DO WHEN TREATING HIV-POSITIVE OR AIDS PATIENTS

- Avoid personal contact.
- Isolate patient from human contact.
- Exhibit personal prejudices.

There is no need to wear protective gloves when directly interacting with these patients. Use gloves only when there is a danger of contact with blood or body fluids, mucous membrane, or non-intact skin. As with any other patient, practicing good hand-washing techniques is always appropriate and in the caregiver's and patient's best interests.

MRSA (METHICILLIN-RESISTANT STAPHYLOCOCCUS AUREUS)

MRSA causes an infection that is resistant to many types of antibiotics. MRSA infections are caused by *Staphylococcus aureus* bacteria—often called "staph." The organism is often sub-categorized as Community-Associated MRSA (CA-MRSA) or Hospital-Associated MRSA (HA-MRSA) depending how the disease is acquired; the two forms of MRSA are distinct strains of the bacterial species. The key to stopping MRSA in hospitals is to practice infection control. CA-MRSA, which is transmitted through close skin-to-skin contact, can be prevented by practicing good hygiene; keeping cuts and scrapes healed and covered; and avoiding sharing personal items, such as towels or washcloths. MRSA infections can be fatal.

MRSA was first discovered in England in 1961, but it is now found worldwide. In recent years the number of MRSA infections in the United States has increased considerably. A 2007 report by the Centers for Disease Control and Prevention (CDC) estimated that the number of MRSA infections treated in hospitals doubled nationwide, from approximately 127,000 in 1999 to 278,000 in 2005, and the number of deaths increased from 11,000 to more than 17,000.

SARS (SEVERE ACUTE RESPIRATORY SYNDROME)

SARS is highly contagious and sometimes fatal respiratory disease. SARS coronavirus, the same family of virus that causes the common cold, is the virus that causes SARS. SARS first appeared in China in 2002. Within six weeks, the virus had spread worldwide. During that outbreak, 8,000 people were infected and 800 died of the disease. In part, the CDC saw the need to develop Respiratory Hygiene/Cough Etiquette as a result of the SARS outbreak.

Patients who can contaminate the area with known communicable diseases should be in a private room or isolated from other patients. There will be a list of diseases which are communicable and the precautions which should be taken available on the patient unit. The Infection Control Department will monitor all admissions and advise staff of any additional precautions that need to be taken.

APPENDIX H: NUTRITION

Without the right fuel, no transportation device can function well. Try putting an inferior grade of gasoline in a car that requires premium fuel, for example, and get ready for the knocks and pings of complaint! It's the same with the human body. Eating right is a major factor in reaching and maintaining good health. In this appendix, you will learn the basics of good nutrition.

Learning Objectives

At the end of the journey through this appendix, you will be able to:

- Explain how the body processes proteins, carbohydrates, and fats in the diet
- Identify the purposes and benefits of vitamins and minerals
- Distinguish between "good" and "bad" cholesterol and identify its effect on the body
- Explain the food pyramid and the advice it provides for daily eating
- Identify the risks and health hazards associated with obesity
- Understand dietary restrictions imposed by some religions
- Explain and differentiate between anorexia nervosa, bulimia, and binge eating disorders

Introduction to Basic Nutrition

When food is taken into the body, it is used in many different ways. The right combination of nutrients work together in the body to provide heat, promote growth, repair tissue, and regulate body processes. Good nutrition promotes a healthier body and mind. It also aids in resistance to illness. When we eat a healthy diet, our energy and vitality are increased. The right foods speed the healing process and help a person feel better and sleep better.

Pronunciation Guide

Correct pronunciation is important in any journey so that you and others are completely understood. Here is a "see and say" Pronunciation Guide for the more difficult terms to pronounce in this chapter.

amino acids (ah MEE no A sidz)

anorexia nervosa
(an uh REK see uh nur VOH suh)

bulimia (buh LEE mee uh)

calorie (KAL uh ree)

cellulose (SEL yo loz)

colitis (kuh LIE tis)

diabetes mellitus
(DYE ah BEE teez mel uh thus)

gastrointestinal
(gas troh in TES tuh nuhl)

hemoglobin (HEE muh gloh bin)

hypertension (hy per TEN shun)

ileitis (il ee EYE tis)

lactation (lak TEY shun)

lipids (LIP idz)

malnutrition (mal noo TRISH un)

metabolic (met uh BOL ik)

metabolism (meh TAH bow liz um)

metabolize (mi TAB e liz)

minerals (MIN er uhlz)

nutrients (NOO tree ents)

protein (PROH teen)

soluble (SOL yuh buhl)

vitamins (VIE tuh minz)

When we do not supply our bodies with the proper nutrients, **malnutrition** occurs, and many things can go wrong. We lose stamina and vitality. An unhealthy diet often results in illness and disease.

Our body gets the energy it needs from food through a process called **metabolism**. Metabolism is a constant process that occurs from the beginning of life to the end of life. Metabolism is the rate of use of nutrients by the body to produce energy. The basal metabolic rate is the amount of energy needed to maintain life when the body is at rest.

Your patients/clients are all from different cultural and religious backgrounds. Each culture and religion has dietary differences. Appetites, food budgets, cultural food preferences, and religious restrictions influence some of these differences. You and your patients need the same basic nutrients provided by a balanced, healthy diet. The health care worker must understand about foods and their effect on the body in order to assess his or her own diet and the patient's diet.

malnutrition *(mal noo TRISH un)*

metabolism *(meh TAH bow liz um)*

BASIC NUTRIENTS

Nutrients are chemical compounds found in food. When the food we eat enters the digestive tract, it is changed into a simple form and absorbed into the blood. The blood carries these nutrients to body cells, where they are used to maintain body functions. Table H-1 lists the essential nutrients.

nutrients *(NOO tree ents)*

TABLE H-1 Essential Nutrients

NUTRIENT	PURPOSE
Proteins	Builds and renews body tissues Generates heat and energy Provides amino acids that aid in body functions
Carbohydrates	Generates heat and energy Excess is converted to fat
Fats (Lipids)	Provides fatty acids for growth and development Generates heat and energy Carries vitamins A and D to cells
Minerals	Regulates the activity of the heart, nerves, and muscles Builds and renews teeth, bones, and other tissues
Vitamins	Assists in metabolism, growth, and body development
Water	Carries other nutrients Regulates body temperature Helps eliminate wastes

PROTEINS

Protein is a source of energy, and also delivers needed enzymes for various body functions. Muscles, hormones, clotting, and antibody generation all depend on proteins. Sources of protein include animal products such as meat and dairy and certain grains and legumes.

protein *(PROH teen)*

amino acids *(ah MEE no AAH sidz)*

Amino acids are the building blocks of proteins. The human body uses 20 amino acids, but many of those are non-essential and are not required by the body. The remaining essential amino acids cannot be created by the body, and must come from dietary sources. These are listed in Table H-2. The body cannot store amino acids, so they must be replenished from food each day.

TABLE H-2 Essential Amino Acids

AMINO ACID	RECOMMENDED DAILY INTAKE (IN MG PER KG OF BODY WEIGHT)
Phenylalanine + Tyrosine*	14 (total)
Leucine	14
Methionine + Cysteine*	14 (total)
Lysine	12
Isoleucine	10
Valine	10
Threonine	7
Tryptophan	3
Histidine	28 in infants, unknown in adults
Arginine	unknown, required for infants, may be required for seniors

* In cases where two amino acids are listed together, the two are interchangeable in the diet; nutrition analysis looks at the total amount of those two combined.

Not all protein sources have all the essential amino acids. Proteins can be divided into two types: complete and incomplete. **Complete proteins** contain all amino acids; **incomplete proteins** do not. Complete proteins are found in animal products; plant-based sources of protein are incomplete. For proper nutrition, vegetarians should carefully select combinations of foods containing incomplete proteins so that they receive all the amino acids they need. For example, combining wheat or rice (which lack lysine) with legumes (which lack tryptophan) results in a complete protein.

Another way of evaluating protein sources is to use their Protein Digestibility Corrected Amino Acid Score (PDCAAS). A PDCAAS value of 1 is the highest and 0 the lowest. Generally speaking, the closer the value is to 1, the more complete the amino acids in the food. Table H-3 lists some of the common PDCAAS scores.

In the United States, overconsumption of protein is a more common problem than underconsumption. Eating too much protein is as much a health concern as eating too little, as it puts an extra burden on the kidneys and liver.

TABLE H-3 Protein Digestibility Corrected Amino Acid Scores

FOOD	SCORE (0 TO 1)
whey	1
egg white	1
casein	1
milk	1
soy protein isolate	1
beef	0.92
soybeans	0.91
kidney beans	0.68
rye	0.68
whole wheat	0.54
lentils	0.52
peanuts	0.52
seitan	0.25

Carbohydrates

Carbohydrates are the most common source of energy in the body. Foods high in carbohydrates include bread, pasta, beans, potatoes, bran, rice, cereals, and sugar.

Some sources differentiate between simple and complex carbohydrates, based on the amount of fiber they contain. **Complex carbohydrates** include starch and dietary fiber (**cellulose**). They are found in grains, fruits, legumes, greens, kale, cabbage, celery, and other vegetables. Fiber adds bulk (roughage) to the diet and helps prevent bowel and colon diseases. It is not digestible by humans and acts as a bulking agent by absorbing water; this helps to prevent constipation. Eating adequate fiber may also lower the risk of heart disease and some cancers. To keep the bowel healthy, a person should eat several servings of fiber each day.

cellulose *(SEL yo loz)*

Simple carbohydrates such as sugar and dairy products do not contain much fiber. They are quickly and easily digested, providing the body with a quick burst of energy. Examples include fruits, fruit juices, milk, yogurt, honey, molasses, and sugar. Simple carbohydrate food is often less nutrient-rich than complex carbohydrate food.

Fats (Lipids)

Fat (**lipids**) is an important part of the diet—although not getting enough fat is very seldom a problem in the diet of the average U.S. citizen! Fats are a source of energy, containing twice as many calories as the same quantity of carbohydrates or protein. Fats also carry fat-soluble vitamins. Stored body fat cushions internal organs and insulates against cold.

lipids *(LIP idz)*

Applied Science

CHOLESTEROL

Cholesterol is a type of lipid (fat) found in food containing animal fats, such as beef, egg yolks, poultry, and shrimp. Excess cholesterol can build up in artery walls, causing atherosclerosis, so people aiming for good circulatory system health often follow a low-cholesterol diet rich in lean muscle meat, fish, skinless poultry, skim milk, vegetables, and fruits. Cholesterol is tested by drawing blood from the patient and measuring the amount of cholesterol in it; the recommended level is under 200 mg/dl.

There are two types of cholesterol: HDL (High-Density Lipoprotein) and LDL (Low-Density Lipoprotein). HDL is also known as "good cholesterol." It removes excess cholesterol from cells to carry back to the liver to be broken down and eliminated. LDL is "bad cholesterol;" it carries fat to cells.

Fats can be categorized as saturated or unsaturated. This has to do with the hydrogen atoms in the fat molecules. A fat is **unsaturated** if there are one or more covalent (double) bonds in the fatty acid chain. A **saturated** fat is saturated with hydrogen atoms, and does not contain covalent bonds. The more unsaturated a fat is, the more vulnerable it is to turning rancid. Therefore a lot of snack food uses saturated fat to extend its shelf life.

Unsaturated fat can be further broken down into **monounsaturated** (containing exactly one covalent bond) and **polyunsaturated** (containing more than one covalent bond). Where covalent bonds are, hydrogen atoms are eliminated. These covalent bonds can be either on the same side of the molecule or on the opposite side. When they are on the same side they are called **cis fats**. When they are on the opposite side they are called **trans fats**.

Generally speaking, unsaturated fats are healthier for your body than saturated ones. However, trans fats are an exception; trans fats can increase one's risk of coronary heart disease by raising levels of LDL cholesterol and lowering levels of HDL cholesterol.

MINERALS

minerals (MIN er uhlz)

Minerals are inorganic compounds needed for human growth and maintenance. The most important minerals for human consumption are:

- **Calcium**, for muscle, hearth, and digestive system health. Calcium builds bone, neutralizes acidity, and supports synthesis and function of blood cells.
- **Potassium** and **sodium**, for regulating adenosine triphosphate (ATP), which transports chemical energy within cells for metabolism

In addition, numerous other minerals are required in trace (very small) amounts, mostly because of their roles in catalysis of enzymes. These include:

- **Iron**: from liver and lean meats, required for many proteins and enzymes, including hemoglobin
- **Fluorine**: for tooth enamel and bones, added to the drinking water in some areas
- **Iodine**: from fish and iodized salt, for biosynthesis of thyroxin

Most of these minerals are present in the average adult diet; however, some people prefer to supplement their mineral intake with a multivitamin that includes minerals.

VITAMINS

Vitamins are active organic compounds that function as coenzymes for normal health and growth. Some vitamins also behave like hormones. Vitamins cannot be created in sufficient quantities by the body, so they must be obtained from the diet. They are distinguished from minerals in that they are organic.

vitamins *(VIE tuh minz)*

Vitamins are named with letters of the alphabet. Vitamins A, D, E, and K are fat soluble, and can be stored in the body. Vitamins B and C are water soluble; they can't be stored, so the excess of these is excreted.

Vitamin deficiency can cause a variety of diseases and disorders. Table H-4 lists the vitamins needed by humans and the Recommended Daily Allowances (RDAs) for each vitamin, along with the diseases associated with getting too much or too little of them.

WATER

Water is essential to the body. It carries nutrients to the body cells and carries waste products away from the body cells. It also lubricates the joints and helps regulate body temperature and body processes. Water makes up 55-65% of body weight. Because water is lost through evaporation, excretion, and respiration, it must be replaced every day.

UNDERSTANDING CALORIES

Food is the source of energy for our bodies. The body **metabolizes** food nutrients to create energy. As the body creates energy, it produces heat. Energy is required to ensure that all of the body systems function. The amount of energy created by the food we eat is measured in **calories**. Calories are the measure of the amount of energy contained within the chemical bonds of the food. Calorie needs vary from person to person. A large, active man needs more calories than a smaller, inactive man does. The number of calories we eat and the amount of exercise we do balance weight. If we eat more calories than we burn, we gain weight. If we burn more calories than we eat, we lose weight. When a patient is inactive and feels ill, she may not want to eat. It is your responsibility to encourage her to take in enough calories to produce the energy needed to heal the body. Some foods are high-calorie but provide few nutrients, like candy. Such foods are sometimes referred to as having "empty calories."

metabolize *(mi TAB e liz)*

calorie *(KAL uh ree)*

The Food Pyramid and a Healthy Diet

Good nutrition and a healthy diet help maintain good health. General guidelines for a healthy diet include:

- Eat a variety of foods
- Maintain a desirable weight, neither too fat nor too thin
- Avoid too much fat, especially saturated fat (animal fat) and cholesterol
- Eat foods that contain adequate fiber
- Avoid too much sugar and salt
- Don't drink alcohol

TABLE H-4 Vitamins

VITAMIN	CHEMICAL NAME	SOLUBILITY	RECOMMENDED DAILY ALLOWANCES	UPPER INTAKE LEVEL	DEFICIENCY DISEASE	OVERDOSE DISEASE
A	Retinoids (retinol, retinoids and carotenoids)	Fat	900 μg	3,000 μg	Night-blindness and Keratomalacia	Hypervitaminosis A
B_1	Thaimine	Water	1.2 mg		Beriberi	
B_2	Riboflavin	Water	1.3 mg		Ariboflavinosis	
B_3	Niacin	Water	16 mg	35 mg	Pellagra	
B_5	Pantothenic acid	Water	5 mg		Paresthesia	
B_6	Pyridoxine, pyridoxamine, pyridoxal	Water	1.3 to 1.7 mg	100 mg	Anemia	Impairment of proprioception, nerve damage
B_7	Biotin	Water	30 μg		Dermatitis, enteritis	
B_9	Folic acid, folinic acid	Water	400 μg	1,000 μg	Deficiency during pregnancy may cause birth defects	Impairment of proprioception, nerve damage
B_{12}	Cyanocobalamin, hydroxycobalamin, methylcobalamin	Water	2.4 μg		Megaloblastic anemia	
C	Ascorbic acid	Water	90 mg	2,000 mg	Scurvy	Indigestion, diarrhea
D	Ergocalciferol, cholecalciferol	Fat	5 to 10 μg	50 μg	Rickets and Osteomalacia	Hypervitaminosis D
E	Tocopherols, tocotrienols	Fat	15 mg	1,000 mg	Mild hemolytic anemia in newborn infants	
K	phylloquinone, menaquinones	Fat	120 μg		Bleeding diathesis	

In 1992, the U.S. Department of Agriculture (USDA) released a now-famous nutrition guide known as the American food pyramid, shown in Figure H-1 ▪, as a guideline for selecting the right proportions of foods. The pyramid illustrates the recommended daily portions of each food group.

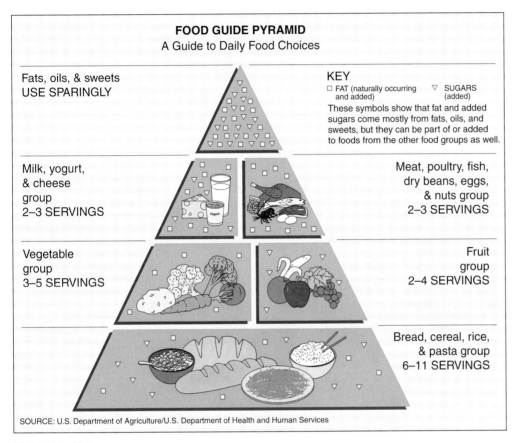

FIGURE ▪ H-1

The American food pyramid.

Many nutritional experts, however, complained that the 1992 pyramid was not accurate based on the newest research on nutrition. They complained that it was too heavy on beef and dairy and did not put enough emphasis on fruit and vegetables. For example, the food pyramid allowed some dietary choices that had been linked to heart disease, such as three cups of whole milk and an eight-ounce serving of hamburger every day. In addition, the pyramid lumped all members of the protein-rich group together (Meat, Poultry, Fish, Dry Beans, Eggs, and Nuts) and made no distinction between whole grains and refined products. Many people complained that the USDA was influenced by corporate interests such as the dairy, meat, and sugar industry and allowed lobbyists to change the wording.

As a result, in 2005, the USDA released a new nutritional guide known as MyPyramid. In the first 72 hours, the MyPyramid Internet sites logged 160 million hits, 20 percent of them from outside the United States. There were no foods pictured on the new MyPyramid image and no text. Instead, the new logo emphasized the importance of physical activity by showing a sort of stick figure climbing the stairs of the pyramid. MyPyramid was designed to be simple; colored vertical bands represent different food groups. Six bands of color run from the top of MyPyramid to the base: orange for grains, green for vegetables, red for fruits, a teeny band of yellow for oils, blue for milk, and purple for meat and beans. Each stripe starts out as the same size, but doesn't end that way at the base. The widths suggest how much food a person should choose from each group. Figure H-2 ■ shows MyPyramid.

Anatomy of MyPyramid

One size doesn't fit all
USDA's new MyPyramid symbolizes a personalized approach to healthy eating and physical activity. The symbol has been designed to be simple. It has been developed to remind consumers to make healthy food choices and to be active every day. The different parts of the symbol are described below.

Activity
Activity is represented by the steps and the person climbing them, as a reminder of the importance of daily physical activity.

Moderation
Moderation is represented by the narrowing of each food group from bottom to top. The wider base stands for foods with little or no solid fats or added sugars. These should be selected more often. The narrower top area stands for foods containing more added sugars and solid fats. The more active you are, the more of these foods you can fit into your diet.

Personalization
Personalization is shown by the person on the steps, the slogan, and the URL. Find the kinds and amounts of food to eat each day at MyPyramid.gov.

Proportionality
Proportionality is shown by the different widths of the food group bands. The widths suggest how much food a person should choose from each group. The widths are just a general guide, not exact proportions. Check the website for how much is right for you.

Variety
Variety is symbolized by the 6 color bands representing the 5 food groups of the Pyramid and oils. This illustrates that foods from all groups are needed each day for good health.

Gradual Improvement
Gradual improvement is encouraged by the slogan. It suggests that individuals can benefit from taking small steps to improve their diet and lifestyle each day.

MyPyramid.gov
STEPS TO A HEALTHIER YOU

USDA U.S. Department of Agriculture
Center for Nutrition Policy and Promotion
April 2005 CNPP-16

USDA is an equal opportunity provider and employer.

| GRAINS | VEGETABLES | FRUITS | OILS | MILK | MEAT & BEANS |

FIGURE ■ **H-2**

MyPyramid.

You can access information about MyPyramid at
http://www.mypyramid.gov/.

One confusing aspect of the new MyPyramid was the absence of recommended serving size. The USDA stated that these amounts differed according to a person's age, weight, gender, and activity level. They suggested that interested Americans search out further details and personalized pyramids at the MyPyramid website. However, this meant that millions of people without access to a computer or the Internet would have trouble getting essential information.

> You can access Harvard's pyramid at
> http://www.hsph.harvard.edu/nutritionsource/pyramids.html

To complicate matters further, the Harvard School of Public Health issued their own "Healthy Eating" pyramid as an alternative to the USDA's suggestions. Harvard's food pyramid included calcium and multi-vitamin supplements as well as moderate amounts of alcohol. Many observers believe that the Harvard pyramid follows the results of scientific nutrition studies more closely that the USDA, which may be influenced by political pressure exerted by food production lobbyists. For example, note that if you combine MyPyramid's fruit and vegetable category into one broad band, that band would have totally dominated the pyramid and sent a very different message to consumers about nutritious choices.

Harvard's nutrition guide is not the only alternative to the USDA's MyPyramid. There are also special nutrition guides/pyramids based on Asian, Latin, Mediterranean, and vegetarian diets. It seems almost certain that the food pyramid wars will continue for the next few years.

Nutrition Labeling

In the United States and the United Kingdom, as well as some other countries, prepared foods are required to contain a Nutrition Facts label, also called a nutrition information panel. This label lists the ingredients in order of most common to least common, the serving size, the number of servings per container, the calories per serving, and the amount of certain key components such as protein, fat, carbohydrates, cholesterol, sodium, and vitamins and minerals per serving, and compares them to the recommended daily allowances.

Pathology Connection: Diseases and Disorders Related to Nutrition

Several medical conditions can be caused or aggravated by the patient's choice of foods. Here is a brief sampling:

Anemia, a decreased number of red blood cells or a decreased amount of **hemoglobin**, results in fatigue, dyspnea (shortness of breath) on exertion, headache, insomnia, paleness, indigestion, and rapid heartbeat. It can be caused by lack of iron.

hemoglobin *(HEE muh gloh bin)*

Constipation is the infrequent, difficult defecation of fecal material. It is commonly caused by lack of activity and by not eating enough vegetables, fruit, or water.

Tooth decay can occur as a result of lack of fluorine.

A **goiter** can be the result of lack of iodine.

Osteoporosis occurs when bones become porous, causing them to break easily. This is caused by inadequate calcium intake or absorption.

Obesity

Obesity is the most common nutritional disorder. Americans, especially young people, are getting fatter. In the last twenty-five years, there has been a dramatic increase in obesity in the United States. In 1980, 25 percent of adults in the United States were overweight. By 1991, this figure had risen to 33 percent. In 2001, about two-thirds of the adult population, more than 120 million people, was classified as overweight. Obesity is defined as being 15% or more over one's optimal weight.

In a way, the numbers are even bleaker for children. The number of overweight children between the ages of 6 and 19 tripled between 1980 and 2004. An estimated 9 million children over the age of six (about 15 percent) are currently obese. About four out of every five of these obese children will remain overweight into adulthood.

So what is so bad about being overweight? The answer is quite simple. Overweight and obese people have an increased risk for many diseases and health conditions such as the following:

Hypertension (high blood pressure)
Diabetes
Coronary heart disease
Stroke
Gallbladder disease
Sleep apnea and respiratory problems
Some cancers (endometrial, breast, and colon)
Osteoarthritis (a degeneration of cartilage and its underlying bone within a joint)

The U.S. government estimates that the cost of obesity in the United States in 2000 was more than $117 billion ($61 billion direct and $56 billion indirect). Obesity is now considered one of the major health crises affecting the United States.

The solution to the problem is simple, but instituting it is not. People have to eat less, eat smarter, and exercise more. Fatty, sugary diets have contributed to growing waistlines. Fast food and junk food are not nutritionally sound. More than half of all Americans do not get the recommended amount of physical activity—thirty minutes a day for adults and sixty minutes daily for children. Kids spend more time staring at television and computer screens they do in physical play; nearly one quarter of American children eight or older spend more than five hours a day watching television or playing computer and video games. Both children and adults have to limit their time in front of the television or computer and follow the recommendations of the new food pyramid.

Religion-Based Diets

When consulting with patients about dietary issues, it is important to remember that other factors than nutrition may influence food habits. For example, many religions follow specific dietary laws. Table H-5 summarizes some of the most common religious restrictions on diet.

TABLE H-5 Religious Dietary Restrictions

RELIGION	RESTRICTIONS
Christian Science Church of Latter-Day Saints (Mormon) Conservative Protestant	Avoid alcohol, coffee, tea, and caffeinated soft drinks
Greek Orthodox	No meat or dairy products on fast days
Muslim	No alcohol, pork, or pork products
Orthodox Jewish	No shellfish, pork, or non-kosher meats No serving milk and milk products with meat No eating leavened bread during Passover Abstain from eating on specific fast days
Roman Catholic	No food one hour before communion and no meat on Ash Wednesday, Good Friday, and all the Fridays during Lent
Buddhist	Generally vegetarian
Hindu	Generally vegetarian

Therapeutic Diets

Therapeutic diets are dietary plans designed to produce a specific result in the patient, such as to:

- Regulate the amount of food in **metabolic** disorders
- Prevent or restrict edema by restricting sodium intake
- Assist body organs to regain and/or maintain normal function
- Aid in digestion by avoiding foods that irritate the digestive tract
- Increase or decrease body weight by adding or eliminating calories

Table H-6 describes various types of therapeutic diets and their purposes, as well as notes on the nutritional adequacy, which becomes increasingly a factor the longer the diet is followed.

metabolic *(met uh BOL ik)*

gastrointestinal *(gas troh in TES tuh nuhl)*
colitis *(kuh LIE tis)*
ileitis *(il ee EYE tis)*
diabetes mellitus *(dahy uh BEE teez mel uh thus)*
soluble *(SOL yuh buhl)*
hypertension *(hy per TEN shun)*
lactation *(lak TEY shun)*

TABLE H-6 Some Therapeutic Diets and Their Purposes

TYPE OF DIET	PURPOSE OF DIET	FOODS	NOTES
Clear liquid	Replaces fluids lost from vomiting, diarrhea, surgery	Plain gelatin, ginger ale, tea, coffee (no cream), fruit or apple juice (no pulp), fat-free broth	Nutritionally inadequate
Full liquid	Trouble chewing or swallowing, gastrointestinal disturbances	All clear liquids, fruit or vegetable juices, strained soup, custard, ice cream, sherbet, milk, cream, eggs, buttermilk, carbonated beverages, eggs, cocoa, eggnog	May be deficient in iron
Soft	For patients who have trouble chewing after surgery	Foods that are soft in consistency, such as fish, ground beef, broth, pureed vegetables, strained cream soup, tender cooked vegetables, fruit juices, cooked fruit, refined cereals, pasta, sherbet, ices, ice cream, custard, plain cookies, angel food cake, tea, coffee, cocoa, carbonated beverages, cheese, cottage cheese	Nutritionally inadequate
Bland	Soothes gastrointestinal tract, avoids irritation in ulcers, colitis	Foods low in fiber and connective tissue that are mild flavored and easy to digest, such as pear (tender, broiled, boiled), prune juice, applesauce, custard, pudding, ice cream, plain cookies, sponge cake, decaffeinated coffee, milk, cheese, yogurt	Nutritionally adequate
Restricted residue	Reduces normal work of the intestine, in cases of rectal diseases, colitis, ileitis	Foods low in fiber and low in bulk, such as milk, buttermilk, cottage cheese, butter, margarine, eggs (not fried), tender poultry, fish, lamb, ground beef (broiled, boiled, baked), broth, refined bread, cereals, pasta, gelatin, angel food or sponge cake, mild-flavored cooked vegetables, lettuce, vegetable and fruit juice, applesauce, canned fruit, citrus fruit without membranes	Nutritionally adequate
Low carbohydrate (diabetic)	Matches food intake with insulin uptake and nutritional requirements, used for patients with hyperinsulism and diabetes mellitus	Foods that supply enough protein, fat, and carbohydrate to maintain health and activities; requires a balance of carbohydrates, protein, and fat to meet the individual need of the patient; restricts sugar, cookies, pies, candies, etc.	Nutritionally adequate

TABLE H-6 Some Therapeutic Diets and Their Purposes (*continued*)

TYPE OF DIET	PURPOSE OF DIET	FOODS	NOTES
Low fat	For patients with gallbladder and liver disease, obesity, and heart conditions	Foods high in carbohydrates and proteins; all fats are limited; skim milk, buttermilk, cottage cheese, lean fish, poultry, meats, fat-free soup broths, cooked vegetables, lettuce, fruit juice, bananas, citrus fruits, gelatin, angel food cake, coffee, tea, carbonated beverages, jelly, honey as desired	Deficient in fat soluble vitamins
Low cholesterol	Regulates amount of cholesterol in the blood for patients with coronary disease and atherosclerosis	Foods low in fat, such as lean muscle meat, fish, poultry without skin or fat, skim milk, vegetables, fruits	Nutritionally adequate
Low calorie (800–2,000 calories)	Reduces number of calories for overweight patients and for clients with arthritis or cardiac conditions	Foods low in fats and calories: skim milk, buttermilk, lean meats, clear soup, vegetables, fresh fruit, coffee, tea, herbs, onions, garlic	Nutritionally adequate
High calorie (2,000+ calories)	For persons 10% or more below desired weight; for patients with anorexia nervosa and hyperthyroidism	All foods with nutritionally balanced proteins, carbohydrates, fats, vitamins, minerals	Nutritionally adequate
Low sodium	Reduces salt intake for patients with kidney disease, cardiovascular disorders, edema, and hypertension	Natural foods prepared without salt, such as fresh fruits, fresh vegetables, foods without salt added	Nutritionally adequate
High protein	For children and adolescents needing additional protein for growth; during pregnancy, lactation; post-surgically; and during illnesses resulting from protein loss	Foods high in protein, such as milk, cheese, eggs, lean meats, fish, and poultry; fruit, cereals, vegetables	Nutritionally adequate

Eating Disorders

Eating disorders are not due to a failure of will. They are real and treatable medical illnesses. The three main types of eating disorders are anorexia nervosa, bulimia, and binge-eating. All three can affect a person's health including causing serious heart conditions, kidney failure, and electrolyte imbalance. The cause of eating disorders is not entirely clear. They seem to have a basis in biology but they are also affected by emotions, genes, and culture.

Eating is controlled by many factors including a person's appetite, the availability of food, family, peer, and cultural practices, and attempts at voluntary control. An eating disorder involves a serious disturbance in eating behavior, such as extreme reduction of food intake, severe overeating, or intentional vomiting. Eating disorders frequently develop during adolescence or early adulthood and can often be found with other problems such as depression, substance abuse, and anxiety disorders.

Women are much more likely than men to develop an eating disorder. Women's magazines, fashion trends, and some activities and professions promote dieting to achieve the "perfect" lean body. This can lead to pressure on women to be thin. Eating disorders are sometimes triggered by the stress of being unable to reach an unattainable goal. Males make up only about 10% of people with eating disorders.

ANOREXIA NERVOSA

anorexia nervosa
(an uh REK see uh   nur VOH suh)

Anorexia is the loss of appetite, for any reason. **Anorexia nervosa** occurs when a person refuses to eat or drastically reduces his or her intake of food as a result of a mental disorder related to an emotional disturbance about image. People with anorexia nervosa see themselves as overweight and have an immense fear of becoming fat even though they are dangerously thin. For them, eating becomes an obsession. They develop unusual eating habits such as avoiding meals, eating only a few foods in small quantities, or carefully weighing food. People with anorexia may repeatedly check their body weight or exercise compulsively.

Anorexia nervosa is seen primarily in teenage girls. Criteria for diagnosis include an intense fear of getting fat, distorted body image, loss of 25% of the original body weight with no accompanying physical illness, amenorrhea (cessation of menstrual periods), and refusal to maintain normal weight.

People with anorexia are often perfectionists. They are driven to succeed but cannot achieve the unattainable standards they set for themselves. When they fail to meet these standards, they often look for a part of their lives they can control such as food or weight.

BULIMIA NERVOSA

bulimia *(buh LEE mee uh)*

Bulimia is diagnosed when a person experiences intervals of food craving and bingeing (eating an excessive amount of food within a short period of time) and then purging (vomiting or overusing laxatives, diuretics, or other medications). The side effects are the same as those of anorexia and result in depression. It is most common in adult women. Bulimics usually feel unable to control their appetite during a bingeing episode. Afterwards, they try desperately to compensate in order to avoid gaining weight.

Because purging follows the binge-eating episodes, people with bulimia usually have a "normal" weight for their age and height. However, like people with anorexia, people with bulimia often feel extremely dissatisfied with their bodies. They might also feel disgusted and ashamed when they binge and perform both bingeing and purging behavior in secret.

BINGE-EATING DISORDER

Binge-eating disorder is similar to bulimia but without the purging. Therefore, many people with the disorder are overweight for their age and height. The out-of-control eating may be associated with eating rapidly, eating until feeling uncomfortably full, or eating large amounts when not feeling hungry. Binge-eaters often feel embarrassed, depressed, or guilty after overeating. This can lead to overeating again, creating a cycle of binge eating.

TREATMENT FOR EATING DISORDERS

People with eating disorders often do not recognize or admit that they are ill. They may strongly resist getting and staying in treatment. Family members or friends can be helpful in making sure that a person with an eating disorder receives the necessary care. Eating disorders can be treated and a healthy weight restored. The sooner a doctor diagnoses and treats these disorders, the better the outcomes are likely to be. Eating disorders often have multiple causes and require a complex treatment plan. This may include medical care, psychological treatment, nutritional counseling, and medication. Ongoing research by scientists continues to advance the understanding and treatment of eating disorders.

GLOSSARY

abdominopelvic cavity (ab dom ih noh PELL vik KAV ih tee) Continuous cavity within the abdomen and pelvis that contains the largest organs of the gastrointestinal system.

abduction (ahb DUK shun) Moving a body part away from the midline of the body.

absorption (ab SORP shun) Process by which digested food nutrients move through villi of the small intestine into the bloodstream or substances move from the kidney tubules into the bloodstream.

accessory muscles Muscles in the neck, chest, and abdomen that can be used, if necessary, to help expand the thoracic cavity on inspiration.

acetylcholine (ass SET ul KOE leen) Neurotransmitter between neurons in the brain and spinal cord or between a neuron and a voluntary skeletal muscle.

acetylcholinesterase (AHS eh till KOH lin ehs tur ace) The enzyme that stops the action of acetylcholine.

acromegaly (ACK row MEG ah lee) A hormonal condition of gradual coarsening and enlargement of bones and facial features.

actin (ak TIN) One of the two proteins in muscle fibers needed for contraction.

action potential The change of the electrical charge of a nerve or muscle fiber when stimulated. Action potentials are all or none.

active transport The movement of cellular material that requires energy.

acute epiglottitis Inflammation of the epiglottis due to infection.

adaptation The adjustment of an organism to changing environments.

adduction (add DUK shun) Moving a body part toward the midline of the body.

adenoid (AD eh noid) Lymphoid tissue in the superior part of the nasopharynx. Also known as the *pharyngeal tonsils*.

adenosine diphosphate (ADP) The compound that can be converted to ATP for energy storage. When ATP is broken down to ADP, energy is released that can be used for cellular energy.

adenosine triphosphate (ATP) Energy storage molecule used to power muscle contraction and other cellular reactions.

adrenal (ad REE nal) Referring to the adrenal gland.

adrenal cortex (ad REE nal KOR teks) Outermost part of the adrenal gland. It produces and secretes three groups of hormones: mineralocorticoids (aldosterone), glucocorticoids (cortisol), and some androgens (male hormones).

adrenal medulla (ad REE nal meh DULL lah) Innermost part of the adrenal gland. It produces and secretes the hormones epinephrine and norepinephrine.

adventitia (add ven TISH ah) The outmost covering of a structure or organ.

afferent arterioles (AFF er ent ahr TEE ree ohlz) Small arteries traveling toward an organ.

agglutinate (ah GLUE tin ate) To clump.

agonist (AG on ist) The muscle that contracts while another muscle relaxes at the same time to cause movement.

albumin (AL byoo men) Most abundant plasma protein.

aldosterone (al DOSS ter ohn) Most abundant and biologically active of the mineralocorticoid hormones secreted by the adrenal cortex. It regulates the balance of electrolytes, keeping sodium (and water) in the blood while excreting potassium in the urine.

alimentary tract (al ah MEN tar ee) Alternate name for the gastrointestinal system.

allergic reaction Relating to any body response stimulated by an allergen.

allergic rhinitis Condition of excessive nasal secretions (runny nose) due to irritations caused by allergic agents.

alopecia (*AL oh PEE she ah*) Absence or loss of hair.

alveoli (*al VEE oh lye*) Hollow spheres of cells in the lungs where oxygen and carbon dioxide are exchanged.

amblyopia (*am blee OH pee ah*) Reduction or dimness of vision. Also known as *lazy eye*.

amenorrhea (*ah MEN oh REE ah*) The absence or lack of menstruation.

amniocentesis A diagnostic test in which a needle and syringe is utilized to withdrawn amniotic fluid that is chemically and cytologically studied for any genetic or biochemical disorders of the fetus. This is usually performed no earlier than 14 weeks gestation.

Amyotrophic Lateral Sclerosis (ALS) (*AA mee o TROF ic*) Neuromuscular disease that destroys both upper and lower motor neurons that leads to a loss of muscle control.

anabolism (*ah NAH bow lizm*) Assembly of new molecules in the body.

anaphylaxis An immediate, usually allergic response characterized by smooth muscle contraction and dilation of capillaries due to various chemical releases by the body.

anastomosis (*ah NASS te MOE sis*) The suturing of one blood vessel to another.

anatomical position (*an ah TOM ih kal*) This is a standard position in which the body is standing erect, the head is up with the eyes looking forward, the arms are by the sides with the palms facing forward, and the legs are straight with the toes pointing forward.

anatomy (*ah NAH tom ee*) The study of the structures of the human body.

anemia (*ah NEE mee ah*) Any condition in which the number of erythrocytes in the blood is decreased.

anesthesia (*an ess THEE zee ah*) Condition in which sensation of any type, including touch, pressure, proprioception, or pain, has been completely lost.

aneurysm (*AN yoo rizm*) Area of dilation and weakness in the wall of an artery. This can be congenital or where arteriosclerosis has damaged the artery. With each heartbeat, the weakened artery wall balloons outward. Aneurysms can rupture without warning. A dissecting aneurysm is one that is enlarging by tunneling between the layers of artery wall.

angina pectoris A constricting pain or uncomfortable heaviness in the chest often radiating to the shoulder and down the left arm due to cardiac muscle ischemia related to coronary disease.

angiogram A radiographic recording of the size, shape and location of the heart and its blood vessels following the injection of a radiopaque medium.

anhidrosis A genetic or disease related condition of diminished or complete absence of sweat. It can be temporary or permanent, depending on the causative agent.

anorexia nervosa An eating disorder exhibiting an unrealistic body image, fear of weight gain, weight loss and emaciation.

antagonists Something that does the opposite of the agonist.

antebrachial The area of the forearm.

antecubital The area in front of the elbow.

anterior (*an TEE ree or*) The front of the body.

anterior commissure (*an TEE ree or KAH mih sure*) A nerve pathway in the anterior part of the brain that allows communication between the right and left halves of the cerebrum.

anteroposterior (AP) A view from the front to the back as in a AP chest X ray.

antibiotics (*AN teh by OT icks*) A natural or synthetic substance that either destroys a microorganism or inhibits its growth.

antibody (*AN teh bahd ee*) Proteins secreted by B lymphocytes (plasma cells) that attack infected cells.

antibody mediated immunity Immunity that results from a formation of antibodies in response to antigens.

antidiuretic hormone (ADH) (*AN tye dye yoo RET ik*) Hormone secreted by the posterior pituitary gland. It stimulates the kidneys to move water back into the blood to increase the volume of blood.

antigen displaying cells (ADC) Cells that engulf antigens and display them as markers.

antigens (*AN tih jenns*) Substance that causes a formation of antibodies. Cell surface markers that help the immune system identify cells.

antiseptic A substance capable of destroying microorganisms.

anus (*AYE nuss*) External opening of rectum, located between the buttocks. The external anal sphincter is under voluntary control. The peri-anal area is around the anus.

anvil One of the three small bones in the ear.

apex The rounded top of each lung and the gently rounded tip of the outer surface of the heart. The ventricles lie beneath the apex.

apocrine glands (*APP oh crin*) Sweat glands found in the pubic and axillary regions which open into the hair follicles.

appendicitis (*ah pen dih SIGH tiss*) Inflammation and infection of the appendix.

appendicular skeleton (*app en DIK yoo lahr SKELL eh ton*) The bones of the shoulders, arms, hips, and legs.

appendix (*ah PEN dicks*) Long, thin pouch on the exterior wall of the cecum. It does not play a role in digestion. It contains lymphatic tissue.

apthous stomatitis (*AFF thess stow meh TIGHT es*) Inflammation of the mouth that results in an ulcer.

aqueous humor (*AY kwee uss HYOO mer*) Clear, watery fluid produced by the ciliary body. It circulates through the posterior and anterior chambers and takes nutrients and oxygen to the cornea and lens.

arachnoid mater (*ah RAK noyd MAY ter*) Web like connective tissues that attach to the pia mater. The space under the arachnoid mater (subarachnoid space) contains cerebral spinal fluid.

arrhythmia (*ah RITH mee ah*) An irregularity or loss of rhythm in the cardiac conduction system.

arteries Blood vessels that carry blood <u>away</u> from the heart.

arteriole (*ahr TEE ree ohl*) Smallest branch of an artery.

arteriosclerosis (*ar tee ree oh skleh ROH sis*) Progressive degenerative changes that produce a narrowed, hardened artery.

arthritis (*ahr THRYE tiss*) Inflammation of the joints.

arthroscopy The endoscopic examination of the interior region of the joint.

articulation (*AHR tick you lay shun*) A joint where two bones come together and join or articulate.

artificial active immunity Vaccination. Intentional exposure to pathogens so patient makes own antibodies.

artificial passive immunity Injection of antibodies to help patient fight infection.

asexual reproduction Reproduction in which sex cells are not utilized.

association neuron Neurons that mediate impulses between a sensory and motor neuron.

asthma (*AZ mah*) Sudden onset of hyperreactivity of the bronchi and bronchioles with bronchospasm (contraction of the smooth muscle). Inflammation and swelling severely narrow the lumen of the airways.

astrocyte (*ASS tre SITE*) Star-shaped cell that provides structural support for neurons, connects them to capillaries, and forms the blood-brain barrier.

ataxia (*ah TAK see ah*) Lack of coordination of the muscles during movement, particularly the gait. Caused by diseases of the brain or spinal cord, cerebral palsy, or an adverse reaction to a drug.

atelectasis (*at ee LEK tah sis*) Incomplete expansion or collapse of part or all of a lung due to mucus, tumor, or a foreign body that blocks the bronchus. The lung is said to be atelectatic. Also known as *collapsed lung*.

atherosclerosis (*ath er oh skleh ROH siss*) Hardening of the arteries as a result of plaque buildup in the lumen of the arteries.

atrial natriuretic peptide (*AY tree al nay tree your RET ick PEP tide*) Hormone released by the atria when blood pressure rises. Cause increased excretion of water by the kidney, thereby decreasing blood pressure.

atrioventricular node (*ay tree oh vehn TRIK yoo lahr*) Small knot of tissue located between the right atrium and right ventricle. The AV node is part of the conduction system of the heart and receives electrical impulses from the SA node.

atrioventricular valve (*ay tree oh vehn TRIK yoo lahr*) The valve situated between the atrium and the ventricle.

atrium (*atria*) (*AY tree um, AY tree ah*) Two upper chambers of the heart. Intra-atrial structures are located within the atria.

atrophy (*AT roh fee*) Loss of muscle bulk in one or more muscles. It can be caused by malnutrition or can occur in any part of the body that is paralyzed and the muscles receive no electrical impulse from the nerves.

auditory canal One of two canals that lead to the ear.

auricle (*AW rih kl*) The visible external ear. Also known as the *pinna*.

autonomic nervous system Division of the peripheral nervous system that carries nerve impulses to the heart, involuntary smooth muscles, and glands. It includes the parasympathetic nervous system (active during sleep and light activity) and the sympathetic nervous system (active during increased activity, danger, and stress).

autoregulation The ability to self control a cardiac impulse.

autorhythmicity (*aw to rith MIH sih tee*) The heart's ability to generate its own stimulus.

avascular (*ay VASS cue ler*) Without blood vessels.

axial skeleton (*AK see al SKELL eh ton*) The bones of the head, chest, and back.

axillary Pertaining to the area of the armpit.

axon (*AK son*) Part of the neuron that is a single, elongated branch at the opposite end from the dendrites. It receives an electrical impulse and releases neurotransmitters into the synapse. Axons are covered by an insulating layer of myelin.

axon terminal The end point of a neuron.

azotemia (uremia) Condition of increased amounts of nitrogenous waste products in the blood.

B lymphocytes (*B LIMF oh sights*) White blood cells which make antibodies to destroy specific pathogens.

bacteria Asexually reproducing cell capable of creating an infection.

basal nuclei (*BAY sal noo KLEE ie*) Clusters of cells (nuclei) deep in the diencephalon, midbrain and cerebrum which help fine tune voluntary movements.

base The bottom of each lung.

basophil (*BAY soh fill*) Least numerous of the leukocytes. It is classified as a granulocyte because granules in its cytoplasm stain dark blue to purple with basic dye. It releases histamine and heparin at the site of tissue damage.

benign (*bee NINE*) Not progressive, non malignant.

benign prostatic hyperplasia (BPH) The non-cancerous enlargement of the prostate gland that often interferes with urine flow.

bicuspid A two leafed structure.

binary fission Asexual reproduction in prokaryotic cells where the cell enlarges, duplicates its chromosomes and creates a transverse septum forming two identical daughter cells.

biopsy Obtaining of tissue samples for microscopic examination to help determine a diagnosis.

Black Lung Disease Also known as Coal Workers Pneumoconiosis, this obstructive lung disease is a result of the chronic inhalation and retention of coal dust.

blood brain barrier A structural and functional barrier that prevents some substances from diffusing from the blood to the brain tissue.

blood platelets Blood cells responsible for clotting. Also known as *thrombocytes*.

bolus (*BOW luss*) A mass of masticated food.

brachial artery Major artery that carries blood to the upper arm.

brain stem Most inferior part of the brain that joins with the spinal cord. It is composed of the midbrain, pons, and medulla oblongata.

breast cancer Malignant tumors located in the breast tissue. Most common malignancy in American women.

bronchi (*BRONG kye*) Tubular air passages that branch off the trachea to the right and left and enter each lung. They carry inhaled and exhaled air to and from the lungs.

bronchioles (*BRONG kee ohlz*) Small tubular air passageways that branch off the bronchi. They carry inhaled and exhaled air to and from the alveoli.

bronchoscope An endoscope utilized to visually examine the airways of the body.

bronchospasm An abnormal narrowing of the bronchi, due to spasms of the airway's smooth muscle, creating partial obstructions of the lumen, often accompanied by coughing and wheezing.

buccal (*BUCK al*) Pertaining to cheek or mouth region.

bulimia An eating disorder marked by recurring episodes of binge eating, self induced vomiting and diarrhea, and other attempts to reduce weight such as excessive exercising, dieting and/or fasting, individual exhibits excessive and unrealistic concerns about body image.

bundle of His (*HISS*) Section of the conduction system of the heart after the AV node. It splits into the right and left bundle branches.

bursa (*BER sah*) Fluid-filled sac that decreases friction where a tendon rubs against a bone near a synovial joint.

bursitis Inflammation of the bursa, resulting in joint swelling and pain.

calcium ion channels Pathways that allow calcium ions to pass through.

cancellous bone (*CAN cell us*) Spongy bone found in the epiphyses of long bones. Its spaces are filled with red bone marrow that makes blood cells.

canine teeth Long, pointed teeth located between the incisors and the premolars. The canines sink deeply into food to hold it. There are four canines: two in the maxilla and two in the mandible. Also known as *cuspids* (because they have one large, pointed cusp) or *eyeteeth*.

capillaries (*KAP ih lair eez*) The smallest blood vessels in the body, connecting blood vessels between arterioles and venules. The exchange of oxygen and carbon dioxide takes place in the capillaries.

capillary bed A network of capillaries.

capsid The protein covering around a virus particle.

cardiac Pertaining to the heart.

cardiac arrhythmias Abnormal conductive and contraction rhythms of the heart.

cardiology Study of the structure, function and pathology of the cardiac system.

cardiopulmonary Pertaining to the heart and lungs.

cardiopulmonary resuscitation (CPR) The actions (both through cardiac compressions and assisted breathing) to create normal heart function in an individual who has cardiac arrest.

cardiovascular system Body system that includes the heart, arteries, veins, and capillaries. It distributes blood throughout the body.

carina (*kuh RINE uh*) A structure with a projecting central ridge such as occurs at the bifurcation of the mainstem bronchi in the lung.

carotene (*CARE eh teen*) A yellow pigment found in plant and animal tissue. The precursor of vitamin A.

carpal bones The eight small bones of the wrist joint.

Carpal Tunnel Syndrome A condition affecting the median nerve distribution of the hand leading to pain and/or numbness. Repetitive motion is the most common cause of this condition.

cartilage (*KAR tih lij*) Smooth, firm, but flexible connective tissue.

cartilaginous Referring to cartilage.

catabolism (*ka TAH bow lizm*) Breaking down of molecules in the body.

cataract (*KAT ah rakt*) Clouding of the lens. Protein molecules in the lens begin to clump together. Caused by aging, sun exposure, eye trauma, smoking, and some medications.

caudal *(KAWD al)* Toward the tailbone, feet, or lower part of the body.

cecum *(SEE kum)* First part of the large intestine. A short, pouchlike area. The appendix is attached to its external wall.

cell mediated immunity Destruction of pathogens by T lymphocytes. T lymphocytes directly attack infected cells, destroying them.

cell membrane Semipermeable barrier that surrounds a cell and holds in the cytoplasm. It allows water and some nutrients to enter and waste products to leave the cell.

cellular respiration The cellular process in which oxygen is used to break down glucose and other nutrients for energy needed for cellular activity.

cellulitis *(sell you LIE tis)* Bacterial infection of skin and subcutaneous tissues.

cementum *(si MEN tum)* Continuous layer of bone-like connective tissue that covers the dentin layer of the tooth and roots below the gum line. It begins at the gum line where the enamel stops. It anchors one side of the periodontal ligaments.

central Located in the middle or equidistant from multiple points.

central nervous system (CNS) Division of the nervous system that includes the brain and the spinal cord.

centrioles An organelle that precedes mitosis.

centrosomes Region of cytoplasm usually near the nucleus that contains 1-2 centrioles.

cephalic *(seh FAL ik)* Toward the head of the body.

cerebellum *(ser eh BELL um)* Small rounded section that is the most posterior part of the brain. Monitors muscle tone and position and coordinates new muscle movements.

cerebral palsy A general term for a non progressive motor impairment due to lesions or abnormalities of the brain that occurred during early stages of its development.

cerebrospinal fluid (CSF) *(ser eh broh SPY nal FLOO id)* The fluid cushion that protects the brain and spinal cord from shock.

cerebrovascular accident (CVA) (Stroke) A sudden loss of neurological function caused by some vascular injury to the brain, either through hemorrhage or blockage of blood flow to the brain.

cerebrum *(SER eh brum)* The largest and most visible part of the brain. Its surface contains gyri and sulci and is divided into two hemispheres.

cerumen *(seh ROO men)* Sticky wax that traps dirt in the external auditory canal.

ceruminous glands *(seh ROO men us)* Gland that produces the wax-like substance in the ear.

cervical Pertaining to the region of the neck and the cervix.

chain of infection A process of infection that must have the following links in sequential order for an infection to occur: infecting agent, reservoir, portal of entry, mode of transmission, susceptible host.

cheilitis *(ky LY tis)* Inflammation of the lips that includes redness and fissures.

chemical synapse *(KEH mih cull SIH naps)* Site of communication between neurons and other excitable cells. Neurotransmitters are released from the neuron (presynaptic cell) and travel to a muscle or gland cell (the post synaptic cell), allowing communication between the two cells.

chickenpox (Varicella) An acute, contagious disease found usually in children, caused by the Varicella Zoster virus, exhibiting papules which become vesicles, and then pustules.

chemotherapy The use of chemical substances or drugs to treat a disease, usually used in reference to cancer treatment.

cholecystitis *(koh lee siss TYE tiss)* Acute or chronic inflammation of the gallbladder because of gallstones.

cholecystokinin (CCK) Hormone released by the duodenum when it receives food from the stomach. It causes the gallbladder to release bile and the pancreas to release its digestive enzymes.

cholelithiasis *(KOH lee lith EYE ah siss)* One or more gallstones in the gallbladder.

cholesterol Lipid-containing compound that is a component of bile (from the gallbladder), sex hormones, neurotransmitters and cell membranes.

chondrosarcoma (*KON droe sar KOE ma*) Cancer of the cartilage.

choroid (*KOH royd*) Spongy membrane of blood vessels that begins at the iris and continues around the eye. In the posterior cavity, it is the middle layer between the sclera and the retina.

chromatin Genetic material found in the nucleus of a cell.

chronic bronchitis (*brong KYE tiss*) Chronic inflammation or infection of the bronchi. Inflammation of the bronchi can be due to pollution or smoking.

chronic obstructive pulmonary disease (COPD) A group of progressive, debilitating and often fatal lung diseases that exhibit difficulty with air movement in and out of the lungs, prolonged expiratory periods, and the loss of normal lung elasticity. COPD includes emphysema, chronic bronchitis and asthma.

chyle (*KILE*) Milk-like substance formed from digested and absorbed fats.

chyme (*KIME*) Mixed food and digestive juices in the stomach and small intestine.

cilia (*SIL ee ah*) Small hairs that flow in waves to move foreign particles away from the lungs toward the nose and the throat where they can be expelled. Also found inside the fallopian tube to propel an ovum toward the uterus.

ciliary muscles (*SILL ee air ee*) Smooth muscle that alters the lens of the eye to accommodate for near vision.

cirrhosis A chronic liver disease that exhibits liver scarring, loss of normal liver architecture and function, often caused by alcoholism and chronic viral hepatitis.

clot A thrombus or coagulated blood.

coagulation The formation of a blood clot by platelets and the clotting factors.

cochlea (*KOHK lee ah*) Structure of the inner ear that is associated with the sense of hearing. It relays information to the brain via the cochlear branch of the vestibulocochlear nerve.

collecting duct Common passageway that collects fluid from many nephrons. The final step of reabsorption takes place there and the fluid is known as urine.

colon The longest part of the large intestine. It has four parts: the ascending colon, the transverse colon, the descending colon, and the S-shaped sigmoid colon.

colonoscopy The visualization of the lower gastrointestinal tract through the use of an endoscope.

coma A state of unconsciousness in which the patient cannot be aroused, even following stimulation.

commissures (*KAH mih sures*) Transverse bands of nerve fibers; carry information from one side of the nervous system to the other.

common cold Usually a communicable viral infection of the upper airways that can spread to other areas of the respiratory tract.

communicable While communicable and contagious may sound like the same thing, there is a fine difference. A communicable disease has the potential to be spread by a variety of ways such as person to person or insect to person. A contagious disease is one that is readily transmitted form one person to another such as the common cold.

compact bone Hard or dense bone forming the superficial layer of all bones.

complement cascade A series of chemical reactions triggered by infection that leads to the destruction of a pathogen.

computed tomography (CT) Imaging diagnostic tool that combines a series of X-rays to give a 3 dimensional view.

conchae (*kong KA*) Shelf-like structures inside the nasal cavity.

concussion An injury resulting from an impact with an object.

cones Light-sensitive cells in the retina that detect colored light. There are three types of cones, each of which responds to either red, green, or blue light.

conjunctiva (*kon JUNK tih vah*) Delicate, transparent mucous membrane that covers the inside of the eyelids and the anterior surface of the eye. It produces clear, watery mucus.

conjunctivitis (*kon JUNK tih VYE tiss*) Inflamed, reddened, and swollen conjunctivae with dilated blood vessels on the sclera. Caused by a foreign substance in the eye or an infection.

constipation Decrease in the frequency of defecation accompanied by difficult or incomplete passage of stool.

contusion A bruise.

cor pulmonale (*KOR pull moh NAY lee*) Failure of the pumping ability of the right ventricle.

corium (*CORE ee um*) Layer of skin immediately under the epidermis. Also known as the *true skin*.

cornea Transparent layer over the anterior part of the eye. It is a continuation of the white sclera.

coronal plane (*kor ROHN al*) Also called *frontal plane*; an imaginary vertical plane that divides the entire body into front and back sections. The coronal plane is named for the coronal suture where the anterior and posterior skull bones meet.

coronary thrombosis A blood clot in the coronary artery, which is the most common cause of a myocardial infarction.

corpus callosum (*KOR pus kah LOH sum*) Thick white band of nerve fibers that connects the two hemispheres of the cerebrum and allows them to communicate and coordinate their activities.

corticobulbar tract (*KOR ti coe BUL bar*) Spinal cord tract that carries impulses to the brainstem from the motor cortex. Carries orders for voluntary movements.

corticospinal tract (*KOR ti coe SPY nal*) Pertaining to the tract between the cerebral cortex and spinal cord. Carries orders for voluntary movements.

counter-current circulation Exchange of substances between two streams on either side of a membrane. Helps control concentration of fluids as in the nephron loop.

cramp Spasmodic muscle contraction.

cranial (*KRAY nee al*) Pertaining to the skull.

cranial nerves Twelve pairs of nerves that originate in the brain. Carry sensory nerve impulses to the brain from the nose, eyes, ears, and tongue for the senses of smell, vision, hearing, and taste. Also carry sensory nerve impulses to the brain from the skin of the face. Also carry motor nerve impulses from the brain to the muscles of the face, mouth, throat; eye, and salivary glands.

cross-sectioning Making slices of a sample for examination purposes.

crown White part of the tooth that is visible above the gum line.

crowning Occurring during delivery, it is the visible presentation of the fetal head as it comes out the vaginal opening.

crural (*CRUR al*) Pertaining to the leg or thigh.

culture and sensitivity test The growth of bacteria and exposing it to a variety of antibiotics to see which drug is most effective.

Cushing's Syndrome A condition usually the result of the usage of higher levels of corticosteroids in which the patient develops physical features that can include: buffalo hump obesity, a rounded face (moon face), weight gain, and increased fragility of the skin.

cuspids Canine teeth.

cutaneous membranes (*cue TAY nee us*) Membranes of the skin.

cuticle Layer of dead skin that arises from the epidermis around the proximal end of the nail. It keeps microorganisms from the nail root.

cyanosis (*sigh ah NOH siss*) Bluish-gray discoloration of the skin from abnormally low levels of oxygen and abnormally high levels of carbon dioxide in the tissues.

cystic fibrosis An inherited disease involving abnormal high sweat chloride levels, pancreatic insufficiency and pus producing pulmonary disease with high levels of thick tenacious mucus.

cytokines (*SIGH tow kines*) Chemicals released by injured body tissues that summon leukocytes and cause them to move to the area.

cytoplasm (*SIGH toh plazm*) Gel-like intracellular substance. Organelles are embedded in it.

cytotoxic T cells (*sigh tow TOX ick*) Type of T lymphocyte that matures in the thymus. Cytotoxic T cells destroy all types of pathogens as well as body cells infected with viruses.

deciduous teeth (*deh SIH jew uss*) Teeth that erupt during childhood from age 6 months to 2 years. Also called the *milk teeth*, *baby teeth*, or *primary teeth*.

decubitus *(dee KYOO bih tus)* Lying down position; on the back.

deep Located underneath or further from a surface.

defecation *(deh fih CAY shun)* Process by which undigested food fiber and water are removed from the body in the form of a bowel movement.

defibrillator An external or internal electronic device used to shock the patient's heart in an attempt to bring it back in to a normal rhythm.

dendrite *(DEN dright)* Multiple branches at the end of a neuron that carry information to the cell body.

dendritic cells *(DEN dright ick)* One of several types of antigen displaying cells that stimulate adaptive immunity.

dental caries Holes or cavities created by bacteria due to poor oral hygiene.

dentin Hard layer of tooth just beneath the enamel layer.

deoxyribonucleic acid (DNA) Sequenced pairs of nucleotides that form a double helix. A segment of DNA makes up a gene.

depolarization Changing of the permeability of the cell membrane of any excitable cell, for example cardiac muscle or a neuron, that leads to a change in charge across the cell membrane.

dermis *(DER miss)* Layer of skin under the epidermis. It is composed of collagen and elastin fibers. It contains arteries, veins, nerves, sebaceous glands, sweat glands, and hair follicles.

diabetes A general term for a group of diseases that are marked by excessive urination.

diabetes insipidus A condition in which there is excessive urination due to either inadequate antidiuretic hormone or and inability of the kidney to respond to the available antidiuretic hormone.

diabetes mellitus (DM) Condition resulting from too little secretion of insulin leading to increased levels of blood glucose.

diagnosis *(dye ahg NOH siss)* A determination as to the cause of the patient's symptoms and signs.

diaphragm *(DYE ah fram)* Muscular sheet that divides the thoracic cavity from the abdominal cavity. Most important ventilation muscle.

diaphysis *(dye AFF ih siss)* The straight shaft of a long bone.

diarrhea Defecation of liquid feces.

diastole *(dye ASS toe lee)* Resting period between contractions. It is when the heart fills with blood.

diencephalon *(dye en SEFF ah lon)* Central part of the brain that contains the thalamus and hypothalamus.

differentiation Process by which embryonic cells assume different shapes and functions in different parts of the body.

diffusion Process of movement of a substance from high concentration to low concentration.

digestion Process of mechanically and chemically breaking food down into nutrients that can be used by the body.

digital Pertaining to the fingers or toes.

dipstick test A general term for a test that involves the application of urine to a chemical impregnated paper for the purpose of analysis. The results are usually read as color changes to the paper.

disease *(dih ZEEZ)* Any change in the normal structure or function of the body.

distal *(DISS tal)* Moving from the body toward the end of a limb (arm or leg).

distal tubule Tubule of the nephron that begins at the loop of Henle. It empties into the collecting duct. Reabsorption takes place there.

diverticulitis An inflammatory condition of the diverticulum of the intestinal tract especially in the colon which can lead to pain, fever, and weight loss.

dorsal *(DOR sal)* Pertaining to the posterior of the body, particularly the back.

dorsal column tract Spinal cord pathway that carries fine touch sensation from the spinal cord to the brain.

dorsal root ganglion *(DOR sal ROOT GANG lee on)* A collection of sensory neurons on the dorsal roots of the spinal cord.

duodenum (*doo ODD eh num*) First part of the small intestine. It secretes cholecystokinin, a hormone that stimulates the gallbladder and pancreas to release bile and digestive enzymes.

dura mater (*DOO ra MAY ter*) Tough, outermost layer of the meninges. The dura mater lies just under the bones of the cranium and vertebrae.

dwarfism A condition of being abnormally undersized. May be due to an endocrine condition, heredity, nutritional deficiency, diseases of the skeleton or others.

dysmenorrhea (*DISS men oh REE ah*) Difficult menstruation.

dyspnea (*DISP nee uh*) Difficulty in breathing.

dysrhythmia See *arrhythmia*.

ecchymosis (*ek ee MOH sis*) Superficial bruising under the skin or mucous membrane due to bleeding.

eccrine glands (*EKK rin*) Sweat glands that cover the entire skin surface.

echocardiography A noninvasive diagnostic procedure using sound waves to visualize cardiac structure and function.

efferent arteriole (*EFF er ent ahr TEE ree ohlz*) Small blood vessels that carry blood away from the glomeruli of the kidney.

ejaculation Sudden discharge of semen from the body.

electrical synapse Gap or junction between two neurons involving electrical charges.

electrocardiogram The graphic record created by echocardiography.

electrolyte Chemical element that carries a positive or negative charge and conducts electricity when dissolved in a solution: Examples include sodium (Na+), potassium (K^+), chloride (Cl^-), calcium (Ca^{++}), and bicarbonate ($HCO3^-$). Electrolytes are carried in the plasma. Excess amounts in the blood are removed by the kidneys.

electromyogram (*ee LEC troh MY ah gram*) A recording or tracing of the muscular activity as a result of electrical stimulation.

electromyography (EMG) (*elec troh my AH graf ee*) Diagnostic procedure to diagnose muscle disease or nerve damage. A needle electrode inserted into a muscle records electrical activity as the muscle contracts and relaxes. The electrical activity is displayed as waveforms on an oscilloscope screen and permanently recorded on paper as an electromyogram.

emboli (*EM bow lie*) Plural of embolus.

embolus (*EM boh luss*) Mass of undissolved matter present in the blood or lymphatic vessels that was brought there by the blood or lymph current.

emphysema (*em fih SEE mah*) Chronic pulmonary disease resulting in destruction of air spaces distal to the terminal bronchiole.

empyema (*em pye EE mah*) Localized collection of purulent material (pus) in the thoracic cavity from an infection in the lungs. Also known as *pyothorax*.

emulsification (*ee mull sih fih KAY shun*) Process performed by bile of breaking down large fat droplets into smaller droplets with more surface area.

enamel Glossy, thick white layer that covers the crown of the tooth. Enamel is the hardest substance in the body.

endemic A condition or disease related to a specific population or region of the world.

endocardium (*ehn doh KAR dee um*) Innermost layer of the heart. It covers the inside of the heart chambers and valves.

endochondral Within a cartilage.

endocrine system (*EHN doh krin*) Body system that includes the testes, ovaries, pancreas, adrenal glands, thymus, thyroid gland, parathyroid glands, pituitary gland, and pineal gland. It produces and releases hormones into the blood to direct the activities of other body organs.

endocrinology (*en doh krin ALL oh jee*) The study of the endocrine system and related disorders.

endocytosis Ingestion of substances by a cell. Substances are taking into the cells after being surrounded by vesicles.

endolymph (*EN doe limf*) Fluid within the labyrinth of the ear.

endometrium (*EHN doh MEE tree um*) The mucous membrane lining the uterus.

endometriosis An inflammation of the mucous membrane lining the uterus.

endoplasmic reticulum Organelle that consists of a network of channels that transport materials within the cell. Also the site of protein, fat, and glycogen synthesis.

endoscopy The inspection of various body organs, structures and cavities by the use of an endoscope.

English System The weights and measuring system commonly used in the United States.

enuresis A condition of involuntary urine discharge after the age at which bladder function becomes under a individual's conscious control.

enzyme Molecules that speed up the rate of chemical reactions in cells. Enzymes are particularly important in the breakdown and synthesis of biological molecules.

eosinophils (*ee oh SIN oh fillz*) Type of leukocyte. It is classified as a granulocyte because it has granules in the cytoplasm. The nucleus has 2 lobes. Eosinophils are involved in allergic reactions and defense against parasites.

ependymal cells Specialized cells that line the walls of the ventricles and spinal canal and produce cerebrospinal fluid.

epidermis (*ep ih DER miss*) Thin, outermost layer of skin. The most superficial part of the epidermis consists of dead cells filled with keratin. The deepest part (basal layer) contains constantly dividing cells and melanocytes.

epidural Located either over or on the dura.

epidural space (*eh pih DURE all*) Area between the dura mater and the vertebral body.

epigastric Above the stomach.

epiglottis (*ep ih glah TISS*) Lid-like structure that seals off the larynx, so that swallowed food goes into the esophagus.

epinephrine (*EP ih NEFF rinn*) Hormone secreted by the adrenal medulla in response to stimulation by nerves of the sympathetic nervous system.

epiphyseal plate (*eh piff ih SEE al*) The growth plate.

epiphysis (*eh PIFF ih siss*) The widened ends of a long bone. Each end contains the epiphysial plate where bone growth takes place.

epithelial tissue (*ep ih THEE lee al*) Layers of cells that form the epidermis of the skin as well as the surface layer of mucous and serous membranes.

erection The stiffing of the penis due to increase blood flow.

erectile dysfunction (ED) An inability to develop a full erection. May be physiological or psychological.

erythrocyte (*eh RITH roh sights*) A red blood cell. Erythrocytes contain hemoglobin and carry oxygen and carbon dioxide to and from the lungs and cells of the body.

erythropoiesis (*eh rith roh poy EE sis*) Formation of red blood cells.

erythropoietin (*eh RITH roh poy EH tin*) In the body, a hormone secreted by the kidneys when the number of red blood cells decreases. It stimulates the bone marrow to make more red blood cells. As a drug, erythropoietin does the same.

esophagus (*eh SOFF ah guss*) Flexible, muscular tube that moves food from the pharynx to the stomach.

etiology (*ee tee ALL oh jee*) The cause or origin of a disease.

eustachian tube (*yoo STAY she ehn*) Tube that connects the middle ear to the nasopharynx and equalizes the air pressure in the middle ear.

excretion (*ik SCREE shun*) Removal of waste matter from the body.

exocytosis Secretion. The expulsion of material from a cell using vesicles.

extension (*eks TEN shun*) Straightening a joint to increase the angle between two bones or body parts.

extensor muscle Muscle that produces extension when it contracts.

external Near or on the outside surface of the body or an organ.

external auditory meatus (*AW dih tor ee mee AY tuss*) Opening at the entrance to the external auditory canal where sound waves enter.

external otitis An inflammation of the outer ear.

external urethral sphincter (*EKS ter nal yoo REE thral SFINK ter*) One of two valves, made of circular muscle, which allows voluntary control of urination.

facilitated diffusion Also known as *carrier mediated passive transport*; the movement of substances into cells via carrier proteins.

familial hypercholesterolemia A condition in which the low density lipoproteins are not being removed from the blood properly by the lipoprotein receptors in the liver.

femoral artery Major artery that carries blood to the upper leg.

fibrin Fiber strands that are the formed by the activation of clotting factors. Fibrin traps erythrocytes and this forms a blood clot.

fibrinogen (*fye BRINN oh jenn*) Blood clotting factor.

fibroblasts Any cell from which connective tissue is created.

fibromyalgia (*fie bro my AL je*) Pain located at specific, small trigger points along the neck, back, and hips. The trigger points are very tender to the touch and feel firm.

filtration Process in which water and substances in the blood are pushed through the pores of the glomerulus. The resulting fluid is known as filtrate.

fissure (*FISH er*) Deep division on the surface of the brain and spinal cord.

flaccid (*FLAH sid*) Limp or without muscle tone.

flagella Hair-like processes on bacteria or protozoon that cause movement.

flexion (*FLEK shun*) Bending of a joint to decrease the angle between two bones or body parts. Opposite of extension.

flexor muscle Muscle that produces flexion when it contracts.

flora (*FLOOR ah*) Plant life occurring in a specific environment.

follicle (*FALL ih kle*) 1. Mass of cells with a hollow center. It holds an oocyte before puberty and a maturing ovum after puberty. The follicle ruptures at the time of ovulation and becomes the corpus luteum. 2. Also a site where a hair is formed. The follicle is located in the dermis.

Forensic Science The application of science and medicine to law.

formed elements Cellular parts of the blood to include red and white blood cells and platelets.

fornix (*FOR niks*) 1. Tract of nerves that joins all the parts of the limbic system. 2. Area of the superior part of the vagina that lies behind and around the cervix.

Fowler's position A semi-sitting position with the torso at a 45–60 degree angle.

frenulum (*FREN you lum*) Structure that attaches the lower side of the tongue to the gum.

frontal lobe Lobe of the cerebrum that is responsible for motor activities, conscious thought, and speech.

frontal plane An imaginary plane parallel with the long axis of the body that divides the body into an anterior and posterior section.

full thickness A more current assessment of the severity of a burn in which the burn has penetrated through all three of the skin layers.

fundus (*FUN duss*) A larger part, base, or body of a hollow organ such as the dome-shaped top of the bladder, uterus above the fallopian tubes, or rounded, most superior part of the stomach.

fungi (*FUN jie*) A plant-like organism that includes mold and yeasts.

furuncle (*FOO rung kle*) A boil.

ganglion (*GANG lee on*) Mass of nervous tissue composed mostly of nerve cell bodies and lying outside the brain and spinal cord.

gastrin (*GAS trin*) Hormone produced by the stomach that stimulates the release of hydrochloric acid and pepsinogen in the stomach.

gastritis An acute or chronic inflammation of the stomach's lining.

gastroenterology (*gas troh en ter ALL oh jee*) The study of diseases related to the gastrointestinal tract.

gastroesophageal reflux disease (GERD) A combination of symptoms caused by a backflow of gastric contents into the esophagus.

gastrointestinal (GI) system Body system that includes the oral cavity, pharynx, stomach esophagus, small and large intestines, and the accessory organs of the liver, gallbladder, and pancreas. Its function is to digest food and remove undigested food from the body. Also known as the *gastrointestinal tract*, *digestive system*, or *digestive tract*.

gel layer The top layer of the mucociliary blanket that entraps inhaled particles.

gene An area on a chromosome that contains all the DNA information needed to produce one type of protein molecule.

genitourinary system (*gen i toe YOUR in air EE*) Combination of two closely related body systems: the female and male genitalia and the urinary system. Also known as the *urogenital system*.

gigantism (Giantism) An excessive development of a body or body part often as a result of an abnormal hormone level.

gingival (*jin jih VAL*) Referring to the gum.

glaucoma (*glaw KOH mah*) Increased intraocular pressure (IOP) because aqueous humor cannot circulate freely. In open-angle glaucoma, the angle where the edges of the iris and cornea touch is normal and open, but the trabecular meshwork is blocked. Open-angle glaucoma is painless but destroys peripheral vision, leaving the patient with tunnel vision. In closed-angle glaucoma, the angle is too small and blocks the aqueous humor. Closed-angle glaucoma causes severe pain, blurred vision, and photophobia. Glaucoma can progress to blindness.

glia (*GLEE ah*) Non-nervous or supporting tissue found in the brain and spinal cord; made of glial cells.

glial cells (*GLEE all sells*) Cells that include astrocytes, oligodendrocytes, ependymal cells, microglia cells, schwann cells, and satellite cells.

glomerular filtrate (*gla MARE you ler FILL trate*) The filtered fluid within the glomerulus.

glomerulonephritis (*gloh MAIR you loh neh FRY tis*) The inflammatory disease of the glomerular capsular membranes of the kidneys.

glomerulosclerosis (*gloh MAIR you loh sklee ROW sis*) Fibrotic or scarring changes of the kidney.

glomerulus (*gloh MAIR yoo luss*) Network of intertwining capillaries within Bowman's capsule in the nephron. Filtration takes place in the glomerulus.

glottis (*GLOT is*) V-shaped structure of mucous membranes and vocal cords within the larynx.

glucagon Hormone secreted by the pancreas that increases the blood glucose by stimulating the liver to change stored glycogen to glucose.

glucose (*GLOO kohs*) 1. A simple sugar found in foods and also the sugar in the blood. 2. Glucose is not normally found in the urine. Its presence (glycosuria) indicates uncontrolled diabetes mellitus with excess glucose in the blood "spilling" over into the urine.

gluteal (*GLOO tee al*) Pertaining to buttocks.

glycogen (*GLIE co jin*) The form that glucose (sugar) takes when it is stored in the liver and skeletal muscles.

goiter Enlargement of the thyroid gland.

golgi apparatus Organelle of the cell that packages cellular material for transport.

gout Disease caused by excessive uric acid in the blood that is deposited in the joints producing inflammation and pain.

Grave's Disease A distinct form of hyperthyroidism that includes an enlarged thyroid gland and various ocular findings such as proptosis (downward displacement), stare, and lid lag.

Guillian-Barré syndrome (GBS) (*gee YAH bar RAY*) A neuromuscular disease that usually leads to ascending flaccid paralysis.

gustatory sense (*GUSS ta tore ee*) Sense of taste.

gynecology (*guy neh KOL oh jee*) The study and treatment of female diseases.

gyri (*JIE rie*) Convolutions of the cerebral hemispheres of the brain.

hammer One of the three small bones of the ear, also known as the *malleus*.

Hashimoto's Thyroiditis An autoimmune disease involving inflammation, then destruction and fibrosis of the thyroid gland, which leads to hypothyroidism.

heart block A blockage of the normal cardiac conduction system pathways.

heart failure (HF) Inability of the heart to pump sufficient amounts of blood. Caused by chronic coronary artery disease or hypertension.

helper T cell Cells that stimulate the production of cytotoxic T cells and B cells. Also known as *CD4 cells*.

hematology *(hee mah TALL oh jee)* The study of blood-related diseases and conditions.

hematoma A swelling as a result of break in a blood vessel(s) usually confined to a specific area (tissue, organ, or space) in the body.

hemisphere One half of the cerebrum, either the right hemisphere or the left hemisphere. The right hemisphere deals with recognizing patterns and three-dimensional structures (including faces) and the emotions of words. The left hemisphere deals with mathematical and logical reasoning, analysis, and interpreting sights, sounds, and sensations. The left hemisphere is active in reading, writing, and speaking.

hemodialysis The use of an artificial kidney to clear waste products, toxins, and excess fluid from the blood.

hemoglobin *(HEE moh GLOH binn)* Substance in an erythrocyte that binds to oxygen and carbon dioxide. Its globin chains give it a round shape. When it is bound to oxygen, it forms the compound oxyhemoglobin.

hemolytic uremic syndrome An acute condition consisting of acute nephropathy, thrombocytopenia, and microangiopathic hemolytic anemia and is caused by eating contaminated or raw meats (esp. hamburger) containing *E. coli 0157:H7*.

hemophilia *(HEE moh FILL ee ah)* Inherited genetic abnormality of a gene on the X chromosome that causes an absence or deficiency of a specific clotting factor. When injured, hemophiliac patients cannot easily form a blood clot and continue to bleed for long periods of time.

hemopoiesis *(HEME ah poy ee sus)* Formation of blood cells.

hemorrhoid Varicose veins in the rectum.

hemostasis *(HEE moh STAY siss)* The cessation of bleeding after the formation of a blood clot.

hemothorax *(HEEM oh THOH raks)* Presence of blood in the thoracic cavity, usually from trauma.

heparin Substance that inhibits coagulation of blood.

hepatic duct *(hepp ah TIC duct)* Duct that carries bile from the liver.

hepatitis Inflammation of the liver most commonly caused by an exposure to an infectious agent, or excessive exposures to drugs (such as acetaminophen), or toxins (such as alcohol).

hernia A weakness in the muscles of the abdominal wall that allows loops of intestine to balloon outward.

herpes Viral infection that can cause a variety of skin conditions to include cold sores and shingles.

hilum *(HIGH lim)* 1. Indentation on the medial side of each lung where the bronchus, pulmonary artery, pulmonary vein, and nerves enter the lung. 2. Indentation in the medial side of each kidney where the renal artery enters and renal vein and the ureter leave.

histamine *(HISS tah meen)* Released by basophils. Histamine dilates blood vessels and increases blood flow to damaged tissue. Allows protein molecules to leak out of blood vessels into the surrounding tissue. This produces redness and swelling.

Hodgkin's Disease A malignant lymphoma in which the early stages of the disease present with no symptoms other than a painless lump or enlarged gland in the neck or armpit. Some patients may develop fevers/night sweats and loss of appetite/weight.

holter monitor A portable device used to record cardiac activity usually during a 24 hour period that may show cardiac arrhythmias that would not be revealed during a regular ECG or stress test.

homeostasis *(hoh mee oh STAY siss)* State of equilibrium of the internal environment of the body, including fluid balance, acid-base balance, temperature, metabolism, and so forth, to keep all the body systems functioning optimally.

horizontal plane Another name for the transverse plane.

hormone *(HOR mohn)* Chemical messenger of the endocrine system that is released by a gland or organ and travels through the blood.

hydrocephalus A general term for the accumulation of excessive amounts of cerebrospinal fluid within the brain's ventricles.

hypercholesterolemia *(HI per koh LESS tur ul ee me ah)* High levels of cholesterol in the blood.

hyperopia *(HIGH per OH pee ah)* Farsightedness. Light rays from a far object focus correctly on the retina, creating a sharp image. However, light rays from a near object come into focus posterior to the retina, creating a blurred image.

hyperpolarized The charge across the cell membrane is more negative than resting.

hypertrophy *(high PER troh fee)* Greater than normal growth.

hypochondriac A person who complains of imaginary conditions or diseases.

hypodermis The fatty tissue layer below the dermis of the skin.

hypopituitarism A condition caused by a less than normal secretion of pituitary hormones.

hyperthyroidism The overproduction of thyroid hormones, often resulting in increased metabolic rate, weight loss, and nervous irritability.

hypothalamus *(high poh THAL ah mus)* Endocrine gland located in the brain just below the thalamus. It produces (but does not secrete) antidiuretic hormone (ADH) and oxytocin. The hypothalamis is in the center of the brain just below the thalamus and coordinates the activities of the pons and medulla oblongata. It also controls heart rate, blood pressure, respiratory rate, body temperature, sensations of hunger and thirst, and the circadian rhythm. It also produces hormones as part of the endocrine system. In addition, the hypothalamus helps control emotions (pleasure, excitement, fear, anger, sexual arousal) and bodily responses to emotions; regulates the sex drive; contains the feeding and satiety centers; and functions as part of the "fight or flight" response of the sympathetic nervous system.

hypothyroidism The underproduction of thyroid hormones, resulting in low metabolism, excessive weight gain, lethargy and depression.

idiopathic A condition with an uncertain or not yet determined cause.

ilium *(ILL ee um)* Most superior hip bone. Bony landmarks include the iliac crest and the anterior-superior iliac spine (ASIS). Posteriorly, each ilium joins one side of the sacrum.

immunology *(im yoo NALL oh jee)* The study of diseases and conditions of the immune system.

impotence The inability of a male to achieve or maintain an erection. The cause of this condition may be a vascular, neurological, or psychological dysfunction.

impulse conduction The propagation of an electrical impulse through the nervous system.

incisors Chisel-shaped teeth in the middle of the dental arch that cut and tear food on their incisal surface. There are 8 incisors, 4 in the maxilla and 4 in the mandible.

incontinence *(in KAH tih nens)* Condition of involuntary urination and defecation.

incus *(ING kuss)* Second bone of the middle ear. It is attached to the malleus on one end and the stapes on the other end. Also known as the *anvil*.

infarct *(in FARKT)* Cellular death due to lack of blood flow (perfusion).

inferior Pertaining to the lower half of the body or a position below an organ or structure.

inferior vena cava The vena cava is the largest vein in the body. The inferior vena cava receives blood from the abdomen, pelvis, and lower extremities and takes it to the heart.

infertility A condition of diminished or an absent ability to produce offspring. It does not necessarily imply sterility.

inflammation *(in flah MAY shun)* Tissue reaction to injury that includes swelling and reddening due to increase blood flow to the area.

inflammatory response The defensive series of actions/events that the body performs in the reaction to trauma or infection.

ingestion Process of taking in material (particularly food).

inguinal region The groin.

innate immunity *(ih NATE im YOO nih tee)* Defense against pathogens that you are born with and that does not improve with experience or remember specific pathogens. The first line of defense.

inotropism *(EYE no TROPE izm)* An influence on the force of muscular contraction.

insula The deep lobes of the cerebral hemispheres.

insulin Hormone secreted by the pancreas that allows for glucose to be utilized for cellular respiration.

intercalated discs *(in ter KUH late ed)* Structures that connect heart tissue cells to facilitate a smooth contraction.

interferon *(in ter FIR on)* Substance released by macrophages that have engulfed a virus. Interferon stimulates body cells to produce an antiviral substance that keeps a virus from entering a cell and reproducing.

interleukin *(in ter LOO kin)* Released by macrophages, it stimulates B cell and T lymphocytes and activates NK cells. It also produces the fever associated with inflammation and infection.

internal Structures deep within the body or an organ.

internal medicine Specialization in the study of function, interrelatedness and treatment of body systems.

interneurons *(in ter NURE ons)* Neurons that facilitate communication between neurons.

iris *(EYE riss)* Colored ring of tissue whose muscles contract or relax to change the size of the pupil in the center of the iris.

ischemia *(iss KEE mee ah)* Tissue injury due to a decrease in blood flow.

jaundice A general term for the yellow staining of body tissues and fluids due to an excessive amount of bilirubin in the blood. There are various causes of jaundice.

jejunum *(jee JOO num)* Second part of the small intestine.

joint Area where two bones come together.

jugular venous distension (JVD) The enlargement of the jugular neck veins due to venous engorgement.

keloid An unusually thick and fibrous scar on the skin.

keratin *(KAIR ah tin)* Hard protein found in the cells of the outermost part of the epidermis and in the nails.

keratinization *(KAIR ah tin eye ZAY shun)* The process of forming a horny growth such as fingernails.

keratinized To become hard.

ketonuria A condition of excessive acetone bodies in the urine.

kidney Organ of the urinary system that filters blood and produces urine, controlling fluid and electrolyte balance.

kyphosis An exaggerated posterior curvature of the thoracic spine, also known as humpback, hunchback, or Pott's curvature.

labia *(LAY bee ah)* An outer pair of vertical fleshy lips covered with pubic hair (the labia majora) and a smaller, thinner, inner pair of lips (the labia minora) that partially cover the clitoris, and urethral and vaginal openings. Part of the external female genitalia.

labyrinth *(LAB ih rinth)* Intricate communicating passage of the inner ear essential for maintaining equilibrium.

labyrinthitis *(LAB ih rinn THYE tiss)* Bacterial or viral infection of the semicircular canals of the inner ear, causing severe vertigo.

lacrimal apparatus *(LAK rim al app ah RA tuss)* Structures involved with the secretion and production of tears.

lacteal *(LACK te al)* Pertaining to milk.

large intestine Organ of absorption between the small intestine and the anal opening to the outside of the body. The large intestine includes the cecum, appendix, colon, rectum, and anus. Also known as the *large bowel*.

laryngitis *(lar in JIGH tis)* Hoarseness or complete loss of the voice, difficulty swallowing, and cough due to swelling and inflammation of the larynx.

laryngopharynx *(lah ring goh FAR inks)* Relates to both the larynx and pharynx.

larynx *(LAR inks)* Triangular-shaped structure in the anterior neck (visible as the laryngeal prominence or Adam's apple) that contains the vocal cords and is a passageway for inhaled and exhaled air. Also known as the *voice box*.

lateral *(LAT er al)* Pertaining to the side of the body or the side of an organ or structure.

lens Clear, hard, disk in the internal eye. The muscles and ligaments of the ciliary body change its shape to focus light rays on the retina.

lesion *(LEE zhun)* General category for any area of visible damage on the skin, whether it is from disease or injury.

leukemia Cancer of leukocytes (white blood cells), including mature lymphocytes, immature lymphoblasts, as well as myeloblasts and myelocytes that mature into neutrophils, eosinophils, or basophils). The malignant leukocytes crowd out the production of other cells in the bone marrow. Leukemia is named according to the type of leukocyte that is the most prevalent and whether the onset of symptoms is acute or chronic. Types of leukemia include acute myelogenous leukemia (AML), chronic myelogeous leukemia (CML), acute lymphocytic leukemia (ALL), and chronic lymphocytic leukemia (CLL).

leukocytes *(LOO koh sights)* White blood cells. There are five different types of mature leukocytes: neutrophils, eosinophils, basophils, lymphocytes, and monocytes.

leukocytosis Increase in the number of white blood cells above normal.

leukopenia Abnormal decrease of white blood cells.

leukoplakia *(loo coe PLAY kee ah)* White patches in the mouth commonly seen in snuff and chewing tobacco users that is a precursor to oral cancer.

lingual *(LING gwal)* Pertaining to the tongue.

ligament Fibrous bands that hold two bone ends together in a synovial joint.

limbic system *(LIM bick)* Processes memories and controls emotions, mood, motivation, and behavior. Links the conscious to the unconscious mind. Limbic system consists of the thalamus, hypothalamus, hippocampus, amygdaloid bodies, and fornix.

lipocyte Cell in the subcutaneous layer that stores fat.

lithotripsy *(LITH oh trip see)* Medical or surgical procedure that uses sound waves to break up a kidney stone.

lobes 1. Large divisions of the lung, visible on the outer surface. 2. Large area of the hemisphere of the cerebrum. Each lobe is named for the bone of the skull that is next to it: frontal lobe, parietal lobe, temporal lobe, and occipital lobe.

local potential Change in the electrical charge across a cell membrane that is proportional to the size of the stimulus.

lordosis An anteroposterior curvature of the spine (backward bending).

lower esophageal sphincter (LES) Ringed muscle leading into the stomach.

lumbar region Two of the nine regions of the abdominopelvic area. The right and left lumbar regions are inferior to the right and left hypochondriac regions. Also refers to the lower back, spinal cord, and spinal column.

lumen Opening in the center of a large tube, for example the center of the tubes in the digestive and respiratory systems and the blood vessels.

lunula *(LOO nyoo lah)* Whitish half-moon visible under the proximal portion of the nail plate. It is the visible tip of the nail root.

luteal *(LOO tee al)* Pertaining to the corpus luteum.

Lyme Disease Named after the town of Lyme, Connecticut, it is the number one vector-borne disease in the US. A multisystem disorder caused by the spirochete, Borrelia burgdorferi and is spread by deer tick bites causing a bull's eye rash at the bite site.

lymph *(LIMF)* Fluid that flows through the lymphatic system.

lymphadenitis An inflammation of lymph nodes, which results from the responsive actions of lymphocytes and phagocytes to invading microorganisms, or cancer cells.

lymph nodes *(LIMF nohdz)* Small, encapsulated pieces of lymphoid tissue located along the lymphatic vessels. Lymph nodes filter and destroy invading microorganisms and cancerous cells present in the lymph.

lymphatic system (*lim FAT ik*) Body system that includes lymphatic vessels, lymph nodes, lymph fluid, and lymphoid tissues (tonsils and adenoids, appendix, Peyer's patches), lymphoid organs (spleen and thymus), and the blood cells, lymphocytes, and macrophages.

lymphatic fluid (*lim FAT ik FLOO id*) Clear and colorless fluid of the lymphatic system.

lymphatic vessels (*lim FAT ik*) Vessels that begin as capillaries carrying lymph, continue through lymph nodes, and empty into the right lymphatic duct and the thoracic duct.

lymphocyte activation Stimulation of lymphocytes, "waking them up" to fight a pathogen.

lymphocyte proliferation Reproduction of activated lymphocytes so there are many copies.

lysis (*LYE siss*) Destruction or breakdown.

lysosome (*LIE so soam*) Organelle that consists of a small sac with digestive enzymes in it. These destroy pathogens that invade the cell.

macrophages (*MAC reh fage ez*) Cells that take fragments of the pathogen they have eaten and present them to a B cell (lymphocyte). This stimulates the B cell to become a plasma cell and make antibodies against that specific pathogen. Macrophages also activate Helper T cells in this way. Macrophages also produce special immune response chemicals: interferon, interleukin, and tumor necrosis factor.

macroscopic anatomy (*MAK roh scop ic ah NAH tom ee*) Study of large structures of the body.

macular degeneration rogressive deterioration of the macula lutea of the retina leading to vision loss.

magnetic resonance imager (MRI) Imaging technique utilizing high intensity magnetism to produce high quality body images for diagnostic purposes.

major calyx, minor calyces (*KAY licks, KAY leh seez*) Tubes in the kidney which carry urine from the nephrons to the renal pelvis.

malignant (*mah LIG nant*) Cancerous, able to spread to distant parts of the body.

malleus (*MALL ee us*) First bone of the middle ear. It is attached to the tympanic membrane on one end and to the incus on the other end. Also known as the *hammer*.

mast cells Connective tissue cells that are important in cellular defense and contain heparin and histamine.

mastication (*MASS tih CAY shun*) Process of chewing, during which the teeth and tongue together tear, crush, and grind food. This is part of the process of mechanical digestion.

mastitis Inflammation or infection of the breasts.

mastoiditis Inflammation of the mastoid sinuses, usually as a result of the spread of infection from acute otitis media.

medial (*MEE dee al*) Pertaining to the middle of the body or the middle of an organ or structure.

mediastinum (*me dee ah STY num*) Central area within the thoracic cavity. It contains the trachea, esophagus, heart, and other structures.

medulla oblongata (*meh DULL lah ob long GALL ah*) Most inferior part of the brainstem that joins to the spinal cord. It relays nerve impulses from the cerebrum to the cerebellum. It contains the respiratory center. Cranial nerves IX through XII originate there.

melanin (*MELL an in*) Dark brown or black pigment that gives color to the skin and hair.

melanocytes (*mell AN oh sights*) Cell that produces melanin.

melanoma A malignant pigmented mole or tumor.

melatonin (*MELL ah TOH ninn*) Hormone secreted by the pineal body. It maintains the 24-hour wake-sleep cycle known as the circadian rhythm.

membrane A thin soft pliable layer of tissue which can line a cavity or cover an organ or structure.

memory B cells Antibody-producing cells that are produced when a pathogen is encountered the first time. Memory B cells are stored until the pathogen comes again. They are responsible for secondary response.

memory T cells Memory cells responsible for cell mediated immunity. They are produced when a pathogen is encountered the first time. Memory T cells are stored until the pathogen comes again. They are responsible for secondary response.

menarche (*men ARE kee*) The onset of the first menstruation.

Meniere's disease (*MEN yerz*) Recurring and progressive disease that includes progressive deafness, ringing ears, dizziness and the feeling of fullness in the ears.

meninges (*men IN jeez*) Three separate membranes that envelope and protect the entire brain and spinal cord. The meninges include the dura mater, arachnoid, and pia mater.

meningitis (*men in JYE tiss*) Inflammation of the meninges of the brain or spinal cord by a bacterial or viral infection. Initial symptoms include fever, headache, nuchal rigidity (stiff neck) lethargy, vomiting, irritability, and photophobia.

menopause The permanent cessation of menses, menstrual activity, usually occurring between ages 35 to 58.

menses (*MEN seez*) The monthly flow of bloody fluid from the endometrium.

menstruation (*MEN stroo AY shun*) The cyclic hormonal shedding of the uterine endometrium.

mesentery (*MEZ in tare ee*) Membranous sheet of peritoneum that supports the jejunum and ileum.

metabolism (*me TAH bow lizm*) Process of using oxygen and glucose to produce energy for cells. Metabolism also produces byproducts like carbon dioxide and other waste products. The ongoing cycle of anabolism and catabolism.

metastasis (*meh TASS tah siss*) Process by which cancerous cells break off from a tumor and move through the blood vessels or lymphatic vessels to other sites in the body.

metastasize To spread throughout the body to distant sites.

metric system System of measurement based on the power of 10.

microglia (*mie crow GLEE ah*) Cells that move, engulf, and destroy pathogens anywhere in the central nervous system.

microscopic anatomy (*MY kroh scop ic ah NAH tom ee*) Study of structures that require the aid of magnification.

midbrain An area that connects the pons and the cerebellum with the hemispheres of the cerebrum.

midsagittal plane (*mid SAJ ih tal*) An imaginary vertical plane that divides the entire body into right and left sides and creates a midline. The midsagittal plane is named for the sagittal suture of the skull.

mitochondria The energy organelle of the cell.

mitotic phase The phase devoted to actual cell division.

mitral Pertaining to the bicuspid or mitral valve of the heart.

mixed nerve A nerve that carries both sensory and motor information.

molars The largest teeth, located posterior to the premolars. They crush and grind food on their large, flat occlusal surfaces.

mononucleosis The presence of higher than normal number of mononuclear leukocytes in the blood, and can be a result of an acute infection of the Epstein Barr virus.

motor neurons Neuron that innervates muscle tissue.

motor system The system responsible for movement.

mucosa 1. Lining throughout the gastrointestinal system that consists of a mucous membrane that produces mucus and an underlying smooth muscle layer that contracts to move food. 2. Mucous membrane that lines the respiratory tract. It warms and humidifies incoming air. It produces mucus to trap foreign particles. 3. Mucous membrane lining the inside of the bladder. 4. Mucous membranes lining the nasal cavity that warm and moisturize the incoming air. They also produce mucus to trap foreign particles.

mucous Pertaining to mucus.

Multiple Sclerosis A chronic disease of the central nervous system involving the destruction of myelin and the nerve axons in various regions of the brain and the spinal cord at different times. This leads to temporary(sometimes sustained) but repetitive episodes of disrupted nerve impulse conduction.

murmur An abnormal heart sound that may or may not be a result of heart disease.

muscle Many muscle fascicles grouped together and surrounded by fascia.

muscular dystrophy *(MUSS kyoo lahr DISS troh fee)* Genetic disease due to a mutation of the gene that makes the muscle protein dystrophin. Without dystrophin, the muscles weaken and then atrophy. Symptoms appear in early childhood as weakness first in the lower extremities and then in the upper extremities. The most common and most severe form is Duchenne's muscular dystrophy; Becker's muscular dystrophy is a milder form.

muscularis externa The outside muscular layer of an organ or tubule.

myalgia *(my AL jee ah)* Pain in one or more muscles due to injury or muscle disease. Polymyalgia is pain in several muscle groups.

myasthenia gravis *(my ass THEE nee ah)* Abnormal and rapid fatigue of the muscles, particularly evident in the muscles of the face; there is ptosis of the eyelids. Symptoms worsen during the day and can be relieved by rest. The body produces antibodies against its own acetylcholine receptors located on muscle fibers. The antibodies destroy many of the receptors. There are normal levels of acetylcholine, but too few receptors remain to produce sustained muscle contractions.

myelin *(MY eh lin)* Fatty sheath around the axon of a neuron. It acts as an insulator to keep the electrical impulse intact. Myelin around the axons of the brain and spinal cord is produced by oligodendrocytes. Myelin around axons of the cranial and spinal nerves is produced by Schwann cells. An axon with myelin is said to be myelinated.

myocardial infarction (MI) The disruption of blood flow to the heart leading to cardiac tissue damage and decreased function, commonly referred to as a heart attack

myofibril *(my ah FIE bril)* Thin filament (actin) and thick filament (myosin) within the muscle fiber that give it its characteristic striated appearance.

myometrium *(MY oh MEE tree um)* Muscle layer in the uterus.

myopathy A general term for muscle disease.

myopia *(my OH pee ah)* Nearsightedness. Light rays from a near object focus correctly on the retina, creating a sharp image. However, light rays from a far object come into focus anterior to the retina, creating a blurred image.

myosin *(MY ah sin)* The thick protein filament found in muscle fibers.

nares *(NAIR eez)* The paired external openings of the nasal cavity.

nasal Pertaining to the nose.

nasal cavity *(NAY zl CAV ih tee)* Hollow area inside the nose that is lined with mucosa or mucous membrane.

nasal polyp Painless non-cancerous rubbery tissue growth that projects from the nasal mucosa.

nasopharynx *(nay zoh FAR inks)* Uppermost portion of the throat where the posterior nares unite. The nasopharynx contains the opening for the eustachian tubes and the adenoids.

natural active immunity Antibodies developed due to exposure to a pathogen.

natural killer (NK) cell Type of lymphocyte that matures in the red marrow and is the body's first cellular defense against invading microorganisms. Without the help of antibodies or complement, an NK cell recognizes a pathogen by the antigens on its cell wall and releases chemicals that penetrate and destroy it.

natural passive immunity Immunity due to the passage of antibodies from mother to child across the placenta or in breast milk.

neck Transitional area between the root and crown where the tooth becomes narrower. The neck of the tooth is located just above and below the gum line.

necrosis *(neh KROH siss)* Gray-to-black discoloration of the skin in areas where the tissue has died.

negative feedback loop Physiological process that works against the trend. Most often brings a variable back to set point. For example, as blood pressure rises, heart rate may decrease to bring blood pressure back to "normal."

negative selection The destruction of lymphocytes which react to "self" antigens. These lymphocytes must be deleted to prevent autoimmunity.

nephrology *(neh FRALL oh jee)* The study of the structure, function, and pathology of the renal system.

nephron *(NEFF rahn)* Microscopic functional unit of the kidney.

nephropathy (*neff ROPP ah thee*) General word for any disease process involving the kidney. Diabetic nephropathy involves progressive damage to the glomeruli because of diabetes mellitus. The tiny arteries of the glomerulus harden (glomerulosclerosis) because of accelerated arteriosclerosis throughout the body.

nerves Bundles of individual axons.

nervous system Body system that includes the brain, cranial nerves, spinal cord, spinal nerves, and neurons. It receives signals from parts of the body and interprets them as pain, touch, temperature, body position, taste, sight, smell, and hearing. It coordinates body movement. It maintains and interprets memory and emotion.

neuroglia (*noo ROH glee ah*) Cells that hold neurons in place and perform specialized tasks. Includes astrocytes, ependymal cells, microglia, oligodendrocytes, satellite cells, and Schwann cells. Also called *glial cells*.

neuromuscular Pertaining to the nervous and muscular system.

neuromuscular junction Area on a single muscle fiber where a nerve connects.

neuron (*NER on*) An individual nerve cell. The functional part of the nervous system.

neurotransmitter (*noo roh TRANS mit ter*) Chemical messenger that travels across the synapse between a neuron and another neuron, muscles fiber or gland.

neutrophils (*NOO troh fill*) A type of leukocyte that perform non-specific phagocytosis.

nociceptors (*NO see SEP torz*) Free nerve endings that are receptors of pain stimuli (pain receptors).

nodes of Ranvier Constriction of the myelin sheath on a myelinated nerve fiber that facilitates nodal transmission of the impulse.

norepinephrine (*NOR ep ih NEFF rinn*) 1. Neurotransmitter for the sympathetic nervous system. It goes between neurons and an involuntary muscle, organ, or gland. Controls the flight or fight response. 2. Hormone secreted by the adrenal medulla in response to stimulation by nerves of the sympathetic nervous system.

normal flora (*FLOOR ah*) The normal microorganisms found in a particular body region.

nosocomial An infection acquired by a patient while at a hospital, nursing home or any health care facility.

nucleolus Round, central region within the nucleus. It makes ribosomes.

nyctalopia (*nick tah LOW pee ah*) The inability to see in dim light, night blindness.

obstetrics (*ob riks*) Branch of medicine concerning managing pregnancy and child birth.

occipital lobe (*ok SIP eh tal*) Lobe of the cerebrum that receives sensory information from the eyes. Contains the visual cortex for the sense of sight.

occult Not readily seen, as in blood in the stool.

oligodendrocytes (*AH li go DEN droe site*) Neuroglial cells which produce myelin in the CNS.

oocyte (*OH oh site*) An immature stage of the female sex egg.

ophthalmology (*off thal MALL oh jee*) The study and treatment of visual conditions.

ophthalmologist (*off thal MALL oh jee*) A doctor who specializes in the eyes.

oral Pertaining to the mouth.

orbit Bony socket in the skull that surrounds all but the anterior part of the eyeball.

organ A part of the body comprised of tissues, that has a specialized function.

organelles Small structures in the cytoplasm that have various specialized functions. Organelles include mitochondria, ribosomes, the endoplasmic reticulum, the Golgi apparatus, and lysosomes.

organism An individual living thing.

oropharynx (*oh roh FAR inks*) Middle portion of the throat just behind the oral cavity. It begins at the level of the soft palate and ends at the epiglottis.

orthopedics (*or thoh PEE diks*) The medical science that deals with structures of the skeleton, joints, muscles, ligaments, and cartilage that create movement.

orthostatic hypotension An abnormal decrease in blood pressure as a result of changing position (such as quickly standing up).

osmosis The passage of the solvent through a semi-permeable membrane to equalize concentrations.

osmotic pressure The pressure which develops when there are two solutions of varying concentrations that are separated by a semi-permeable membrane.

osseous tissue *(AH see us)* Bone, a type of connective tissue.

ossicles *(AH sih kel)* Three tiny bones in the middle ear that function in the process of hearing: malleus, incus, and stapes.

ossification *(AH siff ih cay shun)* Process by which cartilaginous tissue is changed into bone from infancy through puberty. Also known as *osteogenesis.*

osteoarthritis *(OSS tee oh ahr THRYE tiss)* Chronic inflammatory disease of the joints, particularly the large weight-bearing joints of the knees and hips, although it often occurs in the joints that move repeatedly like the shoulders, neck, and hands.

osteoblasts Osteocytes that form new bone.

osteoclasts Osteocytes that break down old or damaged areas of bone.

osteocytes *(OSS tee oh site)* Bone cells. There are two types of osteocytes: osteoclasts and osteoblasts.

osteomalacia *(OSS tee oh mah LAY she ah)* Abnormal softening of the bones due to a deficiency of vitamin D. Chondromalacia is abnormal softening of the cartilage, specifically of the patella.

osteons *(OSS tee ons)* Microscopic unit of compact bone.

osteoporosis *(OSS tee oh por OS sis)* Condition of increased bone porosity that weakens the bones, usually seen in the elderly.

otitis media *(oh TYE tiss MEH dee ah)* Acute or chronic bacterial infection of the middle ear.

otolaryngology *(oh toh lair ing GALL oh jee)* The study of the structure, function and diseases of the ears and throat.

otorhinolaryngology *(oh toh rye noh lair ing GALL oh jee)* The study of the structure, function and diseases of the ears, nose, and throat.

oval window Opening in the temporal bone between the middle ear and the vestibule of the inner ear. The opening is covered by the end of the stapes.

overactive bladder Bladder condition leading to increase urge to urinate.

ovulation *(OV yoo LAY shun)* Periodic ripening and expulsion of female eggs.

oxygen debt The amount of extra oxygen that is required by muscle tissue during its recovery from strenuous exercise.

oxytocin *(AHK see TOH sin)* Hormone secreted by the posterior pituitary gland. It stimulates the uterus to contract and begin labor. It stimulates the "let-down reflex" to get milk flowing for breastfeeding.

pacemaker cells Cells or groups of cells that automatically generate electrical impulses.

palate *(PAHL aht)* Roof of the mouth

palatine tonsils *(PAL ah tighn TAHN sill)* Lymphoid tissue on either side of the throat where the soft palate arches downward in the oropharynx.

pancreas *(PAN kree ass)* A digestive and endocrine organ located in the abdominal cavity that produces digestive enzymes (amylase, lipase, protease, peptidase) and releases them into the duodenum. It also contains the islets of Langerhans (alpha, beta, and delta cells) that produce and secrete the hormones glucagon, insulin, and somatostatin.

pancreatitis *(PAN kree ah TYE tiss)* Inflammation or infection of the pancreas.

pandemic Disease or condition that affects many people world-wide.

paralysis Temporary or permanent loss of muscle function.

paraplegia Paralysis of lower body and of both legs.

parasympathetic nervous system Division of the autonomic nervous system that uses the neurotransmitter acetlycholine and carries nerve impulses to the heart, involuntary smooth muscles, and glands while the body is at rest.

parathyroid glands *(PAIR ah THIGH royd)* Endocrine glands, four of them, on the posterior lobes of the thyroid gland. They produce and secrete parathyroid hormone.

parietal lobe *(pah RYE eh tal)* Lobe of the cerebrum that receives sensory information about temperature, touch, pressure, vibration, and pain from the skin and internal organs.

parietal pleura *(pah RYE eh tal PLOO rah)* One of the two layers of the pleura. It lines the thoracic cavity.

Parkinson's Disease a chronic degenerative disease of the central nervous system that causes impaired movement and changes in mood and cognition. Usually found in individuals older than 65.

parotid salivary gland *(pah RAH tid SAHL ih vair ee gland)* Gland that secretes saliva that helps to lubricate food so that it is easier to chew and swallow.

partial thickness burn A newer classification of the severity of a burn. This level of burn damage includes the entire depth of the epidermis and a portion of the dermis.

passive transport The general term for the transportation of cellular material without the use of energy.

patellar Pertaining to the knee cap.

pathogen *(PATH oh jenn)* Microorganism that causes a disease. Pathogens include bacteria, viruses, protozoa, and other microorganisms, as well as plant cells like fungi or yeast.

pathology *(path ALL oh jee)* The study of disease.

pedal Pertaining to the foot or feet.

pediculosis A condition of lice infestation.

pepsin Digestive enzyme produced by the stomach that breaks down food protein into smaller protein molecules.

peptic ulcer An ulcer occurring in the lower end of the esophagus, stomach or in the duodenum.

perforin Chemical secreted by cytotoxic T cells which makes holes in the cell membranes of pathogens or infected cells, killing them.

perfusion Blood flow to a particular region.

perilymph *(per ih LIMF)* Pale lymph fluid found the labyrinth of the inner ear.

perimetrium *(pair ee MEE tree um)* The serous layer of the uterus.

periodontal *(perr ee oh DAHN til)* Literally means "around the teeth" and may refer to the area of the gum.

periodontal disease Gum disease.

periosteum *(pair ee OSS tee um)* Thick, fibrous membrane that goes around and covers the outside of a bone.

peripheral Referring to "away from center" or the extremities.

peripheral nervous system (PNS) Division of the nervous system that includes the cranial nerves and the spinal nerves.

peripheral neuropathy Condition of the peripheral nerves resulting in pain and dysfunction of these nerves. Often related to diabetes.

peripheral vascular disease (PVD) Any disease of the arteries or veins of the extremities.

peristalsis *(pair ih STALL siss)* Contractions of the smooth muscle of the gastrointestinal tract that propel food through it. Can also be the process of smooth muscle contractions that propel urine through the ureter.

peritoneal dialysis A form of dialysis, in which the lining of the peritoneal cavity is used as the dialyzing membrane.

peritubular capillaries *(per ee TUBE you ler)* Capillaries surrounding the renal tubules.

persistent vegetative state A continuous condition of complete unawareness of the environment with either compete or partial hypothalamic and autonomic brain stem functions. Sleep wake cycles also exist. This condition may be caused by brain trauma, degenerative disease, metabolic disorders, or developmental defects.

pH A test to determine how acidic or alkaline a substance is.

phagocytosis *(fag oh sigh TOH siss)* The process by which a phagocyte destroys a foreign cell or cellular debris; a type of endocytosis.

phantom pain A sensation that an amputated limb is still present and is exhibiting pain.

pharyngitis *(far in JIGH tis)* Bacterial or viral infection of the throat. When the bacteria group *beta-hemolytic streptococcus* causes the infection, it is known as strep throat.

pharynx *(FAR inks)* The throat; contains the passageways for food and for inhaled and exhaled air.

phenylketonuria (PKU) Inherited disease characterized by the excess of phenylketone in the urine due to an accumulation of phenylalanine in the tissues that may lead to brain injury of death.

photopigments *(FOE toe pig ments)* Chemicals in retinal cells that have light sensitivity.

physiology *(fiz ee ALL oh jee)* The study of the function of the body's structures.

pia mater *(PEE ah MAY ter)* Thin, delicate innermost layer of the meninges that covers the surface of the brain and spinal cord. It contains many small blood vessels.

pineal gland *(pih NEE al)* Endocrine gland in the brain that lies posterior to the pituitary gland. It secretes the hormone melatonin.

pinna *(PINN ah)* The auricle of the exterior ear that collects sound waves.

pinocytosis Process in which a cell absorbs fluid material.

pituitary gland *(pih TOO ih tair ee)* Endocrine gland in the brain that is connected by a stalk of tissue to the hypothalamus. Also known as the *hypophysis*. It is known as the master gland of the body. It consists of the anterior and the posterior pituitary gland.

plantar Referring to sole of the foot.

plaque A patch on skin or mucous surface often caused by lipids or calcium. Can build up in atherosclerotic heart disease and block blood vessels.

plasma *(PLAZ mah)* Clear, straw-colored fluid portion of the blood that carries blood cells and contains dissolved substances like proteins, glucose, minerals, electrolytes, clotting factors, complement proteins, hormones, bilirubin, urea, and creatinine.

plasma proteins *(PLAZ mah)* Protein molecules in the plasma. An important one is albumin.

pleural cavity *(PLOO ral)* The space between the parietal and visceral layers of the pleura.

pleural effusion *(PLOO ral eh FYOO zhun)* Accumulation of fluid within the pleural space due to inflammation or infection of the pleura and lungs.

plexus *(PLECK sus)* A network of nerves or vessels.

plicae circulares One of the transverse folds in the small intestine.

pneumonia Infectious lung disease characterized by fever, excessive secretions, shortness of breath, lethargy and cough.

pneumothorax *(NOO moh THOH raks)* Large volume of air that forms in the pleural space and progressively separates the 2 pleural membranes.

polycystic kidney disease everal inherited disorders causing cysts to form in the kidneys and other organs leading to destroyed kidney tissue and function.

polycystic ovarian syndrome *(PCOS)* Formation of multiple cysts on the ovary.

polycythemia *(pall ee sigh THEE mee ah)* Increased number of erythrocytes due to uncontrolled production by the red marrow. The viscosity of the blood increases; it becomes viscous (thick), and the total blood volume is increased.

polymorphonuclear Possessing a nucleus consisting of several parts or lobes connected by fine strands.

polysomnography *(PAUL ee sahm NOG grafee)* The study of sleep patterns and pathology.

pons Area of the brainstem that relays nerve impulses from the body to the cerebellum and back to the body. Area where nerve tracts cross from one side of the body to the opposite side of the cerebrum. Cranial nerves V through VIII originate there.

positive feedback "Vicious cycle." During positive feedback, physiological processes send body chemistry or other attributes further and further away from equilibrium (set point). The trend will continue until something breaks the cycle.

positive selection During lymphocyte development, only those lymphocytes which can actually react to antigens will survive.

postcentral gyrus Ridge on the surface of the cerebrum posterior to the central sulcus in each hemisphere. The postcentral gyrus contains the primary somatic sensory area for your sense of touch.

posterior The back of the body.

posteroanterior (PA) View from the back to the front such as a PA X-ray.

precentral gyrus Ridge on the surface of the cerebrum anterior to the central sulcus in each hemisphere. The postcentral gyrus contains the primary motor cortex for voluntary movements.

premenstrual syndrome (PMS) Syndrome of physical and psychological changes that occurs before menstruation in many women.

premenstrual dysphoric disorder (PMDD) Disorder characterized by severe depression, marked anxiety, and decrease interest in activity.

prepuce (*PREE pewce*) Foreskin.

presbyopia (*PRESS bee OH pee ah*) Loss of flexibility of the lens with blurry near vision and loss of accommodation.

primary motor cortex Located in the precentral gyrus contains a map of the body to control motor function.

primary movers Muscle or muscle group that causes movement. Also known as an *agonist*.

primary somatic sensory cortex Area of the cerebral cortex that allows for understanding and interpretation of somatic sensory information.

proctology (*prok TALL oh jee*) The specialization of medicine dealing with the treatment of diseases of the colon, rectum and anus.

prognosis (*prog NOH siss*) The predicted outcome of a disease.

prolactin (*proh LAK tinn*) Hormone secreted by the anterior pituitary gland. It stimulates milk glands of the breasts to develop during puberty and to produce milk during pregnancy.

proliferative (*pro LIFF er ah TIV*) To increase by rapid reproduction.

prone position Lying with the anterior section of the body down.

proprioception (*PRO pree oh SEP shun*) The sensation of movement and position of the body.

prostate cancer Cancerous condition of the gland that surrounds the upper portion of the urethra in males.

prostrate gland A chestnut size gland that secretes a slightly alkaline fluid that forms part of the seminal fluid.

prothrombin (*pro THROM bin*) Blood clotting factor that is activated just before the thrombus is formed.

ProTime Measure of the time necessary for the coagulation of blood. This is a common term for Prothrombin time.

protozoa Unicellular organisms.

proximal (*PROK sim al*) Referring to "near" a reference point.

proximal tubule The part of the renal tubule closest to the glomerulus. Many substances are secreted or absorbed in this part of the renal tubule.

pseudostratified Apparently composed of layers.

pudendal cleft (*PEW den dall cleft*) Opening between two halves of the labia majora.

pulmonary artery Artery that carries blood from the heart to the lungs. The pulmonary artery is the only artery in the body that carries blood that has low levels of oxygen.

pulmonary edema A potentially life threatening condition in which there is an accumulation of fluid in the interstitium and the alveoli of the lungs that can block the exchange of oxygen and carbon dioxide.

pulmonology (*pull mon ALL oh jee*) The branch of medicine that studies the treatment of lung disorders.

pupil (*PYOO pill*) Round opening in the iris that allows light rays to enter the internal eye.

pustule (*PUS tyool*) A small elevation of skin filled with lymph or pus.

pyloric sphincter (*pye LOR ik SFINK ter*) Muscular ring that keeps food in the stomach from entering the duodenum.

pylorus (*pye LOR uss*) Narrowing canal of the stomach just before it joins the duodenum. It contains the pyloric sphincter.

quadriplegia Paralysis of all four limbs.

quadriplegic One who has paralysis of all four limbs.

rapid eye movement (REM) The dream stage of sleep.

reabsorption Process by which water and substances in the filtrate move out of the renal tubule and into the blood in a nearby capillary.

rectum Final part of the large intestine. It is a short, straight segment that lies between the sigmoid colon and the outside of the body.

referred pain A perceived pain site other than the actual site of the trauma or disease.

reflex Involuntary muscle reaction that is controlled by the spinal cord. In response to pain, the spinal cord immediately sends a command to the muscles of the body to move. All of this takes place without conscious thought or processing by the brain. The entire circuit is also known as a *reflex arc*.

refractory period Short period of time when the myocardium is resting and unresponsive to electrical impulses.

regulatory T cells Type of T cell that shuts down immune response.

renal artery (*REE nal AHR ter ee*) Major artery that carries blood to the kidney.

renal corpuscle (*REE nal KOR puss el*) The filtration apparatus of the kidney, consists of the glomerulus and the glomerular capsule.

renal cortex Tissue layer of kidney just beneath the renal capsule.

renal failure Impaired functioning of the kidneys leading to a build-up of toxic waste products and fluids in the body.

renal nephron Fundamental functional unit of the kidney, consists of the renal corpuscle and renal tubule.

renal pelvis (*REE nal PELL vis*) Funnel-shaped part of the kidney that collects urine.

renal vein Major blood vessels that carries blood away from the kidneys.

renin-angiotensin-aldosterone (*REE nen-an gee oh TEN sen-al DOSS ter ohn*) A complex hormone system that regulates blood volume and blood pressure. The system is triggered when blood flow to the kidney decreases.

repolarization The opposite of depolarization.

respiration (*ress pih RAY shun*) The process of gas exchange at the lungs or tissue sites. Oxygen and carbon dioxide are exchanged in the alveoli during external respiration. Oxygen and carbon dioxide are exchanged at the cellular level during internal respiration.

retina (*RETT in ah*) The inner most membrane lining the eye. It contains rods and cones. Landmarks include the optic disk and macula.

Rh factor A blood group discovered on the surface of erythrocytes of Rhesus monkeys and found to a variable degree in humans. Can be Rh-negative or Rh-positive.

rheumatoid arthritis A chronic, systemic condition affecting synovial joints. Women are three times more affected than men. A common complaint is morning joint stiffness rather than after physical activities.

Rheumatology Division of medicine concerned with rheumatic (inflammation, pain, stiffness in muscles and joints) diseases.

rhinitis An acute temporary, seasonal or year round inflammation of the nasal mucosa with nasal congestion and increased mucus production.

rhinoplasty (*RYE noh plass tee*) Surgical procedure that uses plastic surgery to change the size or shape of the nose.

ribonucleic acid (RNA) Molecule contained in ribosomes and necessary for making proteins.

ribosome Granular organelle located throughout the cytoplasm and on the endoplasmic reticulum. Ribosomes contain RNA and proteins and are the site of protein synthesis.

rigor mortis (*RIG er MORE tiss*) A stiffness that occurs in dead bodies as a result of retained calcium and decreased ATP.

rods Light-sensitive cells in the retina. They detect black and white and function in daytime and nighttime vision.

root Part of the tooth that is hidden below the gum line. The premolars have one or more roots. The molars have multiple roots.

rugae (*ROO guy*) 1. Deep folds in the gastric mucosa. 2. Folds in the mucosa of the bladder that disappear as the bladder fills with urine.

sarcomeres (*SAR koh meerz*) Portion of striated muscle fibril that lies between the two adjacent dark lines.

satellite cells Neuroglia cells that enclose the cell bodies of neurons in the spinal ganglia.

scabies A contagious skin condition caused by the itch mite, Sarcoptes scabiei.

Schwann cell (*SHWAN sells*) Cell that forms the myelin sheaths around axons of the cranial and spinal nerves.

sclera (*SKLAIR ah*) White, tough, fibrous connective tissue that forms the outer layer around most of the eye. Also known as the *white of the eye*.

scoliosis Lateral curvature of the spine.

sebaceous glands (*see BAY shuss*) An exocrine gland of the skin that secretes sebum. Sebaceous glands are located in the dermis. Their ducts join with a hair and sebum coats the hair shaft as it moves toward the surface of the skin. Also known as *oil glands*.

sebum (*SEE bum*) Oily substance secreted by sebaceous glands.

secondary response Increased immune response mediated by memory cells when meeting a pathogen that it recognizes.

secretion (*sih CREE shun*) The movement of a chemical out of a cell or gland.

segmentation Division into similar parts.

self vs. non-self recognition The ability of the immune system to distinguish between the body's cells and cells that do not belong in the body.

semen (*SEE men*) A sperm containing, thick secretion produced by males.

semicircular canals Three canals in the inner ear that are oriented in different planes (horizontally, vertically, obliquely) that help the body keep its balance. It relays information to the brain via the vestibular branch of the vestibulocochlear nerve.

sensations A feeling or awareness as a result of the stimulation of sensory receptors.

septic arthritis A form of joint inflammation due to infection.

septicemia (*sep teh SEE me uh*) Infection by pathogenic organisms in the blood.

septum (*SEHP tum*) 1. Wall of cartilage and bone that divides the nasal cavity into right and left sides. 2. Partitioning wall that divides the right atrium from the left atrium (interatrial septum) and the right ventricle from the left ventricle (interventricular septum).

serosa (*seh ROSE ah*) A serous membrane.

serous membrane (*SEER us*) Double layered membrane lining a serous cavity. The parietal layer lines the wall of the cavity and the visceral layer covers the organs in the cavity. There is a, fluid-filled space between the layers.

set point The normal physiologic range (as in body temperature and blood pressure) to maintain homeostasis.

shin splints A common running induced injury of the extensor muscles and surrounding tissues of the lower leg.

shingles The common name for the skin rash caused by herpes zoster.

silent MI A myocardial infarction that does not exhibit the classic signs of a heart attack, such as chest pain radiating from the left shoulder down the arm.

sinoatrial node (*sigh noh AY tree al*) Pacemaker of the heart. Small knot of tissue located in the posterior wall of the right atrium in a shallow channel near the entrance of the superior vena cava. The SA node dictates the heart rate at 70–80 beats per minute when the body is at rest. It generates the electrical impulse for the entire conduction system of the heart.

sinus Hollow cavity within a bone of the cranium.

sinusitis Inflammation of the sinus.

skeletal muscle (*SKELL eh tal*) One of three types of muscles in the body, but the only muscle that is under voluntary, conscious control. Under the microscope, skeletal muscle has a striated appearance.

smooth muscle One of three types of body muscles that is involuntarily controlled and found in the lining of the airways, blood vessels and uterus.

sol layer The thin, watery layer of mucus in the airways of the lungs.

somatic nervous system (*so MAT ick*) Division of the peripheral nervous system that uses the neurotransmitter acetylcholine and carries nerve impulses to the voluntary skeletal muscles.

spasm Involuntary contraction of a muscle.

spermatogonia (*sper MAT oh GO nee ah*) Large unspecialized germ cells that form primary spermatocytes by mitosis.

spermatocytes (*sper MAT oh sites*) A cell from the spermatogonium that divides to form spermatids.

spermatids (*sper MAT ids*) A cell that is the result of the division of a secondary spermatocyte.

spermatozoa (*sper MAT oh ZOE ah*) The plural form of spermatozoon, the mature male sex cell.

sphincter (*SFING ter*) Muscular ring around a tube; a valve.

Spina Bifida A congenital defect in which the vertebral arches fail to form properly. This leads to varying levels of disability.

spinal block An obstruction of the flow of cerebrospinal fluid through the spinal subarachnoid space, this term is commonly (though inaccurately) used to refer to spinal anesthesia.

spinal cavity (*SPY nal*) A continuation of the cranial cavity as it travels down the midline of the back. The spinal cavity lies within and is protected by the bones (vertebrae) of the spinal column. The spinal cavity contains the spinal cord, the spinal nerves, and spinal fluid.

spinal cord Part of the central nervous system. Continuous with the medulla oblongata of the brain and extends down the back in the spinal cavity. Ends at L2 and separates into individual nerves (cauda equina).

spinal nerves Thirty-one pairs of nerves. Each pair comes out from the spinal cord between two vertebrae. An individual spinal nerve consists of dorsal nerve roots and ventral nerve roots.

spinal roots Axon bundles attached to each spinal cord segment. The dorsal roots are sensory and the ventral roots are motor. The dorsal and ventral roots join to form the spinal nerves.

spinocerebellar tract (*SPY no ser eh BELL ar*) Sensory pathway from the spinal cord to the cerebellum.

spinothalamic tract (*SPY no THAL ah mic*) Sensory pathway from the spinal cord to the thalamus and eventually the primary somatic sensory cortex. Contains pain and crude touch information.

spleen Lymphoid organ located in the abdominal cavity behind the stomach. The spleen destroys old erythrocytes, breaking their hemoglobin into heme and globins. It also acts as a storage area for whole blood. Its white pulp is lymphoid tissue that contains B and T lymphocytes.

spores A protective barrier to allow for future reproduction in a hostile environment.

sprain Overstretching or tearing of a ligament.

squamous cells (*SKWAY muss sells*) A flat, scaly epithelial cell.

stapes (*STAY peez*) Third bone of the middle ear. It is attached to the incus on one end and the oval window on the other end. Also known as the *stirrup*.

stenotic Narrowing of a passage or an orifice

sternal Pertaining to the sternum.

sternum (*STER num*) Vertical bone of the anterior thorax to which the clavicle and ribs are attached. Also known as the *breast bone*.

steroids Ringed lipids that function as extremely powerful hormones.

stomatitis An inflammation of the mouth that can include the lips, tongue, and mucous membranes.

strain Overstretching of a muscle, often due to physical overexertion. This causes inflammation, pain, swelling, and bruising as the capillaries in the muscle tear. There can be small tears in the muscle itself. Also known as a *pulled muscle*.

stratified (*STRAT ih fied*) Having more than one layer.

stratum basale The stem cell layer of the epidermis.

stratum corneum (*STRAY tum core NEE um*) The outer most horny layer of the epidermis.

stress response A general term for an individual's physical, physiological, and psychological reaction to stressors

stress test A diagnostic test that either uses physical activity or a drug to cause the heart to work harder to reveal any possible coronary artery blockage. Commonly called an exercise tolerance test.

striated (*STRY ate ed*) Striped

striated muscle (*STRY ate ed*) Skeletal muscle

stroke (CVA) A condition of sudden neurological function brought about by vascular injury to the brain. Can be caused by a blood blockage due to a clot or by hemorrhage.

subarachnoid space (*SUB ah RACK noyd*) Space beneath the arachnoid layer of the meninges. It is filled with cerebrospinal fluid.

subcutaneous fascia (*sub cue TAY nee us FAY she ah*) Connective tissue layer beneath the skin.

subdural space (*sub DOO ral*) The space between the arachnoid and dura matter.

sublingual salivary glands (*sub LIN gwill SAHL ih vair ee glands*) Smallest of salivary glands found between the tongue and the mandible, one on each side.

submandibular salivary glands (*sub MAN dih bue lar SAHL ih vair ee glands*) Salivary gland beneath the mandible or jaw.

submucosa Layer of connective tissue under a mucous membrane.

sulcus (*SULL cuss*) One of many shallow grooves between the gyri in the cerebrum and cerebellum. Plural: sulci.

superficial Related to or located near the surface.

superior Pertaining to the upper half of the body or a position above an organ or structure.

superior vena cava Largest vein that drains venous blood from the upper portions of the body into the right atrium.

supine position (*sue PINE*) Position of lying on the posterior part of the body. Also known as the *dorsal supine position*.

surfactant (*sir FAC tent*) Protein-fat compound that creates surface tension and keeps the walls of the alveolus from collapsing inward with each inhalation.

sympathetic nervous system Division of the autonomic nervous system that uses the neurotransmitter norepinephrine and carries nerve impulses to the heart, involuntary muscles, and glands during times of increased activity, danger, or stress.

synapse (*SIH naps*) Space between the axon of one neuron and the dendrites of the next neuron.

syndrome (*SIN drohm*) A set of symptoms and signs associated with and characteristic of one particular disease.

synergistic (*sin er GIS tic*) A cooperating action of certain muscles.

synovial fluids (*sin OH vee al*) Clear lubricating fluid that is secreted by the synovial membrane.

synovial membrane (*sin OH vee al*) The membrane lining a capsule of the joint.

Systemic Lupus Erythematosus A chronic autoimmune inflammatory condition that involves multiple organ systems. There are periodic acute episodes of this disease. There is a characteristic "butterfly" rash over the patient's nose and cheeks resembling a wolf's snout.

systems An organized grouping of related structures or parts that perform specific functions.

systole (*SISS toh lee*) Combined contractions of the atria and the ventricles.

T lymphocytes (*T LIMF oh sights*) White blood cells involved in adaptive immunity. There are four types of T lymphocytes: Helper T cells, Cytotoxic T cells, Memory T cells and Regulatory T cells.

tachycardia A more rapid than normal heart rate (greater than 100 beats per minute in adults).

tactile corpuscles (*TAK tle KOR puss els*) Elongated bodies found in nerve ends that act as receptors for slight pressure or touch.

taste buds Sensory end organs that provide us with a sense of taste.

temporal lobe Lobe in the cerebrum that receives sensory information from the auditory cortex for hearing and the olfactory cortex for smelling.

tendonitis Inflammation of any tendon from injury or overuse.

tendon Cordlike white band of non-elastic fibrous connective tissue that attaches a muscle to a bone.

testes (*TESS teez*) Small, egg-shaped glands in the scrotum. Also known as the *testicles*. They contain interstitial cells that secrete testosterone. They also contain the seminiferous tubules that produce spermatozoa.

tetanus *(TETT ah nuss)* An acute infectious disease caused by a bacterium that can lead to severe spasms of voluntary muscles.

thalamus *(THAL ah mus)* Relay station in the brain that receives sensory nerve impulses from the optic nerves and sends them to the visual centers in the occipital lobes of the brain.

thoracic cage *(tho RASS ik)* The portion of the skeleton to include the ribs, sternum and thoracic vertebrae that house and protect the lungs, heart and great vessels.

thoracic cavity *(thoh RASS ik KAV ih tee)* Hollow space within the thorax that is filled with the lungs and structures in the mediastinum.

thoracic duct *(thoh RASS ik)* The main lymph duct of the body.

thoracic surgery *(tho RASS ik)* Manual or operative procedures involving structures or organs located in the thoracic region of the body.

thrombin *(THROM bin)* An enzyme that reacts with fibrinogen , converting it to fibrin which forms a clot.

thrombocyte *(THROM boh sights)* Cell fragment that is flat and does not have a nucleus. It is active in the blood clotting process. Thrombocytes are also known as *blood platelets*.

thrombocytopenia *(THROM boh sigh TOH PEE nee ah)* Deficiency in the number of thrombocytes. This can be due to exposure to radiation or toxic chemicals or drugs that damage the stem cells in the bone marrow.

thrombus *(THROM buss)* A blood clot.

thymus *(THIGH muss)* Lymphoid organ in the thoracic cavity. As an endocrine gland, it releases hormones known as thymosins. Thymosin causes lymphoblasts in the thymus to mature into T lymphocytes.

thyroid gland *(THIGH royd)* Endocrine gland in the neck that produces and secretes the hormones T3, T4, and calcitonin. Its 2 lobes and narrow connecting bridge (isthmus) give it a shield-like shape.

Thyroid Storm A rare but potentially life threatening result of untreated hyperthyroidism that can cause cardiac arrhythmias, congestive heart failure and shock.

tinea Any fungal infection found on the body.

tinnitus *(tinn EYE tuss)* Sounds (buzzing, ringing, hissing, or roaring) that are heard constantly or intermittently in one or both ears, even in a quiet environment.

tissues Collection of similar cells that form a particular function,

tongue Large muscle that fills the oral cavity and assists with eating and talking. It contains taste buds and receptors for the sense of taste.

tonus *(TONE us)* A partial steady contraction of a muscle; firmness.

tonsillitis An inflammation of the tonsil or tonsils.

trabecula *(tra BECK you la)* A fibrous cord of connective tissues that acts as a supporting fiber.

trachea *(TRAY kee ah)* Rigid tubular air pipe between the larynx and the bronchi that is a passageway for inhaled and exhaled air.

transdermal patches A from of drug delivery that allows drug absorption through the skin from a patch, containing the medication, that adheres to the skin.

traumatic brain injury (TBI) Not to be confused with *total body irradiation*, this refers to a severe brain injury that may leave residual impairment.

Trendelenburg *(Trin DELL in berg)* A patient position in which the head is lower than the body and legs.

transitional *(tran ZISH ion al)* Moving from one state to another.

transverse plane *(tranz VERS)* Plane that divides the body into top and bottom sections, superior and inferior.

tricuspid Having three cusps or points as in the tricuspid valve of the right heart.

tuberculosis (TB) *(too ber kew LOH sis)* Lung infection caused by the bacterium mycobacterium tuberculosis and spread by airborne droplets expelled by coughing.

tubular reabsorption Movement of substances out of the renal tubule and back into the blood. The substances will be retained in the body.

tubular secretion Movement of substances from the blood into the renal tubule. These substances will leave the body in the urine.

tumor necrosis factor (TNF) (*neh KROH siss*) Released by macrophages, it destroys endotoxins produced by certain bacteria. It also destroys cancer cells.

tunica externa (*TOO nik ah ex TERN ah*) The outer layer of a blood vessel.

tunica interna (*TOO nik ah in TERN ah*) The inner lining of a blood vessel.

tunica media (*TOO nik ah mee DEE ah*) The middle muscular layer of a blood vessel.

turbinates Three long projections (superior, middle, inferior) of the ethmoid bone that jut into the nasal cavity: superior, middle, and inferior. They break up the stream of air as it enters the nose. Also known as the *nasal conchae*.

tympanic membrane (*tihm PAN ik*) Membrane that divides the external ear from the middle ear. Also known as the *eardrum*.

umbilical Pertaining to the umbilicus ("belly button").

urinary tract infection (UTI) A general term for an infection involving the kidney(s), ureter(s), or bladder.

universal donor A person who has a form of type O blood, which can be transfused to any individual of the ABO blood groups.

universal recipient A person who has type AB blood, and therefore can receive blood from any of the ABO blood groups.

upper respiratory infection (URI) Bacterial or viral infection of the nose that can spread to the throat and ears. The nose is a part of the respiratory system as well as the ENT system. Also known as a *common cold* or *head cold*.

urea The main nitrogenous substance found in urine. The amount excreted in urine is directly related to the amount of protein consumed.

uremia Blood found in urine as a result of a pathologic process.

ureter (*yoo REE ter*) Tube that connects the pelvis of the kidney to the bladder.

urethra (*yoo REE thrah*) Tube that connects the bladder to the outside of the body.

urinary bladder A holding receptacle for urine before it is expelled (voided) from the body.

urology (*yoo RALL oh jee*) A branch of medicine involving the urinary tract of both males and females and the male genital tract.

UTI Urinary Tract infection.

vaginitis (*vaj ih NYE tiss*) An inflammatory process involving the vagina.

vagus nerve Cranial nerve X. Sensation and movement of the throat. Sensory and motor for thoracic and abdominal organs.

valvular insufficiency A pathologic condition in which the valve(s) of the heart do not function normally.

vasculature Network of blood vessels in a particular organ.

vasoconstriction (*vaz oh kon STRIK shun*) Constriction of the smooth muscle in the artery wall causes the artery to become smaller in diameter.

vasodilation (*vaz oh DYE lay shun*) Relaxation of the smooth muscle in the artery wall, causing the artery to become larger in diameter.

vein Blood vessel that carries oxygen-poor blood as well as carbon dioxide and waste products of cellular metabolism away from the cells and back to the heart. Veins have one-way valves that keep blood from flowing backwards, away from the heart. The exception is the pulmonary vein, which carries oxygenated blood from the lungs to the heart.

ventilation (*ven tih LAY shun*) The bulk movement of gas into and out of the lungs.

ventral Pertaining to the anterior of the body particularly the abdomen.

ventricles (*VEN trik lz*) 1. Two lower chambers of the heart. Intraventricular structures are located in the ventricles. 2. Four hollow chambers within the brain that contain cerebrospinal fluid. The 2 lateral ventricles are within the right and left hemispheres of the cerebrum. The third ventricle is small and connects the lateral ventricles to the fourth ventricle, which is at the level of the pons and medulla oblongata.

venules (*VEHN yules*) Smallest branch of a vein.

verruca (*vur ROO kuh*) A wart.

vertebrae (*VER teh bray*) One of 33 irregularly shaped bony segments of the spinal column.

vertigo Sensation of being off balance when the body is not moving. Caused by upper respiratory infection, middle or inner ear infection, head trauma, or degenerative changes of the semicircular canals.

vesicle (*VESS ih kle*) A small bladder or blister; a membrane bound storage sac inside a cell.

vestibule chamber (*VESS tih byool*) A small cavity or space at the beginning of a canal.

vestibulocochlear nerve (*VESS tih byool KOHK lee are*) The eighth cranial nerve responsible for hearing and balance. Sometimes called the *acoustic nerve.*

villi (*VILL eye*) Microscopic projections of the mucosa within the lumen of the small intestine.

virus A parasitic microorganism that depends on other cells for its metabolic and reproductive needs.

visceral (*VISS er al*) Pertaining to organs.

visceral pleura (*VISS er al PLOO rah*) One of the two membranes of the pleura. It covers the surface of the lung.

vital signs Medical procedure during a physical examination in which the temperature, pulse, and respirations (TPR), as well as the blood pressure, are measured.

vitreous humor (*VITT ree uss HYOO mer*) Clear, gel-like substance that fills the posterior cavity of the eye.

vocal cords Connective tissue bands in the larynx that vibrate and produce sounds for speaking and singing.

warts Raised, cutaneous elevations. Also called verruca.

white blood cells (WBCs) Leukocytes; responsible for the immune response.

z lines Lines visible on the surface of skeletal muscle that mark the ends of each sarcomere.

INDEX